ANTIMICROBIALS
Synthetic and Natural Compounds

ANTIMICROBIALS
Synthetic and Natural Compounds

Edited by
Dharumadurai Dhanasekaran
Nooruddin Thajuddin
Annamalai Panneerselvam

CRC Press
Taylor & Francis Group
Boca Raton London New York

CRC Press is an imprint of the
Taylor & Francis Group, an **informa** business

CRC Press
Taylor & Francis Group
6000 Broken Sound Parkway NW, Suite 300
Boca Raton, FL 33487-2742

First issued in paperback 2019

ISBN-13: 978-1-4987-1562-1 (hbk)
ISBN-13: 978-0-367-37715-1 (pbk)

Library of Congress Cataloging-in-Publication Data

Antimicrobials : synthetic and natural compounds / edited by Dharumadurai Dhanasekaran,
Nooruddin Thajuddin, and A. Panneerselvam.
    pages cm
Includes bibliographical references and index.
ISBN 978-1-4987-1562-1
    1. Antibiotics. 2. Anti-infective agents. I. Dhanasekaran, Dharumadurai, ed II. Thajuddin,
Nooruddin. III. Panneerselvam, A.

RM267.A5427 2016
615.7'922--dc23   2015021364

Visit the Taylor & Francis Web site at
http://www.taylorandfrancis.com

and the CRC Press Web site at
http://www.crcpress.com

Contents

Preface

Antimicrobials are secondary metabolites found in microorganisms, plants, and animals. Herbal plants and microorganisms such as bacteria, actinobacteria, cyanobacteria, fungi, and algae attracted more attention in research that led to the discovery of natural antimicrobial compounds. The exploration of natural and synthetic antimicrobial compounds subsequently led to the development of drugs for the treatment of human microbial diseases. Microbial diseases are one of the main causes of morbidity and mortality worldwide. Today, many of such diseases are often caused by multidrug-resistant microorganisms and are very difficult to treat by using conventional antibiotics, and, consequently, lead to substantial increases in health-care costs.

Research in natural and synthetic compounds has attained significant progress with the discovery of novel compounds with antimicrobial potentials. In fact, nature is a wealthy resource of antimicrobial compounds with the potential to treat diseases in plants and animals, including tuberculosis, dermatomycosis, aspergillosis, cancer, and viral and protozoan diseases. Among the known sources of natural and synthetic antimicrobials, we have highlighted plants, sponges, and marine and terrestrial microorganisms, including bacteria and fungi. Nevertheless, there is still a vast fauna and flora that, when systematically explored, can provide additional antimicrobial leads and new drugs.

The book consists of an introductory overview of antimicrobials, which are classified into two main sections: broad spectrum and narrow spectrum antimicrobial compounds. Broad spectrum antimicrobials are further divided into five categories: antimicrobial compounds from microorganisms, animals, plants, rhizosphere microorganisms, and synthetic chemical compounds as antimicrobials. This book provides a comprehensive account of microorganisms and sponge- and plant-derived natural compounds with antibacterial, antifungal, antituberculosis, anticancer, antiviral, antidermatophytic, and antiprotozoan properties. It also discusses synthetic chemical compounds and biogenic nanoparticles. The 25 chapters have been contributed by authors around the world, including Australia, China, Korea, Kenya, the United States, India, Israel, Russia, and Turkey.

The book is considered as necessary reading for microbiologists, biotechnologists, biochemists, pharmacologists, botanists, marine biologists, and those who are doing research in natural and synthetic antimicrobial compounds. It should also be useful to MSc students, MPhil and PhD scholars, scientists, and faculty members of various science disciplines.

We are thankful to all the contributors for the submission of their valuable work. Our special thanks to Hayley Ruggieri for her encouragement in bringing out this book. We also thank Dr. Leong Li-Ming and Dr. Gagandeep Singh for their valuable suggestions during the review of the chapters. We deeply appreciate the technical support from P.M. Gopinath, S. Latha, A. Ranjani, and G. Vinothini, research scholars, Department of Microbiology, Bharathidasan University. We are also grateful to CRC Press for their concern, efforts, and encouragement in the task of publishing this book.

Dharumadurai Dhanasekaran
Nooruddin Thajuddin
Annamalai Panneerselvam

Editors

Dr. Dharumadurai Dhanasekaran, PhD, is an assistant professor, Department of Microbiology, School of Life Sciences, Bharathidasan University, Tiruchirappalli, Tamil Nadu, India. He has experience in the fields of actino-bacteriology and mycology. His current research focus is on actinobacteria, microalgae, fungi, and mushroom for animal and human health improvement. He has deposited around 54 nucleotide sequences in GenBank; published 77 research and review articles and one book, *Fungicides for Plant and Animal Diseases*; guided four PhD candidates; and organized several national-level symposia, conferences, and workshops. Dr. Dhanasekaran has received major research grants from the Department of Biotechnology (DBT), the University Grants Commission (UGC), the Indian Council of Medical Research (ICMR) and the International Foundation for Science (IFS), Sweden. He is a life member of the Mycological Society of India and the National Academy of Biological Sciences, an editorial board member of national and international journals, and a doctoral committee member and board of study member in microbiology. As per the *Indian Journal of Experimental Biology* (Vol. 51, 2013), Dr. Dhanasekaran was ranked in second position in top five institutions in the field of actinobacterial research in India.

Prof. Nooruddin Thajuddin, PhD, is the dean, Faculty of Science, Engineering and Technology, head, Department of Microbiology, School of Life Sciences, Bharathidasan University, Tiruchirappalli, Tamil Nadu, India. He has vast experience in microbial taxonomy, isolation, cultivation, harvesting, and extraction of valuable products. He is an expert in employing molecular tools in the identification and phylogeny of various microorganisms and bioremediation of effluents and bioenergy from microalgae and cyanobacteria. He underwent one-year postdoctoral training in molecular taxonomy and phylogeny of cyanobacteria in the Department of Biology, Rensselaer Polytechnic Institute, Troy, New York, through the Department of Biotechnology (Government of India) overseas fellowship. Professor Thajuddin has deposited around 500 nucleotide sequences in GenBank; published 227 research and review articles and 3 books on microbiology-related topics; developed germplasm of cyanobacteria, microalgae, bacteria, actinobacteria, and fungi in his laboratory; guided 27 PhD candidates; and organized several national-level symposia, workshops, refresher courses, and DST-INSPIRE (Innovation in Science Pursuit for Inspired Research) programs. He has received major research grants from the Department of Biotechnology (DBT), the Department of Science and Technology (DST), the University Grants Commission (UGC) and the Ministry of Earth Sciences (MoES) amounting to $44 million. He is a life member of various academic bodies and an editorial board member of national and international journals. Professor Thajuddin has visited the United States of America, the United Kingdom, Germany, Honk Kong, Malaysia, Kingdom of Saudi Arabia, Republic of Korea and Singapore to disseminate his expertise and to keep abreast of the advanced techniques in the field of microbiology. Recently, the Department of Biotechnology (Gov't of India) sanctioned a major grant to Professor Thajuddin for the establishment of national repository for freshwater microalgae and cyanobacteria. He received the 2014 Dr. G. S. Venkataraman Memorial NABS-Best Scientist Award from the National Academy of Biological Sciences.

 **Prof. Annamalai Panneerselvam, PhD,** is an associate professor and head, Department of Botany and Microbiology, A.V.V.M. Sri Pushpam College (Autonomous) Poondi, Thanjavur District, Tamil Nadu, India. He has more than 32 years of experience in teaching and research. His areas of specialization are mycology, plant pathology, and actinobacteriology. He has visited the National University of Singapore, Nanyang Technical University, TEMSEK Laboratory Singapore, Malaysian Agricultural University, Kuala Lumpur, Malaysia, Bangkok, and also national-level institutes in India. He has received the following awards: INSA visiting fellowship, New Delhi; Member of Research Board of Advisors; Excellent Faculty in Botany; and Man of Year Commemorative model. Professor Panneerselvam has deposited more than 85 nucleotide sequences in GenBank and maintains more than 500 microbial cultures in his laboratory. He is a reviewer for reputed journals and a doctoral committee member for more than 100 research scholars working in Bharathidasan University. He has published more than 300 research and review articles, 23 popular articles, and chapters in more than 12 books. He has guided 30 PhD scholars and organized several symposia, seminars, workshops, DST-INSPIRE programs for plus one students, YSSP program, etc. Professor Panneerselvam has conducted more than 100 training programs on edible mushroom cultivation technology for the welfare of women under a self-help group scheme, Dissemination of Innovative Technology (DIT). He received grants from various funding agencies like the Department of Science and Technology (DST), the University Grants Commission (UGC), the Council of Scientific and Industrial Research (CSIR), the Tamil Nadu State Council for Science and Technology (TNSCST), the National Medicinal Plants Board (NMPB) and the District Rural Development Agency (DRDA) amounting to $5 million. He is a life member of various academic bodies and an editorial board member of reputed journals.

Contributors

Israr Ahmad
Central Institute of Subtropical
  Horticulture
Indian Council of Agricultural Research
Uttar Pradesh, India

Nasim Ahmad
National Research Centre for Integrated
  Pest Management
Indian Council of Agricultural Research
New Delhi, India

Natarajan Amaresan
C.G. Bhakta Institute of Biotechnology
Uka Tarsadia University
Gujarat, India

V. Ambikapathy
Department of Botany and Microbiology
A.V.V.M. Sri Pushpam College
  (Autonomous)
Tamil Nadu, India

Someshwar Bhagat
National Research Centre for Integrated
  Pest Management
Indian Council of Agricultural Research
New Delhi, India

Atanu Bhattacharyya
Department of Biomedical Engineering
Rajiv Gandhi Institute of Technology
Karnataka, India

Ajanta Birah
National Research Centre for Integrated
  Pest Management
Indian Council of Agricultural Research
New Delhi, India

Bhaskar Biswas
Department of Chemistry
Raghunathpur College
West Bengal, India

C. Chattopadhyay
National Research Centre for Integrated
  Pest Management
Indian Council of Agricultural Research
New Delhi, India

Diana R. Cundell
College of Science, Health and
  Liberal Arts
Philadelphia University
Philadelphia, Pennsylvania

Ernest David
Department of Biotechnology
Thiruvalluvar University
Tamil Nadu, India

Dharumadurai Dhanasekaran
Department of Microbiology
Bharathidasan University
Tamil Nadu, India

Christopher M.M. Franco
School of Medicine
Flinders University
Adelaide, South Australia, Australia

P.M. Gopinath
Department of Microbiology
Bharathidasan University
Tamil Nadu, India

Li Han
College of Life and Health Science
Northeastern University
Shenyang, People's Republic of China

Xueshi Huang
College of Life and Health Science
Northeastern University
Shenyang, People's Republic of China

Avilala Janardhan
Department of Virology
Sri Venkateswara University
Andhra Pradesh, India

Velusamy Jayakumar
Division of Crop Protection
Sugarcane Breeding Institute
Tamil Nadu, India

Chenglin Jiang
Yunnan Institute of Microbiology
Yunnan University
Kunming, People's Republic of China

Yi Jiang
Yunnan Institute of Microbiology
Yunnan University
Kunming, People's Republic of China

K. Kannabiran
School of Biosciences and Technology
VIT University
Tamil Nadu, India

L. Karthik
School of Life Sciences and Biotechnology
Shanghai Jiao Tong University
Shanghai, People's Republic of China

Se-Kwon Kim
Department of Marine-Bio Convergence
    Science
Pukyong National University
Busan, Republic of Korea

Ranjith N. Kumavath
Department of Genomic Science
Central University of Kerala
Kerala, India

S. Latha
Department of Microbiology
Bharathidasan University
Tamil Nadu, India

Wen-Jun Li
Ministry of Education
and
Yunnan Institute of Microbiology
Yunnan University
Kunming, People's Republic of China

Deene Manikprabhu
Ministry of Education
and
Yunnan Institute of Microbiology
Yunnan University
Kunming, People's Republic of China

Panchanathan Manivasagan
Department of Marine-Bio Convergence
    Science
Pukyong National University
Busan, Republic of Korea

Mohammad F. Mehbub
School of Medicine
Flinders University
Adelaide, South Australia, Australia

W.M. Muiru
Department of Plant Science and Crop
    Protection
University of Nairobi
Nairobi, Kenya

Golla Narasimha
Department of Virology
Sri Venkateswara University
Andhra Pradesh, India

K. Natarajan
Department of Microbiology
Bharathidasan University
Tamil Nadu, India

M. Nithya
Department of Botany and Microbiology
A.V.V.M. Sri Pushpam College (Autonomous)
Tamil Nadu, India

Didem Deliorman Orhan
Department of Pharmacognosy
Gazi University
Ankara, Turkey

Nilüfer Orhan
Department of Pharmacognosy
Gazi University
Ankara, Turkey

Annamalai Panneerselvam
Department of Botany and Microbiology
A.V.V.M. Sri Pushpam College
   (Autonomous)
Tamil Nadu, India

Michael P. Piechoski
Arts Academy at Benjamin Rush
Philadelphia, Pennsylvania

Arthala Praveen Kumar
Department of Virology
Sri Venkateswara University
Andhra Pradesh, India

K. Rajesh
Department of Microbiology
Bharathidasan University
Tamil Nadu, India

P. Rajesh
Department of Biochemistry
K.S. Rangasamy College of Arts and
   Science
Tamil Nadu, India

V. Rajesh Kannan
Department of Microbiology
Bharathidasan University
Tamil Nadu, India

A. Ranjani
Department of Microbiology
Bharathidasan University
Tamil Nadu, India

Subhasish Saha
South China Sea Institute of Oceanology
Chinese Academy of Sciences
Guangzhou, People's Republic of China

Soumik Sarkar
Department of Microbiology and
   Biotechnology
Al-Ameen College of Arts, Science and
   Commerce
Karnataka, India

Kumar Saurav
Department of Marine Biology
University of Haifa
Haifa, Israel

Gopal Selvakumar
Department of Microbiology
Alagappa University
Tamil Nadu, India

O.P. Sharma
National Research Centre for Integrated
   Pest Management
Indian Council of Agricultural
   Research
New Delhi, India

Manoj Singh
Laboratory of Biomedical
   Nanomaterials
National University of Science and
   Technology
Moscow, Russia

Natesan Sivakumar
School of Biotechnology
Madurai Kamaraj University
Tamil Nadu, India

Nallanchakravarthula Srivathsa
C.G. Bhakta Institute of Biotechnology
Uka Tarsadia University
Gujarat, India

N. Tamilselvan
Department of Zoology
Voorhees College
Tamil Nadu, India

Nooruddin Thajuddin
Department of Microbiology
Bharathidasan University
Tamil Nadu, India

Govindaraj Vaijayanthi
Department of Microbiology
Bharathidasan University Constituent
    College, Kurumbalur, Perambalur
Tamil Nadu, India

Ramasamy Vijayakumar
Department of Microbiology
Bharathidasan University Constituent
    College, Kurumbalur, Perambalur
Tamil Nadu, India

A. Vishnu Kirthi
Department of Biotechnology
C. Abdul Hakeem College
Tamil Nadu, India

Wei Zhang
Centre for Marine Bioproducts
    Development/Medical Biotechnology
Flinders University
Adelaide, South Australia, Australia

chapter one

Antibiotics

From discovery to journey

Deene Manikprabhu and Wen-Jun Li

Contents

1.1 Introduction

An antibiotic is defined as the agent that is capable of killing or inhibiting the growth of a microorganism. The action of an antibiotic is selective in nature: it affects some organisms completely, some to a limited degree, while some not at all (Waksman, 1947).

Though antibiosis was first described by Louis Pasteur and Robert Koch (Landsberg, 1949), the use of the word "antibiosis" dates from the concept first expressed, in 1889, by Vuillemin in the following terms: "one creature destroying the life of another in order to sustain its own" (Vuillemin, 1889). The era of chemotherapy started with Paul Ehrlich who found a toxic dye molecule, a "magic bullet," that specifically binds to pathogens and destroys them without affecting human cells. He further found that the dye, trypan red, was active against the trypanosome that causes African sleeping sickness. Subsequently, Ehrlich and a young Japanese scientist, Sahachiro Hata, discovered that arsphenamine was active against syphilis (Foster and Raoult, 1974).

The story of antibiotics came into existence after the discovery of penicillin by Alexander Fleming. Fleming after returning from a holiday noticed a petri dish containing *Staphylococcus* plate culture was mistakenly left open; the plate was contaminated by blue-green mold *Penicillium*, which formed a visible growth. Fleming noticed that there was a halo of inhibited bacterial growth around the mold. Looking at the plate, Fleming concluded that the mold might release a substance that suppressed the growth or killed the bacteria. He then began his efforts to characterize what he called penicillin. He found that broth from a *Penicillium* culture contained penicillin and that the antibiotic could destroy several pathogenic bacteria (Fleming, 1929). Unfortunately, Fleming's next experiments convinced him that penicillin would not remain active in the body for a long enough time to kill the pathogens. Similarly, another professor, Howard Florey, was testing the bactericidal activity of numerous substances. After reading Fleming's paper, one of Florey's coworkers, Chain, obtained *Penicillum* culture from Fleming, and they both started to obtain pure compound from the *Penicillum*; both were successful in purifying penicillin. When the purified penicillin was injected into the mice infected with *Staphylococcus*, practically all the mice survived. Florey and Chain also had successful trials on humans. The penicillin obtained saved many lives during the world war. For their remarkable contribution, Fleming, Florey, and Chain received the Nobel Prize in 1945 for the discovery and production of penicillin (Lansing et al., 2005).

The great success of penicillin boosted Waksman to conduct experiments for the production of antibiotics. Within a few months of the discovery of penicillin, Waksman discovered streptomycin in 1944 (Schatz et al., 1944; Kingston, 2008). Mayo Clinic found that streptomycin surpassed the activity of inhibiting tuberculosis more than what penicillin did (Kiple, 1993). The journey of antibiotics began then; at present, various antibiotics are available, which are obtained through either a natural or synthetic route. In this chapter, we will discuss a broad classification of antibiotics and some major classes of antibiotics.

1.2   Classification of antibiotics

Antibiotics are classified mainly by the following bases:

1. *Mechanism of action*: An antibiotic may be bactericidal or bacteriostatic. The bactericidal antibiotics are the ones that kill the harmful microorganism, while the bacteriostatic antibiotics stop or slow down bacterial growth.

   The bacterial damage is done by either interrupting cell wall synthesis, protein synthesis, DNA synthesis, RNA synthesis, mycolic acid synthesis, or folic acid synthesis.
2. *Spectrum*: Antibiotics are either narrow or wide spectrum.
3. Route of administration of the drug.
4. *Class*: This includes aminoglycosides, ansamycins, carbacephem (discontinued), carbapenems, cephalosporins, glycopeptides, lincosamides, lipopeptides, macrolides, monobactams, nitrofurans, oxazolidinones, penicillins, polypeptides, quinolones/fluoroquinolones, sulfonamides, tetracyclines, and drugs against mycobacteria (Sunil and Kirti, 2013).

1.2.1   Aminoglycosides

The aminoglycosides are a large, diverse, and key antibiotic, most popular during the 1970s and 1980s (Bosso et al., 2013). The aminoglycoside antibiotics are either derived from *Streptomyces* spp. (streptomycin, neomycin, and tobramycin) or *Micromonospora* spp.

(gentamicin) or synthesized *in vitro* (netilmicin, amikacin, arbekacin, and isepamicin) (Durante et al., 2009). Aminoglycosides derived from bacteria of the *Streptomyces* genus are named with the suffix *mycin*, whereas those that are derived from *Micromonospora* are named with the suffix *micin* (Dewick, 2002; Kroppenstedt et al., 2005).

The first discovered aminoglycoside was streptomycin (Figure 1.1) by Waksman in 1944, which was obtained from *Streptomyces griseus* (Comroe, 1978). Streptomycin became an even more famous antibiotic during that era due to its activity that surpassed that of penicillin in inhibiting tuberculosis (Kiple, 1993). Soon after, another aminoglycoside, neomycin, was discovered by Waksman, which was obtained from *Streptomyces fradiae* in 1949. Neomycin contains two or more amino sugars connected by glycosidic bonds typically used as a topical preparation, such as neosporin. For this extraordinary contribution toward the field of physiology and medicine, Waksman was awarded with the Nobel Prize in 1951 (Schatz et al., 1944; Kingston, 2008). Similarly, kanamycin from *Streptomyces kanamyceticus*, paromomycin from *Streptomyces krestomuceticus* (Garrod et al., 1981), gentamicin from *Micromonospora purpurea* (Weinstein et al., 1963), and sisomicin (Figure 1.2) from *Micromonospora* were obtained (Weinstein et al., 1970).

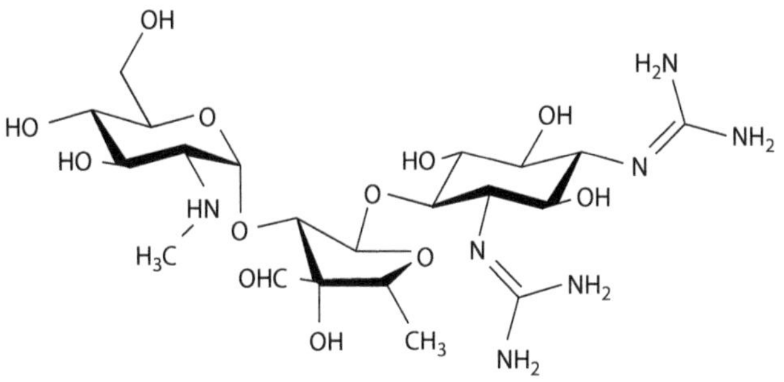

Figure 1.1 Structure of streptomycin.

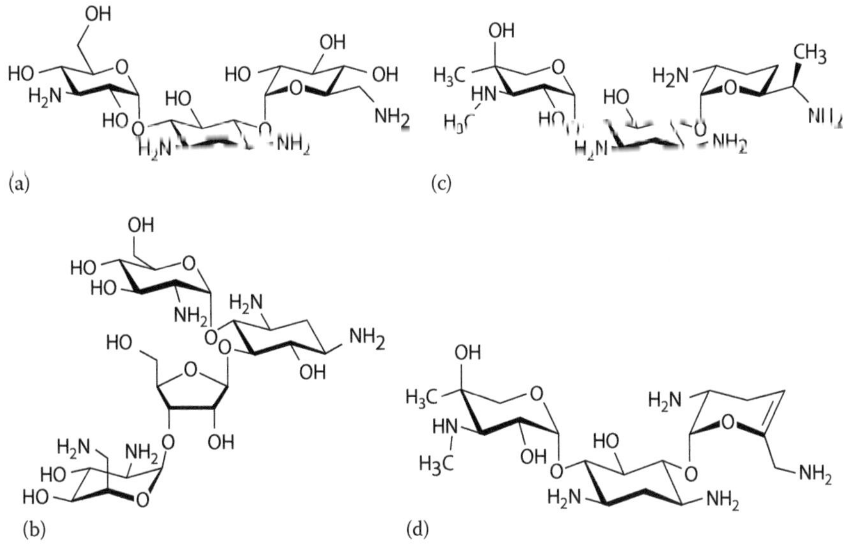

Figure 1.2 Structure of (a) kanamycin, (b) paromomycin, (c) gentamicin, and (d) sisomicin.

Structurally, aminoglycosides contain two or more amino sugars linked by glycosidic bonds to an amino cyclitol component. The cyclitol is 2-deoxystreptamine in most cases, one exception being streptomycin, which has a streptidine moiety streptomycin, which has a streptidine moiety (David, 2000).

The antimicrobial spectrum among aminoglycosides varies, for example, tobramycin is effective against the species of *Pseudomonas*, while kanamycin is effective against *Escherichia coli*, *Proteus* spp., *Serratia marcescens*, and *Klebsiella pneumoniae*. Paromomycin is used in treating intestinal infections such as cryptosporidiosis, amebiasis, and leishmaniasis (Sundar et al., 2007). Gentamicin is more active than tobramycin against *Serratia* spp., while tobramycin has superior activity against *Pseudomonas aeruginosa*. Isepamicin exhibited twofold to fourfold higher *in vitro* activities than amikacin against *Enterobacteriaceae* (Cheng et al., 2005).

Aminoglycosides exhibit several characteristics that make them useful as antimicrobial agents:

1. *They show concentration-dependent bactericidal activity*: The killing potential of aminoglycosides increases with increasing concentrations of the antibiotic (Begg et al., 1992).
2. *They show the postantibiotic effect*: They continue to kill bacteria, even after the aminoglycoside has been removed, following a short incubation with the microorganism (Hessen et al., 1989).
3. *Synergism with other antibiotics*: They show enhanced bactericidal activity in combination with antimicrobial agents. Synergism presumably arises as the result of enhanced intracellular uptake of aminoglycosides caused by the increased permeability of bacteria after incubation with cell wall synthesis inhibitors (Moellering and Weinberg, 1971).

The recent emergence of infections due to gram-negative bacterial strains with advanced patterns of antimicrobial resistance has prompted physicians to reevaluate the use of these antibacterial agents. Current evidence shows that aminoglycosides do retain activity against the majority of gram-negative clinical bacterial isolates in many parts of the world. Still, the relatively frequent occurrence of nephrotoxicity and ototoxicity during aminoglycoside treatment makes physicians reluctant to use these compounds in everyday practice. Recent advances in the understanding of the effect of various dosage schedules of aminoglycosides on toxicity have provided a partial solution to this problem, although more research still needs to be done in order to overcome this problem entirely (Maurin and Raoult, 2001).

1.2.1.1   Aminoglycoside mode of action

The mode of aminoglycoside action is through inhibition of protein synthesis via binding to the bacterial ribosome. Aminoglycosides first cross the bacterial cell walls. A bacterial ribosome complex structure usually comprises three RNA molecules and more than 50 proteins. This complex catalyzes protein synthesis with the assistance of several GTP-hydrolyzing protein factors. The bacterial ribosome consists of two subunits with relative sedimentation rates of 50S and 30S. The large subunit comprises two RNA molecules, referred to as the 5S and 23S RNAs, and 33 proteins, while the small subunit is made up of a single 16S RNA and 20–21 proteins. The ribosome has three functionally important tRNA binding sites, designated A (for aminoacyl), P (for peptidyl), and E (for exit). During protein synthesis, the ribosome decodes information stored in the mRNA and catalyzes sequential incorporation of amino acids into a growing polypeptide chain.

Aminoglycoside antibiotics bind to the 30S ribosomal subunit (some work by binding to the 50S subunit), inhibiting the translocation of the peptidyl-tRNA from the A-site to the P-site and also causing misreading of mRNA, leaving the bacterium unable to synthesize proteins vital to its growth, leading to the death of bacteria (Sergei and Shahriar, 2003).

1.2.2 Penicillin

Penicillin (Figure 1.3) was the first β-lactam antibiotic that was discovered and the most important antibiotic of this group. Although penicillin was announced over eight decades ago, it is of greater use today than it has ever been (Ozcengiz and Demain, 2013).

The parent structure of all penicillins is composed of a β-lactam ring condensed to a thiazoline ring. Connected to this common backbone structure are characteristically different amide-bonded side chains present in every member of this class (Weltzien and Padovan, 1998). Penicillins include natural penicillins, penicillinase-resistant penicillins, aminopenicillins, and beta-lactam–beta-lactamase inhibitor combinations.

Natural penicillin consists of penicillin V, while penicillinase-resistant penicillins include methicillin, oxacillin, and nafcillin (Figure 1.4). Penicillinase-resistant penicillins were developed because of the increasing resistance of *Staphylococci* to natural penicillins. Penicillinase-resistant penicillins are modified penicillins, which have a side chain that inhibits the action of penicillinase.

The aminopenicillins includes ampicillin and amoxicillin (Figure 1.5); these aminopenicillins were the first penicillins discovered to be active against gram-negative rods such as *E. coli* and *Haemophilus influenzae*. Beta-lactam–beta-lactamase inhibitor combinations include clavulanate, sulbactam, and tazobactam; this combination drug provides increased antimicrobial coverage of beta-lactamase–producing strains (Wright and Wilkowske, 1987).

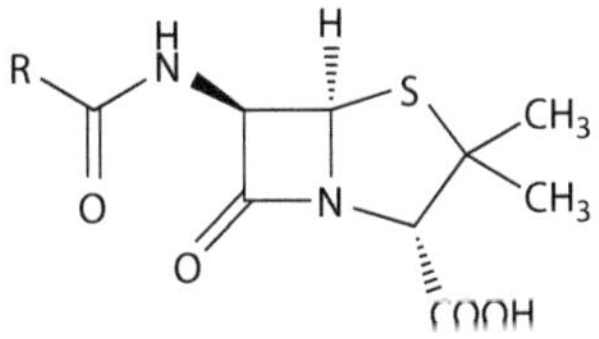

Figure 1.3 Structure of penicillin where "R" is the variable group.

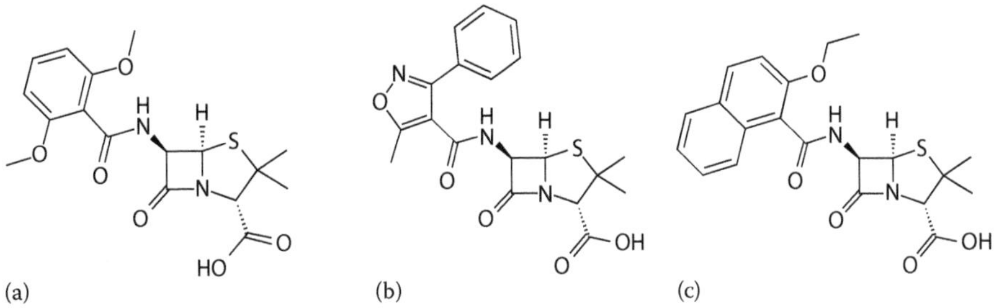

(a)　　　　　(b)　　　　　(c)

Figure 1.4 Structure of (a) methicillin, (b) oxacillin, and (c) nafcillin.

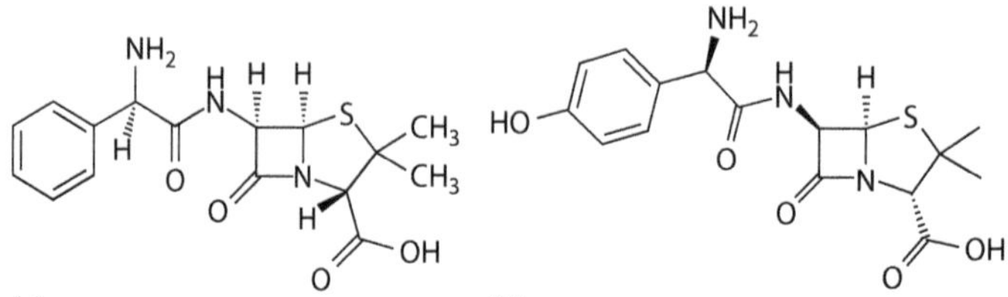

Figure 1.5 Structure of (a) ampicillin and (b) amoxicillin.

1.2.2.1   Penicillin mode of action

Penicillin kills susceptible bacteria by specifically inhibiting the transpeptidase that catalyzes the final step in cell wall biosynthesis, the cross-linking of peptidoglycan. Penicillin inhibits the formation of peptidoglycan cross-links in the bacterial cell wall that is achieved through the binding of the four-membered β-lactam ring of penicillin to the enzyme DD-transpeptidase. As a result, the DD-transpeptidase cannot catalyze formation of these cross-links and an imbalance between cell wall productions. In addition, the buildup of peptidoglycan precursors triggers the activation of bacterial cell wall hydrolases and autolysins, which further digest the bacteria's existing peptidoglycan causing the cell to die rapidly (Yocum et al., 1980).

1.2.3   Cephalosporin

The basic structure of both penicillins and cephalosporins consists of a four-member β-lactam ring. In penicillins, this ring is condensed with a five-member sulfur ring (the thiazolidine ring) and in cephalosporins with a six-member ring (the dihydrothiazine ring). Cephalosporins (Figure 1.6) can be classified according to different criteria, such as their metabolism and stability to the action of beta-lactamases, the substitution of the R_2 side chain, their pharmacokinetic properties, or their microbial properties, mainly related to their antibacterial spectrum.

Group I cephalosporins include cefadroxil, cefacetrile, and cephalexin (Figure 1.7); these antibiotics have greater activity against gram-positive bacteria. Group II cephalosporins include cefaclor, cefonicid, cefprozil, and cefuroxime (Figure 1.8). Unlike Group I cephalosporins, these antibiotics also have greater activity against gram-negative bacteria. Group III cephalosporin includes cefixapene, cefdaloxime, and cefdinir (Figure 1.9). These cephalosporins are active against *P. aeruginosa*. Group IV includes cefepime, which is active against anaerobic bacteria.

From the clinical viewpoint, the most useful classification divides cephalosporins according to their historical development plus their common microbiologic and structural characteristics. Within this classification, according to different generations of

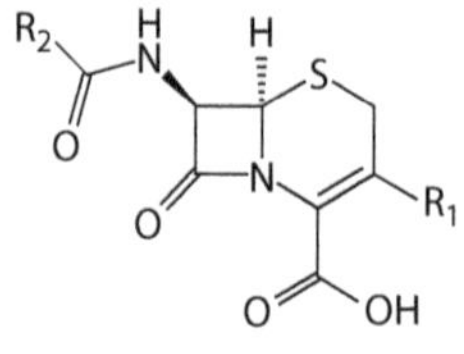

Figure 1.6 Core structure of cephalosporins.

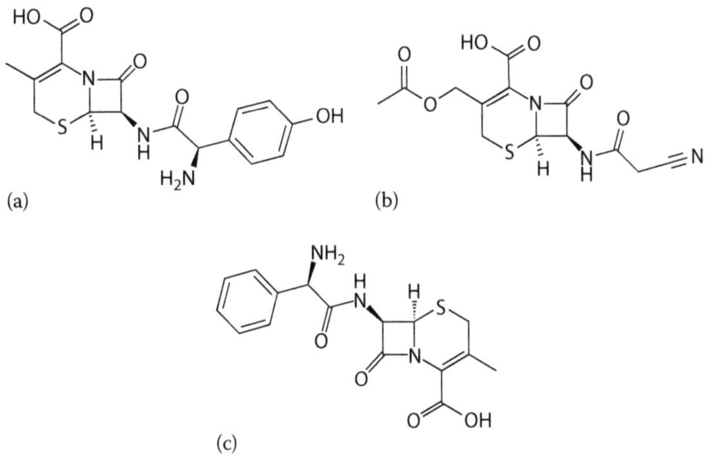

Figure 1.7 Group I cephalosporins: (a) cefadroxil, (b) cefacetrile, and (c) cephalexin.

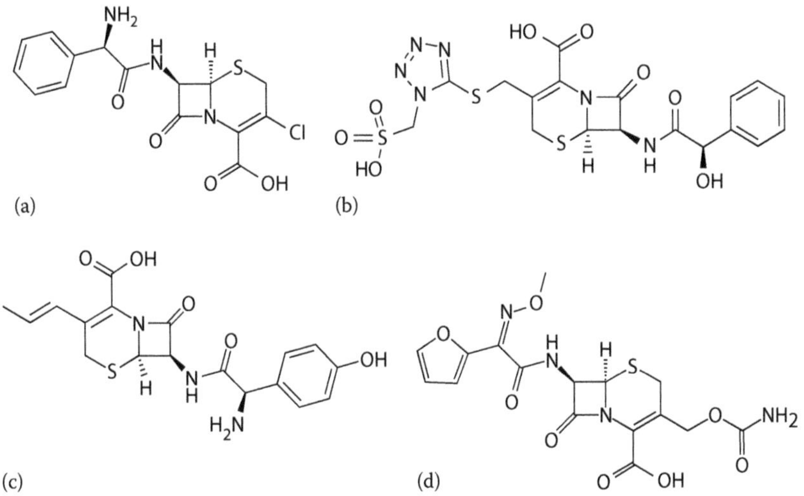

Figure 1.8 Group II cephalosporins: (a) cefaclor, (b) cefonicid, (c) cefprozil, and (d) cefuroxime.

cephalosporins, besides the well-known first-, second-, and third-generation cephalosporins, we now include the fourth generation for which newly developed molecules are continually being added (Perez Inestrosa et al., 2005).

1.2.3.1 Cephalosporin mode of action

Cephalosporins, like other β-lactams, bind to the bacterial penicillin-binding proteins (PBPs). These correspond to the D-ala-D-ala *trans-*, *carboxy-*, and *endo*peptidases responsible for catalyzing the cross-linking of newly formed peptidoglycan. Resistance arises when the PBPs and particularly the transpeptidases are modified or when they are protected by β-lactamases or *permeability barriers*. Target-mediated cephalosporin resistance can involve either the reduced affinity of existing PBPs component or the acquisition of a supplementary β-lactam-insensitive PBP (Livermore, 1987).

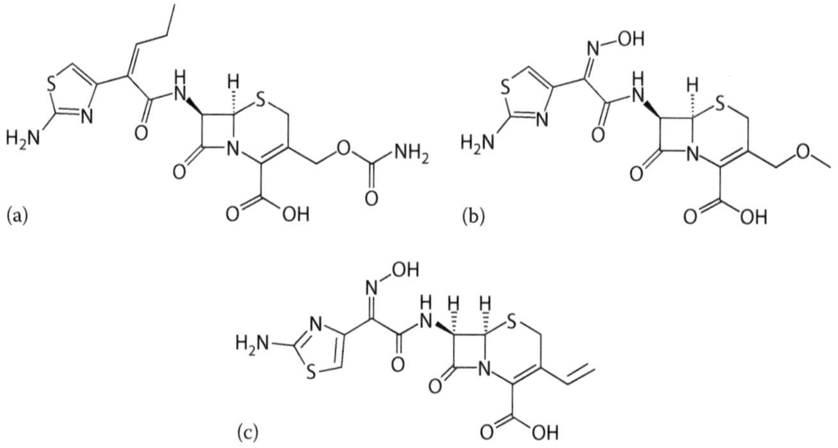

Figure 1.9 Group III cephalosporins: (a) cefcapene, (b) cefdaloxime, and (c) cefdinir.

1.2.4   Carbapenems

Carbapenems (Figure 1.10) are unique among the β-lactam antibiotics because of their extremely broad spectrum of antibacterial activity that encompasses aerobic and anaerobic gram-positive and gram-negative bacteria (Neu, 1985). They have a structure that renders them highly resistant to most β-lactamases (Livermore and Woodford, 2000).

The carbapenems differ from the penicillins in having an unsaturated bond between C2 and C3 and a carbon atom replacing sulfur at position 1 of the thiazolidine ring. Various carbapenems differ primarily in the configuration of the side chains at C2 and C6. Carbapenems include thienamycins, olivanic acids, carpetimycins, asparenomycins, pluracidomycins, and other natural and semisynthetic compounds (Moellering et al., 1989).

Thienamycin was the first carbapenem from *Streptomyces cattleya* obtained and would eventually serve as the parent or model compound for all carbapenems (Krisztina et al., 2011). However, thienamycin could not be marketed due to chemical and biological instabilities. Many carbapenem compounds have long been considered unstable even in neutral conditions, much less in gastric juice and/or intestinal juice. To overcome these problems, many groups started to develop stable compounds and ways of synthesizing thienamycin. Under these circumstances, imipenem (2) containing cilastatin as an inhibitor of renal dehydropeptidase I was marketed by Merck & Co., Inc. Next, the Sankyo group marketed panipenem containing betamipron for the alleviation of nephrotoxicity (Kumagai et al., 2002).

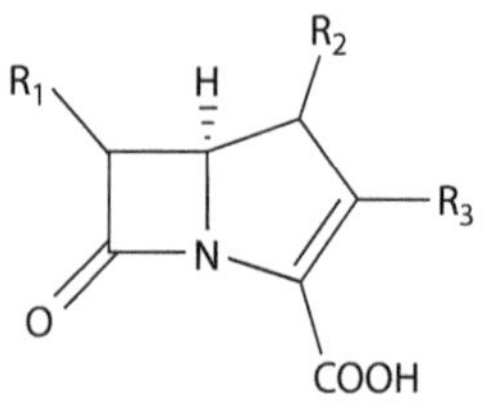

Figure 1.10 Structure of the carbapenem backbone.

1.2.4.1 Carbapenem mode of action

Carbapenems enter gram-negative bacteria through outer membrane proteins, also known as porins. After transversing into the periplasmic space, carbapenems permanently acylate the PBPs. PBPs are enzymes (i.e., transglycosylases, transpeptidases, and carboxypeptidases) that catalyze the formation of peptidoglycan in the cell wall of bacteria. Current insights into this process suggest that the glycan backbone forms a right-handed helix with a periodicity of three per turn of the helix. Carbapenems act as mechanism-based inhibitors of the peptidase domain of PBPs and can inhibit peptide cross-linking as well as other peptidase reactions. A key factor of the efficacy of carbapenems is their ability to bind to multiple different PBPs. Since cell wall formation is a dynamic *3D* process with the formation and autolysis occurring at the same time, when PBPs are inhibited, autolysis continues. Eventually, the peptidoglycan weakens, and the cell bursts due to osmotic pressure (Krisztina et al., 2011).

1.2.5 Glycopeptides

Glycopeptide antibiotics represent a very important group of natural products with regard to medicinal applications as antibacterial and anticancer agents. Glycopeptides include vancomycin, teicoplanin, ramoplanin, bleomycin, mannopeptimycin, and salmochelin (Falko et al., 2007). The first glycopeptide discovered was vancomycin (Figure 1.11) isolated from a strain of *Amycolatopsis orientalis* found in a soil sample collected in Borneo. Similarly, teicoplanin A2 (lipoglycopeptide antibiotic) obtained from *Actinoplanes teichomyceticus* was also introduced for clinical use in the mid-1980s (Grace et al., 2014).

Structurally, glycopeptides are complex molecules of unique structure that has a central, relatively conserved heptapeptide domain in which five of the seven amino acid residues are common to all glycopeptides but differ in the amino acids at positions 1 and 3 and in the substituents of the aromatic amino acid residues. Some of the carbons of the

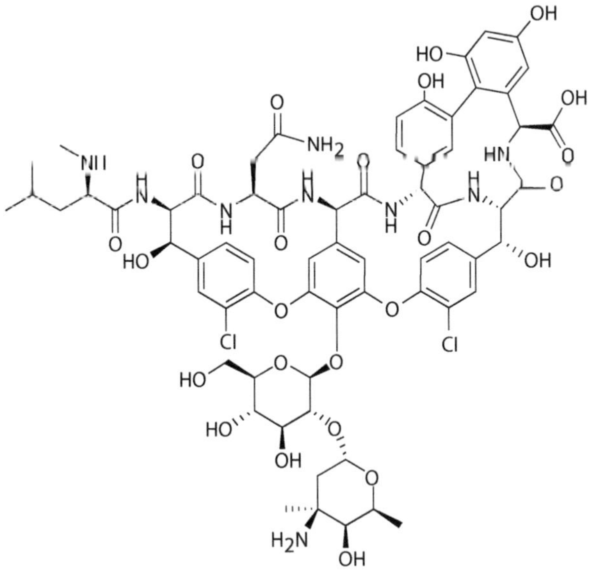

Figure 1.11 Structure of vancomycin.

aromatic residue carry chlorine, hydroxyl, or methyl groups, and some of the hydroxyl groups are substituted with sugar or amino sugar. The presence of phenolic residue permits the formation of two- and three-ring structures in all glycopeptides. Such interaction results in a large group of molecule with very similar rigid 3D structures. Those compounds that have rings 1 and 3 joined in addition to 2, 4 and 6, and 5 and 7 have been shown to adopt a bracelet-like configuration with a substantial cleft in the molecule into which the bacterial site binds with exquisite position. Some glycopeptides, like teicoplanins, have the amino group of amino sugar substituted with a fatty acid chain containing 9–11 carbon atoms (Reynolds, 1989).

1.2.5.1   Glycopeptide mode of action

Biochemical studies indicate that glycopeptides inhibit the late stages of peptidoglycan synthesis (Reynolds, 1989). During cell wall synthesis, muramylpentapeptide forms in the cell cytoplasm, and an N-acetylglucosamine unit is added, together with any amino acids needed for the interpeptide bridge of gram-positive organisms. It is then passed to a lipid carrier molecule, which transfers the whole unit across the cell membrane to be added to the growing end of the peptidoglycan macromolecule. The addition of the new building block is prevented by the glycopeptides vancomycin and teicoplanin, which bind to the acyl-D-alanyl-D-alanine tail of the muramylpentapeptide (Finch et al., 2011). For the rigid polymer that protects bacterial cells from osmotic lysis, by binding to this terminal dipeptide, glycopeptide antibiotics interfere with the proper cell wall formation, which results in eventual death of the cell (Marshall et al., 1998).

1.2.6   Lincosamides

Lincosamides are a class of antibiotics that include clindamycin, lincomycin, and pirlimycin (Figure 1.12). Among them, lincomycin was the first antibiotic to be produced by *Streptomyces lincolnensis* isolated from a soil sample from Lincoln, NE (Chang et al., 1966). Lincomycin proved to be a member of a new class of antibiotics characterized by an alkyl 6-amino-6,8-dideoxy-1-thio-D-erythro-α-D-galacto-octopyranoside joined with a proline moiety by an amide linkage.

Lincomycin has gained clinical acceptance as a major antibiotic for the treatment of diseases caused by gram-positive microbes including pathogenic *Streptococci*, *Staphylococci*, and *Mycoplasma* (Spizek and Rezanka, 2004). While clindamycin is used to treat infections with anaerobic bacteria, it can also be used to treat some protozoal diseases, such as malaria. It is a common topical treatment for acne and can be useful against some

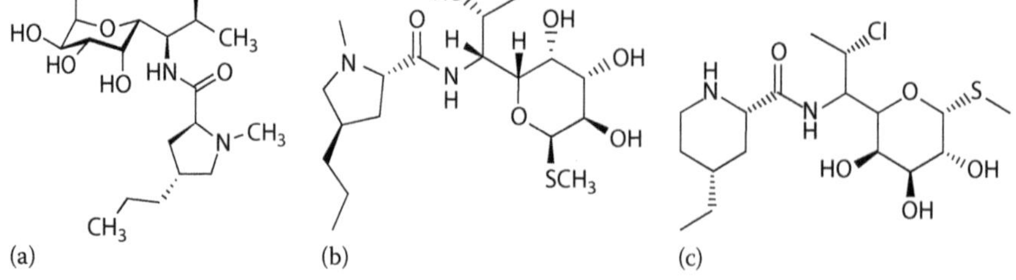

Figure 1.12 Structure of (a) clindamycin, (b) lincomycin, and (c) pirlimycin.

methicillin-resistant *Staphylococcus aureus* infections (Daum, 2007). Pirlimycin is active against gram-positive bacteria, specifically *Staphylococcus aureus* and coagulase negative species of *Staphylococcus* and *Streptococcus*.

1.2.6.1 Lincosamide mode of action

Lincosamides prevent bacteria replication by interfering with the synthesis of proteins. They bind to the 23S portion of the 50S subunit of bacterial ribosomes and cause premature dissociation of the peptidyl-tRNA from the ribosome (Tanel et al., 2003).

1.2.7 Macrolides

Macrolides, which are primarily antibiotics, belong to the polyketide group of natural products. They derive their name from their characteristic structural features, a macrocyclic lactone ring to which various deoxy sugars, most commonly cladinose and desosamine, are attached (Steel et al., 2012). Macrolides are composed of a 12- to 16-member macrolactone ring decorated with various amino sugars that includes azithromycin, clarithromycin, dirithromycin, erythromycin, flurithromycin, josamycin, midecamycin, miocamycin, oleandomycin, rokitamycin, roxithromycin, spiramycin, troleandomycin, tylosin, ketolides, telithromycin, cethromycin, and solithromycin (Alexander, 2008).

Erythromycin was the first macrolide antibiotic discovered from *Streptomyces erythreus* isolated from soil samples (Zuckerman, 2004) and used to treat bacteria responsible for causing infections of the skin and upper respiratory tract, including *Streptococcus*, *Staphylococcus*, and *Haemophilus* genera (Roland, 2002). Similarly, clarithromycin was invented by researchers at the Japanese drug company Taisho Pharmaceutical in the 1970s (Zuckerman, 2004). Animal studies have shown that clarithromycin can induce fetal loss in rabbits and monkeys when used in very low dosages and high dosages, respectively. One observational study concerning pregnant women showed a doubling of the number of miscarriages in women exposed to clarithromycin in early pregnancy compared to a match control group. There is limited knowledge concerning the risk of congenital malformations among women exposed to clarithromycin during pregnancy (Andersen et al., 2013).

Azithromycin is another macrolide antibiotic derived from erythromycin, which was discovered by a team of researchers at the Croatian pharmaceutical company PLIVA (1980). This product has a high bacteriostatic action in front of a wide spectrum of pathogenic bacteria and is used mainly for the treatment of respiratory and dermatological infections (Banic, 2011; Timoumi et al., 2014).

Bacteria resist macrolide antibiotics in three ways: (1) through target-site modification by methylation or mutation that prevents the binding of the antibiotic to its ribosomal target, (2) through efflux of the antibiotic, and (3) by drug inactivation. These mechanisms have been found in the macrolide producers, which often combine several approaches to protect themselves against the antimicrobial that they produce. In pathogenic microorganisms, the impact of the three mechanisms is unequal in terms of incidence and of clinical implications (Roland, 2002).

1.2.7.1 Macrolide mode of action

Macrolides reversibly bind to domain V of 23S ribosomal RNA of the 50S subunit of the bacterial ribosome inhibiting RNA-dependent protein synthesis (Douthwaite and Champney, 2001).

Acknowledgments

This research was supported by the Key Project of International Cooperation of Ministry of Science and Technology (No. 2013DFA31980) and the Key Project of Yunnan Provincial Natural Science Foundation (2013FA004). W.-J. Li was also supported by the *Hundred Talents Program* of the Chinese Academy of Sciences and the Pearl River Scholar Funded Scheme of the Higher Vocational Colleges & Schools of Guangdong Province (2014).

References

Alexander, S.M. 2008. Macrolide myths. *Curr. Opin. Microbiol.*, 11, 414–421.

Andersen, J.T., Petersen, M., Jimenez-Solem, E., Broedbaek, K., Andersen, N.L., Torp-Pedersen, C., Keiding, N., and Poulsen, H.E. 2013. Clarithromycin in early pregnancy and the risk of miscarriage and malformation: A register based nationwide cohort study. *PLoS ONE*, 8, e53327, 1–6.

Banic, T.Z. 2011. The story of azithromycin. *Kemija u Industriji*, 60, 603–617.

Begg, E.J., Peddie, B.A., Chambers S.T., and Boswell, D.R. 1992. Comparison of gentamicin dosing regimens using an *in-vitro* model. *J. Antimicrob. Chemother.*, 29, 427–433.

Bosso, J.A., Haines, L.H., and Gomez, J. 2013. Stable susceptibility to aminoglycosides in an age of low level, institutional use. *Infect. Dis. Ther.*, 2, 209–215.

Chang, F.N., Sih, C.J., and Weisblum, B. 1966. Lincomycin, an inhibitor of aminoacyl sRNA binding to ribosomes. *Proc. Natl. Acad. Sci. USA*, 55, 431–438.

Cheng, N.C., Hsueh, P.R., Liu, Y.C., Shyr, J.M., Huang, W.K., Teng, L.J., and Liu, C.Y. 2005. In vitro activities of tigecycline, ertapenem, isepamicin, and other antimicrobial agents against clinically isolated organisms in Taiwan. *Microb. Drug Resist.*, 11, 330–341.

Comroe J.H. Jr. 1978. Pay dirt: The story of streptomycin. Part I: From Waksman to Waksman. *Am. Rev. Respir. Dis.*, 117, 773–781.

Daum, R.S. 2007. Clinical practice, skin and soft-tissue infections caused by methicillin-resistant *Staphylococcus aureus*. *N. Engl. J. Med.*, 357, 380–390.

David, A.S. 2000. Current methodologies for the analysis of aminoglycosides. *J. Chromatogr. B*, 747, 69–93.

Dewick, P.M. 2002. *Medicinal Natural Products: A Biosynthetic Approach*. England, U.K.: John Wiley & Sons.

Douthwaite, S. and Champney, W.S. 2001. Structures of ketolides and macrolides determine their mode of interaction with the ribosomal target site. *J. Antimicrob. Chemother.*, 48, Topic T1, 1–8.

Durante, M.E., Grammatikos, A., Utili, R., and Falagas, M.E. 2009. Do we still need the aminoglycosides? *Int. J. Antimicrob. Agents.*, 33, 201–205.

Falko, W, Sebastian, S., and Roderich, D.S. 2007. Synopsis of structural, biosynthetic, and chemical aspects of glycopeptide antibiotics. *Top. Curr. Chem.*, 267, 143–185.

Finch, R.G., Greenwood, D., Whitley, R.J., and Norrby, S.R. 2011. *Antibiotic and Chemotherapy: Anti-Infective Agents and Their Use in Therapy*, 9th edn. Edinburgh, U.K.: Churchill-Livingstone, Elsevier Health Sciences.

Fleming, A. 1929. On the antibacterial action of cultures of a *Penicillium*, with special reference to their use in the isolation of *B. influenza*. *Br. J. Exp. Pathol.*, 3, 226–236.

Foster, W. and Raoult, A. 1974. Early descriptions of antibiosis. *J. Roy. Coll. Gen. Pract.*, 24, 889–894.

Garrod, L.P., Lambert, H.P., and O'Grady, F. 1981. *Antibiotic and Chemotherapy*, 5th edn. Edinburgh, U.K.: Churchill-Livingstone, Elsevier Health Sciences.

Grace, Y., Maulik, N.T., Kalinka, K., and Gerard, W. 2014. Glycopeptide antibiotic biosynthesis. *J. Antibiot.*, 67, 31–41.

Hessen, M.T., Pitsakis P.G., and Levison, M.E. 1989. Postantibiotic effect of penicillin plus gentamicin versus *Enterococcus faecalis* in vitro and *in vivo*. *Antimicrob. Agents. Chemother.*, 33, 608–611.

Kingston, W. 2008. Irish contributions to the origins of antibiotics. *Ir. J. Med Sci.*, 177, 87–92.

Kiple, K.F. 1993. *Cambridge World History of Human Disease*. Cambridge, U.K.: Cambridge University Press.

Krisztina, M.P.W., Andrea, E., Magdalena, A.T., and Robert, A.B. 2011. Carbapenems: Past, present, and future. *Antimicrob. Agents. Chemother.*, 55(11), 4943–4960.

Kroppenstedt, R.M., Mayilraj, S., and Wink, J.M. 2005. Eight new species of the genus *Micromonospora*, *Micromonospora citrea* sp. nov., *Micromonospora echinaurantiaca* sp. nov., *Micromonospora echinofusca* sp. nov. *Micromonospora fulviviridis* sp. nov., *Micromonospora inyonensis* sp. nov., *Micromonospora peucetia* sp. nov., *Micromonospora sagamiensis* sp. nov., and *Micromonospora viridifaciens* sp. nov. *Syst. Appl. Microbiol.*, 28, 328–339.

Kumagai, T., Tamai, S., Abe, T., and Hikda, M. 2002. Current status of oral carbapenem development. *Curr. Med. Chem. Anti-Infect. Agents*, 1, 1–14.

Landsberg, H. 1949. Prelude to the discovery of penicillin. *Isis*, 40, 225–257.

Lansing, M.P., John, P.H., and Donald, A. K. 2005. *Microbiology*. New York: McGraw-Hill.

Livermore, D.M. 1987. Mechanisms of resistance to cephalosporin antibiotics. *Drugs*, 34(Suppl. 2), 64–88.

Livermore, D.M. and Woodford, N. 2000. Carbapenemases: A problem in waiting? *Curr. Opin. Microbiol.*, 3, 489–495.

Marshall, C.G., Lessard, I.A., Park, I., and Wright, G.D. 1998. Glycopeptide antibiotic resistance genes in glycopeptide-producing organisms. *Antimicrob. Agents. Chemother.* 42, 2215–2220.

Maurin, M. and Raoult, D. 2001. Use of aminoglycosides in treatment of infections due to intracellular bacteria. *Antimicrob. Agents. Chemother.*, 45, 2977–2986.

Moellering, R.C. Jr., Eliopoulos, G.M., and Sentochnik, D.E. 1989. The carbapenems: New broad spectrum beta-lactam antibiotics. *J. Antimicrob. Chemother.*, 24(Suppl. A), 1–7.

Moellering, R.C. Jr. and Weinberg, A.N. 1971. Studies on antibiotic synergism against enterococci. II. Effect of various antibiotics on the uptake of ^{14}C-labeled streptomycin by enterococci. *J. Clin. Investig.*, 50, 2580–2584.

Neu, H.C. 1985. Carbapenems: Special properties contributing to their activity. *Am. J. Med.*, 78, 33–40.

Ozcengiz, G. and Demain, A.L. 2013. Recent advances in the biosynthesis of penicillins, cephalosporins and clavams and its regulation. *Biotechnol. Adv.*, 31, 287–311.

Perez Inestrosa, E., Suau, R,. Montanez, M.I., Rodriguez, R., Mayorga, C., Torres, M.J., and Blanca, M. 2005. Cephalosporin chemical reactivity and its immunological implications. *Curr. Opin. Allergy Clin. Immunol.*, 5, 323–330.

Reynolds, P.E. 1989. Structure, biochemistry and mechanism of action of glycopeptides antibiotics. *Eur. J. Clin. Microbiol. Infect. Dis.*, 8, 943–950.

Roland, L. 2002. Mechanisms of resistance to macrolides and lincosamides: Nature of the resistance elements and their clinical implications. *Clin. Infect. Dis.*, 34(4), 482–492.

Schatz, A., Bugie, E., and Waksman, S.E. 1944. Streptomycin, a substance exhibiting antibiotic activity against gram-positive and gram negative bacteria. *Proc. Soc. Exp. Biol. Med.*, 55, 65–69.

Sergei, B.V. and Shahriar, M. 2003. Versatility of aminoglycosides and prospects for their future. *Clin. Microbiol. Rev.*, 16, 430–450.

Spizek, J. and Rezanka, T. 2004. Lincomycin, cultivation of producing strains and biosynthesis, *Appl. Microbiol. Biotechnol.*, 63, 510–519.

Steel, H.C., Theron, A.J., Cockeran, R., Anderson, R., and Feldman, C., 2012. Pathogen- and host-directed anti-inflammatory activities of macrolide antibiotics. *Mediat. Inflamm.*, Article ID 584262.

Sundar, S., Jha, T.K., Thakur, C.P., Sinha, P.K., and Bhattacharya, S.K. 2007. Injectable paromomycin for visceral leishmaniasis in India. *N. Engl. J. Med.*, 356, 2571–2581.

Sunil, T.P. and Kirti, M.J. 2013. *Antibiotics—A Manifold Cover Story*, 1st edn. Toronto, Ontario, Canada: Canadian Academic Publishing.

Tanel, T., Martin, L., and Mans, E. 2003. The mechanism of action of macrolides, lincosamides and streptogramin B reveals the nascent peptide exit path in the ribosome. *J. Mol. Biol.*, 330, 1005–1014.

Timoumi, S., Mangin, D., Peczalski, R., Zagrouba, F., and Andrieu, J. 2014. Stability and thermophysical properties of azithromycin dehydrate. *Arabian J. Chem.*, 7(2), 189–195.

Vuillemin, P. 1889. Antibiose et symbiose, Assoc. franc. pour l'Avanc. des Sciences. Paris 2: 525–542.

Waksman, S.A. 1947. What is an antibiotic or an antibiotic substance? *Mycologia*, 39, 565–569.

Weltzien, H.U. and Padovan, E. 1998. Molecular features of penicillin allergy. *J. Invest. Dermatol.*, 110, 203–206.

Weinstein, M.J., Luedemann, G.M., Oden, E.M., Wagman, G.H., Rosselet, J.P., Marquez, J.A., and Black, J. 1963. Gentamicin, a new antimicrobial complex from *Micromonospora*. *J. Med. Chem.*, 6, 463–464.

Weinstein, M.J., Marquez, J.A., Testa, R.T., Wagman, G.H., Oden, E.M., and Waitz, J.A. 1970. Antibiotic 6640, a new *Micromonospora*-produced aminoglycoside antibiotic. *J. Antibiot.*, 23, 551–554.

Wright, A.J. and Wilkowske, C.J. 1987. The penicillins. *Mayo. Clin. Proc.*, 62, 806–820.

Yocum, R.R., Rasmussen, J.R., and Strominger, J.L. 1980. The mechanism of action of penicillin, penicillin acylates the active site of *Bacillus stearothermophilus* D-alanine carboxypeptidase. *J. Biol. Chem.*, 255, 3977–3986.

Zuckerman, J.M. 2004. Macrolides and ketolides: Azithromycin, clarithromycin, telithromycin. *Infect. Dis. Clin. N. Am.*, 18, 621–649.

section one

Broad spectrum antimicrobial compounds from microorganisms

chapter two

Antimicrobial potential of marine actinobacteria
A review

Panchanathan Manivasagan and Se-Kwon Kim

Contents

2.1 Introduction

Marine actinobacteria are the most economically and biotechnologically valuable prokaryotes. They are able to produce a wide range of bioactive secondary metabolites such as antibiotics, antitumor agents, immunosuppressive agents, and enzymes. These metabolites are known to possess antibacterial, antifungal, neuritogenic, anticancer, antialgal, antimalarial, and anti-inflammatory activities (Ravikumar et al., 2011). Actinobacteria have the capacity to synthesize many different biologically active secondary metabolites such as cosmetics, vitamins, nutritional materials, herbicides, antibiotics, pesticides, antiparasitic agents, and enzymes used in waste treatment like cellulase and xylanase (Ogunmwonyi et al., 2010). They are free-living, saprophytic bacteria and a major source for the production of antibiotics (Atta et al., 2009).

    Natural products are a boundless source for important novel compounds having antagonistic activity against pathogenic organisms. The marine environment covers almost 70% of the earth surface (Valli et al., 2012). Organisms present in these environments are extremely rich sources of bioactive compounds (Solanki et al., 2008; Hong et al., 2009). The ocean remains an unexploited source for many drugs and pharmacologically active substances (Sivasubramanian et al., 2011). Actinobacteria are gram-positive, filamentous bacteria that are supreme secondary metabolite producers (Olano et al., 2009). Among the members of the Actinomycetes genus, *Streptomyces* sp. are a dynamic producer of functional, bioeffective metabolites with a broad pharmaceutical range having antimicrobial, antihelminthic, antitumor, and antiviral properties (Ravikumar et al., 2011; Reddy et al., 2011).

Multidrug resistance in microorganisms is an emerging serious problem in the health-care sector. The improper usage of antibiotics plays a major role in drug resistance in pathogenic microbes. Microorganisms acquire resistance toward common antibiotics by altering their metabolism and genetic structure (Raghunath, 2008; Rao, 2012). There is an incessant need to find novel efficient drug molecules against multidrug-resistant microbes. Actinobacteria from terrestrial origin produce hundreds of antibiotics that are widely used at present. Some differences could be expected among organisms existing in marine and terrestrial environments due to variation in physical, chemical, and biological factors (Saurav and Kannabiran, 2010). It is apparent that the marine environment is a potent source for finding new actinobacteria and new antibiotics or biologically active substances (Gärtner et al., 2011). In this chapter, we focus on the antimicrobial potential of marine actinobacteria and classify them in terms of their unique chemical structures (Table 2.1), covering the literature to date.

Table 2.1 Antimicrobial compounds produced by marine actinobacteria

Compound	Species	Other biological activity	References
Antibacterial activity			
Abyssomicins	*Verrucosispora* sp.	—	Riedlinger et al. (2004)
Bonactin	*Streptomyces* sp.	Antifungal	Schumacher et al. (2003)
Chloro-dihydroquinones	*Streptomyces* sp.	Anticancer	Soria-Mercado et al. (2005)
Diazepinomicin	*Micromonospora* sp.	Anticancer; anti-inflammatory	Charan et al. (2004)
Frigocyclinone	*Streptomyces griseus*	—	Bruntner et al. (2005)
Essramycin	*Streptomyces* sp.	—	El-Gendy et al. (2008)
Lynamicins	*Marinispora* sp.	—	McArthur et al. (2008)
Marinopyrroles	*Streptomyces* sp.	Cytotoxic	Hughes et al. (2008)
Caboxamycin	*Streptomyces* sp.	Cytotoxic	Hohmann et al. (2009)
Himalomycins	*Streptomyces* sp.	—	Maskey et al. (2003b)
Marinomycins	*Marinispora*	Antifungal; anticancer	Kwon et al. (2006)
Glyciapyrroles	*Streptomyces* sp.	—	Macherla et al. (2005)
Tirandamycin	*Streptomyces* sp.	—	Carlson et al. (2009)
Bisanthraquinone	*Streptomyces* sp.	—	Socha et al. (2006)
Gutingimycin	*Streptomyces* sp.	—	Maskey et al. (2004)
Helquinoline	*Janibacter limosus*	—	Asolkar et al. (2004)
Lajollamycin	*Streptomyces nodosus*	—	Manam et al. (2005)
TP-1161	*Nocardiopsis* sp.	—	Engelhardt et al. (2010)
Lincomycin	*Streptomyces lincolnensis*	—	Peschke et al. (2006)
Tirandamycins	*Streptomyces* sp.	—	Carlson et al. (2009)
1,4-Dihydroxy-2-(3-hydroxybutyl)-9,10-anthraquinone 9,10-anthrac	*Streptomyces* sp.	—	Ravikumar et al. (2012)
Antifungal activity			
Chandrananimycin	*Actinomadura* sp.	Antialgal; antibacterial; anticancer	Maskey et al. (2003b)
N-(2-hydroxyphenyl)-2-phenazinamine (NHP)	*Nocardia dassonvillei*	Anticancer	Gao et al. (2012)

2.2   Marine microorganisms

Marine microorganisms are increasingly becoming a potential source in the search for industrially important biomolecules. Today both academic and industrial interests in marine microorganisms are on the rise because unique and bioactive metabolites have been reported from marine organisms (Jensen and Fenical, 1994; Imada, 2004; Zhang et al., 2005; Manivasagan et al., 2013, 2014b). Actinobacteria are present in various ecological habitats such as soil, freshwater, backwater, lake, compost, sewage, and marine environments (Goodfellow and Williams, 1983; Manivasaganet al., 2014c). They are considered highly valuable as they produce various antibiotics and other therapeutically useful compounds with diverse biological activities. The vast majority of these metabolites (70%) have been isolated from actinobacteria with the remaining 20% from fungi, 7% from *Bacillus* sp., and 1%–2% from *Pseudomonas* sp. Hence, it is known that the actinobacteria are perhaps the most important group of organisms studied extensively for the discovery of drugs and other bioactive metabolites program (Prabavathy et al., 2006).

The marine environment contains a wide range of distinct microorganisms that are not present in the terrestrial environment. Though some reports are available on antibiotic and enzyme production by marine actinobacteria, the marine environment is still a potential source for new actinobacteria, which can yield novel bioactive compounds and industrially important enzymes (Sharma and Pant, 2001). Since the late 1980s, the number of novel compounds isolated from terrestrial microorganisms has steadily decreased. To cope up with the demand for new pharmaceutical compounds and to combat the antibiotic-resistant pathogens, researchers have been forced to look for novel microorganisms in unusual environments (Ramesh and Mathivanan, 2009).

2.3   Antimicrobial activity

Marine actinobacteria are important sources of new bioactive compounds such as antibiotics and enzymes (Bredholt et al., 2008; Aghamirian and Ghiasian, 2009; Manivasagan et al., 2014a), which have diverse clinical effects and are active against many pathogenic organisms. Actinomycetes and their bioactive compound show antibacterial and antifungal effects against various multidrug-resistant pathogens such as vancomycin-resistant *Enterococci*, methicillin-resistant *Staphylococcus aureus*, *Shigella dysenteriae*, *Klebsiella* sp., and *Pseudomonas aeruginosa* (Saadoun et al., 1998; Bhatnagar and Kim, 2010). The need for new, safe, and effective antimicrobial agent is the major challenge to the pharmaceutical industry today, especially with the obvious increase in opportunistic infections in the immune-compromised host via and multidrug-resistant strains (Bredholt et al., 2008).

2.3.1   Antibacterial activity

An antibacterial substance is an agent that inhibits bacterial growth or kills bacteria. Infectious diseases remain one of the main causes of death due to antibiotic-resistant microorganisms. The frequency of resistance in microbial pathogens continues to grow at an alarming rate throughout the world (Schmitz et al., 1999). Decreased efficacy and resistance of pathogens to antibiotics have necessitated the development of new alternatives (Ravikumar et al., 2010). To overcome this problem, development of effective newer drugs without any side effects is an urgent need.

Caboxamycin (**1**) (Figure 2.1) is a new benzoxazole antibiotic and was detected by HPLC-diode array screening in extracts of *Streptomyces* sp. NTK 937, another strain that was

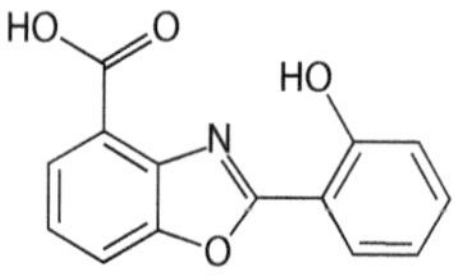

1. Caboxamycin

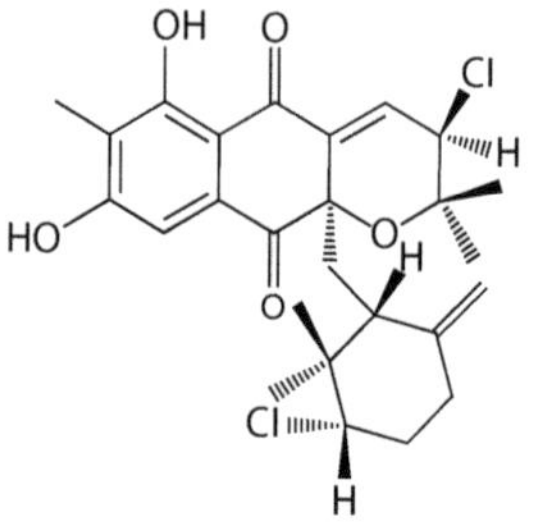

2. Chlorinated dihydroquinones

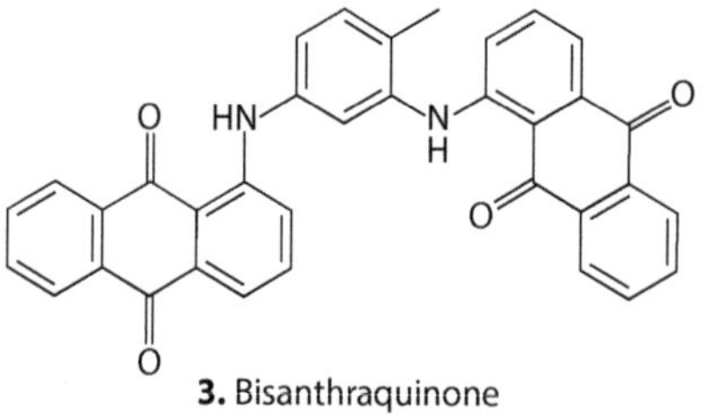

3. Bisanthraquinone

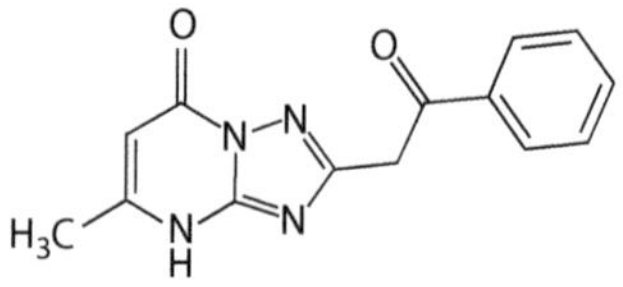

4. Essramycin

Figure 2.1 Chemical structures of caboxamycin, chlorinated dihydroquinones, bisanthraquinone, and essramycin.

isolated from the sediments collected from the Canary Basin. The compound was named after the first letters of the collection site from where the organism was isolated and from the letters drawn from its chemical structure. Caboxamycin showed inhibitory activity against both gram-positive bacteria and the tumor cell lines gastric adenocarcinoma (AGS), hepatocellular carcinoma (Hep G2), and breast carcinoma cells (MCF7). The antibiotic also showed an inhibitory activity against the enzyme phosphodiesterase (Hohmann et al., 2009).

Chlorinated dihydroquinones (**2**) (Figure 2.1) are novel antibiotics produced by a new marine *Streptomyces* sp. (Soria-Mercado et al., 2005). The compounds formally possess new carbon skeletons but are related to several previously reported metabolites of the napyradiomycin class. Structures of the new molecules possess significant antibacterial and cancer cell cytotoxicities. In general, marine actinobacteria are extensively studied for antibacterial activity. Bisanthraquinone (**3**) (Figure 2.1) is a new antibiotic isolated from *Streptomyces* sp. Biological activities were measured against clinically derived isolates of vancomycin-resistant *Enterococcus faecium* (VRE) and methicillin-susceptible, methicillin-resistant, and tetracycline-resistant *Staphylococcus aureus* (MSSA, MRSA, and TRSA, respectively). The most potent antibiotic displayed MIC_{50} values of 0.11, 0.23, and 0.90 μM against a panel ($n = 25$ each) of clinical MSSA, MRSA, and VRE, respectively, was determined to be bactericidal by time–kill analysis (Socha et al., 2006). Essramycin (**4**) (Figure 2.1) is a novel triazolopyrimidine antibiotic isolated from *Streptomyces* sp. The compound is antibacterially active with an minimum inhibitory concentration (MIC) of 2–8 μg/mL against gram-positive and gram-negative bacteria (El-Gendy et al., 2008).

Diazepinomicin (**5**) (Figure 2.2) is a unique farnesylated dibenzodiazepinone produced by a *Micromonospora* strain (Charan et al., 2004). It possesses antibacterial, anti-inflammatory, and antitumor activities. It has a broad spectrum of *in vitro* cytotoxicity and has demonstrated *in vivo* activity against glioma, breast cancer, and prostate cancer in mouse models. Abyssomicin C (**6**) (Figure 2.2) is a novel polycyclic polyketide antibiotic produced by a marine *Verrucosispora* strain (Riedlinger et al., 2004). It is a potent inhibitor of *para*-aminobenzoic acid biosynthesis and, therefore, inhibits the folic acid biosynthesis at an earlier stage than the

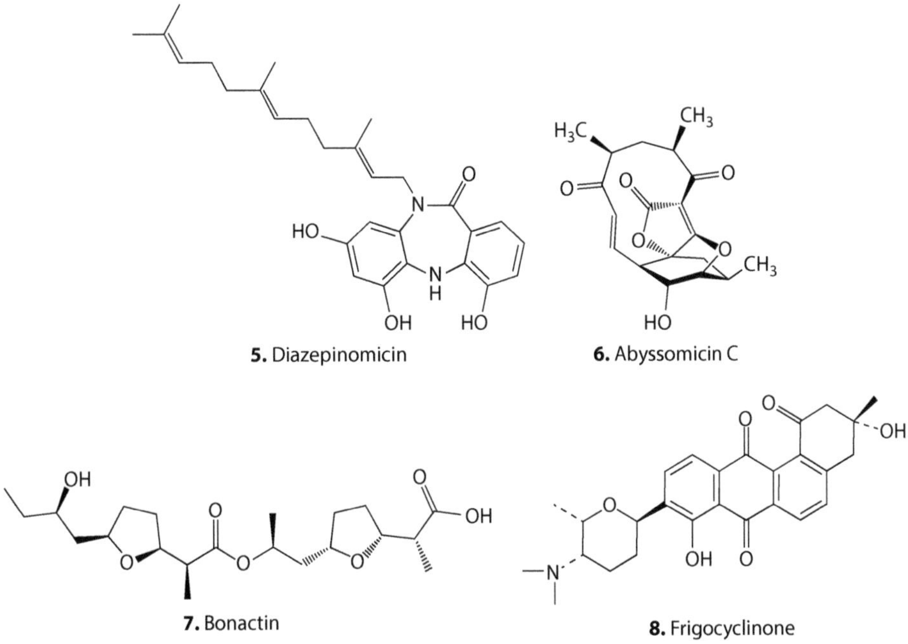

Figure 2.2 Chemical structures of diazepinomicin, abyssomicin C, bonactin, and frigocyclinone.

well-known synthetic sulfa drugs (Bister et al., 2004). Abyssomicin C possesses potent activity against gram-positive bacteria, including clinical isolates of multidrug-resistant and vanco-mycin-resistant *Staphylococcus aureus.* Abyssomicin C or its analog (Rath et al., 2005) has the potential to be developed as an antibacterial agent against drug-resistant pathogens.

A new compound, assigned the trivial name bonactin (**7**) (Figure 2.2), has been iso-lated from the liquid culture of the *Streptomyces* sp. BD21-2, obtained from a shallow-water sediment sample collected at Kailua Beach, Oahu, HI. Bonactin displayed antimicrobial activity against both gram-positive and gram-negative bacteria as well as antifungal activity (Schumacher et al., 2003). Frigocyclinone (**8**) (Figure 2.2) is a new angucyclinone antibiotic isolated from *Streptomyces griseus* strain NTK 97, consisting of a tetrangomy-cin moiety attached through a C glycosidic linkage with the aminodeoxysugar rossamine Frigocyclinone showed antibacterial activities against gram-positive bacteria.

Lynamicins (**9**) (Figure 2.3) are chlorinated bisindole pyrroles, isolated from *Marinispora* sp. The antimicrobial spectrum of these compounds was evaluated against a panel of 11 pathogens, which demonstrated that these substances possess broad-spectrum activity against both gram-positive and gram-negative organisms. Significantly, compounds were active against drug-resistant pathogens such as methicillin-resistant *Staphylococcus aureus* and vancomycin-resistant *Enterococcus faecium* (McArthur et al., 2008). Tirandamycin C (**10**) (Figure 2.3) is a novel dienoyl tetramic acid isolated from *Streptomyces* sp. 307–9. Tirandamycin C showed inhibitory activity against vancomycin-resistant *Enterococcus faecalis* (Carlson et al., 2009). Marinopyrroles (**11**) (Figure 2.3) are densely halogenated and axially chiral metabolites that contain an uncommon bis-pyrrole structure isolated from *Streptomyces* sp. The marinopyrroles possess potent antibiotic activities against methicillin-resistant *Staphylococcus aureus* (Hughes et al., 2008).

Himalomycins A (**12**) and B (**13**) (Figure 2.4) are anthracycline antibiotics. They were obtained from *Streptomyces* sp. 6921, isolated from the marine sediments of Mauritius,

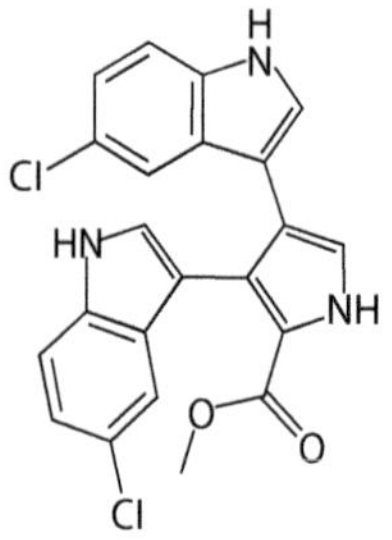

9. Lynamicins

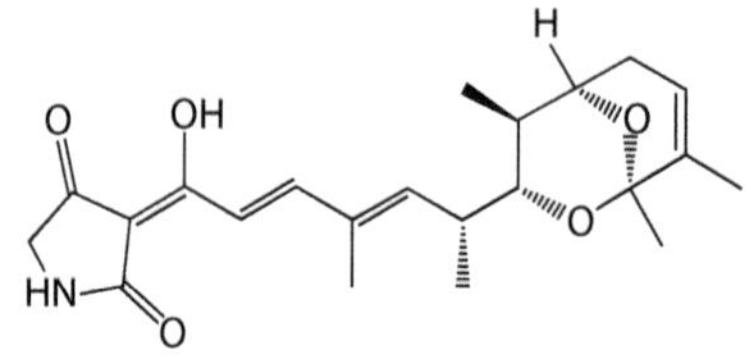

10. Tirandamycin C

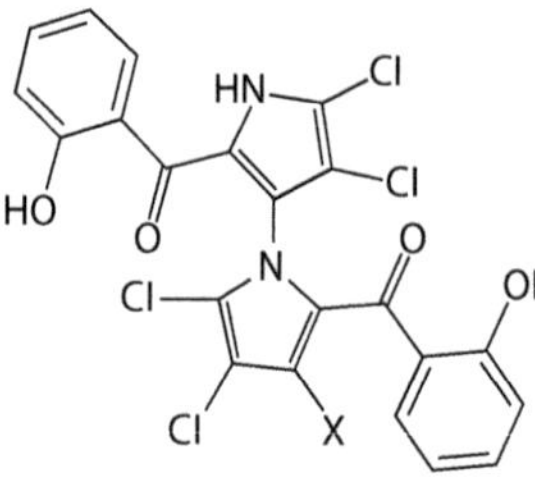

11. Marinopyrroles

Figure 2.3 Chemical structures of lynamicins, tirandamycin C, and marinopyrroles.

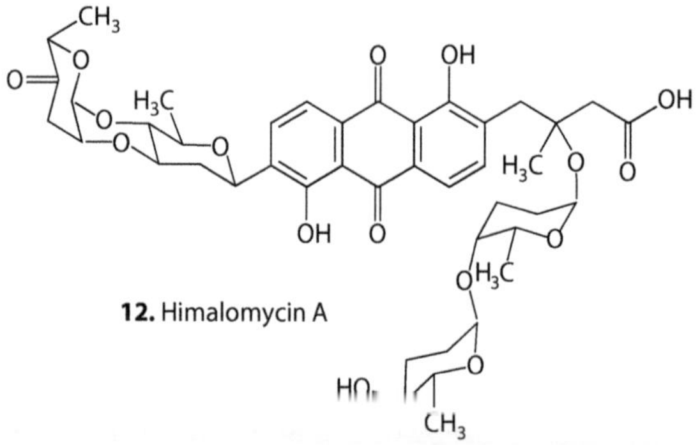

12. Himalomycin A

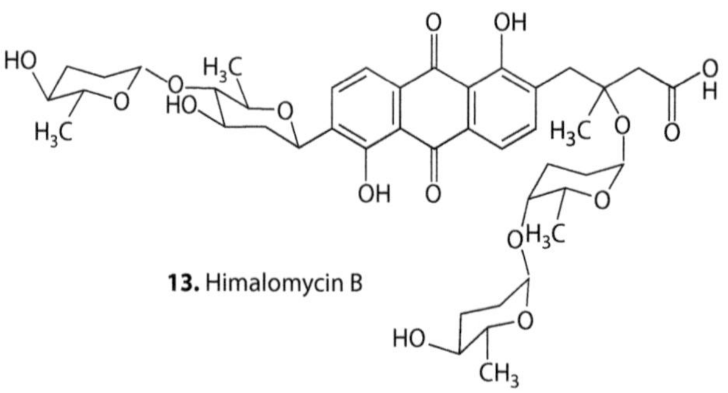

13. Himalomycin B

Figure 2.4 Chemical structures of himalomycins A and B.

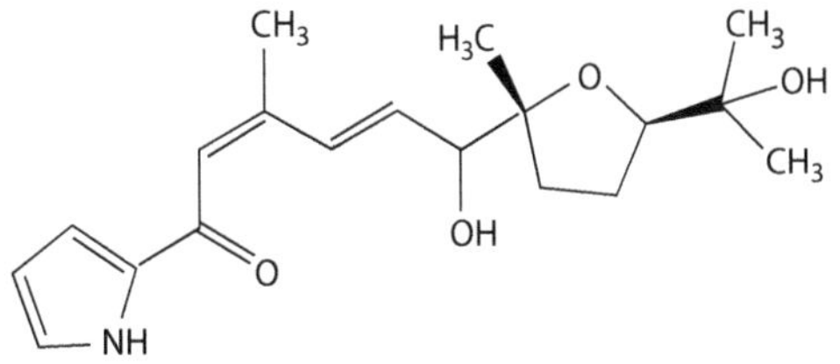

14. Glyciapyrroles A

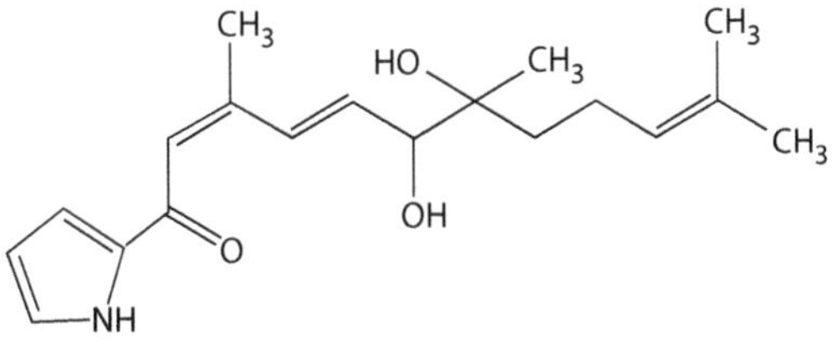

15. Glyciapyrroles B

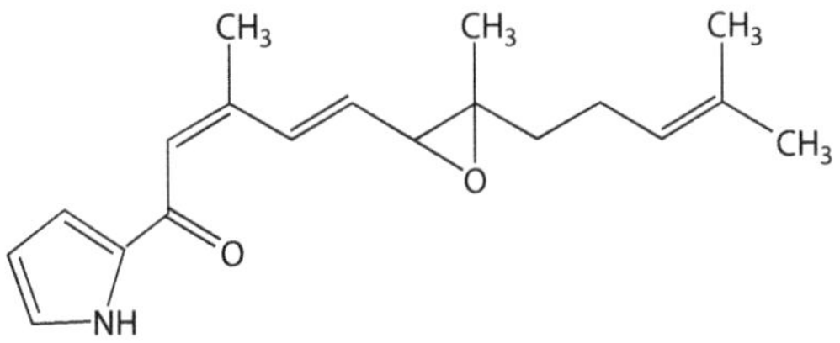

16. Glyciapyrroles C

Figure 2.5 Chemical structures of glyciapyrroles A, B, and C.

and exhibited strong antibacterial activity (Maskey et al., 2003a). Glyciapyrroles A (**14**), B (**15**), and C (**16**) (Figure 2.5) are new pyrrolosesquiterpene antibiotics isolated from the *Streptomyces* sp. NPS008187 (Macherla et al., 2005).

2.3.2   Antifungal activity

Numerous antibiotics have been isolated from a variety of microorganisms; however, studies are still being conducted to identify novel antibiotics effective against pathogenic fungi (Atlas and Bartha, 1986). Marine actinobacteria are useful biological tools for the production of antifungal substances against fungi (Okami and Hotta, 1988). In general, *Streptomyces* species are saprophytic and are commonly associated with soils, where they contribute significantly to the turnover of complex biopolymers and antibiotics (Wanner, 2009). The marine *Streptomyces* sp. DA11 isolated from South China, found to be associated with sponge *Craniella australiensis*, produced the enzyme chitinase and showed antifungal activities against *Aspergillus niger* and *Candida albicans* (Han et al., 2009). Chitin, a linear β 1,4 linked homopolymer of N acetylglucosamine, is one of the three most abundant polysaccharides in nature besides cellulose and starch. The antifungal activity and highly biocompatible quality make chitinase and its derivatives particularly useful for biomedical applications, such as wound healing, cartilage tissue engineering, drug delivery, and nerve generation (Shi et al., 2006; Yan et al., 2006). The biodegradable and antifungal properties of chitinase are also useful for environmental and agricultural uses, food technology, and cosmetics (Rabea et al., 2003; Lin and Lin, 2005).

N-(2-hydroxyphenyl)-2-phenazinamine (NHP) (**17**) (Figure 2.6) is a new antibiotic isolated from *Nocardia dassonvillei*. The new compound showed significant antifungal activity against *Candida albicans*, with an MIC of 64 µg/mL and high cancer cell cytotoxicity against HepG2, A549, HCT-116, and COC1 cells (Gao et al., 2012). Chandrananimycin A (**18**) (Figure 2.6) is a novel antibiotic isolated from *Actinomadura* sp. Chandrananimycin A possesses potent antifungal activity against *Mucor miehei*. It also exhibits antialgal activity against the microalgae *Chlorella vulgaris* and *C. sorokiniana* and antibacterial activity against *Staphylococcus aureus* and *Bacillus subtilis*, along with anticancer activity (Maskey et al., 2003b).

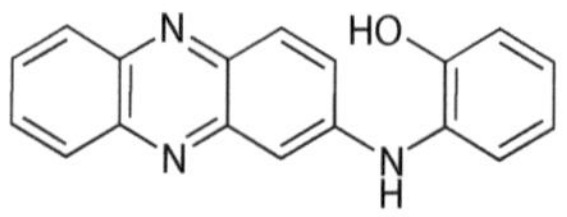

17. *N*-(2-hydroxyphenyl)-2-phenazinamine (NHP) 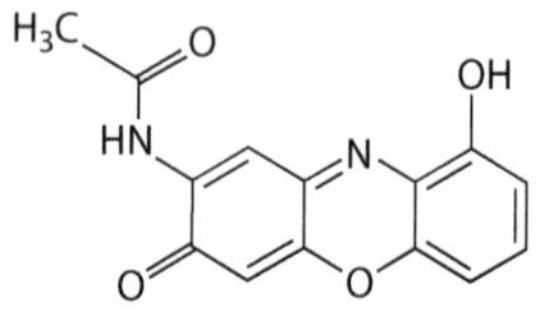   **18.** Chandrananimycin A

Figure 2.6 Chemical structures of *N*-(2-hydroxyphenyl)-2-phenazinamine (NHP) and chandranani-mycin A.

2.4   Conclusion

Marine actinobacteria have a tremendous potential to provide therapeutic leads with distinct chemical structures and biological activities. Actinobacteria, and in particular the genus *Streptomyces*, have the ability to produce a wide variety of secondary metabolites as bioactive compounds, including antibacterial, antifungal, anticancer, antitumor, anti-inflammatory, antimalarial, antiviral, and anti-angiogenesis drugs.

Because of the immense biological diversity in the sea as a whole, it is increasingly recognized that a large number of novel chemical entities exist in the oceans. As marine microorganisms, particularly actinobacteria, have evolved the greatest genomic and metabolic diversity, efforts should be directed toward exploring marine actinobacteria as a potential source for the discovery of novel secondary metabolites and for the development of new antimicrobial drugs.

Acknowledgment

This research was supported by a grant from the Marine Bioprocess Research Center of the Marine Biotechnology Program, funded by the Ministry of Oceans and Fisheries, R and D/2004-6002, Republic of Korea.

References

Aghamirian, M.R. and Ghiasian, S.A. 2009. Isolation and characterization of medically important aerobic actinomycetes in soil of Iran (2006–2007). *Open Microbiology Journal.* 3:53–57.
Asolkar, R.N., Dirk Schroeder, R.H., Siegmund Lang, I.W.D., and Hartmut, L. 2004. Helquinoline, a new tetrahydroquinoline antibiotic from *Janibacter limosus* Hel 1. *Journal of Antibiotics (Tokyo).* 57:17–23.
Atlas, R.M. and Bartha, R. 1986. *Microbial Ecology: Fundamentals and Applications,* 4th edn., New York: Benjamin-Cummings Publishing Company, pp. 174–217.
Atta, H.M., Dabour, S.M., and Desoukey, S.G. 2009. Sparsomycin antibiotic production by *Streptomyces* sp. AZ-NIOFD1: Taxonomy, fermentation, purification and biological activities. *American-Eurasian Journal of Agricultural and Environmental Science* 5(3):368–377.
Bhatnagar, I. and Kim, S.-K. 2010. Immense essence of excellence: Marine microbial bioactive compounds. *Marine Drugs.* 8(10):2673–2701.
Bister, B., Bischoff, D., Stroebele, M., Riedlinger, J., Reicke, A., Wolter, F., and Suessmuth, R.D. 2004. Abyssomicin C—A Polycyclic antibiotic from a marine *Verrucosispora* strain as an inhibitor of the p-aminobenzoic acid/tetrahydrofolate biosynthesis pathway. *Angewandte Chemie International Edition* 43(19):2574–2576.
Bredholt, H., Fjærvik, E., Johnsen, G., and Zotchev, S.B. 2008. Actinomycetes from sediments in the Trondheim fjord, Norway: Diversity and biological activity. *Marine Drugs.* 6(1):12–24.

Bruntner, C., Binder, T., Pathom-aree, W., Goodfellow, M., Bull, A.T., Potterat, O., and Fiedler, H.P. 2005. Frigocyclinone, a novel angucyclinone antibiotic produced by a *Streptomyces griseus* strain from Antarctica. *The Journal of Antibiotics*. 58(5):346–349.

Carlson, J.C., Li, S., Burr, D.A., and Sherman, D.H. 2009. Isolation and characterization of tirandamycins from a marine-derived *Streptomyces* sp. *Journal of Natural Products*. 72(11):2076–2079.

Charan, R.D., Schlingmann, G., Janso, J., Bernan, V., Feng, X., and Carter, G.T. 2004. Diazepinomicin, a new antimicrobial alkaloid from a marine *Micromonospora* sp. *Journal of Natural Products*. 67(8):1431–1433.

El-Gendy, M.M., Shaaban, M., Shaaban, K.A., El-Bondkly, A.M., and Laatsch, H. 2008. Essramycin: A first triazolopyrimidine antibiotic isolated from nature. *Journal of Antibiotics*. 61(3):149–157.

Engelhardt, K., Degnes, K.F., Kemmler, M., Bredholt, H., Fjaervik, E., Klinkenberg, G., and Zotchev, S.B. 2010. Production of a new thiopeptide antibiotic, TP-1161, by a marine *Nocardiopsis* species. *Applied and Environmental Microbiology*. 76(15):4969–4976.

Gao, X., Lu, Y., Xing, Y., Ma, Y., Lu, J., Bao, W., and Xi, T. 2012. A novel anticancer and antifungus phenazine derivative from a marine actinomycete BM-17. *Microbiological Research*. 167:616–622.

Gärtner, A., Ohlendorf, B., Schulz, D., Zinecker, H., Wiese, J., and Imhoff, J.F. 2011. Levantilides A and B, 20-membered macrolides from a *Micromonospora* strain isolated from the Mediterranean deep sea sediment. *Marine Drugs*. 9(1):98–108.

Goodfellow, M. and Williams, S.T. 1983. Ecology of actinomycetes. *Annual Reviews in Microbiology*. 37(1):189–216.

Han, Y., Yang, B., Zhang, F., Miao, X., and Li, Z. 2009. Characterization of antifungal chitinase from marine *Streptomyces* sp. DA11 associated with South China Sea sponge *Craniella australiensis*. *Marine Biotechnology*. 11(1):132–140.

Hohmann, C., Schneider, K., Bruntner, C., Irran, E., Nicholson, G., Bull, A.T., and Fiedler, H.P. 2009. Caboxamycin, a new antibiotic of the benzoxazole family produced by the deep-sea strain *Streptomyces* sp. NTK 937*. *Journal of Antibiotics*. 62(2):99–104.

Hong, K., Gao, A.H., Xie, Q.Y., Gao, H.G., Zhuang, L., Lin, H.P., and Ruan, J.S. 2009. Actinomycetes for marine drug discovery isolated from mangrove soils and plants in China. *Marine Drugs*. 7(1):24–44.

Hughes, C.C., Prieto-Davo, A., Jensen, P.R., and Fenical, W. 2008. The marinopyrroles, antibiotics of an unprecedented structure class from a marine *Streptomyces* sp. *Organic Letters*. 10(4):629–631.

Imada, C. 2004. Enzyme inhibitors of marine microbial origin with pharmaceutical importance. *Marine Biotechnology*. 6(3):193–198.

Jensen, P.R. and Fenical, W. 1994. Strategies for the discovery of secondary metabolites from marine bacteria: Ecological perspectives. *Annual Reviews in Microbiology*. 48(1):559–584.

Kwon, H.C., Christopher, A.K., Paul, R.J., and William, F. 2006. Marinomycins AD, antitumor-antibiotics of a new structure class from a marine actinomycete of the recently discovered genus "*Marinispora*." *Journal of the American Chemical Society*. 128(5):1622–1632.

Lin, C.C. and Lin, H. L. 2005. Remediation of soil contaminated with the heavy metal (Cd²⁺). *Journal of Hazardous Materials*. 122(1):7–15.

Macherla, V.R., Jehnan, L., Christopher, B. et al. 2005. Glaciapyrroles A, B, and C, pyrrolosesquiterpenes from a *Streptomyces* sp. isolated from an Alaskan marine sediment. *Journal of Natural Products*. 68(5):780–783.

Manam, R.R., Teisan, S., White, D.J., Nishino, J.G., Neuteboom, S.T.C., Lam, K.S., Mosca, D.A., Lloyd, G.K., and Potts, B.C.M. 2005. Lajollamycin, a nitro-tetraene spiro-β-lactone-γ-lactam antibiotic from the marine actinomycete *Streptomyces nodosus*. *Journal of Natural Products*. 68:240–243.

Manivasagan, P., Kang, K.H., Sivakumar, K., Li-Chan, E.C., Oh, H.M., and Kim, S.K. 2014a. Marine actinobacteria: An important source of bioactive natural products. *Environmental Toxicology and Pharmacology*. 38(1):172–188.

Manivasagan, P., Venkatesan, J., Sivakumar, K., and Kim, S.K. 2013. Marine actinobacterial metabolites: Current status and future perspectives. *Microbiological Research*. 168(6):311–332.

Manivasagan, P., Venkatesan, J., Sivakumar, K., and Kim, S.K. 2014b. Actinobacterial enzyme inhibitors—A review. *Critical Reviews in Microbiology*. 41(2):261–272.

Manivasagan, P., Venkatesan, J., Sivakumar, K., and Kim, S.K. 2014c. Pharmaceutically active secondary metabolites of marine actinobacteria. *Microbiological Research*. 169:262–278.

Maskey, R.P., Helmke, E., and Laatsch, H. 2003a. Himalomycin A and B: Isolation and structure elucidation of new fridamycin type antibiotics from a marine *Streptomyces* isolate. *Journal of Antibiotics.* 56(11):942–949.

Maskey, R.P., Li, F.C., Qin, S., Fiebig, H.H., and Laatsch, H. 2003b. Chandrananimycins A approximately C: Production of novel anticancer antibiotics from a marine *Actinomadura* sp. isolate M048 by variation of medium composition and growth conditions. *Journal of Antibiotics.* 56(7):622–634.

Maskey, R.P., Sevvana, M., Usón, I., Helmke, E., and Laatsch, H. 2004. Gutingimycin: A highly complex metabolite from a marine streptomycete. *Angewandte Chemie International Edition.* 43(10):1281–1283.

McArthur, K.A., Scott, S.M., and Ginger, T. 2008. Lynamicins A–E, chlorinated bisindole pyrrole antibiotics from a novel marine actinomycete. *Journal of Natural Products.* 71(10):1732–1737.

Ogunmwonyi, I.H., Ntsikelelo, M., and Leonard, M. 2010. Studies on the culturable marine actinomycetes isolated from the Nahoon beach in the eastern Cape Province of South Africa. *African Journal of Microbiology Research.* 4(21):2223–2230.

Okami, Y. and Hotta, K. 1988. Search and discovery of new antibiotics. In: Goodfellow, M., Williams, S.T., and Mordarski, M. (eds), *Actinomycetes in Biotechnology.* San Diego, CA: Academic Press, Inc., pp. 33–67.

Olano, C., Carmen, M., and José, A.S. 2009. Antitumor compounds from marine actinomycetes. *Marine Drugs.* 7(2):210–248.

Peschke, U., Heike, S., Hui-Zhan, Z., and Wolfgang, P. 2006. Molecular characterization of the lincomycin-production gene cluster of *Streptomyces lincolnensis* 78–11. *Molecular Microbiology.* 16(6):1137–1156.

Prabavathy, V.R., Narayanasamy, M., and Kandasamy, M. 2006. Control of blast and sheath blight diseases of rice using antifungal metabolites produced by *Streptomyces* sp. PM5. *Biological Control.* 39(3):313–319.

Rabea, E.I., Mohamed, E.-T.B., Christian, V.S., Guy, S., and Walter, S. 2003. Chitosan as antimicrobial agent: Applications and mode of action. *Biomacromolecules.* 4(6):1457–1465.

Raghunath, D. 2008. Emerging antibiotic resistance in bacteria with special reference to India. *Journal of Biosciences.* 33(4):593–603.

Ramesh, S. and Mathivanan, N. 2009. Screening of marine actinomycetes isolated from the Bay of Bengal, India for antimicrobial activity and industrial enzymes. *World Journal of Microbiology and Biotechnology.* 25(12):2103–2111.

Rao, K.V.B. 2012. *In-vitro* antimicrobial activity of marine actinobacteria against multidrug resistance *Staphylococcus aureus. Asian Pacific Journal of Tropical Biomedicine.* 2(10):787–792.

Rath, J.P., Stephan, K., and Martin, E.M. 2005. Synthesis of the fully functionalized core structure of the antibiotic abyssomicin C. *Organic Letters.* 7(14):3089–3092.

Ravikumar, S., Gnanadesigan, M., Saravanan, A., Monisha, N., Brindha, V., and Muthumari, S. 2012. Antagonistic properties of seagrass associated *Streptomyces* sp. RAUACT 1: A source for anthraquinone rich compound. *Asian Pacific Journal of Tropical Medicine.* 5(11):887–890.

Ravikumar, S., Jacob Inbaneson, S., Uthiraselvam, M., Rajini Priya, S., Ramu, A., and Banerjee, M.B. 2011. Diversity of endophytic actinomycetes from Karangkadu mangrove ecosystem and its antibacterial potential against bacterial pathogens. *Journal of Pharmacy Research.* 4(1):294–296.

Ravikumar, S., Thajuddin, N., Suganthi, P., Jacob Inbaneson, S., and Vinodkumar, T. 2010. Bioactive potential of seagrass bacteria against human bacterial pathogens. *Journal of Environmental Biology.* 31(3):387–389.

Reddy, N.G., Ramakrishna, D.P.N., and Raja Gopal, S.V. 2011. A morphological, physiological and biochemical studies of marine *Streptomyces rochei* (MTCC 10109) showing antagonistic activity against selective human pathogenic microorganisms. *Asian Journal of Biological Sciences.* 4(1):1–14.

Riedlinger, J., Reicke, A., Zähner, H.A.N.S., Krismer, B., Bull, A.T., Maldonado, L.A., and Fiedler, H.P. 2004. Abyssomicins, inhibitors of the para-aminobenzoic acid pathway produced by the marine *Verrucosispora* strain AB-18-032. *Journal of Antibiotics.* 57(4):271–279.

Saadoun, I., Hameed, K.M., and Moussauui, A. 1998. Characterization and analysis of antibiotic activity of some aquatic actinomycetes. *Microbios.* 99(394):173–179.

Saurav, K. and Kannabiran, K. 2010. Diversity and optimization of process parameters for the growth of *Streptomyces* VITSVK9 spp isoled from Bay of Bengal, India. *Journal of Natural and Environmental Sciences*. 1(2):56–65.

Schmitz, F.J., Verhoef, J., and Fluit, A.C. 1999. Prevalence of resistance to MLS antibiotics in 20 European university hospitals participating in the European SENTRY surveillance programme. *Journal of Antimicrobial Chemotherapy* 43(6):783–792.

Schumacher, R.W., Talmage, S.C., Miller, S.A., Sarris, K.E., Davidson, B.S., and Goldberg, A. 2003. Isolation and structure determination of an antimicrobial ester from a marine sediment-derived bacterium. *Journal of Natural Products*. 66(9):1291–1293.

Sharma, S.L. and Pant, A. 2001. Crude oil degradation by a marine actinomycete *Rhodococcus* sp. *Indian Journal of Marine Sciences*. 30(3):146–150.

Shi, C., Zhu, Y., Ran, X., Wang, M., Su, Y., and Cheng, T. 2006. Therapeutic potential of chitosan and its derivatives in regenerative medicine. *Journal of Surgical Research*. 133(2):185–192.

Sivasubramanian, K., Ravichandran, S., and Vijayapriya, M. 2011. Antagonistic activity of marine bacteria *Pseudoalteromonas tunicata* against microbial pathogens. *African Journal of Microbiology Research*. 5(5):562–567.

Socha, A.M., LaPlante, K.L., and Rowley, D.C. 2006. New bisanthraquinone antibiotics and semi-synthetic derivatives with potent activity against clinical *Staphylococcus aureus* and *Enterococcus faecium* isolates. *Bioorganic and Medicinal Chemistry*. 14(24):8446–8454.

Solanki, R., Khanna, M., and Lal, R. 2008. Bioactive compounds from marine actinomycetes. *Indian Journal of Microbiology*. 48(4):410–431.

Soria-Mercado, I.E., Prieto-Davo, A., Jensen, P.R., and Fenical, W. 2005. Antibiotic terpenoid chloro-dihydroquinones from a new marine actinomycete. *Journal of Natural Products*. 68(6):904–910.

Valli, S., Suvathi, S.S., Aysha, O.S., Nirmala, P., Vinoth, K.P., and Reena, A. 2012. Antimicrobial potential of *Actinomycetes* species isolated from marine environment. *Asian Pacific Journal of Tropical Biomedicine*. 2(6):469–473.

Wanner, L.A. 2009. A patchwork of *Streptomyces* species isolated from potato common scab lesions in North America. *American Journal of Potato Research*. 86(4):247–264.

Yan, J., Li, X., Liu, L., Wang, F., Zhu, T.W., and Zhang, Q. 2006. Potential use of collagen-chitosan-hyaluronan tri-copolymer scaffold for cartilage tissue engineering. *Artificial Cells, Blood Substitutes and Biotechnology*. 34(1):27–39.

Zhang, L., An, R., Wang, J., Sun, N., Zhang, S., Hu, J., and Kuai, J. 2005. Exploring novel bioactive compounds from marine microbes. *Current Opinion in Microbiology*. 8(3):276–281.

chapter three

Antimicrobial compounds from microorganisms
Production, characterization, and applications

W.M. Muiru

Contents

3.1   Introduction

Antimicrobials are substances that act against microorganisms by inhibiting their growth or killing them. Antimicrobial metabolites produced by microorganisms are secondary metabolites and are grouped according to the microorganisms they act against, for instance, antibiotics are used against bacteria and antifungals are used for controlling fungal organisms. Antibiotics are the most important antimicrobial metabolites (Basilio et al., 2003). The use of antimicrobial products against microorganisms, either by killing them or inhibiting their growth, has been in place for the past 2000 years (Liras and Martin, 2005). Molds and plant extracts were used by ancient Egyptians and Greeks to treat various infections. Alexander Fleming discovered penicillin, a natural antimicrobial product from *Penicillium rubens*, and this has been successfully used to treat many infections in humans caused by *Streptococcus* bacteria such as pneumonia and gonorrhea. Microbial secondary metabolites have a great variety of chemical structures and are normally formed by microorganisms after the growth phase is complete. Apart from the activity against other microbes, secondary metabolites have different biological activities such as enhancing plant growth and acting as enzyme inhibitors or antitumor agents (Liras and Martin, 2005).

3.2   Sources of antimicrobial metabolites

Many microorganisms have been reported to produce secondary metabolites. Fungi produce a wide range of antimicrobial metabolites, and the production is observed when mycelial growth stops due to some nutrients being limiting while there are still ample carbon sources. Actinomycetes are the most prolific and fruitful group of organisms producing antimicrobial metabolites, and as many as three-quarters of actinomycetes produce antibiotics. Many species of bacteria also produce secondary antagonistic metabolites. Although several hundreds of compounds with antibiotic properties have been isolated and identified, only a few of them have therapeutic applications.

    Some of the microorganisms are known to produce several metabolites with different structures and biological activities, and a single microbial strain may produce more than one secondary metabolite. The compounds produced by one particular strain may have similar chemical structures or completely different structures. Microorganisms contain cluster of genes for the formation of each secondary metabolite. Each cluster has genes encoding structural biosynthetic enzymes and regulatory proteins for the control of metabolite formation. The microorganisms producing the metabolites also contain genes for making the producer organism resistant to the metabolite for those cases where the metabolites have lethal or deleterious biological activity (Liras and Martin, 2005).

3.3 Functions of antimicrobial compounds in nature

The role of secondary metabolites in the organisms that produce them is not very clear, but they are reported to offer ecological advantage to the organism producing it in its natural habitat. The microbial metabolites produced by many microorganisms possess specific or broad-spectrum activities against coexisting microorganisms (Berleman and Kirby, 2009). Some of these metabolites such as antibiotics confer advantage to the producing organisms through direct suppression of other organisms in highly competitive and resource-limited environments (Davies, 1990).

Sethi et al. (2013) postulated that many of these metabolites may act as chemical defense mechanisms in competition for substrates by the various microbes. The same observation has been reported by other workers. Microorganisms that thrive in unusual and extreme habitats have exhibited capabilities of producing unique metabolites. When antibiotics levels are produced at subinhibitory concentrations, they serve other purposes such as in the acquisition of substrates or initiation of changes that enhance survival under stressful conditions (Romero et al., 2011). Antibiotics have also been reported to play a role in virulence on host plants, and they also influence multitrophic interactions that are necessary for adapting to various environments.

3.4 Production of antimicrobial compounds

Different microorganisms vary in the type and quantity of antimicrobial metabolites they produce. The ability of microbes to produce antimicrobial metabolites is greatly influenced by factors such as conditions of nutrition and cultivation (temperature, pH) (Mangamuri et al., 2011). The composition of the media and the intrinsic capacity of the microbe in producing antimicrobial metabolites greatly influence the synthesis of the bioactive metabolites. The medium composition together with the metabolic capacity of the producing microorganism greatly affects synthesis of bioactive metabolites. The environmental factors that influence the production of bioactive metabolites are pH, temperature, incubation media, and duration of incubation (Mangamuri et al., 2011).

Sources of nitrogen and carbon play a critical role in the production of antimicrobial metabolites (Kumar et al., 2012). Carbon is required for growth and production of antimicrobial metabolites, and the choice of carbon sources greatly influences secondary metabolism and, therefore, antibiotic production. Carbon sources such as dextrose, lactose, sucrose, fructose, starch, and glycerol are suitable for production of secondary metabolites for various microorganisms. Glucose may allow the cell to achieve maximum cell growth rates, but it inhibits the production of many secondary metabolites due to intermediates generated from the rapid catabolism of glucose, which interferes with enzymes in the secondary metabolism process. In addition, the duration of fermentation plays a significant role in the growth of culture, and production of antibiotics takes place in the late growth phase of the producing organism. Although the production of antimicrobial metabolites is genetically controlled, it is greatly influenced by environmental manipulations. Variation in fermentation environment results in changes in the yields and compositions or profiles of antimicrobial substances. Limiting the supply of essential nutrients not only restricts the growth but has specific metabolic and regulatory effects (Ripa et al., 2009).

3.4.1 Bacteria as producers of antimicrobial compounds

3.4.1.1 Endophytic bacteria

Endophytes are microbes that colonize living, internal tissues of plants without causing any immediate or overt negative effects. Common genera of bacteria that serve as endophytes are *Serratia, Pseudomonas, Bacillus, Azospirillum, Burkholderia, Azoarcus,* and *Streptomyces.* Endophytic bacteria are known to have such effects as enhancing seed germination, promoting plant growth, and protecting plants from insect pests and pathogens. Currently, only a few secondary metabolites have been isolated from endophytic bacteria since only a few plant species have been studied in depth and their metabolites tested for activity against plant, animal, or human pathogens. There is increasing interest in endophytic bacteria as they are likely to possess new genes for novel bioactive compounds. The bioactive compounds from endophytic bacteria can be used in pharmaceutical industries in the production of drugs, in agriculture as biocontrol agents, and in food industry as food preservatives (Adhikari et al., 2001; Gunatilaka, 2006).

The following antibiotics have been reported from bacterial endophytes: ecomycins from *Pseudomonas viridiflava* and munumbicins from *Streptomyces.* They exhibit antibacterial activities and also have activity against the malarial parasite *Plasmodium falciparum* and the plant-pathogenic oomycete *Pythium ultimum.* Other antibiotics produced by endophytic bacteria are fusaricidins A–D from *Paenibacillus polymyxa* and oocydin, which is a chlorinated macrocyclic lactone produced by *Serratia marcescens* and has been reported to contribute to the natural protection against oomycete pathogens (Menpara and Chanda, 2013). Table 3.1 shows some of the endophytes of host plants and their reported activity against test pathogens.

3.4.1.2 Bacillus

Bacillus produces a wide range of antimicrobial metabolites, mainly polypeptide antibiotics, which have many applications in various fields such as in crop protection, pharmaceutical, and food industry (Sethi et al., 2013). Up to 167 antibiotics are produced by *Bacillus* with *B. subtilis* alone producing 66 antibiotics accounting for more than a third of all the antibiotics produced by the genus. The main antibiotic producers of this genus are *B. subtilis,* which produces polymyxin, mycobacillin, bacitracin, difficidin, and subtilin; *B. cereus,* which produces cerexin and zwittermicin; *B. brevis,* which produces gramicidin and tyrothricin; *B. laterosporus,* which produces laterosporin; *B. circulans,* which produces circulin; *B. polymyxa,* which produces polymyxin and colistin; *B. licheniformis,* which produces bacitracin; and *B. pumilus,* which produces pumulin (Awais et al., 2010).

Most of the peptide antibiotics produced by *Bacillus* are active against gram-positive bacteria; however, some have activity against gram-negative bacteria, whereas others such as bacillomycin, mycobacillin, and fungistatin are effective against molds and yeasts. Two compounds isolated from *Bacillus* and identified as iturin A were shown to have a suppressive effect on common scab disease in a pot assay, decreasing the infection rate from 75% to 35%. This strain also suppressed *Fusarium oxysporum,* the pathogen causing potato dry rot disease.

Bacillus is reported to produce metabolites that suppress *Meloidogyne incognita, Radopholus similis, Ditylenchus dipsaci,* and *Heterodera glycines* through inhibition of egg development and root infection. *Bacillus* is the most numerous producer of peptide antibiotics, producing gramicidins, tyrocidines, and bacitracins.

Table 3.1 Some of the reported bacterial endophytes and their associated
microbial activity to the test organisms

Host plant	Potent endophytes	Activity shown	Tested organisms
Panax ginseng	*Paenibacillus polymyxa GS01, Bacillus* sp. *GS07,* and *Pseudomonas poae JA01*	Antifungal	Phytopathogenic fungi
T. grandiflora, Polyalthia sp., and *Mapania* sp.	*Streptomyces fulvoviolaceus, Streptomyces coelicolor,* and *Streptomyces caelestis*	Antifungal	Phytopathogenic fungi
Scutellaria baicalensis Georgi	*Bacillus amyloliquefaciens*	Antibacterial and antifungal	Phytopathogenic, food-borne pathogenic, and spoilage bacteria and fungi
Panax notoginseng	*Bacillus amyloliquefaciens* subsp. *plantarum* and *Bacillus methylotrophicus*	Antifungal	Phytopathogenic fungi and nematode
Azadirachta indica A. Juss.	*Streptomyces* sp. and *Nocardia* sp.	Antibacterial and antifungal	Phytopathogenic fungi, human pathogenic bacteria, and fungus
Plectranthus tenuiflorus	*Bacillus* sp. and *Pseudomonas* sp.	Antibacterial and antifungal	Human pathogenic bacteria and fungus
Wheat	*Bacillus subtilis*	Antifungal	Phytopathogenic fungi
Anthurium	*B. amyloliquefaciens*	Antibacterial	Phytopathogenic bacteria
Platycodon grandiflorum	*Bacillus licheniformis, Bacillus pumilus,* and *Bacillus* sp.	Antibacterial and antifungal	Phytopathogenic fungi and anti-human food-borne pathogenic organisms
Artemisia annua	*Streptomyces*	Antibacterial and antifungal	Pathogenic bacteria, yeast, and fungal phytopathogens
Centella asiatica	*Bacillus subtilis* and *P. fluorescens*	Antifungal	Phytopathogenic fungi
Panicum virgatum L.	*Bacillus subtilis, C. flaccumfaciens, P. fluorescens,* and *P. ananatis*	Antifungal	Phytopathogenic fungi
Raphanus sativus L.	*Enterobacter* sp. and *B. subtilis*	Antibacterial and antifungal	Phytopathogenic fungi and human pathogenic bacteria
Memecylon edule, Tinospora cordifolia, Phyllodium pulchellum, and *Dipterocarpus tuberculatus*	*Bacillus amyloliquefaciens*	Antibacterial and antifungal	Human pathogenic bacteria and fungus

(Continued)

Table 3.1 (Continued) Some of the reported bacterial endophytes and their associated microbial activity to the test organisms

Host plant	Potent endophytes	Activity shown	Tested organisms
S. lavandulifolia, H. scabrum, and *R. pulcher*	*Bacillus* sp.	Antibacterial and antifungal	Human pathogenic bacteria and saprophytic fungi
Aloe chinensis	*Paenibacillus* species	Antibacterial and antifungal	Pathogenic bacteria and fungi
Epimedium brevicornum Maxim	*Phyllobacterium myrsinacearum*	Antibacterial and antifungal	Phytopathogenic fungi and phytopathogenic bacterium
11 mangrove halophytic plants	*Bacillus thuringiensis* and *Bacillus pumilus*	Antibacterial	Shrimp pathogens
Kandelia candel	*Streptomyces* sp.	Antibacterial	Several pathogenic bacteria
Codonopsis lanceolata	*Bacillus pumilus, B. subtilis,* and *B. licheniformis*	Antifungal	Phytopathogenic fungi
Polygonum cuspidatum	*Streptomyces* sp.	Antifungal	Pathogenic fungi
Manihot esculenta	*Paenibacillus* sp.	Antifungal	Phytopathogenic fungus
Bruguiera gymnorrhiza, Rhizophora stylosa, and *Kandelia candel*	*Bacillus amyloliquefaciens*	Antibacterial and antifungal	Phytopathogenic fungi and phytopathogenic bacteria
Monstera sp.	*Streptomyces* sp.	Antifungal and antimalarial	Pythiaceous fungi and the human fungal pathogen, and malarial parasite
Piper nigrum L.	*P. aeruginosa, P. putida,* and *B. megaterium*	Antifungal	Phytopathogenic fungus
Huperzia serrata	*Burkholderia* sp.	Antifungal	Phytopathogenic fungi
300 plants from upper Amazonian rainforests	*Streptomyces* sp., *Micromonospora* sp., and *Amycolatopsis* sp.	Antibacterial and antifungal	Range of potential fungal and bacterial pathogens
Lycopersicon esculentum	*Streptomyces* sp., *Microbispora* sp., *Micromonospora* sp., and *Nocardia* sp.	Antibacterial and antifungal	Phytopathogenic fungi and phytopathogenic bacteria

Source: Sharma, M., *Int. J. Curr. Microbiol. Appl. Sci.*, 3(2), 801, 2014. Table courtesy of Prof. S. Chanda, Department of Biosciences, Saurashtra University, Rajkot, Gujarat, India.

3.4.1.3 Pseudomonas

Pseudomonas spp. are widely distributed soil inhabitants belonging to the gamma subclass of Proteobacteria, and many of them live in a commensal relationship with plants (Paulsen et al., 2005). Consequently, some of the *Pseudomonas* such as *Pseudomonas fluorescens* are used as plant growth–promoting rhizobacteria. It has the ability to colonize the rhizosphere of host plants and produce a wide range of compounds inhibitory to a number of economically important plant pathogens (Haas and Keel, 2003). Pseudomonads have an exceptional capacity to produce a wide variety of metabolites, including antibiotics that are toxic to plant pathogens.

 P. fluorescens have been reported to produce the following antibiotic compounds: phenazines, pyrrole derivatives, indole derivatives, 2,4-diacetylphloroglucinol (DAPG)

(Ahil et al., 2014; Nowak-Thompson et al., 1994), pyrrolnitrin, and pyoluteorin. Additionally, it produces DAPG, which is toxic to plant juveniles of parasitic nematodes (*Globodera rostochiensis*). A strain of *P. fluorescens* has been observed to be antagonistic to *Meloidogyne javanica* (Raaijmakers et al., 2002). Apart from producing secondary metabolites that suppress plant disease and signal gene expression to neighboring cells inhabiting the rhizosphere, *Pseudomonas* also use siderophores such as pyochelin and pyoverdine from other microorganisms to obtain iron, which increases their survival in iron-limited environments (Paulsen et al., 2005). Production of hydrogen cyanide formed by oxidation of glycine and the siderophores enables the *P. fluorescens* to suppress target pathogens in the rhizosphere through iron competition (Buysens et al., 1996).

3.4.1.4 Lactic acid bacteria

Lactic acid bacteria are a physiologically diverse group of gram-positive, non–spore forming cocci or rods that produce lactic acid as the major end product during carbohydrate fermentation. Lactic acid bacteria (LAB) are able to produce antimicrobial compounds against competing flora, including food-borne spoilage and pathogenic bacteria. Under unfavorable environmental conditions, many species of LAB also produce exopolysaccharides (EPSs), which protect themselves against desiccation, bacteriophage, and protozoan attack (Yang, 2000). LAB are involved in fermentation of a wide range of food products such as vegetable foods, milk, meat, and cereals and have traditionally been used to improve flavor development and ripening of fermented products (Rai and Chikindas, 2011). LAB produces inhibitory substances, namely, metabolic acid products, which are classified as low-molecular-mass compounds. Examples of those antimicrobial products are organic acids, hydrogen peroxide, carbon dioxide, diacetyl, and high-molecular-mass compounds such as bacteriocins.

The production of LAB inhibits spoilage organisms through lowering of pH, thus creating a hostile environment (Yang, 2000). Examples of such inhibitory substances are bacteriocins, which are proteinaceous in nature. They are grouped into four classes, namely small peptides (e.g., nisin), small heat-stable peptides, large heat-labile proteins, and complex bacteriocins. The following genera comprise the LAB: *Lactobacillus, Leuconostoc, Pediococcus, Streptococcus, Vagococcus,* and *Enterococcus* (Yang, 2000).

3.4.2 Actinomycetes as producers of antimicrobial compounds

Actinomycetes comprise an extensive and diverse group of microorganisms, and in addition to the production of biologically active substances, they play a role in soil cycles. The classification of actinomycetes is wrought with controversy due to possession of morphological features similar to fungi and biochemical and physiological features similar to bacteria (Muiru, 2000). This led to describing actinomycetes as bacteria that have the ability to form branching hyphae at some stages of development. The actinomycetes exhibit a very wide range of morphological forms extending from coccus forms through fragmenting hyphal forms to permanent and highly differentiated branched mycelium. Some actinomycetes form spores that include motile zoospores and specialized structures that resist desiccation and mild heat (Dhanasekaran et al., 2009).

Actinomycetes are the most important source of antibiotics, producing over two-thirds of all the known antibiotics. The antibiotics from actinomycetes have high commercial value including medical applications for the treatment of human, animal, and plant diseases. *Streptomyces* and *Micromonospora* are the most commonly isolated genera of actinomycetes, and consequently most of the metabolites identified from screening programs have been identified from the two genera. However, there has been a concerted effort to

screen for bioactive compounds from minor groups of actinomycetes. Other genera in actinomycetes that produce antibiotics are *Nocardiaceae, Pseudonocardiaceae*, and *Actinomadura*.

Most of the soil actinomycetes have preference for neutral and alkaline conditions
for optimal growth; however, some grow under extreme conditions of alkaline and acidic
conditions, and these are the alkalophilic and acidophilic actinomycetes, respectively. The
occurrence of actinomycetes is greatly influenced by environmental conditions such as
temperature, humidity, vegetation, and pH. Halotolerant actinomycetes are adapted to
saline environment and have mostly been isolated from marine habitats. Knowledge on
the requirements for growth of various actinomycetes, especially the rare ones, is essential
to develop isolation conditions that can help in detection of such species. This will help
expand the range of isolated actinomycetes and consequently the bioactive metabolites
produced by these species.

Actinomycetes are known to produce valuable antibiotics such as novobiocin, amphotericin, vancomycin, neomycin, gentamicin, chloramphenicol, tetracycline, erythromycin,
and nystatin, among others (Sharma, 2014). In agriculture, actinomycetes are used as plant
growth–promoting agents in the production of the plant growth hormone indole-3-acetic
acid, as biocontrol tools, as biopesticide agents, as antifungal compounds, and as a source
of agroactive compounds. They are also important in soil biodegradation and humus
formation as they recycle the nutrients associated with recalcitrant polymers, such as
chitin, keratin, and lignocelluloses (Bull and Stach, 2007). Industrially, actinomycetes play
a significant role in the production of various antimicrobial agents, enzymes, and other
industrially important substances (Sharma, 2014).

Other groups of actinomycetes namely Thermonosporaceae and Mycobacteriaceae
and other unclassified species such as *Actinosporangium, Frankia*, and *Sebekia* are reported
to produce bioactive metabolites (Sharma, 2014).

3.4.2.1   Streptomyces

The genus *Streptomyces* is classified in the family Streptomycetaceae, and it is an aerobic
spore-forming actinomycete. The classification is based on the morphological and cell-wall
chemotaxonomic characters. *Streptomyces* includes aerobic, gram-positive bacteria that produce the extensively branched substrate mycelium and aerial hyphae (Liras and Martin,
2005). *Streptomyces* is the most important actinomycete genus in the production of antibiotics, and over two-thirds of the clinically important antibiotics are produced by this genus.

Streptomyces are also some of the most abundant soil microorganisms and are found
in a wide range of habitats under a wide variety of conditions. The genus has more than
500 species and subspecies with around 4000 antibiotics isolated and identified from this
genus. This number of antibiotics represents around 70% of all known antibiotics. Apart
from conventional methods, molecular–systematic methods in the phylogenetic analysis
are increasingly being used in *Streptomyces* systematics (Liras and Martin, 2005).

The production of antibiotics by *Streptomyces* is greatly influenced by factors such as
nutritional source (carbon, nitrogen, and minerals) and environmental factors (incubation
period, pH, and temperature). In artificial conditions, optimization of culture conditions
is essential to obtain high yields of the antimicrobial metabolites (Ozgur et al., 2008).

Abamectin is a macrocyclic lactone metabolite produced by *Streptomyces avermitilis*,
and it is used in seed treatment against nematodes in tomato plants. It kills nematode-
infected larvae and arrests egg hatching and RNA synthesis. Abamectin has been shown
to be highly active against *Pratylenchus* spp. and has significant effect on *Hoplolaimus
galeatus, M. javanica, Radopholus similis, Ditylenchus dipsaci, Aphelenchoides fragariae*, and
Tylenchorhynchus dubius, all of them being plant parasitic nematodes.

3.4.2.2 Micromonosporaceae

The genus has been reported to synthesize a variety of bioactive compounds (Igarashi et al., 2011). It produces a wide range of antibiotics with over 300 being described. This group of actinomycetes produces several antibiotics with varied chemical properties and applications (Carro et al., 2013). *Micromonospora coerulea* strain A058 produces a glutarimide antibiotic named streptimidone, which has been found to be effective in inhibiting the following plant pathogenic organisms: *Didymella bryoniae, Phytophthora capsici, Botrytis cinerea*, and *Magnaporthe grisea*. Greenhouse tests showed high efficacy in the management of *M. grisea, P. capsici*, and *B. cinerea* on rice, pepper, and cucumbers, respectively. The antibiotic was equally as effective as metalaxyl and had no phytotoxicity at relatively high concentrations (Seok et al., 1999). Other antibiotics produced by this genus are gentamicins from *M. purpurea* and *M. echinospora*, fortimicin from *M. olivoasterospora*, rosamicin from *M. rosaria*, and omicin from *M. inyoensis*, among others (Badji et al., 2013).

Two *Micromonospora* strains namely *Micromonospora aurantiaca* and *Micromonospora coriariae* have been reported in actinorhizal nodules from *Casuarina* and *Coriaria* plants, respectively. In addition, *Micromonospora* was reported to be widespread in nodules of legumes such as lupine (Trujillo et al., 2006) and peas.

3.4.2.3 Pseudonocardia

Pseudonocardia have been reported to produce bioactive metabolites that have properties such as antifungal, antibacterial, neuroprotective, and enzyme inhibitors. For instance, phenazostatin, which is a phenazine derivative, acts as a neuroprotective substance (Maskey et al., 2003), azureomycins A and B from *P. azurea* have antimicrobial activity (Omura et al., 1997), and Dekker et al. (1998) reported quinolone compounds from *Pseudonocardia* sp. that have selective and potent activity against *Helicobacter pylori* (Mangamuri et al., 2011).

3.4.2.4 Actinomadura

Actinomadura are slow-growing organisms that take 10–14 days under optimal conditions to form spore-bearing aerial mycelium. Soil is their natural habitat, and conditions for growth are similar to those of other actinomycetes except that chitin and xylan are not utilized. *Actinomadura* is one of the most predominant actinomycete genus in extreme environments and an important target in screening programs for bioactive metabolites (Berdy, 2005). It belongs to the family Thermomonosporaceae and currently has 37 species including 2 subspecies (Zhang et al., 2001). *Actinomadura* are reported to produce over 250 antibiotics, the most common ones being polyether ionophoric antibiotics. Examples of such antibiotics are maduramicins produced by *A. yumanensis* and cationomycin produced by *A. azure*. Some species, such as *A. roseoviolacea*, produce antitumor anthracyclines such as carminomycins. The genus does not produce classical antibacterial macrolides but produces structurally similar products, the macrolactams (Badji et al., 2013).

The genus is reported to produce other bioactive metabolites that are antagonistic to microorganisms. For example, daunomycin, which is an antifungal metabolite and anthracycline type of antibiotic, has activity against *Phytophthora capsici* and *Rhizoctonia solani*, both of which are plant pathogenic organisms. This metabolite also demonstrated activity against *Saccharomyces cerevisiae* and gram-positive bacteria (Beom et al., 2000).

Four active compounds were elucidated from bioactive metabolites from *Actinomadura* species isolated from Algerian Saharan soil. They showed strong antifungal activity against pathogenic and toxinogenic fungi. These compounds were shown to differ from the known antibiotics produced by *Actinomadura* species, and they possessed aromatic rings substituted by aliphatic chains (Badji et al., 2013).

3.4.3   Fungi as producers of antimicrobial compounds

3.4.3.1   Fungal endophytes

Endophytic fungi represent an important and quantifiable component of fungal diversity, with an estimate of hundreds of species. Endophytes spend all or part of their life cycle inter- and intracellularly colonizing healthy tissues of their host plants such as the epidermal cell layers and cause no apparent harm or negative effect to the host (Yu et al., 2010). Endophytes are ubiquitous with over one million species reported and are found in all plant species. They form inconspicuous infections within tissues of healthy plants for all or nearly all their life cycle (Limsuwan et al., 2009). They are known to produce substances that provide protection and ultimate survival of the plant, and different species of plants are a host to one or more endophytes (Strobel, 2003). Endophytes and their host plants share a complex relationship with endophytes indirectly benefiting plant growth by producing special substances, mainly secondary metabolites, to prevent the growth or activity of plant pathogens (Gutierrez et al., 2012).

Endophytes have proven to be a new and potential source of novel natural products for antimicrobial metabolites isolated from endophytes belonging to diverse structural classes, including alkaloids, peptides, steroids, terpenoids, phenols, quinones, flavonoids, lignans, lactones, isocoumarins, and phenylpropanoids (Zhao et al., 2010). Metabolites of endophytes have been reported to inhibit a number of microorganisms, and, therefore, these metabolites or their derivatives have chemotherapeutic value and are used as antifungal and antibacterial products (Yu et al., 2010). Apart from possessing antimicrobial activity, some of the active secondary metabolites have immunosuppressant properties and are used as anticancer compounds. Currently, more than 140 fungal metabolites have been confirmed to possess antitumor activity and have the potential to be used in the treatment of several types of cancer (Wang et al., 2011; Gutierrez et al., 2012).

Examples of endophytes that produce antimicrobial metabolites are *Fusarium oxysporum* NFX06 isolated from the leaf of the medicinal plant *Odulisporium foetida* and *Phomopsis* spp. that produce pyrenocines with good antifungal, antibacterial, and algicidal properties (Hussain et al., 2012a). Other endophytes reported are *Seimatosporium* sp., which produces acaranoic acids, named seimatoporic acids A and B, together with six other compounds (Hussain et al., 2012b), whereas *Pichia guilliermondii* Ppf9 derived from the medicinal plant *Paris polyphylla* var. *yunnanensis* produces three steroids and one nordammarane triterpenoid (Zhao et al., 2010). An endophytic *Penicillium* sp. from the palm tree produced metabolites that showed antimicrobial properties against *Staphylococcus aureus*, *M. luteus*, and *Escherichia coli*. Some of the active metabolite extracted included the rare indole alkaloid glandicoline B (Koolen et al., 2012).

Due to the increasing threats of resistance to drugs by the human, animal, and plant pathogens, it is necessary to investigate more sources of antimicrobial producers, and this means endophytes can increase the chance of finding novel antimicrobial natural products, thus solving the problem of resistance. More research needs to be conducted since endophytes are a rich and reliable source of genetic diversity (Huang et al., 2007).

3.4.3.2   Trichoderma

This is a free-living fungi that are highly interactive in root, soil, and foliar environments with antagonistic properties against plant pathogens. Due to this antagonistic potential, they are being applied commercially as biological control agents against fungal pathogens. The strains commonly used as biological control agents are from the following species: *Trichoderma harzianum, T. viride, T. virens,* and *T. koningii.* They have been effectively used in the management of plant diseases caused by *Sclerotium cepivorum, Pyrenophora tritici-repentis, Sclerotinia sclerotiorum, Pythium ultimum,* and *Rhizoctonia solani.*

Trichoderma uses different modes of action to control the development of plant diseases, and one of these modes is the production of antimicrobial metabolites. In addition, *Trichoderma* spp. isolated from suppressive soil have shown extreme antagonism toward *P. cinnamomi* during the saprophytic stage via antibiosis and mycoparasitism (Keen and Vancov, 2010). *T. koningii* SMF2 produces antimicrobial metabolites with antimicrobial activity against a range of gram-positive bacterial and fungal phytopathogens (Xiao-Yan et al., 2006). Other types of antibiotics produced by *Trichoderma* are viridian, pyrones, gliotoxin, peptaibols, and gliovirin, among others. Trichoderma are also reported to produce over 180 antimicrobial peptides called peptaibols, which have been reported to inhibit spore germination and hyphal elongation of plant pathogenic fungi.

3.4.3.3   Diaporthe

Lignicolous fungi are fungi with the ability to degrade fiber from seaweed, rotten wood, mangrove plants, and marine algae. They live on limited nutritional sources, and they produce bioactive compounds with antimicrobial properties as a competitive means. Phomopsidin, neomangicols A–C, mangicols A, and humicolone are some of the bioactive compounds reportedly produced by *Diaporthe* species. Some of these bioactive compounds have cytotoxic activity against cell lines (Xin et al., 2006).

3.4.3.4   Aspergillus

Aspergillus produces a diverse group of secondary metabolites, organic acids, antibiotics, polyketides, and ribosomal and nonribosomal peptides (Andersen et al., 2013). Some of these are bioactive with antibacterial, antifungal, and antitumor properties. *Aspergillus* species have been reported to produce metabolites with antimicrobial activity against *Candida albicans.* After isolation and purification, structural elucidation of these metabolites yielded three metabolites with the following structures: 5,6-dihydro-5(*S*)-acetoxy-6(*S*)-(1,2-*trans*-epoxypropyl)-2*H*-pyran-2-one (asperline (1), compound 1); 5,6-dihydro-5(*S*)-acetoxy-6(*S*)-(1,2-*trans*-propenyl)-2*H*-pyran-2-one (compound 2); and 5,6-dihydro-5(*R*)-acetoxy-6(*S*)-(1,2-*trans*-epoxypropyl)-*H*-pyran-2-one (compound 3) (Mizuba et al., 1975).

Other metabolites produced by *Aspergillus* are amino acid–derived metabolites such as echinocandins, which are lipopeptide antifungal agents produced by *A. nidulans* and *A. rugulosus* (Badji et al., 2013). *A. terreus* is a prolific producer of secondary metabolites with the following compounds being reportedly produced: geodin, itaconate, lovastatin, questrin, sulochrin, terrecyclic and asterric acids, asterriquinone, butyrolactone I, citrinin, emodin, and aspulvinone.

3.4.3.5   Penicillium

The genus comprises more than 200 species distributed throughout the world in different habitats. Members of this genus are known to produce secondary metabolites with antimicrobial properties (Kang et al., 2007). In addition, they produce immunosuppressants, antitumor drugs, antiviral drugs, and cholesterol-lowering agents (Koolen et al., 2012). Some of the most well-known and economically important metabolites are produced by *P. chrysogenum,* such as penicillins. Penicillins are derived from a tripeptide chain and have been used in the treatment of human diseases caused by various plant pathogenic organisms (Badji et al., 2013). Other metabolites produced by *Penicillium* are compactins produced by *P. solitum* and mycophenolic acid produced by *P. brevicompactum*. In addition, members of this genus are known to produce mycotoxins, and many of these mycotoxins such as citrinin and penicillic acids have antimicrobial activity (Gharaei-Fathabad et al., 2009).

While some species, such as *P. citrinum*, produces citrinin as the only mycotoxin, other species, such as *P. aurantiogriseum*, produce citrinin and penicillic acid simultaneously (Petit et al., 2004). Some species of *Penicillium*, specifically *P. waksmanii*, produce different metabolites namely alkaloids, pyrones, sulfur-containing dioxopiperazines, and griseofulvin (Petit et al., 2004). Two novel tryptoquivaline-like metabolites, tryptoquialanine A (1) and tryptoquialanine B (2), have been isolated from *Penicillium digitatum* (Ariza et al., 2002).

3.4.3.6   Nematophagous fungi

3.4.3.6.1   Arthrobotrys   This nematode-trapping fungi has been reported to produce antimicrobial compounds with nematicidal activities against *Meloidogyne incognita* and *Caenorhabditis elegans*. The nematicidal compound was identified as linoleic acid and was isolated from two species of Arthrobotrys namely *Arthrobotrys conoides* and *Arthrobotrys oligospora* (Jansson and Thiman, 1992). Production of this antimicrobial compound increased with the number of traps formed in both *Arthrobotrys oligospora* and *Arthrobotrys conoides* (Anke et al., 1995).

3.4.3.6.2   Nematoctonus   This group of nematophagous fungi is known to produce antimicrobial metabolites. Two species namely *Nematoctonus concurrens* and *Nematoctonus robustus* produce dihydropleurotinic acid, leucopleurotin, and pleurotin (Anke et al., 1995).

3.4.3.7   Ascomycetous fungi

Some species of ascomycetes produce antimicrobial compounds with nematicidal activities, for instance, *Chlorosplenium* species produces linoleic acid, *Neobulgaria pura* produces 14-epicochlioquinone B, and Daldinia produces naphthalenes derived from the melanin biosynthetic pathway. *Lachnum papyraceum* produces more than 30 metabolites with chlorine or bromine incorporation depending on the culture conditions (Anke et al., 1995).

3.4.3.8   Marine fungi

These have been reported to produce diverse antimicrobial metabolites that have activity against fungal, bacterial, viral, and protozoan infections. Some of these metabolites namely avrainvillamide, sargassamide, and halimide possess anticancer properties. Hypoxysordarin, isolated from *Hypoxylon croceum*, and 1-hydroxy-6-methyl-8-(hydroxymethyl)xanthone, isolated from *Ulocladium botrytis*, have potent antifungal activities. Lactone metabolites from *Phoma* sp. possess antifungal properties against *Cryptococcus neoformans*, *Aspergillus fumigates*, and *Candida albicans*. Some have been reported to produce metabolites that can inhibit protozoans such as the malarial parasite *P. falciparum* (Bhadury et al., 2006).

3.5   Detection/assay of microorganisms producing antagonistic metabolites

The detection of an antimicrobial effect of a crude extract of the culture broth is the first step needed in the discovery of new bioactive compounds. This is followed by the identification of the bioactive compound and finally the elucidation of the structure of potent metabolite (Menpara, and Chanda, 2013). In search for new antimicrobial metab-olites, thousands of microbial strains are isolated, but only a few produce useful metabolites. Microbial antagonism is the basis of selecting organisms that produce such metabolites. Primary screening of isolates is done to ascertain the potential of strains with respect to production of antimicrobial secondary metabolites (Monisha et al., 2011). During primary screening process, a large number of isolates are screened against a range of sensitive strains. On the basis of primary screening results, isolates showing substantial antimicrobial activities are selected for subsequent secondary screening pro-grams; hence, the methods to be used should allow rapid screening of many organisms (Yang, 2000; Ryu et al., 2000).

The commonly used bioassay techniques are "bicultures," "dual cultures," "paired cultures," or "cross cultures" of the potential antagonist and the test pathogen. The potential antagonist produces metabolites that diffuse through the media causing antag-onism to the test pathogen (Sethi et al., 2013). Antagonism is indicated if zone of inhibi-tion develops between the two organisms (Han et al., 2005). The culture media should favor the growth and antibiotic production of potential antagonist and the growth of the test pathogen.

The size of the inhibition zone between the test pathogen and the potential antagonist is taken as a measure of antimicrobial activity, but it should be borne in mind that other factors such as exhaustion of nutrients around a colony or inhibitory pH can result to such zones. The production of the antimicrobial metabolites can be confirmed by testing the cell-free culture filtrates of organisms that cause growth inhibition.

3.6   Production and detection of antimicrobial metabolites in shaken liquid media

The organisms that show antagonistic properties are cultured in liquid media in a mechan-ical shaker or in stationary cultures. Synthetic media or complex organic media can be used. The components of the media, speed of agitation, and duration of incubation are determined empirically. The organism in the shake cultures is grown to obtain sufficient growth. The resultant broth contains a mixture of the antimicrobial metabolites, water, microbial cells, and residues of the nutrients used in the media. Filtration or centrifugation is done to obtain the metabolite from the crude culture filtrate (Vijayakumari et al., 2013), which is then subjected to bioassays.

The evaluation or bioassays of antimicrobial substances depend on factors such as nature of the antibiotic substance, composition of the medium employed, selection of assay organisms, time of action, and the environment conditions for carrying out the tests. The bioassays of these antibiotic substances may be accomplished through biological assay methods or through nonbiological assay methods.

3.7   Factors affecting the bioassay of antibiotic substances

Since the size of the inhibition zones is usually used as an estimation of the quantity of the antimicrobial metabolites produced, the measurement of these zones should be done with a high degree of precision and accuracy. The inhibition zone should be due to the inhibi-tion of the test organism by the antimicrobial metabolite diffusing through the medium, but other factors affect the range, precision, and sensitivity of an assay. The following fac-tors determine the size of the inhibition zone.

3.7.1 Ingredients of the assay medium

The important factors in the media are viscosity and depth in the petri dishes and sub-stances in the media such as peptone, tryptone, yeast extract, and agar. These substances may vary in their mineral content, and this may influence the activity of the antimicro-bials. Calcium, magnesium, and iron affect the sizes of zones produced by gentamicin. Addition of glucose may cause a reduction in the zone of inhibition in some antibiotics, while others such as nitrofurantoin, ampicillin, and carbohydrates may enhance the zones of inhibition. The diffusion of unused nutrients from the zone of the inhibition may cause an enhancement of growth at the edge (Loo et al., 1945).

3.7.2 Choice of the medium and agar thickness

Plates should be poured flat with an even depth of media throughout and all contain-ing the same volume (Collins et al., 1989). This is achieved by supporting the plates so that the plates are level and adding measured volumes of molten agar. Thin layers give larger clear zones with distinct margins, while thick layers give cone-shaped zones with poorly defined margins. Nonuniformity in the thickness causes error in zone size. The use of uninoculated agar below and above the inoculated layer enables the growth layer to be made thin but has the disadvantage of the diffusion of the antimi-crobial metabolites into the uninoculated layer and hence causing modification outside the zone.

3.7.3 Effect of pH

pH has an effect on the size of inhibition zone produced by some antibiotics (Han et al., 2005). Since the agar has very little buffer capacity, it is necessary to control the pH of the samples to be assayed. Activity or zone size is enhanced in alkaline media for aminoglycosides and streptomycin. The activity of tetracyclines is enhanced in acid medium.

3.7.4 Temperature of incubation

Nonuniform incubation temperature is the major cause of inaccurate diffusion assays. The temperature of incubation influences both the zone size and the slope of the dose–response curve. Also, the rate of heating of the contents of the plate influences the zone size and shape. Size of inhibition zone may be increased by delayed incubation. Delayed incubation is achieved by holding plates at a low temperature (4°C) to delay the growth of the test organism while allowing the antibiotic to diffuse into the medium and reach inhibitory concentrations.

3.7.5 Volume of the sample

The volume of the sample applied to the paper disc in the case of paper disc method is of critical importance. The volume applied should be sufficient to saturate the disc but not to cause flooding. Accurate and rapid delivery of this amount of sample can be obtained by using a calibrated pipette.

3.7.6 Nature of inocula propagules and amount of inoculum

The use of spore suspensions in place of vegetative propagules seems to give very reproducible results. This is due to the time required for germination of the other types of inoculum propagules. Higher concentrations of antimicrobial metabolites are needed to prevent growth when small fragments of fungus mycelium are used to seed the media compared to when fungus spores are used. Heavy inocula reduce inhibition zones to some extent, and the ideal inoculum is the one that gives an even dense growth without being confluent.

3.8 Secondary screening for antimicrobial compounds

After the detection of antimicrobial-producing organisms and culturing in the appropriate media, the active metabolites have to be recovered from the crude mixture screened for bioactivity and characterized accordingly. Active metabolites are recovered from the culture broth by fermentation, and production of metabolite can be increased by optimizing the conditions during fermentation. The fermented broth is filtered through a membrane filter (0.20–0.45 µm pore size) or Whatman No. 1 filter to separate the cellular components from the culture filtrate. This can directly be used for the determination of antimicrobial activity against the sensitive organisms or bioactive compounds recovered. Bioactive compounds can be recovered from filtrate by organic solvent extraction method (Badji et al., 2013).

In recovery of active metabolites from filtrate by organic solvent extraction method, culture supernatants are extracted with an equal volume of appropriate organic solvent. The organic fraction obtained is allowed to evaporate under vacuum to dryness using a rotary evaporator to remove all traces of the organic solvent. Stock solution is prepared from the residue and used for antimicrobial analyses to test for bioactivity and characterization of bioactive compounds.

3.9 Identification and characterization of antimicrobial metabolites

Crude culture filtrates containing antimicrobial metabolites are usually mixtures of dissimilar components. Even where the crude culture filtrate contains a single antibiotic, this antibiotic is a very heterogenous group of biologically active compounds. A method that is simple and rapid with applicability to easily prepared samples of crude antimicrobial metabolites is of great importance to help identify new bioactive compounds (Badji et al., 2013). The initial step in identifying unknown antimicrobial metabolites is the determination of movement in specific solvent systems and the nature of microbiological spectrum. Final identification depends on further physical, chemical, and microbiological tests (Jaganathan et al., 2014).

Ideally, to obtain a complete characterization, antimicrobial metabolites should be isolated in pure form and as a single component (Yang, 2000). However, this is not possible especially in a screening program. The following properties and techniques are useful in characterizing antimicrobial metabolites: solubility in different solvents, stability at different temperatures, pH ranges, storage duration, color reactions and fluorescence, light absorption, paper chromatography of the whole antibiotic and of decomposition products, electrophoresis, countercurrent distribution, elementary analysis of physical constant, and mass spectrometry (MS). Characterization of antimicrobial metabolites is important since the ultimate application and the methods of purification are determined by the stability of the metabolites (Lavermicocca et al., 2000).

Bioautography is a method used for the detection of antimicrobial metabolites on paper and thin chromatograms. Bioautography has been used in the classification, search, and identification in search of new antibiotics, development of isolation procedures for unknown antibiotics, preparative chromatography, separation of mixtures of new antibiotics, systematic analysis, and determination of the optimum harvest time.

The paper chromatography is inferior compared to the thin layer in that it has low resolution power resulting in failure to identify macrolide antibiotics and to differentiate closely related antimicrobial metabolites such as antibiotics. In contrast, thin-layer chromatography (TLC) has higher resolution power; thus, it gives an efficient separation and differentiation and also gives results rapidly.

3.9.1 Thin-layer chromatography

TLC is a simple, quick, and inexpensive procedure that tells how many compounds are in a mixture. TLC can also be used to support the identity of a compound in a mixture when the retention factor (Rf) value of a compound is compared with the Rf value of a known compound. Rf value is the retention factor or how far up a plate the compound travels, and it is defined as the distance traveled by the compound divided by the distance traveled by the solvent. Rf values can provide corroborative evidence as to the identity of a compound, and the unknown compounds are spotted together with the standard on a TLC plate side by side or on top of each other. If the two substances have the same Rf value, they are likely (but not necessarily) the same compound. Identity check must be performed on a single plate because it is difficult to duplicate factors that influence Rf values exactly from experiment to experiment.

After spotting a small amount of the mixture to be analyzed near the bottom of the plate, this is placed in a shallow pool of solvent in a developing chamber so that only the very bottom of the plate is in the liquid. This liquid or eluent is the mobile phase, and it slowly rises up the TLC plate by capillary action. As the solvent moves past the spot that was applied, an equilibrium is established for each component of the mixture between the molecules of that component that are adsorbed on the solid and the molecules that are in the solution. Different components differ in solubility and in the strength of their adsorption to the adsorbent, and some components will be carried further up the plate than the others. When the solvent has reached the top of the plate, the plate is removed from the developing chamber, dried, and the separated components of the mixture are visualized using UV lamp.

3.9.2 High-performance liquid chromatography (HPLC)

High-performance liquid chromatography (HPLC) is a chromatographic technique used to separate a mixture of compounds with the purpose of identifying, quantifying, and purifying the individual components of the mixture. HPLC is used in the analysis of unknown compounds against the reference standard. HPLC utilizes special instruments designed to separate, quantify, and analyze components of a chemical mixture. Samples of interest are introduced to a solvent flow path, carried through a column packed with specialized materials for component separation, and component data are obtained through the combination of a detection mechanism coupled with a data-recording system. Chemical separation using HPLC is accomplished by utilizing the fact that certain compounds have different migration rates in a particular column and mobile phase. The extent or degree of separation is mostly determined by the choice of stationery phase and mobile phase. The components being separated or identified using HPLC should have a characteristic peak under certain chromatographic conditions. Chromatographic conditions such as the kind of mobile phase can be adjusted to allow adequate separation and collection of the extract or the desired compound as it elutes from the stationery phase.

In order to identify any compound by HPLC, a detector must first be selected and set to optimal settings. The parameters of separation assay must be developed, and this should allow a clear peak of the known sample to be observed from the chromatograph. The identifying peak should have a reasonable retention time and should be well separated from extraneous peaks at the detection levels that the assay is performed. The retention time of a compound can be altered by manipulating the choice of a column, mobile phase, and the flow rate. To identify an unknown compound by HPLC, a sample of a known compound must be utilized in order to assure identification of the unknown compound.

HPLC can also be used to quantify or determine the unknown concentrations of a compound in a known solution (Grabley and Thiericke, 1999). This involves injecting a series of known concentrations of the standard compound solution onto the HPLC for detection. The chromatographs of these known concentrations give a series of peaks that correlate to the concentration of the compound injected.

3.9.3 Sephadex columns

Sephadex resins are highly specialized gel filtration and chromatographic media that are composed of macroscopic beads synthetically derived from the polysaccharide dextran. The organic chains are cross-linked to give a 3D network having functional ionic groups attached by either linkages to glucose units of the polysaccharide chains. These resins are used to separate (fractionate) a mixture of compounds into its components. When a mixture of molecules and ions are dissolved in a solvent and applied in a column with sephadex, the smaller molecules and ions are distributed through a larger volume of solvent than is available to the large molecules; consequently, the larger molecules move more rapidly through the column enabling the separation of the mixtures.

3.9.4 Other methods of identifying antimicrobial metabolites

Techniques such as nuclear magnetic resonance, spectroscopy, and MS have been used to identify antimicrobial compound from *Lactobacillus* and *Pediococcus* strains where 2-pyrrolidone-5-carboxylic acid has been identified. These metabolites were further separated and purified by chromatographic methods (Yang, 2000). MS is normally used for the determination of molecular mass.

3.9.5 Minimum inhibitory concentration test

Minimum inhibitory concentration (MIC) is the lowest concentration of an antimicrobial compound that inhibits the visible growth of a sensitive strain after the appropriate incubation period. It is the lowest concentration of residue at which there is no visible growth of the pathogenic strain. MIC is determined *in vitro*, by means of agar and broth dilution methods (Kanna et al., 2011; Koolen et al., 2012).

3.10   Utilization of antimicrobial compounds

3.10.1   Use in pharmaceutical industries

Chemical synthetic drugs have been used for a long time in combating human, plant, and animal diseases; however, cases of increased resistance have made it necessary to explore antimicrobial metabolites as an alternative to the use of synthetics (Yu et al., 2010). Pharmacological properties of microbial metabolites have been recognized and exploited for their activity as enzyme inhibitors, receptor agonists, hormone-like regulators, and neurotransmitters (Grabley and Thiericke, 1999). Some antimicrobial metabolites have shown antitumor properties. Antitumor activity of beauvericin to human leukemia cells has been demonstrated (Qinggui and Lijian, 2012). Beauvericin also possesses strong antibacterial activity against human and animal pathogenic bacteria with no selectivity between gram-positive and gram-negative bacteria. It also has antiviral activity against HIV (Rattanachaikunsopon and Phumkhachorn, 2010).

Reutericyclin, a lactic acid antibiotic produced by *Lactobacillus reuteri*, has a broad-spectrum activity against fungi, protozoa, and a wide range of bacteria including both gram-positive and gram-negative bacteria. *E. coli*, *S. aureus*, *S. epidermidis*, and *Candida albicans* (Li et al., 2008) are strongly inhibited by endophytes belonging to the *Streptomyces* genus. These metabolites possessing activity against various organisms have been used to develop and formulate pharmaceutical products to combat human, animal, and plant pathogenic diseases.

3.10.2   Food industry

Food spoilage and contamination with microorganisms and their associated mycotoxins are a major problem in food industry. Production of high-quality and safe food free from pathogens relies on antimicrobial metabolites that inhibit spoilage microorganism. Antimicrobial metabolites from bacteria, algae, and fungi are in use in food industry (Rai and Chikindas, 2011). Naturally produced antimicrobial metabolites are used as biocides to kill and nullify the effects of contaminants. Biocides are used as disinfectants, antiseptics, and food preservatives.

A range of antimicrobial metabolites such as lactic, acetic, and propionic acids produced during the fermentation process by the LAB have a preservative action since they create unfavorable environment for the growth of many spoilage and pathogenic organisms. Acids act against microbes by interfering with the cell membrane, inhibiting active transport, and inhibiting a variety of other metabolic functions. Some acids, such as propionic acid produced by propionic acid bacteria, have antimicrobial activity against yeast and moulds (Yang, 2000).

3.10.3   Crop protection

Hundreds of antimicrobial metabolites have demonstrated antagonistic activity against phytopathogenic fungi and bacteria, and consequently a number of them have been used to formulate products for use in crop protection. Extracts from *Streptomyces* sp. collected from *Allium fistulosum* showed the potential to suppress infection of *Alternaria brassicicola* on Chinese cabbage seedlings. Some actinomycete strains of plant origin produce herbicidal antibiotics, for instance, herbicidin H is a metabolite of *Streptomyces* sp. strain, SANK 63997, that was isolated from the leaves of *Setaria viridis* var. *pachystachys*.

Another strain of *Microbispora* sp., SANK 62597, recovered from *Carex kobomugi* produced γ-glutamylmethionine sulfoximine in culture broth. A strain of *Dactylosporangium* sp. isolated from *Cucubalus* sp. was found to produce streptol and two plant growth inhibitors that inhibit germination of *Brassica rapa* (Hasegawa et al., 2006).

Beauvericin produced by fungi, such as *Beauveria bassiana* and *Fusarium* spp., is a hexadepsipeptide mycotoxin and has a strong insecticidal activity against a broad spectrum of insect pests (Qinggui and Lijian, 2012). In addition, it has antiviral and cytotoxic activities. Antimicrobial metabolites from bacterial endophytes have exhibited activity against various plant pathogenic organisms. Endophytes from *Bacillus subtilis* from wheat roots showed strong activity against *Fusarium graminearum*, *Rhizoctonia cerealis*, *Botrytis cinerea*, and *Macrophomina kuwatsukai*, among others. Endophytes from *B. pumilus* and *B. licheniformis* from balloon flower showed antifungal activity against *Pythium ultimum*, *Rhizoctonia solani*, *Fusarium oxysporum*, and *Phytophthora capsici* (Menpara and Chanda, 2013).

3.10.4 Other applications

Apart from production of antimicrobial metabolites, *Bacillus* also produces industrial enzymes such as subtilisins, cellulases, and amylases used in laundry. Antimicrobial compounds and the organisms producing them have been used in promoting plant growth. Endophytes colonizing inside plants usually get nutrition and protection from the host plants. In return, they confer profoundly enhanced fitness to the host plants by producing a variety of bioactive metabolites. Among other mechanisms, they stimulate growth of plants through biocontrol of phytopathogens through production of antibiotics or siderophores, nutrient competition, and induction of systemic disease resistance. Some of these metabolites directly affect physiology of the host plants, but others do so indirectly by affecting the microbial population by antibiosis and/or competition (Raajmakers and Mazzola, 2012).

Pteridic acids A and B from the fermentation broth of an endophytic *Streptomyces hygroscopicus* TP-A045 (isolated from *Pteridium aquilinum*) as plant growth promoters with auxin-like activity. These compounds accelerated formation of adventitious roots in hypocotyls of kidney beans at 1 nM as effectively as indole acetic acid (Hasegawa et al., 2006).

Actinomycetes play a role in ecological balance and are responsible for much of the digestion of resistant carbohydrates such as chitin and cellulose. Some of them are renowned as degraders of toxic materials and have been shown to degrade hydrocarbons, explosives, chlorinated solvents, and plastics and, as a result, are used in bioremediation (Sharma, 2014). *Pseudonocardia* can utilize hydrocarbons, methyl sulfides, and tetrahydrofuran as growth substrates (Mahendra and Alvarez-Cohen, 2005). Adaptation to harsh environments such as high temperatures allows actinomycetes to be used for composting purposes (Sharma, 2014).

3.11 Challenges in the utilization of antimicrobial metabolites and future prospects

Due to the emergence of drug-resistant infections, there is a need to identify and develop new antibiotics. Despite the critical need for new antibiotics, very few new antibiotics are being developed. Although several hundreds of compounds with antibiotic activity have been isolated from microorganisms over the years, at present only 1% of the microbial world have been explored, so there is still a need to tap the vast reservoir of microbial community for their antimicrobial potential.

Other factors that limit utilization of antimicrobial metabolites are that although some microbes produce antimicrobial metabolites, some of these have unspecific toxicity and they are toxic to human beings along with pathogenic organisms; thus, they cannot be used, and some have moderate antimicrobial activity and cannot be used as medicine (Menpara and Chanda, 2013). Hence, more work needs to be done to come up with bioactive substances without any side effects to humans and plants. Studies of regulatory gene in the synthesis of antimicrobial compounds can be manipulated to increase the yield of these metabolites, and by modification of the structure, one can enhance antimicrobial activity, thereby improving the efficacy and reducing the toxicity, thus decreasing the side effects (Menpara and Chanda, 2013; Lancini and Lorenzetti, 1993).

References

Adhikari, T.B., Joseph, C.M., Yang, G., Phillips, D.A., and Nelson, L.M. 2001. Evaluation of bacteria isolated from rice for plant growth promotion and biological control of seedling disease of rice. *Can. J. Microbiol.* 47(10), 916–924.

Ahil, S.B., Ameer, B., Govardhanam, R., Mallela, V., Nagesh, K., Yukthi, S., Jagannath, V.P., Yuhei, T., and Yoshinori, F. 2014. Isolation and characterization of antimicrobial cyclic dipeptides from *Pseudomonas fluorescens* and their efficacy on sorghum grain mold fungi. *Chem. Biodivers.* 11(1), 92–100.

Andersen, M.R., Jakob Nielsen, J.B., Andreas, K., Lene, M.P., Zachariasen, M., Tilde, J.H., and Blicher, L.H. 2013. Accurate prediction of secondary metabolite gene clusters in filamentous fungi. *Proc. Natl. Acad. Sci.* 110(1), E99–E107.

Anke, H., Stadler, M., Mayer, A., and Sterner, O. 1995. Secondary metabolites with nematicidal and antimicrobial activity from nematophagous fungi and Ascomycetes. *Can. J. Bot.* 73(S1), 932–939. doi:10.1139/b95-341.

Ariza, M.R., Larsen, T.O., Petersen, B.O., Duus, J.Q., and Barrero, B.F. 2002. *Penicillium digitatum* metabolites on synthetic media and citrus fruits. *J. Agric. Food Chem.* 50, 6361–6365.

Awais, M., Pervez, A., Asim, Y., and Shah, M.M. 2010. Production of antimicrobial metabolites by *Bacillus subtilis* immobilized in polyacrylamide gel. *Pak. J. Zool.* 42(3), 267–275.

Badji, B., Mostefaoui, A., Sabaou, N., Mathieu, F., and Lebrihi, A. 2013. Identification of a new strain of *Actinomadura* isolated from Saharan soil and partial characterization of its antifungal compounds. *Afr. J. Biotechnol.* 10, 13878–13886.

Basilio, A., Gonzalez, I., Vicente, M.F., Gorrochategui, J., Cabello, A., Gonzalez, A., and Genilloud, O. (2003). Patterns of antimicrobial activities from soil actinomycetes isolated under different conditions of pH and salinity. *J. Appl. Microbiol.* 95(4), 814–823.

Beom, S., Surk, S.M., and Byung, K.H. 2000. Structure elucidation and antifungal activity of an anthracycline antibiotic, daunomycin, Isolated from *Actinomadura roseola*. *J. Agric. Food Chem.* 48(5), 1875–1881.

Berdy, J. 2005. Bioactive microbial metabolites. *J. Antibiot.* 58(1), 1–26.

Berleman, J.E. and Kirby, J.R. 2009. Deciphering the hunting strategy of a bacterial wolfpack. *FEMS Microbiol. Rev.* 33, 942–957.

Bhadury, P., Mohammad, B.T., and Wright, P.C. 2006. The current status of natural products from marine fungi and their potential as anti-infective agents. *J. Ind. Microbiol. Biotechnol.* 33, 325–337.

Bull, A.T. and Stach, J.E. 2007. Marine actinobacteria: New opportunities for natural product search and discovery. *Trends Microbiol.* 15, 491–499.

Buysens, S., Heungens, K., Poppe, J., and Hofte, M. 1996. Involvement of pyochelin and pyoverdin in suppression of *Pythium*-induced damping-off of tomato by *Pseudomonas aeruginosa* 7NSK2. *Appl. Environ. Microbiol.* 62, 865–871.

Carro, L., Pujic, P., Trujillo, M.E., and Normand, P. 2013. *Micromonospora* is a normal occupant of actinorhizal nodules. *J. Biosci.* 38, 685–693.

Collins, C.H., Lyne, P.M., and Grenge, J.M. 1989. *Microbiological Methods*. Butterworth and Company Publishers Ltd., London, U.K.

Davies, J. 1990. What are antibiotics? Archaic functions for modern activities. *Mol. Microbiol.* 4, 1227–1232.

Dekker, K.A., Inagaki, T., Gootz, T.D., and Huang, L.H. 1998. New quinolone compounds from *Pseudonocardia* sp. with selective and potent anti-*Helicobacter pylori* activity. *J. Antibiot.* 51, 145–152.

Dhanasekaran, D., Selvamani, S., Panneerselvam, A., and Thajuddin, N. 2009. Isolation and characterization of actinomycetes in Vellar estuary, Annagkoil, Tamil Nadu. *Afr. J. Biotechnol.* 8(17), 4159–4162.

Gharaei-Fathabad, E., Tajick-Ghanbary, M.A., and Shahrokhi, N. 2009. Antimicrobial properties of *Penicillium* species isolated from agricultural soils of Northern Iran. *Res. J. Toxins* 6(1), 1–7.

Grabley, S. and Thiericke, R. (Eds.). 1999. *Drug Discovery in Nature*. Springer-Verlag, Berlin, Germany.

Gunatilaka, A. 2006. Natural products from plant-associated microorganisms: Distribution, structural diversity, bioactivity, and implications of their occurrence. *J. Nat. Products* 69, 509–526.

Gutierrez, R.M., Gonzalez, A.M., and Ramirez, A.M. 2012. Compounds derived from endophytes: A review of phytochemistry and pharmacology. *Curr. Med. Chem.* 19(18), 2992–3030.

Haas, D. and Keel, C. 2003. Regulation of antibiotic production in root-colonizing *Pseudomonas* spp. and relevance for biological control of plant disease. *Annu. Rev. Phytopathol.* 41, 117–153.

Han, J.S., Cheng, J.H., Yoon, T.M., Song, J., Rajkarnikar, A., Kim, W.G., Yoo, I.D., Yang, Y.Y., and Suh, J.W. 2005. Biological control agent of common scab disease by antagonistic strain *Bacillus* sp. sunhua. *J. Appl. Microbiol.* 99, 213–221.

Hasegawa, S., Akane, M., Masafumi, S., Tomio, N., and Hitoshi, K. 2006. Endophytic *Actinomycetes* and their interactions with host plants. *Actinomycetologica* 20, 72–81.

Huang, W.Y., Cai, Y.Z., Hyde, K.D., Corke, H., and Sun, M. 2007. *World J. Microbiol. Biotechnol.* 23(9), 1253–1263.

Hussain, H., Ahmed, I., Schulz, B., Draeger, S., and Krohn, K. 2012a. Pyrenocines J-M: Four new pyrenocines from the endophytic fungus. *Fitoterapia* 83(3), 523–526.

Hussain, H., Krohn, K., Schulz, B., Draeger, S., Nazir, M., and Saleem, M. 2012b. Two new antimicrobial metabolites from the endophytic fungus, *Seimatosporium* sp. *Nat. Prod. Commun.* 7(3), 293–294.

Igarashi, Y., Yanase, S., Sugimoto, K., Enomoto, M., Miyanaga, S., Trujillo, M.E., Saiki, I., and Kuwahara, S. 2011. Lupinacidin C, an inhibitor of tumor cell invasion from *Micromonospora lupini*. *J. Nat. Prod.* 74, 862–865.

Jansson, H.B. and Thiman, L. 1992. A preliminary study of chemotaxis of zoospores of the nematode parasitic fungus *Catenaria anguillulae*. *Mycologia* 84, 109–112.

Jeganathan, P., Rajasekaran, K.M., and Asha Devi, N.K. 2014. Antimicrobial activity and characterization of marine bacteria in coastal sea water. *Sch. Acad. J. Pharm.* 3(1), 73–78.

Kang, SW, Park, C.H., Hong, S.I., and Kim, S.W. 2007. Production of penicillic acid by *Aspergillus sclerotiorum* CGF. *Bioresour. Technol.* 98, 191–197.

Kanna, M., Solanki, R., and Lal, R. 2011. Selective isolation of rare actinomycetes producing novel antimicrobial compounds. *Int. J. Adv. Biotechnol. Res.* 2(3), 357–375.

Keen, B. and Vancov, T. 2010. Phytophthora cinnamomi suppressive soils. In: Méndez, A., ed., *Current Research Technology and Education Topics in Applied Microbiology and Microbial Biotechnology*, Vol. 1, Formatex Microbiology Series Publication, Formatex Research Center, Badajoz, Spain, pp. 239–250.

Koolen, H.H.F., Soares, E.R., da Silva, F.M.A., de Almeida, R.A., and de Souza, A.D.L. 2012. An antimicrobial alkaloid and other metabolites produced by *Penicillium* sp. an endophytic fungus isolated from *Mauritia Flexuosa* L. f. *Quim. Nova* 35(4), 771–774.

Kumar, S.N., Siji, J.V., Ramya, R., Nambisan, B., and Mohandas, C. 2012. Improvement of antimicrobial activity of compounds produced by *Bacillus* sp. associated with a *Rhabditid* sp. (entomopathogenic nematode) by changing carbon and nitrogen sources in fermentation media. *J. Microbiol. Biotechnol. Food Sci.* 1(6), 1424–1438.

Lancini, G. and Lorenzetti, R. 1993. *Biotechnology of Antibiotics, and Other Bioactive Microbial Metabolites*. Plenum Press, New York.

Lavermicocca, P., Valerio, F., Evidente, A., Lazzaroni, S., Corsetti, A., and Gobbetti, M. 2000. Purification and characterization of novel antifungal compounds from the sourdough *Lactobacillus plantarum* strain 21B. *Appl. Environ. Microbiol.* 66(9), 4084–4090.

Li, J., Zhao, G., Chen, H., Wang, H., Qin, S., Zhu, W., Xu, L., Jiang, C., and Li, W. 2008. Antitumour and antimicrobial activities of endophytic *Streptomycetes* from pharmaceutical plants in rainforest. *Lett. Appl. Microbiol.* 47, 574–580.

Limsuwan, S., Trip, E.N., Kouwenc, T., Piersmac, S., Hiranrat, A., and Mahabusarakam, W. 2009. Rhodomyrtone: A new candidate as natural antibacterial drug from *Rhodomyrtus tomentosa*. *Phytomedicine* 16, 645–651.

Liras, P. and Martin, J.F. 2005. Assay methods for detection and quantification of antimicrobial metabolites produced by *Streptomyces clvuligerus*. *Methods Biotechnol.* 18, 149–164.

Loo, Y.H., Skell, P.S., Thornberry, H.H., Ehrlich, J., McGuire, J.M., Savage, G.M., and Sylvester, J.C. 1945. Assay of streptomycin by the paper disc plate method. *J. Bact.* 50, 701–709.

Mahendra, S. and Alvarez-Cohen, L. 2005. *Pseudonocardia dioxanivorans* sp. nov., a novel actinomycete that grows on 1,4-dioxane. *Int. J. System. Evolution. Microbiol.* 55, 593–598.

Mangamuri, U.K., Poda, S., Kamma, S., and Muvva, V. 2011. Optimization of culturing conditions for improved production of bioactive metabolites by *Pseudonocardia* sp. VUK-10. *Mycobiology* 39(3), 174–181.

Maskey, R.P., Kock, I., Helmke, E., and Laatsch, H. 2003. Isolation and structure determination of phenazostatin D, a new phenazine derivative from a marine actinomycete isolate *Pseudonocardia* sp. B6273. *Z. Naturforsch. B. J. Chem. Sci.* 58, 692–694.

Menpara, D. and Chanda, S. 2013. Endophytic bacteria—Unexplored reservoir of antimicrobials for combating microbial pathogens. *Microbial Pathogens and Strategies for Combating Them: Science, Technology and Education* (Méndez-Vilas, A., Ed.), pp. 1095–1103.

Mizuba, S., Lee, K., and Jiu, J. 1975. Three antimicrobial metabolites from *Aspergillus caespitosus*. *Can. J. Microbiol.* 21(11), 1781–1787.

Monisha, K., Renu, S., and Rup, L. 2011. Selective isolation of rare *Actinomycetes* producing novel antimicrobial compounds. *Int. J. Adv. Biotechnol. Res.* 3, 357–375.

Muiru, W.M. 2000. Isolation of soil actinomycetes, characterization and screening of their antibiotics against economically important plant pathogens. MSc thesis, University of Nairobi, Nairobi, Kenya.

Nowak-Thompson, B., Gould, S.J., Kraus, J., and Loper, J.E. 1994. Production of 2,4-diacetylphloroglucinol by the biocontrol agent *Pseudomonas fluorescens* Pf-5. *Can. J. Microbiol.* 40, 1064–1066.

Omura, S., Tanaka, H., Tanaka, Y., Spiri-nakagawa, P., Oiwa, R., Takahashi, Y., Matsuyama, K., and Iwai, Y. 1997. Studies on bacterial cell wall inhibitors. VII. Azureomycins A and B new antibiotics produced by *Pseudonocardia azurea* nov. sp. taxonomy of the producing organism, isolation, characterization and biological properties. *J. Antibiot.* 32, 985–994.

Ozgur, C., Gulten, O., and Aysel, U. 2008. Isolation of soil *Streptomyces* as source antibiotics active against antibiotic-resistant bacteria. *Eurasia. J. Biol. Sci.* 2, 73–82.

Paulsen, I.I., Press, C.M., and Ravel, J. 2005. Complete genome sequence of the plant commensal *Pseudomonas fluorescens* Pf-5. *Nat. Biotechnol.* 23, 873–878.

Petit, K.E., Mondeguer, F., Roquebert, M.F., Biard, J.F., and Pouchus, Y.F. 2004. Detection of griseofulvin in a marine strain of *Penicillium waksmanii* by ion traps spectrometry. *J. Microbiol. Methods* 58, 59–65.

Qinggui, W. and Lijian, X. 2012. Beauvericin, a bioactive compound produced by fungi: A short review. *Molecules* 17, 2367–2377.

Raaijmakers, J.M., Vlami, M. and de Souza, J.T. 2002. Antibiotic production by bacterial biocontrol agents. *Antonie Van Leeuwenhoek.* 81, 537–547.

Raaijmakers, J.M. and Mazzola, M. 2012. Diversity and natural functions of antibiotics produced by beneficial and plant pathogenic bacteria. *Ann. Rev. Phytopathol.* 50, 403–424.

Rai, M. and Chikindas, M. 2011. *Natural Microbials in Food Safety and Quality*. CABI International, London, U.K.

Rattanachaikunsopon, P. and Phumkhachorn, P. 2010. Lactic acid bacteria: Their antimicrobial compounds and their uses in food production. *Ann. Biol. Res.* 1(4), 218–228.

Ripa, F.A., Nikkon, F., Zaman, S., and Khond, P. 2009. Optimal conditions for antimicrobial metabolites production from a new *Streptomyces* sp. RUPA-08PR isolated from Bangladeshi Soil. *Mycobiology* 37, 211–214.

Romero, D., Traxler, M.F., Lopez, D., and Kolter, R. 2011. Antibiotics as signal molecules. *Chem. Rev.* 111, 5492–5505.

Ryu, J., Lee, S., Lee, Y., Lee, S., Kim, D., Cho, S., Ba, D., Park, K., and Yun, H. 2000. Screening and identification of antifungal *Pseudomonas* sp. that suppresses balloon flower root rot caused by *Rhizoctonia solani*. *J Microbiol. Biotechnol*. 10, 435–440.

Seok, K.B., Surk, S.M., and Byung, K.H. 1999. Isolation, antifungal activity, and structure elucidation of the glutarimide antibiotic, streptimidone, produced by *Micromonospora coerulea*. *J. Agric. Food Chem*. 47, 3372–3380.

Sethi, S., Ravi, K., and Saksham, G. 2013. Antibiotic production by microbes isolated from soil. *Int. J. Pharm. Res*. 4, 2967–2973.

Sharma, M. 2014. Actinomycetes: Source, identification, and their applications. *Int. J. Curr. Microbiol. Appl. Sci*. 3(2), 801–832.

Strobel, G. 2003. Endophytes as sources of bioactive products. *Microbes Infect*. 5, 535–544.

Trujillo, M., Kroppenstedt, R., Schumann, P., Carro, L., and Martínez-Molina, E. 2006. *Micromonospora coriariae* sp. nov., isolated from root nodules of *Coriaria myrtifolia*. *Int. J. Syst. Evol. Microbiol*. 56, 2381–2385.

Vijayakumari, S.J., Sasidharannair, N.K., and Nambisan, B. 2013. Optimization of media and temperature for enhanced antimicrobial production by bacteria associated with *Rhabditis* sp. *Iran. J. Microbiol*. 5, 136–141.

Wang, L.W., Zhang, Y.L., Lin, F.C., Hu, Y.Z., and Zhang, C.L. 2011. Natural products with antitumor activity from endophytic fungi. *Mini Rev. Med. Chem*. 11, 1056–1074.

Xiao-Yan, S., Qing-Tao, S., Shu-Tao, X., Xiu-Lan, C., Cai-Yun, S., and Yu-Zhong, Z. 2006. Broad-spectrum antimicrobial activity and high stability of trichokonins from *Trichoderma koningii* SMF2 against plant pathogens. *FEMS Microbiol. Lett*. 260(1), 119–125.

Xin, L., Yaojian, H., Meijuan, F., Jianfeng, W., Zhonghui, Z., and Wenjin, S. 2006. Cytotoxic and antimicrobial metabolites from marine lignicolous fungi, *Diaporthe* sp. *FEMS Microbiol. Lett*. 251(1), 53–58.

Yang, Z. 2000. Antimicrobial compounds and extracellular polysaccharides produced by lactic acid bacteria: Structures and properties. Academic dissertation, Department of Food Technology, University of Helsinki, Helsinki, Finland.

Yu, H., Lei, Z., Lin, L., Zheng, C., Lei, G., Wenchao, L., Peixin, S., and Luping, Q. 2010. Recent developments and future prospects of antimicrobial metabolites produced by endophytes. *Microbiol. Res*. 165, 437–449.

Zhao, J., Yan, M., Tijiang, S., Yan, L., Ligang, Z., Mingan, W., and Jingguo, W. 2010. Antimicrobial metabolites from the endophytic fungus *Pichia guilliermondii* isolated from *Paris polyphylla* var. yunnanensis. *Molecules* 15, 7961–7970.

Zhang, Z., Kudo, T., Nakajima, Y., and Wang, Y. 2001. Clarification of the relationship between the members of the family *Thermomonosporaceae* on the basis of 16S rDNA, 16S–23S rRNA internal transcribed spacer and 23S rDNA sequences and chemotaxonomic analyses. *Int. J. Syst. Evol. Microbiol*. 51, 373–383.

chapter four

Animal fecal actinomycetes
A new source for the discovery of drug leads

Yi Jiang, Li Han, Xueshi Huang, and Chenglin Jiang

Contents

4.1   Introduction

The clinical demand for new drugs worldwide is substantial and extremely urgent, reflecting the rapid expansion of dangerous diseases, including influenza, SARS, helopyra, tuberculosis, and HIV, the frequency of multiple common ailments (cancer, hypertension, diabetes, hyperlipidemia, etc.), the emergence of new diseases without known causes, and the evolution and spread of antibiotic-resistant pathogens (Payne et al., 2007; Jiang et al., 2008; Goodfellow et al., 2010).

Based on the most recent statistics obtained in 2012, the success rate for the development of new drugs from synthetic compounds is only 0.005%, while the rate for the development of whole natural products is 0.6% and that for the development of natural microbial products is 1.6% (Bérdy, 2012). Most of the drugs currently available on the market are those obtained from the synthesis or semisynthesis of natural products. Thus, microorganisms remain an important source for the development of new drugs. The bioactive compounds recovered from microbes can be modified using gene manipulation to obtain more superior or ideal drugs, which are subsequently produced on a large scale.

Actinomycetes (Actinobacteria) have recently received much attention as these bacteria produce a variety of natural drugs and other bioactive metabolites, including antibiotics, enzyme inhibitors, and enzymes. More than 22,000 bioactive secondary metabolites (including antibiotics) from microorganisms have been identified and published in the scientific and patent literature, and approximately half of these compounds are produced by actinomycetes. Currently, approximately 160 antibiotics have been used in human therapy and agriculture, and 100–120 of these compounds, including streptomycin, erythromycin, gentamicin, and avermectin, are produced by actinomycetes (Bérdy, 2005). However, the use of general approaches to develop new drugs from actinomycetes growing in common habitats is extremely difficult (Jiang et al., 2009). Although several microorganisms have been identified, described, screened, and used in many applications, more than 90% of all microorganisms remain unknown (Jennifer et al., 2001; Kaeberlein et al., 2002; Zengler et al., 2002; Joseph et al., 2003; Chiao, 2004; Handelsman 2004; Patrick and Handelsman, 2005; Lior 2007; Ibrahim et al., 2009), and these unknown microbes might offer hope for the development of new drugs.

To overcome the challenges of drug development from microbial metabolites, new concepts based on genomics have been described, that is, "new habitats, new methods, new species, new gene clusters, new products, and new uses" (Jiang et al., 2009; Jensen, 2010; Xu et al., 2010). Thus, novel microbes should contain new gene clusters that synthesize novel secondary metabolites to obtain new bioactive compounds (Jiang et al., 2009). Many laboratories and companies have focused on new actinomycete resources from new habitats, such as oceans (Bull and Stach, 2007; Abdelmohsen et al., 2010; Blunt, 2011; Jiang et al., 2011; Blunt et al., 2013; Subramani and Aalbersberg, 2013), extreme environments (Jiang et al., 2006; Javad et al., 2013), and plants, for the development of new drugs. Generating pure cultured microorganisms from uncultured sources might provide information concerning new drug leads. Actinomycetes remain an important source for the development of new natural drugs. Baltz (2008) proposed a "Renaissance in antibacterial discovery from actinomycetes."

Animal intestinal and fecal microorganisms (fecal microbiota) have been studied for decades (Savage, 1977). A large number of microbes exist in the gastrointestinal tract and feces of animals. The intestinal microbial community, comprising 10^{13} to 10^{14} microorganisms, outnumbers somatic and germ cells by at least one order of magnitude (Simpson et al., 2002). However, most of these microorganisms have not been cultured (Daly et al., 2001; Greetham et al., 2002; Simpson et al., 2002; Suchodolski et al., 2004, 2008; Ritchie et al., 2008; Durso et al., 2010; Run-chi et al., 2010; Zhang and Chen, 2010). Exploring and utilizing the enormous beneficial resources from fecal microbiota are a tempting challenge. For example, probiotics as dietary supplements containing friendly bacteria have been widely used to recover microbial balance, improve intestinal and overall health, and protect against disease (Falagas et al., 2006; Meyer et al., 2006).

Every species of animal possesses a specific intestinal microbial community (intestinal or fecal microbiota), reflecting coevolution and natural selection between microbes and hosts. However, the relationship between microorganisms and hosts remains elusive due to the complexity of the internal ecological system (Hooper and Gorden, 2001; Curtis and Sperandio, 2011). Notably, animals are sterile before birth. However, after birth, the combined effects of the living environment (food, air, water, climate, etc.) contribute to the establishment of relationships with beneficial, harmful, anaerobic, and aerobic microorganisms; subsequent adaptation between the microorganism and its host results in the gradual formation of an extremely complex and relatively stable microbial flora in these animals. The microbial flora gradually changes with increasing age and changing habitat. For example, the intestinal microbiota changes when animals are treated with antibiotics and subsequently slowly returns to normal after the treatment is completed. Indeed, the relationships among hosts and microbes and between different microbes are extremely complex and continuously evolving.

Notably, all animals eat *dirty* and *raw* foods every day but rarely become sick. This phenomenon is quite common, suggesting that animals have strong antibacterial and immune mechanisms, and the microorganisms in the intestinal tract are important components of these immune mechanisms. With respect to the host, microorganisms play an important role in food digestion, absorption, immunity, antimicrobial defenses, and health maintenance, producing a variety of bioactive substances (such as antibiotics, inhibitor agents, immune inhibitors, vitamins, various enzymes, and enzyme inhibitors). These roles reflect the coevolution and natural selection between microbes and hosts. Indeed, these microorganisms and their bioactive products are nontoxic to the host, generated through a long-term toxic test involving the symbiosis between microorganisms and hosts. Thus, nontoxic microbes provide an important advantage for the development of medicines, pesticides, and health products.

Actinomycetes, as human and animal pathogens, have been widely studied (Beman, 1983). However, until recently, there have been few studies concerning the use of fecal actinomycetes as a source for the discovery of novel drug. To obtain more unknown actinomycetes for the discovery of new bioactive metabolites, fecal samples from 48 species of carnivorous, herbivorous, and omnivorous animals, including primates, mammals (perissodactyl, artiodactyl, and ruminant), birds, amphibians, fishes, and insects, were collected. The fecal actinomycetes were isolated, cultivated, and identified. The antimicrobial and enzymatic activities and synthesis of five antibiotics from selected strains were determined. The metabolites produced from selected highly bioactive strains were studied. A mixture of microbial manure containing fecal streptomycetes was applied for the preventive treatment of soil-borne notoginseng diseases in the field. Some results are reported herein.

4.2 Isolation methods of animal fecal actinomycetes

4.2.1 Collection and pretreatment of samples

Fresh fecal samples were collected from 48 selected animal species in the Yunnan Wild Animal Park, Kunming, and South China Sea, China. Some samples were collected from the original animal habitats. The samples were immediately transferred to sterile dishes and dried for 10 days at 28°C. Two grams of each dried sample was pretreated at 80°C for 1 h and subsequently dissolved in 18 mL of sterile water containing 0.1% $Na_4P_2O_5$, followed by shaking at 220 rpm for 60 min. The suspension was treated with ultrasound waves for 40 s at 150 W before coating (Jiang et al., 2010). The suspension was diluted from 10^{-1} to 10^{-8}, and three dilutions, 10^{-5}, 10^{-6}, and 10^{-7}, were used for isolating actinomycetes.

4.2.2 Isolation medium for actinobacteria (per liter)

4.2.2.1 HV medium
(Hayakawa and Nonomura, 1987)

4.2.2.2 YIM 171
Glycerol 10 g, asparagine 1 g, $K_2HPO_4·H_2O$ 1 g, $MgSO_4·7H_2O$ 0.5 g, $CaCO_3$ 0.3 g, vit mixture of HV medium 3.7 mg, and agar 15 g; pH 7.2

4.2.2.3 YIM 212
Mycose 5 g, proline 1 g, $(NH4)2SO_4$ 1 g, NaCl 1 g, $CaCl_2$ 2 g, K_2HPO_4 1 g, $MgSO_4·7H_2O$ 1 g, vit mixture of HV medium 3.7 mg, and agar 15 g; pH 7.2

4.2.2.4 YIM 47
Na_2HPO_4 0.5 g, KCl 1.7 g, $MgSO_4·7H_2O$ 0.05 g; $FeSO_4·7H_2O$ 0.01 g, $CaCl_2$ 1 g, soy bean flour 0.2 g, lignin 1 g, vit mixture of HV medium 3.7 mg, soil extract 100 mL, and water 900 mL; pH 7.5

4.2.2.5 YIM 601
Solution starch 10 g, casein 0.3 g, KNO_3 2 g, $MgSO_4·7H_2O$ 0.05 g, NaCl 2 g, K_2HPO_4 2 g, $CaCO_3$ 0.02 g, $FeSO_4$ 10 mg, vit mixture of HV medium 3.7 mg, and agar 15 g; pH 7.2 ~ 7.4

4.2.3 Inhibitors

All media were supplemented with four filter-sterilized mixtures or single solutions containing inhibitors against fungi and gram-negative bacteria (per liter): (1) 50 mg cycloheximide, 50 mg nystatin, 20 mg nalidixic acid, and 3 mg penicillin; (2) 100 mg cycloheximide, 100 mg nystatin, 40 mg nalidixic acid, and 5 mg penicillin; (3) 50 mg $K_2Cr_2O_7$ and 5 mg penicillin; and (4) 75 mg $K_2Cr_2O_7$ and 5 mg penicillin.

The plate dilution method was used to isolate actinobacteria from the sample suspension. Approximately 0.1 mL of each sample (10^{-5}, 10^{-6}, and 10^{-7} dilutions) was used to coat the plates and cultivated for 7–35 days at 28°C. Subsequently, the colonies were counted, and a single *Actinobacteria* colony was picked to inoculate a slant with the same isolation medium. The pure strains were cultured at 4°C and in 20% of glycerol at –20°C.

4.2.4 Effect of isolation media

The quantity of cultivable actinomycetes in mixed samples containing eight species of animal feces was 7–21 × 10^9, and that of other bacteria was 6 × 10^9; mostly, the growth of gram-negative bacteria was suppressed using inhibitors. The optimum fecal suspension dilutions for isolating actinobacteria were 10^{-6} and 10^{-7}, in which approximately 22–133 colonies were observed on the isolation plates (Table 4.1). However, the optimum concentration for each animal fecal sample should be determined in advance.

Table 4.2 shows the results of the selective isolation of actinobacteria with five media from fecal samples obtained from 13 species of animals. In Table 4.1, about 123 pure

Table 4.1 Cfu/g^a of actinobacteria on YIM 171 isolation medium at different dilutions

Dilution times	Actinobacteria		Other bacteria	Fungi
	Mixture fecal samples of seven species of animal	Fecal sample of *Vicugna pacos*		
4th	1634 × 10^5	1324 × 10^5	132 × 10^5	0
5th	204 × 10^6	188 × 10^6	87 × 10^6	0
6th	133 × 10^7	98 × 10^7	32 × 10^7	0
7th	66 × 10^8	22 × 10^8	11 × 10^8	0
8th	21 × 10^9	7 × 10^9	6 × 10^9	0
CK^b	17 × 10^8		486 × 10^8	11 × 10^7

^{a} cfu/g = colony-forming units.
^{b} CK = 7th dilution without inhibitors, and single actinomycete colonies were not obtained.

Table 4.2 Effect of the selective isolation of actinobacteria from the fecal samples of 17 animal species using five different media (number of strains obtained)

Sample source	YIM medium No.					Total
	HV	47	171	212	601	
Hylobates hoolock	19	6	15	11	16	67
Panthera tigris	12	9	22	23	32	98
Panthera tigris altaica	11	6	15	21	19	72
Ailuropoda melanoleuca	15	16	18	21	17	87
Viverra zibetha	12	10	12	19	9	62
Cavnlvara zlrsidae	19	9	18	33	12	91
Vicugna pacos	38	39	27	36	16	156
Rhinoceros sondaicus	43	8	34	33	22	140
Buceros bicornis	11	7	20	14	12	64
Aceros undulatus	12	4	15	6	20	57
Testudo elephantopus	3	7	7	3	0	20
Ursus thibetanus	19	9	18	33	12	91
Cervus nippon	33	22	18	36	19	128
Total	247	152	239	289	206	1123

cultivated strains of actinomycetes were isolated. Among all the strains, 156 and 140 were isolated from *Vicugna pacos* and *Rhinoceros sondaicus*, respectively. Only 20 strains were isolated from *Testudo elephantopus*. YIM 212, HV, and YIM 171 media were better for isolating actinobacteria, resulting in the identification of 289, 247, and 239 strains of actinobacteria, respectively.

4.2.5   Key points for isolating actinobacteria from animal feces

The abundance of gram-negative bacteria in animal feces presents a major challenge for the isolation of fecal actinobacteria. To eliminate gram-negative bacteria and fungi and to obtain more unknown actinobacteria for discovering novel lead compounds, some key points for sampling and isolation should be considered.

First, based on the results of previous experiments, it is best to collect fresh fecal samples from wild animals living in original habitats; second, the fresh samples should be dried at 25°C–28°C for 7–10 days; third, the dried samples should be treated for 60 min at 80°C, and the fecal suspension should be treated with ultrasound waves for 40 s at 150 W before coating (Jiang et al., 2010); fourth, potassium bichromate 50 and 5 mg penicillin or nystatin 50 mg, nalidixic acid 20 mg, and 5 mg penicillin should be added per liter of isolation medium to inhibit the growth of fungi and gram-negative bacteria; fifth, the samples should be diluted to 10^{-5}, 10^{-6}, and 10^{-7}, and the optimum dilution concentration for each animal fecal sample should be determined in advance; sixth, YIM 212, YIM 171, and HV media are better for the isolation of fecal actinobacteria, and these media should be improved and constantly updated with respect to different samples; and seventh, all experiments should be performed under strict sterile conditions for avoiding spread of pathogen.

Animal fecal actinomycetes represent a new field of study. The physiological features of these bacteria are not understood. Therefore, the method for the selective isolation of actinomycetes (including the isolation media, pH, inhibitors, sample pretreatment, and culture temperatures) should be different from those used to isolate bacteria from soil, sea, and plant samples, and this method should be continually improved and updated for isolation from different fecal samples.

Some actinobacteria were the pathogen of humans and animals (Beman, 1983). For example, *Cellulosimicrobium funkei*, which was first discovered in a clinic, was an opportunistic pathogen, close to *Oerskovia* in taxonomy, and mainly infected people who have immune dysfunction or inflammation (Brown et al., 2006; Petkar et al., 2011); this was found in our study, too. Therefore, the fresh fecal samples have to be collected from health bodies. The whole length of research work should be carried under strict sterile conditions for avoiding pathogens to interference researchers.

4.3   Diversity of animal fecal actinomycetes

4.3.1   Identification of pure cultivated actinobacteria

A total of 3049 pure strains were isolated from the feces samples obtained from 48 animal species; 1869 strains were obtained after eliminating duplicate strains based on morphological and cultural characteristics. The DNA was extracted from pure strains for 16S rDNA analysis (Orsini and Romano-Spica, 2001). PCR amplification of the 16S rDNA, followed by purification and sequencing of the PCR products, was performed as previously described (Cui et al., 2001). The forward primer F8 (8 ± 27; 5′-GAG AGT TTG ATC CTG GCT CAG-3′) and the reverse primer (1510 ± 1492; 5′-GGT TAC CTT GTT ACG ACT T-3′) were used. The resulting sequences were manually aligned using the sequences from available, public databases. Phylogenetic trees were inferred using neighbor-joining (Saitou and Nei, 1987) and maximum-likelihood methods (Felsenstein, 1981). All pure cultivated strains were identified at a genus and species level.

4.3.2   Diversity of actinobacteria

The actinomycete communities in 19 of the 48 fecal samples are described.

4.3.2.1   Hoolock gibbon (Hylobates hoolock)

The hoolock gibbon is a member of the *Hylobatidae* primates. This primate is classified as a class I protected species according to the Law of the People's Republic of China on the Protection of the Wildlife (LPW), the Convention on International Trade in Endangered Species of Wild Fauna and Flora (CITES), and the International Union for Conservation of Nature and Natural Resources (IUCN). Fresh fecal samples from two individual primates were collected at three different times, and a total of 108 pure cultured strains were isolated. After eliminating several duplicate strains based on morphological and cultural characteristics, 76 strains were selected, and the 16S rDNA sequences were determined. A phylogenetic analysis was performed. The strains were identified at the genus and species levels. The isolated strains comprised 12 genera of actinobacteria: *Arthrobacter, Cellulosimicrobium, Corynebacterium, Dietzia, Gulosibacter, Kocuria, Microbacterium, Nocardia, Oerskovia, Rhodococcus, Streptomyces,* and *Zimmermannella.* The molinate-degrading actinomycete *Gulosibacter,* initially identified by Célia et al. (2004), currently contains three species. Ten other bacteria, *Acinetobacter, Bacillus, Jeotgalicoccus, Kurthia, Leuconostoc, Planococcus, Psychrobacter, Pseudomonas, Psychrobacillus,* and *Rummeliibacillus,* were also identified.

4.3.2.2   Yunnan snub-nosed monkey (Rhinopithecus bieti)

The rare Yunnan snub-nosed monkey belongs to *Cercopithecidae* primates. This monkey is classified as class I protected animal of China according to the CITES and the IUCN. Fresh fecal samples from three individuals were collected at three different times. A total of 122 strains were isolated, and 66 strains were identified, belonging to 13 genera of actinobacteria: *Agrococcus, Arthrobacter, Cellulosimicrobium, Citricoccus, Corynebacterium, Gordonia, Jiangella, Kocuria, Microbacterium, Mobilicoccus, Oerskovia, Rhodococcus,* and *Streptomyces. Jiangella* are distributed in various habitats including deserts, alkaline soils, caves, and plants (Song et al., 2005; Lee, 2008; Qin et al., 2009; Kampfer et al., 2011; Tang et al., 2011) Other bacteria, namely *Bacillus, Enterococcus, Leuconostoc, Paenibacillus, Calellibacterium,* and *Myroides,* were also identified.

4.3.2.3   Assamese macaque (Macaca assamensis)

The Assamese macaque is a member of the *Cercopithecidae* primates. This primate is a class I protected animal in China. Ninety-three selected strains were identified. These Actinobacteria belonged to eight genera, including *Acinetobacter, Corynebacterium, Kocuria, Luteococcus, Microbacterium, Nocardiopsis, Rhodococcus,* and *Streptomyces.* About 11 other bacteria, namely, *Achromobacter, Acinetobacter, Bacillus, Bordetella, Enterococcus, Escherichia, Jeotgalicoccus, Methylobacterium, Planococcus, Pseudomonas,* and *Shigella,* were also identified.

4.3.2.4   Bengal tiger (Panthera tigris)

The Bengal tiger is a member of the family *Felidae* of the order *Carnivore* and has been classified as a class I protected animal of China according to the CITES and the IUCN. These animals inhabit the same type of forest as the Manchurian tiger. Around 258 strains of actinomycetes were isolated from fresh fecal samples, and 177 of these strains, belonging to 13 genera, namely, *Arthrobacter, Corynebacterium, Dietzia, Enteractinococcus, Kocuria, Microbacterium, Nocardia, Nocardiopsis, Oerskovia, Promicromonospora, Saccharomonospora, Streptomyces,* and *Yaniella,* were identified. The novel genus *Enteractinococcus* was described in the International Journal of Systematic and Evolutionary Microbiology (Cao et al., 2012). The genus *Yaniella* is typically found in saline soil (Li et al., 2004, 2008). *Streptomycetaceae* occupied 64% of actinobacteria observed, representing the predominant strain, followed by *Micrococcaceae* with 7% (Figures 4.1 and 4.2).

4.3.2.5   *Manchurian tiger* (Panthera tigris altaica)

The Manchurian tiger is classified as a class I protected animal in China according to the CITES and the IUCN, and this animal is a member of the carnivorous *Felidae* family. Fresh fecal samples from four Manchurian tigers were collected, and 302 pure cultured actinomycete strains were isolated. Among these, 107 strains were identified. Nine genera of actinobacteria were identified from these fecal samples, namely, *Arthrobacter, Enteractinococcus, Microbacterium, Nocardia, Oerskovia, Promicromonospora, Saccharomonospora, Streptomyces,* and *Yaniella.* The fecal actinomycete component in the Manchurian tiger fecal samples was less four genera than that of the Bengal tigers, although the nine identified genera were the same.

4.3.2.6   *Giant panda* (Ailuropoda melanoleuca)

The giant panda is a rare, national treasure in China, classified as a class I protected animal according to the IUCN and the CITES. This panda only inhabits a limited area of northern Sichuan, China, and is a member of the family *Ailuridae* (*Ailuropodidae*) of the order *Carnivora*; its main food is bamboo. Five individual feces were sampled. A total of 330 pure cultured actinomycete strains were isolated, and 133 of these strains were identified through phylogenetic analysis of the 16S rRNA gene sequences. These strains belonged to 13 genera of actinobacteria: *Agrococcus, Arthrobacter, Cellulomonas, Cellulosimicrobium, Janibacter, Micrococcus, Micromonospora, Mycobacterium, Oerskovia, Patulibacter, Rhodococcus, Streptomyces,* and *Verrucosispora.* The genus *Janibacter* was identified in 1997 (Martin et al., 1997). *Verrucosispora* are members of the family *Micromonosporaceae,* and these bacteria were isolated from a

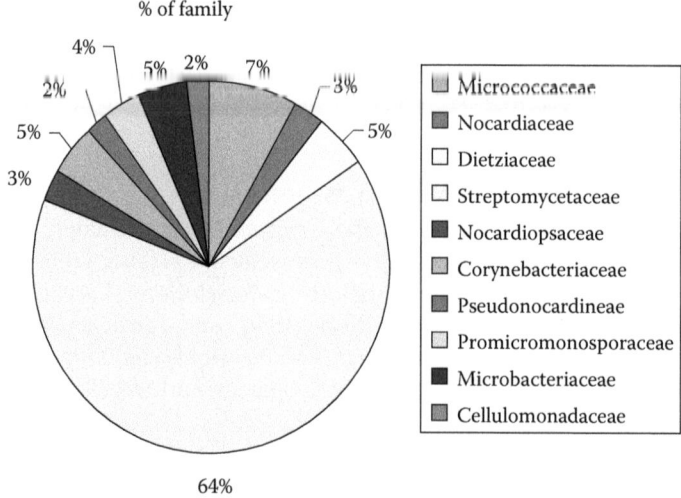

Figure 4.1 Composition of actinobacteria in *Panthera tigris* feces.

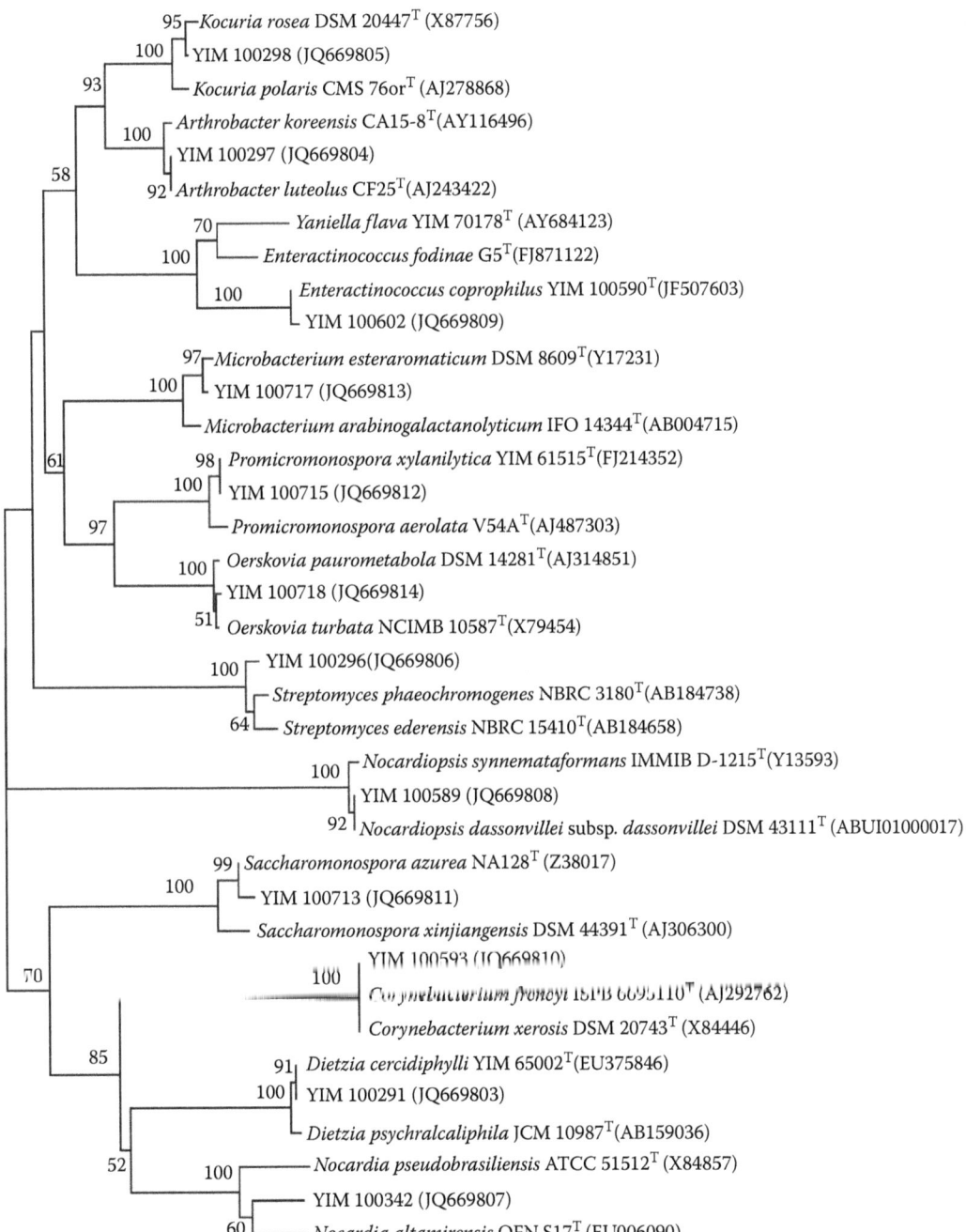

Figure 4.2 Neighbor-joining tree showing the phylogenetic relationships based on the 16S rRNA gene sequences of culturable actinomycetes isolated from the fecal samples of *Panthera tigris*. The sequences identified in the present study are bolded. The bootstrap values (expressed as percentages of 1000 replications) > 50% are indicated at the nodes. Bar = 1 nt substitution per 100 nt.

peat bog near Gifhorn, Germany, and they typically inhabit peat bogs, lakes, and oceans (Goodfellow et al., 2012b). *Patulibacter* belongs to the order *Solirubrobacterales* (Gundlapally and Ferran, 2009).

4.3.2.7   Red panda (Ailurus fulgens)

The red panda belongs to the family *Procyonidae* of the order *Carnivora*. This panda is classified as a class II protected animal in China. About 66 selected strains were identified, belonging to 10 genera of actinobacteria, namely, *Agrococcus, Arthrobacter, Corynebacterium, Gulosibacter, Leucobacter, Microbacterium, Micrococcus, Pseudonocardia, Rhodococcus,* and *Streptomyces*. Six genera of other bacteria, namely *Aerococcus, Brevundimonas, Jeotgalicoccus, Planomicrobium, Pseudomonas,* and *Psychrobacter*, were identified.

4.3.2.8   Civet (Viverra zibetha)

Civet (small Indian civet, *Viverra indica, Viverricula malaccensis thai, Viverricula hanensis, Viverricula pallida taivana,* and *Viverra pallida*) lives in tropical and subtropical rain forests and evergreen broadleaf forests. This omnivorous species are classified as class II protected animals in China according to the IUCN and the CITES. The civet is named for producing musk. Civet fecal samples were collected from three individuals in a wild animal park in Malaysia twice. A total of 88 strains of Actinobacteria were isolated using five different media. Fifty-eight strains were identified, comprising of 15 genera: *Arthrobacter, Cellulomonas, Cellulosimicrobium, Corynebacterium, Curtobacterium, Enteroactinococcus, Isoptericola, Kocuria, Leucobacter, Microbacterium, Micrococcus, Rhodococcus, Saccharopolyspora, Sanguibacter,* and *Streptomyces*. *Curtobacterium* were rarely observed in the feces of these animals. *Cellulosimicrobium* were identified as inhabitants of termites (Bakalidou et al., 2002; Stackebrantd et al., 2004). Other bacteria, including *Bacillus, Enterococcus, Flavobacterium, Hansschlegelia, Methylobacterium, Ochrobactrum, Pseudomonas, Sporosarcina,* and *Stenotrophomonas*, were also identified from the civet fecal samples.

4.3.2.9   Asiatic black bear (Ursus thibetanus)

Asiatic black bears (moon bear) are omnivorous animals widely distributed throughout the forests of broad areas from north to south in the eastern half sphere of China, Russia, Iran, Pakistan, Laos, Japan, and India. These bears are classified as class II protected animals in China according to the IUCN and the CITES. Fecal samples of three individuals were collected twice. One hundred twelve strains of actinobacteria were isolated, and 32 strains were identified. They belong to 11 genera: *Cellulosimicrobium, Dietzia, Kribbella, Gordonia, Microlunatus, Microbacterium, Nocardiopsis, Promicromonospora, Rhodococcus, Saccharomonospora,* and *Streptomyces*. Five genera of other bacteria, namely, *Massilia, Myroides, Methylobacterium, Rhizobium,* and *Stenotrophomonas*, were identified.

4.3.2.10   Shanxi sika (Cervus nippon)

Shanxi sika (sika, sika deer, spotted deer) is a class I protected animal in China according to the IUCN. These phytophagous animals are widely distributed throughout forests and grasslands. Notably, the prevalence of these animals is rapidly decreasing. Fecal samples were obtained from 12 individuals in Kunming and Shenyang, China, and 337 pure strains were isolated. Among these, 117 strains were identified, belonging to 16 genera of actinobacteria: *Actinocorallia, Agrococcus, Arthrobacter, Cellulosimicrobium, Citricoccus, Isoptericola, Kocuria, Leucobacter, Microbacterium, Mycobacterium, Promicromonospora, Nocardiopsis, Rhodococcus, Salinibacterium, Streptomyces,* and *Tsukamurella*. Only two other bacteria, namely *Bosea* and *Stenotrophomonas*, were identified.

4.3.2.11 Sambar (Rusa unicolor)

Sambars are members of the family *Cervidae* of the order *Artiodactyla*, and these animals are classified as class II protected animal in China according to the IUCN. A total of 33 selected strains were identified, including eight actinobacteria, namely, *Actinotalea, Arthrobacter, Corynebacterium, Dietzia, Kocuria, Micrococcus, Streptomyces*, and *Tessaracoccus*, and seven other bacteria, including *Bacillus, Desemzia, Paracoccus, Planococcus, Planomicrobium, Psychrobacillus*, and *Sphingomonas*.

4.3.2.12 Vicuna (Vicugna pacos)

Vicunas were originally identified in frigid zones from 3650 to 4800 m above sea level in the Andes Mountains. Humans have long reared these camels, for it produce high-grade hair. Vicunas belong to the family *Camelidae* of the order *Artiodactyla*. Three individuals were imported from the Republic of Chile into the Yunnan Wild Animal Park. Fecal samples were obtained from these animals, and 87 strains of actinomycetes were isolated. Among these, 66 strains were identified, which belong to 15 genera, namely, *Arthrobacter, Brevundimonas, Cellulosimicrobium, Corynebacterium, Dietzia, Enteractinococcus, Gordonia, Isoptericola, Kocuria, Microbacterium, Micrococcus, Nocardiopsis, Rhodococcus, Saccharomonospora*, and *Streptomyces*. The number of *Arthrobacter arilaitensis* was 80×10^5/g in the fresh fecal sample. Eleven other bacteria, including *Achromobacter, Advenella, Ancylobacter, Jeotgalicoccus, Kurthia, Lysobacter, Methylobacterium, Ornithinibacillus, Psychrobacillus, Shigella*, and *Solibacillus*, were also isolated and identified.

4.3.2.13 Common wildebeest (Connochaetes taurinus)

Common wildebeests are typically found living in the grasslands in Africa and were imported to the Yunnan Wild Animal Park. The wildebeest belongs to the family *Bovidae* of the order *Artiodactyla*. Fecal samples were obtained from four healthy individuals, and 96 purified strains were isolated. Among these, 43 strains were selected for identification. Only five genera of actinobacteria, namely, *Citricoccus, Microbacterium, Micrococcus, Rhodococcus*, and *Streptomyces*, were identified. These actinomycetes are commonly found in nature. Other bacteria, including *Stenotrophomonas* and *Methylobacterium*, were also identified.

4.3.2.14 Rhino (Rhinoceros sondaicus)

The rhino is a rare, nationally treasured animal in China, classified as a class I protected animal according to the IUCN. The rhino belongs to both *Rhinocerotidae* and *Perissodactyla*. Fecal samples were collected from two individuals in Indonesia. About 202 strains were isolated, and 112 strains were identified. Fourteen genera were identified: *Arthrobacter, Brevundimonas, Corynebacterium, Dietzia, Gulosibacter, Kocuria, Microbacterium, Micromonospora, Nocardiopsis, Promicromonospora, Pseudoclavibacter, Rhodococcus, Streptomyces*, and *Tessaracoccus. Gulosibacter* and *Tessaracoccus* are rarely observed in nature. One strain (YIM 100770), identified as a novel genus, was characterized using polyphasic taxonomic procedures (Xu et al., 2007). Nineteen other bacteria, including *Achromobacter, Alcaligenes, Ancylobacter, Bacillus, Hansschlegelia, Kurthia, Luteimonas, Methylobacterium, Massilia, Novosphingobium, Paracoccus, Pseudomonas, Psychrobacillus, Rhizobium, Shigella, Solibacillus, Sphingobacterium, Staphylococcus*, and *Stenotrophomonas*, were also identified.

4.3.2.15 Indian elephant (Elephas maximus)

The Indian elephant is typically found in south China and Asia, and the prevalence of this rare, nationally treasured animal is rapidly decreasing. Thus, this elephant is classified as a class I protected animal in China according to the IUCN. The Indian elephant belongs to the family *Elephantidae* of the order *Proboscidea*. Fresh fecal samples were obtained from four individuals in the Xiaomemgyang National Natural Protect area and Yunnan Wild Animal Park. One hundred twenty-one strains were isolated, and 68 strains were identified, comprising 15 genera of actinobacteria, including *Arthrobacter, Cellulomonas, Cellulosimicrobium, Citricoccus, Janibacter, Kocuria, Leucobacter, Microbacterium, Micrococcus, Micromonospora, Promicromonospora, Rhodococcus, Sphaerobacter, Streptomyces,* and *Verrucosispora*. Three genera of gram-negative bacteria, namely, *Bacillus, Devosia,* and *Planococcus*, were identified.

4.3.2.16   *Chinese bamboo rat* (Rhizomys sinensis)

Humans have long reared the Chinese bamboo rat for producing high protein. The bamboo rat is a member of the family *Rhizomyidae* of the order *Rodentia*. Fresh fecal samples were obtained from six individuals, and actinomycetes were isolated from these samples at two different times. A total of 306 strains were isolated, and 104 strains were identified, belonging to 13 genera, namely, *Agrococcus, Arthrobacter, Brachybacterium, Corynebacterium, Dietzia, Gordonia, Labedella, Microbacterium, Oerskovia, Rhodococcus, Sanguibacter, Streptomyces,* and *Williamsia*. Three genera of gram-negative bacteria, namely *Comamonas, Flavobacterium,* and *Psychrobacter*, were identified.

4.3.2.17   *Peacock* (Pavo cristatus)

The peacock, a member of the family *Phasianidae* of the order *Galliformes*, is classified as a class I protected animal according to the LPW, the IUCN, and the CITES. Fecal samples were obtained from 12 individuals, and the actinomycetes were isolated for 3 times. One hundred eighty-eight strains were isolated, and 118 selected strains were identified, comprising 18 genera of actinobacteria, namely, *Arthrobacter, Brevibacterium, Cellulosimicrobium, Curtobacterium, Dietzia, Gordonia, Isoptericola, Janibacter, Kineococcus, Kocuria, Leucobacter, Microbacterium, Nocardiopsis, Oerskovia, Rhodococcus, Pseudonocardia, Sanguibacter,* and *Streptomyces*, representing the largest genus observed in the present study. Other bacteria, including *Bacillus, Devosia, Lysinibacillus, Methylobacterium, Planococcus, Planomicrobium, Shigella,* and *Staphylococcus*, were also identified.

4.3.2.18   *Common black-headed gull* (Larus ridibundus)

The common black-headed gull belongs to the family *Laridae* of the order *Ciconiiformes*. It is listed in the directory by LPW and IUCN. The common black-headed gull, a typical migrant bird, is typically found in Siberia, Russia. Over 30,000 *Larus ridibundus* migrate to Dian Lake, Kunming, from Siberia between December and March every year. A total of 37 selected strains were identified from the fecal samples of these birds, belonging to 10 genera of actinobacteria, namely, *Arthrobacter, Blastococcus, Devosia, Microbacterium, Oerskovia, Paracoccus, Plantibacter, Promicromonospora, Pseudoclavibacter,* and *Streptomyces*.

4.3.2.19   *Red-crowned crane* (Grus japonensis)

The red-crowned crane belongs to the family *Gruidae* of the order *Gruiformes*. This bird has been classified as a class I protected animal according to the LPW, the IUCN, and the CITES. Forty-six selected strains were identified from the feces of these birds. Nine genera of actinobacteria were identified: *Arthrobacter, Blastococcus, Cellulosimicrobium, Microbacterium, Mycobacterium, Nocardia, Rhodococcus, Streptomyces,* and *Yaniella*. Actinomycete compositions of a total of 48 species of animal feces were studied in our laboratories recent years. Related results for rest of the 29 species of animals are not shown here.

4.3.3 Diverse features of animal fecal actinomycetes

A total of 222 genera, 53 families, and 23 orders have been collected and described in *Bergey's Manual of Systematic Bacteriology* (Goodfellow et al., 2012a). Fifty-one genera of pure culturable *Actinobacteria* were isolated and identified in fecal samples collected from only 48 animal species. These *Actinobacteria* belong to 27 families of 12 orders, representing 24% of 222 genera, 51% of 53 families, and 52% of 23 orders. An incertae sedis, *Sphaerobacter*, was isolated from Indian elephant feces. Thus, these results showed that the diversity of animal fecal actinobacteria is very rich. The actinobacteria identified herein are listed in Table 4.3.

Forty-eight other bacteria, including *Achromobacter, Acinetobacter, Advenella, Aerococcus, Alcaligenes, Ancylobacter, Aurantimonas, Bacillus, Bordetella, Bosea, Brevundimonas, Brochothrix, Catellibacterium, Desemzia, Devosia, Enterococcus, Escherichia, Flavobacterium, Hansschlegelia,*

Table 4.3 Component of the phylum Actinobacteria in the fecal samples of 48 animal species

	Family	Genus
Class I *Actinobacteria*		
Corynebacteriales	*Corynebacteriaceae*	*Corynebacterium*
	Dietziaceae	*Dietzia*
	Mycobacteriaceae	*Mycobacterium*
	Nocardiaceae	*Gordonia, Nocardia, Rhodococcus, Williamsia*
	Tsukamurellaceae	*Tsukamurella*
Frankiales	*Frankiaceae*	*Blastococcus*
	Geodermatophilaceae	*Mobilicoccus*
Jiangellales	*Jiangellaceae*	*Jiangelle*
Kineosporiales	*Kineosporiaceae*	*Kineococcus*
Micrococcales	*Micrococcaceae*	*Arthrobacter, Citricoccus, Enteractinococcus, Kocuria, Micrococcus, Yaniella*
	Brevibacteriaceae	*Brevibacterium*
	Cellulomonadaceae	*Cellulomonas*
	Dermabacteraceae	*Brachybacterium*
	Intrasporangiaceae	*Janibacter*
	Promicromonosporaceae	*Cellulosimicrobium, Isoptericola, Promicromonospora*
	Microbacteriaceae	*Agrococcus, Curtobacterium, Gulosibacter, Labedella, Leucobacter, Microbacterium, Plantibacter, Pseudoclavibacter, Salinibacterium, Zimmermannella*
	Cellulomonadaceae	*Oerskovia*
	Sanguibacteraceae	*Sanguibacter*
	Incertae sedis	*Actinotalea*
Micromonosporales	*Micromonosporaceae*	*Micromonospora, Verrucosispora*
Propionibacteriales	*Propionibacteriaceae*	*Luteococcus, Microlunatus, Tessaracoccu*
Pseudonocardiales	*Pseudonocardiaceae*	*Pseudonocardia, Saccharomonospora, Saccharopolyspora*
Streptomycetales	*Streptomycetaceae*	*Streptomyces*
Streptosporangiales	*Thermomonosporaceae*	*Actinocorallia*
	Nocardiopsaceae	*Nocardiopsis*
Class II *Thermoleophilia*		
Solirubrobacterales	*Patulibacteriaceae*	*Patulibacter*
Incertae sedis	*Incertae sedis*	*Sphaerobacter*

Jeotgalicoccus, Kurthia, Lactococcus, Luteimonas, Leuconostoc, Lysinibacillus, Lysobacter, Massilia, Methylobacterium, Myroides, Novosphingobium, Ochrobactrum, Ornithinibacillus, Paenalc ligenes, Paenibacillus, Paracoccus, Planococcus, Planomicrobium, Pseudomonas, Psychrobacter, Psychrobacillus, Shigella, Solibacillus, Sphingobacterium, Sphingomonas, Sporosarcina, Staphylococcus, Stenotrophomon as, and *Rhizobium,* were also identified.

Eighteen genera of actinobacteria were identified from fecal samples of peacock, and the actinomycete component was the complexest; the second was 16 genera in Shanxi sika deers; the third was 15 genera in civet, vicuna, and Indian elephants; and the least was only 3 genera from *Tragelaphus buxtoni* and *Python reticulates* (unknown).

Members of the genus *Streptomyces* were isolated from 100% samples, representing the most predominant microbes, and the colony-forming units (cfu)/g fresh sample ranged from 2×10^5 to 176×10^7 in different fecal samples. *Streptomyces albus, S. albidoflavus, S. griseus, S. hygroscopicus, S. rutgersensis, S. tendae,* and *S. violaceoruber* were observed at a high frequency.

Members of *Rhodococcus* were isolated and identified in the fecal samples from all animal species, representing the second most prevalent and widely distributed genus. *Rhodococcus coprophilus, Rh. corynebacterioides, Rh. equi, Rh. pyridinivorans,* and *Rh. zopfii* were most frequently observed.

The genome sizes of *Streptomyces* and *Rhodococcus,* up to 9×10^7 base pairs, are the largest genomes in actinobacteria, and some species contain 20 or more natural product biosynthetic gene clusters (Ōmura et al., 2001; Bentley et al., 2002; Mcleod et al., 2006). These results are consistent with the idea that the function of actinomycetes in the host intestinal tract is primarily influenced through bioactive substances produced by the members in these two genera.

Members of *Arthrobacter, Microbacterium,* and *Micrococcus* were identified from most of the animal feces examined. Twenty-six genera belong to the order *Micrococcales* and eight genera belong to the order *Corynebacteriales* in the 48 tested animal feces.

Based on these results, two conclusions can be drawn: first, members of both *Streptomyces* and *Rhodococcus* exhibited the widest distribution and contained the largest numbers, and second, the composition of actinobacteria with chemotypes IV–IX (Lechevalier and Lechevalier, 1970, 1980), globose, and bacilliform shapes, particularly the order *Micrococcales,* exhibited the richest diversity and were detected at a high frequency in most of animal fecal samples. These distinct features of the fecal actinobacterium community differ from those of the soil, marine, and plant communities.

Some members of rare actinobacteria, such as *Yaniella* (Li et al., 2004, 2008), were identified from the feces of two species of tigers. A strain (YIM 100708) of *Jiangella* (Tang et al., 2011), a genus widely distributed in saline and alkaline soil, desert, indoor wall material, caves, and plant stems, was also isolated from the feces of *Rhinopithecus bieti*. The members of a novel genus, *Enteractinococcus,* belonging to *Micrococcaceae,* were isolated and characterized from three species of tigers (Cao et al., 2012).

Members of *Micromonospora* were isolated from the feces of *Ailuropoda melanoleuca, Rhinoceros sondaicus, Elephas maximus,* and *Cygnus melancoryphus* (not shown). Interestingly, *Nocardia* was only isolated from *Panthera tigris altaica* and *Grus japonensis; Saccharopolyspora* was only isolated from *Viverra zibetha;* and *Verrucosispora* was only isolated from *Ailuropoda melanoleuca*. Actinomycetes possessing cell wall chemotypes II (e.g., *Actinoplanes* and *Dactylosporangium*) and III (*Actinomadura* and *Streptosporangium*) are commonly found in soil and lakes, but these genera are not isolated from the animal fecal samples examined herein.

It is worth to notice that among the 1869 sequenced strains, 16S rDNA sequence similarities of 106 strains with valid published species were below 98.5%. Thus, nearly 6% pure cultivated strains were unknown, potentially representing novel species (Stackebrandt and Gorbel, 1994; Xu et al., 2007).

4.3.4   Comparison of actinomycete diversity in different habitats

In previous studies of the actinomycetes diversity, 17 genera were isolated from soil samples collected from primeval forests in Grand Shangri-La, southwest China (Cao et al., 2009); 13 genera were isolated from soil samples from subtropical every green forest in Gulin, Sichuan (Cao et al., 2010); 26 genera were isolated from soil samples from tropical rain forests in Xishuangbanna, southwest China (Jiang et al., 2013); and 16 genera were isolated from soil samples from hypersaline soil in Qinghai, west China (Jiang et al., 2006, 2011) (Table 4.4). The aerobic mycelium of these soil actinomycetes was abundant. Fifteen genera of actinomycetes were identified from samples collected from the Baltic Sea, and the members of *Micromonospraceae* were the most (Jiang et al., 2011). Until recently, approximately 48 genera of actinobacteria were isolated and identified from marine habitats worldwide (Goodfellow and Fiedler, 2010). In the present study,

Table 4.4 Comparison of actinomycete diversity in different habitats

Habitat	Composition of actinobacteria	References
Primeval forest soil in Grand Shangri-La	*Actinomadura, Actinopolymorpha, Agromyces, Arthrobacter, Dactylosporangium, Kocuria, Lentzea, Mycetocola, Nocardia, Nocardioides, Oerskovia, Promicromonospora, Pseudonocardia, Rhodococcus, Streptomyces, Streptosporangium, Tsukamurella*	Cao et al. (2009)
Subtropical evergreen forest soil in Sichuan	*Actinomadura, Actinopolymorpha, Micromonospora, Mycobacterium, Nocardia, Nocardioides, Nonomuraea, Promicromonospora, Pseudonocardia, Rhodococcus, Saccharomonospora, Streptomyces, Verrucosispora*	Cao et al. (2010)
Tropical rain forest soil in Xishuangbanna	*Actinomadura, Actinoplanes, Actinopolymorpha, Agrococcus, Agromyces, Arthrobacter, Citricoccus, Dactylosporangium, Friedmanniella, Kribbella, Lentzea, Microbacterium, Micromonospora, Mycobacterium, Nocardia, Nocardioides, Nonomuraea, Oerskovia, Planosporangium, Promicromonospora, Pseudonocardia, Rhodococcus, Saccharopolyspora, Sphaerisporangium, Streptomyces, Streptosporangium*	Jiang et al. (2013)
Hypersaline soil in Qinghai	*Citricoccus, Corynebacterium, Isoptericola, Jiangella, Marinococcus, Myceligererans, Nesterenkonia, Nocardiopsis, Prauserella, Rhodococcus, Saccharomonospora, Salinimicrobium, Streptomonospora, Streptomyces, Yaniella, Zhihengliuella*	Jiang et al. (2006)
Baltic Sea	*Actinomadura, Actinoplanes, Amycolatopsis, Arthrobacter, Cellulomonas, Isoptericola, Kocuria, Micromonospora, Microbacterium, Myceligenerans, Mycobacterium, Nocardiopsis, Promicromonospora, Rhodococcus, Streptomyces*	Jiang et al. (2011)
48 species of animal feces	*Actinocorallia, Actinotalea, Agrococcus, Arthrobacter, Brachybacterium, Brevibacterium, Brevundimonas, Cellulomonas, Cellulosimicrobium, Citricoccus, Corynebacterium, Curtobacterium, Dietzia, Enteractinococcus, Gordonia, Gulosibacter, Isoptericola, Janibacter, Jiangella, Kineococcus, Kocuria, Labedella, Leucobacter, Luteococcus, Microbacterium, Micrococcus, Micromonospora, Mobilicoccus, Mycobacterium, Nocardia, Nocardiopsis, Oerskovia, Patulibacter, Plantibacter, Promicromonospora, Pseudoclavibacter, Pseudonocardia, Rhodococcus, Saccharomonospora, Saccharopolyspora, Salinibacterium, Sanguibacter, Sphaerobacter, Streptomyces, Tessaracoccus, Tsukamurella, Verrucosispora, Williamsia, Yaniella, Zimmermannella*	This study

51 genera of actinobacteria were isolated from only 48 species of animal feces, and 23% of 222 genera, 51% of 53 families, and 52% of 23 orders were classified in Bergey's Manual. Therefore, actinomycete communities of animal feces are not only complex but also different from those in soil, extreme environments, and marine environments. More than one million animal species have been identified worldwide. Thus, animal fecal actinomycetes are tremendous natural resources.

4.4   Bioactivities of animal fecal actinomycetes

4.4.1   Antimicrobial activity

The strains were fermented in nutrient broth (YIM 61, soybean meal 20 g, glucose 10 g, peptone 4 g, K_2HPO_4 1 g, $MgSO_4 \cdot 7H_2O$ 0.5 g, NaCl 1 g, $CaCO_3$ 2 g, water 1000 mL, pH 7.8), shaking for 7 days at 28°C. The agar diffusion method was used to determine the antimicrobial activities against *Bacillus subtilis* (DSM 3258[T]), *Staphylococcus aureus* (DSM 30501[T]), *Mycobacterium tuberculosis avium* (unpathogen from Dr. Lixin Zhang), *Candida albicans* (DSM 24506[T]), and *Aspergillus niger* (IAM 190). The results from the experiments performed with animal fecal actinomycetes are shown in Table 4.5.

Table 4.5 Antimicrobial activities of some fecal actinobacteria

Source of Strains	Number of test strains	*Bacillus subtilis*	*Staphylococcus aureus*	*Mycobacterium tuberculosis*	*Candida albicans*	*Aspergillus niger*
Hylobates hoolock	26	4	1	1	0	11
Rhinopithecus bieti	29	4	1	1	3	14
Panthera tigris altaica	21	4	0	0	2	1
Panthera tigris tigris	13	1	0	0	2	0
Ailuropoda melanoleuca	28	2	3	2	2	0
Viverricula indica	7	5	4	1	0	4
Helarctos malayanus	3	0	0	0	0	0
Elephas maximus	32	5	8	0	12	6
Cervus Nippon	43	5	3	0	7	1
Cervus elaphus	32	3	6	0	3	10
Elaphurus davidianus	18	4	1	0	6	0
Giraffa camelopardalis	7	0	0	0	3	0
Connochaetes taurinus	6	0	1	0	1	0
Vicugna pacos	18	4	1	0	7	1
Oryx leucoryx	9	4	2	0	4	0
Tragelaphus buxtoni	7	2	3	0	5	0
Aceros undulatus	3	0	0	0	0	0
Pavo cristatus	3	0	0	0	0	0
Cygnus cygnus	16	8	2	3	4	2
Struthio camelus	5	1	1	1	1	3
Indotestudo elongata	11	1	0	0	1	2
Python reticulates	47	27	15	14	13	6
Total	384	84	52	23	76	61
%		22	13	6	20	16

The antimicrobial activities of 384 strains isolated from 22 species of animal feces were determined using the agar diffusion method (Table 4.5). Eighty-four strains (22%) showed inhibitory activity against *Bacillus subtilis*, 52 strains (13%) showed activity against *Staphylococcus aureus*, 23 strains (6%) showed activity against *Mycobacterium tuberculosis*, 76 strains (20%) showed activity against *Candida albicans*, and 61 strains (16%) showed activity against *Aspergillus niger*. Notably, 13–55% of the 47 strains isolated from *Python reticulates*, which primarily included streptomycetes, exhibited higher inhibitory activities against all five tested microbes. Fewer anti–*Mycobacterium tuberculosis* strains (6%) exhibited antimicrobial activity. Three strains isolated from *Helarctos malayanus, Aceros undulatus,* and *Pavo cristatus* feces did not exhibit antimicrobial activities. These results showed that actinobacteria from animal feces have wide antimicrobial activities.

4.4.2   Enzymatic activities

The activities of 19 enzymes were determined using the API ZYM Kit (Biomèrieux). The hydrolysis of cellulose and chicken hair were determined using the methods of Shirling and Gottlieb.

The enzymatic activities of 233 strains from feces were determined using the API ZYM Kit (Figure 4.3). More than 90% of the tested strains showed activity for five enzymes, namely, alkaline phosphatase, acid phosphatase, leucine arylamidase, naphthol-AS-BI-phosphohydrolase, and α-glucosidase. A total of 10%–90% of strains showed activity for 12 other enzymes. Less than 10% of the strains produced α-mannosidase and α-fucosidase. No strains showed β-glucuronidase activity (Figure 4.3). The strains isolated from 13 species of animal feces were able to hydrolyze cellulose and chicken hair; however, those isolated from *Panthera tigris altaica*, a carnivorous animal, were not able to hydrolyze cellulose. More than 90% of strains from *Cervus elaphus, Giraffa camelopardalis, Vicugna pacos, Elaphurus davidianus,* and *Oryx leucoryx* were able to hydrolyze chicken hair. Approximately 80% of strains from *Giraffa camelopardalis, Equus burchelli,* and *Oryx leucoryx* were able to hydrolyze cellulose. Only a few strains isolated from *Panthera tigris altaica* and *Ailuropoda melanoleuca* showed activity for the hydrolysis of chicken hair, and a few strains from

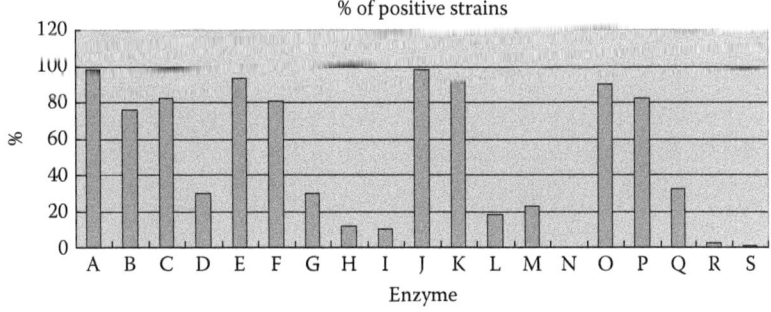

Figure 4.3 Enzyme activities of some actinomycetes isolated from animal feces: A = alkaline phosphatase, B = esterase (C4), C = esterase lipase (C8), D = lipase (C14), E = leucine arylamidase, F = valine arylamidase, G = cysteine arylamidase, H = trypsinase, I = α-chymotrypsinase, J = acid phosphatase, K = naphthol-AS-BI-phosphohydrolase, L = α-galactosidase, M = β-galactosidase, N = β-glucuronidase, O = α-glucosidase, P = β-glucosidase, Q = *N*-acetyl-β-glucosaminidase, R = α-mannosidase, and S = α-fucosidase.

Hypoblasts hoolock, Rhinopithecus bieti, Ailuropoda melanoleuca, and *Elephas maximus* showed activity for the hydrolysis of cellulose (Figure 4.4).

The activity of 233 strains for the hydrolysis of cellulose and chicken hair was determined (Figure 4.4). The results showed that all strains were able to hydrolyze cellulose and chicken hair, except those isolated from the carnivorous animal *Panthera tigris altaica.* More than 90% of strains isolated from *Cervus elaphus, Giraffa camelopardalis, Vicugna pacos, Elaphurus davidianus,* and *Oryx leucoryx* were able to hydrolyze chicken hair. Approximately 80% of strains from *Giraffa camelopardalis, Equus burchelli,* and *Oryx leucoryx* were able to hydrolyze cellulose. The strains isolated from *Panthera tigris altaica* were unable hydrolyze cellulose, and 20% of the strains isolated from *Panthera tigris altaica* and *Ailuropoda melanoleuca* were able to hydrolyze chicken hair. Approximately 20% of the strains isolated from *Hypoblasts hoolock, Rhinopithecus bieti, Ailuropoda melanoleuca,* and *Elephas maximus* were able to hydrolyze cellulose (Figure 4.4). Two strains, YIM 100110 and YIM 100135, were able to completely hydrolyze chicken hair at 28°C after 1 week (Figure 4.5). YIM 100118 was also

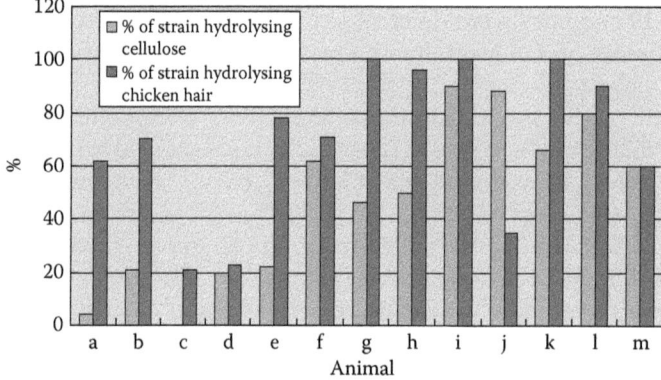

Figure 4.4 Hydrolytic activity of some actinomycetes for cellulose and chicken hair: a = *Hypoblasts hoolock,* b = *Rhinopithecus bieti,* c = *Panthera tigris altaica,* d = *Ailuropoda melanoleuca,* e = *Elephas maximus,* f = *Cervus Nippon,* g = *Cervus elaphus,* h = *Elaphurus davidianus,* i = *Giraffa camelopardalis,* j = *Equus burchelli,* k = *Vicugna pacos,* l = *Oryx leucoryx,* and m = *Tragelaphus buxtoni.*

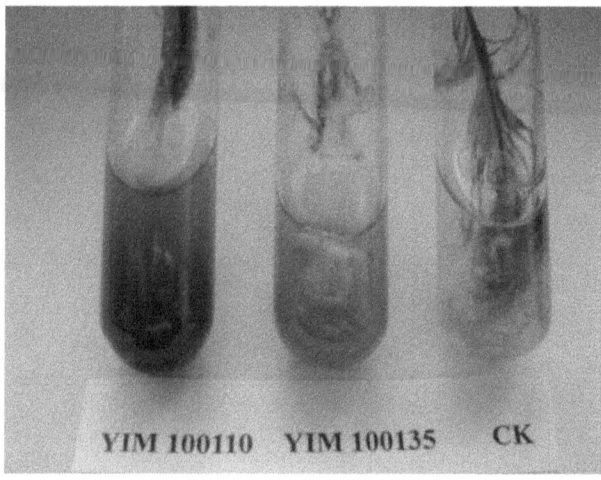

Figure 4.5 Hydrolysis of chicken hair by YIM 100110, YIM 100135, and control (CK).

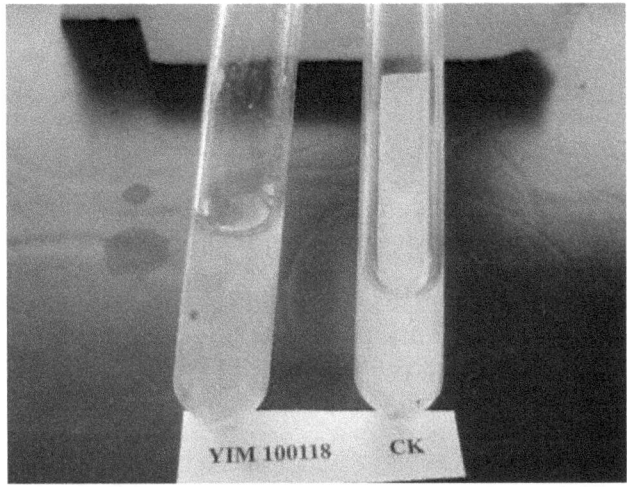

Figure 4.6 Hydrolysis of cellulose by YIM 100118 and control (CK).

able to completely hydrolyze cellulose under the same conditions (Figure 4.6). Thus, the enzyme preparation developed from animal fecal actinomycetes was significant.

4.4.3   Antitumor activities

The antitumor activities of 238 fecal actinobacterial strains were determined using HL60, HepG-2, Skov-3, A431, and K562 cell lines in vitro. The results are shown in Table 4.6. Approximately 33% and 30% of the tested strains showed K562 and HL60 cell line inhibition activity, respectively, and more than 50% of the strains from *Rhinopithecus bieti*

Table 4.6 Antitumor activities of some fecal actinomycete strains

Source of strains	K562 Cell line		HL60 Cell line	
	Number of test strains	Number of strains >90% inhibition	Number of test strains	Number of strains >60% inhibition
Hylobates hoolock	19	12	16	4
Rhinopithecus bieti	24	12	24	13
Panthera tigris altaica	13	6	12	2
Ailuropoda melanoleuca	23	12	18	0
Elephas maximus	37	13	24	11
Cervus Nippon	25	4	42	14
Cervus elaphus	19	0	20	5
Giraffa camelopardalis	0	0	12	0
Elaphurus davidianus	19	5	23	11
Connochaetes taurinus	14	0	5	2
Vicugna pacos	22	7	11	0
Oryx leucoryx	13	5	25	9
Total	238	76	232	69
%		33		30

could inhibit K562 and HL60. The IC_{50} of the crude extracts from some strains was below 4 µg/mL. Several of the tested strains showing antitumor activities exhibited the distinct features of actinomycetes from animal feces.

4.4.4　Genes encoding the biosynthetic enzymes of five antibiotics

The genes encoding the biosynthetic enzyme of five antibiotics developed from 201 strains from 15 species of animal feces were analyzed. The genes encoding the biosynthetic enzymes of type I and II polyketide synthases (PKSs), nonribosomal peptide synthetase (NRPS), and polyene cytochrome P450 (CYP) hydroxylase were determined through PCR using specific primers (Cao et al., 2010). The 3-amino-5-hydroxybenzoic acid (AHBA) gene was determined using the method of Hui-tu et al. (2009), and the results are shown in Figure 4.7.

The results from Figure 4.7 shows that 19%, 15%, 34%, and 22% of strains contained PKS I, PKS II, NRPS, and CYP genes, respectively. Many of the strains isolated from *Rhinopithecus bieti*, *Panthera tigris altaica*, *Cavnlvara zlrsidae*, *Vicugna pacos*, *Rhinoceros sondaicus*, and *Pavo cristatus* possess these four genes. None of these genes were detected in the strains from *Cervus nippon* and *Equus burchelli*. The gene encoding the biosynthetic enzyme AHBA, involved in ansamycin biosynthesis, was determined; however, no positive strains were detected. These results showed that fecal actinobacteria contained various genes encoding biosynthetic enzymes, implicating these species as a new source for developing novel bioactive metabolites (Figure 4.5). The AHBA gene was detected in only one strain (YIM 100801) isolated from the feces of *Viverra zibetha*, suggesting that this gene is not well expressed in animal feces.

In previous studies of the antimicrobial activities of actinomycetes, we used the five test strains described earlier. The study of soil actinomycetes from three areas in Yunnan and Sichuan, southwest China (Cao et al., 2010), showed that 12%–15% streptomycetes and 5%–12% rare actinomycetes presented antimicrobial activities and, in addition, 4.3%–21% actinomycetes from Baltic Sea presented antimicrobial activities. Fecal actinomycetes showed a wide spectrum of antimicrobial activities. The antimicrobial activities of 486 strains isolated from animal feces were detected (data not shown). The results show that 15%–27% of strains could inhibit *Bacillus subtilis*, 8%–30% of strains inhibited *Staphylococcus aureus*, 2%–9% of strains inhibited *Mycobacterium tuberculosis avium*, 11%–20% of strains inhibited *Candida albicans*, and 2%–18% of strains inhibited *Aspergillus niger*. Some strains generated large zones of inhibition, up to 60 mm.

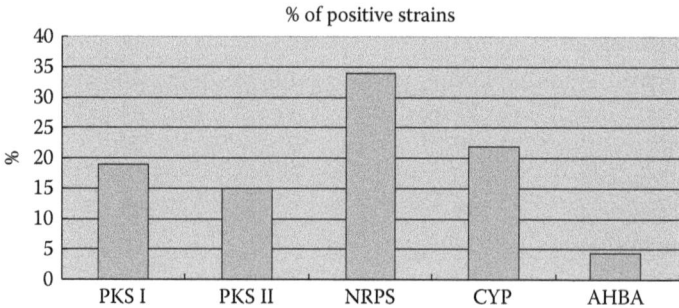

Figure 4.7 The genes encoding the biosynthetic enzyme of five compounds produced from fecal actinomycetes.

Fecal actinomycetes also showed a wide spectrum of antitumor abilities. At least 30% of fecal actinomycete strains could inhibit two types of tumor cell lines, and more than 50% of the strains isolated from *Rhinopithecus bieti* could inhibit K562 and HL60. Moreover, the fermented crude extracts from some of these strains showed high inhibition activities against tumor cells, with an IC_{50} below 4 µg/mL. Fecal actinomycetes also contained various enzymes showing high activity for the degradation of difficult substances, such as cellulose and chicken hair. These active symbiotic fecal actinomycetes could provide enormous benefits to hosts, such as to improve food digestion and absorption, maintain the microbial balance in the intestinal tract, provide resistance against pathogens and tumors, and improve the overall health of hosts.

4.4.5   Toxins of two streptomycetes to mice

Fifteen mice were fed with two *streptomycetes*, T005 and T019, containing living cell 10×25.2^8/mL and 10×36.6^8/mL, respectively, for 15 days. All tested mice did not die, and their body weights were not reduced, suggesting that streptomycetes are nontoxic to mice.

4.5   Discovery of bioactive compounds

More than 80 bioactive secondary metabolites have been isolated and characterized from some fecal actinomycete strains, including 6-*N*,*N*-dimethyladenosine (Li et al., 2012), abkhazomycin, actinomycin, AI 77B, akashin A, alazopeptin, apigenin, candicidin, cosmomycin, cyclo(4-Hyp-Phe) (Li et al., 2012), desertomycin, desferrioxamine E, discodermolide, emodin, enopetin A/B, erythromycins, favofungin, geldanamycin, isostreptazolin (Zheng et al., 2012), kasugamycin, kidamycins, leucomycin, longestin, panosialin-wA, puromycin, rutamycin, rhodomycinone, sannaphenol (Zheng et al., 2012), sannastatin (Yang et al., 2011), tirandamycin, vicenistatin, violapyrones A–G (Zhang et al., 2013), and polyene macrolides. Some of these compounds are described in Figure 4.8.

These compounds showed complex and different structures and exhibited various activities. Several novel compounds, such as sannastatin, a novel macrolactam polyketide glycoside, produced by an unidentified *Streptomyces* sp. YIM 100282, had been identified (Yang et al., 2011). The compounds displayed significant cytotoxicity against brine shrimp nauplii (*Artemia salina*). A new compound with high activity to some targets of senile dementia was obtained (unshown), and related study was carried out

4.6   Effects of microbial manure on notoginseng disease

Notoginseng (*Panax pseudoginseng* var. *Notoginseng*) is a local and rare medicinal herb from Wenshan, Yunnan, a main raw material used in many traditional Chinese medicines. Annual yield of it in Yunnan is 90% in China. However, the cultivation of notoginseng is severely affected by soil-borne diseases causing continuous cropping. Nine streptomyces strains, which showed the strongest inhibitory activity against many pathogens of notoginseng, were selected from actinomycete strains isolated from animal feces. The nine strains were fermented, mixed with manure, and used for the preventative treatment of soil-borne diseases affecting notoginseng fields in Wenshan, Yunnan for 3 years. The morbidity of the plants at a dosage of 20 kg/Chinese Mu (666 m²) of microbial manure was 81% lower than that observed using agricultural chemicals (Figure 4.9). These results suggest that microbial manure can be widely used in large tracts of land.

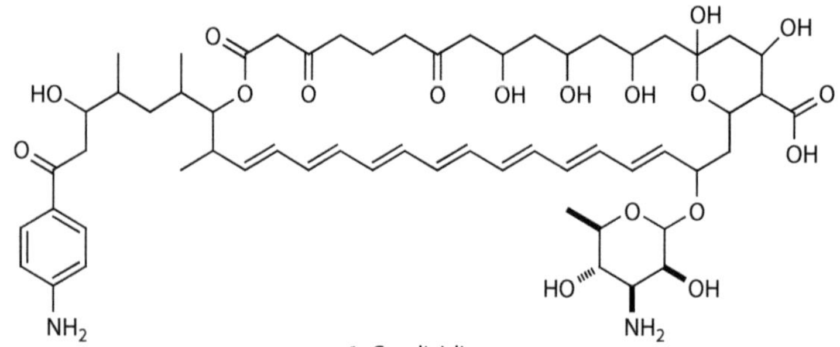

1. Candicidin

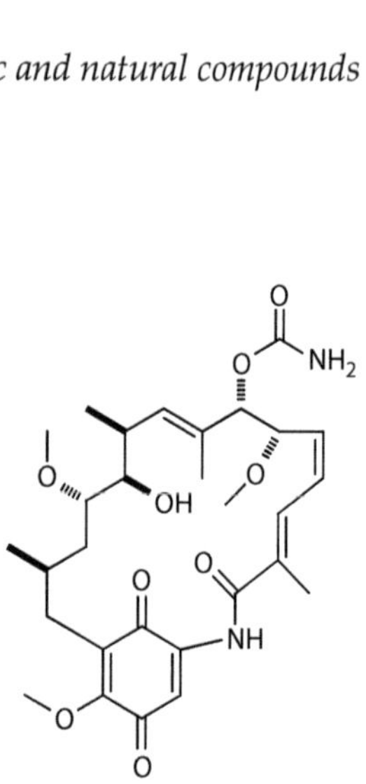

2. Geldanamycin

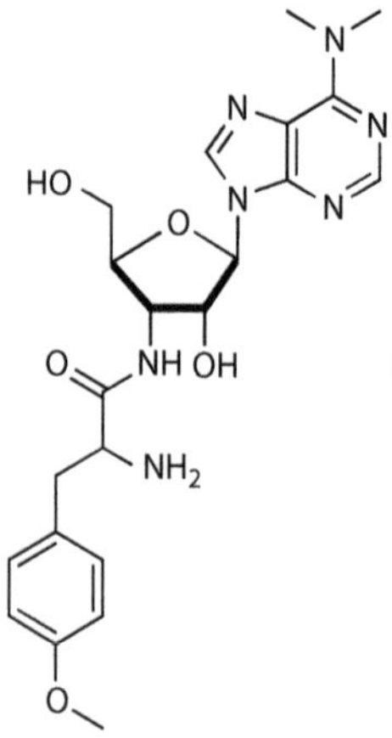

3. Puromycin

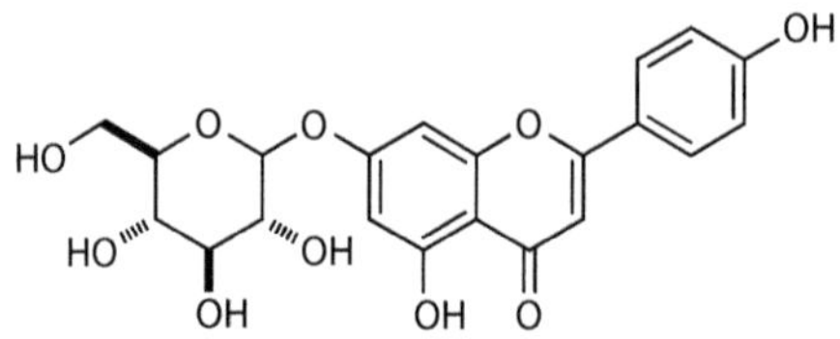

4. Apigenin

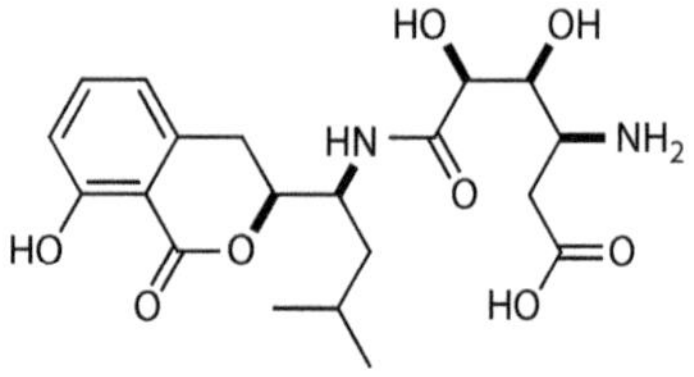

5. AI 77B

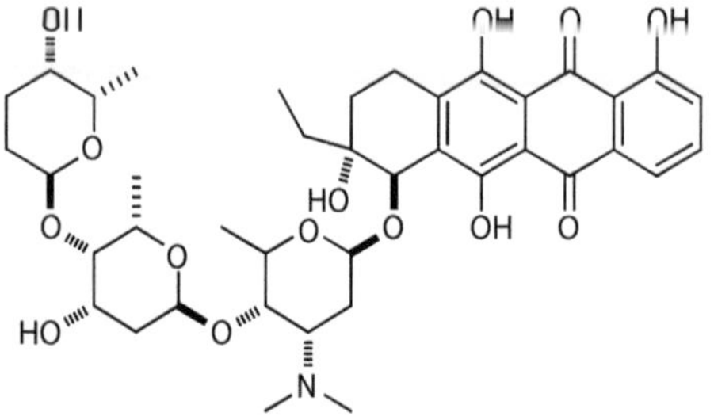

6. Cosmomycin

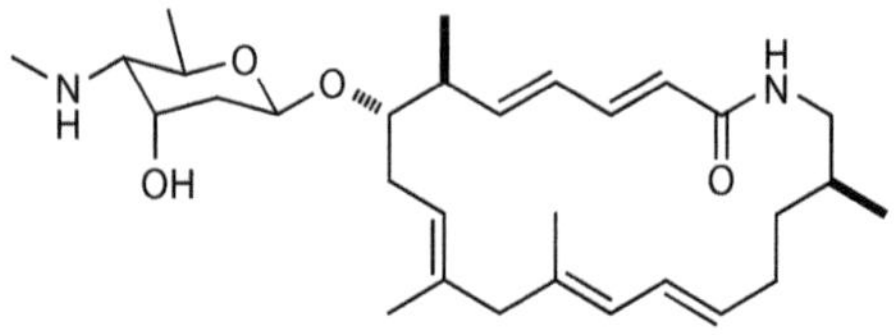

7. Vicenistatin

(a)

Figure 4.8 Bioactive compounds produced from actinomycete strains from animal feces.

(Continued)

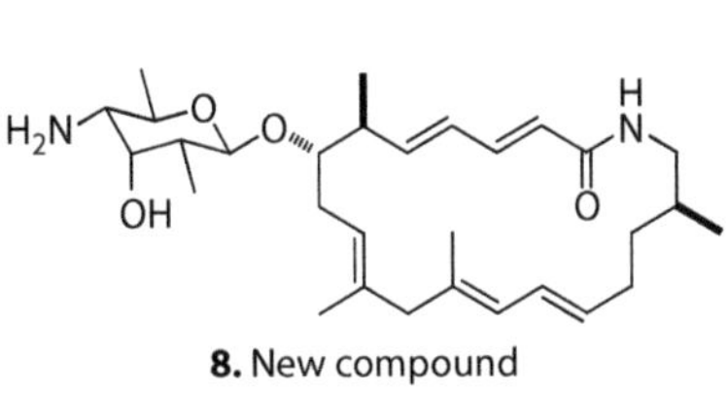

8. New compound

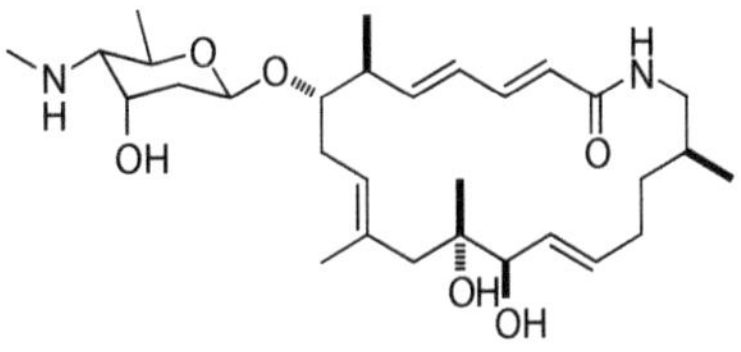

9. Sannastatin (new compound)

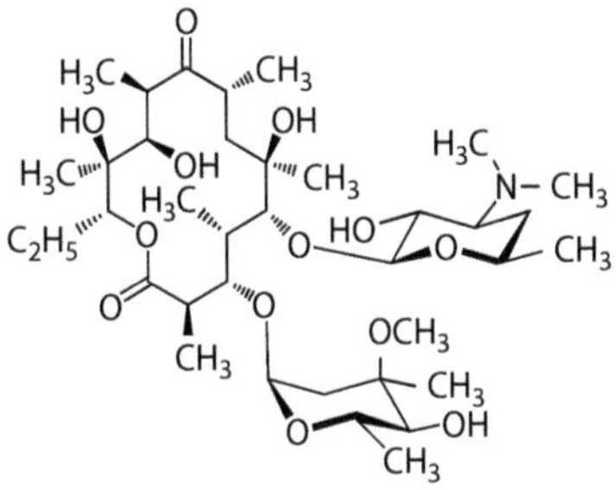

10. Erythromycins

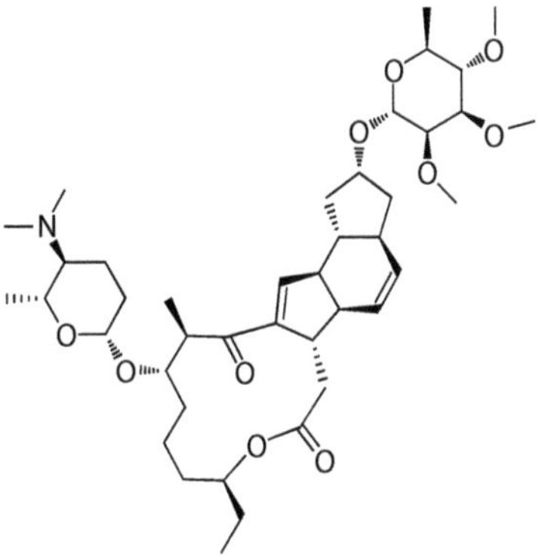

11. Spinosyns

(b)

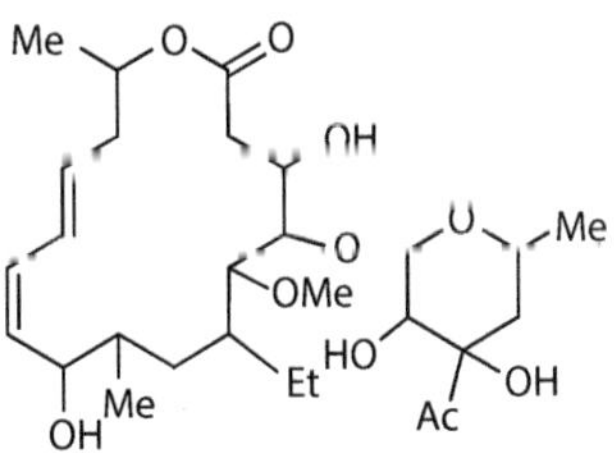

12. Leuconolides

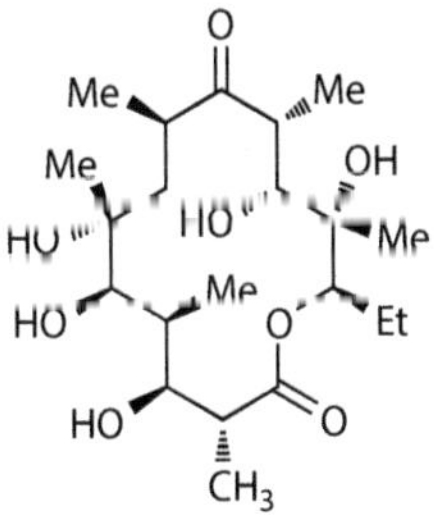

13. Erythronolide

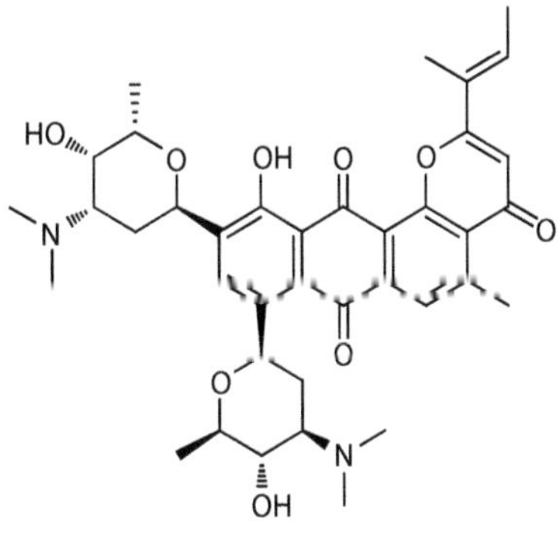

14. Kidamycin

(c)

Figure 4.8 (Continued) Bioactive compounds produced from actinomycete strains from animal feces.   *(Continued)*

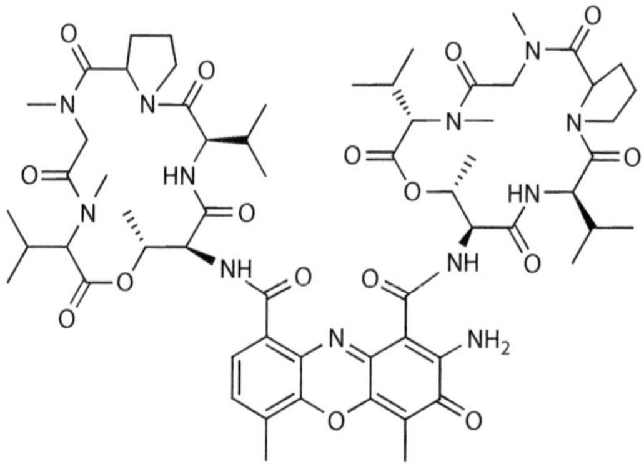

15. Actinomycins

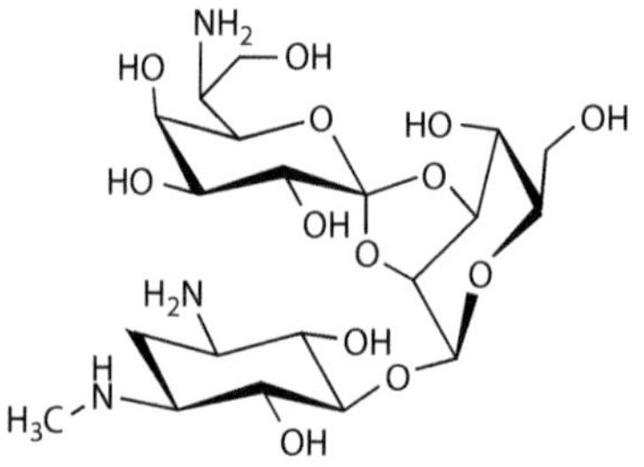

16. Desertomycin 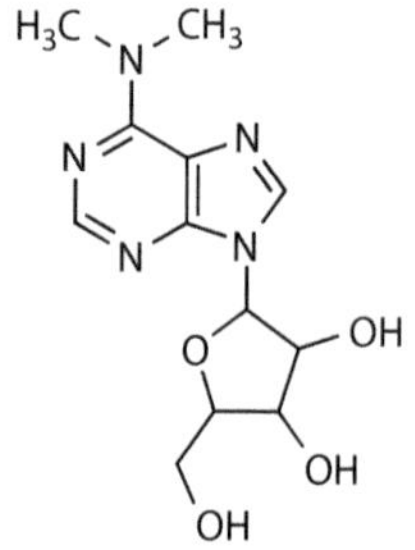   **17.** 6-*N*, *N*-dimethyladenosine

(d)

Figure 4.8 (Continued) Bioactive compounds produced from actinomycete strains from animal feces.

(a)   (b)   (c)

Figure 4.9 Effectiveness of using microbial manure to treat soil-borne notoginseng diseases: (a) agricultural chemicals control, (b) 15 g/m² microbial manure, and (c) 30 g/m² microbial manure.

4.7 Conclusion

The results obtained from the 8-year study of animal fecal actinomycetes should be highlighted with several key points:

1. The diversity of animal fecal actinomycetes is rich. Fifty-one genera were identified from only 48 animal fecal samples. The members of the genera *Streptomyces, Rhodococcus, Microbacterium,* and *Micrococcineae* (including the genus *Arthrobacter*) were predominant. However, many of the actinomycetes present in animal feces remain unknown.

2. The antimicrobial and antitumor activities of actinomycetes were strong and widespread. The activities of many enzymes were also strong. Actinomycetes play important roles in health, including improving food digestion and absorption, maintaining the balance of microbial ecological system in intestinal tract, providing resistance to various pathogens and tumors, and improving the overall health of the host.

3. Fecal actinomycetes produce multifarious secondary metabolites with bioactivities, and these metabolites exhibit kaleidoscope structures. We further propose that the metabolites produced by nonpathogen fecal actinobacteria should be not toxic or less toxic to the hosts. This is a very important excellence comparing with the metabolites of other microorganisms from other habitats. Thus, discovering new compounds now should focus on unknown streptomycetes.

4. There are millions of species of animals on the earth. Similar to the actinomycetes in other habitats (soil, sea, plant), animal fecal actinomycetes represent a tremendous resource for the development of drug leads.

5. To discover much more ideal drug leads from animal fecal actinomycetes, first, the isolation methods should be innovated, improved, and updated constantly to make uncultivable into pure cultured actinomycete strains; second, fermentation is an important prerequisite to produce bioactive substances, and thus studies on the improvement of fermentation approaches are needed; third, new and unknown isolates should be screened using new and highly individualized models; and fourth, a rapid, simple, and accurate remove–repeat system in the early stage should be established.

Acknowledgments

This research was supported by the National Natural Science Foundation of China (Nos 30900002, 31270001, and 21062028), the National Major Scientific and Technology Special Projects (2009ZX09302–003), and the National Institutes of Health in the United States (1P41GM086184–01A1). We thank Professor H. Laatsch and Dr. S. X. Yang (University of Göttingen, Germany); Professor G. L. Challis and Dr. L. J. Song (University of Warwick, United Kingdom) for the analysis of bioactive metabolites; Dr. Yanru Cao, Dr. Xiu Chen, Dr. Dan Zheng, Dr. Jiang Liu, and Ms. Chunhua Yang for taking part in some works; and Mr. Li Youlong and Ms. Shumei Qiu for collecting the test samples.

References

Abdelmohsen, U. R., Pimentel-Elardo, S. M., Hanora, A., Radwan, M., Abou-El-Ela, S. H., Ahmed, S., and Hentschel, U. 2010. Isolation, phylogenetic analysis and anti-infective activity screening of marine sponge-associated *Actinomycetes. Mar. Drugs*, 8, 399–412.

Bakalidou, A., Kämpfer, P., Berchtold, M., and Kuhnigk, T. 2002. *Cellulosimicrobium variable* sp. *nov.*, a cellulolytic bacterium from the hindgut of the termite *Mastotermes darwiniensis. Int. J. Syst. Evol. Microbiol.*, 52, 1185–1192.

Baltz, R. H. 2008. Renaissance in antibacterial discovery from actinomycetes. *Curr. Opin. Pharmacol.*, 8, 1–7.

Beman, B. L. 1983. Actinomycete pathogen. In: Goodfellow, M., Mordarski, M., and Williams, S. T. (eds.), *The Biology of the Actinomycetes*. Academic Press, London, U.K., pp. 457–480.

Bentley, S. D., Chater, K. F., Cerdeno-Tarraga, A. M., Challis, G. L., Thomson, N. R., James, K. D., and Hopwood, D. A. 2002. Complete genome sequence of the model actinomycetes *Streptomyces coelicolor* A3(2). *Nature*, 417, 141–147.

Bérdy, J. 2005. Bioactive microbial metabolites. *J. Antibiot. Tokyo*, 58, 1–26.

Bérdy, J. 2012. Thoughts and facts about antibiotics: Where we are now and where we are heading. *J. Antibiot.*, 65, 385–395.

Blunt, J. W. 2011. Marine natural products. *Nat. Prod. Rep.*, 28, 196–268.

Blunt, J. W., Copp, B. R., Keyzers, R. A., Munro, M. H., and Prinsep, M. R. 2013. Marine natural products. *Nat. Prod. Rep.*, 30, 237–323.

Brown, J. M., Steigerwalt, A. G., Morey, R. E., Daneshvar, M. I., Romero, L. J., and McNeil, M. M. 2006. Characterization of clinical isolates previously identified as *Oerskovia turbata*: Proposal of *Cellulosimicrobium funkei* sp. nov. and emended description of the genus *Cellulosimicrobium*. *Int. J. Syst. Evol. Microbiol.*, 56, 801–804.

Bull, A. T. and Stach, J. E. M. 2007. Marine actinobacteria: New opportunities for natural produce search and discovery. *Trends Microbiol.*, 15, 491–499.

Cao, Y., Jiang, Y., Wang, Q., Zhao, L., Jin, R., and Jiang, C. 2010. Diversity and some bioactivities of cultured actinomycetes in four areas in Sichuan and Yunnan. *Acta Microbiol. Sinica*, 50(8), 995–1000.

Cao, Y. R., Jiang, Y., Jin, R. X., Han, L., He, W. X., Li, Y. L., and Xue, Q. H. 2012. *Enteractinococcus coprophilus* gen. *nov.*, sp. *nov.*, of the family Micrococcaceae isolated from *Panthera tigris* amoyensis feces, transfer of *Yaniella fodinae* Dhanjal et al. 2011 to the genus Enteractinococcus as *Enteractinococcus fodinae* comb. *nov. Int. J. Syst. Evol. Microbiol.*, 62, 2710–2716.

Cao, Y. R., Jiang, Y., and Xu, L. H. 2009. Actinomycetes composition and bioactivities in Grand Shangri-La. *Acta Microbiol. Sinica*, 49, 105–109.

Célia, M. M., Balbina, N., Norbert, W., and Olga, C. N. 2004. *Gulosibacter molinativorax* gen. *nov.*, sp. *nov.*, a molinate-degrading bacterium, and classification of 'Brevibacterium helvolum' DSM 20419 as *Pseudoclavibacter helvolus* gen. *nov.*, sp. *nov. Int. J. Syst. Evol. Microbiol.* 54, 783–789.

Chiao, J. S. 2004. An important mission for microbiologist in the new century—Cultivation of the un-culturable microorganism. *Chin. J. Biotechnol.*, 20, 641–645.

Cui, X. L., Mao, P. H., Zeng, M., Li, W. J., Zhang, L. P., Xu, L. H., and Jiang, C. L. 2001. *Streptomonospora salina* gen. *nov.*, sp. *nov.*, a new member of the family Nocardiopsaceae. *Int. J. Syst. Evol. Microbiol.*, 51, 357–363.

Curtis, M. M. and Sperandio, V. 2011. A complex relationship: The interaction among symbiotic microbes, invading pathogens, and their mammalian host. *Mucosal Immunol.*, 4, 133–138.

Daly, K., Stewart, C. S., Flint, H. J., and Shirazi-Beechey, S. P. 2001. Bacterial diversity within the equine large intestine as revealed by molecular analysis of cloned 16S rRNA genes. *FEMS Microbiol. Ecol.*, 38, 141–151.

Durso, L. M., Harhay, G. P., Smith, T. P., Bono, J. L., DeSantis, T. Z., Harhay, D. M., and Clawson, M. L. 2010. Animal-to-animal variation in fecal microbial diversity among beef cattle. *Appl. Environ. Microbiol.*, 76, 4858–4862.

Falagas, M., Betsi, G., and Athanasiou, S. 2006. Probiotics for prevention of recurrent vulvovaginal candidiasis: A review. *J. Antimicrob. Chemother.*, 58(2), 266–272.

Felsenstein, J. 1981. Evolutionary trees from DNA sequences: A maximum likelihood approach. *J. Mol. Evol.*, 17, 368–376.

Goodfellow, M. and Fiedler, H. P. 2010. A guide to successful bioprospecting: Informed by actinobacterial systematics. *Antonie van Leeuwenhoek*, 98, 119–142.

Goodfellow, M., Kämpfer, P., Busse, H. J., Trujillo, M. E., Suzuki, K., Ludwig, W., and Whitman, W. B. 2012a. *Bergey's Manual of Systematic Bacteriology*, 2nd eds, vol. 5, The Actinobacteria, Part A, B. Springer, New York.

Goodfellow, M., Stach, J. E., Brown, R., Bonda, A. N. V., Jones, A. L., Mexson, J., and Bull, A. T. 2012b. *Verrucosispora maris* sp. *nov.*, a novel deep-sea actinomycete isolated from a marine sediment which produces abyssomicins. *Antonie van Leeuwenhoek*, 101(1), 185–193.

Greetham, H. L., Giffard, C., Hutson, R. A., Collins, M. D., and Gibson, G. R. 2002. Bacteriology of the labrador dog gut: A cultural and genotypic approach. *J. Appl. Microbiol.*, 93, 640–646.

Gundlapally, S. N. R. and Ferran, G. P. 2009. Description of *Patulibacter americanus* sp. nov., isolated from biological soil crusts, emended description of the genus *Patulibacter* Takahashi et al. 2006 and proposal of *Solirubrobacterales* ord. nov. and Thermoleophilales ord. nov. *Int. J. Syst. Evol. Microbiol.*, 59, 87–94.

Handelsman, J. 2004. Metagenomics: Application of genomics to uncultured microorganisms. *Microbiol. Molecular Biol.*, 68, 669–685.

Hayakawa, M. and Nonomura, H. 1987. Humic acid-vitamin agar, a new medium for the selective isolation of soil actinomycetes. *J. Ferment. Technol.*, 65, 501–509.

Hooper, L. V. and Gorden, J. I. 2001. Commensal host-bacyrtial relationship in the gut. *Science*, 292, 1115–1118.

Hui-tu, Z., Ai-ming, L., Lin-zhuan, W., Gui-zhi, S., Feng, H., Qun-jie, G., and Yi-guang, W. 2009. Screening of AHBA synthase gene and discovery of actinomycin D production in *Streptomyces violaceusniger* 4353. *J. Chin. Antibiotics*, 34, 30–33.

Ibrahim, A. S. S., Al-Salamah, A. A., Hatamleh, A. A., El-Shiekh, M. S., and Ibrahim, S. S. S. 2012. Tapping uncultured microorganisms through metagenomics for drug discovery. *Afr. J. Biotechnol.*, 11(92), 15823–15834.

Javad, H., Fatemeh, M., and Antonio, V. 2013. Systematic and biotechnological aspects of halophilic and halotolerant actinomycetes. *Extremophiles*, 17, 1–13.

Jennifer, B. H., Jessica, J. H., Taylor, H. R., and Brendan, J. M. B. 2001. Counting the uncountable: Statistical approaches to estimating microbial diversity. *Appl. Environ. Microbiol.*, 67(10), 4399–4406.

Jensen, P. R. 2010. Linking species concepts to natural product discovery in the post-genomic era. *J. Ind. Microbiol. Biotechnol.*, 37, 219–224.

Jiang, Y. 2011. Diversity of cultured actinomycetes in the Baltic Sea. *Acta Microbiol. Sinica*, 51(11), 1461–1457.

Jiang, Y., Cao, Y. R., Wiese, J., Lou, K., Zhao, L. X., Imhoff, J. F., and Jiang, C. L. 2009. A new approach of research and development on pharmaceuticals from actinomycetes. *J. Life Science US*, 3(7), 52–56.

Jiang, Y., Cao, Y. R., Zhao, L., Tang, S., Wang, Y., Li, W., and Xu, L. 2011. Large numbers of new bacterial taxa found by Yunnan Institute of Microbiology. *Chin. Sci. Bull.*, 56(8), 709–712.

Jiang, Y., Cao, Y. R., Zhao, L., Wang, Q., Jin, R., He, W., and Xue, Q. 2010. Treatment of ultrasonic to soil sample for increase of the kind of rare actinomycetes. *Acta Microbiol. Sinica*, 50(8), 1094–1097.

Jiang, Y., Chen, X., Cao, Y. R., and Ren, Z. 2013. Diversity of cultivable actinomycetes in tropical rainy forest of Xishuangbanna, China. *Open J. Soil Sci.*, 3, 9–14.

Jiang, Y., Li, W. J., Xu, P., Tang, S. K., and Xu, L. H. 2006. Study on actinomycete diversity under salt and alkaline environments. *Acta Microbiol. Sinica*, 46, 191–195.

Jiang, Y., Xu, P., Lou, K., Xu, L., and Liu, Z. 2008. Problem and countermeasure on development of pharmaceuticals from actinomycete resources. *Microbiology*, 35(2), 272–274.

Joseph, S. J., Hugenholtz, P., Sangwan, P., Osborne, C. A., and Janssen, P. H. 2003. Laboratory cultivation of widespread and previously uncultured soil bacteria. *Appl. Environ. Microbiol.*, 69(12), 7210–7215.

Kaeberlein, T., Lewis, K., and Epstein, S. S. 2002. Isolating "unculturable microorganisms" in pure culture in a simulated natural environment. *Science*, 296, 1127–1129.

Kampfer, P., Schafer. J., Lodders, N., and Martin, K. 2011. *Jiangella muralis* sp. *nov.*, from an indoor environment. *Int. J. Syst. Evol. Microbiol.*, 61, 128–131.

Lechevalier, M. P. and Lechevalier, H. A. 1970. Chemical composition as a criterion in the classification of aerobic *Actinomycetes*. *Int. J. Syst. Bacteriol.*, 20, 435–443.

Lechevalier, M. P. and Lechevalier, H. A. 1900. The chemotaxonomy of *Actinomycetes*. In: Dietz, A. and Thayer, D. W. (eds.), *Actinomycete Taxonomy*, special publications No. 6. Society for Industrial Microbiology, Arlington, VA, pp. 227–291.

Lee, S. D. 2008. *Jiangella alkaliphila* sp. *nov.*, an actinobacterium isolated from a cave. *Int. J. Syst. Evol. Microbiol.*, 58, 1176–1179.

Li, W. J., Chen, H. H., Xu, P., Zhang, Y. Q., Schumann, P., Tang, S. K., and Jiang, C. L. 2004. *Yania halotolerans* gen. *nov.*, sp. *nov.*, a novel member of the suborder Micrococcineae isolated from saline soil in China. *Int. J. Syst. Evol. Microbiol.*, 54, 525–531.

Li, W. J., Zhi, X. Y., and Euzéby, J. P. 2008. Proposal of *Yaniellaceae fam. nov.*, *Yaniella* gen. *nov.* and *Sinobaca* gen. *nov.* as replacements for the illegitimate prokaryotic names Yaniaceae Li et al. 2005, Yania Li et al. 2004, emend Li et al. 2005 and Sinococcus Li et al. 2006, respectively. *Int. J. Syst. Evol. Microbiol.*, 58, 525–527.

Li, X., Han, L., Cao, Y. R., Jiang, Y., Huang, X. S. 2012. Separation and identification of chemical constituent from a *Streptomyces* sp. isolated from Sika deer feces. *Chemistry & Bioengineering*, 39(6), 31–49.

Lior, P. 2007. Interpreting the unculturable majority. *Nat. Methods*, 4, 479–480.

Martin, K., Schumann, P., Rainey, F. A., and Schuetze, B. 1997. *Janibacter limosus* gen. *nov.*, sp. *nov.*, a New *Actinomycete* with meso-uiaminopimelic acid in the cell wall. *Int. J. Syst. Bacteriol.*, 47, 529–534.

McLeod, M. P., Warren, R. L., Hsiao, W. W., Araki, N., Myhre, M., Fernandes, C., and Eltis, L. D. 2006. The complete genome of *Rhodococcus* sp. RHA1 provides insights into a catabolic powerhouse. *Proc. Natl Acad. Sci. USA*, 103(42), 15582–15587.

Meyer, A., Micksche, M., Herbacek, I., and Elmadfa, I. 2006. Daily intake of probiotic as well as conventional yogurt has a stimulating effect on cellular immunity in young healthy women. *Ann. Nutr. Metab.*, 50(3), 282–289.

Ōmura, S., Ikeda, H., Ishikawa, J., Hanamoto, A., Takahashi, C., Shinose, M., and Hattori, M. 2001. Genome sequence of an industrial microorganism *Streptomyces avermitilis*: Deducing the ability of producing secondary metabolites. *Proc. Natl Acad. Sci. USA*, 98(21), 12215–12220.

Orsini, M. and Romano-Spica, V. 2001. A microwave-based method for nucleic acid isolation from environmental samples. *Lett. Appl. Microbiol.*, 33, 17–20.

Patrick, D. S. and Handelsman, J. 2005. Metagenomics for studying unculturable microorganisms: Cutting the Gordian knot. *Genome Biol.*, 2005, 6, 229–302.

Payne, D. J., Gwynn, M. N., Holmes, D. J., and Pompliano, D. L. 2007. Drugs from bad bugs: Confronting the challenges of antibacterial discovery. *Nat. Rev. Drug Discov.*, 6, 29–40.

Petkar, H., Li, A., Bunce, N., Duffy, K., Malnick, H., and Shah, J. J. 2011. *Cellulosimicrobium funkei*: First report of infection in a nonimmunocompromised patient and useful phenotypic tests for differentiation from *Cellulosimicrobium cellulans* and *Cellulosimicrobium terreum*. *J. Clin. Microbiol.*, 49(3), 1175–1178.

Qin, S., Zhao, G. Z., Li, J., Zhu, W. Y., Xu, L. H., and Li, W. J. 2009. *Jiangella alba* sp. nov., an endophytic actinomycete isolated from the stem of *Maytenus austroyunnanensis*. *Int. J. Syst. Evol. Microbiol.*, 59, 2162–2165.

Ritchie, L. E., Steiner, J. M., and Suchodolski, J. S. 2008. Assessment of microbial diversity along the feline intestinal tract using 16S rRNA gene analysis. *FEMS Microbiol. Ecol.*, 66, 590–598.

Run-chi, G., Zun-xi, H., Chang-fei, W., Bo, X., and Xiao-yan, W. 2010. Culture-independent analysis of microflora in gayals (Bos frontalis) feces. *Afr. J. Biotechnol.*, 9, 2774–2788.

Saitou, N. and Nei, M. 1987. The neighbor-joining method: A new method for reconstructing phylogenetic trees. *Mol. Biol. Evol.*, 4, 406–425.

Savage, D. C. 1977. Microbial ecology of gastrointestinal tract. *Annu. Rev. Microbiol.*, 31, 107–133.

Simpson, J. M., Martineau, B., Jones, W. E., Ballam, J. M., and Mackie, R. I. 2002. Characterization of fecal bacterial population in canines: Effects of age, breed and dietary fiber. *Microb. Ecol.*, 44, 186–197.

Song, L., Li, W. J., Wang, Q. L., Chen, G. Z., Zhang, Y. S., and Xu, L. H. 2005. *Jiangella gansuensis* gen. nov., sp. nov., a novel actinomycete from a desert soil in north-west China. *Int. J. Syst. Evol. Microbiol.*, 55, 881–884.

Stackebrandt, E., Schumann, P., and Cui, X. L. 2004. Reclassication of *Cellulosimicrobium variabile* Bakalidou et al. 2002 as *Isoptericola variabilis* gen. nov., comb. nov. *Int. J. Syst. Evol. Microbiol.*, 54, 685–688.

Subramani, R. and Aalbersberg, W. 2013. Culturable rare *Actinomycetes*: Diversity, isolation and marine natural product discovery. *Appl. Microbiol. Biotechnol.*, 97, 9291–9321.

Suchodolski, J. S., Camacho, J., and Steiner, J. M. 2008. Analysis of bacterial diversity in the canine, duodenum, jejunum, ileum, and colon by comparative 16S rRNA gene analysis. *FEMS Microbiol. Ecol.*, 66, 567–578.

Suchodolski, J. S., Ruaux, C. G., Steiner, J. M., Fetz, K., and Williams, D. A. 2004. Application of molecular fingerprinting for qualitative assessment of small-intestinal bacterial diversity in dogs. *J. Clin. Microbiol.*, 42, 4702–4708.

Tang, S. K., Zhi, X. Y., Wang, Y., Shi, R., Lou, K., Xu, L. H., and Li, W. J. 2011. *Haloactinopolyspora alba* gen. nov., sp. nov., a halophilic filamentous actinomycete isolated from a salt lake, with proposal of *Jiangellaceae* fam. nov. and *Jiangellineae* subord. nov. *Int. J. Syst. Evol. Microbiol.*, 61, 194–200.

Xu, L. H., Li, W. J., Liu, Z. H.,and Jiang, C. L. 2007. *Actinomycete Taxonomy*. Academic Press, Beijing, China, pp. 202–208.

Xu, L. H., Zhang, H., Zhang, L. P., Xue, Q. H., Zhang, L. X., and Xiong, Z. 2010. *Microbial Resource Science*, 2nd ed. Academic Press, Beijing, China.

Yang, S. X., Gao, J. M., Zhang, A. L., and Laatsch, H. 2011. Sannastatin, a novel toxic macrolactam polyketide glycoside produced by actinomycete *Streptomyces sannanensis*. *Bioorg. Medical Chem.*, 21, 3905–3908.

Zengler, K., Toledo, G., Rappé, M., Elkins, J., Mathur, E. J., Short, J. M., and Keller, M. 2002. Cultivating the uncultured. *Proc. Natl Acad. Sci. USA*, 99(24), 15681–15686.

Zhang, H. H. and Chen, L. 2010. Phylogenetic analysis of 16S rRNA gen sequences reveals distal gut bacterial diversity in wild wolves (*Canis lupus*). *Mol. Biol. Res.*, 37, 4013–4023.

Zhang, J., Jiang, Y., Cao, Y., Liu, J., Zheng, D., Chen, X., and Huang, X. 2013. Violapyrones A–G, α-pyrone derivatives from streptomyces violascens isolated from *Hylobates hoolock* feces. *J. Natural Products.*, 76(11), 2126–2130.

Zheng, D., Han, L., Li, Y., Li, J., Rong, H., Leng, Q., and Huang, X. 2012. Isostreptazolin and sannaphenol, two new metabolites from *Streptomyces sannanensis. Molecules*, 17, 836–842.

chapter five

Potentially novel Actinobacteria-derived antibiotics from unique microenvironments

Diana R. Cundell and Michael P. Piechoski

Contents

5.1   Introduction

Antibiotics are natural or synthetic molecules able to inhibit the growth of bacteria and display a diversity of targets including the cell wall, the cell membrane, protein synthesis, and nucleic acid synthesis (Singh et al., 2012). Their production is not constant and is typically induced by a change in the environment such as lowered nutrient supply or temperature changes. Currently, some 75% of the world's antibiotics are derived from terrestrial *Streptomyces* bacteria, with the phyla to which they belong, the Actinobacteria, is thought to produce as many as 12,000 bioactive secondary molecules (Raja and Prabakaran, 2011). The remaining bacterial sources of antibiotics are from terrestrial species of *Bacilli* and *Pseudomonas* (Singh et al., 2012).

Studies have shown that antibiotics are produced by these species in small amounts when the growth of the colony begins to slow due to nutrient exhaustion (Challis and Hopwood, 2003). Although possibly originally produced as signaling molecules able to induce gene expression (Chater et al., 2010), antibiotics provide a key to bacterial survival and dominance in natural environments since they remove competitors. This is, however, not without cost to the bacteria producing them since antibiotics often require large numbers of enzymatic proteins involved in their assembly (Challis and Hopwood, 2003). A good example is the bacterium *Saccharomyces erythrea* (formerly *Streptomyces erythrea*) in which 60 kb of its DNA is involved in making the complex macrolide erythromycin (Donadio et al., 1991).

First discovered to produce "chemicals able to kill other organisms" by Gasperini in 1892 (Waksman and Lechevalier, 1962), soil Actinobacteria are dominant species in most microenvironments where they play a vital role as decomposers of organic substances including chitin and cellulose (Hogan, 2010). Their ability to thrive in diverse habitats is a function of several attributes. First, Actinobacteria produce resistant spores, which allow them to survive changes in nutrient and water levels in many diverse habitats (Schaechter, 2010). Second, researchers have shown that although Actinobacteria prefer a neutral to acid pH, they are resistant to desiccation, allowing their colonization of desert and sandy soils (Schaechter, 2010). They are especially important to the continued fertility of soils, being the second group to appear after pigmented bacteria and are the commonest bacteria in desert crusts (Kieft, 1991). Third, Actinobacteria have been shown to exist in an extreme diversity of climates and are tolerant to heat (Kurapova et al., 2012), cold (Augustine et al., 2012), as well as alkaline and saline soils (Zvyagintsev and Zenova, 2007).

Actinobacteria species already produce an impressive group of bioactive molecules, but it is apparent that they probably represent the mere tip of the biological iceberg. First, studies have shown that between 20% and 50% of all terrestrial and aquatic species, Actinobacteria are able to generate antibacterial and antifungal factors. Second, it is probable that only around 10% of the Actinobacteria species in existence on our planet have actually been isolated, despite intensive screening of soils by the pharmaceutical industry for over 50 years (Enright, 2003). Given these parameters, *Streptomyces* species that come from unique, biologically challenging locations such as tropical soils and arid desert environments are also likely to be a potential source for novel antimicrobials. These locations may also be seeded with the so-called "rare Actinomycetes," most of which are still uncharacterized as to their antimicrobial activity (Lazzarini et al., 2001). One group of interest in this category is the genus *Micromonospora*, which is already a source of a variety of macrolides (Lancini and Lorenzetti, 1993).

Given the current and also putative future capabilities of the Actinobacteria phylum, the remaining goal of this chapter is twofold. First is to describe the hurdles that need to be overcome in order to successfully identify and launch new Actinobacteria antibiotics. Second is to survey the scientific literature as to the potential of antibiotics isolated from two novel environments namely sandy (including temperate and tropical) and tropical locales with other soil types, particularly the unique environment of Jamaica.

5.2   Hurdles in commercialization of Actinobacteria antibiotics

Although Actinobacteria have been studied since the 1940s, their methods for producing antibiotics are still often poorly understood. The path from discovery to commercialization is also a difficult one, with several major barriers along the way. First is the ability to both successfully culture and stabilize a novel antibiotic; second is the testing procedure, which is often lengthy and may prove unsuccessful in terms of the overall, that is, nonbacterial, toxicity of the novel therapeutic; and third is the clinical testing (Silver, 2011). The next sections explore each of these issues with a view to how each is being addressed.

5.2.1 Successful culture methodologies for Actinobacteria

The issue with recovery of antibiotics in culture is twofold. First, current culture methods tend to be primarily a "one size fits all" model with the bacterium being separated out of its natural environment. For over 40 years, since the first observations of Waksman in 1952, primary methods of Actinobacteria isolation have involved separating the individual viable colonies from their habitat by plating diluted soil samples onto an albumen agar medium and incubating at 30°C for 1 week (Boisvert-Bertrand et al., 2004). Although many viable producers can be recovered this way, it tends to select for the more rapidly growing colonies and, interestingly, researchers have developed new methods for recovery of new types of antibiotics (Li, 1989). For example, air drying and heating soil, followed by culture on streptomycin- and rifamycin-supplemented isolation media allowed slower growing strains of *Actinomadura* and *Microtetraspora* Actinobacteria to grow and produce antibiotics (Li, 1989). This has since paved the way for the discovery of over 200 antibiotics from these genera (Lazzarini et al., 2001).

Another issue in generating antibiotics from Actinobacteria is that some produce antibiotics only in response to soil competition and not when grown in isolation. A notable example of this was *Streptomyces coelicor*, which was originally believed not to be capable of producing significant antibiotic activity (Marinelli, 2009). Genomic sequencing of the bacterium uncovered 23 gene clusters capable of producing secondary metabolites with only six ever having been recovered from growth media of the microbe (Marinelli, 2009). With the recognition that the majority of *Streptomyces* species produce polyketide or polypeptide antibiotics, genes responsible for their production can be screened for, thus also facilitating the identification of novel antibiotic-producing microbes (Busti et al., 2006).

More recently, it has also been recognized that horizontal gene transfer may also occur when antibiotic-producing and non-antibiotic-producing Actinobacteria species are cocultured (Kurosawa et al., 2008). Research on the bacterium *Rhodococcus fascians* demonstrated that it possessed many genes involved in antibiotic synthesis, although the bacterium had never been shown to release any under normal growth conditions (Kurosawa et al., 2008). Only when cocultured with a second Actinobacteria species, *Streptomyces padanus*, did *R. fascians* generate antibiotics (Kurosawa et al., 2008). Furthermore, the rhodostreptomycins produced by *R. fascians* were found to belong to a novel class of aminoglycosides that were distinct from the actinomycin that *S. padanus* generated (Kurosawa et al., 2008).

A second hurdle is that many of the Actinobacteria recovered grow only slowly in culture and produce small amounts of viable product. This is particularly the case for Actinobacteria recovered on humic acid agar, which may take as much as a month to produce viable isolates (Hayakawa, 2008). Researchers identifying glycopeptide antibiotics also need to wrestle with the problems of low yield, mixtures of molecules at varying stages of glycosylation, and lack of extrusion of the product from the mycelium (Nicolaou et al., 1999). Being aware of the bacteria's culture needs becomes even more important when trying to isolate and identify antibiotic-producing Actinobacteria from unusual environments. Our own studies of Presque Isle's temperate, sandy loam (Boisvert-Bertrand et al., 2004) and Jamaica's tropical and varied soils (Piechoski et al., 2012) were successful in identifying some effective colonies but may well have missed many other organisms requiring nonclassical isolation strategies.

Others have also struggled in culturing Actinobacteria from unique environments. Indeed, a recent study by Hozzein and colleagues (2008) demonstrated that desert Actinobacteria are quite fastidious and instead of a rich medium require a special minimal medium that includes glucose, yeast extract, and mineral salts. By using this medium,

Hozzein and his team (2008) were able to increase the recovery of Actinobacteria from the Western and Eastern Egyptian deserts by as much as 10-fold and also expanded the diversity of the population recovered. This study alone underlines the fact that many species may be discovered from these locations given optimal isolation methodologies.

5.2.2   Drug discovery and development of antibiotics

Early drug discovery of natural products (1940s–1960s) typically involved empirical screening of bacterial cultures for antibiotic activity and secondary testing for cytotoxicity (Silver, 2011). Classic, standard qualitative procedures involved overlaying the colonies on agar embedded with a bacterial pathogen and then defining antibiotic activity as a zone of "no growth" of >3 mm (Atlas et al., 1995). Quantitative analysis of actual antibiotic activity typically then involved a microassay in which extracts of growth inhibitory activity from growth media were compared with the minimum inhibitory concentration of antibiotics as to their ability to kill a panel of standard, clinical strains of bacteria (Zgoda and Porter, 2001). To ensure consistency, bacterial strains to be tested were usually obtained from reference libraries such as the American Type Culture Collection (ATCC) and then cultured in rich media with continuous agitation to allow consistent oxygenation and maximum growth. Secondary cytotoxicity testing then employed exposing brine shrimp to extracts of the culture medium and estimating the LD_{50} of the bioactive moiety (Solis et al., 1993).

Unfortunately, the initial empirical-based screening process of identification of antibiotic activity did not distinguish between already-described antibiotic activities, and to avoid replication, additional screening methods were next developed by the pharmaceutical industry to identify the cellular target for the putative novel therapeutic (Hancock, 2007; Silver, 2011). The first screens addressed inhibitors of peptidoglycan (cell wall synthesis) as this is an effective strategy to limit bacterial growth, and utilization of this procedure identified several classes of antibiotics including the monobactams and carbapenems (Gadebusch et al., 1992). In addition, it was also apparent that most successful individually used antibiotics targeted multiple components within a bacterial cell, for example, macrolides target bacteria with multiple copies of rRNA genes and so much focus was made in their discovery (Silver, 2011). Finally, the arrival of genetic cloning and overexpression technologies, as well as the use of microarray gene signatures, allowed researchers to identify genetic signatures and, in turn, to also understand more fully the mechanisms by which antibiotics are effective (Hancock, 2007). Genetic analyses now estimate that there are around 160 enzyme targets that are useful to bacteria and absent from human beings (Payne et al., 2007), and Lange et al. (2007) list 16 enzyme classes that are already targets for commercialized antibiotics. Such screenings, although increasing understanding of how antibiotics work have, however, been notable in not yielding significant numbers of new antibiotics other than a few like platensin and the fluoroquinolones (Silver, 2011).

Interestingly, despite the advances in genomic research and high-throughput screening of potential antibiotics, many companies lost faith in these rather generic methods and current screening technologies typically involve either searching for novel natural products or reevaluating underexploited natural products (Brötz-Oesterhelt and Sass, 2010). It seems that in the search for new marketable pharmaceuticals, there is a definite split with at least five major companies (Wyeth, Sanofi-Aventis, Hoechst, Cubist, and Merck) electing to choose this route (Brötz-Oesterhelt and Sass, 2010). Both strategies have proven successful with the discovery of the natural products mannopeptimycins, friulimicin, and plectasin and the reevaluation of usage for daptomycin and ramoplanin (Brötz-Oesterhelt and Sass, 2010). This latter strategy was very successful for Bayer who used functional genomics in combination with mode of action studies to identify two novel classes of bacterial translation inhibitors, dihydropyrimidinones and biphenomycins (Bandow et al., 2003; von Nussbaum et al., 2006), and a new target in fatty acid biosynthesis, the pseudopeptide pyrrolidine diones (Freiberg et al., 2004). Dihydropyrimidinones are derived from a natural product, Tan-1057, and bind to the large ribosomal subunit, thus inhibiting the peptidyl transferase reaction (Bandow et al., 2003), while biphenomycins affect bacterial protein synthesis (von Nussbaum et al., 2006). New derivatives of both these compounds were able to increase the selectivity for gram-positive bacteria and remain well in the range of efficacy reported for existing antibiotics (von Nussbaum et al., 2006).

Another technique used by pharmaceutical companies, anxious to optimize antibiotic efficacy by decreasing toxicity, circumventing existing resistance, and increasing bioavailability (as well as minimize costs), has turned to the modification of existing antibiotics rather than the search for novel ones. Older Actinobacteria antibiotics, such as the aminoglycoside kanamycin, now have semisynthetic forms (amikacin and arbekacin) whose altered chemistry has decreased the toxicity of these agents while maintaining their efficacy against multidrug-resistant bacteria including methicillin-resistant *Staphylococcus aureus* (MRSA), *Pseudomonas aeruginosa*, and enteric bacteria (Cordeiro et al., 2001; Falagas et al., 2008).

For some Actinobacteria antibiotics, the generation of semisynthetic variants did not just reduce toxicity, but it allowed their usage. This was the case for oxytetracycline (Terramycin) initially discovered by Dr. Robert Woodward in 1950 and deemed highly since it prevented protein synthesis by blocking aminoacyl tRNA entry (Nelson and Levy, 2011). Although Terramycin was short-lived due to its many side effects, the discovery was to be important because structural modifications to their polyketide structure resulted in the semisynthetic antibiotics doxycycline and minocycline (Fuoco, 2012). Both doxycycline and minocycline have been found to be effective against most gram-positive and gram-negative bacteria, intracellular bacteria, mycoplasma, and spirochetes (Fuoco, 2012). Their use is widespread, and revenues from their prescription as doxycycline hyclate netted $264 million in 2011 (Mylan Receives Final FDA Approval for First Generic Version of Doryx® Tablets, 150 mg).

Having obtained a potentially useful antibiotic, in terms of *in vitro* performance, it must survive several phases of clinical trials as well as receive FDA approval. For one successful drug to reach marketability, it is estimated that 10,000 compounds must be discovered (Raja and Prabakarana, 2011). Indeed, current studies suggest that for around 60% of the therapeutics that received FDA approval, their journey will end at or before phase III trials when they are found to have toxic side effects in human subjects or in animal models (Raja and Prabakarana, 2011). Many also cannot be effectively stabilized and will need to be administered intravenously, which will limit their usage. At this point, pharmaceutical companies have invested large sums of money, and although each new antibiotic can yield annual profits, which may be in the several-hundred-million-dollar range, each may cost as much as a billion dollars to produce (Raja and Prabakarana, 2011).

The costs are, however, worth the benefits. According to the most recent report by the IMS Institute for Healthcare Informatics in 2011, azithromycin was ranked 8 of the 10 most popular drugs prescribed in the United States accounting for 56 million written prescriptions (IMS, 2011). This represents an increase of nearly three million prescriptions from 2010, with consumers enjoying the ease of the Z-Pak form of the antibiotic in which the 5-day course is numbered for ease of memory. Savvy marketing has helped make this antibiotic a household name, and a study in the *Journal of Family Practice* suggests that as many as four out of five patients with upper respiratory infections will ask for the drug directly by name (Hickner, 2007).

Despite the promise of an excellent market return on a successful product, many biotech and pharmaceutical companies left the market to focus instead on the more lucrative chronic diseases (Brötz-Oesterhelt and Sass, 2010). Consequently, the number of patent applications has declined (Katz et al., 2006) along with the number of approved antibiotics (Powers, 2004). This has led some scientists to believe that current trends will result in there being no effective antibiotics for some microbes in the next 10 years (de Lima Procópio et al., 2012).

The good news is that those biotechnology companies that remain investigated in antibiotic drug discovery are demonstrating recent, renewed vigor in research (Brötz-Oesterhelt and Sass, 2010). Armed with well-established protocols for identification of antibiotic activity, as well as modeling the metabolism and pharmacokinetics of the active moieties in highly predictive animal models, it can be anticipated that more drugs will be developed (Brötz-Oesterhelt and Sass, 2010). Indeed, success rates from Investigational New Drug applications (those associated with human trials) are now 30% for anti-infective therapeutics (Powers, 2004). The majority (70%) of these novel antibiotics appear to be coming from natural products, as suggested in a review of the literature by Brötz-Oesterhelt and Sass (2010).

Natural products typically strongly surpass their synthetic counterparts in activity and structural diversity and in addition usually have more complex modes of action (von Nussbaum et al., 2006). This is not surprising since antibiotic-producing strains have coevolved to occupy and compete for an environmental niche, thus honing their products accordingly. In addition to broad-spectrum antibiotics, many companies have also begun searching for novel narrow-spectrum antibiotics, with concomitantly fewer side effects on normal human flora (Brötz-Oesterhelt and Sass, 2010; de Lima Procópio et al., 2012). For many researchers, these new narrow- and broad-spectrum antibiotics are to be sought from the Actinobacteria indigenous to unique, diverse environments (Baltz, 2008).

5.3   Novel sources of Actinobacteria species and antibiotics

Although pharmaceutical companies have been harvesting natural Actinobacteria for over 50 years, it is likely that only a fraction of the terrestrial species have been identified, with still fewer from nutrient-poor environments, such as sandy locations or very diverse tropical environments. Tropical environments are also home to many Actinobacteria, which have only recently begun to be even described (Araragi, 1979; Janso and Carter, 2010; Kumar et al., 2012a; Piechoski et al., 2012; Ambrose et al., 2013). Thus, in terms of trying to identify novel, natural products, terrestrial locations with unique flora and soil structures clearly represent the next leap forward in the natural product world. In the next two sections, two such biomes will be discussed: first, sandy locations where nutrients and water are limited and second, tropical soils with soil types that are rich in humus and possess a diverse microflora.

5.3.1   Sandy locations (tropical and temperate)

Sandy soils and deserts provide an extreme terrestrial environment where few microbial species can survive, but they have been shown to be habitats for bacteria able to form desiccation-resistant structures such as spores (Kieft, 1991). Actinomycetes in these locations often use the cyanobacterium *Microcoleus* as a food source and are also typically associated with any plant life especially plant roots (Skujins, 1984). These species tend to migrate further down into the subsoils in harsher climates and, therefore, be more frequently isolated 2–15 cm down rather than at the surface (Cameron et al., 1970). Although most Actinobacteria are primarily mesophilic species needing an adequate supply of water, thermophilic and thermotolerant *Streptomyces, Micromonospora, Actinomadura,* and *Streptosporangium* (Kurapova et al., 2012) as well as alkaliphiles (Ali et al., 2009; Osman et al., 2011) have also been described from desert environments.

Several reports exist of antibiotic-secreting Actinobacteria in tropical and subtropical desert locations (Table 5.1). Though some are known *Streptomyces,* at least one new species, *S. atacamicus,* has emerged and is capable of secreting a novel class of antibiotics, the chaxamycins (Chaxamycins, 2011). Chaxamycins act on bacteria, particularly MRSA, by attaching to the ATP-binding domain (Chaxamycins, 2011).

Three alkaline desert locations, two in Egypt and one in Bangladesh, have also yielded successful novel antibiotics (Table 5.1). Alkaline deserts typically consist of dry, compacted clay soils, which, unlike moist clay that is typically nutrient rich, are typically nutrient poor (Chhabra, 1996). Life in these environments is, therefore, very extreme and competitive, and many microbes may theoretically only exist in dormant forms (Skujins, 1984). In the case of the Egyptian deserts, the first, Wadi El Natrun, is a saline, alkaline desert, which yielded meroparamycin in 2006 (El-Naggar et al., 2006). This novel therapeutic comes from a previously unidentified *Streptomyces* species and is highly effective against several gram-positive and gram-negative bacterial species (El-Naggar et al., 2006). The second,

Table 5.1 Antibiotic-producing Actinobacteria from desert sand or sandy/loam areas

Location/soil type	Species	Antibiotic	Antibacterial efficacy
Atacama, Chile[1] (hyperarid)	*Streptomyces* sp.	Chaxamycins	MRSA, *E. coli*
Cairo, Egypt[2] (sand)	*S. bikiniensis*	Unknown aminoglycoside	Broad-spectrum antibacterial
Falmouth, MA[3] (sandy, temperate)	*Nocardia*	Neocitreamicin	MRSA and VRE
Iran (Eastern desert)[4]	Strong similarities to several *Streptomyces* strains	Unknown	Broad spectrum
Karnataka, India[5] (black alkaline)	Actinomycetes (unknown)	Unknown	Broad-spectrum antibacterial
Keffi Metropolis, Nigeria[6] (sandy, loam)	Various *Streptomyces* and *Actinomycetes* species	Various, unknown	Broad spectrum
Presque Isle, PA (sandy, loam, temperate)[7]	Actinomycetes (unknown)	Unknown	Narrow spectrum; *S. aureus, P. aeruginosa,* and *B. megaterium*
Thar, Rajasthan[8] (subtropical desert)	*S. hygroscopicus* subsp. *ossamyceticus*	Unknown	Vancomycin-resistant *S. aureus* and *Klebsiella*
Wadi El Natrun, Egypt[9] (saline, alkaline desert)	*S. coelicolor, S. flavius, S. griseoruber,* and *S. plicatus*	Unknown	Broad-spectrum antibacterial and antifungal
Wadi Araba, Egypt[10] (alkaline desert)	*Nocardiopsis dansonvilli* WA52	WA52 Macrolide	Broad-spectrum antibacterial

[1] Chaxamycins (2011), [2] El-Khagawa and Megahed (2012), [3] Peoples et al. (2008), [4] Tabrizi et al. (2013), [5] Basavaraj et al. (2010), [6] Makut and Abdalazeez (2012), [7] Boisvert-Bertrand et al. (2004), [8] Selvameena et al. (2009), [9] Osman et al. (2011), and [10] Ali et al. (2009).

Wadi Araba, which is also an alkaline desert, yielded a macrolide, WA52, from a novel species, *Nocardiopsis dansonvilli* (Osman et al., 2011). The Thar desert contained a black, alkaline soil and yielded an *S. hygroscopicus* species able to secrete a broad-spectrum antibiotic activity (Selvameena et al., 2009). The isolation of classes of antibiotics from previously unidentified species from the Bangladesh and Egyptian desert alkaline *Streptomyces* may therefore herald a whole series of new microbes and novel antibiotics.

Temperate locations have also proven relatively fruitful for the isolation of novel Actinobacteria (Table 5.1). Neocitreamicins were isolated from *Streptomyces* species growing in sandy soil from Falmouth, Massachusetts (Peoples et al., 2008). They are chemically closely related to citreamicins, which are polycyclic xanthones first isolated in the 1980s and possess antibiotic activity against MRSA and vancomycin-resistant *Enterococci* (VRE) (Carter et al., 1990). Although their exact antibacterial mode of action remains unclear, they are known to be toxic for tumor cells by inhibiting cell division at G1 phase (Liu et al., 2013). Our team was also able to identify narrow-spectrum Actinomycetes from sandy, loam soil in Presque Isle, Pennsylvania, which were able to selectively inhibit *Pseudomonas aeruginosa*, *Bacillus* species, and *S. aureus* (Boisvert-Bertrand et al., 2004). In this study, the number of Actinobacteria species closely mirrored those of fungal isolates and *S. aureus* in the soils, and these species thus have these microbes as their competitors for the microenvironment (Boisvert-Bertrand et al., 2004). Temperate sandy soils with their narrow complement of microflora may therefore well provide suitable locations from which to obtain novel antistaphylococcal and antifungal agents.

5.3.2   Tropical locations with other soil types

Tropical terrestrial environments are known to comprise many novel species (Janso and Carter, 2010), and these organically nutrient-rich locations are home for thousands of bacterial species, each of which must compete for territory (Hawksworth and Colwell, 1992). Studies by Araragi (1979) on fertile tropical Thai soil being used as farmland have suggested that although many of these species are *Streptomyces* and *Nocardia* species, there is also a differential distribution based on soil type. For example, soils with high water retention ability (grumusols) tend to have more *Monospora* species, whereas humus, bedrock (rendzina), and iron-rich, forest soils (grey podzolic) tend to have more *Nocardia* (Araragi, 1979). More recent studies performed on the granite, loam, acidic soils in the tropical forest in Western Ghats, India, have similarly suggested that *Actinomycete* species are typically *Streptomyces*, *Nocardia*, and *Micromonospora* (Panaiyadiyan and Chellaia, 2011). In addition, as is the case for all soil ecosystems, the species and dynamics of Actinobacteria growing in these locations may be influenced by many biotic factors, including the plant species growing on the soils (Ambrose et al., 2013) and the animal species (ants, earthworms, etc.) living within it (Kumar et al., 2012b). The next section will deal with antibiotics that have been identified in each of these three very separate tropical environments.

5.3.2.1   Terrestrial and tropical Actinobacteria antibiotics

Many reports of novel terrestrial tropical Actinobacteria exist including those harvested from rich soils in Bangladesh (Al-Bari et al., 2005) and Jamaica (Piechoski et al., 2012). Some of the diverse locations are shown in Table 5.2. Jamaica is a particularly diverse location from which to sample in this respect since it possesses a tropical, maritime climate conducive to bacterial proliferation and diversity (Borneman and Triplett, 1997). The primary bedrock soil is limestone, but the soil diversity ranges from erosion-resistant upland plateaus to lowland sand and gravel. This variety in tropical soil construct

Table 5.2 Antibiotic-producing Actinobacteria from tropical, rich soils

Location/soil type	Species	Antibiotic	Antibacterial efficacy
Jordanian red soils (clay)[1]	Unknown	Unknown	*Micrococcus luteus* and *S. aureus*
Karnataka, India[2] (rhizosphere, college campus, rich soil)	*Streptomyces* sp.	Unknown	MRSA
Kutumsar Cave, India[3] (Limestone)	*S. prasinosporus, S. roseus, S. aurantiacus,* and *S. longisporoflavus*	Unknown	*Escherichia coli, Staphylococcus aureus,* and *Pseudomonas aeruginosa*
Natore, Bangladesh[4]	*S. bangladeshensis*	bis-2-Ethylhexyl phthalate	Broad spectrum
Port Antonio, Jamaica[5] (clay soil)	Unknown Actinobacteria	Unknown	*Mycobacterium tuberculosis*
Tamil Nadu, India[6] (Manakudy mangrove swamp)	*Streptomyces* sp. PVRK-1	Unknown	Narrow-spectrum MRSA
Shola Soils, Kerala, India[7] (Tropical Forest)	*Streptomyces* sp.	Unknown	Broad spectrum
Samed Island, Indonesia[8] (forest)	*S. sporoclivatus*	Geldanamycin	MRSA

[1] Falkingham et al. (2009), [2] Prashith Kekuda et al. (2010), [3] Rajput et al. (2012), [4] Al-Bari et al. (2005), [5] Piechoski et al. (2012), [6] Kannan et al. (2011), [7] Varghese et al. (2012), and [8] Anansiriwattana et al. (2006).

has been previously shown to be previously associated with extreme microbial diversity (Juo and Franzluebbers, 2003). In addition, the northwestern area of the island, around the St. Ann's district, is associated with the rare red soils (*terra rossa*) that are currently being mined for copper, iron, silver, and gold ores (Wall Street Journal Market Watch, 2013). Red soils may be further sites for novel Actinobacteria, as evidenced by the studies of Falkingham and colleagues (2009). Falgingham et al. (2009) sampled the Jordanian red soils and demonstrated that they contained Actinobacteria with potent antibacterial activities. These soils have a long history of being used as topical treatments for diaper rash and wounds, which led weight to their therapeutic content (Falkingham et al., 2009). Finally, the coast, especially to the east, is surrounded by coral reefs, which are also known sources of novel Actinobacteria. Our own studies showed the presence of Actinobacteria producing narrow-spectrum antibiotic activity against tuberculosis bacteria in soils from three locations: Sunset Cove, where limestone is the predominant rock and bats are common; Bonnie view, a former cholera burial site where clay soil predominates; and the sands of Folly Ruin, under a piece of coral (Piechoski et al., 2012). In this particular case, soil abiotic components and animal activity in the Jamaican soil isolates have not been delineated, but this is not the case for all tropical, terrestrial environments. The next sections will therefore discuss the currently available information for the roles played by animals and plants in Actinobacteria growth and production of antibiotics.

5.3.2.2   Role of animals in Actinobacteria growth and production of antibiotics

Environmentally, in both terrestrial and aquatic locations, the majority of the Actinobacteria behave as strict saprophytes involved in maintaining the carbon cycle, but some may be parasitic or even form symbiotic associations with animals (Barke et al., 2010). An interesting symbiosis of this nature was recently demonstrated in the fungus farming ant *Acromyrmex octospinosus* in which the mutualist Actinobacteria species *Streptomyces* and *Pseudonocardia* were able to generate antibiotic and antifungal agents, presumably to protect the colony against microbial infection (Barke et al., 2010). Ants are common inhabitants of tropical soils where they may encourage or even "farm" bacteria and fungi able to release antibiotic substances (Barke et al., 2010; Mendes et al., 2013). Actinobacteria have been successfully isolated from attine ants and include the species *Pseudonocardia* and *Streptomyces* (vander Meer, 2012). Both have been shown to produce the antibiotic 6-deoxy-8-*O*-methylrabelomycin, active against *B. subtilis* as well as anti-Candida urauchimicins from *Streptomyces* species (Mendes et al., 2013). Our studies of Jamaican tropical soil also isolated *Streptomyces* able to secrete effective anti-Candida activity from a location with heavy ant contamination (Folly View; Piechoski et al., 2012).

Studies by Kumar and colleagues (2012b) have also demonstrated the presence of various antibiotic-secreting Actinobacteria in earthworm castings of Doon Valley, India, soils. Earthworms have long been studied by scientists as they are known to be important vectors for the transport of beneficial bacteria in soils (Doube et al., 1994), and these species have now been found to include a variety of Actinobacteria species of *Streptomyces* were found to be the dominant organisms and Actinobacteria were found to be more frequent in worm casts originating in forest soils as compared to agricultural land (Kumar et al., 2012b). Antibiotic production was seen in many *Streptomyces* species with *S. rochei* generating excellent antibacterial and antifungal activities (Kumar et al., 2012b).

Another tropical location, which has yielded novel and unique microbes, is limestone caves including Kotumsar and Magura (Rajput et al., 2012; Tomova et al., 2013). Caves typically possess higher humidity and more stable temperatures than outside locations, and their floors are covered with bacteria derived from bat guano (*E. coli*, *S. aureus*, and *Pseudomonas*) as well as Actinobacteria able to generate antibiotic activities to inhibit these microbes (Ikner et al., 2007). As with other terrestrial locations, *Streptomyces* often predominate (Niyomvong et al., 2012), but for the Magura cave, they contain novel *Arthrobacter* and *Myroides* species (Yucel and Yamac, 2010; Tomova et al., 2013). Interestingly, one of the Jamaican locations that proved fruitful in our own studies was associated with bat guano (Sunset Cove) and was the site of *Streptomyces* isolates with activity against both tuberculosis and *Candida albicans* (Piechoski et al., 2012).

Finally, animal associations with Actinobacteria are not limited to terrestrial locations. Two major studies have demonstrated the association of novel antibiotic-secreting Actinobacteria associated with soft (Fu et al., 2013) and hard (scleractinian; Nithyand et al., 2010) corals. Strepchloritides A and B, with activity against MRSA, were identified from South China Sea soft corals (Fu et al., 2013) and *S. akiyoshinensis* associated with Gulf of Mannar, Tamil Nadu scleractinian corals inhibited *S. pyogenes* biofilm formation associated with atopic dermatitis and impetigo (Nithyanand et al., 2010). Jamaica harbors more than 60 species of coral including elk horn and stag horn species, and the reefs fringe up to a mile out in the northern coasts, where they are commonest and more sporadically in the south (Idjadi et al., 2006). Like many of the coral reef environments around the world, Jamaican coral is endangered (Broad, 1994). Studies demonstrated that amounts decreased from 1970s from 52% to 3% due to a combination of intense fishing for parrotfish, which are the major grazers on algae that smothers coral, and hurricane damage. Some attempts to reseed the coral seem successful as shown in Dairy Bull east of Discovery Bay, which is again dominated by scleractinian corals (Idjadi et al., 2006). Given that these corals may also have potential for novel, narrow-spectrum antibiotics, the preservation of these unique locations is clearly of paramount importance on several levels.

5.3.2.3   Role of plants in Actinobacteria growth
and production of antibiotics: Endophytic species

Researchers have also recently identified several tropical endophyte Actinobacteria able to secrete novel antibiotics (Ambrose et al., 2013). Endophyte bacteria are unusual in that they spend their lives associated with plant species in a beneficially mutualistic relationship that may also now extend to secreting antibiotics to protect them from infection (Ambrose et al., 2013). At least three important classes of antibiotics have been described from endophyte Actinobacteria: the munumbicins, the kakadumycins, and the xiamycins (Ambrose et al., 2013).

Munumbicins are secreted by *Streptomyces* growing on snakevine plants in the tropical, Northern Territory of Australia and have their primary activity against gram-positive bacteria such as *Bacillus cereus*, MRSA, VRE, and *Mycobacterium tuberculosis* (Castillo et al., 2002). In this latter respect, munumbicin B antibiotic is particularly interesting since it has activity against multidrug resistant (MDR) forms of tuberculosis (Ambrose et al., 2013) and may therefore represent a new therapeutic for the treatment of this bacterium. Kakadumycins have been identified in cultures of endophytic *Streptomyces* (NRRL30566) bacteria isolated from the grevillea tree (*Grevillea pteridifolia*, synonym: *Grevillea chrysodendron* R.Br.) also native to the Northern Territory of Australia (Castillo et al., 2003). Two kakadumycins have been identified: kakadumycin A is similar to echinomycin, a quinoxaline antibiotic, and kakadumycin B, which is similar in spectrum to the munumbicins (Castillo et al., 2003). Xiamycins represent a third class of novel antibiotics and are novel pentacyclicindolosesquiterpenes, obtained from *Streptomyces* sp. strain GT2002/1503, an endophyte from the mangrove plant *Bruguiera gymnorrhiza* (Ding et al., 2010). In addition to potent activity against MRSA and VRE, similar to that demonstrated by the munumbicins, xiamycins also demonstrate selective effectiveness against the HIV (Ding et al., 2010). These compounds, therefore, have promise in several treatment areas.

5.4   Conclusions and future directions

Even though drug discovery and development aim to maintain effective therapy by replacing antibiotics to which bacteria have become resistant with new ones, few new natural drugs of significance have made their appearance in the past 20 years (Raja and Prabakarana, 2011). There are two major reasons for this shortfall: failure to match growth conditions with antibiotic product and cost of therapeutic development. Since it is likely that many of the more dominant Actinobacteria have already been isolated and identified as antibiotic producers, many researchers are now beginning to scour the globe for more remote, as yet untapped environments where novel species might yield novel drugs. Given that many terrestrial Actinobacteria produce their antibiotics in response to external pressures including both biotic (competitors, food) and abiotic (changes in temperature and pH) factors, it would be anticipated that both hostile and nutrient-rich locales would yield effective antibiotic-producing species. For example, in addition to the terrestrial, tropical Actinobacteria, several researchers have identified tropical marine antibiotic-producing Actinobacteria, which seem to represent a separate ecosystem rather than a subset of terrestrial washings (Kumar et al., 2012a). Aquatic antibiotic-secreting bacteria belonging primarily to the genera *Streptomyces* and *Rhodococcus* have been isolated from Indian sediments and the Arabian Sea as well as the Bay of Bengal, Andhra Pradesh (Kumar et al., 2012a). The majority of these antibiotics have not yet been characterized, and it is not known if they are novel or represent rediscoveries of already known isolates.

Antibiotic-resistant bacteria are now our constant companions, and new strains are emerging all the time, despite effective measures to limit the use of therapeutics (Levy, 1998). MDR tuberculosis is a case in point, affecting 500,000 annually, which requires 2 years of treatment and is effectively cured in only 50% of patients (de Lima Procópio et al., 2012). Even more virulent is extensively drug-resistant tuberculosis (XDR-TB), which has been reported in 58 countries (Mlambo et al., 2008). More worrying are genotyping studies that suggest between 63% and 75% of MDR-TB strains can progress to becoming XDR-TB (Mlambo et al., 2008). Tropical Actinobacteria may be capable of generating antibiotics effective against this bacterium, as suggested by the work of researchers in the Amazon Biotechnology Center (CBA) (de Lima Procópio et al., 2012) and in our own study (Piechoski et al., 2012). Lack of toxicity of antibiotics from any of these sources could result in new hope for many across the world.

Finally, a return to natural antibiotics may help allay mistrust by many in the ability of "artificial medicines" to effectively treat disease. Synthetic and semisynthetic antibiotics have received bad press recently with the limitations placed on telithromycin (Ketek®; Echols, 2011) as well as dangerous side effects from trimethoprim/sulfa drugs (Antoniou et al., 2011). Even the well-loved semisynthetic azithromycin (Z-Pak®) was suggested by one set of authors to increase cardiovascular death, although others labeled the studies inconclusive azithromycin and the risk of cardiovascular death (Ray et al., 2012). These observations, published in reputable journals, reduce public confidence in the safety of antibiotics and may have even been a contributory factor in the increased use of natural, herbal medicines, particularly in the United States (Bauer, 2003). Natural antibiotics and their derivatives typically possess better bioavailability and capacity to bind to cellular targets than their synthetic counterparts, again lending support for new therapeutics to come from natural sources (Wright, 2010). Given that many natural microbial metabolites have been and will continue to be derived from Actinobacteria species, these resilient bacteria are destined to remain unchallenged as primary industrial sources of our antibiotics.

Acknowledgments

The authors thank Dr. Anne Bower and Dr. John Porter for their valuable scientific insights into the Jamaican soil project and also Catherine Boisvert-Bertrand and Stefanie Emanuel for their bench skills.

References

Al-Bari, M. A. A., Bhuiyan, M. S. A., Flores, M. E., Petrosyan, P., Garcia-Varela, M., and Islam, M. A. U. 2005. *Streptomyces bandladeshensis* sp. nov., isolated from soil, which produces bis-(2-ethylhexyl) phthalate. *Int. J. Syst. Evol. Microbiol.*, 55(5), 1973–1977.

Ali, M. I., Ahmad, M. S., and Hozzein, M. N. 2009. WA52 a macrolide produced by alkalophile *Nocardiensis dassonvillei* WA52. *Aus. J. Basic Appl. Sci.*, 3, 607–616.

Ambrose, C., Varghese, C., and Bhore, S. J. 2013. Endophytic bacteria as a source of novel antibiotics— An overview. *Pharmacogn. Rev.*, 7(13), 11–16.

Antoniou, T., Gomes, T., Mamdani, M. M., Yao, Z., Hellings, C., Garg, A. X., and Juurlink, D. N. 2011. Trimethoprim-sulfamethoxazole induced hyperkalaemia in elderly patients receiving spironolactone: Nested case-control study. *BMJ*, 343, d5228.

Araragi, M. 1979. Actinomycete flora of tropical, upland farmland soil on the basis of genus composition and antagonistic property. *Soil Sci. Plant. Nutr.* 25, 514–523.

Atlas, R. M., Parks, L. C., and Brown, A. E. 1995. Screening for the isolation of an antibiotic producer. In: Atlas, R. M., Parks, L. C., and Brown, A. E., eds., *Laboratory Manual of Experimental Microbiology*, exercise 52, 2nd edn. Mosby Publishers, St. Louis, MO, pp. 425–426.

Augustine, N., Wilson, P. A., Kerkar, S., and Thomas, S. 2012. Arctic actinomycetes as potential inhibitors of *Vibrio cholera* biofilm. *Curr. Microbiol.*, 64, 338–342.

Baltz, R. H. 2008. Renaissance in antibacterial discovery from Actinomycetes. *Curr. Opin. Pharmacol.*, 8, 557–563.

Bandow, J. E., Brötz, H., Leichert, L. I. O., Labischinski, H., and Hecker, M. 2003. Proteomic approach to understanding antibiotic action. *Antimicrob. Agents Chemother.*, 50, 519–526.

Barke, J., Seipke, R. F., Grüschow, S., Heavens, D., Drou, N., Bibb, M. J., and Hutchings, M. I. 2010. A mixed community of actinomycetes produce multiple antibiotics for the fungus farming ant *Acromyrmex octospinus*. *BMC Biology*, 8, 109–119.

Bauer, B. A. 2003. Herbal medicine the good, the bad and the ugly. *SoCRA SOURCE*, pp. 27–30.

Boisvert-Bertrand, C., Brendley, B., and Cundell, D. 2004. Searching for narrow spectrum antibiotics from microbes in soil from Presque Isle, Pennsylvania. *J. Young Invest.*, 10(4). Accessed January 13, 2014. http://legacy.jyi.org/volumes/volume10/issue4/articles/boisvert.html.

Borneman, E. W. and Triplett, J. 1997. Molecular microbial diversity in soils from eastern Amazonia: Evidence for unusual microoorganisms and microbial population shifts associated with deforestation. *Appl. Environ. Microbiol.*, 63, 2647–2453.

Broad, W. J. 1994. Coral reefs endangered in Jamaica. *The New York Times*, September 9.

Brötz-Oesterhelt, H. and Sass, P. 2010. Postegenomic strategies in antibacterial drug therapy. *Future Microbiol.*, 5, 1553–1579.

Busti, E., Monciardini, P., Cavaletti, L., Bamonte, R., Lazzarini, A., Sosio, M., and Donadio, S. 2006. Antibiotic-producing ability by representatives of a newly discovered lineage of actinomycetes. *Microbiology*, 152, 675–683.

Cameron, R. E., King, J., and David, C. N. 1970. Soil microbial ecology of Wheeler Valley, Antarctica. *Soil Sci.*, 109, 110–120.

Carter, G. T., Nietsche, J. A., Williams, D. R., and Borders, D. B. 1990. Citreamicins, novel antibiotics from *Micromonospora citrea*: Isolation, characterization and structure determination. *J. Antibiot.*, 43, 504–512.

Castillo, U., Harper, J. K., Strobel, G. A., Sears, J., Alesi, K., Ford, E., and Teplow, D. 2003. Kakadumycins, novel antibiotics from *Streptomyces* sp. NRRL 30566, an endophyte of *Grevillea pteridifolia*. *FEMS Microbiol. Lett.*, 224, 183–190.

Castillo, U. F., Strobel, G. A., Ford, E. J., Hess, W. M., Porter, H., Jensen, J. B., and Yaver, D. 2002. Munumbicins, wide-spectrum antibiotics produced by *Streptomyces* NRRL 30562, endophytic on *Kennedia nigriscans*. *Microbiology*, 148, 2675–2685.

Challis, G. L. and Hopwood, D. A. 2003. Synergy and contingency as driving forces for the evolution and multiple secondary metabolite production by *Streptomyces* species. *Proc. Natl. Acad. Sci.*, 100, 14555–14561.

Chater, K. F., Biró, S., Lee, K. J., Palmer, T., and Schrempf, H. 2010. The complex extracellular biology of *Streptomyces*. *FEMS Microbiol. Rev.*, 34, 171–198.

Chaxamycins, A. D. 2011. Bioactive ansamycins from a hyper-arid desert *Streptomyces* sp. *J. Nat. Prod.*, 74(6), 1491–1499.

Chhabra, R. 1996. *Soil Salinity and Water Quality*. Oxford & IBH Publishing Co. Pvt. Ltd., New Delhi, India (South Asian edition), p. 284.

Cordeiro, J. C. R., Reis, A. O., Miranda, E. A., and Sader, H. S. 2001. In vitro antimicrobial activity of the aminoglycoside arbekacin tested against oxacillin-resistant *Staphylococcus aureus* isolated in Brazilian hospitals. *Braz. J. Infect. Dis.*, 5(3), 130–135.

de Lima Procópio, R. E., da Silva, I. R., Martins, M. K., de Azevedo, J. L., and de Araújo, J. M. 2012. Antibiotics produced by *Streptomyces*. *Braz. J. Infect. Dis.*, 16(5), 466–471.

Ding, L., Münch, J., Goerls, H., Maier, A., Fiebig, H. H., Lin, W. H., and Hertweck, C. 2010. Xiamycin, a pentacyclic indolosesquiterpene with selective anti-HIV activity from a bacterial mangrove endophyte. *Bioorg. Med. Chem. Lett.*, 20(22), 6685–6687.

Donadio, S., Staver, M. J., McAlpine, J. B., Swanson, S. J., and Katz, L. 1991. Modular organization of genes required for complex polyketide biosynthesis. *Science*, 252, 675–679.

Doube, B. M., Stephens, P. M., Davoren, C. W., and Ryder, M. H. 1994. Interaction between earth-worms, beneficial microbes and root pathogens. *Appl. Soil. Ecol.*, 1, 3–10.

Echols, R. M. 2011. Understanding the regulatory hurdles for antibacterial drug development in the post-Ketek world. *Ann. NY Acad. Sci.*, 1241, 153–161.

El-Khagawa, M. A. and Megahed, M. M. M. 2012. Antibacterial and anti-insecticidal activity of acti-nomycetes isolated from sandy soil of Cairo Egypt. *Egypt Acad. J. Biolog. Sci.*, 4, 53–67.

El-Naggar, M. Y., El-Assar, S. A., and Abdul-Gawad, S. M. 2006. Meroparamycin production by newly isolated *Streptomyces* sp. strain MAR01: Taxonomy, fermentation, purification and struc-tural elucidation. *J. Microbiol.*, 44, 432–438.

Enright, M. C. 2003. The evolution of a resistant pathogen—The case of MRSA. *Curr. Op. Pharmacol.*, 3, 474–479.

Falagas, M. E., Grammatikos, A. P., and Michalopoulos, A. 2008. Potential of old-generation antibiot-ics to address current need for new antibiotics. *Expert Rev. Anti Infect. Ther.*, 6(5), 593–600.

Falkinham, J. O., Wall, T. E., Tanner, J. R., Tawaha, K., Alali, F. Q., Li, C., and Oberlies, N. H. 2009. Proliferation of antibiotic-producing bacteria and concomitant antibiotic production as the basis for the antibiotic activity of Jordan's red soils. *Appl. Environ. Microbiol.*, 75(9), 2735–2741.

Freiberg, C., Brunner, N. A., Schiffer, G., Lampe, T., Pohlmann, J., Brands, M., and Ziegelbauer, K. 2004. Identification and characterization of the first class of potent bacterial acetyl-CoA car-boxylase inhibitors with antibacterial activity. *J. Biol. Chem.*, 279(25), 26066–26073.

Fu, P., Kong, F., Wang, Y., Wang, Y., Liu, P., Zuo, G., and Zhu, W. 2013. Antibiotic metabolites from the coral-associated *Streptomyces* species OUCMDZ-1703. *Chin. J. Chem.*, 31(1), 100–104.

Fuoco, D. 2012. Classification framework and chemical biology of tetracycline-structure-based drugs. *Antibiotics*, 1, 1–13.

Gadebusch, H. H., Stapley, E. O., and Zimmerman, S. B. 1992. The discovery of cell wall active anti-bacterial antibiotics. *Crit. Rev. Biotechnol.*, 12, 225–243.

Hancock, R. E. W. 2007. The complexities of antibiotic action. *Mol. Syst. Biol.*, 3, 142.

Hawksworth, D. L. and Colwell, R. R. 1992. Biodiversity amongst microorganisms and its relevance. *Biodiver. Conserv.*, 1, 221–345.

Hayakawa, M. 2008. Studies of the isolation and distribution of rare Actinomycetes in soil. *Actinomycetologica*, 22, 12–19.

Hickner, J. 2007. Can you call in a Z-Pak for me? *J. Family Pract.*, 56(4), 262.

Hogan, C. M. 2010. Bacteria. In: Draggan, S. and Cleveland, C. J., eds., *Encyclopedia of Earth*. National Council for Science and the Environment, Washington, DC.

Hozzein, W. M., Ali, M. I. A., and Rabie, W. 2008. A new preferential medium for enumeration and isolation of desert Actinomycetes. *World J. Microbiol. Biotechnol.*, 24(8), 1547–1552.

Idjadi, J. A., Lee, S. C., Bruno, J. F., Precht, W. F., Allen-Requa, L., and Edmunds, P. J. 2006. Rapid phase shift reversal on a Jamaican coral reef. *Coral Reefs.*, 25, 209–211.

Ikner, L. A., Toomey, R. S., Nolan, G., Neilson, J. W., Pryor, B. M., and Maier, R. M. 2007. Culturable microdiversity and the impact of tourism in Kartchner Caverns, Arizona. *Microb. Ecol.*, 53, 30–42.

IMS Institute for Healthcare Informatics. 2011. The use of medicines in the United States: Review of 2011. Accessed January 14, 2014, http://www.imshealth.com/ims/Global/Content/Insights/IMS%20Institute%20for%20Healthcare%20Informatics/IHII_Medicines_in_U.S_Report_2011.pdf.

Janso, J. E. and Carter, G. T. 2010. Phylogenetically unique endophytic actinomycetes from tropical plants possess great biosynthetic potential. *Appl. Environ. Microbiol.*, 77, 4377–4386.

Juo, A. S. and Franzluebbers, K. 2003. *Tropical Soils: Properties and Management for Sustainable Agriculture* (Topics in Sustainable Agronomy), 1st edn. Oxford University Press, New York, p. 78.

Kannan, R.R., Iniyan, M.M., and Prakash, V.S.G. 2011. Isolation of a small molecule with anti-MRSA activity from a mangrove symbiont *Streptomyces* sp PVRK-1 and its biomedical studies in zebrafish embryos. *Asian Pac. J. Trop. Biomed.*, 5, 341–347.

Katz, M. L., Mueller, L. V., Polyakov, M., and Weinstock, S. F. 2006. Where have all the antibiotic patients gone? *Nat. Biotechnol.*, 24, 1529–1531.

Kieft, T. L. 1991. Soil microbiology in reclamation of arid and semi-arid lands. In: Skujins, J., ed., *Semiarid Lands and Deserts: Soil Resource and Reclamation*. Marcel Dekker, New York, pp. 209–256.

Kumar, K. S., Haritha, R., Mohan, Y. S. Y. V. J., and Ramana, T. 2012a. Screening of marine Actinobacteria for antimicrobial compounds. *Res. J. Microbiol.*, 6(4), 385–393.

Kumar, V., Bharti, A., Negi, Y. K., Gusain, O., Pandey, P., and Bisht, G. S. 2012b. Screening of actinomycetes from earthworm castings for their antimicrobial activity and industrial enzymes. *Braz. J. Microbiol.*, 43(1), 205–214.

Kurapova, A. I., Zenova, G. M., Sudnitsyn, I. I., Kizilova, A. K., Manucharova, N. A., Norovsuren, Z., and Zvyagintsev, D. G. 2012. Thermotolerant and thermophilic Actinomycetes from soils of Mongolia desert steppe zone. *Microbiology*, 81, 105–116.

Kurosawa, K., Ghivriga, I., Sambandan, T. G., Lessard, P. A., Barbara, J. E., Rha, C., and Sinskey, A. J. 2008. Rhodostreptomycins, antibiotics biosynthesized following horizontal gene transfer from *Streptomyces padanus* to *Rhodococcus fascians. JACS Commun.*, 130, 1126–1127.

Lancini, G., and Lorenzetti, R. 1993. *Biotechnology of Antibiotics and Other Bioactive Microbial Metabolites*. Plenum Press, New York, pp. 49–57.

Lange, R. P. Locher, H. H., Wyss, P. S., and Then, R. L. 2007. The targets of currently used antibiotic agents: Lessons for drug discovery. *Curr. Pharm. Des.*, 13, 3140–3154.

Lazzarini, A., Cavaletti, L., Toppo, G., and Marinelli, F. 2001. Rare genera of actinomycetes as potential producers of new antibiotics. *Antonie van Leeuwenhoek*, 78, 399–405.

Levy, S. B. 1998. The challenge of antibiotic resistance. *Sci Am.*, 398, 451–456.

Li, G. P. 1989. Isolation of actinomycetes for antibiotic screening. *Chin. J. Antibiotic*, 14, 452–465.

Liu, L. L., He, L. S., Xu, Y., Han, Z., Li, Y. X., Zhong, J. L., and Qian, P. Y. 2013. Caspase-3-dependent apoptosis of citreamicin epsilon-induced HeLa cells is associated with reactive oxygen species generation. *Chem. Res. Toxicol.*, 26(7), 1055–1063.

Makut, M. D. and Abdalazeez, U. S. 2012. A survey of antibiotic-producing actinomycetes in the soil environment of Keffi metropolis, Nasarawa State, Nigeria. *J. Microbiol. Antimicrobial.*, 4(5), 74–78.

Marinelli, F. 2009. Antibiotics and *Streptomyces*: The future of antibiotic discovery. *Microbiology Today*, pp. 20–23.

Mendes, T. D., Borges, W. S., Rodrigues, A., Solomon, S. E., Vieira, P. C., Duarte, M. C., and Pagnocca, F. C. 2013. Anti-Candida properties of urauchimycins isolated from Actinobacteria associated with *Trachymyrmex* ants. *Biomed. Res. Int.* 8. Article ID 835081. Accessed January 20, 2014. http://www.hindawi.com/journals/bmri/2013/835081/cta/.

Mlambo, C. K., Warren, R. M., Poswa, X., Victor, T. C., Duse, A. G., and Marais, E. 2008. Genotypic diversity of extensively drug-resistant tuberculosis (XDR-TB) in South Africa. *Int. J. Tuberc. Lung. Dis.*, 12(1), 99–104.

Nelson, L. M. and Levy, S. B. 2011. The history of the tetracyclines. *Ann. NY Acad. Sci.*, 1241, 17–32.

Nicolaou, K. C., Boddy, C. N. C., Bräse, S., and Winssinger, N. 1999. Review: Chemistry, biology, and medicine of the glycopeptide antibiotics. *Angew. Chem. Int. Ed.*, 38, 2096–2152.

Nithyanand, P. Thenmozhi, R., Rathna, J., and Pandian, S. K. 2010. Inhibition of *S. pyogenes* biofilm formation by coral associated Actinomycetes. *Curr. Microbiol.*, 60, 454–460.

Niyomvong, N., Pathom-aree, W., Thamchaipenet, A., and Duangmal, K. 2012. Actinomycetes from tropical limesone caves. *Chiang Mai J. Sci.*, 39(3), 373–388.

Osman, M. E., Ahmed, F. A. H., and Abd El All, W. S. M. 2011. Antibiotic production from local *Streptomyces* isolates from Egyptian soil at Wady El Natron: Isolation, identification and optimization. *Aust. J. Basic Appl. Sci.*, 5(9), 782–792.

Panaiyadiyan, P. and Chellaia, S. R. 2011. Biodiversity of microorganisms isolated from rhizosphere soils of Pachamalai hills, Tamilnadu, India. *Res. J. Forestry*, 5, 27–35.

Payne, D. J., Gwynn, M. N., Holmes, D. J., and Pompliano, D. L. 2007. Drugs for bad bugs: Confronting the challenge for antibacterial discovery. *Nat. Rev. Drug Discov.*, 6, 29–40.

Peoples, A. J., Zhang, Q., Millett, W. P., Rothfeder, M. T., Pescatore, B. C., Madden, A. A., and Moore, C. M. 2008. Neocitreamicins I and II, novel antibiotics with activity against methicillin-resistant *Staphylococcus aureus* and vancomycin-resistant *Enterococci. J. Antibiot. (Tokyo)*, 61, 457–463.

Piechoski, M., Cundell, D., Bower, A., and Porter, J. 2012. Bioassay and antibiotic activity of Jamaican *Actinomyces* isolates. *J. Young Investigat.* Accessed January 13, 2014. http://www.jyi.org/issue/bioassay-and-antibiotic-activity-of-jamaican-actinomycetes-isolates/.

Powers, J. H. 2004. Antimicrobial drug development—The past, the present and the future. *Clin. Microbiol. Infect.*, 10, 23–31.

Prashith Kekuda, T. R., Shobha, K. S., and Onkarappa, R. 2010. Studies on antioxidant and anthelmintic activity of two *Streptomyces* species isolated from Western Ghat soils of Agumbe, Karnataka. *J. Pharm. Res.*, 3, 26–29.

Raja, A. and Prabakarana, P. 2011. Actinomycetes and drug—An overview. *Am. J. Drug Discov. Dev.*, 1(2), 75–84.

Rajput, Y., Rai, V., and Biswas, J. 2012. Screening of bacterial isolates from various microhabitat sediments of Kotumsar Cave: A cogitation on their respective benefits and expected threats for complete biosphere and tourists. *Res. J. Environ. Toxicol.*, 6, 13–24.

Ray, W. A., Murray, K. T., Hall, K., Arbogast, P. G., and Stein, C. M. 2012. Azithromycin and the risk of cardiovascular death. *N Eng. J. Med.*, 366(20), 1881–1890.

Schaechter, M., ed. 2010. *Desk Encyclopedia of Microbiology.* Academic Press, San Diego, CA.

Selvameenal, L., Radhakrishnan, M., and Balagurunathan, R. 2009. Antibiotic pigment from desert soil actinomycetes; Biological activity, purification and chemical screening. *Indian J. Pharm. Sci.*, 71(5), 499–504.

Silver, L. 2011. Challenges of antibacterial discovery. *Clin. Microbiol. Rev.*, 24, 71–109.

Singh, A. B., Singh, R. B., and Mishra, S. 2012. Microbial and biochemical aspects of antibiotic producing microorganisms from soil samples of certain industrial area of India—An overview. *Open Nutraceut. J.*, 5, 107–112.

Skujins, J. 1984. Microbial ecology of desert soils. In: Marshal, C. C., ed., *Advances in Microbial Biology.* Plenum Press, New York, pp. 49–51.

Solis, P. N., Wright, C. W., Anderson, M. M., Gupta, M. P., and Phillipson, J. D. 1993. A microwell cytotoxicity assay using *Artemia salina. Plant Med.*, 59, 250–252.

Tabrizi, S. G., Hamedi, J., and Mohammadipanah, F. 2013. Screening of soil actinomycetes against *Salmonella* serovar NCTC 5761 and characterization of the prominent active strains. *Iran J. Microbiol.*, 5, 356–365.

Tomova, I., Lazarkevich, I., Tomova, A., Kambourova, M., and Vasileva-Tonkova, E. 2013. Diversity and biosynthetic potential of culturable aerobic heterotrophic bacteria isolated from Magura cave, Bulgaria. *Int. J. Spelol.*, 42(1), 65–76.

Vander Meer, R. V. 2012. Ant interactions with soil organisms and associated semiochemicals. *J. Chem. Ecol.*, 38(6), 728–745.

Varghese, R., Nishamol, S., Suchithra, R., Jyothy, S., and Hatha, A. M. 2012. Distribution and antibacterial activity of Actinomycetes from shola soils of tropical montane forest in Kerala, South India. *J. Environ.*, 1, 93–99.

Von Nussbaum, F., Brands, M., Hinzen, B., Weigand, S., and Häbich, D. 2006. Antibacterial natural products in medicinal chemistry—Exodus or revival? *Angew. Chem. Int. Ed. Engl.*, 45, 5072–5129.

Waksman, S. A. and Lechevalier, H. A. 1962. *The Actinomycetes, Volume III, Antibiotics of Actinomycetes.* The Williams and Wilkins Company, Baltimore, MD.

Wall Street Journal Market Watch. Vela Minerals acquires 100% interest in Mavis Bank and Port Antonio properties Jamaica. *The Wall Street Journal*, June 6, 2013. Accessed January 22, 2014. www.marketwatch.com/story/vela-minerals-acquires-100-interest-in-mavis-bank-and-port-antonio-properties-jamaica-2013-06-06.

Wright, G. D. 2010. Q&A: Antibiotic resistance: Where does it come from and what can we do about it? *BMC Biol.*, 8, 123.

Yucel, S. and Yamac, M. 2010. Selection of *Streptomyces* isolates from Turkish karstic caves against antibiotic-resistant organisms. *Pak. J. Pharm. Sci.*, 23, 1–6.

Zgoda, J. R. and Porter, J. R. 2001. A convenient microdilution method for screening natural products against bacteria and fungi. *Pharm. Biol.*, 39, 221–225.

Zvyagintsev, D. G. and Zenova, G. M. 2007. *Aktinomitsety zasolennykh i schchelochnykh pochv (Actinomycetes of Saline and Alkaline Soils).* Knizhnyi Som Universitet, Moscow, Russia.

chapter six

Antimicrobial agents from actinomycetes
Chemistry and applications

Deene Manikprabhu and Wen-Jun Li

Contents

6.1 Introduction

The antimicrobial era started after the discovery of penicillin by Alexander Fleming (1929). Back from a holiday, Fleming noticed that a petri dish containing *Staphylococcus* plate culture, which was mistakenly left open, was contaminated by a blue-green mould (*Penicillium*), which formed a visible growth. Fleming noticed that there was a halo of inhibited bacterial growth around the mould. Watching the plate Fleming concluded that the mould might release a substance that suppressed the growth or killed the bacteria. The substance was nothing but penicillin (Figure 6.1a), which had an antibacterial effect that inhibited the growth of *Staphylococcus*.

Although the discovery was accidental and a lucky one, let us imagine for an instance that after returning from the holiday, Fleming noticed something "funny" about this plate and just discarded it as a culture that had been made useless as it was contaminated. Perhaps the best known of Pasteur's obiter is that "in matters of observation, fortune favours only the prepared mind" (Kingston, 2008). Fleming discovered the penicillin because his mind was prepared; in his earlier study, he discovered lysozyme when he had a heavy cold and mucus from his nose fell on the bacterial plate. After a few days, he noticed no growth around the mucus. He had hoped that this mucus would have antibacterial activity; from this experience, he discovered the penicillin.

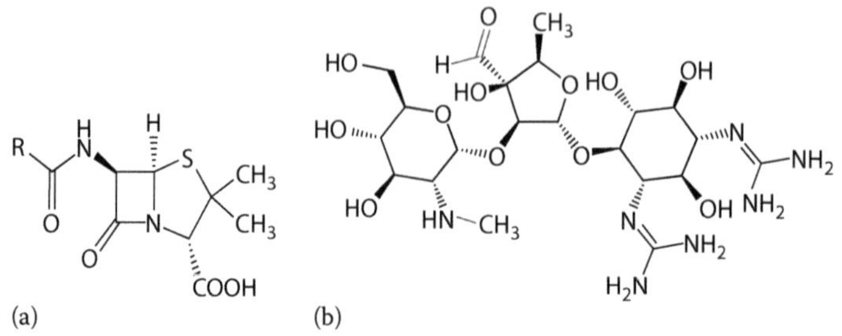

Figure 6.1 The two important compounds of antimicrobial era. (a) Structure of penicillin where R is the variable group. (b) Structure of streptomycin.

This remarkable contribution of Fleming to medicinal science saved many lives during World War II. Though penicillin was a very effective antimicrobial agent at that time, unfortunately, it had one drawback: it has negligible effect on tuberculosis. The great success of penicillin made Waksman rush into his laboratory and shout to his colleagues, "see what these Englishmen have discovered a mould can do. I know the actinomycetes will do better!" Within a few months, Waksman discovered streptomycin (Figure 6.1b), and it was announced in a 1944 paper (Schatz et al., 1944; Kingston, 2008). Furthermore, the Mayo Institute found that "streptomycin" surpassed the activity of penicillin in inhibiting tuberculosis (Kiple, 1993). Since then, the golden era of actinomycetes started, and today actinomycetes are well known for producing antimicrobial agents.

Actinomycetes are gram-positive, filamentous bacteria that have a proven capacity to produce bioactive secondary metabolites. Particularly, it has been estimated that approximately two-thirds of natural antibiotics have been isolated from actinomycetes, and about 75% of them are produced by members of the genus *Streptomyces* (Table 6.1) (Franco-Correa et al., 2010; Wang et al., 2013a). The golden age of antimicrobial agents continues. Everything seemed to be very well controlled, but because of the widespread and uncontrolled use of antimicrobial agents, the golden age came to an end as microorganisms

Table 6.1 Some antimicrobial agents obtained from *Streptomyces*

Antibiotic/ molecular formula	Source	Application	References
Paromomycin $C_{23}H_4N_5O_{18}S$	*Streptomyces krestomuceticus*	Used to treat intestinal infections such as cryptosporidiosis, amoebiasis, and other diseases like leishmaniasis	Davidson et al. (2009)
Carbomycin $C_{42}H_{67}NO_{16}$	*Streptomyces halstedii*	Used as a mild antibiotic and used to treat granuloma inguinale	Wagner et al. (1953)
Kanamycin $C_{18}H_{36}N_4O_{11}$	*Streptomyces kanamyceticus*	Used as an antibacterial agent	Garrod et al. (1981)
Lincomycin $C_{18}H_{34}N_2O_6S$	*Streptomyces lincolnensis*	Used against mycoplasma and some species of *Plasmodium*	Spizek and Rezanka (2004)
Daptomycin $C_{72}H_{101}N_{17}O_{26}$	*Streptomyces roseosporus*	Used to treat skin infections mainly caused by gram-positive bacteria	Woodworth et al. (1992)
Tylosin $C_{46}H_{77}NO_{17}$	*Streptomyces fradiae*	Effective against gram-positive bacteria but has limited range on gram-negative bacteria	Giguere et al. (2007)

became resistant to them (Deene and Lingappa, 2013). The biggest advantage of microbes against antibacterial agents is their incredible adaptability to change in the environment, which is mainly due to their flexible metabolic power (Bérdy, 2012). The increasing emergence of multidrug-resistant bacteria throughout the world and the lack of antibiotics to combat such pathogenic agents continue to be the major concern of the medical community (Kitouni et al., 2005). To combat multidrug-resistant bacteria, many different antimicrobial agents from actinomycetes isolated from various sources have been reported. In this chapter, we focus on the chemistry and applications of antimicrobial agents from actinomycetes isolated from different sources.

6.2 Source of antimicrobial agent–producing actinomycetes

6.2.1 Soil actinomycetes

Soil is not just a mass of dead debris that resulted simply from the physical and chemical weathering of rocks, but it teems with life. Every small particle of soil contains numerous types of living organisms so small that they cannot be recognized with the naked eye. The living organisms comprise numerous types of bacteria, fungi, algae, protozoa, nematodes, and other invertebrates that vary considerably in their structure, size, mode of living, and relationship to soil processes (Waksman and Starkey, 1931). Soils contain by far the greatest diversity of organisms (Bonkowski and Roy, 2005), for instance, it has been estimated that 1 g of soil contains as many as 10^{10}–10^{11} bacteria (Marcel et al., 2008), of which 10^7 is actinomycetes (Steffan et al., 1988) mainly belonging to the genus *Streptomyces*.

Since the discovery of streptomycin, various scientists started finding new antibacterial agents mainly isolated from soil. Eli Lilly's research team in 1949 discovered erythromycin, a macrolide antibiotic having the molecular formula $C_{37}H_{67}NO_{13}$ and a molecular mass of 733.93 g/mol, mainly obtained from *Saccharopolyspora erythraea* isolated from soil samples (McGuire et al., 1952). Erythromycin (Figure 6.2a) is used to treat skin and upper respiratory tract infections caused by *Streptococcus*, *Staphylococcus*, and *Haemophilus* genera (Sarbani, 2006).

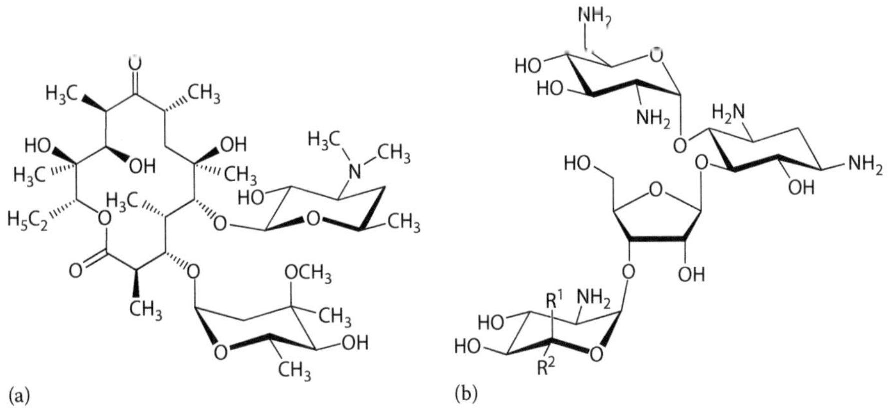

(a)   (b)

Figure 6.2 Antimicrobial agent from actinomycetes isolated from soil: (a) structure of erythromycin and (b) structure of neomycin.

In the same year, neomycin was discovered by Selman Waksman who was later awarded the Nobel Prize in 1951 for his comprehensive contribution toward the field of physiology and medicine (Schatz et al., 1944; Kingston, 2008). Neomycin (Figure 6.2b) is an amino glycoside antibiotic having the molecular formula $C_{23}H_{46}N_6O_{13}$ and a molecular mass of 614.644 g/mol. Naturally produced by *Streptomyces fradiae*, neomycin works by binding to the bacterial 30S ribosomal subunit inhibiting the translocation of the peptidyl-tRNA from the A-site to the P-site and also causing misreading of mRNA, leaving the bacterium unable to synthesize protein vital to its growth, leading to its death. *Streptomyces* was not the only center of attraction, but other actinomycetes were also explored.

Edmund Kornfeld isolated vancomycin in 1953 from *Amycolatopsis orientalis* from soil samples collected from the interior jungles of Borneo (Michael and Plotkin, 2003). Vancomycin, a glycopeptide antibiotic (Figure 6.3a), has the molecular formula $C_{66}H_{75}Cl_2N_9O_{24}$ and a molecular mass of 1449.3 g/mol. It is used in the prophylaxis and treatment of infections caused by gram-positive bacteria like *Staphylococcus aureus* (Levine, 2006). The minimum inhibitory concentration (MIC) of vancomycin against *S. aureus* is found to be ≤2 µg/mL according to the Clinical and Laboratory Standards Institute (CLSI) guidelines (2012).

Similarly, another aminoglycoside antibiotic, gentamicin, was isolated from *Micromonospora purpurea* by Weinstein in 1963. Gentamicin (Figure 6.3b) has the molecular formula $C_{21}H_{43}N_5O_7$ and a molecular mass of 477.596 g/mol (Weinstein et al., 1967), and it is active against gram-negative bacteria.

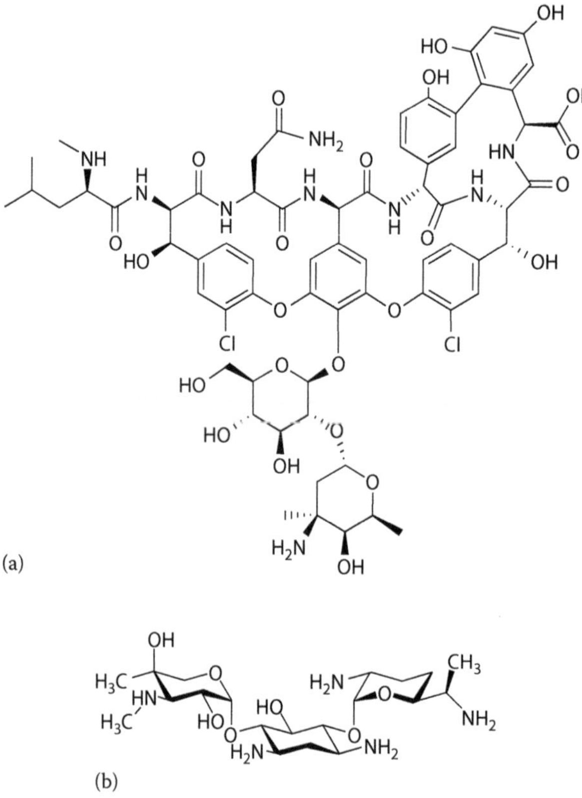

(a)

(b)

Figure 6.3 Structure of (a) vancomycin and (b) gentamicin.

The antimicrobial agent produced by actinomycetes not only helped us to overcome human pathogens but is also used as biological control agents for many plant pathogens and also solved many environmental problems.

Cavity spot disease in the plant caused by the pathogen *Pythium coloratum* was controlled by the antimicrobial agent produced by *Actinoplanes philippinensis* isolated from the soil collected from carrot rhizosphere (El-Tarabily et al., 1997). The antimicrobial agent from actinomycetes was used to solve many environmental problems like cyanobacteria bloom, which posed a serious health threat to animals and humans. Many cyanobacteria produce cyanotoxins (Sivonen and Jones, 1999). To overcome this toxin many compounds from actinomycetes were isolated, such as a novel actinomycete, *Streptomyces aurantiogriseus*, which showed algicidal activity against the toxic cyanobacterium *Microcystis aeruginosa* (Somdee et al., 2013). Though many antimicrobial agents produced from actinomycetes have been isolated (Figure 6.4a through f), it is just a "tip of the iceberg" that has been explored (Baltz, 2005).

In the past few decades, researchers have made many attempts to isolate novel antimicrobial agents not only from the soil but also from the plants generally regarded as endophytes. Attempts were also made to isolate actinomycetes from extreme conditions like alkaliphilic actinomycetes (actinomycetes that can grow at high pH), halophilic actinomycetes (actinomycetes that can grow at high salt concentration), and also actinomycetes from the marine ecosystem. In this chapter, we discuss antimicrobial agents from actinomycetes isolated from the aforementioned environments.

6.2.2 Endophytic actinomycetes

Endophytes are microorganisms that live for the whole or part of their life history inside plant tissues. As a result of these long-held associations, endophytic microorganisms and plants have developed good information transfer (Zhao et al., 2011a,b). The antimicrobial activity of endophytes was reported more than 50 years ago when Smith (1957) succeeded to isolate *Micromonospora* from the tomato plant, which had an antagonistic activity. Since then, many endophytic actinomycete compounds were isolated and have found application not only as an antimicrobial agent but also in many fields.

Many endophytic actinomycete compounds were used as biocontrol agents like compounds from *Nocardia globerula* used to control the *Helminthosporium solani* pathogen, which caused the silver scurf disease in potatoes (Elson et al., 1997). Furthermore, compounds like chitinase produced by the endophyte *Actinoplanes missouriensis* are used against root rot disease of lupine caused by *Plectosporium tabacinum* (El-Tarabily, 2003).

Besides biocontrol agents, various compounds from endophytic actinomycetes helped to protect plants from fungal pathogens. The compound from *Streptomyces* sp. isolated from *Rhododendron* showed antifungal activity against seven fungal pathogens, namely *Phytophthora cinnamomi* Rands, *Pythium aphanidermatum* (Edson) Fitzpatrick, *Rhizoctonia solani* Kühn, *Fusarium avenaceum* (Corda: Fries) Saccardo, *Sclerotinia homoeocarpa* Bennett, *Botrytis cinerea* Persoon, and *Pestalotiopsis sydowiana* Bresadola (Shimizu et al., 2000). Similarly, two white amorphous compounds, 5,7-dimethoxy-4-*p*-methoxyphenylcoumarin ($C_{18}H_{16}O_5$) and 5,7-dimethoxy-4-phenylcoumarin ($C_{17}H_{14}O_5$), obtained from *Streptomyces aureofaciens* CMUAc130 isolated from the root tissue of *Zingiber officinale* Rosc. (Zingiberaceae) helped to suppress the soil fungal pathogens (Taechowisan et al., 2005).

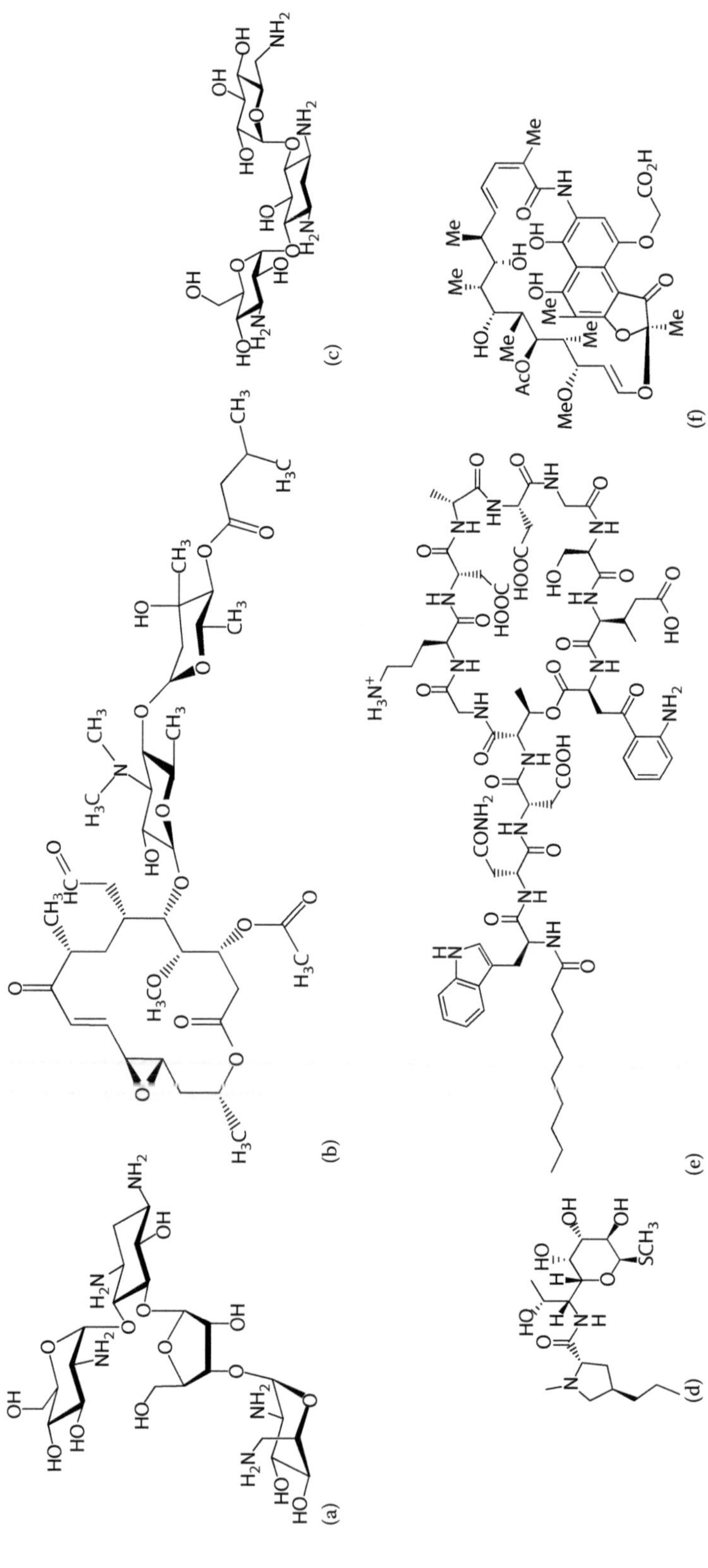

Figure 6.4 Structure of (a) paromomycin, (b) carbomycin, (c) kanamycin, (d) lincomycin, (e) daptomycin, and (f) rifamycin B.

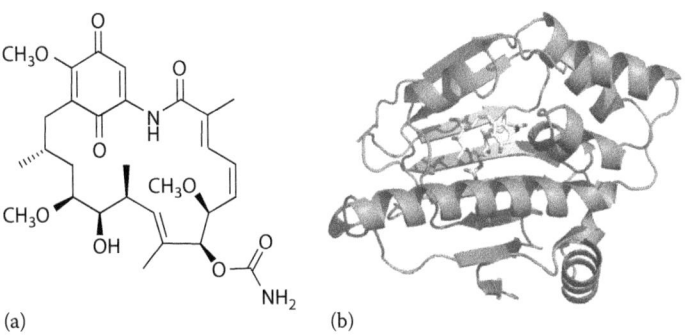

Figure 6.5 Structure of (a) geldanamycin and (b) Hsp90–geldanamycin complex.

A macrolide antibiotic obtained from *Streptomyces* sp. CS isolated from *Maytenus hookeri* was not only a good antimicrobial agent but also a good antitumor agent (Lu and Shen, 2003). Celastramycins A and B isolated from *Streptomyces* sp. MaB-QuH-8 isolated from the *Maytenus aquifolium* plant showed excellent antimicrobial activity (Pullen et al., 2002).

Compounds like "siderophore," an iron-chelating compound isolated from endophytic actinomycetes, not only help in plant growth but also act as an indirect antimicrobial agent (Franco-Correa et al., 2010). Many structures similar to siderophore like "spoxazomicins A" (molecular formula $C_{16}H_{21}N_3O_3S$ and molecular weight 335 g/mol), pyochelin, a siderophore isolated from the orchid endophyte *Streptosporangium oxazolinicum* (Inahashi et al., 2011), and geldanamycin, a type of benzoquinone ansamycin antibiotic ($C_{29}H_{40}N_2O_9$) isolated from endophytic actinomycetes, also showed antimicrobial activity (Dhanya and Padmavathy, 2014). Geldanamycin (Figure 6.5a and b) inhibits the function of heat shock protein 90 (Hsp90) by binding to the unusual ADP/ATP-binding pocket of the protein. Hsp90 client proteins play important roles in the regulation of the cell cycle, cell growth, cell survival, apoptosis, angiogenesis, and oncogenesis (Michael and Plotkin, 2003). Similarly, many antimicrobial agents from endophytic actinomycetes were isolated. Table 6.2 lists some of the endophytic actinomycetes producing antimicrobial agents.

6.2.3 Alkaliphilic actinomycetes

Sato's team was the first to report the production of antibiotics in alkaline medium (pH of the medium, 10.5) (Sato et al., 1980). Since then, many attempts were made to isolate antimicrobial compounds from alkaliphilic actinomycetes like phenazine and its derivatives.

Phenazine is a water-soluble antibiotic obtained as a yellow-to-brown crystal powder (Figure 6.6a), isolated from the alkaliphile *Nocardiopsis dassonvillei* OPC-15. It has the molecular formula $C_{12}H_8N_2$, a molecular mass of 180.21 g/mol, and a melting point of 177°C and is reported to have an antimicrobial activity at pH 10 (Tsujibo et al., 1988; McDonald et al., 2001).

Similarly, pyrocoll (Figure 6.6b), a constituent of cigarette smoke isolated from the alkaliphilic *Streptomyces* sp. AK, has biological activity against various *Arthrobacter* strains, filamentous fungi, several pathogenic protozoa, and some human tumor cell lines (Dietera et al., 2003). Furthermore, an antifungal compound, naphthospironone A, produced by the

Table 6.2 List of some endophytic actinomycetes, origin and their compound applications

Name of endophytic actinomycetes	Origin	Compound application	References
Microbispora	Roots and leaves of maize	Antibacterial agent against *Bacillus subtilis, S. aureus,* and *Micrococcus luteus*	Araujo et al. (2000)
Streptomyces sp.	Roots of *Zingiber officinale*	Antifungal activity against *Colletotrichum musae* and *Fusarium oxysporum*	Taechowisan and Lumyong (2003)
Streptomyces griseorubiginosus	Banana roots	Biocontrol agent against wilt disease of banana, caused by the fungal pathogen *Fusarium oxysporum*	Cao et al. (2005)
Actinoplanes philippinensis	Cucumber	Biocontrol agent against damping-off disease caused by the pathogen *Pythium aphanidermatum*	El-Tarabily (2006)
Microbispora, Streptomyces, and *Micromonospora*	Chinese cabbage	Antagonists to *Plasmodiophora brassicae*	Lee et al. (2008)
Streptomyces, Streptosporangium, Microbispora, Streptoverticillium, Saccharomonospora, and *Nocardia*	From various parts of *Azadirachta indica*	Antibacterial and antifungal agents	Verma et al. (2009)
Nocardioides sp.	*Triticum aestivum*	Used to treat wheat take-all disease caused by *Gaeumannomyces graminis* var. *tritici*	Coombs et al. (2004)

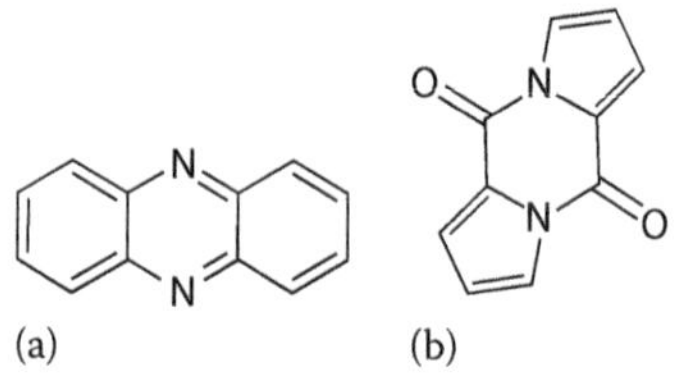

(a)   (b)

Figure 6.6 Structure of (a) phenazine and (b) pyrocoll.

Nocardiopsis sp. YIM DT266 isolated from the soil sample (pH 10), is a colorless solid with the molecular formula $C_{20}H_{18}O_9$, a molecular mass of 403.1023 g/mol, and a melting point of 173°C. It shows maximum absorption at 227 nm in methanol and has antibacterial, antifungal, and moderate cytotoxic activities (Ding et al., 2010). Similarly, the two pyran-2-one compounds, namely nocardiopyrones A and B (Figure 6.7a and b), isolated from the novel alkaliphilic actinomycete species *Nocardiopsis alkaliphila* YIM-80379 and having the molecular formula $C_{14}H_{20}O_4$ and $C_{12}H_{18}O_3$ and a molecular mass of 275.1255 and 233.1150 g/mol, respectively, showed antimicrobial activity against *Pseudomonas aeruginosa, Enterobacter aerogenes, Escherichia coli, S. aureus,* and *Candida albicans* (Wang et al., 2013b).

There are only scanty reports on antimicrobial activity produced by alkaliphilic actinomycetes. This may be due to the fact that antibiotics produced were not stable under higher pH (Horikoshi, 1999). Thus, in future attempts, isolation of stable alkaliphilic antibiotics should be made so that they can be applied in many applications.

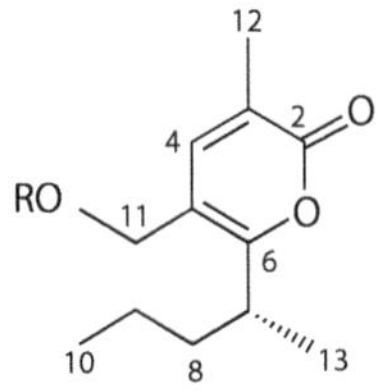

(a) R = Ac
(b) R = H

Figure 6.7 Structure of (a) nocardiopyrone A and (b) nocardiopyrone B.

6.2.4   Halophilic actinomycetes

The actinomycetes growing in saline environments are probably most metabolically and physiologically different from their type strains that thrive in other environments. *Actinopolyspora halophila* is the first halophilic actinomycete isolated as a contaminant of a culture medium containing 25% NaCl (Johnson et al., 1986). Since then, many compounds from halophilic actinomycetes were isolated, characterized, and used in many fields as a biological control agent, and as antibacterial, antifungal, and anticancer agents.

The biological control agents spinosyns A and D, isolated from the halophilic actinomycete *Saccharopolyspora spinosa*, prevent acetylcholine from binding to appropriate receptor sites, which causes paralysis and death in insects, mostly caterpillars, by both contact and ingestion (Kirst et al., 1992). Himalomycins (Table 6.3) isolated from *Streptomyces* sp. have antibacterial activity (Maskey et al., 2003).

The new *p*-terphenyl 6′-hydroxy-4,2′,3′,4″-tetramethoxy-*p*-terphenyl obtained as colorless needles from the halophilic actinomycete *Nocardiopsis gilva* YIM 90087 is not only a good antibacterial and antifungal agent but also acts as an antioxidant (Tian et al., 2013). Furthermore, Table 6.3 presents some of the halophilic actinomycetes and their products.

Due to the diverse application that halophilic actinomycete compounds possess, our team have made attempts and successfully isolated many novel actinomycetes (listed in Table 6.4), and furthermore, we believe that our team can isolate many unique antimicrobial agents from them in the future.

6.2.5   Marine actinomycetes

More than 70% of our planet's surface is covered by oceans, and life on Earth is believed to have originated from the sea. Experts estimate that the biological diversity in marine ecosystems is much higher than in the tropical rain forests (Haefner, 2003). The early evidence supporting the existence of marine actinomycetes came from the description of *Rhodococcus marinonascene*, the first marine actinomycete species to be characterized (Helmke and Weyland, 1984). Since then, many researchers have switched over to this environment to find actinomycetes that produce novel antimicrobial agents mainly due to the belief that marine environmental conditions are extremely different from terrestrial conditions. Thus, the marine ecosystem may harbor many unique microorganisms that produce novel antimicrobial agents and obtain positive results. The compounds isolated from marine actinomycetes have many diverse applications like control of many multidrug-resistant pathogens, for example, the polyketide antibiotic SBR-22, isolated from marine streptomycete BT-408, showed antibacterial activity against methicillin-resistant

Table 6.3 List of some antimicrobial agents obtained from halophilic actinomycetes

Halophilic actinomycetes and compounds name	Structure	References
Streptomyces sp. Himalomycins		Maskey et al. (2003)
Actinopolyspora sp. YIM90600 Erythronolides H and I		Huang et al. (2009)
Actinopolyspora sp. YIM90600 Actinopolysporins A–C		Zhao et al. (2011b)

Table 6.4 List of some novel halophilic actinomycetes

Halophilic actinomycetes	Salt concentration range	References
Streptomonospora alba	5–25	Li et al. (2003b)
Prauserella halophila	5–25	Li et al. (2003a)
Nesterenkonia xinjiangensis	0–25	Li et al. (2004a)
Nocardiopsis salina	3–20	Li et al. (2004b)
Nesterenkonia halotolerans	0–25	Li et al. (2004a)
Nocardiopsis rhodophaea	0–18	Li et al. (2006)
Nocardiopsis chromatogenez	0–18	Li et al. (2006)
Saccharomonospora saliphila	5–20	Syed et al. (2008)
Marinactinospora thermotolerans	0–5	Tian et al. (2009)
Amycolatopsis halophila	1–15	Tang et al. (2010)
Actinopolyspora erythraea	10–25	Tang et al. (2011)
Streptomyces oceani	2.5–12.5	Tian et al. (2012)

S. aureus (Sujatha et al., 2005). Similarly, the compound abyssomicin C, a novel polycyclic polyketide antibiotic produced by the marine actinomycete *Verrucosispora* sp., possesses potent activity against vancomycin-resistant *S. aureus* (Rath et al., 2005; Riedlinger et al., 2004).

Another compound, diazepinomicin (for the structure, refer to Table 6.5), isolated from the *Micromonospora* strain DPJ12 and having the molecular formula $C_{28}H_{34}N_2O_4$, not only possesses antibacterial activity but also has anti-inflammatory and antitumor activities (Charan et al., 2004). Furthermore, *Pseudonocardia* sp. SCSIO 01299, isolated from marine sediments collected from a depth of 3258 m, produced four compounds: deoxynyboquinone and pseudonocardians A–C. These four compounds were extracted from the biomass using acetone and butanone as a solvent, respectively. The compound deoxynyboquinone (Figure 6.8a) was obtained as red needles having a molecular weight of 284. The compound, having an MIC of 1 µg/mL, showed antibacterial activity against *Bacillus thuringiensis*, *S. aureus*, and *Enterococcus faecalis*. The compound was tested for in vitro cytotoxic activities against three human tumor cell lines, namely SF-268 (human glioma cell line), MCF-7 (human breast adenocarcinoma cell line), and NCI-H460 (human non–small cell lung cancer cell line), and the IC_{50} (µM) values were estimated to be 0.022, 0.015, and 0.080, respectively. The compound pseudonocardian A (Figure 6.8b) was obtained as a white solid having the molecular formula $C_{18}H_{18}N_2O_5$ with a molecular mass of 343.1305 g/mol. The compound, having an MIC of 4 µg/mL, showed antibacterial activity against *B. thuringiensis*, *S. aureus*, and *E. faecalis*. The estimated IC_{50} (µM) against SF-268, MCF-7, and NCI-H460 were 0.028, 0.027, and 0.209, respectively.

Similarly, pseudonocardian B (Figure 6.8c) was obtained as a white solid having the molecular formula $C_{19}H_{20}N_2O_5$, with a molecular mass of 357.1456. The compound, having an MIC of 2 µg/mL, showed antibacterial activity against *B. thuringiensis*, *S. aureus*, and *E. faecalis*. The estimated IC_{50} (µM) against SF-268, MCF-7, and NCI-H460 were 0.022, 0.021, and 0.177, respectively. The compound pseudonocardian C (Figure 6.8d) was obtained as a red-brown powder having the molecular formula $C_{21}H_{24}N_2O_8$, with a molecular mass of 433.1609 g/mol. The compound, having an MIC of >128 µg/mL, showed antibacterial activity against *B. thuringiensis*, *S. aureus*, and *E. faecalis*. The estimated IC_{50} (µM) against SF-268, MCF-7, and NCI-H460 were 6.70, 8.02, and 43.28, respectively (Li et al., 2011). Furthermore, Tables 6.5 and 6.6 list some of the important antimicrobial agents isolated from the marine ecosystem. In our point of view, the marine ecosystem hinders many novel antimicrobial agents that can be used against multidrug-resistant pathogens. Thus, in the future, further research in finding novel compounds is required in order to compete with harmful pathogens.

6.3   Future developments

Since antiquity, humans have been fighting with pathogenic microorganisms. Many antimicrobial agents were discovered, but still we are unable to eradicate them. Since the discovery of penicillin, many antimicrobial agents were discovered and we think we are ready for the battle, but we always forget that the microorganisms are also developing exponentially, though very tiny but still giving equal competition. So, we should not stop and keep discovering. Actinomycetes are the largest hub for antimicrobial agents. Though many antimicrobial agents from actinomycetes were reported, it is just the "tip of the iceberg" that has been explored. Various unique and virgin areas should be explored, and different isolation techniques and isolation media should be developed in order to isolate novel actinomycetes having an extraordinary antimicrobial agent to triumph against pathogenic microorganisms.

Table 6.5 List of compounds from marine actinomycetes and their applications

Compound name/origin	Compound applications	References
Diazepinomicin from *Micromonospora* sp.	Antibacterial, anticancer, and anti-inflammatory	Charan et al. (2004)
Sporolides A and B from *Salinispora tropica*	Antibacterial	Buchanan et al. (2005)
Ammosamide D from *Streptomyces variabilis*	Cytotoxic activity	Pan et al. (2012)
Lajollamycin from *Streptomyces nodosus*	Antibacterial	Manam et al. (2005)
Helquinoline from *Janibacter limosus*	Antibacterial	Asolkar et al. (2004)

(a) $R_1 = Cl$ $R_2 = H$
(b) $R_1 = H$ $R_2 = Cl$

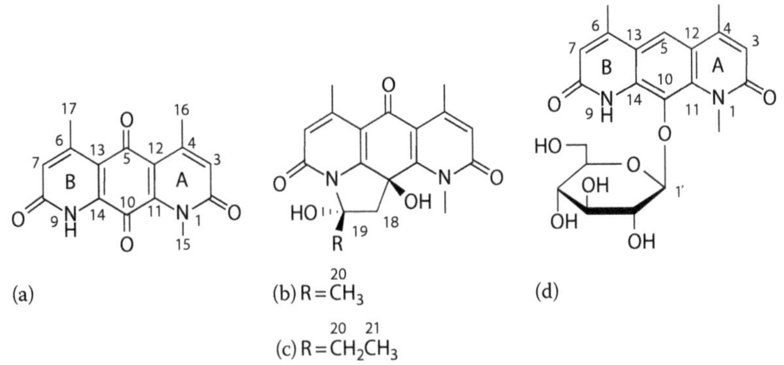

Figure 6.8 Structure of (a) deoxynyboquinone, (b) pseudonocardian A, (c) pseudonocardian B, and (d) pseudonocardian C.

Table 6.6 List of compounds from marine actinomycetes and their applications

Origin	Compound name	Compound applications	References
Streptomyces griseus	Frigocyclinone	Antibacterial	Bruntner et al. (2005)
Streptomyces sp.	Glaciapyrroles	Antibacterial	Macherla et al. (2005)
Streptomyces sp.	Resistoflavin methyl ether	Antibacterial and antioxidative	Kock et al. (2005)
Verrucosispora sp.	Proximicins	Antibacterial and anticancer	Fiedler et al. (2008)
Streptomyces sp.	Caboxamycin	Antibacterial and anticancer	Hohmann et al. (2009)
Streptomyces sp.	2-Allyloxyphenol	Antimicrobial, food preservative, and oral disinfectant	Arumugam et al. (2010)
Salinispora arenicola	Arenimycin	Antibacterial and anticancer	Asolkar et al. (2010)
Streptomyces sp.	Tirandamycins	Antibacterial	Carlson et al. (2009)

Acknowledgments

This research was supported by the Key Project of International Cooperation of Ministry of Science & Technology (MOST) (No. 2013DFA31980) and the Key Project of Yunnan Provincial Natural Science Foundation (2013FA004). W.-J. Li was also supported by the "Hundred Talents Program" of the Chinese Academy of Sciences and Guangdong Province Higher Vocational Colleges & Schools Pearl River Scholar Funded Scheme (2014).

References

Araujo, J.M., Silva, A.C., and Azevedo, J.L. 2000. Isolation of endophytic actinomycetes from roots and leaves of maize (*Zea mays L.*). *Braz. Arch. Biol. Technol.* 43, 447–451.

Arumugam, M., Mitra, A., Jaisankar, P., Dasgupta, S., Sen, T., Gachhui, R., and Mukherjee, J. 2010. Isolation of an unusual metabolite 2-allyloxyphenol from a marine actinobacterium, its biological activities and applications. *Appl. Microbiol. Biotechnol.* 86, 109–117.

Asolkar, R.N., Kirkland, T.N., Jensen, P.R., and Fenical, W. 2010. Arenimycin, an antibiotic effective against rifampin- and methicillin-resistant *Staphylococcus aureus* from the marine actinomycete *Salinispora arenicola*. *J. Antibiot.* 63, 37–39.

Asolkar, R.N., Schroeder, D., Heckmann, R., Lang, S., Wagner-Doebler, I., and Laatsch, H. 2004. Helquinoline, a new tetrahydroquinoline antibiotic from *Janibacter limosus* Hel 1+. *J. Antibiot.* 57, 17–23.

Baltz, R.H. 2005. Antibiotic discovery from actinomycetes: Will a renaissance follow the decline and fall? *SIM News* 55, 186–196,

Bérdy, J. 2012. Thoughts and facts about antibiotics: Where we are now and where we are heading. *J. Antibiot.* 65, 385–395.

Bonkowski, M. and Roy, J. 2005. Soil microbial diversity and soil functioning affect competition among grasses in experimental microcosms. *Oecologia* 143, 232–240.

Bruntner, C., Binder, T., Pathom-aree, W., Goodfellow, M., Bull, A.T., Potterat, O., and Fiedler, H.P. 2005. Frigocyclinone, a novel angucyclinone antibiotic produced by a *Streptomyces griseus* strain from Antarctica. *J. Antibiot.* 58, 346–349.

Buchanan, G.O., Williams, P.G., Feling, R.H., Kauffman, C.A., Jensen, P.R., and Fenical, W. 2005. Sporolides A and B: Structurally unprecedented halogenated macrolides from the marine actinomycete *Salinispora tropica*. *Org. Lett.* 7, 2731–2734.

Cao, L., Qiu, Z., You, J., Tan, H., and Zhou, S. 2005. Isolation and characterization of endophytic *Streptomycete* antagonists of *Fusarium* wilt pathogen from surface-sterilized banana roots. *FEMS Microbiol. Lett.* 247, 147–152.

Carlson, J.C., Li, S., Burr, D.A., and Sherman, D.H. 2009. Isolation and characterization of tirandamycins from a marine-derived *Streptomyces* sp. *J. Nat. Prod.* 72, 2076–2079.

Charan, R.D., Schlingmann, G., Janso, J., Bernan, V., Feng, X., and Carter, G.T. 2004. Diazepinomicin, a new antimicrobial alkaloid from marine *Micromonospora* sp. *J. Nat. Prod.* 67, 1431–1433.

CLSI. 2012. *Performing Standards for Antimicrobial Susceptibility Testing: Twenty-second Informational Supplement*. CLSI document M100-S22. Wayne, PA: Clinical and Laboratory Standards Institute.

Coombs, J.T., Michelsen, P.P., and Franco, M.M. 2004. Evaluation of endophytic actinobacteria as antagonists of *Gaeumannomyces graminis var. tritici* in wheat. *Biol. Control* 29, 359–366.

Davidson, R.N., den Boer, M., and Ritmeijer, K. 2009. Paromomycin. *Trans. R. Soc. Trop. Med. Hyg.* 103, 653–660.

Deene, M. and Lingappa, K. 2013. Antibacterial activity of silver nanoparticles against methicillin-resistant *Staphylococcus aureus* synthesized using model *Streptomyces* sp. pigment by photo-irradiation method. *J. Pharma. Res.* 6, 255–260.

Dhanya, N.N. and Padmavathy, S. 2014. Impact of endophytic microorganisms on plants, environment and humans. *Sci. World J.*, 2011, 1–11.

Dietera, A., Hamm, A., Fiedler, H.P., Goodfellow, M., Müller, W.E., Brun, R., and Bringmann, G. 2003. Pyrocoll, an antibiotic, antiparasitic and antitumor compound produced by a novel alkaliphilic *Streptomyces* strain. *J. Antibiotics* 56, 639–646.

Ding, Z.G., Li, M.G., Zhao, J.Y., Ren, J., Huang, R., Xie, M.J., and Wen, M.L. 2010. Naphthospironone A: An unprecedented and highly functionalized polycyclic metabolite from an alkaline mine waste extremophile. *Chem. Eur. J.* 16, 3902–3905.

Elson, M.K., Schisler, D.A., and Bothast, R.J. 1997. Selection of microorganisms for biological control of silver scurf (*Helminthosporium solani*) of potato tubers. *Plant Dis.* 81, 647–652.

El-Tarabily, K.A. 2003. An endophytic chitinase-producing isolate of *Actinoplanes missouriensis*, with potential for biological control of root rot of lupine caused by *Plectosporium tabacinum*. *Aus. J. Bot.* 51, 257–266.

El-Tarabily, K.A. 2006. Rhizosphere-competent isolates of streptomycete and non-streptomycete actinomycetes capable of producing cell-wall degrading enzymes to control *Pythium aphanidermatum* damping-off disease of cucumber. *Can. J. Bot.* 84, 211–222.

El-Tarabily, K.A., Hardy, G.E., Sivasithamparam, K., Hussein, A.M., and Kurtböke, D. 1997. The potential for the biological control of cavity spot disease of carrots caused by *Pythium coloratum* by streptomycete and non-streptomycete actinomycetes in Western Australia. *New Phytol.* 137, 495–507.

Fiedler, H.P., Bruntner, C., Riedlinger, J., Bull, A.T., Knutsen, G., Goodfellow, M., and Sussmuth, R.D. 2008. Proximicin A, B and C, novel aminofuran antibiotic and anticancer compounds isolated from marine strains of the actinomycete *Verrucosispora*. *J. Antibiot.* 61, 158–163.

Fleming, A. 1929. On the antibacterial action of cultures of a penicillium, with special reference to their use in the isolation of *B. influenzae*. *Br. J. Exp. Pathol.* 10(3), 226.

Franco-Correa, M., Quintana, A., Duque, C., Suarez, C., Rodríguez, M.X., and Barea, J.M. 2010. Evaluation of actinomycete strains for key traits related with plant growth promotion and mycorrhiza helping activities. *Appl. Soil Ecol.* 45, 209–217.

Garrod, L.P., Lambert, H.P., and O'Grady, F. 1981. *Antibiotic and Chemotherapy.* Churchill Livingstone, New York, p. 131.

Giguere, S., Prescott, J.F., Baggot D., Walker R.D., and Dowling, P.M. 2007. *Antimicrobial Therapy in Veterinary Medicine*, 4th edn. Blackwell Publishing, Ames, IA.

Haefner, B. 2003. Drugs from the deep: Marine natural products as drug candidates. *Drug. Discov. Today* 8, 536–544.

Helmke, E. and Weyland, H. 1984. *Rhodococcus marinonascens* sp. Nov., an actinomycete from the sea. *Int. J. Syst. Bacteriol.* 34, 127–138.

Hohmann, C., Schneider, K., Bruntner, C., Irran, E., Nicholson, G., Bull, A.T., and Fiedler, H.P. 2009. Caboxamycin, a new antibiotic of the benzoxazole family produced by the deep-sea strain *Streptomyces* sp. NTK 937. *J. Antibiot.* 62, 99–104.

Horikoshi, K. 1999. Alkaliphiles: Some applications of their products for biotechnology. *Microbiol. Mol. Biol. Rev.* 63, 735–750.

Huang, S.X., Zhao, L.X., Tang, S.K., Jiang, C.L., Duan, Y., and Shen, B. 2009. Erythronolides H and I, new erythromycin congeners from a new halophilic actinomycete *Actinopolyspora* sp. YIM90600. *Org. Lett.* 11, 1353–1356.

Inahashi, Y., Iwatsuki, M., Ishiyama, A., Namatame, M., Nishihara-Tsukashima, A., Matsumoto, A., and Shiomi, K. 2011. Spoxazomicins A-C, novel antitrypanosomal alkaloids produced by an endophytic actinomycete, *Streptosporangium oxazolinicum* K07–0450T. *J. Antibiot.* 64, 303–307.

Johnson, K.G., Lanthier, P.H., and Gochnauer, M.B. 1986. Studies of two strains of *Actinopolyspora halophila*, an extremely halophilic actinomycete. *Arch. Microbiol.* 143, 370–378.

Kingston, W. 2008. Irish contributions to the origins of antibiotics. *Ir. J. Med. Sci.* 177, 87–92.

Kiple, K.F. (ed.). 1993. *Cambridge World History of Human Disease.* Cambridge University Press, Cambridge, U.K.

Kirst, H.A., Michel, K.H., Mynderase, J.S., Chio, E.H., Yao, R.C., Nakasukasa, W.M., and Deeter, J.B. 1992. Discovery, isolation, and structure elucidation of a family of structurally unique, fermentation derived tetracyclic macrolides. In: Baker, D.R., Fenyes, J.G., Steffens, J.J. (eds.) *Synthesis and Chemistry of Agrochemicals III.* American Chemical Society, Washington, DC, pp. 214–225.

Kitouni, M., Boudemagh, A., Oulmi, L., Reghioua, S., Boughachiche, F., Zerizer, H., and Boiron, P. 2005. Isolation of actinomycetes producing bioactive substances from water, soil and tree bark samples of the north–east of Algeria. *J. Mycol. Med.* 15, 45–51.

Kock, I., Maskey, R.P., Biabani, M.A., Helmke, E., and Laatsch, H. 2005. 1-Hydroxy-1-norresistomycin and resistoflavin methyl ether: New antibiotics from marine-derived *Streptomycetes. J. Antibiot.* 58, 530–534.

Lee, O.O., Choi, G.J., Choi, Y.H., Jang, K.S., Park, D.J., Kim, C.J., and Kim, J.C. 2008. Isolation and characterization of endophytic actinomycetes from Chinese cabbage roots as antagonists to *Plasmodiophora brassicae. J. Microbiol. Biotechnol.* 18, 1741–1746.

Levine, D.P. 2006. Vancomycin: A history. *Clin. Infect. Dis.* 42(Suppl. 1), S5–S12.

Li, S., Tian, X., Niu, S., Zhang, W., Chen, Y., Zhang, H., and Zhang, C. 2011. *Pseudonocardians* A-C, new diazaanthraquinone derivatives from a deep-sea actinomycete *Pseudonocardia* sp. SCSIO 01299. *Mar. Drugs* 9, 1428–1439.

Li, W.J., Chen, H.H., Zhang, Y.Q., Schumann, P., Stackebrandt, E., Xu, L.H., and Jiang, C.L. 2004a. *Nesterenkonia halotolerans* sp. nov. and *Nesterenkonia xinjiangensis* sp. nov., actinobacteria from saline soils in the west of China. *Int. J. Syst. Evol. Microbiol.*54, 837–841.

Li, W.J., Kroppenstedt, R.M., Wang, D., Tang, S.K., Lee, J.C., Park, D.J., and Jiang, C.L. 2006. Five novel species of the genus *Nocardiopsis* isolated from hypersaline soils and emended description of *Nocardiopsis salina. Int. J. Syst. Evol. Microbiol.* 56, 1089–1096.

Li, W.J., Park, D.J., Tang, S.K., Wang, D., Lee, J.C., Xu, L.H., and Jiang, C.L. 2004b. *Nocardiopsis salina* sp. nov., a novel halophilic actinomycete isolated from saline soil in China. *Int. J. Syst. Evol. Microbiol.* 54, 1805–1809.

Li, W.J., Xu, P., Tang, S.K., Xu, L.H., Kroppenstedt, R.M., Stackebrandt, E., and Jiang, C.L. 2003a. *Prauserella halophila* sp. nov. and *Prauserella alba* sp. nov., moderately halophilic actinomycetes from saline soil. *Int. J. Syst. Evol. Microbiol.* 53, 1545–1549.

Li, W.J., Xu, P., Zhang, L.P., Tang, S.K., Cui, X.L., Mao, P.H., and Jiang, C.L. 2003b. *Streptomonospora alba* sp. nov., a novel halophilic actinomycete, and emended description of the genus *Streptomonospora*. *Int. J. Syst. Evol. Microbiol.* 53, 1421–1425.

Lu, C.H. and Shen, Y.M. 2003. A new macrolide antibiotics with antitumor activity produced by *Streptomyces* sp. CS, a commensal microbe of *Maytenus hookeri*. *J. Antibiot.* 56, 415–418.

Macherla, V.R., Liu, J., Bellows, C., Teisan, S., Lam, K.S., and Potts, B.C.M. 2005. Glaciapyrroles A, B and C, pyrrolosesquiterpenes from a *Streptomyces* sp. isolated from an Alaskan marine sediment. *J. Nat. Prod.* 68, 780–783.

Manam, R.R., Teisan, S., White, D.J., Nicholson, B., Grodberg, J., Neuteboom, S.T.C., Lam, K.S., Mosca, D.A., Lloyd, G.K., and Potts, B.C. 2005. Lajollamycin, a nitro-tetraene spiro-b-lactone-c-lactam antibiotic from the marine actinomycete *Streptomyces nodosus*. *J. Nat. Prod.* 68, 240–243.

Maskey, R.P., Helmke, E., and Laastsch, H. 2003. Himalomycin A and B: Isolation and structure elucidation of new fridamycin type antibiotics from a marine *Streptomyces* isolate. *J. Antibiot.* 56, 942–949.

McDonald, M., Mavrodi, D.V., Thomashow, L.S., and Floss, H.G. 2001. Phenazine biosynthesis in *Pseudomonas fluorescens*: Branchpoint from the primary shikimate biosynthetic pathway and role of phenazine-1,6-dicarboxylic acid. *J. Am. Chem. Soc.* 123, 9459–9460.

McGuire, J.M., Bunch, R.L., Anderson, R.C., Boaz, H.E., Flynn, E.H., Powell, H.M., and Smith, J.W. 1952. Ilotycin, a new antibiotic. *Antibiot. Chemother.* 2, 281–283.

Michael, S. and Plotkin, M.J. 2003. *The Killers Within: The Deadly Rise of Drug-Resistant Bacteria.* New York: Little Brown & Company. ISBN 978-0-316-73566-7.

Pan, E., Jamison, M., Yousufuddin, M., and MacMillan, J.B. 2012. Ammosamide D, an oxidatively ring opened ammosamide analog from a marine-derived *Streptomyces variabilis*. *Org. Lett.* 14, 2390–2393.

Pullen, C., Schmitz, P., Meurer, K., Bamberg, D.D.V., Lohmann, S., Franca, S.D.C., and Leistner, E. 2002. New and bioactive compounds from *Streptomyces* strains residing in the wood of *Celastraceae*. *Planta* 216, 162–167.

Rath, J.P., Kinast, S., and Maier, M.E. 2005. Synthesis of the full functionalized core structure of the antibiotic abyssomicin C. *Org. Lett.* 7, 3089–3092.

Riedlinger, J., Reicke, A., Zähner, H.A.N.S., Krismer, B., Bull, A.T., Maldonado, L.A., and Fiedler, H.P. 2004. Abyssomicins, inhibitors of the para-aminobenzoic acid pathway produced by the marine *Verrucosispora* strain AB-18-032. *J. Antibiot.* 57, 271–279.

Sarbani, P. 2006. A journey across the sequential development of macrolides and ketolides related to erythromycin. *Tetrahedron* 62, 3171–3200.

Sato, M., Beppu, T., and Arima, K. 1980. Properties and structure of a novel peptide antibiotic no. 1970. *Agric. Biol. Chem.* 44, 3037–3040.

Schatz, A., Bugie, E., and Waksman, S.E. 1944. *Streptomycin*, a substance exhibiting antibiotic activity against gram-positive and gram negative bacteria. *Proc. Soc. Exp. Biol. Med.* 55, 65–69.

Shimizu, M., Nakagawa, Y., Yukio, S., Furumai, T., Igarashi, Y., Onaka, H., and Kunoh, H. 2000. Studies on endophytic actinomycetes (I) *Streptomyces* sp. Isolated from *Rhododendron* and its antifungal activity. *J. Gen. Plant Pathol.* 66, 360–366.

Sivonen, K. and Jones, G. 1999. Cyanobacterial toxins. In: Chorus, I. and Bartran, J. (eds.) *Toxic Cyanobacteria in Water: A Guide to Their Public Health Consequences. Monitoring and Management.* E&FN Spon., London, U.K., pp. 41–111.

Smith, G.E. 1957. Inhibition of *Fusarium oxysporum* f.sp. *lycopersici* by a species of *Micromonospora* isolated from tomato. *Phytopathology* 47, 429–432.

Somdee, T., Sumalai, N., and Somdee, A. 2013. A novel actinomycete *Streptomyces aurantiogriseus* with algicidal activity against the toxic cyanobacterium *Microcystis aeruginosa*. *J. Appl. Phycol.* 25, 1587–1594.

Spizek, J. and Rezanka, T. 2004. Lincomycin, cultivation of producing strains and biosynthesis. *Appl. Microbiol. Biotechnol.* 63, 510–519.

Steffan, R.J., Goskoyr, J., Bej, A.K., and Atlas, R.M. 1988. Recovery of DNA from soil and sediments. *Appl. Environ. Microbiol.* 54, 2908–2915.

Sujatha, P., Bapi Raju, K.V.V.S.N., and Ramana, T. 2005. Studies on a new marine streptomycete BT-408 producing polyketide antibiotic SBR-22 effective against methicillin resistant *Staphylococcus aureus. Microbiol. Res.* 160, 119–126.

Syed, D.G., Tang, S.K., Cai, M., Zhi, X.Y., Agasar, D., Lee, J.C., and Li, W.J. 2008. *Saccharomonospora saliphila* sp. nov., a halophilic actinomycete from an Indian soil. *Int. J. Syst. Evol. Microbiol.* 58, 570–573.

Taechowisan, T., Lu, C., Shen, Y., and Lumyong, S. 2005. Secondary metabolites from endophytic *Streptomyces aureofaciens* CMUAc130 and their antifungal activity. *Microbiology* 151, 1691–1695.

Taechowisan, T. and Lumyong, S. 2003. Activity of endophytic actinomycetes from roots of *Zingiber officinale* and *Alpinia galanga* against phytopathogenic fungi, *Ann. Microbiol.* 53, 291–298.

Tang, S.K., Wang, Y., Guan, T.W., Lee, J.C., Kim, C.J., and Li, W.J. 2010. *Amycolatopsis halophila* sp. nov. a halophilic actinomycete isolated from a salt lake. *Int. J. Syst. Evol. Microbiol.* 60, 1073–1078.

Tang, S.K., Wang, Y., Klenk, H.P., Shi, R., Lou, K., Zhang, Y.J., and Li, W.J. 2011. *Actinopolyspora alba* sp. nov. and *Actinopolyspora erythraea* sp. nov., isolated from a salt field, and reclassification of *Actinopolyspora iraqiensis* as a heterotypic synonym of *Saccharomonospora halophile. Int. J. Syst. Evol. Microbiol.* 61, 1693–1698.

Tian, S.Z., Pu, X., Luo, G., Zhao, L.X., Xu, L.H., Li, W.J., and Luo Y. 2013. Isolation and characterization of new *p*-terphenyls with antifungal, antibacterial, and antioxidant activities from a halophilic actinomycete *Nocardiopsis gilva* YIM 90087. *J. Agric. Food. Chem.* 27, 3006–3012.

Tian, X.P., Tang, S.K., Dong, J.D., Zhang, Y.Q., Xu, L.H., Zhang, S., and Li, W.J. 2009. *Marinactinospora thermotolerans* gen. nov., sp. nov., a marine actinomycete isolated from a sediment in the northern South China Sea. *Int. J. Syst. Evol. Microbiol.* 59, 948–952.

Tian, X.P., Xu, Y., Zhang, J., Li, J., Chen, Z., Kim, C.J., and Zhang, S. 2012. *Streptomyces oceani* sp. nov., a new obligate marine actinomycete isolated from a deep-sea sample of seep authigenic carbonate nodule in South China Sea. *Antonie Van Leeuwenhoek* 102, 335–343.

Tsujibo, H., Sato, T., Inui, M., Yamamoto, H., and Inamori, Y. 1988. Intracellular accumulation of phenazine antibiotics production by an alkalophilic actinomycete. *Agric. Biol. Chem.* 52, 301–306.

Van der Heijden, M.G.A., Bardgett, R.D., and van Straalen, N.M. 2008. The unseen majority: Soil microbes as drivers of plant diversity and productivity in terrestrial ecosystems. *Ecol. Lett.* 11, 296–310.

Verma, V.C., Gond, S.K., Kumar, A., Mishra, A., Kharwar, R.N., and Gange, A.C. 2009. Endophytic actinomycetes from *Azadirachta indica* A. Juss.: Isolation, diversity, and anti-microbial activity. *Microb. Ecol.* 57, 749–756.

Wagner, R.L., Hochstein, F.A., Murai, K., Messina, N., and Regna, P.P. 1953. Magnamycin, a new antibiotic. *J. Am. Chem. Soc.* 75, 4684–4687.

Waksman, S.A. and Starkey, R.L. 1931. *The Soil and the Microbe. An Introduction to the Study of the Microscopic Population of the Soil and Its Role in Soil Processes and Plant Growth.* John Wiley & Sons, Inc., New York.

Wang, X., Tabudravu, J., Jaspars, M., and Deng, H. 2013a. Tianchimycins, A-B, 16-membered macrolides from the rare actinomycete *Saccharothrix xinjiangensis. Tetrahedron* 69, 6060–6064.

Wang, Z., Fu, P., Liu, P., Wang, P., Hou, J., Li, W., and Zhu, W. 2013b. New pyran-2-ones from alkalophilic actinomycete, *Nocardiopsis alkaliphila* sp. nov. YIM-80379. *Chem. Biodivers.* 10, 281–287.

Weinstein, M.J., Wagman, G.H, Oden, E.M., and Marquez, J.A. 1967. Biological activity of the antibiotic components of the gentamicin complex. *J. Bacteriol.* 94, 789–790.

Woodworth, J.R., Nyhart, E.H., Brier, G.L., Wolny, J.D., and Black, H.R. 1992. Single-dose pharmacokinetics and antibacterial activity of daptomycin, a new lipopeptide antibiotic, in healthy volunteers. *Antimicrob. Agents. Chemother.* 36, 318–325.

Zhao, K., Penttinen, P., Guan, T., Xiao, J., Chen, Q., Xu, J., and Strobel, G.A. 2011a. The diversity and anti-microbial activity of endophytic actinomycetes isolated from medicinal plants in Panxi Plateau, China. *Curr. Microbiol.* 62, 182–190.

Zhao, L.X., Huang, S.X., Tang, S.K., Jiang, C.L., Duan, Y., Beutler, J.A., and Shen, B. 2011b. Actinopolysporins A-C and tubercidin as a Pdcd4 stabilizer from the halophilic actinomycete *Actinopolyspora erythraea* YIM 90600. *J. Nat. Prod.* 74, 1990–1995.

chapter seven

Actinobacteria

A predominant source of antimicrobial compounds

**Ramasamy Vijayakumar, Govindaraj Vaijayanthi,
Annamalai Panneerselvam, and Nooruddin Thajuddin**

Contents

7.1   Antibiotics: An introduction

An antibiotic is an agent that either kills or inhibits the growth of a microorganism (Dorlands, 2010). The term antibiotic was first used in 1942 by Selman Waksman and his collaborators and they described it as any substance produced by a microorganism that is antagonistic to the growth of other microorganisms (Waksman, 1947). Most of the recent antibacterials are semisynthetic and are modifications of various natural compounds (Von Nussbaum et al., 2006). These include the beta-lactam antibiotics, which include the penicillins (produced by fungi in the genus *Penicillium*), cephalosporins, and carbapenems. Compounds that are still isolated from living organisms are the aminoglycosides, whereas other antibacterials, for example, the sulfonamides, quinolones, and oxazolidinones, are produced solely by chemical synthesis. In accordance with this, many antimicrobial compounds are classified on the basis of chemical/biosynthetic origin into natural, semisynthetic, and synthetic. Another classification system is based on biological activity; in this classification, antibacterials are divided into two broad groups according to their biological effect on microorganisms. Bactericidal agents kill bacteria, and bacteriostatic agents slow down or stall bacterial growth.

Antibiotics are produced industrially by a process of fermentation, where the source microorganism is grown in large containers (100,000–150,000 L or more) containing a liquid growth medium. Oxygen concentration, temperature, pH, and nutrient levels must be optimal and are closely monitored and adjusted if necessary. As antibiotics are secondary metabolites, the population size must be controlled very carefully to ensure that maximum yield is obtained before the cells die. Once the process is complete, the antibiotic must be extracted and purified to a crystalline product. This is simpler to achieve if the antibiotic is soluble in organic solvent. Otherwise, it must first be removed by ion exchange, adsorption, or chemical precipitation.

Marine organisms represent a promising source for natural antibiotics of the future due to the incredible diversity of chemical compounds that were isolated particularly from the marine microorganisms. The oceans cover almost 70% of the earth's surface and over 90% of volume of its crust (Aghighi et al., 2005). From the entire microbial flora, actinobacteria from the marine environment have been traditionally a rich source for biologically active metabolites. Although heavily studied over the past three decades, actinobacteria continue to prove themselves as reliable sources of novel antimicrobial compounds. Among the well-characterized pharmaceutically relevant microorganisms, actinobacteria remain major sources of novel, therapeutically relevant natural products (Arasu et al., 2009). The majority of these compounds demonstrate one or more bioactivities, many of which developed into antibiotics for treatment of a wide range of diseases in human, veterinary, and agriculture sectors (Atalan et al., 2000).

7.2   Origin of antibiotics

After accidental discovery of penicillin (first antibiotic) by Sir Alexander Fleming from *Penicillium notatum*, a fungus in 1929 that inhibited the growth of *Staphylococcus aureus*, a dangerous/major microbial killer from the early twentieth century to our days, has been a fascinating, exciting, continuously changing, and developing adventure. Following the discovery of penicillin is the second most important antibiotic, streptomycin, which was discovered by Waksman from *Streptomyces griseus*, an actinobacterium. Interests toward the field were generally increasing, although sometimes declining; the interest and the whole story show some cyclic features with successes and failures and evolved around changing clinical needs and new enabling technology.

Generally, the microbial natural products appear as the most promising source of the future antibiotics that society is expecting (Fernando, 2006). Antibiotics are produced by bacteria, fungi, actinobacteria, algae, lichens, and green plants. Since the isolation of actinomycin and streptomycin, the actinobacteria have received tremendous attention (Waksman and Schatz, 1946). Members of *streptomycetes* are a rich source of bioactive compounds, notably antibiotics, enzymes, enzyme inhibitors, and pharmacologically active agents, and about 75% of the known commercially and medically useful antibiotics are produced by streptomycetes (Sujatha et al., 2005). Beijerinck (1900) and Eriko (2002) established that actinobacteria occur in great abundance in the soil and have a great role in the management of microbial stability with the production of antibiotic substances. Actinobacteria live in soil and decompose organic matter such as cellulose, hemicellulose, pectin, and chitin. In the drug discovery, these microorganisms are widely recognized for their ability to produce secondary metabolites with commercially viable antibiotic activity. Actinobacteria were first recognized as a common and important group of soil microorganisms (Waksman et al., 1941; Berger et al., 1949). Streptomycin, the first antibiotic for tuberculosis, was derived from the largest genus of actinobacteria, *Streptomyces*; erythromycin and tetracycline are two other common medicines derived originally from these microorganisms.

Most streptomycetes and other actinobacteria produce a diverse array of antibiotics, including aminoglycosides, macrolides, peptides, polyenes, polyether, and tetracyclines. In searching for new antibiotics, over 1000 different bacteria (including actinobacteria), fungi, and algae have been investigated. To prevent exponential emergence of microorganisms becoming resistant to the clinically available antibiotics already marketed, the periodic replacement of the existing antibiotics is necessary. Thus, the microbiologists/pharmacologists have regularly screened/searched the novel antibiotics particularly from microorganisms.

7.3   Types of antibiotics

On the basis of inhibitory action against different types of pathogenic microorganisms, the antibiotics are grouped as antibacterial, antifungal, antiparasitic, antiviral, anticancer, etc. They also classified as broad-spectrum and narrow-spectrum antibiotics as they possess inhibitory action on several pathogenic microorganisms as well as single pathogens, respectively.

7.3.1   Antibacterial compounds

The emergence of multidrug-resistant bacteria is a phenomenon of concern to the clinician and the pharmaceutical industry, as it is the major cause of failure in the treatment of infectious diseases. The most common resistance mechanism of pathogenic bacteria to the antibiotics, namely, aminoglycoside, beta-lactam (penicillins and cephalosporins), and chloramphenicol types, involves the enzyme inactivation of the antibiotic by hydrolysis or by formation of inactive derivatives. Such resistance determinants most probably were acquired by pathogenic bacteria from a pool of resistance genes in other microbial genera, including antibiotic-producing organisms. Certain antibiotics produced by actinobacteria have a high activity upon bacteria and fungi. Some antibiotics are used by the geneticists to select a mutant bacteria. Others are used by the biochemists as specific inhibitors of metabolic reaction, such as chloramphenicol (inhibition of protein synthesis), antimycin A (inhibition of cytochrome oxidase), and actinomycin (activity upon nucleic acids). Among the various uses, the treatment of various infectious diseases of human and animals is the remarkable role of antibiotics (Waksman, 1943).

7.3.2   Antifungal compounds

Antifungal compounds have been overshadowed by antibacterials in research interest and applications due to the greater impact of bacterial infections on health. Resistance to antibacterial drugs and the resultant clinical impact are of widespread concern regarding public health (Parks and Casey, 1996). However, resistance by pathogenic fungal infections to drug treatment has become more common in the past two decades. Certain antibiotics produced by actinobacteria have a high activity upon fungi. Some of them like nystatin, candicidin, candidin, trichomycin, hamycin, and amphotericin are polyenes in nature. Some of these antibiotics, notably actidione, and some of the polyenes have found practical applications.

7.3.3   Antiviral compounds

The term "antiviral agents" has been defined in very broad terms as substances other than a virus or virus-containing vaccine or specific antibody that can produce either a protective or a therapeutic effect to the clear detectable advantage of the virus-infected host (Swallow, 1977). Unlike the search for antibiotics, which took root from the discovery of penicillin during the late 1930s, the search for antiviral agents began in the 1950s (Kinchington et al., 1995) but had a breakthrough in 1964 (Kucera and Hermann, 1965). Early success in this direction included the use of methisazone for the prophylaxis of small pox and the use of idoxuridine for the treatment of herpes keratitis. Two major obstacles to the development and use of effective antiviral chemotherapy are the close relationship that exists between the multiplying virus and the host cell and that viral diseases can only be diagnosed and

Table 7.1 Chemical class and action of antiviral compounds

Antiviral compound	Chemical class	Main action
Phytolacca American protein	Glycoprotein (amphoteric)	Polio and influenza
Antiviral factor	Interferon 1	Antiviral
Lycoricidinol	N_2-containing heterocyclic antibiotic (acidic)	Antiviral
Chelerythrine	Alkaloid (basic)	Antiphage
5 – X	Pyran derivative	Herpes and polio
Fisetin	Flavanone (acidic)	Antiviral

Source: Data from Abonyi, D.O. et al., *Afr. J. Biotechnol.*, 8(17), 3989, 2009.

recognized after it is too late for effective treatment. In the first case, an effective antiviral agent must prevent the completion of the viral growth cycle in the infected cell without being toxic to the surrounding normal cells (Kinchington et al., 1995). One encouraging development is the discovery that some virus-specific enzymes are synthesized during multiplication of the viral particles, and this may be a point of attack by a specific inhibitor (Berdy, 1982). Some of the antiviral compounds currently used for therapy are given in Table 7.1.

7.3.4   Antiparasitic compounds

During the evolution of humans, a broad set of parasites have evolved, which inhabits with us as a host organism. Usually, a parasite will not kill its host (at least not immediately), as this would by an evolutionary dead end for a parasite. However, most parasites are either unpleasant for us (lice and fleas) or weaken our health (most internal parasites). However, some parasitic infections such as malaria, trypanosomiasis, or chagas can be deadly if the patients are not treated with adequate therapeutics. Because humans usually live in close proximity and often without good hygienic conditions, the transmission of parasites within a human population is often facilitated. It is very likely that humans have always tried to get rid or minimize the impact of parasites. External parasites (ectoparasites) could be reduced or eliminated mechanically. This could be done individually or in groups. Grooming is a common behavior in primates and monkeys, delousing each other in sequence. Humans probably did the same. More complicated to treat were internal parasites (endoparasites). We know that humans have used medicinal plants for several thousands of years to treat illness and health disorders (Van Wyk and Wink, 2004). Whereas the availability of effective drugs for the treatment of parasitic infections is scanty, only azole compounds and very few other drugs are effective for these parasitic infections.

7.3.5   Anticancer compounds

Anticancer, or antineoplastic, drugs are used to treat malignancies, or cancerous growths. Drug therapy may be used alone or in combination with other treatments such as surgery or radiation therapy. Cancer is commonly defined as the uncontrolled growth of cells, with loss of differentiation and commonly, with metastasis, spread of the cancer to other tissues and organs. In contrast, benign growths remain encapsulated and grow within a well-defined area. Although benign tumors may be fatal if untreated, due to pressure on essential organs, as in the case of a benign brain tumor, surgery and radiation are the preferred methods of treating growths that have a well-defined location. Drug therapy is used when the tumor has spread, or may spread, to all areas of the body (Canetta and

Eisenhauer Rozencsweig, 1994). Several classes of drugs may be used in cancer treatment, depending on the nature of the organ involved. Newer methods of antineoplastic drug therapy have taken different approaches, including angiogenesis—the inhibition of formation of blood vessels feeding the tumor and contributing to tumor growth. Although these approaches hold promise, they are not yet in common use. Developing new anticancer drugs is the work of ongoing research. In 2003, a new technique was developed to streamline the search for effective drugs. Researchers pumped more than 23,000 chemical compounds through a screening technique to identify those that help to fight cancer while leaving healthy cells unharmed. The system identified nine compounds matching the profile, including one previously unidentified drug for fighting cancer (U.S. Administration of Food and Drug, 2008). The chemical classes and usage of anticancer compounds are listed in Table 7.2.

7.4   Sources of antibiotics

7.4.1   Plants

The use and search of drugs and dietary supplements derived from plants have accelerated in recent years. Ethnopharmacologists, botanists, microbiologists, and natural product chemists are combing the earth for phytochemicals and "leads" that could be developed for the treatment of infectious diseases. While 25%–50% of current pharmaceuticals are derived from plants, none are used as antimicrobials. Traditional healers have long been used plants to prevent or cure infectious conditions; western medicine is trying to duplicate their successes. Plants are rich in a wide variety of secondary metabolites, such as tannins, terpenoids, alkaloids, and flavonoids, which have been found *in vitro* to have antimicrobial properties. Since many of the bioactive compounds are currently available as unregulated botanical preparations and their use by the public is increasing rapidly, clinicians need to consider the consequences of patients self-medicating with these preparations. Plants have an almost limitless ability to synthesize aromatic substances, most of which are phenols or their oxygen-substituted derivatives (Elvin-Lewis and Lewis, 1995). Most are belonging to secondary metabolites, of which at least 12,000 have been isolated from plants, a number estimated to be less than 10% of the total metabolites (Borkada, 1978). In many cases, these substances serve as plant defense mechanisms against predation by microorganisms, insects, and herbivores. Some, such as terpenoids, give plants their odors, and others (quinones and tannins) are responsible for plant pigment. Many compounds are responsible for plant flavor (e.g., the terpenoid capsaicin from chili peppers), and some of the same herbs and spices used by humans to season food yield useful medicinal compounds (Savoia, 2012). The approximate number of the known natural products derived from the main types of plant and animal organisms is summarized in Table 7.3.

Mainstream medicine is increasingly receptive to the use of antimicrobial and other drugs derived from plants, as traditional antibiotics become ineffective and as new, particularly viral, diseases remain intractable to this type of drug. Another driving factor for the renewed interest in plant antimicrobials in the past two decades has been the rapid rate of plant species extinction (Georges and Pandelai, 1949). Unfortunately, there is an alike feeling among natural products, chemists, and microbiologists that the multitude of potentially useful phytochemical structures that could be synthesized chemically is at risk of being lost irretrievably (Rojas et al., 1992). There is a scientific discipline known as

Table 7.2 Chemical class and usage of anticancer compounds

Generic (brand name)	Clinical uses
Altretamine (Hexalen)	Treatment of advanced ovarian cancer
Asparaginase (Elspar)	Commonly used in combination with other drugs; refractory acute lymphocytic leukemia
Bleomycin (Blenoxane)	Lymphomas, Hodgkin's disease, testicular cancer
Busulfan (Myleran)	Chronic granulocytic leukemia
Carboplatin (Paraplatin)	Palliation of ovarian cancer
Carmustine	Hodgkin's disease, brain tumors, multiple myeloma, malignant melanoma
Chlorambucil (Leukeran)	Chronic lymphocytic leukemia, non-Hodgkin's disease
Cisplatin (Platinol)	Treatment of the bladder, ovarian, uterine, testicular, and head and neck cancers
Cladribine (Leustatin)	Hairy cell leukemia
Cyclophosphamide (Cytoxan)	Hodgkin's disease, non-Hodgkin's lymphomas, neuroblastoma; breast, ovarian, and lung cancers; acute lymphoblastic leukemia in children; multiple myeloma
Cytarabine (Cytosar-U)	Leukemias occurring in adults and children
Dacarbazine (DTIC-Dome)	Hodgkin's disease, malignant melanoma
Diethylstilbestrol (DES)	Breast cancer in postmenopausal women, prostate cancer
Ethinyl estradiol (Estinyl)	Advanced breast cancer in postmenopausal women, prostate cancer
Etoposide (VePesid)	Acute leukemias, lymphomas, testicular cancer
Mitomycin (Mutamycin)	Bladder, breast, colon, lung, pancreas, rectum, and head and neck cancers, malignant melanoma
Mitotane (Lysodren)	Cancer of the adrenal cortex (inoperable)
Mitoxantrone (Novantrone)	Acute nonlymphocytic leukemia
Paclitaxel (Taxol)	Advanced ovarian cancer
Pentostatin (Nipent)	Hairy cell leukemia unresponsive to alpha-interferon
Pipobroman	Chronic granulocytic leukemia
Plicamycin (Mithracin)	Testicular tumors
Prednisone (Meticorten)	Used in combined therapy for palliation of symptoms in lymphomas, acute leukemia, and Hodgkin's disease
Procarbazine (Matulane)	Hodgkin's disease
Streptozocin (Zanosar)	Islet cell carcinoma of pancreas
Tamoxifen (Nolvadex)	Advanced breast cancer in post menopausal
Teniposide (Vumon)	Acute lymphocytic leukemia in children
Vinblastine (Velban)	Breast cancer, Hodgkin's disease, metastatic testicular cancer
Vincristine (Oncovin)	Acute leukemia, Hodgkin's disease, lymphomas

Source: Data from Kelecom, A., *Ann. Braz. Acad. Sci.,* 74(1), 151, 2002.

ethnobotany/ethnopharmacology, whose goal is to utilize the impressive array of knowledge assembled by indigenous peoples about the plant and animal products they have used to maintain health (Silva et al., 1996). Ultimately, the ascendancy of the human immunodeficiency virus has spurred intensive investigation into the plant derivatives that may be effective, especially for use in underdeveloped nations with little access to expensive western medicines.

Table 7.3 Approximate number of known natural products

Source	All known compounds	Bioactives	Antibiotics
Natural products	Over one million	200,000–250,000	25,000–30,000
Plant kingdom	600,000–700,000	150,000–200,000	~25,000
Microbes	Over 50,000	22,000–23,000	~17,000
Algae, lichens	3,000–5,000	1,500–2,000	~1,000
Higher plants	500,000–600,000	~100,000	10,000–12,000
Animal kingdom	300,000–400,000	50,000–100,000	~5,000
Protozoa	Several hundreds	100–200	~50
Invertebrates	~100,000	NA	~500
Marine animals	20,000–25,000	7,000–8,000	3,000–4,000
Insects/worms	8,000–10,000	800–1,000	150–200
Vertebrates (mammals, fishes, amphibians, etc.)	200,000–250,000	50,000–70,000	~1,000

Source: Data from Berdy, J., *J. Antibiot.*, 58(1), 1, 2005.

7.4.2   Animals

Louis Pasteur, for instance, began to study rabies in animals in 1881, and over a number of years he developed methods of producing attenuated virus preparations by progressively drying spinal cord of rabbits experimentally infected with the agent. Also through experimental transmission to mice, in 1900, Walter Reed demonstrated that yellow fever was caused by a virus and spread by mosquitoes. This discovery eventually enabled Max Theiler in 1937 to propagate the virus in chick embryo and to produce an attenuated vaccine from the 17 D strain that is still in use today. The success of this approach had led to increased use of animal systems to identify and propagate pathogenic viruses even with the adoption/perfection of tissue culture technique. In the area of antiviral chemotherapeutic research, animal models have been used either primarily as screening tools or applied in testing the efficacy of the test compound when it had been identified as effective/potent using any other method (Likar and Japelj, 1977; Sloan et al., 1977).

7.4.3   Bacteria

Among the many challenges facing drug discovery research in recent years, the search for new antibacterial agents has proved to be among the most unproductive. Many factors have contributed to this problem, but one of the key areas for improvement is the need to test compounds that are more appropriate, as it has become obvious that the screening of randomly assembled, diverse compound libraries has produced extremely low hit rates (Fernando, 2006). Besides, *in vitro* screening often delivers nondrug-like and non-target-specific structures that tend to face serious efficacy issues in *in vivo* experiments. To address the major challenge of antibacterial drug discovery, it is critical to access compound libraries that are capable of delivering excellent chemical starting points for completely new classes of antibacterials. A large proportion of known antibacterials have derived from natural products, and these compounds clearly have structures and properties that have made them a particularly rich source. The discovery of novel antibiotics and other potential molecules of pharmaceutical interest through microbial secondary metabolite screening is becoming increasingly fruitful. There is wide acceptance that

Table 7.4 Approximate number of bioactive microbial products according to their producers

Source	Antibiotics	Other bioactive metabolites	Total bioactive metabolites	Practically used (in human therapy)	Inactive metabolites
Bacteria	2,900	900	3,800	10 ~ 12 (8 ~ 10)	3,000–5,000
Actinobacteria	8,700	1,400	10,100	100 ~ 120 (70 ~ 75)	5,000–10,000
Fungi	4,900	3,700	8,600	30 ~ 35 (13 ~ 15)	2,000–15,000
Total	16,500	6,000	22,500	140 ~ 160 (~100)	20,000–25,000

Source: Data from Berdy, J., *J. Antibiot.*, 58(1), 1, 2005.

microorganisms are virtually unlimited sources of novel substances with many therapeutic applications. Among them, actinobacteria hold a predominant position due to their diversity and had proven their ability to provide new and novel substances (Gayathri et al., 2011). As the best approximate, the total number of additional *inactive* microbial products is about 20,000–25,000; therefore, today close to 50,000 microbial metabolites may be known (Vijayakumar et al., 2013). According to the main types of microbial producers, the numbers of compounds, including both antibiotics and *other bioactive* metabolites, practically used compounds, and the approximate numbers of the inactive microbial metabolites are summarized in Table 7.4.

7.4.4 Fungi

Fungi make an extraordinarily important contribution to managing disease in humans and other animals. At the beginning of the twenty-first century, fungi were involved in the industrial processing of more than 10 of the 20 most profitable products used in human medicine. In 1941, penicillin from the fungus *Penicillium chrysogenum* was first used successfully to treat an infection caused by a bacterium. The use of penicillin revolutionized the treatment of pathogenic disease. Many formally fatal diseases caused by bacteria became treatable, and new forms of medical intervention were possible. When penicillin was first produced, the concentration of active ingredient was approximately 1 µL/mL of broth solution. Today, improved strains and highly developed fermentation technologies produce more than 700 µL/mL of active ingredient. The natural penicillins have a number of disadvantages. They are destroyed in the acid stomach and so cannot be used orally; they are sensitive to beta lactamases, which are produced by resistant bacteria, thus reducing their effectiveness. Also, they only act on gram-positive bacteria. Modifications to manufacturing conditions have resulted in the development of oral forms. However, antibiotic resistance among bacteria is becoming an extremely important aspect determining the long-term use of all antibiotics (Zabriskie and Jackson, 2000). Cephalosporins also contain the beta-lactam ring. The original fungus found to produce the compounds was *Cephalosporium*. As with penicillin, the cephalosporin antibiotics have a number of disadvantages. Industrial modification of the active ingredients has reduced these problems. The only broadly useful antifungal agent from fungi is griseofulvin. The original source was *Penicillium griseofulvum*. Griseofulvin is fungistatic, rather than fungicidal. It is used for the treatment of dermatophytes, as it accumulates in the hair and skin following topical application (Erturk, 2006).

More recently, several new groups have been developed. Strobilurins target the ubihydroquinone oxidation center, and in mammals, the compound from fungi is immediately excreted. Basidiomycetes, especially from tropical regions, produce an enormous diversity of these compounds. Sordarins are structurally complex molecules that show a remarkably narrow range of action against yeasts and yeastlike fungi. The compounds inhibit protein biosynthesis and so may become important agents against a number of fungal pathogens of humans. Echinocandins are cyclic peptides with a long fatty acid side chain. They target cell wall formation. Semisynthetic members of the group of compounds include pneumo-candins that are in use in humans.

7.4.5   *Actinobacteria*

A rapid increase in discoveries of new antibiotics occurred from the late 1910s to 1960s, followed by a gradual fall until 1968 and then again raised in the 1970s. But most of the major antibiotics had been discovered by the 1960s. The second increase in the isolation of new antibiotics in recent years may reflect the development of new technique that has enabled us to isolate and characterize antibiotic of closely related chemicals structures, which are antibiotics present in the culture as minor components. The recent increase in the description of new antibiotics might also be attributed to the rediscoveries of old antibiotics, which were once isolated in crude state and completely described in the early stage of antibiotic era. Since more than 2000 antibiotics have already been isolated and described, the chances of finding new antibiotic have now become poor. Under such a situation, the importance of the preliminary identification of antibiotic producer in the first step of screening cannot be stressed too much. If one could identify the antibiotic producer simply by taxonomic studies on the antagonistic actinobacteria under study, it would be of a great help in the search for new antibiotics produced simply by taxonomic placement of actinobacteria and their antibiotic production. Krasilnikov (1960) reported that the pro-duction of specific antibiotics by actinobacteria can be used as one of the criteria for their classification. On the other hand, Baldacci et al. (1954) claimed that the uses of antibiotic production in taxonomy present staggering problems since actinobacteria produced sev-eral hundred antibiotics in various mixtures. Lechavalier et al. (1971) suggested that the antibiotic production helped in the speciation of actinobacteria.

Generally, the productivity and activity of antimicrobial compounds are not uniform, which may be varied between the producing organisms where they have isolated and against test pathogenic microorganisms, respectively. In a study, among the 24 antibac-terial antagonistic actinobacteria, 20 isolates inhibited the growth of gram-positive bac-teria, 17 inhibited the growth of gram-negative bacteria, and 13 inhibited the growth of both gram-positive and gram-negative bacteria. The maximum percentage of antibacterial compound producing actinobacteria was found in saltpan soil (47.4%), followed by sea-shore soil (33.33%) and mangrove soil (28.57%) (Vijayakumar, 2006). Correspondingly, out of 50 isolates, 29 (58%) isolates exhibited antimicrobial activity. Among them, 22 (75.89%) isolates had activity against gram-positive bacteria, 26 (89.66%) against gram-negative bac-teria, and 20 (68.97%) against both gram-positive and gram-negative bacteria (Cholarajan, 2014). Several reports have been published by various researchers, and they bring out the potentials of actinobacteria as producer of commercially valuable antibiotics and other bioactive compounds. The actinobacteria producing some antibiotics are given in Table 7.5.

7.4.6   *Other sources*

Numerous compounds isolated in the past few years from lichens proved to be active in certain physiological and pharmacological terms, and a number of them were confirmed to be effective antimicrobial agents. These lichen products belong to several distinct

Table 7.5 Some examples for actinobacteria-producing antibiotics

Organism	Antibiotic
Micromonospora galeriensis	Primycin
Actinoplanes ianthinogenes	Purpuromycin
Streptosporangium albidum	Sporaviridin
Planomonospora parontospora	Sporangiomycin
Actinosporangium griseoroseum	Hepcin
Nocardia acidophilus	Mycomycin
Micromonospora chalcea	Chalacidin
Thermoactinopolyspora coremialis	Thermothiocin
Streptomyces venezuelae	Chloramphenicol
Streptomyces roseosporus	Daptomycin
Streptomyces fradiae	Fosfomycin
Streptomyces lincolnensis	Lincomycin
Streptomyces fradiae	Neomycin
Streptomyces alboniger	Puromycin
Streptomyces griseus	Streptomycin
Streptomyces rimosus, Streptomyces aureofaciens	Tetracycline
Streptomyces avermitilis	Avermectin

Source: Data from Kelecom, A., *Ann. Braz. Acad. Sci.,* 74(1), 151, 2002.

chemical classes including polysaccharides, coumarone derivatives, lactone derivatives, orcinol and orcinol-type despsides and depsidones, and long-chain aliphatic oligocarboxy hydroxy acids (Esimone, 1997). Most of the lichen products are primarily active against gram-positive bacteria and mycobacteria, though the extracts of *Parmelia caperata, Evernia prunastri,* and *Usnea* spp. have been shown to be active against gram-negative bacteria (Rowe et al., 1989). Some lichen polysaccharides and other lichen products such as psoronic acid had antitumor activity, and usnic acid, the most widely distributed and best-known lichen antibiotics, has been used in several countries as a topical antibacterial agent for human skin diseases (Berdy, 1982) and is also reported to exert some antimitotic action, but at low concentrations, the acid displays a capacity to stimulate cell metabolism in some biological systems tested (Cardarelli et al., 1997). In Table 7.6, some of the clinically important antibiotics are given.

7.5 Isolation of actinobacteria

7.5.1 Selection of samples for the isolation of actinobacteria

Many thousands of actinobacteria have been isolated from the environment until now, but little information only is available about the geographical or ecological distribution of these microbes. So, it is generally impossible to predict the sites in which a particular actinobacterial taxon or strain will occur. Thus, the selection of macro or microenvironments as a source of useful isolates remains largely a matter of chance and hopeful initiative. According to the results of Takahashi et al. (1993), most of the actinobacteria occur within one meter below the ground. Compared to terrestrial soils, the marine sediments have proved to be the best source for the isolation of antagonistic actinobacteria by various workers not only from marine soils and sediments (Vijayakumar et al., 2007, 2012a–c) but also from salt mining samples (Yang et al., 2008); mangrove environments (Dhanasekaran

Table 7.6 Some clinically important antibiotics

Antibiotic	Producer organism	Activity	Site or mode of action
Penicillin	*Penicillium chrysogenum*	Gram-positive bacteria	Wall synthesis
Cephalosporin	*Cephalosporium acremonium*	Broad spectrum	Wall synthesis
Griseofulvin	*Penicillium griseofulvum*	Dermatophytic fungi	Microtubules
Bacitracin	*Bacillus subtilis*	Gram-positive bacteria	Wall synthesis
Polymyxin B	*Bacillus polymyxa*	Gram-negative bacteria	Cell membrane
Amphotericin B	*Streptomyces nodosus*	Fungi	Cell membrane
Erythromycin	*Streptomyces erythreus*	Gram-positive bacteria	Protein synthesis
Neomycin	*Streptomyces fradiae*	Broad spectrum	Protein synthesis
Streptomycin	*Streptomyces griseus*	Gram-negative bacteria	Protein synthesis
Tetracycline	*Streptomyces rimosus*	Broad spectrum	Protein synthesis
Vancomycin	*Streptomyces orientalis*	Gram-positive bacteria	Protein synthesis
Gentamicin	*Micromonospora purpurea*	Broad spectrum	Protein synthesis
Rifamycin	*Streptomyces mediterranei*	Tuberculosis	Protein synthesis

Source: Data from Kelecom, A., *Ann. Braz. Acad. Sci.*, 74(1), 151, 2002, modified and reprinted with permission from Prescott, L.M. et al., *Microbiology*, 5th edn., Tata McGraw-Hill Company, New Delhi, India, 2002.

et al., 2008, 2009, 2011; Remya and Vijayakumar, 2008), estuaries, sand dunes and industrially polluted coast soil, and salt marsh soil (Al-Zarban et al., 2002; Kathiresan et al., 2005); coral reefs (Lam, 2006); marine sediments (Olano et al., 2009); salt pan environment (Vijayakumar et al., 2012c); sea anemone (Chen et al., 2009); marine sponge (Gandhimathi et al., 2009); beach soil (Ogunmwonyi et al., 2010); endophytic actinobacteria (Ravikumar et al., 2010); seawater (Reddy et al., 2011); and saltern (Chun et al., 2000).

7.5.2   Pretreatment of the samples

The collected samples were processed and inoculated onto various culture media using one method, or in some cases (especially for the deeper sediments), as many as eight methods described in the following are used.

7.5.2.1   Method 1 (dry/stamp)

Sediment was dried overnight in a laminar flow hood and, when clumping occurred, ground lightly with an alcohol-sterilized mortar and pestle. An autoclaved foam plug (2 cm in diameter) was pressed onto the sediment and then repeatedly onto the surface of an agar plate in a clockwise direction creating a serial dilution effect.

7.5.2.2   Method 2 (sterile dry/scrape)

The method was used for small rocks that had been dried overnight in a laminar air flow hood and then scraped with a spatula generating a powder that was processed as per Method 1. In some cases, the powder was collected with a wet cotton-tipped applicator, or the rock was rubbed directly with the applicator that was then used to inoculate the surface of an agar plate.

7.5.2.3   Method 3 (dry/dilute)

Dried sediment (0.5 g) was diluted with 5 mL of sterile seawater (SSW). The diluted sample was vortex mixed and allowed to settle for a few minutes, and 50 µL of the resulting solution was inoculated onto the agar surface and spread with sterile glass rod.

7.5.2.4   Method 4 (dilute/heat)

Dried sediment was volumetrically added to 3 mL of SWW (dilutions 1:3 or 1:6) and heated to 55°C for 6 min, and 50–75 µL of the resulting suspension was inoculated onto agar media as per Method 3.

7.5.2.5   Method 5 (dilute/heat/2)

Dried sediment was treated as per Method 4 (dilution 1:6) with the addition of a second heat treatment at 60°C for 10 min.

7.5.2.6   Method 6 (dry/stamp + dilute/heat)

The surface of an agar medium was inoculated using a sample treated as Method 1. The dried sediments were then processed using Method 4, and the same agar plate was inoculated a second time with the heat-treated samples.

7.5.2.7   Method 7 (freeze/dilute)

Wet sediment was frozen at –20°C for at least 24 h, thawed, and volumetrically diluted in SSW (1:3 to 1:120 depending on particle size), and 50 µL of the resulting suspension was inoculated onto the surface of an agar plate as per Method 3.

7.5.2.8   Method 8 (freeze/dilute/2)

Wet sediment was treated as per Method 7 except that the thawed and diluted samples were incubated at room temperature for 48 h before inoculation onto the surface on agar plate (Jensen et al., 2005).

7.5.3   Isolation and culture conditions

First, the isolates were isolated on starch casein agar (SCA) medium (g/L, starch 10, casein 0.3, KNO_3 2, NaCl 2, K_2HPO_4 2, $MgSO_4$ 2, $7H_2O$ 0.05, $CaCO_3$ 0.02, $FeSO_4$ $7H_2O$ 0.01, and agar 18); supplemented with griseofulvin and cycloheximide (Himedia, Mumbai) of 25 and 10 µg/mL density (Vijayakumar et al., 2007). The plates were incubated for 14 days in an incubator of darkness at 28°C ± 1°C. Colonies of morphologically distinct isolates were purified by using pure culture technology (streak plate method) (Figures 7.1 and 7.2). The studies on colony morphological characteristics of the isolated marine actinobacteria were carried out following the methods recommended by the International *Streptomyces* Project (ISP) (Waksman, 1943; Shirling and Gottlieb, 1966). The population

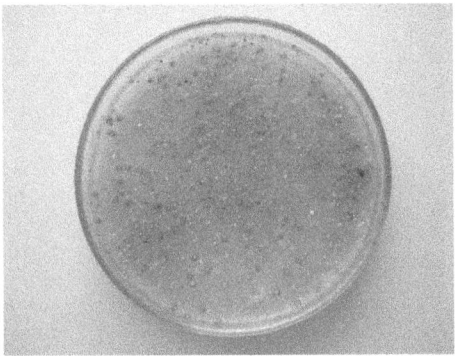

Figure 7.1 Isolated colonies of actinobacteria on a SCA medium.

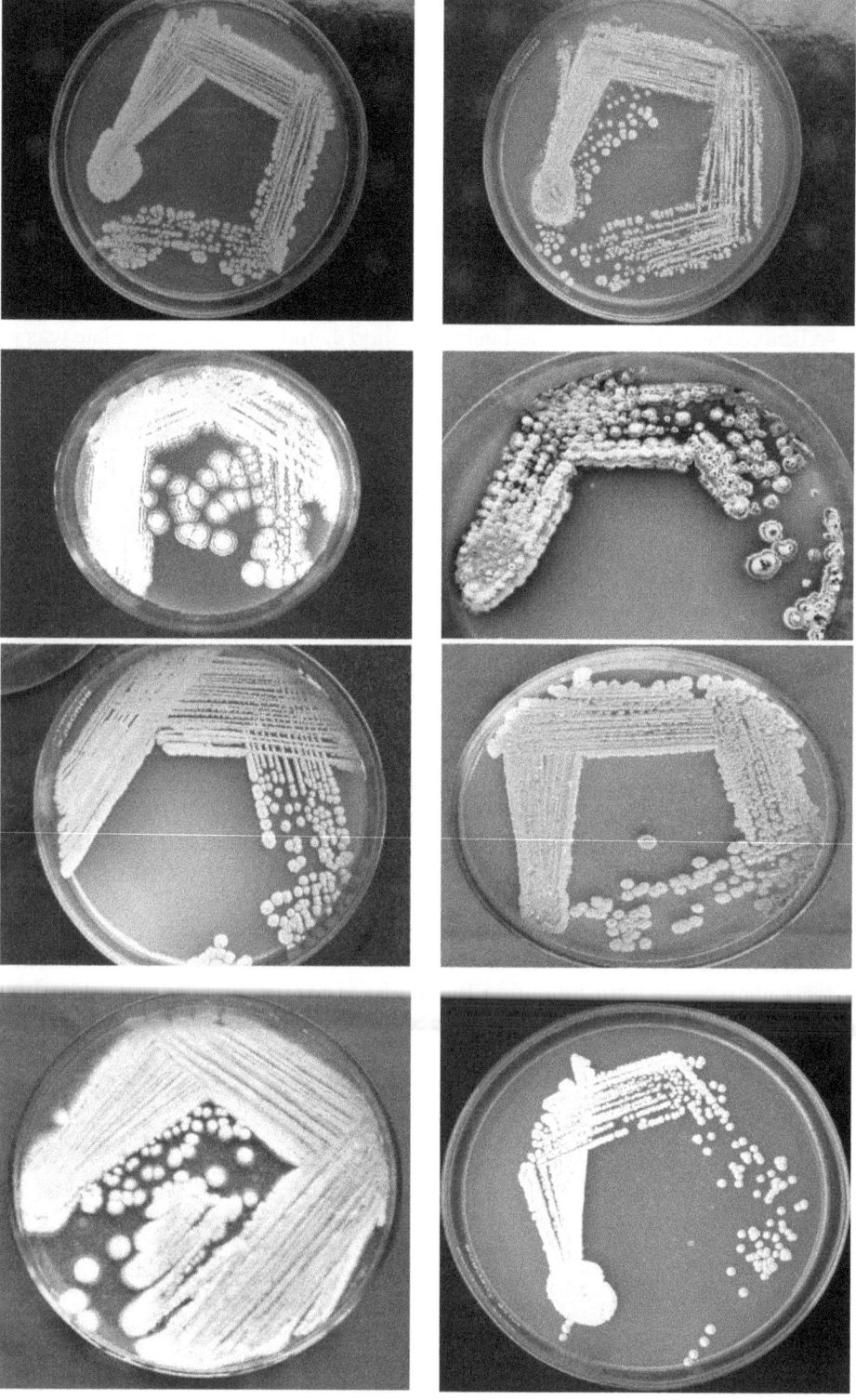

Figure 7.2 Various colonial morphologies of some actinobacteria on a SCA medium.

of actinobacteria will be varied depending on the samples and their collection location as well as media with and without antibiotics (inhibit other bacterial and fungal population) used for cultivation. Thus, usage of different culture media for the isolation of actinobacteria from different sampling environments will give the complete picture (occurrence) of them (Table 7.7).

7.6 Identification of actinobacteria

7.6.1 Phenotypic characterization

7.6.1.1 Microscopic characterization

Purified isolates of actinobacteria were initially characterized by using morphological properties followed by the cultural, biochemical, and physiological characteristics by the methods of ISP. The morphology of the spore-bearing hyphae with the entire spore chain, the structure and arrangement of the spore chain with the substrate, and the aerial mycelium of the actinobacteria were examined and identified using slide culture technique. After growth, the microscopic morphology including the formation of aerial and substrate mycelia and sporophore nature was observed by light microscopy. The spore and mycelial sizes and spore surface nature were observed under scanning electron microscopy.

7.6.1.2 Colony morphological/cultural characterization

Aerial spore mass color and reverse side pigments of the actinobacterial isolates were recorded in various culture media such as oat meal agar (ISP 3), yeast extract–malt extract agar (ISP 2), inorganic salt starch agar (ISP 4), glycerol–asparagine agar (ISP 5), tyrosine agar base (ISP 7), starch–casein agar, and Czapek–Dox agar. The reverse side pigments of the colony, namely, distinctive (+) and not distinctive (–), were tested using peptone–yeast extract–iron agar (ISP 6) (Das et al., 2008). The production of melanoid pigments was tested on tryptone–yeast extract agar (ISP 1) and ISP 7 medium. Colors of aerial spore mass and reverse side colors were determined with the ISCC-NBS centroid color charts (Kenneth et al., 1976). Generally, it is difficult to characterize all the isolated actinobacteria with respect to colony morphology on various culture media and biochemically and physiologically. Hence, bioactive actinobacteria alone studied for their phenotypic and genotypic characteristics will allow us to consume the time duration.

7.6.1.3 Biochemical characterization

The biochemical characteristics are used to differentiate the microorganisms at species/strain level, in which indole production; methyl red and Voges–Proskauer tests, triple sugar iron test; citrate utilization; nitrate reduction; production of urease, catalase, and oxidase; hydrolyses of gelatin, starch, lipid, and casein; degradation of xanthine; fermentation of carbohydrates; tolerance to pH; and temperature and salt tests are performed and used to identify the actinobacteria (Holt, 1989). Methodologies for all the described examinations were performed with reference to Cappuccino and Sherman (2005) and Bergey and Holt (1994). In addition, the analysis of cell wall amino acids and whole cell sugars of the actinobacteria has also provided valuable information for generic-level identification.

7.6.1.4 Physiological characterization

The ability of the isolates to utilize various carbon and nitrogen sources was studied as per the methods recommended by ISP. Carbon sources like glucose, mannitol, fructose, xylose, sucrose, raffinose, inositol, arabinose, and rhamnose were tested on carbon utilization agar

Table 7.7 Different sources and media for isolation of actinobacteria

Source	Media	References
From soil		
Forest soil	Starch–casein medium	Kuster and Williams (1964)
Humus layer of forest soil	Humic acid–vitamin agar	Choi and Park (1993)
	Starch–casein–nitrate (SCN) agar	Hayakawa and Nonomura (1987)
	Hair hydrolysate–vitamin agar	
	Bennett's agar	Seong et al. (2001)
Cornfield, cow's barnyard, and forest	Arginine–glycerol salt medium	Porter et al. (1960)
	Chitin medium	Lingappa and Lockwood (1961)
	Modified Benedict's medium	Porter et al. (1960)
	Soybean meal–glucose medium	Tsao et al. (1960)
	Gauze's agar medium	Rehacek (1959)
	Czapek's agar medium	Waksman (1961)
	Egg albumen medium	
	Glucose–asparagine medium	
	Glycerol–asparagine agar 2	
Lake soil	Chitin agar	Hsu and Lockwood (1975)
Soil	Coal–vitamin agar (CVA)	Wakisaka et al. (1982)
Antarctic soil	Mineral salt (MS) medium	Kosmachev (1954)
Mitidja plain (Algeria)	Yeast extract–malt extract agar	Shirling and Gottlieb (1966)
Marine soil	SCN agar medium	Amorso et al. (1998)
From water		
Stream sediments and lake muds	Chitin agar media	Lingappa and Lockwood (1961, 1962)
	M3 agar medium	Jones (1949)
	Bennett's medium	
Marine sediments	Starch–casein agar	Grein and Meyers (1958)
	Asparagine agar	
	Glycerol–glycine agar	Lindenbein (1952)
Marine sediments (South China)	Actinomycete isolation agar (AIA) medium	You et al. (2007)
From root and stem samples of four plants		
Cinnamomum zeylanicum, Zingiber spectabile, Elettariopsis curtisii, and *Labisia pumila*	Starch–yeast–casein–agar, AIA, humic acid–vitamin–gellan gum, tap water–yeast extract agar, CVA	Zin et al. (2007)
Mangrove sediments	Asparagine–glucose agar medium	Shirling and Gottlieb (1966)

Source: Data from Sharma, M., *Int. J. Curr. Microbiol. App. Sci.*, 3(2), 801, 2014.

(ISP 9) supplemented with 1% carbon sources (Nonomura, 1974). The ability of the isolate to utilize various nitrogen sources like leucine, histidine, tryptophan, serine, glutamic acid, lysine, arginine, methionine, and tyrosine for growth was also tested.

7.6.2   Molecular characterization

Sequence-based identification is becoming an increasingly important tool of identification. Based on the molecular characterization of phylogenetic relatedness, 16S rRNA genes provide enough information for the species-level identification of actinobacteria. The genomic DNA isolation of actinobacteria was done by the method described by Sambrook et al. (1989) and amplified by polymerase chain reaction (PCR) using master mix (Kit Medox Mix). PCR conditions, the primers and the methodology for sequencing, were adapted from Magarvey et al. (2004) and Mincer et al. (2002). The sequencing was carried out in both sense and antisense directions. The similarity and homology of the 16S rDNA partial gene sequence were analyzed with the similar existing sequences available in the data bank of NCBI using BLAST search. The DNA sequences were aligned and phylogenetic tree was constructed by neighbor joining (NJ tree) method using neighbor tree software. A bootstrap analysis of 100 replicates was carried out.

7.7   Screening of antibiotic production

7.7.1   Primary screening: Cross streak plate technique

The search for novel metabolites especially from actinobacteria requires a large number of isolates (over thousands) in order to discover actinobacterial population with novel compound of pharmaceutical interest. Because of this, the research will be more promising if diverse and more actinobacteria are sampled and screened. Such attempts need to be continued both in the sample collection area and from the adjacent places during various climatic conditions as to screen more isolates for novel antimicrobial compounds. Antimicrobial producing property of the actinobacteria was initially screened by cross streak method (Egorov, 1985). The single streak of the actinobacteria was made on the surface of the modified nutrient agar medium and incubated at 28°C ± 2°C. After observing a good ribbon-like growth of the actinobacteria on the Petri plates, the test pathogens were streaked at right angles to the original streak of actinobacteria and incubated at 27°C. The inhibition zone was measured after 24 and 48 h. By this cross streak plate method, deserved candidates (antibiotic-producing actinobacteria) were selected for further studies. The isolates with no activity may also be cultured and maintained for identification so as to understand the actinobacteria with or without biological activities.

7.7.2   Secondary screening: Well diffusion assay

The molten sterile nutrient agar was prepared and inoculated with the test organisms. Wells were made using sterile borer; 50 µL of clear broth (supernatant) was added to each well. The plates were kept in a refrigerator for about 2 h to allow the diffusion of the bioactive metabolite. After 2 h, plates were incubated at 37°C in an incubator. The inhibition zones were measured after 24 h using an antibiotic zone reader (Grove and Randall, 1955).

7.8   Extraction and separation of antimicrobial compounds

The selected antagonistic actinobacterial isolates were inoculated into liquid starch–casein medium and incubated at 28°C on a shaker (200–250 rpm) for 7 days. After incubation, the medium was filtered through Whatman No. 1 filter paper and then through 0.45 µm Millipore filter (Millipore Millex-HV Hydrophilic PVDF). The filtrate was transferred aseptically into a conical flask and stored at 4°C for further assay. To the culture filtrate, equal volume of various solvents (*viz.*, chloroform, ethyl acetate, petroleum ether, and methanol) was added separately and centrifuged at 8000 rpm for 10 min at 4°C to extract the antimicrobial compound (Sambamurthy and Ellaiah, 1974). The supernatant was collected and maintained in the refrigerator for further use. Extraction of an antimicrobial compound is a crucial and vital process with as much as organic solvents. Therefore, it is essential to keep in mind that the solvent selection is crucial step not only for the extraction of antimicrobial compounds but also for the minimization of the total production cost of an antibiotic.

7.8.1   Thin-layer chromatography

The crude extract of the selected actinobacteria was applied manually on a preparative thin-layer chromatography (TLC) glass plate (20 × 20 cm; 1.5 cm thickness) with inorganic fluorescent indicator binder. After air drying, the plate was developed, using the same mobile phase as used in the analytical TLC, in a presaturated glass chamber. In each experiment, two plates were used in parallel. One plate from each set of experiment was sprayed with 2,3,5-triphenyl tetrazolium chloride solution, as described earlier, and the bands with antimicrobial activity determined by bioautography were scraped off carefully from the second plate of each set of experiment. The scratched sample was dissolved in high-performance liquid chromatography (HPLC) grade methanol and centrifuged at 12,000 rpm for 15 min in order to remove debris of silica gel. The supernatant was collected, filtered from 0.22 µm filter, and dried under reduced pressure. Further, all the dried samples were passed under nitrogen gas for 5 min and then dissolved in methanol for further characterization and bioactivity analysis. The entire purification process was carried out under dark or dim light conditions (Rajauria and Abu-Ghannam, 2013).

7.8.2   High-performance liquid chromatography

HPLC is a technique used to separate a mixture of compounds in analytical chemistry and biochemistry with the purpose of identifying, quantifying, and purifying the individual components of the mixture. The chromatographic separation of antibacterial compound was carried out on a LC-10 AT vp model HPLC using 250 × 4.60 mm Rheodyne column (C-18). The solvent system used was methanol (HPLC grade) and water (HPLC grade) in the ratio of 88:12. The operating pressure was 114 kgf, at a flow rate 0.8 mL/min, and the temperature was set at 30°C. The ultraviolet–visible (UV-Vis) (SPD-10 A vp) detector was set at 210 nm. The sample was mixed with the solvent in the ratio of 50:50 and filtered using Millipore filter before injection. About 25 µL of the sample filtrate was injected into the column. The sample was run for 10 min and the retention time was noted. The elution time was compared with the standard, and the compound was determined (Swami et al., 1983; Sethi, 2001).

7.9 Characterization and identification of antibiotics

7.9.1 pH stability test

Five milliliters of actinobacterial culture filtrates were put in vials using 0.1 N HCl or 0.1 N NaOH adjusted to the pH levels 2, 4, 6, 8, 10, and 12. These antibiotic-containing media were held for 3–4 h at room temperature (20°C–24°C) and then readjusted to pH 6.0 (Muiru et al., 2012). The experiment was replicated three times and arranged in a completely randomized design as outlined earlier. The activity of these treated antibiotic culture filtrates was determined against pathogenic microorganisms by the measurements of the inhibition zones using the paper disc method as described earlier (Grove and Randall, 1955).

7.9.2 Temperature stability test

The culture filtrates of actinobacteria were adjusted to pH 6.0 using 0.1 N HCl or 0.1 N NaOH, and 5 mL of each culture filtrate was put in sterile universal bottles and then subjected to different temperatures in a water bath for 10 min. The temperature range from 30°C to 121°C was employed for the determination of the stability of antimicrobial compounds. Autoclaving was done for the temperature 121°C at one bar pressure for 15 min. The stability of the various culture filtrates to these temperatures was determined by measuring the size of the inhibition zones using the paper disc method as described earlier. For each temperature, three replicates (plates) were maintained (Muiru et al., 2012).

7.9.3 Melting point

The melting point of the antimicrobial compound was determined on electrically heated oil bath. The dried antimicrobial compound was taken in the glass capillary tubes. The tubes were kept in the electrical heat oil bath. Thermometer was monitored regularly to notice the melting point of the secondary metabolite (Gandhi et al., 1976).

7.9.4 Quality test

Quality assurance is essential to ensure the quality of antimicrobial susceptibility test by diffusion methods. Routine internal quality control testing with a range of control strains is a major part of the quality assurance process, as it facilitates monitoring of the performance of the test. Most standardized methods include tables of acceptance zone size ranges for control strains, and in addition to checking that control zone diameters are within the published ranges, rules or statistical approaches may be applied to indicate deviations from acceptable performance (King and Brown, 2001).

7.10 Prediction of antibiotic structures

7.10.1 Ultraviolet spectrum

UV–visible (UV–Vis or UV/Vis) spectrophotometry refers to absorption spectroscopy or reflectance spectroscopy in the UV–Vis spectral region. This means that it uses light in the visible and adjacent (near-UV) and near-infrared ranges. The absorption or reflectance in the visible range directly affects the perceived color of the chemicals involved. In this region of the electromagnetic spectrum, molecules undergo electronic transitions. This technique is complementary to fluorescence spectroscopy, in that fluorescence deals with transitions from the excited state to the ground state, while absorption measures transitions from the ground state to the excited state. Molecules containing π-electrons or nonbonding electrons (n-electrons) can absorb the energy in the form of UV or visible light to excite these electrons to higher antibonding molecular orbitals (Douglas et al., 2007). The more easily excited the electrons (i.e., lower energy gap between the HOMO and the LUMO), the longer the wavelength of light it can absorb.

7.10.2 Fourier transform infrared spectrum

The powdered antimicrobial compound samples were mixed with potassium bromide (KBr pellet) and subjected to a pressure of about 5.10^6 Pa in an evacuated to produce a clear transparent disc of diameter 13 mm and thickness 1 mm. IR spectra in frequency region 4000–400 cm^{-1} were recorded at room temperature on a PerkinElmer fourier transform spectrometer equipped an air-cooled deuterated triglycine sulfate detector. For each spectrum, 100 scans were coadded at a spectral resolution of 4 cm. The frequencies for all sharp bands were accurate to 0.01 cm. All the spectral values were expressed in (%) transmittance.

7.10.3 Mass spectrum

Mass spectroscopy is an important physicochemical tool applied for structural elucidation of compounds from natural product including medicinal plants. The fundamental principle of mass spectrum (MS) is the use of different physical means for sample ionization and separation of the ions generated based on their mass (m) to change (z) ratio (m/z) (De Rijke et al., 2006), for example, ionization (APCI), electron ionization, fast atom bombardment, and matrix-assisted laser desorption ionization. Mass spectroscopy has high sensitivity with detection limit of fentogram compared to nuclear magnetic resonance (NMR) with sensitivity limit of nanogram range and above. The high sensitivity and the flexibility for hyphenation with other chromatographic techniques made MS a versatile analytical instrument.

7.10.4 Nuclear magnetic resonance spectrum

One-dimensional (^{1}H and ^{13}C) and 2D NMR spectra (^{1}H, ^{1}HCOSY, HMQC, and HMBC) were recorded on a varian VNMRS 600 spectrometer with tetramethylsilane as internal standard. Standard pulse sequence was used for homo- and heteronuclear correlation experiments. ^{1}H NMR spectra were measured at 599 MHz, whereas ^{13}C NMR spectra were run at 150 MHz multiplicities of ^{13}C NMR resonances determined by DEPT experiments. All NMR experiments were performed at constant temperature (27°C) using software supplied by the manufacturer, employing deuteriochloroform, deuteriomethanol, and deuteriosulfoxide as solvent on the basis of solubility of the sample and literature data (Figure 7.3).

7.11 Conclusion

Around 23,000 bioactive secondary metabolites produced by microorganisms have been reported, and over 10,000 of these compounds are produced by actinobacteria. Their metabolic potential offers a strong area for research. For novel drug delivery, scientists still exploit the chemical and biological diversity from diverse actinobacterial group to

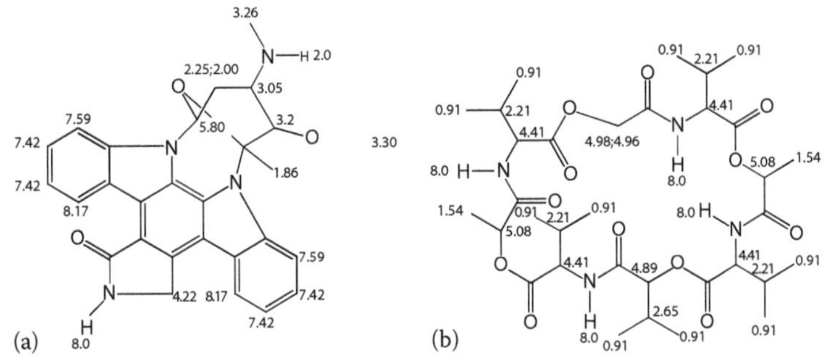

Figure 7.3 Structure of antimicrobial compounds: (a) staurosporine and (b) octa-valinomycin.

maximize the possibility of successful discovery of novel strain. However, further characterization of actinobacteria and their product for utilization in plant biotechnology, environmental biotechnology, urban waste management, and some other applications is yet to be done. The potential numbers of metabolites from actinobacteria may be discovered in the future.

References

Abonyi, D.O., Adikwu, M.U., Esimone, C.O., and Ibezim, E.C. 2009. Plants as sources of antiviral agents. *Afr. J. Biotechnol.* 8(17):3989–3994.

Aghighi, S.G.H., Bonjar, S., and Saadoun, I. 2005. First report of antifungal properties of a new strain of *Streptomyces plicatus* (strain 101) against four Iranian phytopathogenic isolate of *Verticillium dahliae*, a new horizon in biocontrol agents. *Biotechnology* 3(1):90–97.

Al-Zarban, S.S., Al-Musallam, A.A., Abbas, I.H., and Fasasi, Y.A. 2002. Noteworthy salt-loving actinomycetes from Kuwait. *Kuwait J. Sci. Eng.* 29(1):99–109.

Amorso, M.J., Castro, G.R., Carlino, F.J., Romero, N.C., Hill, R.T., and Oliver, G. 1998. Screening of heavy metal-tolerant actinomycetes isolated from the Salí River. *J. Gen. Appl. Microbiol.* 44:129–132.

Arasu, M.V., Duraipandiyan, V., Agastian, P., and Ignacimuthu, S. 2009. *In vitro* antimicrobial activity of *Streptomyces* spp. ERI-3 isolated from Western Ghats rock soil (India). *J. Mycol. Méd.* 19:22–28.

Atalan, E., Manfio, G.P., Ward, A.C., Kroppenstedt, R.M., and Goodfellow, M. 2000. Biosystematic studies on novel *Streptomycetes* from soil. *Anton van Leeuwhoek* 77: 337–353.

Balducci, E., Opalla, C., and Grein, A. 1954. The classification of the *Actinomyces* (*Streptomyces*) species. *Arch. Microbiol.* 20.047–357.

Beijerinck, M.W. 1900. *Streptothrix chromogena. Proc. Natl Acad. USA* 100(12):4555–4561.

Berdy, J. 1982. *Handbook of Antibiotics Compounds, Vol. IX. Antibiotics from Higher Forms of Life: Lichens, Algae and Animal Organisms*. Boca Raton, FL: CRC Press, pp. 39–227.

Berdy, J. 2005. Bioactive microbial metabolites. *J. Antibiot.* 58(1):1–26.

Berger, J., Jampolsky, L.M., and Goldberg, M.W. 1949. Borrelidin, a new antibiotic with antiborrelia activity and penicillin-enhancement properties. *Arch. Biochem.* 22:476–478.

Bergey, D.H. and Holt, J.G. 1994. *Bergey's Manual of Determinative Bacteriology*, 9th edn. Philadelphia, PA: Lippincott Williams and Wilkins.

Berkada, B. 1978. Preliminary report on warfarin for the treatment of herpes simplex. *J. Irish Coll. Phys. Surg.* 22:56.

Canetta, R.M. and Eisenhauer Rozencsweig, E.M. 1994. *Methods for Administration of Taxol for Cancer Treatment with Reduced Toxicity*. New York: Bristol-Myers Squibb Co., Application, p. 38.

Cappuccino, J.G. and Sherman, N. 2005. *Microbiology: A Laboratory Manual*, 7th edn. San Francisco, CA: The Benjamin Cummings.

Cardarelli, M., Serino, G., Campanella, L., Ercole, P., Nardone, F.D.C., Alesiani, O., and Rossiello, F. 1997. Antimitotic effects of usnic acid on different biological systems. *Cell. Mol. Life Sci.* 53:667–672.

Chen, Y.G., Wang, Y.X., Zhang, Y.Q., Tang, S.K., Liu, Z.X., Xiao, H.D., and Li, W.J. 2009. *Nocardiopsis litoralis* sp. nov., a halophilic marine actinomycete isolated from a sea anemone. *J. Syst. Evol. Microbiol.* 59:2708–2713.

Choi, J.D. and Park, U.Y. 1993. Identification of the marine microorganisms producing bioactives.1. Isolation and cultural conditions of the marine actinomycetes no. 101 producing antimicrobial compounds. *Bull. Kor. Fish. Soc.* 26(4):305–311.

Cholarajan, A. 2014. Diversity, characterization and antimicrobial compounds from actinobacteria in terrestrial soil of Thanjavur District, Tamilnadu, India. PhD thesis, Bharathidasan University, Tiruchirappalli, Tamil Nadu, India.

Chun, J., Bae, K.S., Moon, E.Y., Jung, S.O., Lee, H.K., and Kim, S.J. 2000. *Nocardiopsis kunsanensis* sp. nov., a moderately halophilic actinomycete isolated from a saltern. *Int. J. Syst. Evol. Microbiol.* 50:1909–1913.

Das, S., Lyla, P.S., and Khan, S.A. 2008. Characterization and identification of marine actinomycetes existing systems, complexities and future directions. *Natl. Acad. Sci. Lett.* 31:149–160.

De Rijke, E., Out, P., Niessen, W.M.A., Ariese, F., Gooijer, C., and Brinkman, U.A. 2006. Analytical separation and detection methods for flavonoids. *J. Chromatograp.* 1112(1–2):31–63.

Dhanasekaran, D., Selvamani, S., Panneerselvam, A., and Thajuddin, N. 2009. Isolation and characterization of actinomycetes in Vellar Estuary, Annagkoil, Tamil Nadu. *Afr. J. Biotechnol.* 8(17):4159–4162.

Dhanasekaran, D., Thajuddin, N., and Panneerselvam, A. 2008. Distribution and ecobiology of antagonistic *Streptomycetes* from agriculture and coastal soil in Tamilnadu, India. *J. Cul. Coll.* 6:10–20.

Dhanasekaran, D., Thajuddin, N., and Panneerselvam, A. 2011. Applications of actinobacterial fungicides in agriculture and medicine. In *Fungicides for Plant and Animal Diseases*, Dhanasekaran, D., Thajuddin, N., and Panneerselvam, A., eds. Rijeka, Croatia: InTech, pp. 29–54.

Dorlands Medical Dictionary: Antibacterial. Archived from the original on 2010–11–17. Accessed October 29, 2010.

Douglas, A., Skoog, F., Holler, J., Stanley, R. Crouch. 2007. *Principles of Instrumental Analysis*, 6th edn., Belmont, CA: Thomson Brooks/Cole, pp. 169–173.

Egorov, N.S. 1985. *Antibiotics: A Scientific Approach*. Moscow, Russia: Mir Publishers.

Elvin-Lewis, M. and Lewis, W.H. 1995. New concepts and medical and dental ethnobotany. In *Ethnobotany Evolution of a Discipline*, Schultes, R. and Von Reis, S., eds. Portland, OR: Discords Press, pp. 303–310.

Eriko, T. 2002. g-Butyrolactones: *Streptomyces* signalling molecules regulating antibiotic production and differentiation. *Opin. Microbiol.* 9:287–294.

Erturk, O. 2006. Antibacterial and antifungal activity of ethanolic extracts from eleven spice plants. *Biologia. (Bratisl)* 61(3):275–278.

Esimone, C.O. 1997. Antimicrobial properties of extracts from the lichen *Ramalina farinacea*. Masters Degree Dissertation, Department of Pharmaceutics, University of Nigeria, Nsukka, Nigeria.

Fernando, P. 2006. The history delivery of antibiotics from microbial nature products. *Biochem Pharmacol.* 71:981–990.

Food and Drug Administration U.S. Approval statistics of oncology drugs. 2008. http://www.accessdata.fda.gov/scripts/cder/onctools/statistics.cfm. Accessed November 5, 2002.

Gandhi, N., Sawant, S., and Joshi, J. 1976. Studies on the lipozyme-catalyzed synthesis of butyl laurate. *Biotechnol. Bioeng.* 46:1–12.

Gandhimathi, R., Seghal Kiran, G., Hema, T.A., Selvin, J., Rajeetha Raviji, T., and Shanmughapriya, S. 2009. Production and characterization of lipopeptide biosurfactant by a sponge-associated marine actinomycetes *Nocardiopsis alba* MSA10. *Bioprocess Biosyst. Eng.* 32:825–835.

Gayathri, V., Madhanraj, P., and Panneerselvam, A. 2011. Diversity, antibacterial activity and molecular characterization of actinomycetes isolated from saltpan region of Kodiakarai, Nagapattinam district. *Asian J. Pharm. Tech.* 1(3):79–81.

Georges, M. and Pandelai, K.M. 1949. Investigations on plant antibiotics. IV. Further search for antibiotic substances in Indian medicinal plants. *Ind. J. Med. Res.* 37:169–181.

Grein, A. and Meyers, S.P. 1958. Growth characteristics and antibiotic production of actinomycetes isolated from littoral sediments and materials suspended in seawater. *J. Bacteriol.* 76:457–463.

Grove, D.C. and Randall, W.A. 1955. *Assay Methods of Antibiotics: A Laboratory Manual* [Antibiotics Monographs, 02]. New York: Medical Encyclopedia, Inc., p. 80.

Hayakawa, M. and Nonomura, H. 1987. Humic acid-vitamin agar, a new medium for the selective isolation of soil actinomycetes. *J. Ferment. Technol.* 65:501–509.

Holler, F.J., Skoog, D.A., and Crouch, S.R. 2007. *Principles of Instrumental Analysis*, 6th edn. Belmont, CA: Thomson Brooks/Cole, pp. 169–173.

Holt, J.G. 1989. Streptomycetes and related genera. In *Bergey's Manual of Systematic Bacteriology*, Williams, S.T. and Sharpe, M.E., eds. Baltimore, MD: Williams and Wilkins, p. 747.

Hsu, S.C. and Lockwood, J.L. 1975. Powdered chitin agar as a selective medium for enumeration of actinomycetes in water and soil. *Appl. Microbiol.* 29(3):422–426.

Jensen, P.R., Gontang, E., Mafnas, C., Mincer, T.J., and Fenical, W. 2005. Culturable marine actinomycetes diversity from tropical Pacific Ocean sediments. *Environ. Microbiol.* 7(7):1039–1048.

Jones, K.L. 1949. Fresh isolates of actinomycetes in which the presence of sporogenous aerial mycelia is a fluctuating characteristic. *J. Bacteriol.* 57:141–145.

Kathiresan, K., Balagurunathan, R., and Masilamani Selvam, M. 2005. Fungicidal activity of marine actinomycetes against phytopathogenic fungi. *Ind. J. Biotechnol.* 4:271–276.

Kelecom, A. 2002. Secondary metabolites from marine microorganisms. *Ann. Braz. Acad. Sci.* 74(1):151–170.

Kelly, K.L. and Judd, D.B. 1976. *National Bureau of Standards*, Spec. Publ. 440, NBS/ISCC Centroids. U.S. Government Printing Office, Washington, DC, pp. 189.

Kinchington, D., Kangro, H., and Jeffries, K.J. 1995. Design and testing of antiviral compounds. In *Medical Virology: A Practical Approach*, Desselberger, U., ed. New York: Oxford University Press, pp. 147–171.

King, A. and Brown, D.F.J. 2001. Quality assurance of antimicrobial susceptibility testing by disc diffusion. *J. Antimicrob. Chemother.* 48:71–76.

Kosmachev, A.E. 1954. Thermophilic actinomycetes and their antagonistic properties. PhD thesis, Institute of Microbiology, Moscow, Russia (in Russian).

Krasilnikov, N.A. 1960. Taxonomic principles of the actinomycetes. *J. Bacteriol.* 79:65–74.

Kucera, L.S. and Hermann, E.E. 1965. *Annual Report in Medicinal Chemistry*. London, U.K.: Academic Press, pp. 129.

Kuster, E. and Williams, S.T. 1964. Production of hydrogen sulphide by *Streptomyces* and methods for its detection. *Appl. Microbiol.* 12:46–52.

Lam, K.S. 2006. Discovery of novel metabolites from marine actinomycetes. *Curr. Opin. Microbiol.* 9:245–251.

Lechavalier, H.A., Lechavalier, M.P., and Gerber, N.N. 1971. Chemical composition as a criterion in the classification of actinomycetes. *Adv. Appl. Microbiol.* 14:47–72.

Likar, M. and Japelj, M. 1977. Animal corneas as tools for the testing of antiviral compounds. *Ann. N.Y. Acad. Sci.* 284:182–189.

Lindenbein, W. 1952. Uber einige chemisch interressante Actinomycetenstamme und ihre Klassifizierung. *Arch. Mikrobiol.* 17:361–383.

Lingappa, Y. and Lockwood, J.L. 1961. A chitin medium for isolation, growth and maintenance of actinomycetes. *Nature* 189:158–159.

Lingappa, Y. and Lockwood, J.L. 1962. Chitin media for selective isolation and culture of actinomycetes. *Phytopathology* 52:317–323.

Magarvey, N.A., Keller, J.M., Bernan, V., Dworkin, M., and Sherman, D.H. 2004. Isolation and characterization of novel marine-derived actinomycete taxa rich in bioactive metabolite. *Appl. Environ. Microbiol.* 70(12):7520–7529.

Mincer, T.J., Jensen, P.R., Kauffman, C.A., and Fenical, W. 2002. Widespread and persistent populations of a major new marine actinomycete taxon in ocean sediments. *Appl. Environ. Microbiol.* 68(10):5005–5011.

Muiru, W.M., Mutitu, E.W., and Mukunya, D.M. 2012. Characterization of antibiotic metabolites from actinomycete isolates. *Afr. Crop Sci. Conf. Proc.* 8:2103–2107.

Nonomura, H. 1974. Key for classification and identification of 458 species of the *Streptomyces* included in ISP. *J. Ferment. Technol.* 52:78–92.

Ogunmwonyi, I.H., Mazomba, N., Mabinya, L., Ngwenya, E., Green, E., Akinpelu, D.A., and Okoh, A.I. 2010. Studies on the culturable marine actinomycetes isolated from the Nahoon beach in the Eastern Cape Province of South Africa. *Afr. J. Microbiol. Res.* 4(21):2223–2230.

Olano, C., Méndez, C., and Salas, J.A. 2009. Antitumor compounds from marine actinomycetes. *Mar. Drugs.* 7:210–248.

Parks, L.W. and Casey, W.M. 1996. Fungal sterols. In *Lipids of Pathogenic Fungi*, Prasad, R. and Ghannoum, M., ed. Boca Raton, FL: CRC Press, pp. 63–82.

Porter, J.N., Wilhelm, J.J., and Tresner, H.D. 1960. Method for the preferential isolation of actinomycetes from soils. *Appl. Microbiol.* 8:174–178.

Prescott, L.M., Harley, J.P., and Klein, D.A. 2002. *Microbiology*, 5th edn. New Delhi, India: Tata McGraw-Hill Company.

Rajauria, G. and Abu-Ghannam, N. 2013. Isolation and partial characterization of bioactive fucoxanthin from *Himanthalia elongata* brown seaweed: A TLC-based approach. *Int. J. Anal. Chem.* Article ID 802573: 1–6. doi: 10.1155/2013/802573.

Ravikumar, S., Gnanadesigan, M., Thajuddin, N., Deepan Chakkaravarthi, V.S., and Beula Banerjee, M. 2010. Anticancer property of sponge associated actinomycetes along Palk Strait. *J. Pharm. Res.* 3(10):2415–2417.

Reddy, N.G., Ramakrishna, D.P.N., and Raja Gopal, S.V. 2011. A morphological, physiological and biochemical studies of marine *Streptomyces rochei* (MTCC 10109) showing antagonistic activity against selective human pathogenic microorganisms. *Asian J. Biol. Sci.* 4(1):1–14.

Rehacek, Z. 1959. Isolation of actinomycetes and determination of the number of their spores in soil. *Microbiology* 28:220–225.

Remya, M. and Vijayakumar, R. 2008. Isolation and characterization of marine antagonistic actinomycetes from West Coast of India. *Facta Universitatis: Med. Biol.* 15(1):13–19.

Rojas, A., Hernandez, L., Pereda-Miranda, R., and Mata, R. 1992. Screening for antimicrobial activity of crude drug extracts and pure natural products from Mexican medicinal plants. *J. Ethnopharmacol.* 35:275–283.

Rowe, J.G., Saenz, M.T., and Garcia, M.D. 1989. Antibacterial activity of some lichens from Southern Spain. *Ann. Pharm. Francaises.* 47:89–94.

Sambamurthy, K. and Ellaiah, P. 1974. A new Streptomycin producing Neomycin (B and C) complex, *S. marinensis* (part-1). *Hind. Antibiot. Bull.* 17:24–28.

Sambrook, J., Fritsch, E.F., and Maniatis, T. 1989. *Molecular Cloning: A Laboratory Manual*, 2nd edn. Cold Spring Harbor, NY: Cold Spring Harbor Laboratory Press.

Savoia, D. 2012. Plant-derived antimicrobial compounds: Alternatives to antibiotics. *Future Microbiol.* 7(8):979–990.

Seong, C.N., Choi, J.H., and Baik, K.S. 2001. An improved selective isolation of rare actinomycetes from forest soil. *J. Microbiol.* 39(1):17–23.

Sethi, P.D. 2001. *High Performance Liquid Chromatography: Quantitative Analysis of Pharmaceutical Formulations*, 1st edn. New Delhi, India: CBS Publishers and Distributors.

Sharma, M. 2014. Actinomycetes: Source, identification and their applications. *Int. J. Curr. Microbiol. App. Sci.* 3(2):801–832.

Shirling, E.B. and Gottlieb, D. 1966. Methods for characterization for *Streptomyces* species. *Int. J. Syst. Bacteriol.* 16:313–340.

Silva, O., Duarte, A., Cabrita, J., Pimentel, M., Diniz, A., and Gomes, E. 1996. Antimicrobial activity of Guinea-Bissau traditional remedies. *J. Ethnopharmacol.* 59:55–59.

Sloan, B.J., Kielty, J.K., and Miller, F.A. 1977. Effect of a novel adenosine deaminase inhibitor upon the antiviral activity *in vitro* and *in vivo* of vidarabine for DNA virus replication. *Ann. N. Y. Acad. Sci.* 284:60–80.

Sujatha, P., Bapi Raju, K.V.V.S.N., and Ramana, T. 2005. Studies on a new marine Streptomycete BT-408 producing polyketide antibiotic SBR-22 effective against methicillin resistant *Staphylococcus aureus*. *Microbiol. Res.* 160(2):119–126.

Swallow, D.L. 1977. Progress in medicinal chemistry, Ellis, G.P. and West, G.B. eds. London, U.K.: Butterworth Group, p. 120.

Swami, M.B., Sastry, M.K., Nigudkar, A.G., and Nanda, R.K. 1983. Correlation of HPLC retention time with structure and functional group of macrolide polyene. *Hind. Antibiot. Bull.* 25:81–99.

Takahashi, S., Miyaoka, H., Tanaka, K., Enokita, R., and Okazaki, T. 1993. Milbemycins alpha11, alpha12, alpha13, alpha14 and alpha15: A new family of milbemycins from *Streptomyces hygroscopicus* sub sp. *aureolacrimosus. J. Antibiot.* 46(9):1364–1371.

Tsao, P.H., Leben, C., and Keitt, G.W. 1960. An enrichment method for isolating actinomycetes that produce diffusible antifungal antibiotics. *Phytopathology* 50:88–89.

Van Wyk, B.E. and Wink, M. 2004. *Medicinal Plants of the World: An Illustrated Scientific Guide to Important Medicinal Plants and Their Uses.* Portland, OR: Timber Press.

Vijayakumar, R. 2006. Studies on actinomycetes from Palk Strait region of Tamilnadu coast with reference to antibiotic production. PhD thesis, Bharathidasan University, Tiruchirappally, Tamilnadu, India, p. 66.

Vijayakumar, R., Gopika, G., Dhanasekaran, D., and Saravanamuthu, R., 2012a. Marine actinomycetes: A potential biocontrol agent against *Colletotrichum vulgatum, Thielaviopsis paradoxa, Trichoderma virida* and *Fusarium semitectum. Arch. Phytopathol. Pl. Prot.* 45(9):1010–1025.

Vijayakumar, R., Muthukumar, C., Thajuddin, N., Panneerselvam, A., and Saravanamuthu, R. 2007. Studies on the diversity of actinomycetes in the Palk Strait region of Bay of Bengal, India. *Actinomycetologica* 21(2):59–65.

Vijayakumar, R., Panneer Selvam, K., Muthukumar, C., Thajuddin, N., Panneerselvam, A., and Saravanamuthu, R. 2012b. Optimization of antimicrobial production by the marine actinomycetes *Streptomyces afghaniensis* VPTS3–1 isolated from Palk Strait, East Coast of India. *Ind. J. Microbiol.* 52(2):230–239.

Vijayakumar, R., Panneer Selvam, K., Muthukumar, C., Thajuddin, N., Panneerselvam, A., and Saravanamuthu, R. 2012c. Antimicrobial potentiality of a halophilic strain of *Streptomyces* sp. VPTSA18 isolated from the saltpan environment of Vedaranyam, India. *Ann. Microbiol.* 62(3):1039–1047.

Vijayakumar, R., Panneerselvam, A., and Thajuddin, N. 2013. Marine actinobacteria—A potential source for antifungal compounds. In *Marine Pharmacognosy: Trends and Applications,* Kim, S.K., ed. Busan, South Korea: CRC Press, pp. 231–252.

Von Nussbaum, F., Brands, M., Hinzen, B., Weigand, S., and Häbich, D. 2006. Medicinal chemistry of antibacterial natural products—Exodus or revival. *Angew. Chem. Int. Ed.* 45(31):5072–5129.

Wakisaka, Y., Kawamura, Y., Yasuda, Y. Koizumi, K., and Nishimoto, Y. 1982. A selective isolation procedure for *Micromonospora. J. Antibiot.* 35:822–836.

Waksman, S.A. 1943. Production and activity of streptothricin. *J. Bacteriol.* 45:299–310.

Waksman, S.A. 1947. What is an antibiotic or an antibiotic substance? *Mycologia* 39(5):565–569. doi:10.2307/3755196. JSTOR 3755196. PMID 20264541.

Waksman, S.A. 1961. *The Actinomycetes: Classification, Identification and Descriptions of Genera and Species,* Vol. II. Baltimore, MD: Williams and Wilkins Co., p. 363.

Waksman, S.A. and Henrici, A.T. 1943. The nomenclature and classification of actinomycetes. *J. Bacteriol.* 10.007 041.

Waksman, S.A., Horning, E.S., Welsch, M., and Woodruff, H.G. 1941. The distribution of antagonistic actinomycetes in nature. *Soil. Sci.* 54:281–296.

Waksman, S.A. and Schatz, A. 1946. Production of antibiotic substances by actinomycetes. *Ann. N. Y. Acad. Sci.* 48:73–86.

Yang, R., Zhang, L.P., Guo, L.G., Shi, N., Lu, Z., and Zhang, X. 2008. *Nocardiopsis valliformis* sp. nov., an alkaliphilic actinomycete isolated from alkali lake soil in China. *Int. J. Syst. Evol. Microbiol.* 58:1542–1546.

You, J., Xue, X., Cao, L., Lu, X., Wang, J., Zhang, L., and Zhou, S. 2007. Inhibition of *Vibrio* biofilm formation by a marine actinomycete strain A66. *Appl. Microbiol. Biotechnol.* 76(5):1137–1144.

Zabriskie, T.M. and Jackson, M.D. 2000. Lysine biosynthesis and metabolism in fungi. *Nat. Prod. Rep.* 17(1):85–97.

Zin, N.M., Sarmin, N.I., Ghadin, N., Basri, D.F., Sidik, N.M., Hess, W.M., and Strobel, G.A. 2007. Bioactive endophytic streptomycetes from the Malay Peninsula. *FEMS Microbiol. Lett.* 274:83–88.

chapter eight

Novel antimicrobial and anticancer drugs from bacteria

Ranjith N. Kumavath

Contents

8.1 Introduction

8.1.1 Production of secondary metabolites from aromatic amino acids

Microorganisms such as bacteria and fungi are promising sources for structurally diverse and potent bioactive compounds (Laatsch, 2006; Lebar et al., 2007). Actinomycetes are mostly responsible for the production of half of the discovered secondary metabolites (Bull, 2004; Berdy, 2005), such as antibiotics (Strohl, 2004), antitumor agents (Olano et al., 2009), immunosuppressive agents (Mann, 2001), and enzymes (Pecznska-Czoch and Mordarski, 1988; Oldfield et al., 1998). The studies on metabolism of aromatic amino acids in microorganisms show that they evolved secondary metabolic pathways with the capacity to produce compounds displaying an effective array of pharmacological applications including pigments, toxins, enzyme inhibitors, pesticides, herbicides, antiparasitics, myco-toxins, antitumor agents, antibiotics, cytotoxicity activities, and growth promoters of ani-mals and plants. Structurally diverse classes of secondary metabolites were produced by the marine *Bacillus* species, such as lipopeptides, polypeptides, macrolactones, fatty acids, polyketides, lipoamides, and isocoumarins (Hamdache et al., 2011). These compounds show a wide range of biological activities such as antimicrobial, anticancer, antialgal, and antiperonosporomycetal (Brauzzi et al., 2011).

These bioactive compounds had a wide range of applications such as chemothera-peutic agents for the treatment of human and animal diseases (Schwartsmann et al., 2001; Jha and Zi-rong, 2004). There is a continuing demand for novel bioactive compounds for the treatment of drug-resistant human and animal pathogens (Nathan, 2004; von Nussbaum et al., 2006) and the management of destructive pathogens of crops, which are insensitive to existing chemical pesticides (Islam, 2005; Islam et al., 2011; Islam and Hossain, 2013). Another application for the bioactive compounds is quorum sensing (QS), which helps the bacteria for cell-to-cell communication using intermediary sub-stances, which are excreted from bacterial cells into the environment. The metabolite concentration reaches the threshold level in the bacteria; results in certain types of gene expression resulting in bioluminescence, biofilm formation, swarming motility, anti-biotic biosynthesis, and virulence factor production (Falcao et al., 2004; Antunes et al., 2010); and is triggered by the metabolite (Kazuhiro et al., 2013). The strong impact of classical natural compounds with advanced microbial genetics and bioinformatics will help to overcome supply and sustainability issues from the past and to promote bioac-tive substances from microbial species.

8.2   *Indole terpenoid ethers and esters*

The novel indole terpenoids, namely, sphestrin (Sunayana et al., 2005a) and rhodestrin (Sunayana et al., 2005b), were isolated from the culture supernatants of *Rhodobacter sphaeroides* OU5 grown in the presence of 2-aminobenzoate. Indole terpenoid esters/ethers are recent discoveries of biomolecules produced by a purple nonsulfur bacterium *Rhodobacter sphaeroides* OU5 having phytohormonal activity. They were produced in the presence of a processor like 2-aminobenzoate (Nanda et al., 2000) or aniline (Vijay et al., 2006). There are a few reports of indole ester biosynthesis in plant systems and micro-organisms. 4-Chloroindole-3-acetic acid and its esters were chemically synthesized from 2-chloro-6-nitrotoluene as the base material (Katayama, 2000). Esters of indole-3-acetic acid (IAA) were extracted and purified from the liquid endosperm of immature fruits of various species of the horse chestrint (*Aesculus parviflora, Aesculus baumannii, Aesculus pavia rubra*, and *Aesculus pavia humulis*). The liquid endosperm contained at least 12 chro-matographically distinct esters. One of these compounds were purified and characterized as an ester of IAA and myoinositol (Domagaski et al., 1987). Indole-3-acetyl-myo-inositol esters have been demonstrated as an endogenous component of etiolated *Zea mays* shoot tissue. The amount of indole-3-acetyl-myo-inositol esters in the shoots was determined to be 74 nmoles/kg fresh weight (Chisnell, 1984).

Indoles are extensively produced by the chemical industry for a variety of appli-cations including pharmaceuticals, pesticides, and dyes, and they are widely used as analgesics (Magniez et al., 1995), anti-inflammatory agents (Verma et al., 1994), antihyper-tensive (Frishman, 1983), anti-HIV compound (Britcher et al., 1995), and phytohormones (Elsorra et al., 2003). Many indole esters were found as COX-2 selective enzyme inhibitors (Olgen and Nebioglu, 2002; Olgen et al., 2007); they were also found to have anti–lipid peroxidation (LP) activity and antisuperoxide formation (SOD) (Olgen and Coban, 2003, Olgen et al., 2007). Indole esters were also found to have more phytohormonal activity than their corresponding acids in the auxin bioassay (Katayama, 2000).

8.2.1   Rhodethrin

Rhodethrin, which is a terpenoid metabolite, is isolated from the culture supernatants of *Rhodobacter sphaeroides* OU5 when grown on L-tryptophan as a sole source of nitrogen under photoheterotrophic conditions. The International Union of Pure and Applied Chemistry (IUPAC) name of the compound is 3-hydroxy-6-(1H-indol-3-yloxy)-4-methylhexanoic acid. This compound better acts as a phytohormone, along with its cytotoxic activity against SUP-T1 lymphoma and Colo-125 cancer cell lines at 10 nM (Ranjith et al., 2007).

8.2.2   Rubrivivaxin

The *Rubrivivax benzoatilyticus* JA2 bacterial species produces a phenol terpenoid called rubrivivaxin. The IUPAC name of the compound is 3,4-dihy-droxy-5-carboxy-3-methylpentyl ester. The significance of cyclooxygenase-I inhibitory, antimicrobial, cytotoxic activities against U937 (human leukemic monocyte lymphoma), and Jurkat (T lymphocyte) cell lines is conferred by rubrivivaxin (Ranjith et al., 2011).

8.2.3   Rhodestrin

Rhodestrin is a phytohormone isolated from the metabolite of anthranilate photobiotransformation by *Rhodobacter sphaeroides* OU5. The IUPAC name of the compound is 24-hydroxy-2,6,10,14,19 pentamethyl tetrecosa-2,4,6,8,10,12,14,16,18 nonenyl-2(hydroxymethyl)-1H-indole-3-carboxylate. This molecule shows antitumor and antimicrobial activities (Sunayana et al., 2005).

m/z 595 [MH]+

8.2.4   2.4.3,6-Disubstituted indoles A, B, and C

The rich sources of compounds for the development of pharmaceutical agents were found in the marine microorganisms (Fenical, 1993). One of them is marine *Streptomyces* sp. strain BL-49–58–005 that was identified for the production of the 3,6-disubstituted indole compounds, which are cytotoxic in nature. These compounds are in three forms based upon the functional group substituted at the first position of the indole ring of the pyrimidine structure (Sanchez Lopez et al., 2003).

3,6-Disubstituted indole A   **3,6-Disubstituted indole B**   **3,6-Disubstituted indole C**

8.3   Indole terpenoid ethers and esters

8.3.1   Indole-3-acetic acid

Light/horseradish peroxidase (HRP) activation has been suggested as a new photodynamic cancer therapy by forming free radicals (Dong et al., 2006) such as indolyl and peroxyl radicals (Folkes et al., 2002), which can cause LP. The combination of IAA and HRP shows cytotoxic to mammalian cells including G361 human melanoma cells (Domagalski et al., 1987; Dong et al., 2006) and human pancreatic cancer BXPC-3 cells (Chen et al., 2005).

8.3.2   Indigo

The blue dye indigo has been known since prehistoric times and is still one of the most economical important textile dyes; the first report of microbial indigo production was recorded in 1928 (Gray, 1928). It is biosynthesized in bacteria via the oxidation of indole by a naphthalene dioxygenase and subsequent oxidation and dimerization (Ensley et al., 1983). The desire to achieve a competitive alternative to the chemical production of indigo rejuvenated interest in microbial indigo production (Murdock et al., 1993) since many microorganisms expressing both monooxygenase (Allen et al., 1997) and dioxygenase (Murdock et al., 1993) during the growth of aromatic hydrocarbons have been shown to transform indole to indigo, and the production of these has focused on the naphthalene dioxygenase from *Pseudomonas putida* PpG7 expressed in *Escherichia coli*. Some of the genes of indigo biosynthetic pathway have been cloned and used to construct "engineering bacteria" with this kind of bacteria; more efficient fermentation systems for indigo production have been exploited (Han et al., 2008).

8.3.3   Violacein

Chromobacterium violaceum was first reported as an isolate from wet rice paste; one of the characteristics of this microorganism is the ability to produce a purple (deep violet) pigment known as violacein under aerobic conditions. The biological role of violacein in *Chromobacterium violaceum*, as well as its biosynthesis pathway, was well reported (DeMoss and Evans, 1959), in addition to the role of tryptophan and other indole derivatives (DeMoss and Evans, 1960; Hoshino et al., 1987; Hoshino and Ogasawara, 1990; Duran et al., 1994; Momen and Hoshino, 2000). Tryptophan appears to be the only precursor molecule in violacein biosynthesis; its production is apparently essential for pigment production in *Chromobacterium violaceum* (Vasconcelos et al., 2003; Regina and Creczynski-Pasa, 2004). The IUPAC name and molecular mass of violacein are (3-[1,2-dihydro-5-(5-hydroxy-1H-indol-3-yl)-2-oxo-3H-pyrrol-3-ylidene]-1,3-dihydro-2H-indol-2-one) and [m/z 343.34], respectively. Violacein has attracted interest owing to its important multiple biological activities and pharmacological potentials such as antibiotic, bactericide (Duran et al., 1983), trypanocide (Duran et al., 1994), antitumoral (Melo et al., 2003; Ferreira et al., 2004; Saraiva et al., 2004), antiviral (Kodach et al., 2006), and genotoxic (Andrighetti et al., 2003) properties, as well as its antioxidant efficiency against oxygen- and nitrogen-reactive species as a scavenger of hydroxyl, superoxide, and nitric oxide radicals (Konzen et al., 2006). In addition, it is capable of inducing apoptosis in cancer cell cultures (Duran and Menck, 2001) and effective against a panel of neoplastic cell lines, including leukemia lineage cancer diseases (Melo et al., 2003).

8.3.4   Indolmycin

Indolmycin is a secondary metabolite produced by *Streptomyces griseus* ATCC 1248 (formally *Streptomyces albus* BA 3972A), which was isolated from a sample of African soil. Indolmycin completely inhibits bacterial tryptophanyl-tRNA synthetase enzyme (Makoto et al., 2002), and it exhibits antimicrobial activity against gram-positive and gram-negative bacteria. Recently, researchers have shown that indolmycin is active against *Mycobacteria* and *Helicobacter pylori*, which is known as a major causative agent of chronic active gastritis.

8.4   Phenols and its derivatives

8.4.1   Alkyl esters of gallic acid

The chief source for obtaining gallic acid is through the hydrolysis of plant-based products like tannins (Inoue et al., 1995). Microbial production of gallic acid was also reported using tannic acid as substrate (Kar et al., 1999). The esters iso-amyl-(iAG), *n*-amyl-(nAG), iso-butyl-(iBG), *n*-butyl-(nBG), and isopropyl gallate (iPG) were chemically synthesized from gallic acid (Christian et al., 2007). Gallic acid (3,4,5-trihydroxybenzoic acid) is an industrially important phenol and finds its applications in various fields (Kar et al., 1999).

Alkyl esters of gallate are an important group of biogenic molecules reported from plants (Yang et al., 2003), bee propolis (Ahn et al., 2004), and yeasts like *Candida* (Stevenson et al., 2007). These molecules are of biotechnological significance since they are known to have antioxidant (Chen and Ho, 1997), anticancer (Samaha et al., 1997; Li et al., 2003), anti-HIV (Burke et al., 1995), and antifungal/microbial (Tawata et al., 1996) activities. Alkyl esters of gallic acid have antiviral, antibacterial, and antifungal properties (Fujita and Kubo, 2002) specifically against gram-positive bacteria (Kubo et al., 2004).

8.5   Polyketides

Polyketides are the products of repetitive condensations of the small carboxylic acid units similar to fatty acid synthesis. These are synthesized by the polyketide synthase, which is a mono- or bifunctional enzyme (Staunton and Weissman, 2001). These compounds are preferentially isolated from the marine *Streptomyces* sp. JP95 strains (Li and Piel, 2002). These compounds show functional and structural variabilities; these molecules may be of simple to complex structures.

8.5.1   Daryamides A, B, and C

Daryamides are the class of polyketides that act as antibiotic and antitumor agents, isolated from marine-derived actinobacteria (Ratnakar et al., 2006). These compounds are of different types based upon the complex structures designated: A = 1, B = 2, and C = 3.

8.6   Piericidin and derivatives

Piericidin are the group of molecules found in the marine actinomycete and *Streptomyces* sp. These molecules mostly act as anti-QS, antimicrobial (Kazuhiro et al., 2013), and antitumor agents. Among these compounds, piericidin C7 and C8 are mostly distributed in the marine microorganisms (Yoichi et al., 2007; Selvakumar, 2010).

8.6.1 Piericidins C_7 and C_8

The marine *Actinomycete* sp. produces these antitumor compounds against retinoblastoma, and these molecules act as antimicrobial agents for the screening against marine microorganisms. These molecules are selectively cytotoxic against rat glial cells (Yoichi et al., 2007).

Piericidin C_7 R = H
Piericidin C_8 R = CH_3

Piericidin A_1 R = H
Piericidin A_2 R = CH_3

8.7 Conclusion

Microbes are the large repositories of bioactive compounds with different structural and functional forms, namely, sugars, acids, esters, ethers, terpenoids, peptides, proteins, and nucleopeptides, which are meant for the immense applications in the treatment of the wide range of disease types, such as anticancer, anti-inflammatory, antimicrobial, antiviral, and antifungal. As long as the microorganisms are treated with the classical antibiotics, they are supposed to be developing the nature of resistance to the specific disease types. Now, it is time to prepare the new drugs and to make modifications in the structural characteristic of the target-based drug design for the specific disease. Hence, the utilization of these vast resources is poorly understood in microorganisms. However, the microbial screening provides an opportunity to manipulate new-generation drugs for efficient antimicrobial and anticancer approaches.

Acknowledgment

The authors would like to thank SERB, Govt. of India.

References

Ahn, M.R., Kumazawa, S., Hamasaka, T., Bang, K.S., and Nakayama, T. 2004. Antioxidant activity and constituents of propolis collected in various areas of Korea. *J Agric Food Chem.* 52: 7286–7292.

Allen, C.C.R., Boyd, D.R., Larkin, M.J., Reid, K.A., Sharma, N.D., and Wilson, K. 1997. Metabolism of naphthalene, 1-naphthol, indene, and indole by *Rhodococcus* sp. strain NCIMB12038. *Appl Environ Microbial.* 63: 151–155.

Andrighetti, F.C.R., Antonio, R.V., Creczynski, P.T.B., Barandi, C.R.M., and Simoes, C.M.O. 2003. The antiviral and cytotoxic activities of violacein using three different methods. *MemInst Oswaldo Cruz.* 98: 834–848.

Antunes, L.C., Ferreira, R.B., Buckner, M.M., and Finlay, B.B. 2010. Quorum sensing in bacterial virulence. *Microbiology.* 156(8): 2271–2282.

Baruzzi, F., Quintieri, L., Morea, M., and Caputo, L. 2011. Antimicrobial compounds produced by *Bacillus* spp. and applications in food. In: *Science Against Microbial Pathogens: Communicating Current Research and Technological Advances*, Vilas, A.M., Ed. Formatex, Badajoz, Spain, pp. 1102–1111.

Bassler, B.L. and Losick, R. 2006. Bacterially speaking. *Cell.* 125(2): 237–246.

Berdy, J. 2005. Bioactive microbial metabolites. *J. Antibiotics (Tokyo).* 58: 1–26.

Britcher, S.F., Lymna, W.C., Young, S.D., Grey, V.E., and Tran, L.D. 1995. 3-Substituted heterocyclic indoles as inhibitors of HIV reverse transcriptase. *Briton UK Pat Appl GB2.* 282: 808.

Bull, A.T. 2004. *Microbial Diversity and Bioprospecting.* ASM Press, Washington, DC.

Chen, H., Li, Y.L., Tu, S.S., Lei, N., Ling, Y., Xiao, Y.H., Jing, S.H., Li, P.S., Yu, L., and Lu, S.S. 2005. Apoptosis of pancreatic cancer BXPC-3 cells induced by indole-3-acetic acid in combination with horseradish peroxidase. *World J Gastroenterol.* 11(29): 4519–4523.

Chen, J.H. and Ho, C.T. 1997. Antioxidant activities of caffeic acid and its related hydroxycinnamic acid compounds. *J Agric Food Chem.* 45: 2374–2378.

Chisnell, J.R. 1984. Myo-inositol esters of indole-3-acetic acid are endogenous component of *Zea mays* L. shoot tissue. *Plant Physiol.* 74: 278–283.

Christian, F., Mario, P., Gianni, C., Sergio, M., Enrique, R., Jorge, M., Antonio, M., Juan, D.M., and Jorge, F. 2007. Comparative cytotoxicity of alkyl gallates on mouse tumor cell lines and isolated rat hepatocytes. *Comp Biochem Physiol A.* 146: 520–527.

DeMoss, R.D. and Evans, N.R. 1959. Physiological aspects of violacein biosynthesis in nonproliferating cells. *J Bacteriol.* 78: 583–586.

DeMoss, R.D. and Evans, N.R. 1960. Incorporation of C14-labeled substrates into violacein. *J Bacteriol.* 79: 729–735.

Domagalski, W., Schulze, A., and Bandurski, R.S. 1987. Isolation and characterization of esters of indole-3-acetic acid from the liquid endosperm of the horse chestnut *(Aesculus* species). *Plant Physiol.* 84: 1107–1113.

Dong, S.K., So, Y.K., Yun, M.J., Sang, E.J., Myo, K.K., Sun, B.K., Jung, I.N., and Kyoung, C.P. 2006. Light-activated indole-3-acetic acid induces apoptosis in G361 human melanoma cells. *Biol Pharm Bull.* 29(12): 2404–2409.

Duran, N., Antonio, R.V., Haun, M., and Pilli, R.A. 1994. Biosynthesis of a trypanocide by *Chromobacterium violaceum. World J Microbiol Biotechnol.* 10: 686–690.

Duran, N., Erazo, S., and Campos, V. 1983. Bacterial chemistry-II: Antimicrobial photoproduct from pigment of *Chromobacterium violaceum. An Acad Bras Cienc.* 55: 231–234.

Duran, N. and Menck, C.F. 2001. *Chromobacterium violaceum*: A review of pharmacological and industrial perspective. *Critical Rev Microbiol.* 27: 201–222.

Elsorra, E.I., Domingo, J.I., Manuel, T., and Rainer, B. 2003. Tryptophan dependent production of indole-3-acetic acid (IAA) affect level of plant growth promotion by *Bacillus amyloliquefaciens* FZB42. *Mol Plant Microbe Interact.* 20(6): 619–626.

Ensley, B.D., Ratzkin, B.J., Osslund, T.D., Simon, M.J., Wackett, L.P., and Gibson, D.T. 1983. Expression of naphthalene oxidation genes in *Escherichia coli* results in biosynthesis of indigo. *Science.* 222: 167–168.

Falcao, J.P., Sharp, F., and Perandio, S.V. 2004. Cell-to-cell signaling in intestinal pathogens. *CIIM.* 5(1): 9–17.

Fenical, W. 1993. Chemical studies of marine bacteria: Developing a new resource. *Chem. Rev.* 93:1673–1683.

Ferreira, C.V., Bos, C.L., Versteeg, H.H., Justo, G.Z., Duran, N., and Peppelenbosch, M.P. 2004. Molecular mechanism of violacein-mediated human leukemia cell death. *Blood.* 104: 1459–1464.

Fineran, P.C., Slater, H., Everson, L., Hughes, K., and Salmond, G.P. 2005. Biosynthesis of Tripyrrole and Beta-Lactam secondary metabolites in *Serratia*: Integration of quorum sensing with multiple new regulatory components in the control of prodigiosin and carbapenem antibiotic production. *Mol. Microbiol.* 56(6): 1495–1517.

Folkes, L.K., Greco, O., Dachs, G.U., Stratford, M.R., and Wardman, P. 2002. 5-Floroindole-3-acetic acid: A prodrug activated by a peroxidase with potential for use in target therapy. *Biochem Pharmacol.* 63: 265–272.

Frishman, W.H. 1983. Pindolol: A new β-adrenoceptor antagonist with partial activity. *N Engl J Med.* 308: 940–994.

Fujita, K. and Kubo, I. 2002. Antifungal activity of octyl gallate. *Int J Food Microbial.* 79: 193–201.

Gray, P.M.M. 1928. Indigo formation by aromatic hydrocarbon-degrading bacteria. *Proc Royal Soc London Ser B.* 102: 2263–2279.

Hamdache, A., Lamarti, A., Aleu, J., and Collado, I.G. 2011. Non-peptide metabolites from the genus *Bacillus. J Nat Products.* 74: 893–899.

Han, X., Wang, W., and Xiao, X. 2008. Microbial biosynthesis and biotransformation of indigo and indigo like pigments. *Chin J Biotech.* 24(6): 921–926.

Hoshino, T., Kondo, T., Uchiyama, T., and Ogasawara, N. 1987. Biosynthesis of violacein: A novel rearrangement in tryptophan metabolism with a 1,2-shift of the indole ring. *Agri Biol Chem.* 51: 965–970.

Hoshino, T. and Ogasawara, N. 1990. Biosynthesis of violacein: Evidence for the intermediary of 5-hydroxy-L-tryptophan and the structure of a new pigment, oxyviolacein, produced by the metabolism of 5-hydroxytryptophan. *Agri Biol Chem.* 64: 2339–2345.

Inoue, M., Suzuki, R., Sakaguchi, N., Li, Z., Takeda, T., Ogihara, Y., Jiang, B.Y., and Chen, Y. 1995. Selective induction of cell death in cancer cells by gallic acid. *Bio Pharma Bull.* 18: 1526–1530.

Islam, M.T. 2011. Potentials for biological control of plant disease by *Lysobacter* spp., with special reference to strain SB-K88. In: *Bacteria in Agrobiology: Plant Growth Responses*, Maheshwari, D.K., ed. Springer-Verlag, Berlin/Heidelberg, Germany, pp. 355–364.

Islam, M.T. and Hossain, M.M. 2013. Biological control of *Peronosporomycete* phytopathogens by antagonistic bacteria. In: *Bacteria in Agrobiology: Plant Disease Management*, Maheshwari, D.K., ed. Springer-Verlag, Berlin/Heidelberg, Germany, pp. 167–218.

Islam, M.T., von Tiedemann, A., and Laatsch, H. 2011. Protein kinase C is likely to be involved in zoosporogenesis and maintenance of flagellar motility in the peronosporomycete zoospores. *MPMI.* 24: 938–947.

Jha, R.K. and Zi-rong, X. 2004. Biomedical compounds from marine organisms. *Marine Drugs* 2: 123–146.

Kar, B., Banerjee, R., and Bhattacharya, B.C. 1999. Microbial production of gallic acid by modified solid-state fermentation. *J Ind Micro Biotechnol.* 23: 173–177.

Katayama, M. 2000. Synthesis and biological activities of 4-chloroindole-3-acetic acid and its esters. *Biosci Biotechnol Biochem.* 64: 808–815.

Kazuhiro, O., Atsushi, F., Tomoe, Y., Kanako, S., Ryo, I., YojiroAnzai, P., and Fumio, K. 2013. Novel quorum-sensing Inhibitors against *Chromobacterium violaceum* CV026, from *Streptomyces* sp. TOHO-Y209 and TOHO-O348. *Open J Med Chem.* 3: 93–99.

Kodach, L.L., Bos, C.L., Duran, N., Peppelenbosch, M.P., Ferreira, C.V., and Hardwich, J.C. 2006. Violacein synergistically increases 5-fluorouracil cytotoxicity, induces apoptosis and inhibits Akt-mediated signal transduction in human colorectal cancer cells. *Carcinogenesis.* 27: 508–516.

Konzen, M., De Marco, D., Clarissa, A.S. C., Tiago, O.V., Regina, V.A., and Tania, B.C.P. 2006. Antioxidant properties of violacein: Possible relation on its biological function. *Bioorg Med Chem.* 14(24): 8307–8313.

Kubo, I., Fujita, K., Nihei, K., and Nihei, A. 2004. Antibacterial activity of alkyl gallates against *Bacillus subtilis. J Agric Food Chem.* 52: 1072–1076.

Laatsch, H. 2006. Marine bacterial metabolites. In: *Frontiers in Marine Biotechnology*, Proksch, P. and Muller, W.E.G., eds. Horizon Bioscience, Norfolk, U.K., pp. 225–288.

Lebar, M.D., Heimbegner, J.L., and Baker, B.J. 2007. Cold-water marine natural products. *Nat Prod Rep.* 24: 774–797.

Li, A. and Piel, J. 2002. A gene cluster from a marine *Streptomyces* encoding the biosynthesis of the aromatic spiroketal polyketide Griseorhodin A. *Chem. Biol.* 9: 1017–1026.

Makoto, K., Ali, K., Arnold, D., Sakamoto, K., Yokoyama, S., and Soll, D. 2002. Indolmycin resistance of *Streptomyces coelicolor* A3 by induced expression of one of its two tryptophanyl-tRNA synthetases. *J Biol Chem.* 277(26): 23882–23887.

Mann, J. 2001. Natural products as immunosuppressive agents. *Nat Prod Rep.* 18: 417–430.

McLean, R.J.C., Whiteley, M., Stickler, D.J., and Fuqua, W.C. 1997. Evidence of autoinducer activity in naturally occurring biofilms. *FEMS Microbiol Lett.* 154(2): 259–263.

Nanda Devi, P., Sasikala, C., and Ramana, C.V. 2000. Light-dependent transformation of anthranilate to indole by *Rhodobacter sphaeroides* OU5. *J Ind Microbiol Biotechnol.* 24: 219–221.

Nathan, C. 2004. Antibiotics at the crossroads. *Nature.* 431: 899–902.

Nealson, K.H. 2006. Autoinduction of bacterial luciferase occurrence, mechanism and significance. *Arch Microbiol.* 112(1): 73–79.

Olano, C., Mendez, C., and Salas, J.A. 2009. Antitumour compounds from marine actinomycetes. *Marine Drugs.* 7: 210–248.

Oldfield, C., Wood, N.T., Gilbert, S.C., Murray, F.D., and Faure, F.R. 1998. Desulphurisation of benzothiophene and dibenzothiophene by actinomycete organisms belonging to the genus *Rhodococcus*, and related taxa. *Antonie Van Leeuwenhoek.* 74: 119–132.

Olgen, S. and Nebioglu, D. 2002. Synthesis and biological evolution of N-substituted indole esters as inhibitors of cyclo-oxygenase-2 (COX-2). *Farmaco.* 57: 677–683.

Olgen, S., Zuhal, K., Ada, A., and Coban, T. 2007. Synthesis and anti-oxidant properties of novel N-H and N-substituted propanamide derivatives. *Archv der Pharmazi.* 340(3): 140–146.

Pecznska-Czoch, W. and Mordarski, M. 1988. Actinomycete enzymes. In: *Actinomycetes in Biotechnology.* Academic Press, London, U.K., pp. 219–283.

Ranjith, N.K., Ramana, C.V., and Sasikala, C. 2007. Rhodethrin: A novel indole terpenoid ether produced by *Rhodobacter sphaeroides* has cytotoxic and phytohormonal activities. *Biotechnol Lett.* 29: 1399–1402, doi: 10.1007/s10529–007–9413–7.

Ranjith, N.K., Ramana, C.V., and Sasikala, C. 2011. Rubrivivaxin, a new cytotoxic and cyclooxygenase-I inhibitory metabolite from *Rubrivivax benzoatilyticus* JA2. *World J Microbiol Biotechnol.* 27: 11–16, doi: 10.1007/s11274–010–0420–9.

Ratnakar, N.A., Paul, R.J., Christopher, A.K., and William, F. 2006. Daryamides A–C, weakly cytotoxic polyketides from a marine-derived actinomycete of the genus *Streptomyces* strain CNQ-085. *J Nat Prod.* 69: 1756–1759.

Regina, V.A. and Creczynski-Pasa, T.B. 2004. Genetic analysis of violacein biosynthesis by *Chromobacterium violaceum. Genet. Mol. Res.* 3(1): 85–91.

Samaha, H.S., Kelloff, G.J., Steele, V., Rao, C.V., and Reddy, B.S. 1997. Modulation of apoptosis by sulindac, curcumin, phenylethyl 3-methylcaffeate and 6-phenylhexyl isothiocyanate: Apoptotic index as a biomarker in colon cancer chemoprevention and promotion. *Cancer Res.* 57: 1301–1305.

Sanchez Lopez, J.M., Martinez Insua, M., Perez Baz, J., Fernandez Puentes, J.L., and Canedo Hernandez, L.M. 2003. New cytotoxic indolic metabolites from a marine *Streptomyces. J Nat Products.* 66: 863–864.

Saraiva, V.S., Marshall, J.C., Cools-Lartigue, J., and Burnier Jr., M.N. 2004. Cytotoxic effects of violacein in human uveal melanoma cell lines. *Melanoma Res.* 14: 421–424.

Schwartsmann, G., da Rocha, A.B., Berlinck, R.G.S., and Jimeno, J. 2001. Marine organisms as a source of new anticancer agents. *Lancet Oncol.* 2: 221–225.

Selvakumar, D. 2010. Marine *Streptomyces* as a novel source of bioactive substances. *World J Microbiol Biotechnol.* 26: 2123–2139.

Slater, H., Crow, M., Everson, L., and Salmond, G.P. 2003. Phosphate availability regulates biosynthesis of two antibiotics, prodigiosin and carbapenem, in *Serratia* via both quorum-sensing-dependent and -independent path-ways. *Mol. Microbiol.* 47(2): 303–320.

Staunton, J. and Weissman, K.J. 2001. Polyketide biosynthesis: A millennium review. *Nat Prod Rep.* 18: 380–416.

Stevenson, D.V., Parkar, S.G., Zhang, J., Stanley, R.A., Jenson, D.J., and Cooney, J.M. 2007. Combinatorial enzymatic synthesis for functional testing of phenolic acid ester catalyzed by *Candida Antarctica* lipase B (Novozym 435®). *Enzyme Microbial Technol.* 40: 1078–1086.

Strohl, W.R. 2004. Antimicrobials. In: *Microbial Diversity and Bioprospecting*, Bull, A.T. ed. ASM Press, Washington, DC, pp. 336–355.

Sunayana, M.R., Sasikala, C., and Ramana, V.C. 2005a. Production of novel indole ester from 2-amonobenzoate by *Rhodobacter sphaeroides* OU5. *J Ind Microbiol Biotech.* 32: 41–45.

Sunayana, M.R., Sasikala, C., and Ramana, V.C. 2005b. Rhodestrin: A novel indole terpenoid phytohormones from *Rhodobacter sphaeroides* OU5. *Biotech Lett.* 27: 1897–1900.
Tawata, S., Taira, S., Kobamoto, N., Nhu, J., Ishihara, M., and Toyama, S. 1996. Synthesis and antifungal activity of cinnamic acid ester. *Biosci Biotechnol Biochem.* 60: 909–910.
Vasconcelos, A.T.R., Almeida, D.F., Hungria, M. et al. 2003. The complete genome of *Chromobacterium
violaceum* reveals remarkable and exploitable bacterial adaptability. *Proc Natl Acad Sci USA.*
100: 11660–11665.
Verma, M., Tripathi, M., Saxena, A.K., and Shanker, K. 1994. Anti-inflammatory activity of novel
indole derivatives. *Eur J Med Chem.* 29: 941–946.
Vijay, S., Sunayana, M.R., Ranjith, N.K., Sasikala, C., and Ramana, V.C. 2006. Light dependent transformation of aniline to indole esters by a purple bacterium *Rhodobacter sphaeroides* OU5. *Curr
Microbiol.* 52: 413–415.
Von Nussbaum, F., Brands, M., Hinzen, B., Weigand, S., and Habich, D. 2006. Antibacterial natural
products in medicinal chemistry—Exodus or revival. *Angew Chem Int Ed.* 45: 5072–5129.
Yang, G., Song, L., Li, K., and Hu, C. 2003. Studies on chemical constituents of *Polygonum
orientale*. *ZhongguoYaoxue Zazhi.* 38: 338–350.

chapter nine

Bacteriocin

A natural alternative to synthetic antibacterial antibiotics

S. Latha and Dharumadurai Dhanasekaran

Contents

9.1 Introduction

Antimicrobials are unarguably one of the most important medical discoveries of the twentieth century and function as a vital medicine for the treatment of bacterial infections in both humans and animals. Moreover, they have played a major role in the growth and development of food-producing animals (poultry, goat, sheep, beef, and dairy cattle) mainly by improving their efficiency in growth rate, feed utilization, mortality reduction, etc. However, the continuous use of antibiotics led to the emergence of microbial resistance, dissemination of resistant bacteria, and resistance genes to pathogenic bacteria in both humans and animals. A major issue is that antimicrobial resistance (AMR) not only occurs among disease-causing organisms but has also become an issue for other resident organisms in the host. In addition, consumers are also becoming increasingly concerned about the accumulation of drug residues in meat products of food animals (Thacker, 2013).

This alarming spread of resistance to classic antimicrobial agents has driven the search for new antimicrobials that are broadly effective and less likely to induce AMR.

Bacteriophages (phages), bacterial cell wall hydrolases (BCWHs), and antimicrobial peptides (AMPs) are some of the promising antimicrobial alternatives (Parisien et al., 2008). Bacteriophages are highly specific and can be active against a single strain of bacteria. Therefore, using bacteriophage against infecting strains was suggested to control undesirable bacterial species in mucosal systems. This approach was first developed in the last century and showed much potential; however, it also aroused much controversy and concern (Gillor et al., 2008). BCWHs are enzymes that degrade peptidoglycan (major bacterial cell wall component) due to their lytic enzymatic activities, that is, attacking specific sites in the peptidoglycan network, leading to hydrolysis and consequently bacteriolysis (Masschalck and Michiels, 2003). Though they are well established, safe, and efficient against antibiotic-resistant bacteria, the main obstacle to use them for clinical application is the high production costs and not being effective on most gram-negative bacteria due to the presence of outer membrane (Parisien et al., 2008).

AMP is another major group of prospective novel substitute to antibiotics based on their effectiveness, safety, and enormous diversity. The bactericidal mechanism of AMP includes formation of ion channels or pores across the cytoplasmic membrane; inhibition of cell wall biosynthesis, ribonuclease, or deoxyribonuclease (DNase) activities; and induction of cytoplasmic membrane perturbations by depolarization and perforation of the membrane (Damasko et al., 2005; Parisien et al., 2008). AMPs are a large family of naturally occurring peptides from various sources, having diverse structures and functionalities. Based on their origins, they were classified into three types, namely, eukaryotic AMPs, phage-encoded AMPs, and bacteriocins. This review summarizes the natural AMPs, bacteriocins, by highlighting their potential as alternative to conventional antibiotics.

9.2   Bacteriocins: A class of antimicrobial peptides

Bacteriocins are proteinaceous compounds (natural AMPs) produced by a bacterium, which are active against other bacteria. They are generally ribosomally synthesized, although some are extensively posttranslationally modified. Bacteriocins usually have low molecular weight (rarely over 10 kDa) and can be easily degraded by proteolytic enzymes especially by the proteases of the mammalian gastrointestinal tract, which makes them safe for human consumption. In general, they are cationic, amphipathic molecules as they contain an excess of lysyl and arginyl residues. They are usually unstructured when they are incorporated in aqueous solutions; however, they form a helical structure when exposed to structure-promoting solvents such as trifluoroethanol and anionic phospholipids membranes (Zacharof and Lovitt, 2012).

Bacteriocins are produced by all major lineages of bacteria, archaea, and constitute a heterogeneous group of peptides with respect to the size, structure, mode of action, antimicrobial potency, immunity mechanisms, and target cell receptors (Dobson et al., 2012). Bacteriocin-producing bacteria (BPB) belong to different systematic groups and occupy various ecological niches such as soil, dairy and meat products, fermented plant products, and gastrointestinal tract of warm-blooded animals. Bacteriocin provides them with a competitive advantage in their environment, eliminating competitors to gain resources. They are mainly known for their bactericidal and bacteriostatic activities against various human and animal pathogens (Hammami et al., 2007). The inhibitory spectrum of some bacteriocins also includes food spoilage and food-borne pathogenic microorganisms.

Bacteriocin genes are either chromosomally or plasmid encoded with resulting toxins employing a variety of killing mechanisms, including cytoplasmic membrane pore formation, cell wall interference, and nuclease activity (Gillor et al., 2005; Borges et al., 2014). Additional roles have been proposed for some bacteriocins produced by gram-positive bacteria, such as chemical mediators in quorum sensing and communication molecules in bacterial consortia.

9.3 Bacteriocins versus conventional antibiotics

All organisms produce AMPs that represent part of the natural and innate immune systems that protect them against invading organisms. In the microbial world, the AMPs are an important part of the defense system of bacteria, and they are referred to as bacteriocins. They are often confused with the antibiotics that would limit their use in food and medical applications from a legal standpoint. However, bacteriocins are clearly distinguishable from clinical antibiotics, and also they are safe and more effective in the control of target pathogens.

An important criterion of being a bacteriocin is that they are ribosomally synthesized, while antibiotics are made by multienzyme complexes (Nes et al., 2007). Besides, they are produced during the primary phase of growth, whereas antibiotics are usually secondary metabolites. The major difference between them is that bacteriocins restrict their activity to strains of species related to the producing species and particularly to strains of the same species; antibiotics, on the other hand, have a wider activity spectrum, and even if their activity is restricted, this does not show any preferential effect on closely related strains (Zacharof and Lovitt, 2012). Moreover, bacteriocins are more potent against their target bacteria, while higher concentrations of traditional antibiotics are needed to kill the target bacteria (Table 9.1). In contrast to the currently used antibiotics, bacteriocins are often considered more natural because they are thought to have been present in many of the foods eaten since ancient times (Cleveland et al., 2001).

Table 9.1 Distinctive features of bacteriocins and conventional antibiotics

S. no.	Characteristics	Bacteriocins	Conventional antibiotics
1.	Example	Penicillin	Defensin
2.	Chemical composition	Proteinaceous	Complex ring structure
3.	Synthesis	Ribosomal	Secondary metabolite
4.	Molecular characteristics	Small amphipathic peptides < 10 kDa, feasible to synthesize with higher production cost	Small compounds <2000 Da, easy to synthesize with lower production cost
5.	Activity	Narrow spectrum	Varying spectrum
6.	Application	Food	Clinical
7.	Interaction requirements	Sometimes docking molecules	Specific target
8.	Host cell immunity	Yes	No
9.	Mechanism of target cell resistance or tolerance	Usually adaptation affecting cell membrane composition	Usually a genetically transferable determinant affecting different sites depending on the mode of action
10.	Mode of action	Mostly pore formation, but in a few cases possibly cell wall biosynthesis	Cell membrane or intracellular targets
11.	Toxicity side effects	Not known	Yes

9.4   Classification of bacteriocins

It has been hypothesized that 99% of bacteria produce at least one bacteriocin, which provides the basis for an optimistic view that it is only a matter of directing enough research resources to this area to identify more bacteriocins for different bacteria (Klaenhammer, 1988). Due to the extensive focus on bacteriocins, a number of classification schemes have been proposed, which are largely applicable to gram-positive and gram-negative bacteriocins. Methods of classification include producing strain (bacteria, archaea), method of production (ribosomal, postribosomal modifications, nonribosomal), mechanism of killing (pore forming, DNase, nuclease, murein production inhibition, etc.), genetics (large plasmids, small plasmids, chromosomal), molecular weight and chemistry (large protein, polypeptide, with/without sugar moiety, containing atypical amino acids like lanthionine), and resistance mechanisms (Blinkova et al., 2003). According to their origin, bacteriocins are classified as gram-negative bacteriocin, archaebacterial bacteriocin, and gram-positive bacteriocin (Figure 9.1).

9.4.1   Bacteriocins from gram-negative bacteria

Most bacteriocins produced by gram-negative bacteria are relatively large and range in size from <10 to 20 kDa (roughly 400 to 700 amino acids). They differ from gram-positive bacteriocins in two fundamental ways: (1) usually released through cell lysis and (2) dependent on host regulatory pathways, like SOS regulation. Colicins are the most studied gram-negative bacteriocins and were first discovered by Gratia (2000). They are plasmid-encoded high-molecular-weight proteins (over 20 kDa) that are active against *Escherichia coli* strains and other closely related bacteria, such as *Salmonella*. Over 30 types of colicins have been identified, based on killing activity and immunity specificity (Gillor et al., 2005).

Colicins produced by *E. coli* can be pore-forming toxins or nuclease-type toxins. Their operons are on plasmids and consist of colicin toxin gene, lysis gene, and immunity gene. Target specificity is governed by a receptor domain on the colicin protein that binds a specific cell surface receptor found only on certain strains of *Escherichia* genus or other enterobacteria. Microcins (Figure 9.2a), which are also produced by *E. coli*, have been treated as a

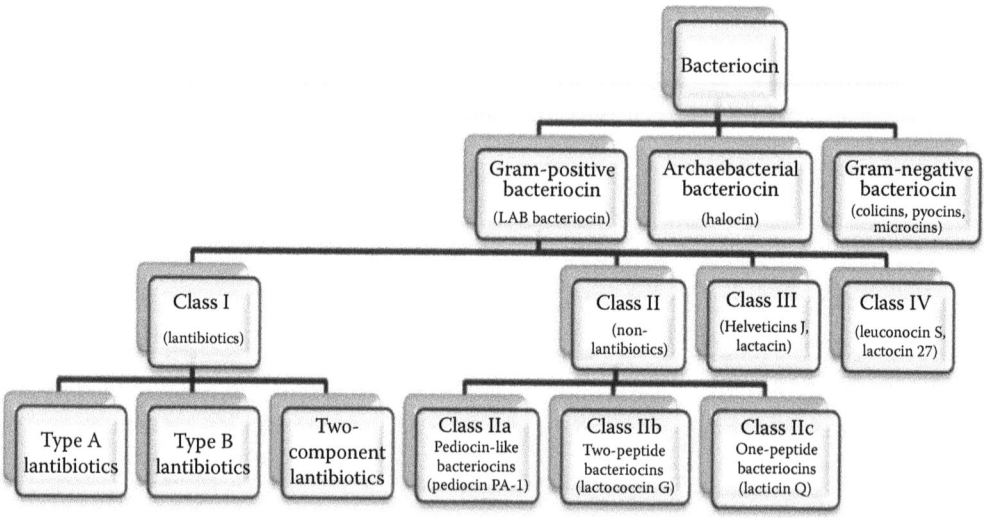

Figure 9.1 Classification of bacteriocin based on producers.

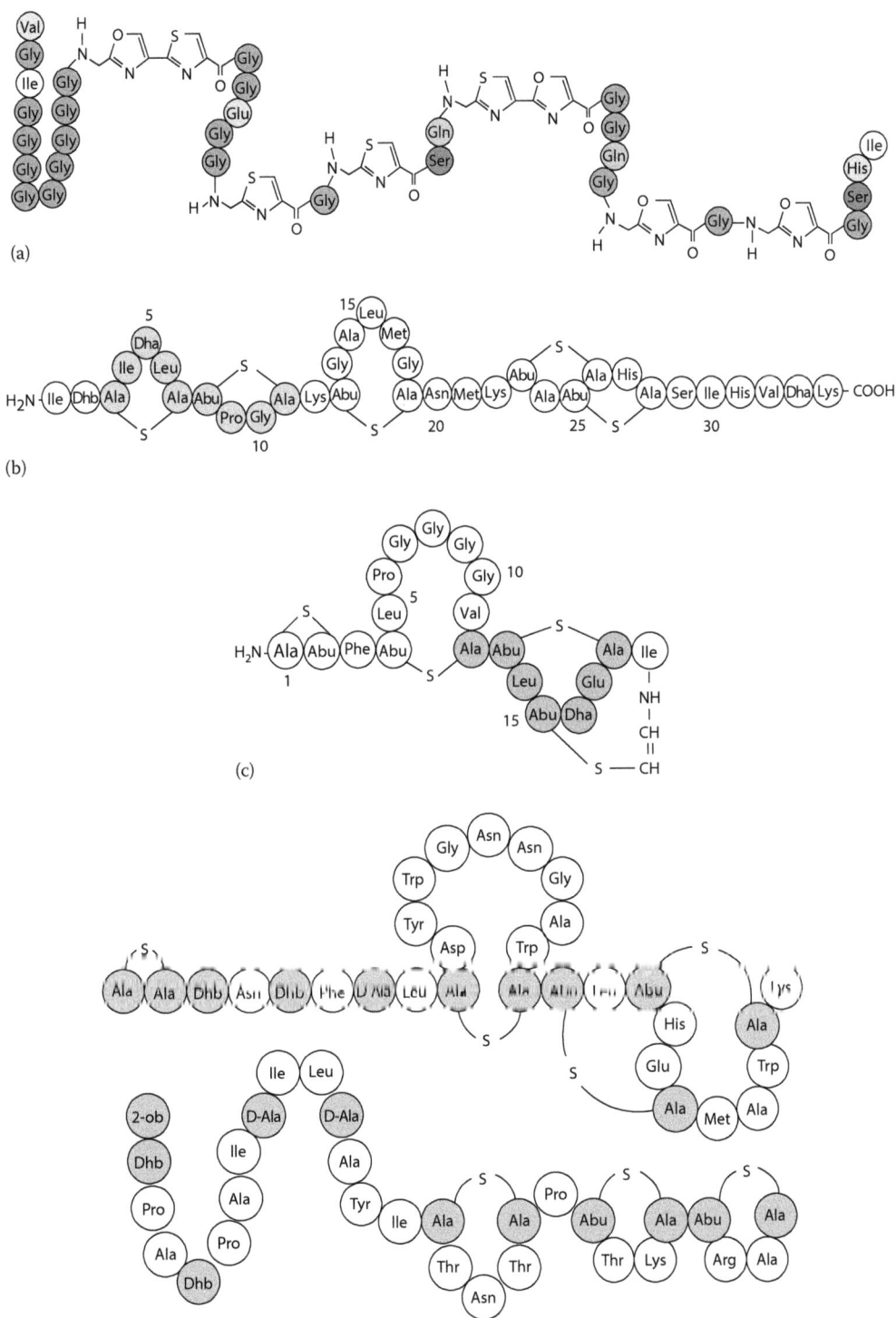

Figure 9.2 Covalent structure of peptide bacteriocins: (a) microcin B17, (b) nisin, (c) mersacidin, (d) lacticin 3147. *(Continued)*

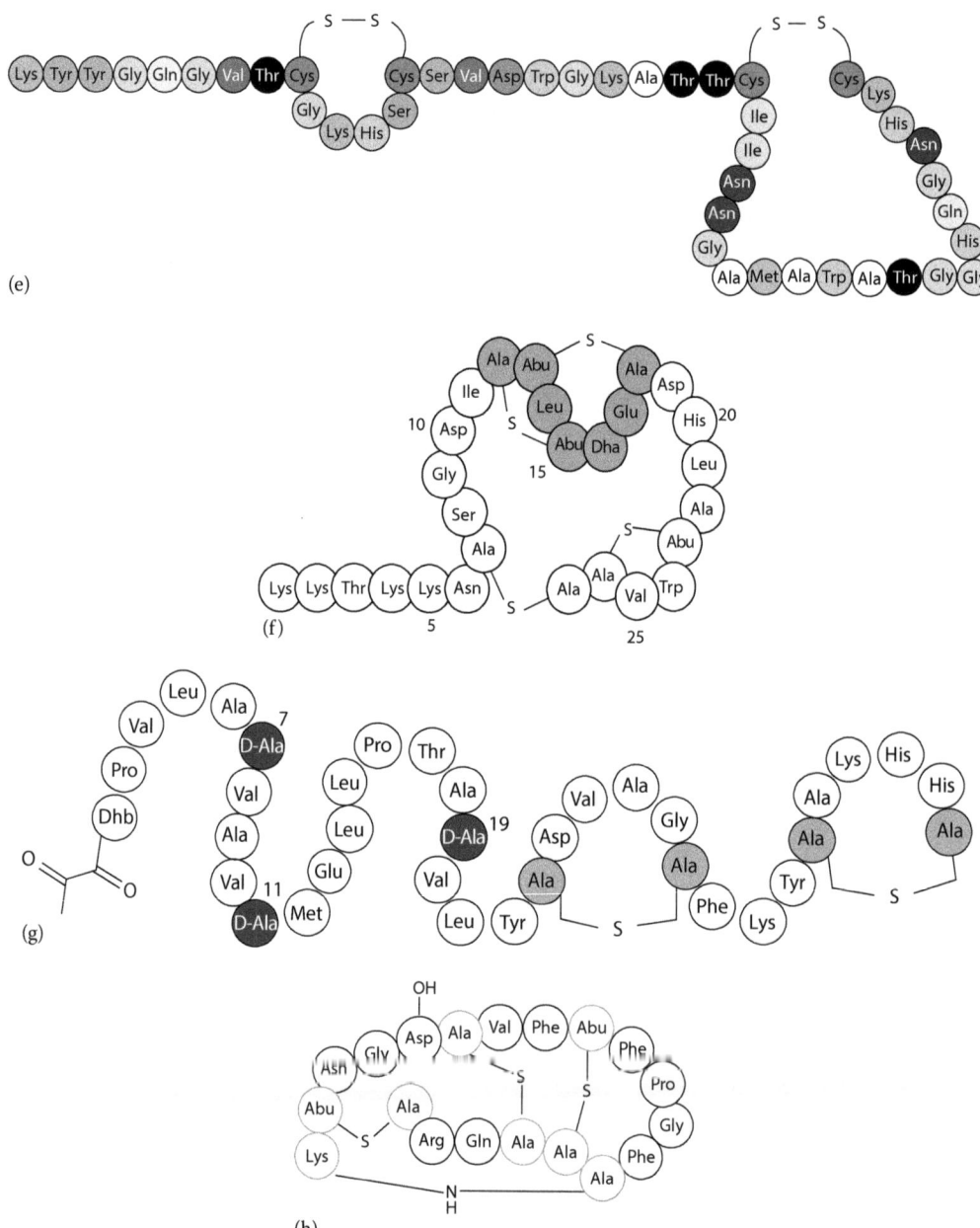

Figure 9.2 (Continued) Covalent structure of peptide bacteriocins: (e) pediocin PA-1, (f) plantaricin C, (g) lactocin S, and (h) cinnamycin.

separate class because of their significantly smaller molecular weight. Only nine microcins have been identified so far (Moreno et al., 2002), and unlike colicins, few have been characterized at the level of protein structure or mode of action. The killing spectrum of microcins is broad compared to that of colicins, but they are primarily directed against genera of Enterobacteriaceae. Microcins kill their target cells by forming pores or by disrupting the cell membrane potential (Destoumieux-Garzón et al., 2002).

Pyocins are the nuclease proteins produced by *Pseudomonas aeruginosa*, having a sequence similar to that of colicin. The colicin and related bacteriocins have typical domain structure where there are generally a receptor recognition domain, a translocation domain, and a toxin domain, all of which have sequence similar to their counterparts in other gram-negative species. Most of the enteric bacteria have the ability to produce bacteriocins. Recent surveys of *Hafnia alvei, Citrobacter freundii, Klebsiella oxytoca, K. pneumoniae*, and *Enterobacter cloacae* reveal levels of bacteriocin production ranging from 3% to 26% of environmental enteric isolates (Riley et al., 2003). Molecular investigations reveal that enteric bacteriocins employ similar killing mechanisms, although they often utilize novel receptor recognition and translocation functions, ensuring their narrow killing specificities (Riley et al., 2001; Wertz and Riley, 2004; Gillor et al., 2005).

9.4.2   Bacteriocins from gram-positive bacteria

Bacteriocins from gram-positive bacteria are even more diverse than those of gram-negative bacteria. The majority of gram-positive bacteriocins are relatively small consisting of 30–70 amino acids and seem to possess a broader range of susceptible organisms. They are generally secreted by the producer cell and thus do not require death of the producer cell to release mature bacteriocin.

9.4.2.1   Lactic acid bacterial bacteriocin

The best studied members of the gram-positive group of molecules are derived from lactic acid bacteria (LAB), and they are prolific in their use in cultured food products as natural preservatives (Lüders et al., 2003). The uses of LAB and their metabolic products are generally considered as safe (GRAS, Grade One). Many bacteriocin-producing LAB (*Lactobacillus, Lactococcus, Enterococcus, Streptococcus, Pediococcus, Leuconostoc*) and *Bifidobacterium* have been isolated from different food matrices, such as fermented dairy products, vegetables, fruits, meat, and fish, and also from the human and animal gastrointestinal tracts (Todorov et al., 2010).

Bacteriocins produced by LAB are small, ribosomally synthesized AMPs or proteins that possess activity towards closely related gram-positive bacteria, whereas producer cells are immune to their own bacteriocin(s). The application of the produced antimicrobial compounds as a natural barrier against pathogens and food spoilage caused by bacterial agents has been proven to be efficient (Leroy, 2007). LAB bacteriocins are categorized into different classes based on their biochemical and genetic properties such as mode of action, heat tolerance, biological activity, presence of modified amino acids, and secretion mechanism.

Class I peptides are the lantibiotics, which are small, posttranslationally modified peptides that contain unusual amino acids such as lanthionine, β-methyl lanthionine, dehydroalanine, dehydrobutyrine, and several dehydrated amino acids. They target a broad range of gram-positive bacteria and are subdivided into three groups on the basis of their structure and mode of action. Type A lantibiotics (2–5 kDa), such as nisin (Figure 9.2b), are small, elongated, screw-shaped proteins that contain positively charged molecules, which kill via the formation of pores, leading to the dissipation of membrane potential and the efflux of small metabolites from the sensitive cells (Nagao et al., 2006). Type B lantibiotics (about 2 kDa), such as mersacidin (Figure 9.2c), kill by interfering with cellular enzymatic reactions, for example, cell wall synthesis (Pag and Sahl, 2002). Another subgroup is composed of two-component lantibiotics, such as lacticin 3147 (LtnA1, LtnA2) consisting of two lantibiotic peptides (Figure 9.2d) that synergistically display antimicrobial activity (Wiedemann et al., 2006; Gillor et al., 2008).

Class II includes unmodified, small, heat-stable, non-lanthionine-containing peptides (molecular masses of <10 kDa) that share a conserved N-terminal sequence and pore formers and have anti-*Listeria* activity. The majority of bacteriocins in this group kill the target bacteria by inducing membrane permeabilization and subsequent leakage of molecules (Oppegård et al., 2007). These bacteriocins are organized into three subgroups, namely, class IIa pediocin-like bacteriocins (Figure 9.2e), class IIb two-peptide bacteriocins (Figure 9.2f), and class IIc others, that is, non-pediocin-like one peptide-bacteriocins (Borges et al., 2014). Class III peptides are large, thermosensitive proteins (molecular masses of >10 kDa) such as bacteriolysin, helveticins J, and lactacin B (Dobson et al., 2007). The best studied bacteriolysin is lysostaphin, a 27 kDa peptide that hydrolyses several *Staphylococcus* sp. cell walls, principally *S. aureus*. Class IV bacteriocins are a group of complex proteins, associated with other lipid or carbohydrate moieties, which appear to be required for activity. They are relatively hydrophobic and heat stable (Mahrous et al., 2013). Little is known about the structure and function of class IV bacteriocins. Examples include leuconocin S, lactocin 27, and lactocin S (Figure 9.2g) (Choi et al., 1999; Vermeiren et al., 2006).

9.4.2.2 Actinobacterial bacteriocin

Actinobacteria are a group of gram-positive bacteria that constitute a significant component of the microbial population in most soils. Traditionally, actinobacteria have been a rich source of biotechnological products like antibiotics, industrial enzymes, and other bioactive molecules. Followed by LAB, actinobacteria is known to be a potential source for different kinds of bacteriocins.

Among human intestinal microbiota, *Bifidobacterium* is one of the most familiar genus that constitutes up to 25% of the total population in the intestinal tract in adults and 95% in newborns. The positive effect of *Bifidobacterium* in human is the production of antimicrobial compounds other than organic acids, such as bacteriocins. Although some reports have been suggested that the production of organic acids (via the heterofermentative pathway) is partially responsible for the inhibitory activity of bifidobacteria (Ibrahim and Salameh, 2001; Bruno and Shah, 2002), it is well accepted that some genera of this group also produce the bacteriocins (Cheikhyoussef et al., 2010).

To date, some bacteriocins such as bifidin I (Cheikhyoussef et al., 2010), bifidocin B (Yildirim et al., 1999), and bacteriocin-like inhibitory substances (BLISs) (Cheikhyoussef et al., 2009) have been reported to be produced by bifidobacteria. Zoulti et al. (2011) reported that *Bifidobacterium* sp. RBL 68 and RBL 85 isolated from newborn feces were found to produce two BLISs with inhibitory activities against a wide range of gram-positive and gram-negative bacteria that makes them potentially useful as antimicrobial agents in foods. A list of known *Bifidobacterium*-associated bacteriocins and putative bacteriocins as well as their main characteristics is represented in Table 9.2.

Apart from *Bifidobacterium*, other actinobacterial genera such as *Streptomyces* and *Actinomyces* also have the ability to produce bacteriocin and inhibit bacterial pathogens. A strain of *A. odontolyticus*, originally isolated from human dental plaque, produced a non-dialyzable, trypsin-sensitive substance called odontolyticin that was bactericidal for certain strains of bifidobacteria at 42°C but not at 37°C (Franker et al., 1977). Bacteriocin production within the genus *Streptomyces* has been previously reported, with bactericidal spectra described as species specific (Zhang et al., 2003) or genus specific (Roelants and Naudts, 1964). *S. scopuliridis* sp. nov., RB72ᵀ isolated from woodland bluff soil in northern Alabama, United States, produces a broad-spectrum bacteriocin with inhibitory activity against gram-positive and gram-negative bacteria (Farris et al., 2011).

Table 9.2 Bifidobacterium-associated bacteriocins and their inhibitory spectrum

S. no.	Bacteriocin	Species and strain	Inhibitory spectrum	References
1.	Bifidin	*B. bifidum* NCDC 1452	Gram-positive and gram-negative bacteria	Anand et al. (1985)
2.	Bifidocin B	*B. bifidum* NCFB 1454	*Bacillus cereus, E. faecalis, S. faecalis, P. acidilactici, L. monocytogenes*	Yildirim et al. (1999)
3.	Bifilong	*B. longum*	Gram-positive and gram-negative bacteria	Kang et al. (1989)
4.	Bifilact Bb-46	*B. longum* Bb-46	*S. aureus, S. typhimurium, B. cereus, E. coli*	Saleh et al. (2004)
5.	Bifilact Bb-12	*B. lactis* Bb-12	*S. aureus, S. typhimurium, B. cereus, E. coli*	Saleh et al. (2004)
6.	Thermophilicin B67	*B. thermophilum* RBL67	*Listeria* sp., *L. acidophilus*	Von Ah (2006)
7.	Bifidin I	*B. infantis* BCRC 14602	LAB strains, *Staphylococcus, Streptococcus, Salmonella, Shigella, Bacillus, E. coli*	Cheikhyoussef et al. (2009, 2010)
8.	Lantibiotic (bisin)	*B. longum* DJO10A	*S. thermophilus* ST403, *S. epidermidis, B. subtilis, Serratia marcescens, E. coli* DH5a, *Clostridium perfringens*	Lee et al. (2011)

In the same way, bacteriocin of *Streptomyces* sp. JD9 (KF878075) isolated from the feces of indigenous chicken (Latha and Dhanasekaran, 2013) exhibited inhibitory activity against human and animal bacterial pathogens, namely, *E. coli* MTCC 9537, *S. enterica* MTCC 3224, *S. flexneri* MTCC 9543, *K. pneumoniae* MTCC 109, *E. faecalis* MTCC 439, and *S. aureus* MTCC 96 and *E. coli* AP1, *S. typhimurium* AP2, *Pasteurella multocida* AP3, and *S. aureus* AP4, respectively (Figures 9.3 and 9.4).

Cinnamycin (Figure 9.2h) is one of the peptide antibiotics closely related to type B lantibiotics duramycin, duramycin B, duramycin C, and ancovenin. These compounds are derived from 19-aa propeptides and have lanthionine residues in similar positions. They are all produced by actinobacteria, particularly the duramycins and cinnamycins that are exclusively produced by streptomycete strains. Despite their antimicrobial activities, these compounds also have other potentially useful pharmaceutical properties, including

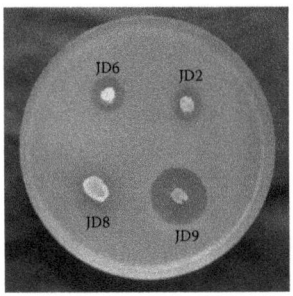
S. typhimurium AP2

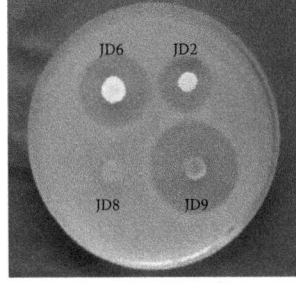
P. multocida AP3

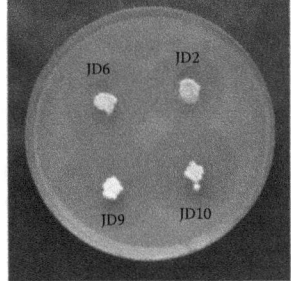
S. aureus MTCC 96

Figure 9.3 Determination of bacteriocin activity by spot agar test.

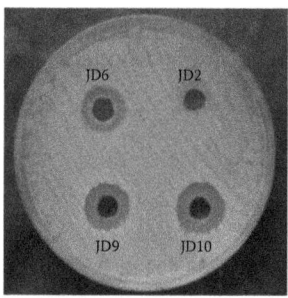 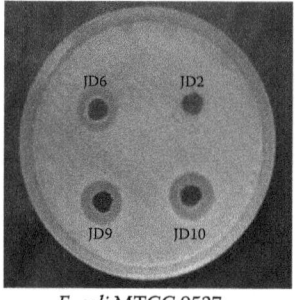 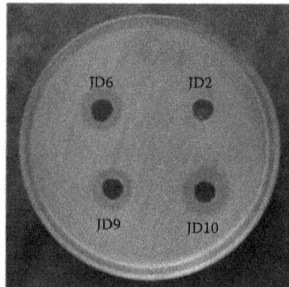

E. faecalis MTCC 439	*E. coli* MTCC 9537	*K. pneumoniae* MTCC 109

Figure 9.4 Determination of bacteriocin activity by well diffusion assay.

inhibition of angiotensin-converting enzyme, phospholipase A2 and prostaglandin, and leukotriene biosynthesis (Widdick et al., 2003).

9.4.3   Bacteriocins from archaea

The archaea produce their own distinct family of bacteriocin-like antimicrobials, known as archaeocins. The only well-characterized member of archaeocin is the halocin, produced by the Halobacteriaceae family (Platas et al., 2002; Sun et al., 2005). Halocin production is recognized as a nearly universal feature of halobacteria. However, only the halocins like H1, H4, H6, S8, and R1 have been purified and described in detail (Parisien et al., 2008). The limited number of known halocins exhibits substantial diversity in size, ranging from proteins as large as 35 kDa (e.g., halocin H4) to peptides as small as 3.6 kDa (e.g., halocin S8). The first discovered halocin is S8, a short hydrophobic peptide of 36 amino acids, which is processed from a much larger proprotein of 34 kDa (Price and Shand, 2000). It is encoded on a megaplasmid and extremely hardy; it can be desalted, boiled, subjected to organic solvents, and stored at 4°C for extended periods without losing activity. Similar characteristics are also found in microhalocins, and they are usually quite resistant to acids and bases (Li et al., 2003). Archaeocins are generally produced during the stationary phase of the cell growth. When resources are limited, producing cells lyse sensitive cells and enrich the nutrient content of the local environment. As stable proteins, they may remain in the environment long enough to reduce competition during subsequent phases of nutrient flux (Riley and Wertz, 2002).

9.5   Potential applications of bacteriocin

Bacteriocins have been the focus of an extensive number of studies for the past 60 years due to their important role in nature and, more recently, their potential use as probiotics and therapeutics. The antimicrobial activity of this group of natural substances against food-borne pathogens, as well as spoilage bacteria, has raised considerable interest for their application in food preservation. The application of bacteriocins reduces the use of chemical preservatives and/or the intensity of heat and other physical treatments, satisfying the demands of consumers for foods that are fresh-tasting, ready to eat, and lightly preserved. In addition, bacteriocins are of interest in medicine because they are made by nonpathogenic bacteria that normally colonize the human body. The other most important contributions of bacteriocins include animal health improvement and biocontrol of aquatic and plant bacterial pathogens (Figure 9.5).

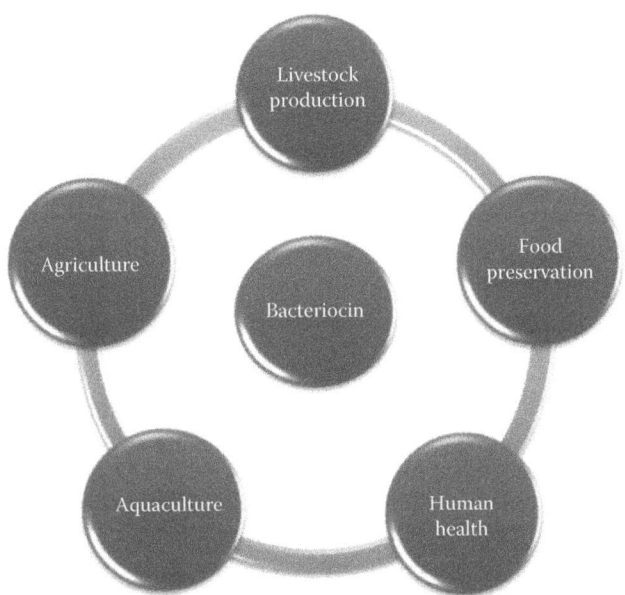

Figure 9.5 Applications of bacteriocin in diverse field.

9.5.1 Bacteriocins in food biopreservation

Nowadays, bacteriocins have been widely utilized in the field of food preservation. The uses of bacteriocins in food industry especially on dairy, egg, vegetable, and meat products have been extensively investigated. They can be applied in a purified or crude form and the product previously fermented with a bacteriocin-producing strain as an ingredient during food processing or incorporated as a starter culture. Furthermore, bacteriocins could be combined with other antimicrobial compounds such as sodium acetate and sodium lactate resulting in enhanced inactivation of bacteria. They can also be used to improve the food quality and sensory properties, for example, increasing the rate of proteolysis or in the prevention of gas blowing defect in cheese. Another application of bacteriocins is bioactive packaging, a process that can protect the food from external contaminants (Ross et al., 2002; Deegan et al., 2006; Leroy, 2007; Zacharof and Lovitt, 2012).

The potential use of lantibiotics in food, human, and animal health applications has been well documented. There are several features making them attractive for such applications such as a relatively broad killing spectrum, an autoregulation system, and stability and cost-effective production processes. Moreover, several lantibiotics are produced by food-grade bacteria that have been safely consumed by humans for centuries. The best studied lantibiotic is undoubtedly nisin, produced by *L. lactis*. It is thus far the only bacteriocin that has been approved by Food and Drug Administration as a food preservative, and it is being used for this purpose in more than 50 countries. Maisnier-Patin et al. (1992) demonstrated the potential of using nisin-producing starters for the inhibition of *L. monocytogenes* in camembert cheese. Daeschel et al. (1991) and Radler (1990a,b) indicated the possibility of using nisin for combating contamination in wines as most strains of *Leuconostoc* and *Pediococcus* are nisin sensitive.

Bacteriocin-producing strains of LAB have also been shown to improve the flavor and quality of meat products when used as starter adjuncts. For example, *L. lactis* DPC4275, a transconjugant strain producing lacticin 3147, was used as a starter for the manufacture of salami and was compared to salami manufactured with a conventional starter (*L. sake* and *S. carnosus*) in terms of pH development, water activity (a_w) value, weight loss, color development, and sensory characteristics. Salami produced with *L. lactis* DPC4275 exhibited pH values below 5.1 and an a_w value of 0.9 that is favorable for preservation and hygienic stability. In addition, these salamis had good sensory and colorimetric qualities (Coffey et al., 1998). Improvements in the flavor of fermented meat were also observed by Vogel et al. (1993) when curvacin A–producing *L. curvatus* LTH1174 was used as a starter culture in fermented sausages.

9.5.2 Bacteriocins in human health

Bacteriocins produced by several food-grade bacteria have been safely consumed by humans for centuries. Since the mode of action of bacteriocins is remarkably different from conventional antibiotics, they considered as a novel source for the control of microbial pathogens. Besides being used as food preservative, nisin (nontoxic bacteriocin) was also used as an antibacterial wipe for the udder prior to milking, and it has been suggested as a contraceptive agent (Gillor et al., 2005). It has a potential in treating peptic ulcer disease by inhibiting *Helicobacter pylori* growth and colonization (Delves-Broughton et al., 1996).

Nisin was additionally used to inhibit the growth of multidrug-resistant pathogens such as *Staphylococcus* and *Streptococcus* sp. Bower and his colleagues (2002) treated catheters and tracheotomy tubes with nisin, which had a protective effect against infection by gram-positive bacteria, albeit for only a short period (5–12 h), and produced no systematic or adverse effects. It also has the ability to control respiratory tract infections caused by *Staphylococcus* sp. De Kwaadsteniet et al. (2009) reported that nisin F inhibited the growth of *S. aureus* in the respiratory tract of rats when administered intranasally.

Bacteriocins of LAB are usually not active against gram-negative bacteria, and only a few works referred to activity against *Salmonella* sp. Bacteriocin AS-48, produced by *E. faecalis*, inhibited the growth of *S. choleraesuis* at pH 4.0 (Abriouel et al., 1998). Similarly, a bacteriocin-like substance produced by a strain of *L. plantarum* inhibited the growth of *Salmonella* sp. isolated from mango (Ragazzo-Sanchez et al., 2009). Bacteriocins appear to be capable of displacing or suppressing the growth of other bacteria and perhaps provide an advantage to microorganisms in fermenting ecosystems, including the female genital tract. As bacteriocins do not induce vaginal irritation, they are suitable for human use.

The inhibitory activity of bacteriocin SB83 produced by *P. pentosaceus* SB83 against *L. monocytogenes* (serotypes 1/2a, 1/2b, and 4b) revealed that the isolate can be used as a potential vaginal probiotic (Borges et al., 2014). The bacteriocin, subtilosin has proven antimicrobial activity against a wide variety of human pathogens, including *L. monocytogenes*, *Gardnerella vaginalis*, *S. agalactiae*, and *Micrococcus luteus*. Its activity against *G. vaginalis*, combined with its lack of effect on probiotic vaginal *Lactobacillus* isolates, indicates that subtilosin could target the vaginal pathogen while leaving the healthy vaginal microflora intact (Sutyak et al., 2008).

9.5.3 Bacteriocins in livestock production

Antibiotic therapy has been a valuable tool used in animal research, as growth promoters or therapeutic agents, and their efficacy and cost-effectiveness contribute to their popularity. However, in recent years, bacteriocins and probiotic microbes replaced the antibiotic growth promoters and reduced the antibiotic-associated problems such as occurrence of antibiotic residues in the environment and veterinary products as well as an increase in the frequency of resistance among bacterial species (Ochoa-Zarzosa et al., 2008).

Bacteriocins produced by different gram-positive bacteria have been tested both *in vitro* and *in vivo* for the improved livestock production. The peptides that have been tested differ in their physicochemical characteristics and spectrum of activity, but preliminary studies indicated that they might be a potential and effective alternative to classical antibiotics used in animal husbandry. However, due to rapid degradation of the bacteriocins in the digestive tract of mammals, feeding or applying bacteriocins alone to livestock is less common. On the other hand, a variety of probiotic bacteria have been tested to control animal pathogenic bacteria and to improve the production of livestock. The pathogen inhibition mechanism of probiotic microorganisms is primarily mediated by the production of bacteriocins. Accordingly, the method of feeding BPB is generally preferred as an alternative antimicrobial strategy in animal husbandry.

In poultry, the use of BPB has been mainly targeted for the control of *Salmonella* sp. Administration of bacteriocin-producing *E. faecium* strain J96 after hatching decreased the poultry pathogen *S. pullorum* population and increased the survival rate of young broiler chicks (Audisio et al., 2000). Microcins produced by *E. coli* hold promise in reducing the abundance of *S. typhimurium* in adult chickens (Portrait et al., 1999; Gillor et al., 2004). BLIS produced by *B. amyloliquefaciens* CECT 5940 was also used as a probiotic in poultry systems that showed the reduction of pathogenic bacteria, such as *C. perfringens*, *E. coli*, and *Yersinia* sp. (Diaz, 2007).

In cattle, the cow's rumen serves as a major reservoir for *E. coli* O157:H7, a pathogen that is exceedingly difficult to control using antibiotics (Hussein and Bollinger, 2005; Hussein, 2007). In fact, studies have shown that antibiotic treatment increases the amount of Shiga toxin released by this pathogen, resulting in higher levels of bacterial virulence. Recently, there have been reports that administration of colicin-producing bacteria into the rumen of cows can reduce the level of enteric pathogens in the animal. In the same way, the two-peptide lantibiotic, lacticin 3147, produced by *L. lactis* was found to be active against mastitis-causing bacteria *Streptococci* and *Staphylococci* in dairy cattle. Significant increase in cure rates of infections caused by *S. agalactiae*, *S. aureus*, and other mastitis pathogens (90.1%, 50%, and 65.2%) was observed when cows were treated with nisin Z, via intramammary administration (Wu et al., 2007). The application of BPB for improvements in productivity has not been limited to poultry and cattle as several researchers have explored the use of probiotic strains capable of producing bacteriocins to increase the growth rate of swine (Rodriguez et al., 2003).

9.5.4   *Bacteriocins in aquaculture*

Bacteriocinogenic bacterial strains emerged as an excellent antibiotic substitute for disease control of aquatic animals. Administration of probiotic bacteria was reported to competitively exclude pathogenic bacteria through the production of inhibitory compounds, improvement of water quality, enhancement of immune response, and nutrition of host species through the production of supplemental digestive enzymes (Verschuere et al., 2000). Bacteriocins produced by marine bacteria have generated a great deal of excitement due to their potential to serve as probiotics and antibiotics in the seafood industry (Gálvez et al., 2008; Garcia et al., 2010). A recent antimicrobial screening of 258 bacterial strains isolated from water and sediment in the Yucatan Peninsula revealed that 46 strains belonging to the genera *Aeromonas*, *Bacillus*, *Burkholderia*, *Photobacterium*,

Pseudomonas, Serratia, and *Stenotrophomonas* possessed antimicrobial activity. Approximately 50% of this antimicrobial activity was due to bacteriocins or BLISs (De la Rosa-Garcia et al., 2007).

The first bacteriocin isolated from marine microorganism was detected in *Vibrio harveyi* (formerly *Beneckea harveyi*). McCall and Sizemore (1979) screened a total of 795 *Vibrio* sp. isolated from Galveston Island, Texas, for bacteriocin production. About 5% of the *Vibrio* sp. possessed a high-molecular-weight bacteriocin-like killing agent named as harveyicin. Piscicocins V1a and V1b, divercin V41, piscicocin CS526, divergicin M35, carnocin U149, and carnobacteriocin B2 are some of the bacteriocins isolated from marine *Carnobacterium* species (Stoffels et al., 1992; Bhugaloo-Vial et al., 1996; Metivier et al., 1998; Duffes et al., 1999; Tahiri et al., 2004; Suzuki et al., 2005). These bacteriocins share similar characteristics with class II bacteriocins of gram-positive bacteria.

A. media strain A199 was found to produce several BLISs and was shown to control infection by *V. tubiashii* in pacific oyster larvae (Gibson et al., 1998) and reduce saprolegniosis-related mortality in eels (Lategan et al., 2004). Irianto and Austin (2002) reported that cultures of *A. hydrophila* and *V. fluvialis* were effective at controlling infections by *A. salmonicida* in rainbow trout. In addition, Ruiz-Ponte et al. (1999) found that BLIS-producing *Roseobacter* sp. strain BS107 inhibits the pathogenic effect of *Vibrio* sp. resulting in enhanced survival of scallop larvae (Gillor et al., 2008).

9.5.5 Bacteriocins in agricultural settings

Bacteria occupying plant niches exhibit diverse lifestyles ranging from epiphytic colonization of phyllospheres or rhizospheres to infection of plant tissues as endophytes, symbionts, or pathogens. A number of studies indicate that bacteriocin production plays a major role in the competitive colonization of the plant environment by phytobacteria. Owing to their highly selective killing spectrum and potent cytotoxicity, bacteriocins show potential as targeted next-generation antibiotics for medical and agricultural uses. One possible application of bacteriocinogenic strains in agriculture includes their use in biological control of soilborne or phyllosphere-inhabiting bacterial pathogens.

A number of studies have shown that bacteriocins can be used as an effective agent in the treatment of bacterial diseases in plants. In this way, heterologous production of the peptide bacteriocin trifolitoxin by an avirulent *Agrobacterium* strain effectively enhances the biological control of *A. vitis* crown gall (Herlache and Triplett, 2002). Expression of the trifolitoxin genes from *Rhizobium leguminosarum* bv. *trifolii* T24 in *R. etli* CE3 increases bean nodulation competitiveness of the recombinant strain in the presence of indigenous rhizobia under agricultural conditions (Robleto et al., 1998; Parret et al., 2005). Dipping plants in a suspension of a bacteriocin-producing avirulent strain of *Ralstonia solanacearum* prevented tobacco wilt infection (Chen and Echandi, 1984). The incidence and severity of bacterial blight infection that causes leaf streak in rice were reduced by treatment with a nonpathogenic bacteriocin-producing strain of *Xanthomonas campestris* pv. *oryzae* (Sakthivel and Mew, 1991).

Lavermicocca et al. (2002) reported the use of pyocin, an uncharacterized bacteriocin from *P. syringae* pv. *ciccarone* in the prevention of olive knot disease. The results showed 60%–80% reduction in knot formation when stem wounds were pretreated with crude bacteriocin before infection with the causal agent of the disease, *P. syringae* pv. *savastanoi*. Additionally, a 350- to 400-fold decrease in the epiphytic numbers of the pathogen was observed when unwounded olive plants were treated with a crude preparation of the bacteriocin (Grinter et al., 2012).

Glycinicin A, the best described bacteriocin produced by *X. campestris* pv. *glycines*, is a heterodimer of two polypeptides. It was found to be active against most tested *Xanthomonas* phytopathogenic bacterial strains (Heu et al., 2001; Gillor et al., 2005). A bacteriocin-based strategy to control fire blight of apple (devastating necrotic disease) was proposed by Jabrane et al. (1996), based on the inhibitory activity of *S. plymuthica* J7 culture supernatant against several strains of γ-proteobacterial species, including *Erwinia amylovora*. The antagonistic activity was exerted by a phage-tail-like bacteriocin named serracin P (Jabrane et al., 2002).

9.6 Conclusion

Bacteriocins produced by various bacteria have the potential to cover a broad field of applications, including food industry and medical sector. In the food industry, bacteriocin-producing starter or cocultures have been successfully applied in pilot-scale experiments yielding food quality and food safety advantages. With respect to medical applications, bacteriocins are antagonistic to many important clinical and veterinary pathogens. In particular, bacteriocins of probiotic microbes play a major role during the *in vivo* interactions occurring in the human and animal gastrointestinal tracts, hence contributing to gut health. They have the ability to target a relatively narrow range of bacteria without affecting much of the natural microbiota, which is an important advantage compared to other antibiotics. Although they do not target many pathogens like antibiotics, they have the potential to perform a very specific role. However, future studies should turn these bacteriocins into practical clinical substitutes to antibiotics and prove their anticipated efficacy, safety, and affordability.

Acknowledgment

The author is grateful to acknowledge the Department of Science and Technology (DST), New Delhi, India, for the award of INSPIRE fellowship (DST Award Letter No. IF110317/DST/INSPIRE Fellowship/2011/Dt. 29.06.2011).

References

Abriouel, H., Valdivia, E., Gálvez, A., and Maqueda, M. 1998. Response of *Salmonella choleraesuis* LT2 spheroplasts and permeabilized cells to the bacteriocin AS 48. *Appl. Environ. Microbiol.* 64(11): 4623–4626.

Anand, S. K., Srinivasan, R. A., and Rao, L. K. 1985. Antibacterial activity associated with *Bifidobacterium bifidum*-II. *Cult. Dairy Prod. J.* 20: 21–23.

Audisio, M. C., Oliver, G., and Apella, M. C. 2000. Protective effect of *Enterococcus faecium* J96, a potential probiotic strain, on chicks infected with *Salmonella pullorum*. *J. Food Prot.* 63(10): 1333–1337.

Bhugaloo-Vial, P., Dousset, X., Metivier, A., Sorokine, O., Anglade, P., Boyaval, P., and Marion, D. 1996. Purification and amino acid sequences of piscicocins V1a and V1b, two class IIa bacteriocins secreted by *Carnobacterium piscicola* V1 that display significantly different levels of specific inhibitory activity. *Appl. Environ. Microbiol.* 62(12): 4410–4416.

Blinkova, L. P., Butova, L. G., Sergeev, V. V., Elkina, S. I., Al'tshuler, M. L., and Kalina, N. G. 2003. Effectiveness of the oral administration of tomicide in experimental infection. *Zh. Mikrobiol. Epidemiol. Immunobiol.* 1: 74–77.

Borges, S., Barbosa, J., Silva, J., and Teixeira, P. 2014. Characterization of a bacteriocin of *Pediococcus pentosaceus* SB83 and its potential for vaginal application. *Antiinfect. Agents* 12(1): 68–74.

Bower, C. K., Parker, J. E., Higgins, A. Z., Oest, M. E., Wilson, J. T., Valentine, B. A., Bothwell, M. K., and McGuire, J. 2002. Protein antimicrobial barriers to bacterial adhesion: *In vitro* and *in vivo* evaluation of nisin-treated implantable materials. *Colloids Surf. B.* 25(1): 81–90.

Bruno, F. and Shah, N. P. 2002. Inhibition of pathogenic and putrefactive microorganisms by 541 *Bifidobacterium* sp. *Milchwissenschaft* 57(11/12): 617–621.

Cheikhyoussef, A., Cheikhyoussef, N., Chen, H., Zhao, J., Tang, J., Zhang, H., and Chen, W. 2010. Bifidin I–A new bacteriocin produced by *Bifidobacterium infantis* BCRC 14602: Purification and partial amino acid sequence. *Food Control* 21(5): 746–753.

Cheikhyoussef, A., Pogori, N., Chen, H., Tian, F., Chen, W., and Tang, J. 2009. Antimicrobial activity 550 and partial characterization of bacteriocin-like inhibitory substances (BLIS) produced by *Bifidobacterium infantis* BCRC 14602. *Food Control* 20: 553–559.

Chen, W. Y. and Echandi, E. 1984. Effects of avirulent bacteriocin-producing strains of *Pseudomonas solanacearum* on the control of bacterial wilt of tobacco. *Plant Pathol.* 33(2): 245–253.

Choi, H. J., Lee, H. S., Her, S., Oh, D. H., and Yoon, S. S. 1999. Partial characterization and cloning of leuconocin J, a bacteriocin produced by *Leuconostoc* sp. J2 isolated from the Korean fermented vegetable kimchi. *J. Appl. Microbiol.* 86(2): 175–181.

Cleveland, J., Montville, T. J., Nes, I. F., and Chikindas, M. L. 2001. Bacteriocins: Safe, natural antimicrobials for food preservation. *Int. J. Food Microbiol.* 71(1): 1–20.

Coffey, A., Ryan, M., Ross, R. P., Hill, C., Arendt, E., and Schwarz, G. 1998. Use of a broad host range bacteriocin-producing *Lactococcus lactis* transconjugant as an alternative starter for salami manufacture. *Int. J. Food Microbiol.* 43(3): 231–235.

Daeschel, M. A., Jung, D. S., and Watson, B. T. 1991. Controlling wine malolactic fermentation with nisin and nisin-resistant strains of *Leuconostoc oenos. Appl. Environ. Microbiol.* 57(2): 601–603.

Damasko, C., Konietzny, A., Kaspar, H., Appel, B., Dersch, P., and Strauch, E. 2005. Studies of the efficacy of enterocoliticin, a phage-tail like bacteriocin, as antimicrobial agent against *Yersinia enterocolitica* Serotype O3 in a cell culture system and in mice. *J. Vet. Med. B.* 52(4): 171–179.

Deegan, L. H., Cotter, P. D., Hill, C., and Ross, P. 2006. Bacteriocins: Biological tools for bio-preservation and shelf-life extension. *Int. Dairy J.* 16(9): 1058–1071.

De Kwaadsteniet, M., Doeschate, K. T., and Dicks, L. M. T. 2009. Nisin F in the treatment of respiratory tract infections caused by *Staphylococcus aureus. Lett. Appl. Microbiol.* 48(1): 65–70.

De la Rosa-Garcia, S. C., Muñoz-García, A. A., Barahona-Pérez, L. F., and Gamboa-Angulo, M. M. 2007. Antimicrobial properties of moderately halotolerant bacteria from cenotes of the Yucatan Peninsula. *Lett. Appl. Microbiol.* 45(3): 289–294.

Delves-Broughton, J., Blackburn, P., Evans, R. J., and Hugenholtz, J. 1996. Applications of the bacteriocin, nisin. *Antonie van Leeuwenhoek* 69(2): 193–202.

Destoumieux-Garzón, D., Peduzzi, J., and Rebuffat, S. 2002. Focus on modified microcins: Structural features and mechanisms of action. *Biochimie* 84(5): 511–519.

Diaz, D. 2007. Effect of *Bacillus amyloliquefaciens* CECT-5940 spores on broiler performance and digestibility. Accessed November 2007. http://en.engormix.com/MA-poultry-industry/articles/effect-bacillus-amyloliquefaciens-cect5940-t795/p0.htm.

Dobson, A., Cotter, P. D., Ross, R. P., and Hill, C. 2012. Bacteriocin production: A probiotic trait?. *Appl. Environ. Microbiol.* 78(1). 1–6.

Dobson, A. E., Sanozky-Dawes, R. B., and Klaenhammer, T. R. 2007. Identification of an operon and inducing peptide involved in the production of lactacin B by *Lactobacillus acidophilus. J. Appl. Microbiol.* 103(5): 1766–1778.

Duffes, F., Corre, C., Leroi, F., Dousset, X., and Boyaval, P. 1999. Inhibition of *Listeria monocytogenes* by *in situ* produced and semipurified bacteriocins of *Carnobacterium* spp. on vacuum-packed, refrigerated cold-smoked salmon. *J. Food Protect.* 62(12): 1394–1403.

Farris, M. H., Duffy, C., Findlay, R. H., and Olson, J. B. 2011. *Streptomyces scopuliridis* sp. nov., a bacteriocin-producing soil streptomycete. *Int. J. Syst. Evol. Microbiol.* 61(9): 2112–2116.

Franker, C. K., Herbert, C. A., and Ueda, S. 1977. Bacteriocin from *Actinomyces odontolyticus* with temperature-dependent killing properties. *Antimicrob. Agents Chemother.* 12(3): 410–417.

Gálvez, A., López, R. L., Abriouel, H., Valdivia, E., and Omar, N. B. 2008. Application of bacteriocins in the control of foodborne pathogenic and spoilage bacteria. *Crit. Rev. Biotechnol.* 28(2): 125–152.

Garcia, P., Rodriguez, L., Rodriguez, A., and Martinez, B. 2010. Food biopreservation: Promising strategies using bacteriocins, bacteriophages and endolysins. *Trends Food Sci. Tech.* 21(8): 373–382.

Gibson, L. F., Woodworth, J., and George, A. M. 1998. Probiotic activity of *Aeromonas media* on the Pacific oyster, *Crassostrea gigas*, when challenged with *Vibrio tubiashii. Aquaculture* 169(1): 111–120.

Gillor, O., Etzion, A., and Riley, M. A. 2008. The dual role of bacteriocins as anti-and probiotics. *Appl. Microbiol. Biotechnol.* 81(4): 591–606.

Gillor, O., Kirkup, B. C., and Riley, M. A. 2004. Colicins and microcins: The next generation antimicrobials. *Adv. Appl. Microbiol.* 54: 129–146.

Gillor, O., Nigro, L. M., and Riley, M. A. 2005. Genetically engineered bacteriocins and their potential as the next generation of antimicrobials. *Curr. Pharm. Des.* 11(8): 1067–1075.

Gratia, J. P. 2000. Andre Gratia: A forerunner in microbial and viral genetics. *Genetics* 156(2): 471–476.

Grinter, R., Milner, J., and Walker, D. 2012. Bacteriocins active against plant pathogenic bacteria. *Biochem. Soc. Trans.* 40(6): 1498.

Hammami, R., Zouhir, A., Hamida, J. B., and Fliss, I. 2007. BACTIBASE: A new web-accessible database for bacteriocin characterization. *BMC Microbiol.* 7(1): 89.

Herlache, T. C. and Triplett, E. W. 2002. Expression of a crown gall biological control phenotype in an avirulent strain of *Agrobacterium vitis* by addition of the trifolitoxin production and resistance genes. *BMC Biotechnol.* 2(1): 2.

Heu, S., Oh, J., Kang, Y., Ryu, S., Cho, S. K., Cho, Y., and Cho, M. 2001. gly gene cloning and expression and purification of glycinecin A, a bacteriocin produced by *Xanthomonas campestris* pv. *glycines* 8ra. *Appl. Environ. Microbiol.* 67(9): 4105–4110.

Hussein, H. S. 2007. Prevalence and pathogenicity of shiga toxin-producing *Escherichia coli* in beef cattle and their products. *J. Anim. Sci.* 85(Suppl 13): E63–E72.

Hussein, H. S. and Bollinger, L. M. 2005. Prevalence of shiga toxin–producing *Escherichia coli* in beef cattle. *J. Food Protect.* 68(10): 2224–2241.

Ibrahim, S. A. and Salameh, M. M. 2001. Simple and rapid method for screening antimicrobial activities of *Bifidobacterium* species of human isolates. *J. Rapid Meth. Aut. Mic.* 9(1): 53–62.

Irianto, A. and Austin, B. 2002. Use of probiotics to control furunculosis in rainbow trout, *Oncorhynchus mykiss* (Walbaum). *J. Fish Dis.* 25(6): 333–342.

Jabrane, A., Ledoux, L., Thonart, P., Deckers, T., and Lepoivre, P. 1996. The efficacy in vitro and in vivo of bacteriocin against *Erwinia amylovora*: Comparison of biological and chemical control of fire blight. *Acta Hortic.* 411: 355–359.

Jabrane, A., Sabri, A., Compère, P., Jacques, P., Vandenberghe, I., Van Beeumen, J., and Thonart, P. 2002. Characterization of serracin P, a phage-tail-like bacteriocin, and its activity against *Erwinia amylovora*, the fire blight pathogen. *Appl. Environ. Microbiol.* 68(11): 5704–5710.

Kang, K. H., Shin, H. J., Park, Y. H., and Lee, T. S. 1989. Studies on the antibacterial substances produced by lactic acid bacteria: Purification and some properties of antibacterial substance "Bifilong" produced by *B. longum. Kor. J. Dairy Sci.* 11: 204–216.

Klaenhammer, T. R. 1988. Bacteriocins of lactic acid bacteria. *Biochimie* 70(3): 337–349.

Lategan, M. J., Torpy, F. R., and Gibson, L. F. 2004. Control of saprolegniosis in the eel *Anguilla australis* Richardson, by *Aeromonas media* strain A199. *Aquaculture* 240(1): 19–27.

Latha, S. and Dhanasekaran, D. 2013. Antibacterial and extracellular enzyme activities of gut actinobacteria isolated from *Gallus gallus domesticus* and *Capra hircus. J. Chem. Pharm. Res.* 5(11): 379–385.

Lavermicocca, P., Lonigro, S. L., Valerio, F., Evidente, A., and Visconti, A. 2002. Reduction of olive knot disease by a bacteriocin from *Pseudomonas syringae* pv. *ciccarone. Appl. Environ. Microbiol.* 68(3): 1403–1407.

Lee, J. H., Li, X., and O'Sullivan, D. J. 2011. Transcription analysis of a lantibiotic gene cluster from *Bifidobacterium longum* DJO10A. *Appl. Environ. Microbiol.* 77(17): 5879–5887.

Leroy, L. D. V. F. 2007. Bacteriocins from lactic acid bacteria: Production, purification, and food applications. *J. Mol. Microbiol. Biotechnol.* 13: 194–199.

Li, Y., Xiang, H., Liu, J., Zhou, M., and Tan, H. 2003. Purification and biological characterization of halocin C8, a novel peptide antibiotic from *Halobacterium* strain AS7092. *Extremophiles* 7(5): 401–407.

Lüders, T., Birkemo, G. A., Fimland, G., Nissen-Meyer, J., and Nes, I. F. 2003. Strong synergy between a eukaryotic antimicrobial peptide and bacteriocins from lactic acid bacteria. *Appl. Environ. Microbiol.* 69(3): 1797–1799.

Mahrous, H., Mohamed, A., El-Mongy, M. A., El-Batal, A. I., and Hamza, H. A. 2013. Study bacteriocin production and optimization using new isolates of *Lactobacillus* spp. isolated from some dairy products under different culture conditions. *Food Nutri. Sci.* 4(3): 342–356.

Maisnier-Patin, S., Deschamps, N., Tatini, S. R., and Richard, J. 1992. Inhibition of *Listeria monocytogenes* in Camembert cheese made with a nisin-producing starter. *Le Lait* 72(3): 249–263.

Masschalck, B. and Michiels, C. W. 2003. Antimicrobial properties of lysozyme in relation to foodborne vegetative bacteria. *Crit. Rev. Microbiol.* 29(3): 191–214.

McCall, J. O. and Sizemore, R. K. 1979. Description of a bacteriocinogenic plasmid in *Beneckea harveyi. Appl. Environ. Microbiol.* 38(5): 974–979.

Metivier, A., Pilet, M. F., Dousset, X., Sorokine, O., Anglade, P., Zagorec, M., Piard, J., Marlon, D., Cenatiempo, Y., and Fremaux, C. 1998. Divercin V41, a new bacteriocin with two disulphide bonds produced by *Carnobacterium divergens* V41: Primary structure and genomic organization. *Microbiology* 144(10): 2837–2844.

Moreno, F., Gónzalez-Pastor, J. E., Baquero, M. R., and Bravo, D. 2002. The regulation of microcin B, C and J operons. *Biochimie* 84(5): 521–529.

Nagao, J. I., Asaduzzaman, S. M., Aso, Y., Okuda, K. I., Nakayama, J., and Sonomoto, K. 2006. Lantibiotics: Insight and foresight for new paradigm. *J. Biosci. Bioeng.* 102(3): 139–149.

Nes, I. F., Yoon, S., and Diep, D. B. 2007. Ribosomally synthesized antimicrobial peptides (bacteriocins) in lactic acid bacteria: A review. *Food Sci. Biotechnol.* 16(5): 675.

Ochoa-Zarzosa, A., Loeza-Lara, P. D., Torres-Rodríguez, F., Loeza-Ángeles, H., Mascot-Chiquito, N., Sánchez-Baca, S., and López-Meza, J. E. 2008. Antimicrobial susceptibility and invasive ability of *Staphylococcus aureus* isolates from mastitis from dairy backyard systems. *Antonie van Leeuwenhoek*, 94(2): 199–206.

Oppegård, C., Rogne, P., Emanuelsen, L., Kristiansen, P. E., Fimland, G., and Nissen-Meyer, J. 2007. The two-peptide class II bacteriocins: Structure, production, and mode of action. *J. Mol. Microbiol. Biotechnol.* 13(4): 210–219.

Pag, U. and Sahl, H. G. 2002. Multiple activities in lantibiotics-models for the design of novel antibiotics? *Curr. Pharm. Des.* 8(9): 815–833.

Parisien, A., Allain, B., Zhang, J., Mandeville, R., and Lan, C. Q. 2008. Novel alternatives to antibiotics: Bacteriophages, bacterial cell wall hydrolases, and antimicrobial peptides. *J. Appl. Microbiol.* 104(1): 1–13.

Parret, A. H., Temmerman, K., and De Mot, R. 2005. Novel lectin-like bacteriocins of biocontrol strain *Pseudomonas fluorescens* Pf-5. *Appl. Environ. Microbiol.* 71(9): 5197–5207.

Platas, G., Meseguer, I., and Amils, R. 2002. Purification and biological characterization of halocin H1 from *Haloferax mediterranei* M2a. *Int. Microbiol.* 5(1): 15–19.

Portrait, V., Gendron-Gaillard, S., Cottenceau, G., and Pons, A. M. 1999. Inhibition of pathogenic *Salmonella enteritidis* growth mediated by *Escherichia coli* microcin J25 producing strains. *Can. J. Microbiol.* 45(12): 988–994.

Price, L. B. and Shand, R. F. 2000. Halocin S8: A 36-amino-acid microhalocin from the haloarchaeal strain S8a. *J. Bacteriol.* 182(17): 4951–4958.

Radler, F. 1990a. Possible use of nisin in winemaking. I. Action of nisin against lactic acid bacteria and wine yeasts In solid and liquid media. *Am. J. Enol. Vitic.* 41(1): 1–6.

Radler, F. 1990b. Possible use of nisin in winemaking. II. Experiments to control lactic acid bacteria in the production of wine. *Am. J. Enol. Vitic.* 41(1): 7–11.

Ragazzo-Sanchez, J. A., Sanchez-Prado, L., Gutiérrez-Martínez, P., Luna-Solano, G., Gomez-Gil, B., and Calderon-Santoyo, M. 2009. Inhibition of *Salmonella* spp. isolated from mango using bacteriocin-like produced by lactobacilli. *Cyta J. Food* 7(3): 181–187.

Riley, M. A., Goldstone, C. M., Wertz, J. E., and Gordon, D. 2003. A phylogenetic approach to assessing the targets of microbial warfare. *J. Evol. Biol.* 16(4): 690–697.

Riley, M. A., Pinou, T., Wertz, J. E., Tan, Y., and Valletta, C. M. 2001. Molecular characterization of the klebicin B plasmid of *Klebsiella pneumoniae. Plasmid* 45(3): 209–221.

Riley, M. A. and Wertz, J. E. 2002. Bacteriocins: Evolution, ecology, and application. *Annu. Rev. Microbiol.* 56(1): 117–137.

Robleto, E. A., Kmiecik, K., Oplinger, E. S., Nienhuis, J., and Triplett, E. W. 1998. Trifolitoxin production increases nodulation competitiveness of *Rhizobium etli* CE3 under agricultural conditions. *Appl. Environ. Microbiol.* 64(7): 2630–2633.

Rodriguez, E., Arques, J. L., Rodriguez, R., Nunez, M., and Medina, M. 2003. Reuterin production by lactobacilli isolated from pig faeces and evaluation of probiotic traits. *Lett. Appl. Microbiol.* 37(3): 259–263.

Roelants, P. and Naudts, F. 1964. Properties of a bacteriocin-like substance produced by *Streptomyces virginiae*. *Antonie van Leeuwenhoek* 30(1): 45–53.

Ross, R. P., Morgan, S., and Hill, C. 2002. Preservation and fermentation: Past, present and future. *Int. J. Food Microbiol.* 79(1): 3–16.

Ruiz-Ponte, C., Samain, J. F., Sanchez, J. L., and Nicolas, J. L. 1999. The benefit of a *Roseobacter* species on the survival of scallop larvae. *Mar. Biotechnol.* 1(1): 52–59.

Sakthivel, N. and Mew, T. W. 1991. Efficacy of bacteriocinogenic strains of *Xanthomonas oryzae* pv. *oryzae* on the incidence of bacterial blight disease of rice (*Oryza sativa* L.). *Can. J. Microbiol.* 37(10): 764–768.

Saleh, F. A., Kholif, A. M., El-Sayed, E. M., Abdou, S. M., and El-Shibiny, S. 2004. Isolation and characterization of bacteriocins produced by *Bifidobacterium lactis* BB-12 and *Bifidobacterium longum* BB-46. 9ᵗʰ *Egyptian Conference for Dairy Science and Technology*, Cairo, Egypt, pp. 9–11.

Stoffels, G., Nes, I. F., and Guomundsdóttir, A. 1992. Isolation and properties of a bacteriocin-producing *Carnobacterium piscicola* isolated from fish. *J. Appl. Bacteriol.* 73(4): 309–316.

Sun, C., Li, Y., Mei, S., Lu, Q., Zhou, L., and Xiang, H. 2005. A single gene directs both production and immunity of halocin C8 in a haloarchaeal strain AS7092. *Mol. Microbiol.* 57(2): 537–549.

Sutyak, K. E., Wirawan, R. E., Aroutcheva, A. A., and Chikindas, M. L. 2008. Isolation of the *Bacillus subtilis* antimicrobial peptide subtilosin from the dairy product-derived *Bacillus amyloliquefaciens*. *J. Appl. Microbiol.* 104(4): 1067–1074.

Suzuki, M., Yamamoto, T., Kawai, Y., Inoue, N., and Yamazaki, K. 2005. Mode of action of piscicocin CS526 produced by *Carnobacterium piscicola* CS526. *J. Appl. Microbiol.* 98(5): 1146–1151.

Tahiri, I., Desbiens, M., Benech, R., Kheadr, E., Lacroix, C., Thibault, S., Ouellet, D., and Fliss, I. 2004. Purification, characterization and amino acid sequencing of divergicin M35: A novel class IIa bacteriocin produced by *Carnobacterium divergens* M35. *Int. J. Food Microbiol.* 97(2): 123–136.

Thacker, P. A. 2013. Alternatives to antibiotics as growth promoters for use in swine production: A review. *J. Anim. Sci. Biotechnol.* 4(35): 1–12.

Todorov, S. D., Wachsman, M., Tomé, E., Dousset, X., Destro, M. T., Dicks, L. M. T., Franco, B. D., Vaz-Velho, M., and Drider, D. 2010. Characterisation of an antiviral pediocin-like bacteriocin produced by *Enterococcus faecium*. *Food Microbiol.* 27(7): 869–879.

Vermeiren, L., Devlieghere, F., Vandekinderen, I., and Debevere, J. 2006. The interaction of the non-bacteriocinogenic *Lactobacillus sakei* 10A and lactocin S producing *Lactobacillus sakei* 148 towards *Listeria monocytogenes* on a model cooked ham. *Food Microbiol.* 23(6): 511–518.

Verschuere, L., Rombaut, G., Sorgeloos, P., and Verstraete, W. 2000. Probiotic bacteria as biological control agents in aquaculture. *Microbiol. Mol. Biol. Rev.* 64(4): 655–671.

Vogel, R. F., Pohle, B, S., Tichaczek, P. S., and Hammes, W. P. 1993. The competitive advantage of *Lactobacillus curvatus* LTH 1174 in sausage fermentations is caused by formation of curvacin A. *Syst. Appl. Microbiol.* 16(3): 457–462.

Von Ah, U. 2006. Identification of *Bifidobacterium thermophilum* RBL67 isolated from baby faeces and partial purification of its bacteriocin, Doctoral dissertation, Technische Wissenschaften, Eidgenössische Technische Hochschule ETH Zürich, Nr. 16927.

Wertz, J. E. and Riley, M. A. 2004. Chimeric nature of two plasmids of *Hafnia alvei* encoding the bacteriocins alveicins A and B. *J. Bacteriol.* 186(6): 1598–1605.

Widdick, D. A., Dodd, H. M., Barraille, P., White, J., Stein, T. H., Chater, K. F., Gasson, M. J., and Bibb, M. J. 2003. Cloning and engineering of the cinnamycin biosynthetic gene cluster from *Streptomyces cinnamoneus cinnamoneus* DSM 40005. *Proc. Natl. Acad. Sci.* 100(7): 4316–4321.

Wiedemann, I., Böttiger, T., Bonelli, R. R., Wiese, A., Hagge, S. O., Gutsmann, T., Seydel, U. et al. 2006. The mode of action of the lantibiotic lacticin 3147–a complex mechanism involving specific interaction of two peptides and the cell wall precursor lipid II. *Mol. Microbiol.* 61(2): 285–296.

Wu, J., Hu, S., and Cao, L. 2007. Therapeutic effect of nisin Z on subclinical mastitis in lactating cows. *Antimicrob. Agents Chemother.* 51(9): 3131–3135.

Yildirim, Z., Winters, D. K., and Johnson, M. G. 1999. Purification, amino acid sequence and mode of action of bifidocin B produced by *Bifidobacterium bifidum* NCFB 1454. *J. Appl. Microbiol.* 86(1): 45–54.

Zacharof, M. P. and Lovitt, R. W. 2012. Bacteriocins produced by lactic acid bacteria: A review article. *APCBEE Procedia* 2: 50–56.

Zhang, X., Clark, C. A., and Pettis, G. S. 2003. Interstrain inhibition in the sweet potato pathogen *Streptomyces ipomoeae*: Purification and characterization of a highly specific bacteriocin and cloning of its structural gene. *Appl. Environ. Microbiol.* 69(4): 2201–2208.

Zouhir, A., Kheadr, E., Fliss, I., and Hamida, J. B. 2011. Partial purification and characterization of two bacteriocin-like inhibitory substances produced by bifidobacteria. *Afr. J. Microbiol. Res.* 5(4): 411–418.

chapter ten

Protease inhibitors from marine organisms

L. Karthik and A. Vishnu Kirthi

Contents

10.1 Introduction

10.1.1 Protease enzymes

Proteases are essential constituents found in prokaryotes, fungi, plants, and animals. Proteases are enzymes that are involved in the breakdown of proteins in a process called proteolysis. These enzymes are involved in a multitude of physiological reactions ranging from simple digestion of food proteins to highly regulated cascades. Proteases determine the lifetime of other proteins by playing an important physiological role like hormones, antibodies, or other enzymes. This is one of the fastest *switching on* and *switching off* regulatory mechanisms in the physiology of organisms. Several bacteria secrete proteases to hydrolyze the protein peptide bond into simple monomer units. Some of these bacterial proteases also act as an exotoxin, which will destroy the extracellular structures. Proteases play a critical role in many complex physiological and pathological processes such as protein catabolism, blood coagulation, cell growth and migration, tissue arrangement, morphogenesis in development, inflammation, tumor growth and metastasis, activation of zymogens, release of hormones and pharmacologically active peptides from precursor proteins, and transport of secretory proteins across membranes (Chambers et al., 2001). In general, extracellular proteases catalyze the hydrolysis of large proteins to smaller molecules for subsequent absorption by the cell, whereas intracellular proteases play a critical role in the regulation of metabolism. Since proteases are physiologically necessary for living organisms, they are ubiquitous, being found in a wide diversity of sources such as plants, animals, and microorganisms. Besides being necessary from the physiological point of view, proteases are potentially hazardous to their proteinaceous environment and the respective cell or organism must precisely control their activity. When uncontrolled, proteases can be responsible for serious diseases. The control of proteases is generally achieved by regulated expression/secretion and/or activation of proproteases, by degradation of mature enzymes, and by the inhibition of their proteolytic activity (Fitzpatrick and O'Kennedy, 2004).

Moreover, protease enzymes are used for a long time in various forms of clinical therapies. Their use in medicine is gaining more and more attention as several clinical studies are indicating their benefits in oncology, inflammatory conditions, blood rheology control, and immune regulation. The pathogenesis of many parasites include their involvement in invasion of a host by parasite migration through host tissue barriers, degradation of hemoglobin and blood proteins, immune invasions, and activation of inflammation. Thus, proteases play a foremost role in pathogenesis. However, uncontrolled action causes deleterious effects in the body. Protease enzyme that is located inside the cancer cell is capable of breaking through other healthy cell walls and membranes, thereby spreading and growing into other cell organs and areas of the body, thus causing metastasis (Glinsky, 1993; Kim et al., 1998; Bellail et al., 2004).

10.1.2   Classification of proteases

Proteases are currently classified into six groups:

1. Serine proteases
2. Threonine proteases
3. Cysteine proteases
4. Aspartate proteases
5. Metalloproteases
6. Glutamic acid proteases

During catalysis, serine, aspartate, threonine, and cysteine groups as well as metal ions will play a major role. All these types of enzymes are present in bacteria. Among these, glutamic proteases are found only in fungi. Serine, cysteine, and metalloproteases are widely spread in many pathogenic bacteria, where they play critical functions related to evasion of host immune defenses, acquisition of nutrient for growth and proliferation, facilitation of dissemination, or tissue damage during infection (Drag and Salvesen, 2010).

10.2   Protease inhibitors

Protease inhibitors are usually proteins with domains that enter or block the protease active site to prevent substrate access. Activation of proteases causes alteration of a number of specific proteins leading to subcellular remodeling and cardiac dysfunction. Most of the protease inhibitors have been isolated from terrestrial animals, plants, fungi, and actinomycetes. The presence of protease inhibitors in microorganisms came into existence from the studies on antibiotics as they act as inhibitors of the enzymes, which are involved in the growth and multiplication of microorganisms. Proteolytic enzymes outside of microbial cells hydrolyze organic nitrogen compounds in the medium, so they are thought to be harmful to cells. The production of inhibitors of the proteolytic enzymes by microorganisms is probably a mechanism to provide cellular protection. In contrast to the inhibitors of proteolytic enzymes obtained from animals and plants, the inhibitors from microorganisms are of smaller molecules in nature. Specific inhibitors of microbial origin have been used as useful tools in biochemical analysis of biological functions and diseases (Fear et al., 2007). Natural protease inhibitors include the family lipocalin proteins, which play a major role in cell regulation and differentiation. The status of protease inhibitor is summarized in Table 10.1.

Table 10.1 Protease used in structure-based drug design

Peptidase	Biological function	Disease
Cysteine peptidases		
Cathepsin B	Antigen processing	Acute pancreatitis, cancer
Cathepsins L and S	Lysosomal proteolysis	Inflammation
FP-2	Hemoglobin degradation	Malaria
Caspase 1	Maturation of interleukin 1-β	Ameliorate inflammation, endotoxic shock
Caspases 3 and 7	Executioner caspases in apoptosis	Neuronal and cardiac ischemic injuries
Calpains 1 and 2	Degradation of cytoskeletal proteins	Stroke, neural injuries
Picornain cysteine peptidases	Processing of viral proprotein	Virus infection
Serine peptidases		
Thrombin	Proteolysis of fibrinogen	Thrombosis
Factor Xa	Conversion of prothrombin to thrombin	Thrombosis
Factor VIIa	Activation of factors IX and X	Thrombosis
Urokinase	Activation of plasminogen	Cancer
Flavivirus peptidases	Processing of polyprotein	Viral infection
DPP-4	Processing of hormone precursors	Type 2 diabetes mellitus
20S Proteasome	Ubiquitin-dependent protein degradation	Cancer
Aspartic peptidases		
HIV peptidase	Processing of viral proprotein	HIV infection
Renin	Processing of angiotensinogen	Blood pressure
Memapsin 2	β Secretase activity	Alzheimer's disease
Plasmepsin	Hemoglobin degradation	Malaria
Metallopeptidases		
Angiotensin-converting enzyme	Conversion of angiotensin	Hypertension
Botulinum neurotoxin	Cleavage of SNAP proteins	*Clostridium* and tetanus infection
Anthrax lethal factor	Cleavage of MAPKK	*Bacillus anthracis* infection
Matrix metallopeptidase 1	Degradation of connective tissue	Tissue damage in tumor invasion
FtsH	Elimination of misfolded proteins	Neurological diseases
Carboxypeptidases B and U	Cleavage of tissue plasminogen activator	Blood coagulation
PSMA	Liberates glutamate from Ac-Asp-Glu in the brain	Marker for prostate cancer

Source: Mittl, P.R. and Grütter, M.G., *Curr. Opin. Struct. Biol.*, 16(6), 769, 2006.

10.3   Marine microorganisms as a source of metabolites

The biological and chemical diversity of the marine environment has been the source of unique chemical compounds with the potential for industrial development as pharmaceuticals, cosmetics, nutritional supplements, molecular probes, enzymes, fine chemicals, and agrochemicals (Ireland et al., 1994). The oceans represent a virtual untapped resource for the discovery of even more novel compounds with useful activity. Even though the commercial success stories in biotechnology are familiar, such stories in marine biotechnology are far less familiar and far fewer (Zilinskas and Hill, 1995). In the past two decades, there has been a continuous effort to learn more about the still largely unexplored realm of marine-related products.

Besides microorganisms like bacteria, fungi, and actinobacteria, many other marine organisms such as fishes, prawns, crabs, snakes, plants, and algae have also been studied to tap the arsenal of the marine world. Properties like high salt tolerance, hyperthermostability, barophilicity, cold adaptivity, and ease in large-scale cultivation are the key interests of scientists. These properties may not be expected in terrestrial sources as marine organisms thrive in habitats such as hydrothermal vents, oceanic caves, and some areas where high pressure and the absence of light are obvious (Debashish et al., 2005).

Marine bioactive compounds are organic compounds produced by both prokaryotes and eukaryotes. These compounds generally help the host organism to protect themselves and to maintain homeostasis in their environment. So far, less than 1.0% of the total marine organisms producing bioactive metabolites are revealed. Based on the review of literature, the seawater has bactericidal properties and it endorsed the production of antibiotics by planktonic algae and bacteria, respectively (Lipton, 2003).

10.3.1   Marine microorganism source of protease inhibitor

Imada et al. (1985) isolated the first protease inhibitor–producing marine bacteria from coastal seawater at the Aburatsubo Inlet of Sagami Bay in Kanagawa Prefecture, Japan. Based on the amino acid sequences, it is identified as a marinostatin. It is capable to inhibit the serine protease (Imada et al., 1986). Monastatin was the second protease inhibitor that had an inhibitory activity against thiol protease (Imada et al., 1985). Leupeptin was the third protease inhibitor that had an inhibitory activity against thiol and serine proteases (Hamato et al., 1992). These all are produced by *Alteromonas* sp. metalloprotease inhibitor from *Spirulina platensis* capable to inhibit the melanoma, carcinoma, and fibrosarcoma cell lines (Mishima et al., 1998). The antioxidants extracted from *Chlorella vulgaris* have proven inhibiting MMP-mediated cancer proliferation and progression (Wang et al., 2010).

Enzyme inhibitors are the third important product of marine actinobacteria. Although it is used for the study of enzyme structures and reaction mechanisms, their use in pharmacology was started late (Bode and Huber, 1993). These selective inhibitors can be used as a powerful tool for inactivating target proteases in the pathogenic processes of human diseases such as malaria, emphysema, arthritis, pancreatitis, thrombosis, high blood pressure, muscular dystrophy, cancer, and AIDS (Demuth, 1990). Enzyme inhibitors from marine microorganisms were less studied. However, *Streptomyces* isolated from terrestrial environment is a one of the potential producers of enzyme inhibitors (Umezawa et al., 1972). Some of the secondary metabolite from actinobacteria was acting as an enzyme inhibitor such as streptomycin (Martin, 1989), actinomycin (Katz and Weissbach, 1962), chloramphenicol (Jones and Westlake, 1974), and candicidin (Liras et al., 1977). The isolation of novel enzyme inhibitor from terrestrial sources is rare; hence, marine actinobacteria will provide new potential inhibitors.

10.3.2 Marine animal source of protease inhibitor

In general, marine animals, particularly marine invertebrates, are the promising sources of novel bioactive compounds. This is an adaptation strategy to thrive in the extreme environmental conditions of the sea and as a defense strategy to escape from predators by the marine invertebrates especially soft-bodied animals like sponges (Muller et al., 2004). Focusing on sponges, a conceptual progress occurred with the study of Thakur et al. (2003), who suggested that marine animals and their symbiotic microorganisms (bacteria and fungi) produce an array of bioactive compounds against foreign attackers. Pharmaceutical interest in sponges was aroused in the early 1950s by the discovery of a number of unknown nucleosides: spongothymidine and spongouridine in the marine sponge *Cryptotethia crypta* (Bergmann and Burke, 1956, Bergmann et al., 1957).

Marine organisms are rich sources of structurally diverse bioactive compounds with various biological activities. The importance of marine organisms as a source of novel bioactive substances is growing rapidly. With marine species comprising approximately one-half of the total global biodiversity, the sea offers an enormous resource for novel compounds. Proteases are enzymes that catalyze the hydrolysis of the peptide bonds forming the primary structure of proteins (Dixon and Webb, 1979). Evolved over millions of years, they are a goldmine of genetic diversity and novel secondary metabolites. They are common to organisms, from microorganism to plants and animals. In higher animals, proteases are involved in several processes including digestion, proenzyme activation, and defense mechanisms such as blood clotting and complement activation. Due to several reasons, namely, inaccessibility of their habitats and very low yield of bioactive metabolites, systematic chemical and pharmacological investigations of these organisms were not popular until a few decades ago. However, the recent advances in underwater exploration, natural product chemistry, genome mining, and bioassays have led to a great surge in the search for novel biomolecules from this rather underexploited habitat. A cursory review of the literature indicates that more than 70% of marine metabolites are obtained from marine sponges, corals, and microorganisms. The contribution from other organisms like molluscs, ascidians, and algae adds up to only 30% (Blunt et al., 2007). More than 90% of marine bacteria are gram-negative psychrophiles, and nearly half of them require high Na^+ concentration for growth (Macleod, 1968). Their microbial growth and the products of their metabolic activities differ considerably from those of terrestrial microorganisms. Maeda and Taga (1976) reported that the activity of deoxyribonuclease isolated from a marine *Vibrio* sp. was activated strongly by Mg^{2+} and stabilized by Ca^{2+}. Kobori and Taga (1980) isolated a marine phosphatase-producing bacterium. At 1000 atm hydrostatic pressure, the enzyme activity of this microbe was more than three times higher than that at 1 atm. This implies that enzymes of marine microorganisms may also differ considerably from those of terrestrial organisms. Therefore, inhibitors of these enzymes can be expected to show different characteristics than those of terrestrial inhibitors. Enzyme inhibitors have received increasing attention as useful tools, not only for the study of enzyme structures and reaction mechanisms but also for potential utilization in pharmacology (Bode and Huber, 1993). Specific and selective protease inhibitors are potentially powerful tools for inactivating target proteases in the pathogenic processes of human diseases such as emphysema, arthritis, pancreatitis, thrombosis, high blood pressure, muscular dystrophy, cancer, and AIDS.

Protease inhibitors can be further classified into five groups (serine, threonine, cysteine, aspartyl, and metalloprotease inhibitors) according to the mechanism employed at the active site of proteases they inhibit. Some protease inhibitors interfere with more than one type of protease. For example, the serine family of protease inhibitors (serpins) is generally thought of as active against serine proteases, yet contains several important inhibitors of cysteine proteases as well. Proteolytic inhibition by protease inhibitors can occur via two mechanisms: irreversible trapping reactions and reversible tight binding reactions (Rawlings et al., 2004). Inhibitors that bind through a trapping mechanism change conformation after cleaving an internal peptide bond and *trap* the enzyme molecule covalently; neither the inhibitor nor the protease can participate in further reactions. In tight binding reactions, the inhibitor binds directly to the active site of the protease; these reactions are reversible, and the inhibitor can dissociate from the enzyme in either the virgin state or after modification by the protease (Fear et al., 2007).

10.3.3   Classification of protease inhibitors

Protease inhibitors are classifieds based on the type of protease they inhibit.

10.3.3.1   Cysteine protease inhibitors

Nature continues to be one of the most significant sources of pharmacologically active compounds in the quest for drugs. *Penicillium* species are widely distributed in nature and often found living on foods and in indoor environments. They are the source of several β-lactam antibiotics, most significantly penicillin. Other secondary metabolites of *Penicillium chrysogenum* include various different penicillins, roquefortine C, meleagrin, chrysogine, xanthocillins, secalonic acids, sorrentanone, sorbicillin, and PR toxin (de Hoog et al., 2000). Several undescribed, marine-derived *Penicillium* sp. have been recently isolated from a variety of substrates such as mollusks, sponges, algae, and sands. These *Penicillium* species are important producers of new metabolites such as sculezonones A and B (Komatsu et al., 2000).

Serine and cysteine protease inhibitors named marinostatins and monastatins, respectively, were isolated from a marine bacterial strain collected from a neritic Japanese marine environment. The isolated strain, *Alteromonas* sp. (Imada et al., 1985), is a gram-negative rod-shaped microorganism with DNA of low G + C contents. It requires seawater for growth and produces both serine and cysteine protease inhibitors (Imada et al., 1985). Marine microorganisms have become an important source of pharmacologically active metabolites. Published reviews show the importance of these organisms as potential sources of pharmaceutical leads. More specifically, fungi from the marine environment have shown great potential as suggested by the diversity of secondary metabolites. Equistatin, a protease inhibitor isolated from the hydrophilic extract of the whole body of *Actinia equine* (Lenarcic et al., 1998), is an acidic protein composed of three thyroglobulin type 1 domains (Galea et al., 2000). It was further shown that the N-terminal domain alone acts as a Cys protease inhibitor (K_i of 0.6 nM for papain) and the second domain acts as an aspartic protease inhibitor (K_i of 0.3 for cathepsin D). The function of the third domain remains unknown (Štrukelj et al., 2000). Sea anemones contain a variety of polypeptide toxins, presumably in specialized stinging organelles (nematocysts). The sea anemone polypeptide toxins are considered to serve as chemical weapons to capture prey animals and as defensive substances against predators. A number of protease inhibitors have already been isolated from *Actinia equine* (Ishida et al., 1997), *Anemonia sulcata* (Wunderer et al., 1981; Schweitz et al., 1995), *Anthopleura aff. xanthogrammica* (Minagawa et al., 1997, 1998), *Radianthus macrodactylus* (Zykova, 1985), *Stichodactyla helianthus* (Antuch et al., 1993), and *Stichodactyla haddoni* (Honma et al., 2008), although their physiological function is not fully understood yet. Interestingly, three protease inhibitors (AsKC 1–3 or kalicludines 1–3) from *A. sulcata* (Schweitz et al., 1995) and one inhibitor (SHTX III) from *S. haddoni* (Honma et al., 2008) have been demonstrated to exhibit blocking activity against Kv1 potassium channels as well as protease inhibitory activity.

10.3.3.2   Serine protease inhibitors

Marine microorganisms are widely recognized as new frontiers in natural product research. In particular, the actinomycete bacteria and higher fungi from diverse marine environments are very prolific sources of structurally unique and biologically active metabolites. Since the 1990s, more than a few hundred novel compounds have been annually isolated from these organisms, which significantly contribute to both the chemical and biomedical aspects of natural product research. The bioactivities of these marine-derived microbial metabolites are highly diverse but with a focus on cytotoxic and antimicrobial activities. Among the antimicrobial-related bioactivities, the sortase enzymes are regarded as being a promising target for therapy because gram-positive bacteria proteins are displayed on the cell surface with these enzymes (Bugni and Ireland, 2004; Cragg et al., 2009; Clancy et al., 2010; Julianti et al., 2013).

Serine protease inhibitors are the largest and most widely distributed superfamily of protease inhibitors (Hedstrom, 2002; Krowarsch et al., 2003; Di Cera, 2009), and based on their possession of conserved functional motifs, they can be subdivided into many classes, being the Kunitz-type inhibitors, the best characterized of them, probably due to their abundance in several organisms (Lingaraju and Gowda, 2008; Yuan et al., 2008; Zhao et al., 2011; Isaeva et al., 2012). The Kunitz-type motif consists of a polypeptide chain of ~60 amino acid residues stabilized by three disulfide bridges (C_I–C_{VI}, C_{II}–C_{IV}, C_{III}–C_V). Marine cyanobacteria have emerged as a rich source of biologically and ecologically active secondary metabolites, in particular peptides and depsipeptides (Gunasekera et al., 2009). The Kunitz-type protease inhibitors are the best characterized family of serine protease inhibitors, probably due to their abundance in several organisms. These inhibitors consist of a chain of ~60 amino acid residues stabilized by three disulfide bridges and were first observed in the bovine pancreatic trypsin inhibitor (BPTI) like protease inhibitors, which strongly inhibit trypsin and chymotrypsin. Canonical serine protease inhibitors have been the most extensively studied so far, and 18 families are currently recognized (Laskowski et al., 2000). Among them, the BPTI Kunitz family (PFAM: PF00014), which comprises extremely potent inhibitors of serine proteases, is composed of more than 1000 different sequences isolated from a variety of animals ranging from invertebrates to mammals (Kunitz and Northrop, 1936; Delfín et al., 1996). De Almeida Nogueira et al. (2013) investigated the interference of the *S. helianthus* Kunitz-type serine protease inhibitor (ShPI-I), a 55–amino acid peptide, in *Trypanosoma cruzi* serine peptidase activities, parasite viabilities, and parasite morphologies. Potent biological activity is often found in products isolated from marine organisms because of their novel molecular structures (Donia and Hamann, 2003; Molinski et al., 2009). Trabectedin (Yondelis), cytarabine (Ara-C), and eribulin (Halaven), which are known as antitumor drugs, were developed from compounds found in marine organisms (Mayer et al., 2010). Two new potent serine protease inhibitors, cyclotheonamides E2 and E3, have been isolated from a marine sponge of the genus *Theonella* (Nakao et al., 1998). The effects of classical serine protease inhibitors and a Kunitz-type inhibitor, obtained from sea anemone *S. helianthus* (ShPI-I), on the viability and morphology of parasites in *Leishmania* culture were studied. The *N*-tosyl-L-phenylalanine chloromethyl ketone (TPCK) and benzamidine (Bza) inhibitors, which are potential *Leishmania* protease inhibitors, in all experimental conditions reduced the parasite viability, with regard to time dependence. On the other hand, *N*-tosyllysine chloromethyl ketone did not significantly affect the parasite viability as it was a poor *Leishmania* enzyme inhibitor. Ultrastructural analysis demonstrated that both Bza and TPCK induced changes in the flagellar pocket region with membrane alteration, including bleb formation. However, TPCK effects were more pronounced than those of Bza in *Leishmania* flagellar pocket in plasma membrane, and intracellular vesicular bodies were visualized. ShPI-I proved to be a powerful inhibitor of *L. amazonensis* serine proteases and the parasite viability (Silva-Lopez et al., 2007).

Also, there are reports of isolation of protease inhibitors, for example, the bioassay-guided isolation of serine protease inhibitors from a marine sponge, *Coscinoderma mathewsi*, has yielded 1-methylherbipoline salts of halisulfate 1 and of suvanine (Kimura et al., 1998). Two cyclodepsipeptides named kempopeptins A and B were isolated from a collection of a Floridian marine cyanobacterium *Lyngbya* sp. that had previously afforded the structurally related potent elastase inhibitors lyngbyastatin 7 and somamide B (Taori et al., 2008).

A serine protease inhibitor, CvSI-1, was recently identified from the plasma of the eastern oyster, *Crassostrea virginica* (Xue et al., 2006). CvSI-1 is a 7609.6 Da protein consisting of a 71–amino acid sequence with no homology to any known protein sequence in databases. CvSI-1 inhibited subtilisin A, trypsin, and perkinsin, the major serine protease secreted by the oyster protozoan parasite *Perkinsus marinus*, in a slow tight binding manner (Xue et al., 2006).

The Ras family of guanine nucleotide–binding proteins plays important roles in signal transduction and the regulation of cell differentiation and proliferation. The Ras protein requires posttranslational processing in order to associate with the plasma membrane and to function in signal transduction or cellular transformation. The first processing step is catalyzed by farnesyl protein transferase (FPT), which adds a farnesyl group to a cysteine residue near the carboxy terminus of the Ras protein. Inhibition of FPT is a potential therapeutic target for novel anticancer agents. Cembranolide diterpene, isolated from the soft coral *Lobophytum cristagalli*, showed potent inhibitory activity against FPT with an IC_{50} value of 0.15 µM.

Serine, cysteine, and threonine proteases have many common active site features including an active site nucleophile and a general base, which are often the target of irreversible inhibitors. So far, this group includes the majority of proteolytic enzymes and many significant enzymes with involvement in human diseases. We will cover inhibitors commonly considered to be irreversible. This includes inhibitors that form *stable* covalent bonds with the enzyme. Covalent irreversible inhibitors of cysteine, serine, and threonine proteases are capable of either alkylating or acylating their target enzymes. Alkylating agents are very effective cysteine protease inhibitors, and they include chloromethyl ketones, fluoromethyl ketones, diazomethyl ketones, acyloxymethyl ketones, epoxides, and vinyl sulfones. The nucleophilic active site cysteine attacks an activated carbon and forms an irreversible carbon–sulfur bond. Because the active site serine of serine proteases is generally less nucleophilic than the corresponding catalytic cysteine, alkylating agents primarily target cysteine proteases. An exception is the chloromethyl ketones, which are capable of inhibiting serine proteases, although they have a slightly different mechanism of inhibition. After a nucleophilic attack of the carbonyl of the inhibitor, the inhibitor alkylates the catalytic histidine. The enzyme can then be deacylated, but the alkylation of the catalytic histidine results in a covalently bound inhibitor (Farady and Craik, 2010).

10.3.3.3 *Threonine protease inhibitors*

Threonine proteases are part of a multicomponent proteasome complex in microbial cells. The archaebacterial proteasome has 14 active sites in the inner channel, one on each β-subunit. The hydrolytic sites are spatially separated from the intracellular components. Recent reports have indicated that the active site nucleophile is the hydroxyl group on the threonine at the

N-terminus of the β-subunit. The replacements of the terminal threonine by serine in archae-bacterial proteasomes allow complete proteolytic activity. Therefore, the conservation of the threonines in the active sites of all threonine proteases from bacteria to eukaryotes is unclear. Looking at the diverse functions of the threonine proteases in bacteria and mammals, it is evident that the phylogenetically ancient proteasome has undergone adaptations that favor different functions in different physiological situations (Menon and Rao, 2012).

10.3.3.4 Aspartic protease inhibitors

Proteases are responsible either directly or indirectly for all bodily functions, including cell growth, differentiation, and death (apoptosis), cell nutrition, intra- and extracellular protein turnover (housekeeping and repair), cell migration and invasion, and fertilization and implantation. These functions extend from the cellular level to the organ and organism levels to produce cascade systems such as homeostasis and inflammation and complex processes at all levels of physiology and pathophysiology. Any system that encompasses normal and abnormal bodily functions must have effective regulatory counterparts, that is, protease inhibitors. Hence, the research interest in protease inhibitors has evoked tremendous attention in many disciplines. Multicellular organisms possess endogenous protein protease inhibitors to control proteolytic activity. Most of these inhibitory proteins are directed against serine proteases, although some are known to target cysteine, aspartyl, or metalloproteases. Indeed, inhibitors of serine, cysteine, and metalloproteases are distributed ubiquitously throughout the biological world (Menon and Rao, 2012). Xylanases (1,4-β-D-xylan xylanohydrolase) are glycosidases that catalyze the hydrolytic cleavage of β-1,4-linked polymers of D-xylose. They have raised enormous interest in the past decade in view of their application in clarification of juices and wines, conversion of renewable biomass into liquid fuels, and development of environmentally sound prebleaching processes in the paper and pulp industry (Kulkarni et al., 1999). Although extensive studies have been carried out on the industrial applications of xylanases, there is a paucity of reports on their molecular enzymology and clinical implications. Recently, glycosidases have been studied with a clinical perspective of locating enzyme allergens (Tarvainen et al., 1991). Some of these enzymes, including xylanases and cellulases, have been found to cause occupational and nonoccupational allergies, such as respiratory and irritant contact dermatitis. Therefore, from the biomedical point of view, inhibitors of this class of enzymes will have tremendous importance in the near future. In addition, the inhibition of cellulolytic and hemicellulolytic enzymes has potential applications to prevent the degradation of wood and cloth by the action of the hydrolytic enzymes present in the gut of termites (Menon and Rao, 2012). A bifunctional low-molecular-weight, linear, peptidic inhibitor API, from thermotolerant *Bacillus licheniformis*, is reported to exhibit a slow tight binding inhibition against ChiA. The bifunctionality of the inhibitor can be defined as it was previously reported to inhibit an aspartic protease, pepsin (Kumar and Rao, 2006).

Class-specific transition state analogs have been developed to interfere specifically with the catalytic residues of each class of proteases. Aspartic protease inhibitors have long been designed around substrate polypeptides, with a replacement of the scissile amide bond with a noncleavable, transition-state isostere. The first specific inhibitor for aspartic proteases, pepstatin A, was discovered from *Actinomyces*, as an inhibitor for pepsin. It also showed strong inhibitory activity against several other aspartic proteases. Pepstatin A is a peptide, but the scissile bond is replaced with a statin group [(3S,4S)-4-amino-3-hydroxyl-6-methylheptanoic acid]. Instead of a trigonal carbonyl, statins have a chiral hydroxyl group, giving it the ability to mimic the tetrahedral state of the substrate transition state (Eder et al., 2007).

10.4   Conclusion

The family of protease inhibitors, although rather small, contains a number of validated and potential drug targets making drug discovery efforts in this area very fruitful and exciting. There have been substantive advances in our understanding of the use of protease inhibitors as therapeutic agents. Several synthetic protease inhibitors have been approved by the FDA for therapy of HIV and hypertension. These drugs represent prime examples of structure-based drug design. Moreover, the inhibitory principles and compounds, which have been established and discovered, now enable mechanism-based drug discovery across the whole family. The sequencing of the human genome and the resulting knowledge on all human aspartic proteases allow an exhaustive profiling of inhibitors for specificity toward all family members, thereby reducing the risk of unwanted side effects. A number of natural and peptidomimetic inhibitors performed well in different phases of clinical testing to treat other human disorders, including cancer, inflammation, cardiovascular, neurodegenerative, and various infectious diseases. Despite this impressive progress, there is much to learn about the cross talk between signal transduction pathways and protease activation cascades. Additionally, the development of successful protease inhibitors for clinical use is reliant on maximizing bioavailability, specificity, and potency of inhibition of the target enzyme. Ideally, localizing protease inhibitors to a single target area of the body may also help minimize the potential for complications and detrimental side effects. There is the further issue of the development of drug resistance to protease inhibitors in the face of a buildup of substrate pressure and selection of catalytically active mutant or other salvage proteases that do not have complementarity for carefully designed inhibitors of wild-type proteases. The future appears to still hold considerable promise for protease inhibitors. We can anticipate new, overexpressed proteases from genomic/biochemical comparisons made between normal/diseased cells, host/pathogen, healthy/unhealthy subjects leading to more effective and efficient validation of proteases as drug targets. New advances in protein chemistry will lead to faster production and greater quantities of pure recombinant proteases, and advances in structural biology (crystallography, NMR spectroscopy) will produce faster and more accurate inhibitor protease structures. Inhibitors (naturally occurring and synthetic) have permitted detailed biochemical and crystallographic investigations to be made, but an understanding of the selectivity of such inhibitors may be of just as much importance for the design and synthesis of specific inhibitors for use therapeutically in controlling individual aspartic proteases. The discovery of novel selective inhibitors can proceed only through the combination of screening of chemical libraries, rational design, computational technology, and exploration of natural compounds. The exploitation of vast microbial diversity will also generate a large amount of biologic aspartic protease inhibitors. Furthermore, future research into the synergistic capabilities of inhibitors will help elucidate the most effective combination therapies. These advances, together with more careful attention to inhibitor conformation, mechanism of action, and drug-like composition, are expected to result in more potent, more selective, and more bioavailable inhibitors with a higher probability of success in the clinic.

References

Antuch, W., Berndt, K. D., Chavez, M. A., Delfin, J., and Wuthrich, K. 1993. The NMR solution structure of a Kunitz type proteinase inhibitor from the sea anemone *Stichodactyla helianthus*. *Eur J Biochem* 212(3):675–684.

Bellail, A. C., Hunter, S. B., Brat, D. J., Tan, C., and Van Meir, E. G. 2004. Microregional extracellular matrix heterogeneity in brain modulates glioma cell invasion. *Int J Biochem Cell Biol* 36(6):1046–1069.

Bergmann, W. and Burke, D. C. 1956. Contributions to the study of marine products. XL. The nucleosides of sponges. 1 IV. Spongosine 2. *J Org Chem* 21(2):226–228.

Bergmann, W., Watkins, J. C., and Stempien Jr, M. F. 1957. Contributions to the study of marine products. XLV. Sponge nucleic acids, 1. *J Org Chem* 22(11):1308–1313.

Blunt, J. W., Copp, B. R., Hu, W. P., Munro, M. H., Northcote, P. T., and Prinsep, M. R. 2007. Marine natural products. *Nat Prod Rep* 24(1):31–86.

Bode, W. and Huber, R. 1993. Natural protein proteinase inhibitors and their interaction with proteinases. *Eur J Biochem Rev* 204:43–61.

Bugni, T. S. and Ireland, C. M. 2004. Marine-derived fungi: A chemically and biologically diverse group of microorganisms. *Nat Prod Rep* 21(1):143–163.

Chambers, L. S., Black, J. L., Poronnik, P., and Johnson, P. R. 2001. Functional effects of protease-activated receptor-2 stimulation on human airway smooth muscle. *Am J Physiol* 281:L1369–L1378.

Clancy, K. W., Melvin, J. A., and McCafferty, D. G. 2010. Sortase transpeptidases: Insights into mechanism, substrate specificity, and inhibition. *Biopolymers* 94(4):385–396.

Cragg, G. M., Grothaus, P. G., and Newman, D. J. 2009. Impact of natural products on developing new anti-cancer agents. *Chem Rev* 109(7):3012–3043.

De Almeida Nogueira, N. P., Morgado-Diaz, J. A., Menna-Barreto, R. F., Paes, M. C., and da Silva-Lopez, R. E. 2013. Effects of a marine serine protease inhibitor on viability and morphology of *Trypanosoma cruzi*, the agent of chagas disease. *Acta Trop* 128(1):27–35.

De Hoog, G. S., Guarro, J., Gene, J., and Figueras, M. J. 2000. *Atlas of Clinical Fungi*. Centraalbureau voor Schimmelcultures (CBS).

Debashish, G., Malay, S., Barindra, S., and Joydeep, M. 2005. Marine enzymes. *Mar Biotechnol* 1, 189–218.

Delfin, J., Martinez, I., Antuch, W., Morera, V., Gonzalez, Y., Rodriguez, R., Marquez, M. et al. 1996. Purification, characterization and immobilization of proteinase inhibitors from *Stichodactyla helianthus*. *Toxicon* 34(11–12):1367–1376.

Demuth, H. U. 1990. Recent developments in inhibiting cysteine and serine proteases. *J Enzyme Inhib* 3(4):249–278.

Di Cera, E. 2009. Serine proteases. *IUBMB Life* 61(5):510–515.

Dixon, M. and Webb, E. C. 1979. *Enzymes*. Longam Group Limited, London, U.K.

Donia, M. and Hamann, M. T. 2003. Marine natural products and their potential applications as anti-infective agents. *Lancet Infect Dis* 3(6):338–348.

Drag, M. and Salvesen G. S. 2010. Emerging principles in protease-based drug discovery. *Nat Rev Drug Discov* 9(9):690–701.

Eder, J., Hommel, U., Cumin, F., Martoglio, B., and Gerhartz, B. 2007. Aspartic proteases in drug discovery. *Curr Pharm Des* 13(3):271–285.

Farady, C. J. and Craik, C. S. 2010. Mechanisms of macromolecular protease inhibitors. *Chembiochem* 11(17):2341–2346.

Fear, G., Komarnytsky, S., and Raskin, I. 2007. Protease inhibitors and their peptidomimetic derivatives as potential drugs. *Pharmacol Ther* 113(2):354–368.

Fitzpatrick, B. and O'Kennedy, R. 2004. The development and application of a surface plasmon resonance-based inhibition immunoassay for the determination of warfarin in plasma ultrafiltrate. *J Immunol Methods* 291(1–2):11–25.

Galea, K., Bavec, S., Turk, V., and Lenarcic, B. 2000. Cloning and expression of functional equistatin. *Biol Chem* 381(1):85–88.

Glinsky, G. E. 1993. Cell adhesion and metastasis: Is the site specificity of cancer metastasis determined by leukocyte-endothelial cell recognition and adhesion? *Crit Rev Oncol Hematol* 14(3):229–278.

Gunasekera, S. P., Miller, M. W., Kwan, J. C., Luesch, H., and Paul, V. J. 2009. Molassamide, a depsipeptide serine protease inhibitor from the marine cyanobacterium *Dichothrix utahensis*. *J Nat Prod* 73(3):459–462.

Hamato, N., Takano, R., Kamei-Hayashi, T., Imada, C., and Hara, S. 1992. Leupeptins produced by the marine *Alteromonas* sp. B-10–31. *Biosci Biotech Biochem* 56:1316–1318.

Hedstrom, L. 2002. Serine protease mechanism and specificity. *Chem Rev* 102(12):4501–4524.

Honma, T., Kawahata, S., Ishida, M., Nagai, H., Nagashima, Y., and Shiomi, K. 2008. Novel peptide toxins from the sea anemone *Stichodactyla haddoni*. *Peptides* 29(4):536–544.

Imada, C., Maeda, M., Hara, S., Taga, N., and Simidu, U. 1986. Purification and characterization of subtilisin inhibitors 'Marinostatin' produced by marine *Alteromonas* sp. *J Appl Microbiol* 60(6):469–476.

Imada, C., Nobuo, T., and Masachika, M. 1985. Cultivation conditions for subtilisin inhibitor-producing bacterium and general properties of the inhibitor marinostatin. *Bull Japan Soc Sci Fish* 505–810.

Ireland, P., Jolley, D., Giles, G., O'Dea, K., Powles, J., Rutishauser, I., and Williams, J. 1994. Development of the Melbourne FFQ: A food frequency questionnaire for use in an Australian prospective study involving an ethnically diverse cohort. *Asia Pac J Clin Nutr* 3(1):19–31.

Isaeva, M. P., Chausova, V. E., Zelepuga, E. A., Guzev, K. V., Tabakmakher, V. M., Monastyrnaya, M. M., and Kozlovskaya, E. P. 2012. A new multigene superfamily of Kunitz-type protease inhibitors from sea anemone *Heteractis crispa*. *Peptides* 34(1):88–97.

Ishida, M., Minagawa, S., Miyauchi, K., Shimakura, K., Nagashima, Y., and Shiomi, K. 1997. Amino acid sequences of Kunitz-type protease inhibitors from the sea anemone *Actinia equina*. *Fish Sci* 63(5):794–798.

Jones, A. and Westlake, D. W. S. 1974. Regulation of chloramphenicol synthesis in *Streptomyces* sp. 3022 a. properties of arylamine synthetase, an enzyme involved in antibiotic biosynthesis. *Can J Microbiol* 20(11):1599–1611.

Julianti, E., Lee, J. H., Liao, L., Park, W., Park, S., Oh, D. C., Oh, K. B., and Shin, J. 2013. New polyaromatic metabolites from a marine-derived fungus *Penicillium* sp. *Org Lett* 15(6):1286–1289.

Katz, E. and Weissbach, H. 1962. Biosynthesis of the actinomycin chromophore; enzymatic conversion of 4-methyl-3-hydroxyanthranilic acid to actinocin. *J Biol Chem* 237(3):882–886.

Kim, J, Yu, W., Kovalski, K., and Ossowski, L. 1998. Requirement for specific proteases in cancer cell intravasation as revealed by a novel semiquantitative PCR-based assay. *Cell* 94(3):353–362.

Kimura, J., Ishizuka, E., Nakao, Y., Yoshida, W. Y., Scheuer, P. J., and Kelly-Borges, M. 1998. Isolation of 1-methylherbipoline salts of halisulfate-1 and of suvanine as serine protease inhibitors from a marine sponge, *Coscinoderma mathewsi*. *J Nat Prod* 61(2):248–250.

Kobori, H. and Taga, N. 1980. Extracellular alkaline phosphatase from marine bacteria: Purification and properties of extracellular phosphatase from a marine *Pseudomonas* sp. *Can J Microbiol* 26(7):833–838.

Komatsu, K., Shigemori, H., Mikami, Y., and Kobayashi, J. 2000. Sculezonones A and B, two metabolites possessing a phenalenone skeleton from a marine-derived fungus *Penicillium* species. *J Nat Prod* 63(3):408–409.

Krowarsch, D., Cierpicki, T., Jelen, F., and Otlewski, J. 2003. Canonical protein inhibitors of serine proteases. *Cell Mol Life Sci* 60(11):2427–2444.

Kulkarni, N., Shendye, A., and Rao, M. 1999. Molecular and biotechnological aspects of xylanases. *FEMS Microbiol Rev* 23(4):411–456.

Kumar, A. and Rao, M. 2006. Biochemical characterization of a low molecular weight aspartic protease inhibitor from thermo-tolerant *Bacillus licheniformis*: Kinetic interactions with pepsin. *Biochim Biophys Acta* 1760(12):1845–1856.

Kunitz, M. and Northrop, J. H. 1936. Isolation from beef pancreas of crystalline trypsinogen, trypsin, a trypsin inhibitor, and an inhibitor-trypsin compound. *J Gen Physiol* 19(6):991–1007.

Laskowski Jr, M. I. C. H. A. E. L., Qasim, M. A., and Lu, S. M. 2000. Interaction of standard mechanism, canonical protein inhibitors with serine proteinases. *Protein–Protein Recognition* (Kleanthous, C., ed.), Oxford University Press, Oxford, U.K. pp. 228–279.

Lenarcic, B., Ritonja, A., Strukelj, B., Turk, B., and Turk, V. 1998. Equistatin, a new inhibitor of cysteine proteinases from *Actinia equina*, is structurally related to thyroglobulin type-1 domain. *J Biol Chem* 273(20):12682.

Lingaraju, M. H. and Gowda, L. R. 2008. A Kunitz trypsin inhibitor of *Entada scandens* seeds: Another member with single disulfide bridge. *Biochim Biophys Acta* 1784(5):850–855.

Lipton, A. P. 2003. Isolation and application of marine natural products. 1–8.

Liras, P., Villanueva, J. R., and Martin, J. F. 1977. Sequential expression of macromolecule biosynthesis and candicidin formation in *Streptomyces griseus*. *J Gen Microbiol* 102(2):269–277.

Macleod, R. A. 1968. On the role of inorganic ions in the physiology of marine bacteria. *Adv Microbiol Sea* 1:95–126.

Maeda, M. and Taga, N. 1976. Extracellular nuclease produced by a marine bacterium. II. Purification and properties of extracellular nuclease from a marine *Vibrio* sp. *Can J Microbiol* 22(10):1443–1452.

Martin, J. F. 1989. Molecular mechanisms for the control by phosphate of the biosynthesis of antibiotics and other secondary metabolites. *Regulation of Secondary Metabolism in Actinomycetes*: 213–237.

Mayer, A. M., Glaser, K. B., Cuevas, C., Jacobs, R. S., Kem, W., Little, R. D., McIntosh, J. M., Newman, D. J., Potts, B. C., and Shuster, D. E. 2010. The odyssey of marine pharmaceuticals: A current pipeline perspective. *Trends Pharmacol Sci* 31(6):255–265.

Menon, V. and Rao, M. 2012. Microbial aspartic protease inhibitors. 1–30.

Minagawa, S., Ishida, M., Shimakura, K., Nagashima, Y., and Shiomi, K. 1997. Isolation and amino acid sequences of two Kunitz-type protease inhibitors from the sea anemone *Anthopleura* aff. *xanthogrammica*. *Comp Biochem Physiol B Biochem Mol Biol* 118(2):381–386.

Minagawa, S., Ishida, M., Shimakura, K., Nagashima, Y., and Shiomi, K. 1998. Amino acid sequence and biological activities of another Kunitz-type protease inhibitor from the sea anemone *Anthopleura* aff. *xanthogrammica*. *Fish Sci* 64:157–161.

Mishima, T., Murata, J., Toyoshima, M., Fujii, H., Nakajima, M., Hayashi, T., and Saiki, I. 1998. Inhibition of tumor invasion and metastasis by calcium spirulan (Ca-SP), a novel sulfated polysaccharide derived from a blue-green alga, *Spirulina platensis*. *Clin Exp Metastasis* 16(6):541–550.

Mittl, P. R. and Grütter, M. G. 2006. Opportunities for structure-based design of protease-directed drugs. *Curr Opin Struct Biol* 16(6):769–775.

Molinski, T. F., Dalisay, D. S., Lievens, S. L., and Saludes, J. P. 2009. Drug development from marine natural products. *Nat Rev Drug Discov* 8(1):69–85.

Muller, W. E., Schroder, H. C., Wiens, M., Perovic-Ottstadt, S., Batel, R., and Muller, I. M. 2004. Traditional and modern biomedical prospecting: Part II-the benefits: Approaches for a sustainable exploitation of biodiversity (secondary metabolites and biomaterials from sponges). *Evid Based Complement Alternat Med* 1(2):133–144.

Nakao, Y., Oku, N., Matsunaga, S., and Fusetani, N. 1998. Cyclotheonamides E2 and E3, new potent serine protease inhibitors from the marine sponge of the genus *Theonella*. *J Nat Prod* 61(5):667–670.

Rawlings, N. D., Tolle, D. P., and Barrett, A. J. 2004. Evolutionary families of peptidase inhibitors. *Biochem J* 378(Pt 3):705–716.

Schweitz, H., Bruhn, T., Guillemare, E., Moinier, D., Lancelin, J. M., Beress, L., and Lazdunski, M. 1995. Kalicludines and kaliseptine two different classes of sea anemone toxins for voltage sensitive K⁺ channels. *J Biol Chem* 270(42):25121–25126.

Silva-Lopez, R. E., Morgado-Diaz, J. A., Chavez, M. A., and Giovanni-De-Simone, S. 2007. Effects of serine protease inhibitors on viability and morphology of *Leishmania amazonensis* promastigotes. *Parasitol Res* 101(6):1627–1635.

Štrukelj, B., Lenarčič, B., Gruden, K., Pungerčar, J., Rogelj, B., Turk, V., and Jongsma, M. A. 2000. Equistatin, a protease inhibitor from the sea anemone *Actinia equina*, is composed of three structural and functional domains. *Biochem Biophys Res Commun* 269(3):732–736.

Taori, K., Paul, V. J., and Luesch, H. 2008. Kempopeptins A and B, serine protease inhibitors with different selectivity profiles from a marine cyanobacterium, *Lyngbya* sp. *J Nat Prod* 71(9):1625–1629.

Tarvainen, K., Kanerva, L., Tupasela, O., Grenquist-Norden, B., Jolanki, R., Estlander, T., and Keskinen, H. 1991. Allergy from cellulase and xylanase enzymes. *Clin Exp Allergy* 21(5):609–615.

Thakur, N. L., Hentschel, U., Krasko, A., Pabel, C. T., Anil, A. C., and Müller, W. E. 2003. Antibacterial activity of the sponge *Suberites domuncula* and its primmorphs: Potential basis for epibacterial chemical defense. *Aquatic Microb Ecol* 31(1):77–83.

Umezawa, S., Tatsuta, K., Fujimoto, K., Tsuchiya, T., and Umezawa, H. 1972. Structure of antipain, a new sakaguchi positive product of *Streptomyces. J Antibiot* 25(4):267–270.

Wang, H. M., Pan, J. L., Chen, C. Y., Chiu, C. C., Yang, M. H., Chang, H. W., and Chang, J. S. 2010. Identification of anti-lung cancer extract from *Chlorella vulgaris* CC by antioxidant property using supercritical carbon dioxide extraction. *Process Biochem* 45(12):1865–1872.

Wunderer, G., Machleidt, W., and Fritz, H. 1981. The broad-specificity proteinase inhibitor 5 II from the sea anemone *Anemonia sulcata. Methods Enzymol* 80:816–820.

Xue, Q. G., Waldrop, G. L., Schey, K. L., Itoh, N., Ogawa, M., Cooper, R. K., Losso, J. N., and La Peyre, J. F. 2006. A novel slow-tight binding serine protease inhibitor from eastern oyster (*Crassostrea virginica*) plasma inhibits perkinsin, the major extracellular protease of the oyster protozoan parasite *Perkinsus marinus. Comp Biochem Physiol B Biochem Mol Biol* 145(1):16–26.

Yuan, C. H., He, Q. Y., Peng, K., Diao, J. B., Jiang, L. P., Tang, X., and Liang, S. P. 2008. Discovery of a distinct superfamily of Kunitz-type toxin (KTT) from tarantulas. *PLoS One* 3(10):e3414.

Zhao, R., Dai, H., Qiu, S., Li, T., He, Y., Ma, Y., Chen, Wu, Z., Li, Y. W., and Cao, Z. 2011. SdPI, the first functionally characterized Kunitz-type trypsin inhibitor from scorpion venom. *PLoS One* 6(11):e27548.

Zilinskas, R. A. and Hill, R. T. 1995. The global challenge of marine biotechnology: A status report on the United States, Japan, Australia, and Norway. Maryland Sea Grant College.

Zykova, T. A. 1985. Amino-acid sequence of trypsin inhibitor IV from *Radianthus macrodactylus. Bioorg Khim* 11:293–301.

chapter eleven

Ganoderma

A bioresource of antimicrobials

K. Rajesh and Dharumadurai Dhanasekaran

Contents

11.1 Introduction

"Mushroom" is not a taxonomic category. The term "mushroom" should be used here according to the definition of Chang and Miles (1992) as "a macro fungus with a distinctive fruiting body, which can be hypogeous or epigeous, large enough to be seen with the naked eye and to be picked by hand." From a taxonomic point of view, mainly basidiomycetes but also some species of ascomycetes belong to mushrooms. Mushrooms constitute at least 14,000 and perhaps as many as 22,000 known species.

The natural products and herbal medicine industry have become increasingly popular over the past three decades (Hamburger and Hostettmann, 1991; Shu, 1998). The recognition of the value of traditional medical systems, particularly of Asian origin, and identification of indigenous medicinal plants that have shown to have healing power (Elvin Lewis, 2001) are factors that have had significant influence in the expansion of the natural product industry. Furthermore, there is a constant search for new and effective drugs, in order to overcome a number of pathogenic organism which are resistant to many therapeutic products available in market.

Mushrooms have been a major focus of investigations for novel biologically active compounds from natural resources, and in recent years, pharmaceutical companies have spent a lot of time developing these natural products to produce more affordable and cost-effective remedies (Farnsworth, 1994). In recent years, more mushrooms have been isolated and identified, and the number of mushrooms being cultivated for food or medicinal purposes has been increasing rapidly (Chang, 1995). Mushroom "nutriceuticals" are bioactive compounds that are extractable from mushrooms, and they have nutritional and medicinal features that may be used in the prevention and treatment of disease (Chang and Buswell, 1996).

Ganoderma has been used in folk medicine in China and Japan for over 2000 years for a wide range of ailments. *Ganoderma*, commonly known as Reishi mushroom, is one of the most important medicinal mushrooms widely used in various countries. *Ganoderma* has been regarded as a cure for all types of disease perhaps due to its demonstrated efficacy as a popular remedy to treat a large number of diseases, namely, chronic hepatitis, arthritis, hypertension, hyperlipidemia, bronchitis, asthma, gastic ulcer, diabetes, etc.; almost all medicinal properties have been attributed to *Ganoderma lucidum*, and thus, it is known as mushroom of immortality. Due to its ability to cure many different diseases, it received names like "elixir of life" and food of god. Its intracellular and extracellular polysaccharide shows inhibition of growth of several types of cancer cells. The different kinds of polysaccharide derived from mushrooms and their effects on host immune system are very important today.

Antimicrobial activity including antibacterial, antifungal, antiparasitic, and antiviral agents is the third widespread therapeutic effect reported in mushrooms (Kettering et al., 2005; Ngai and Ng, 2003; Obuchi et al., 1990; Okamoto and Li, 1993). Indeed, it is thought that over 200 higher fungus species showed antimicrobial properties (Anke and Sterner, 1991; Gianetti et al., 1986; Wang et al., 1993; Yoon et al., 1995). The combined biological properties of different mushroom genera such as *G. lucidum*, *G. frondosa*, *Lentinus edodes*, *Omphalotus illudens (Schwein.)*, *P. betulinus*, *P. ostreatus*, and *Rozites caperatus* were reported by Wasser and Weis (1999b), and Ying et al. (1987) generated a broad spectrum of antimicrobial activities including antibacterial effects (Beltran-Garcia et al., 1997; Cherqui et al., 1999; Donnelly et al., 1985; Dornberger et al., 1986; Min et al., 2000; Piraino, 2006; Wasser and Weis, 1999; Ying et al., 1987). *G. lucidum*, through a laccase, showed a potent inhibitory activity against HIV-1 reverse transcriptase. *G. frondosa* also demonstrates anti-HIV properties. Its action is simultaneously general and topical.

The limitations of traditional identification techniques indicate that alternative methods need to be developed for the identification of these fungi. With the development of molecular biology, some new techniques have been applied to fungal classical taxonomy. DNA fingerprinting techniques, however, would be allowed to identify the *Ganoderma* species and cultivars, indicating that it is a useful tool for the valid protection of newly bred cultivars.

11.2   Classification of Ganoderma

Ganoderma species belong to the kingdom of fungi, the division of Basidiomycota, the class of Homobasidiomycetes, the order of Aphyllophorales, the family of Polyporaceae, (Ganodermataceae), and the genus of *Ganoderma* (Chang, 1995; Wassser and Weis, 1999). Fungi from the family of Polyporaceae are classified as they have many tiny holes on the underside of the fruiting body, which are pores that contain the reproductive spores. They have a woody or leathery feel, and the presence of these pores is an obvious characteristic that distinguishes polypores from other common types of mushrooms. Polypore does, like other fungi, grow on wood as an expensive network of microscopic tubes known as mycelium. They degrade the wood over time and produce a fruiting body (or conk) on the surface of the wood. *Ganoderma* species are among those fungi that can thrive under hot and humid conditions and are usually found in subtropical and tropical regions (Moncalvo and Ryvarden, 1998).

Ganoderma species are not classified as edible, as the fruiting bodies are always thick, corky, and tough and do not have the fleshy texture characteristic of true edible mushrooms such as the common white button mushroom *Agaricus bisporus*. Although they are not classified as edible, several types of *Ganoderma* products are available on the market including ground fruiting bodies or mycelium processed into capsule or tablet form, extracts from fruiting body or mycelium dried and processed into capsule or tablet form, *Ganoderma* beer, and *Ganoderma* hair tonics (Jong and Birmingham, 1992).

Within the genus *Ganoderma*, over 250 taxonomic names have been reported worldwide (Moncalvo et al., 1995) including *G. adspersum*, *G. applanatum*, *G. australe*, *G. boninense*, *G. cupreum*, *G. incrassatum*, *G. lipsience*, *G. lobatum*, *G. lucidum*, *G. oerstedii*, *G. oregonense*, *G. pfeifferi*, *G. platense*, *G. resinaceum*, *G. sessile*, *G. sinense*, *G. tornatum*, *G. tsugae*, and *G. webrianum*.

11.3   History *of* Ganoderma

Ganoderma lucidum has been treasured in China and Japan for many thousands of years (Willard, 1990). In Chinese, the mushroom is called "lingzhi"; in Japanese "Reishi, Mannentake, or Sachitake"; and "Youngzhi" in Korean. Chinese tradition proclaims that *Ganoderma* is also called "miraculous zhi" or "aunicious herb" and is usually considered to "symbolize happy augury, and to bespeak good fortune, good health and longevity, even immortality" (Wasson, 1968).

As early as 800 years ago in the Yuan Dynasty (1280–1368 A.D.), *G. lucidum* has been represented in paintings, carvings, furniture, carpet design, jewellery, perfume bottles, and many more creative artworks (Wasser and Weis, 1999). According to the two famous Chinese herb medical books, *Shen Nong Ben Cao Jing* (25–220 A.D., Eastern Han Dynasty) and *Ben Cao Gang Mu* (1590 A.D., Ming Dynasty), there were six known species of *Ganoderma* (lingzhi) in China at that time, whereas now more than 250 species have been described (Moncalvo et al., 1995).

11.4   Medicinal Ganoderma

G. lucidum (lingzhi) was the favorite species within the Ganodermataceae family as it was believed to be the only mushroom containing therapeutic properties (Willard, 1990). In the literature today, there is much confusion as to which is the true *Ganoderma* species. The Japanese believed that the true *Ganoderma* was red and that a *Ganoderma* species with different colors was a red *Ganoderma* that had become discolored due to changes in environmental conditions such as temperature, humidity, and light. In China, they believed that the true *Ganoderma* was black as there were reports of a black *Ganoderma* that had unusual medicinal benefits not produced by the red mushroom (Mayzumi et al., 1997).

Chang (1995) suggested that *Ganoderma* (lingzhi) encompassed several *Ganoderma* species, although most investigations and therapeutic practices refer to the species *G. lucidum*. More recently, other species such as *G. tsugae*, *G. boninense*, *G. capense*, *G. sinense*, *G. japonicum*, *G. applanatum*, *G. tropicum*, *G. tenue*, and *G. luteum* have become increasingly popular for the investigation of medicinal properties. A number of reviews have described the bioactive substances, medicinal effects, and health benefits of *Ganoderma* species (Chang, 1995; Chang and Buswell, 1996; Chang and Miles, 1996; Jong and Birmingham, 1992; Mizuno et al., 1995). It is also noted that the majority of the studies concerning the Ganodermataceae family relate to the antitumour and antiviral effects, while the antioxidant properties associated with this fungus have only recently become apparent (Mau et al., 2002; Yen and Wu, 1999; Zhu et al., 1999). There appears to be limited information available that reports the antimicrobial properties of *Ganoderma* species.

11.5   Bioactive substances in Ganoderma *sp.*

Many bioactive compounds have been found in mushroom, some of which inhibit the growth of cancer cell used under *in vitro* condition. These substances may be useful as starting materials for the development of chemical therapeutic agents in cancer treatment and other ailments (Mizuno, 1995). In addition, the modes of action of these compounds are being investigated (Gao et al., 2002; Lei and Lin, 1992; Zhang et al., 2002). The fruiting body, mycelia, and spores of G. *lucidum* contain approximately 400 different bioactive compounds, which mainly include triterpenoids, polysaccharides, nucleotides, sterols, steroids, fatty acids, proteins, peptides, and trace elements.

11.5.1   Terpenoid compounds

A new terpenoid, named ganosporeric acid A, was recently isolated from the ether-soluble fraction of the spores. Min et al. (2000) reported the isolation of six new lanostane-type triterpenes and also from the spores (ganodericacids g, d, e, z, Z, and y). Preliminary studies indicate that the spores contain considerably higher contents of ganoderic acids than other parts of the fungus, and triterpene composition of the fruit body varies; spores also contain triterpene lactones. The mode of action of triterpenes has been of interest to many researchers (Gonzalez et al., 2002; Kimura et al., 2002; Lin et al., 2003). The triterpenes are used to enhance the immune system, as similar to polysaccharides; triterpenes have been shown to have direct cytotoxicity against tumor cells (Gonzalez et al., 2002). A list of the important terpenoid bioactive components structure and their biological functions found in *Ganoderma* species is given in Table 11.1 and Figure 11.1.

Ha et al. (2000) also isolated two lanosteroids from the basidiocarpe (fruiting body) of this mushroom, one of which markedly increased the activity of NAD(P)H:quinone oxidoreductase. Since this enzyme takes part in xenobiotic metabolism, the determination of its activity can be used to detect the antitumor chemopreventive potential of the product.

11.5.2   Proteins and polysaccharides

One of the most important proteins isolated from the mycelium of G. *lucidum* is lingzhi (LZ-8) (Kino et al., 1989). LZ-8 is a polypeptide consisting of 110 amino acid residues with an acetylated amino terminus and has a molecular mass of 12 kDa (Tanaka et al., 1989). The native form of LZ-8, with a molecular mass of 24 kDa, is a homodimer of the LZ-8 polypeptide (Tanaka et al., 1989). This protein has been shown to have mitogenic activity *in vitro* and immunomodulating activity *in vivo* (Haak-Frendscho et al., 1993; Kino et al., 1989; Van der Hem et al., 1995).

More than 100 types of polysaccharides have been isolated from the fruiting body, spores, and mycelia or separated from the broth of a submerged liquid culture of G. *lucidum*. These polysaccharides are the major sources of G. *lucidum*'s pharmacologically active compounds.

Table 11.1 Major triterpenoid bioactive constituents in *Ganoderma* species and their function

S. no.	Name of the species	Effects	Compound	References
1.	*Ganoderma lucidum, G. applanatum*	Cytotoxicity	Lucidenic acid N, methyl lucidenate F, lucialdehydes A–C, and ganoderic alcohol	Gao et al. (2002), Gonzalez et al. (2002), Kimura et al. (2002), Lin et al. (2003), Min et al. (2000), Su (1991)
2.	*Ganoderma pfeiferri*	Antiviral	Ganoderic acid β, ganoderiol F, ganodermanotriol	El-Mekkawy et al. (1998), Min et al. (1998)
3.	*Ganoderma lucidum*	Anticomplement activity	Lucialdehydes	Min et al. (2001)
4.	*Ganoderma lucidum*	Hypolipidemic (cholesterol inhibitors)	Ganoderan B	Komoda et al. (1989), Shiao (1992)
5.	*Ganoderma lucidum* and *G. japonicum*	Antiplatelet aggregation activity	Ganodermic acid S	Shiao (1992), Wang et al. (1993)
6.	*Ganoderma lucidum*	Antioxidant	Ganosporeric acid A, lucidenic acid A, and ganoderic acids B	Zhu et al. (1999)
7.	*Ganoderma lucidum, Ganoderma pfeifferi, Ganoderma resinaceum, Ganoderma* sp. DKR1	Antibacterial	Ganomycin A, ganomycin B, and lanostanoid terpenoids	Smania et al. (1999), Rajesh and Dhanasekaran (2014)
8.	*Ganoderma lucidum* and *G. tsugae*	Antiinflammatory	N-acetylglucosamine	Giner-Larza et al. (2000)
9.	*Ganoderma lucidum*	Antiprotozoal	Ganoderic acid, ganodermanondiol, 23-hydroxyganoderic acid S, and ganofuran B	Adams et al. (2006)
10.	*Ganoderma* sp. DKR1	Antifungal	Terpenoids	Rajesh and Dhanasekaran (2014)

G. lucidum polysaccharides such as beta-D-glucans, heteropolysaccharides, and glycoprotein have been isolated and characterized and are considered the major contributors of bioactivity of the mushroom.

11.5.3 *Antimicrobial compounds*

In recent years, there have been a significant number of human pathogenic bacteria becoming resistant to antimicrobial drugs (Davis, 1994; Donadio et al., 2002). Antimicrobial drug resistance is of major economic concern having an impact on physicians, patients, health-care administrators, pharmaceutical producers, and the public (McGowan, 2001). Therefore, bacterial and fungal pathogenic diseases are often difficult to treat (Baratta et al., 1998).

For the past two decades, attention has turned to the extracts and biologically active compounds used in traditional herbal medicine to expose their remedial effects and to seek new lead compounds for development into therapeutic drugs (Cragg, 1997). In addition

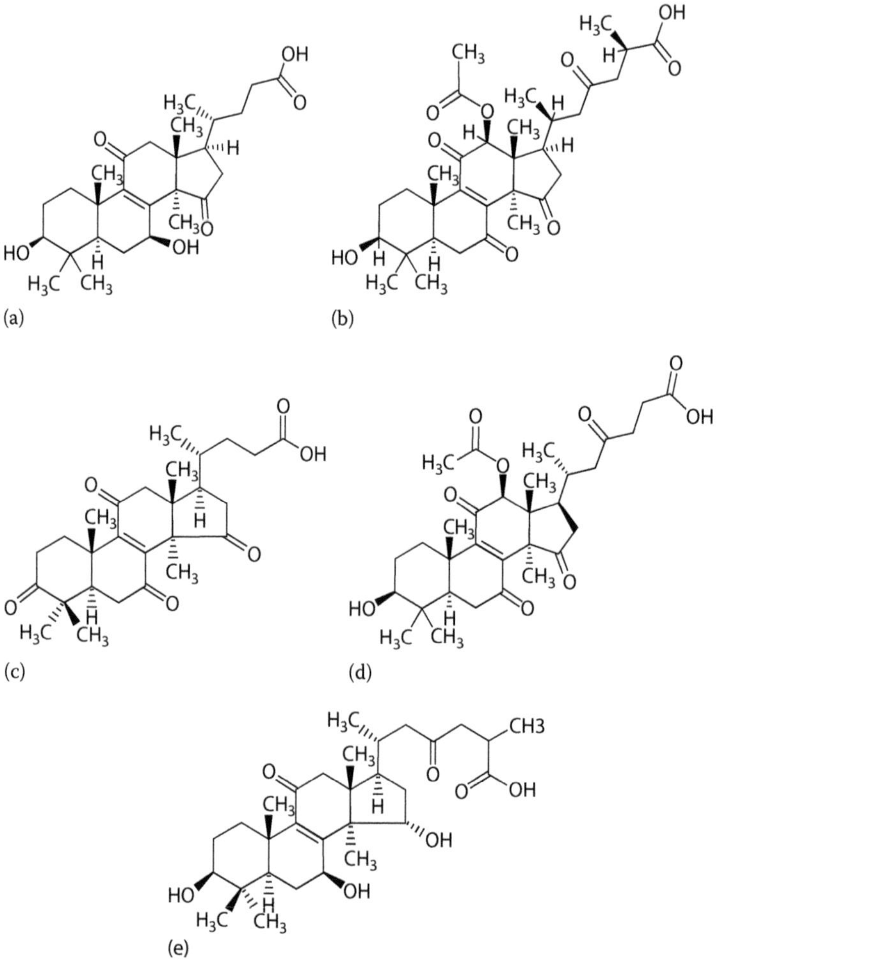

Figure 11.1 Structure of triterpenoids from *Ganoderma lucidum*: (a) lucidenic acid N, (b) ganoderic acid H, (c) lucidenic acid E, (d) ganoderic acid AM, and (e) ganoderic acid C_6.

to plants extracts as sources of antimicrobial agents, research is being performed on fungi for their ability to prevent bacterial, viral, or fungal pathogens that are resistant to current therapeutic agents (Wasser and Weis, 1999a). Fungi are well known for the production of important antibiotic compounds such as penicillin; however, the occurrence of antibiotics in the class of fungi known as basidiomycetes (the mushrooms) is less well documented, and there are only few reviews that summarize the antibacterial activity from this type of mushrooms (Gao et al., 2003; Wasser and Weis, 1999a,b; Zjawiony, 2004).

11.6   *Antibacterial activity*

Antibacterial activity has been observed against gram-positive bacteria from the fruiting body extracts of *G. lucidum* (Kim et al., 1993), *Ganoderma* sp. DKR1 (Rajesh et al., 2014), and *G. orgonense*. Furthermore, observed that seven Indonesian *Ganoderma* species inhibited the growth of *B. subtilis*. Yoon et al. (1995) investigated the additive effect on the activity of an aqueous extract of *G. lucidum* with four known antibiotics and observed that the antibacterial activity increased. The basidiomycetous mushrooms have been shown to possess antibacterial activity against this group of bacteria. The triterpenes have a great antibacterial effect (Pinducciu et al., 1995; Wilkens et al., 2002), and it is well documented that triterpenes are one of the major constituents isolated from *Ganoderma*.

A few triterpenoids isolated from the fruit body of *G. applanatum* exhibited inhibitory activity against gram-positive and gram-negative bacteria such as *Bacillus cereus*, *Corynebacterium diphtheriae*, *Escherichia coli*, *Pseudomonas aeruginosa*, *Staphylococcus aureus*, *Staphylococcus saprophyticus*, and *Streptococcus pyogenes* (Smania et al., 1999). The antibacterial activity of these compounds was determined by MIC and minimal bactericidal concentration (MBC). Among the seven bacterial species tested, the gram positives were more sensitive (Rajesh and Dhanasekaran, 2014) (MIC: 0.003–2.0 mg/mL; MBC: 0.06–4.0 mg/mL) than were the gram negatives (MIC: 1.0–4.0 mg/mL; MBC: 2.0–4.0 mg/mL).

11.7   Antifungal activity

The antifungal drugs available today are not always successful in treating immune compromised patients due to the ineffectiveness or toxicity that many of them have on the host (Seltrennikoff, 1995), and hence, there is a need for the identification of novel antifungal agents. The basidiomycetes belong to the kingdom of fungi; they are thought to have weak antifungal activities (Mizuno, 1995) and, therefore, have rarely been investigated for their bioactivity as antifungal agents. It is only recently that they have become of interest due to their secondary metabolites exhibiting a wide range of antimicrobial activities.

11.8   Antiviral activity

Recently, a number of reviews by Gao et al. (2003) were published on the antiviral value of the genus *Ganoderma*, which summarized the major biologically active constituents and their effect or mode of action toward a number of viruses such as herpes simplex virus (HSV) and human immunodeficiency virus (HIV). Polysaccharide fractions from *G. lucidum* exhibited activity against HSV-1 and HSV-2 (Eo et al., 1999). Lanostane-type triterpenes from *G. pfeifferi* have also been shown to exhibit activity against HSV-1 as well as inhibit the influenza A virus.

The fruiting bodies of *Ganoderma lucidum* are the source of antiviral triterpenoids. Ganoderic acid β (a) isolated from the spores of *G. lucidum* showed significant anti-HIV-1 protease activity, with an IC_{50} value of 20 μM (Min et al., 1998). The same species also produced ganoderiol F (b) and ganodermanontriol (c), which have anti-HIV-1 activity (El-Mekkawy et al., 1998) (Figure 11.2).

Herbal medicines may have potential activity against Epstein-Barr virus (EBV) infection or EBV-mediated tumor promotion. A few polyoxygenated lanostanoid triterpenes isolated from *G. applanatum* inhibited the 12-O-tetradecanoyl phorbol-13-acetate–induced EBV early antigen expression in Raji cells (Chairul and Hayashi, 1994). Similar effects have been observed with Zingiberaceae rhizomes, a commonly used traditional medicine in Malaysia (Vimala et al., 1999). These results indicate that herbal medicines like as *Ganoderma* may behave as antitumor agents.

11.9   Antiprotozoal effect of Ganoderma sp.

Protozoa are single-celled creatures with nuclei that show some characteristics usually associated with animals, most notably mobility and heterotrophy. According to the World Health

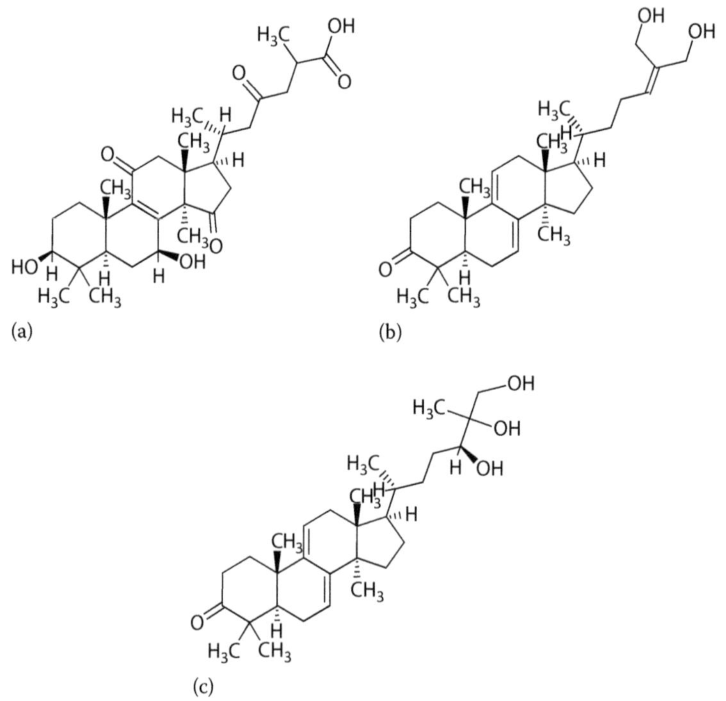

Figure 11.2 Structure of antiviral compound from *Ganoderma* species: (a) ganoderic acid β, (b) ganoderiol F, and (c) ganodermanotriol.

Organization (WHO) estimates, approximately 300–500 million individuals are infected with malaria, with death totals ranging from 1.5 to 3.5 million annually (Stanley, 1997). It is caused by infection with protozoa of the genus *Plasmodium* consisting of four species of obligate intracellular sporozoans: *P. malariae, P. vivax, P. ovale,* and *P. falciparum.* Of the four species, *P. falciparum* is the predominant species with about 400 million new cases and one million deaths per year globally. *P. falciparum* is the most dangerous of these infections as *P. falciparum* malaria has the highest rates of complications and mortality. Drug therapy, in particular compounds arising from natural resources, is the major approach to manage malaria. The fight against *P. falciparum* species has proven to be formidable with an increasing rise in drug-resistant strains being isolated. Thus, novel compounds are sought for controlling malaria (Figure 11.3).

Interestingly, the extract from *G. lucidum* showed inhibitory activity against *Plasmodium falciparum* (a pyrimethamine-resistant malarial parasite) (Lovy et al., 1999). Other mushrooms including *Polyporus umbellatus, Russula xerampelina, Trametes versicolor, Lentinula aurantiacum, Laetiporus sulphureus, Boletus variipes, Boletus queletii, Grifola frondosa,* and *Lentinula edodes* also demonstrated activity against *Plasmodium falciparum* (Lovy et al., 1999).

11.10   Current biomedical applications

A number of biomedical applications of medicinal mushrooms (Borchers et al., 1999; Mizuno, 1995; Wasser and Weis, 1999) and *Ganoderma* species were reported earlier (Su, 1991). Recently, a review was published on the immunomodulating effects of *G. lucidum,* which outlined the mode of actions of a number of biologically extracted compounds (Zhou and Gao, 2002). *Ganoderma* is best known for its immune stimulatory effects in aiding cancer treatment and for

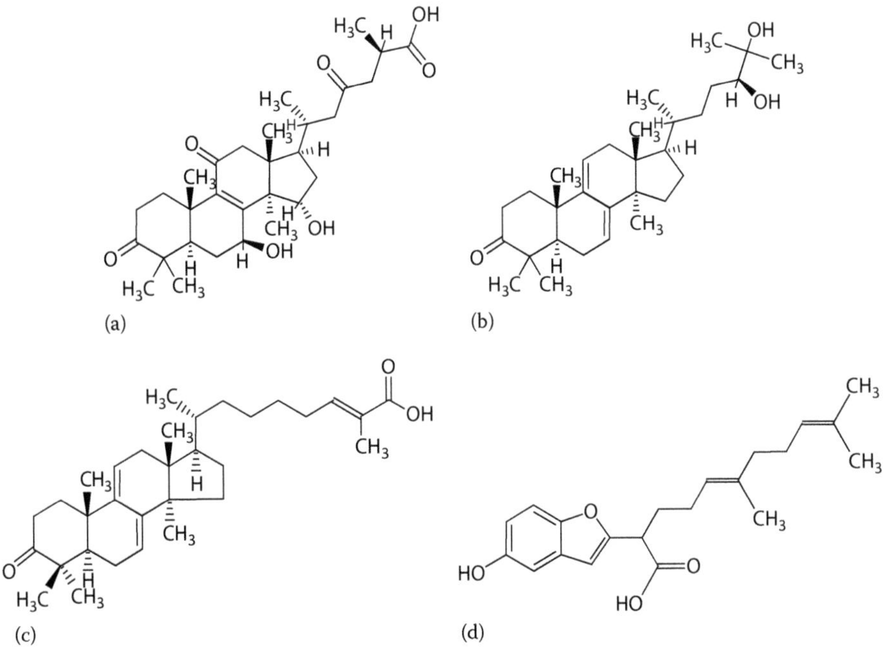

Figure 11.3 Structure of antiprotozoval compound from *Ganoderma lucidum*: (a) ganoderic acid, (b) ganodermanondiol, (c) 23-hydroxyganoderic acid S, and (d) ganofuran B.

its anti-HIV activity. Over the past decade, there has been an increasing amount of research to investigate *Ganoderma* species for new biomedical applications. Some of the applications for which the mushroom extracts and constituents have been found to play important roles include the following: hepatoprotective activity, hypoglycaemic activity, hypolipidimic activity, antihistamine release, anti-inflammatory properties, antiplatelet aggregation activity, anti-complement activity, antiviral activity, enzyme inhibition, and in the healing of open skin wounds.

11.11   Future perspective

In recent years, more varieties of mushrooms have been isolated and identified, and the number of mushrooms being cultivated for food or medicinal purposes has been increasing rapidly (Chang, 1995). Mushroom "nutriceuticals" are bioactive compounds that are extractable from mushrooms, and they have nutritional and medicinal features that may be used in the prevention and treatment of disease (Chang and Buswell, 1996). Several nutriceutical products have been isolated from medicinal mushrooms, and three of these, which are carcinostatic polysaccharides drugs, have been developed from mushroom in many countries. With particular focus on *Ganoderma* species, it is apparent that most of the available data on active extracts and compounds relate to the pharmacological effects on tumor cells, which appear to be based on the enhancement of the host's immune system.

Acknowledgement

The valuable support of Department of Microbiology, Bharathidasan University, Tiruchirappalli is greatly acknowledged.

References

Adams, M., Ettl, S., Kunert, O., Wube, A.A., Haslinger, E., Bucar, F., and Bauer, R. 2006. Antimycobacterial activity of geranylated furocoumarins from *Tetradium daniellii*. *Planta Med* 72: 1132–1135.

Anke, H. and Sterner, O. 1991. Comparison of the antimicrobial and cytotoxic activities of twenty unsaturated sesquiterpene dialdehydes from plants and mushrooms. *Planta Medica* 57: 344–346.

Baratta, M.T., Dorman, H.J.D., Deans, S.G., Figueiredo, A.C., Barroso, J.G., and Ruberto, G. 1998. Antimicrobial and antioxidant properties of some commercial essential oils. *Flav Frag J* 13: 235–244.

Beltran-Garcia, M.J., Estarron-Espinosa, M., and Ogura, T. 1997. Volatile compounds secreted by the Oyster mushroom and their antibacterial activities. *J Agric Food Chem* 45: 4049–4052.

Borchers, A.T., Stern, J.S., Hackman, R.M., Keen, C.L., and Gershwin, M.E. 1999. Mushrooms, tumors, and immunity. *Proc Soc Exp Biol Med* 221: 281–293.

Chairul, C.S.M. and Hayashi, Y. 1994. Lanostanoid triterpenes from *Ganoderma applanatum*. *Phytochemistry* 35: 1305–1308.

Chang, A.W. and Miles, P.G. 1996. Biomedical research and the application of mushroom nutriceuticals from *Ganoderma lucidum*. *Proceedings of the 2nd International Conference on Mushroom Biology and Mushroom Products*, pp. 161–175. University Park, PA.

Chang, S.T. 1995. Ganoderma—The leader in production and technology of mushroom neutraceuticals. In: Kim BK, Kim IH, Kim YS, eds., *Proceedings of the 6th International Symposium. Recent Advances in Ganoderma lucidum Research*, pp. 43–52.

Chang, S.T. and Buswell, J.A. 1996. Mushroom nutriceuticals. *World J Microbiol Biotechnol* 12: 473–476.

Chang, S.T. and Miles, P.G. 1992. Mushroom biology—A new discipline. *Mycologist* 6: 64–65.

Cherqui, F., Rapior, S., and Cuq, P. 1999. Les activités biologiques de *Lepista nebularis* (Batsch:Fr.) Harm. *Ann Soc Hortic Hist Nat Hérault* 139(3): 75–86.

Cragg, W.J. 1997. Phytochemicals-guardians of our health. *J Am Diet Assoc* 97: S199–S204.

Davis, J. 1994. Inactivation of antibiotic and the dissemination of resistance genes. *Science* 264: 375–382.

Donadio, S., Carrano, L., Brandi, L., Serina, S., Soffientini, A., Raimondi, E., Montanini, N., Sosio, M., and Gualerzi, C.O. 2002. Targets and assays for discovering novel antibacterial agents. *J Biotechnol* 99: 175–185.

Donnelly, D.M.X., Abe, F., Coveney, D., and Fukuda, N. 1985. Antibacterial sesquiterpene aryl esters from *Armillaria mellea*. *J Nat Prod* 48(1): 10–16.

Dornberger, K., Ihn, W., Schade, W., and Tresselt, D. 1986. Antibiotics from Basidiomycetes evidence for the occurrence of the 4-hydroxybenzenediazonium ion in the extracts of *Agaricus xanthodermus* Genevier (Agaricales). *Tetrahedron Lett* 27(5): 559–560.

El-Mekkawy, S., Meselhy, M.R., Nakamura, N., Tezuka, Y., Hattori, M., Kakiuchi, N., and Otake, T. 1998. Anti-HIV-1 and anti-HIV-1-protease substances from *Ganoderma lucidum*. *Phytochemistry* 49(6), 1651–1657.

Eo, S.K., Kim, Y.S., Lee, C.K., and Han, S.S. 1999. Antitheraputic activates of various protein bound polysaccharides isolated from *Ganoderma lucidum*. *J Ethnopharmacol* 68: 175–181.

Farnsworth, N.R. 1994. Ethnopharmacology and drug development. In *Ethnobotany and the Search for New Drugs*, Vol. 185, (ed. G.T. Prance), pp. 42–59. John Wiley & Sons, Chichester, U.K.

Gao, Y., Zhou, S., Wen, J., Huang, M., and Xu, A. 2002. Mechanism of the antiulcerogenic effect of *Ganoderma lucidum* polysaccharides on indomethacin induced lesions in the rat. *Life Sci* 72: 731–745.

Gao, Y., Zhou, S., Huang, M., and Xu, A., 2003. Antibacterial and antiviral value of the genus *Ganoderma P. Karst.* species (aphyllophoromycetideae): A review. *Int J Med Mushroom* 5: 235–246.

Gianetti, B.M., Steffan, B., Steglich, W., Kupka, J., and Anke, T. 1986. Antibiotics from basidiomycetes. Part 24. Antibiotics with a rearranged hirsutane skeleton from *Pleurotus hypnophilus* (Agaricales). *Tetrahedron* 42(13): 3587–3593.

Giner-Larza, E.M., Máñez, S., Giner, R.M., Recio, M.C., Prieto, J.M., Cerdá-Nicolás, M., and Rios, J.L. 2002. Anti-inflammatory triterpenes from *Pistacia terebinthus* galls. *Planta Med* 68(4): 311–315.

Gonzalez, A.G., Leon, F., Rivera, A., Padran, J.I., Quintana, J., Estevez, F., and Bermejo, J. 2002. New lanostanoids from the fungus *Ganoderma concinna*. *J Nat Prod* 65: 417–421.

Ha, T.B., Gerhäuser, C., Zhang, W.D., Ho-Chong-Line, N., and Fouraste, I. 2000. New lanostanoids from *Ganoderma lucidum* that induce NAD(P) H:quinone oxidoreductase in cultured hepalcic7 murine hepatoma cells. *Planta Med* 66: 681–684.

Haak-frendscho, M., Kino, K., Sone, T., and Jardieu, P. 1993. Ling Zhi-8: A novel T cell mitogen induces cytokine production and upregulation of ICAM-1 expression. *Cell Immunol* 150: 101–113.

Hamburger, M. and Hostettmann, K. 1991. Bioactivity in plants: The link between phytochemistry and medicine. *Phytochemistry* 30: 3864–3874.

Jong, S.C. and Birmingham, J.M. 1992. Medicinal benefits of the mushroom *Ganoderma*. *Adv Appl Microbiol* 37: 101–134.

Kettering, M., Valdivia, C., Sterner, O., Anke, H., and Thines, E. 2005. Heptemerones A-G, seven novel diterpenoids from *Coprinus heptemerus*: Producing organism, fermentation, isolation and biological activities. *J Antibio* 58(6): 390–396.

Kim, B.K., Cho, H.Y., Kim, J.S., Kim, H.W., and Choi, E.C. 1993. Studies on constituents of higher fungi of Korea. Antitumor components of the cultured mycelia of *Ganoderma lucidum*. *Kor J Pharmacol* 24: 203–212.

Kimura, Y., Taniguchi, M., and Baba, K. 2002. Antitumor and antimetastatic effects on liver of triterpenoid fractions of *Ganoderma lucidum*: Mechanism of action and isolation of an active substance. *Anticancer Res* 22: 3309–3318.

Kino, K., Yamashita, A., Yamaoka, K., Watanabe, J., Tanaka, S., Ko, K., Shimizu, K., and Tsunoo, H. 1989. Isolation and characterization of new immunomodulatory protein, (LZ-8), from *Ganoderma lucidum*. *J Biol Chem* 264: 472–478.

Komoda, Y., M. Shimizu, Sonoda, Y., and Sato, Y. 1989. Ganoderic acid and derivatives as cholesterol synthesis inhibitors. *Chem Pharm Bull (Tokyo)* 37: 531–533.

Lei, L.S. and Lin, Z.B. 1992. The effect of *Ganoderma* polysaccharides on T cell subpopulations and production of interleukin 2 in mixed lymphocyte response. *Acta Pharmaceut Sin* 27: 331–335.

Lin, S.B., Li, C.H., Lee, S.S., and Kan, L.S. 2003. Triterpene-enriched extracts from *Ganoderma lucidum* inhibit growth of hepatoma cells via suppressing protein kinase C, activating mitogen-activated protein kinase and G2-phase cell cycle arrest. *Life Sci* 72: 2381–2390.

Lovy, A., Knowles, B., Labbe, R., and Nolan, L. 1999. Activity of edible mushrooms against the growth of human T4 leukemic cancer cells, HeLa cervical cancer cells, and *Plasmodium falciparum*. *J Herb Spice Med Plant* 6: 49–57.

Mau, J.L., Lin, H.C., and Chen, C.C. 2002. Antioxidant properties of several medicinal mushrooms. *J Agric Food Chem* 50: 6072–6077.

Mayzumi, F., Okamoto, H., and Mizuno, T. 1997. Cultivation of *Ganoderma lucidum*. *Food Rev Int* 13: 365–382.

McGowan, J., J.E. 2001. Economic impact of antimicrobial resistance. *Emerg Infect Dis* 7(2): 286.

Min, B.S., Gao, J.J., Nakamura, N., and Hattori, M. 2000. Triterpenes from the spores of *Ganoderma lucidum* and their cytotoxicity against meth-A and LLC tumor cells. *Chem Pharm Bull (Tokyo)* 48: 1026–1033.

Min, B.S., Nakamura, N., Miyashiro, H., Bae, K.W., and Hattori, M. 1998. Triterpenes from the spores of *Ganoderma lucidum* and their inhibitory activity against HIV-1 protease. *Chem Pharm Bull (Tokyo)* 46(10): 1607–1612.

Min, B.S., Gao, J.J., Hattori, M., Lee, H.K., and Kim, Y.H. 2001. Anticomplement activity of terpenoids from the spores of *Ganoderma lucidum*. *Planta Med* 67: 811–814.

Mizuno, T. 1995. Bioactive biomolecules of mushrooms: Food function and medicinal effect of mushroom fungi. *Food Rev Int* 11: 7–21.

Mizuno, T., Wang, G., Zhang, J., Kawagishi, H., Nishitoba, T., and Li, J. 1995. Reishi *Ganoderma lucidum* and *Ganoderma tsugae*: Bioactive substances and medicinal effects. *Food Rev Int* 11: 151–166.

Moncalvo, J.F. and Ryvarden, F. 1998. Nomenclature of ganodermataceae. *Syn Fungorum* 11: 1–109.

Moncalvo, J.M., Wang, H., and Hseu, R.S. 1995. Gene phylogeny of the *Ganoderma lucidum* complex based on ribosomal DNA sequences. Comparison with traditional taxonomic characters. *Mycol Res* 99: 1489–1499.

Ngai, P.H. and Ng, T.B. 2003. Lectin, a novel and potent antifungal protein from shitake mushroom with inhibitory effects on activity of human immunodeficiency virus-1 reverse transcriptase and proliferation of leukemia cells. *Life Sci* 73: 63–74.

Obuchi, T., Kondoh, H., Watanabe, N., Tamai, M., Omura, S., Yang, J.S., and Liang, X.T. 1990. Armillaric acid, a new antibiotic produced by *Armillaria mellea*. *Planta Medica* 56: 198–201.

Okamoto, H. and Li, J. 1993. Antitumor active polysaccharides from the Chinese mushroom Songshan lingzhi, the fruiting body of *Ganoderma tsugae*. *Biosci Biotechnol Biochem* 57: 894–900.

Pinducciu, G., Serra, C., Cagetti, M.G., Cotti, M., Deidda, D., Pinza, M., and Pompei, R. 1995. Selective antibacterial activity of triterpene derivatives. *Med Microbiol Lett* 4: 83–90.

Piraino, F.F. 2006. Emerging antiviral drugs from medicinal mushrooms. *Int J Med Mush* 8(2): 101–114.

Rajesh, K. and Dhanasekaran, D. 2014. Phytochemical screening and biological activity of medicinal mushroom *Ganoderma* species. *Malaya J Biosci* 1(2): 67–75.

Rajesh, K., Dhanasekaran, D., and Panneerselvam, A. 2014. Isolation and taxonomic characterization of medicinal mushroom *Ganoderma* sp. *Acad J Microbiol Res* 2(2): 061–071.

Seltrennikoff, C. 1995. *Antifungal Drugs: (1,3)-Beta-Glucan-Synthase Inhibitors*, pp. 6–8. Springer-Verlag, Heidelberg, Germany.

Shiao, M.S. 1992. Triterpenoid natural products in the fungus *Ganoderma lucidum*. *J Chin Chem Soc* 39: 669–674.

Shu, Y.Z. 1998. Recent natural products based drug development: A pharmaceutical industry perspective. *J Nat Prod* 61: 1053–1071.

Smania, A., Delle Monache, F., Smania, E.F., and Cuneo, R.S. 1999. Antibacterial activity of steroidal compounds isolated from *Ganoderma applanatum* (Pers.) Pat. (Aphyllophoromycetideae) fruit body. *Int J Med Mushr* 1: 325–330.

Stanley, J. 1997. Malaria. *Emerg Med Clin North Am* 15: 113–155.

Su, C.H. 1991. Taxonomy and physiology active compounds of *Ganoderma*—A review. *J Taipei Med Coll* 20: 1–16.

Tanaka, S., Ko, K., Tsuchiya, K., Yamashita, A., Murasugi, A., Sakuma, S., and Tsunoo, H. 1989. Complete amino acid sequence of an immunomodulatory protein, (LZ-8). An immuno-modulator from a fungus, *Ganoderma lucidum* having similarity to immunoglobin variable region. *J Biol Chem* 264: 16372–16377.

Van der Hem, L.G., Van der Vliet, J.A., Bocken, C.F., Kino, K., Hoitsma, A.J., and Tax, W.J. 1995. Ling Zhi-8: Studies of a new immunomodulating agent. *Transplantation* 60: 438–443.

Vimala, S., Norhanom, A.W., and Yadav, M. 1999. Anti-tumour promoter activity in Malaysian ginger rhizobia used in traditional medicine. *Br J Cancer* 80: 110–116.

Wang, G., Zhang, J., Mizuno, T., Zhuang, C., Ito, H., Mayuzumi, H., Okamoto, H., and Li, J. 1993. Antitumor active polysaccharides from the Chinese mushroom Songshan lingzhi, the fruiting body of *Ganoderma tsugae*. *Biosci Biotechnol Biochem* 57: 894–900.

Wasser, S.P. and Weis, A.L. 1999a. Therapeutic effects of substances occurring in higher basidio-mycete's mushrooms: A modern perspective. *Crit Rev Immunol* 19: 65–96.

Wasser, S.P. and Weis, A.L. 1999b. General description of the most important medicinal higher basidiomycetes mushroom. *Int J Med Mush* 1: 351–370.

Wasson, R.G. 1968. *Soma—Divine Mushroom of Immortality*, 381 p. Harcourt, Brace and World, Inc., New York.

Wilkens, M., Alarcon, C., Urzua, A., and Mendoza, A. 2002. Characterization of the bactericidal activity of natural diterpene kaurenoic acid. *Plata Medica* 68: 452–454.

Willard, T. 1990. *Reishi Mushroom. Herb of Spiritual Potency and Medicinal Wonder*, 167 p. Sylvan Press, Seattle, WA.

Yen, G.C. and Wu, J.Y. 1999. Antioxidant and radical scavenging properties of extracts from *Ganoderma tsugae*. *Food Chem* 65: 375–379.

Ying, J., Mao, X., Ma, Q., Zong, Z., and Wen, H. 1987. *Icons of Medicinal Fungi from China* (ed. X. Yuehan) 575 p. Science Press, Beijing, China.

Yoon, J.O., Min, T.J., and Yoon, H. 1995. An antibacterial lectin from *Lampteromyces japonicus* "LJAP" (*Lampteromyces japonicus* antibiotic protein). *Han'guk Kyunhakhoechi* 23(1): 46–52.

Zhang, G.L., Wang, Y.H., Ni, W., Teng, H.L., and Lin, Z.B. 2002. hepatoprotective role of *Ganoderma lucidum* polysaccharides against BCG- induced immune liver injury in mice. *World J Gastroenterol* 8: 728–733.

Zhou, S. and Gao, Y. 2002. The immunomodulating effects of *Ganoderma lucidum* (Curt.:Fr.) P. Karst. (Ling Zhi, Reishi mushroom) (Aphyllophoromycetideae). *Int J Med Mush* 4: 1–11.

Zhu, M., Chang, Q., Wong, L.K., Chong, F.S., and Li, R.C. 1999. Triterpene antioxidents from *Ganoderma lucidum*. *Phytother Res* 13: 529–531.

Zjawiony, J.K. 2004. Biologically active compounds from aphyllophorales (polypore) fungi. *J Nat Prod* 67: 300–310.

chapter twelve

Marine cyanobacteria
A prolific source of antimicrobial natural products

Natesan Sivakumar and Gopal Selvakumar

Contents

12.1   Introduction

Antibiotics are essential for the treatment of various microbial infectious diseases. However, emergence of antibiotic-resistant microorganisms is one of the major problems in recent years. For example, penicillin resistance in *Staphylococcus* spp. emerged rapidly. Due to increasing antibiotic-resistant microorganisms, a new antibiotic is required, which is active against antibiotic-resistant pathogens. In response to bacterial resistance, the pharmaceutical industry has produced a remarkable range of antibiotics (Coates and Hu, 2007). In addition, currently available drugs are effective against only one-third of the diseases as a result of increased antibiotic resistance in pathogens. Therefore, identification of new metabolites is urgently required for the development of new drugs. In recent decades, the main focus on natural products from marine microbial sources has been increased. The majority of new chemical compounds introduced as drugs during the period of 1981–2002 are of natural origin (Newman et al., 2003).

Recently, high-throughput screening like genome-based bioinformatics tools is used to design new bioactive compounds. In spite of these technological advances, the number of new entities reaching the market has declined from 53 to only 26 within a time span (1996–2005) of 9 years (Singh et al., 2011). As a result, insufficient numbers of new therapeutic drugs are available in the market. To fulfill the demand for new therapeutic drugs, scientists have been searching novel bioactive compounds from microbial origin. Among the producers of commercially important metabolites, cyanobacteria have proven to be a prolific source for the vast majority of compounds discovered until now (Burja et al., 2001; Kumar et al., 2013). In this chapter, we have made an attempt to consolidate the recent researches in the field of cyanobacterial bioactive metabolites as antibiotic, antifungal, antimalarial, antiviral, and anti–quorum sensing activities.

12.2 Cyanobacteria

Cyanobacteria are highly diverse groups of gram-negative oxygenic photoautotrophic prokaryotic microorganisms also known as blue-green algae. The name "cyanobacteria" comes from the color of the bacteria (Adams, 2002). These organisms can inhabit a wide range of habitats including freshwater, marine, and soil environments, as well as extreme habitats (Taton et al., 2006). The morphology of cyanobacteria varies from unicellular to filamentous or colonial forms (Figure 12.1). The colonies are often surrounded by a

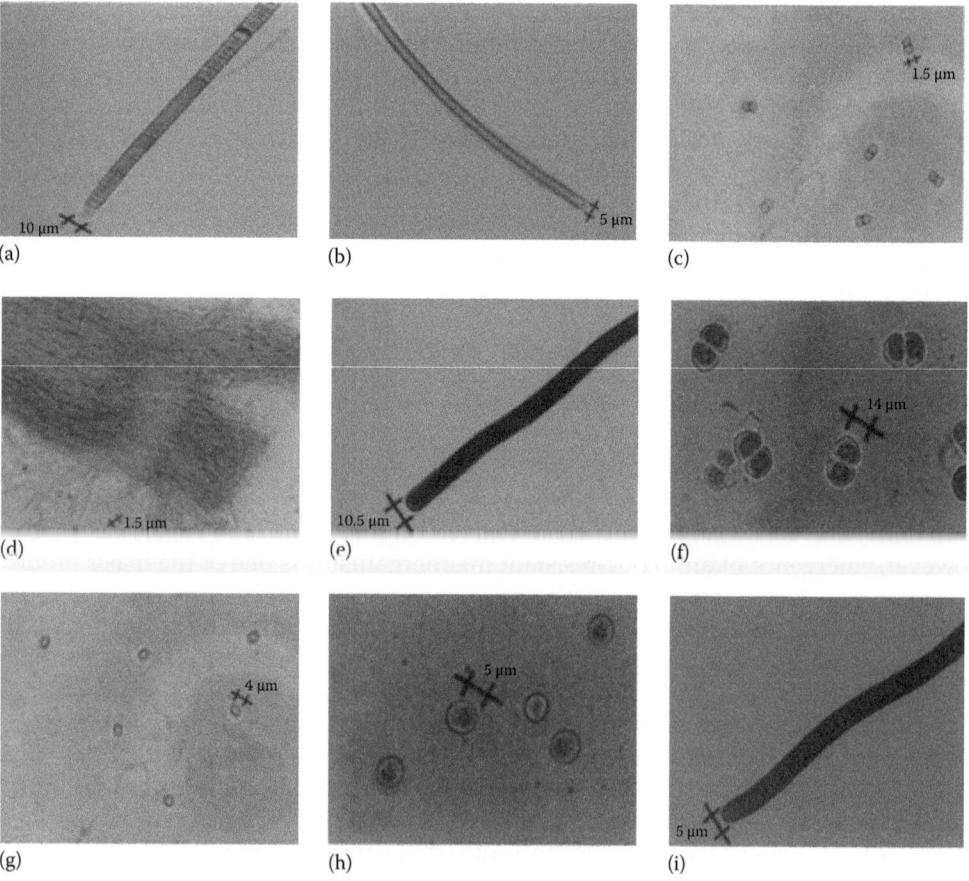

Figure 12.1 Photomicrographs of cyanobacterial species: (a) *Lyngbya confervoides* AU0045, (b) *Phormidium ambiguum* AU0015, (c) *Gloeocapsa* sp. AU0033, (d) *Phormidium tenue* AU0028, (e) *Oscillatoria subbrevis* AU0026, (f) *Chroococcus* sp. AU0052, (g) *Microcystis* sp. AU0036, (h) *Aphanocapsa littoralis* AU0055, and (i) *Oscillatoria princeps* AU0012.

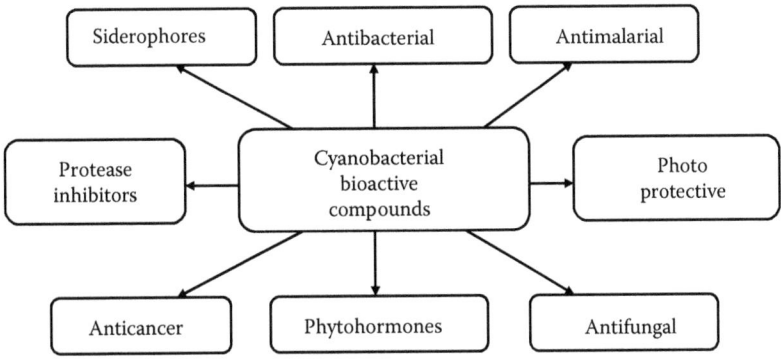

Figure 12.2 Bioactivity of cyanobacterial compounds.

mucilaginous or gelatinous sheath, depending on environmental conditions. Some of the filamentous cyanobacteria have three types of cell morphology, namely, vegetative cells, akinetes, and thick-walled heterocysts. They are either free living or associated with other living organisms to form symbiosis or communalistic relationship. The cyanobacteria can form bloom under certain environmental conditions.

In recent years, significant importance has been given to cyanobacterial biotechnology (Thajuddin and Subramanian, 2005). Marine cyanobacteria are one of the richest sources of known and novel bioactive compounds including toxins with wide pharmaceutical applications (Browitzka, 1995; Singh et al., 2011). During recent decades, attention is drawn toward cyanobacterial bioactive metabolites (Gademann and Portmann, 2008), probably secondary metabolites (Figure 12.2). The secondary metabolites from cyanobacteria include a range of compounds showing toxicity on animals and antibacterial, antifungal, antiinflammatory, antimalarial, antiprotozoal, antituberculous, antiviral, and antitumor activities (Mayer et al., 2009). More than 800 secondary metabolic compounds have been isolated from cyanobacteria and named on the basis of chemical structure, bioassay method, and their toxicological targets (Pearson et al., 2010).

Cyanobacteria can grow in extreme environmental conditions containing highly diverse bioactive compounds with a broad range of activity. In general, cyanobacteria do not produce specific class of compounds; its secondary metabolite spectrum includes 10.2% lipopeptides, 5.6% amino acids, 4.2% fatty acids, 4.2% macrolides, and 9.4% amides and others (Figure 12.3; Burja et al., 2001). The bioactivity spectrum (antibacterial, antifungal, antiviral, and cytotoxic) of the metabolites can be explained on the basis of their ubiquitous occurrence and long evolutionary history. Broad activity spectrum of the secondary metabolites may clearly indicate the pharmaceutical potential of cyanobacteria (Devillers et al., 2007).

12.3   *Marine cyanobacteria as a source of natural products*

Marine cyanobacteria have been considered a well-known source of structurally diverse and biologically active natural products (Nunnery et al., 2010) such as phenols, fatty acids, terpenes, lipopeptides, sterols, and alkaloids. Many cyanobacteria produce compounds that are generally considered as secondary metabolites. Wide diversity of the natural product is a result of the cyanobacterial biosynthetic genes, having the ability to synthesize various classes of compounds such as linear peptides, linear lipopeptides, depsipeptides (Han et al., 2005), cyclic peptides, glicomacrolides (Teruya et al., 2009), cyclic depsipeptides

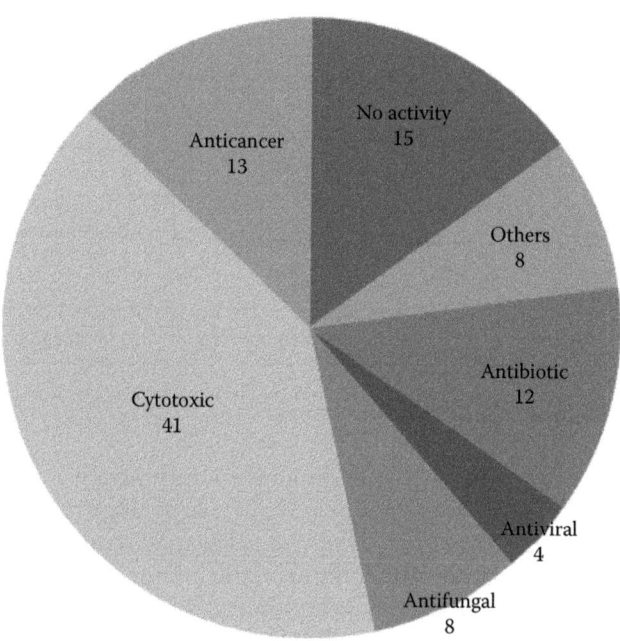

Figure 12.3 Bioactivity of marine cyanobacterial compounds. (From Burja, A.M. et al., *Tetrahedron*, 57, 9347, 2001.)

(Soria-Mercado et al., 2009), and fatty acid amides (Chang et al., 2011). They have potentially useful biological activities such as cytotoxic, antiviral, antibiotic, antimycotic, multidrug resistance reversing, immunosuppressive, anti-inflammatory, and enzyme inhibition properties (Priyadarshani and Rath, 2012; Mukherjee et al., 2013). Cyanobacterial species accumulate high amounts of these bioactive compounds in their biomass, which leach into their surrounding environment (Mukherjee et al., 2013). Most of the cyanobacterial metabolites have cytotoxic activity, for example, lyngbyatoxins B and C produced by *Lyngbya majuscula* (Aimi et al., 1990), microcystins (Dawson, 1998), hermitamides A and B from *L. majuscula* (Tan et al., 2000), apratoxin D produced by *Lyngbya* sp. (Pérez Gutiérrez et al., 2008a,b), cymplocamide A isolated from *Symploca* sp. (Linington et al., 2008), malyngamide 2 (Malloy et al., 2011), and antillatoxin (Okura et al., 2013). Organic solvent extracts exhibit antimicrobial activity, for example, extracellular diterpenoids from *Nostoc commune* (Jaki et al., 1999, 2000) have antibacterial activity. Ether- and water-soluble fractions obtained from *L. majuscula* have been reported to have antibacterial action against *Pseudomonas fluorescens*, *Micrococcus pyogenes*, and *Mycobacterium smegmatis* (Starr et al., 1962). Extract from *Phormidium fragile* has antagonistic effect against *Bacillus subtilis* (Palaniselvam, 1998). *Phormidium corium* has exhibited inhibition zone (46 mm) against *Staphylococcus aureus* (Madhumathi et al., 2011). Extract from *Tolypothrix* sp. reported to have inhibition zone (17 and 21 mm) against *Bacillus cereus* and *Staphylococcus epidermidis* (Zeeshan et al., 2010), and *Plectonema boryanum* and *Anabaena variabilis* inhibited (17 and 12 mm) *S. epidermidis* (Suhail et al., 2011).

More than 50% of the marine cyanobacteria are potentially exploitable for extracting bioactive substances that are effective in killing cancer cells by inducing apoptotic death (Tidgewell et al., 2010). Extensive studies revealed that cyanobacterial species, especially those belonging to the genus *Lyngbya*, are a prolific source of unique bioactive natural

Table 12.1 Different classes of cyanobacterial secondary metabolites

Organism	Compound	Class of compound
Lyngbya semiplena	Wewakpeptins	Cyclic depsipeptides
Lyngbya bouillonii	Alotamide A	
L. majuscula	Pitipeptolides C and D, lagunamides A and B, apratoxin D, homodolastatin, lagunamide C, malyngolide dimmer, palmyramide A, pitipeptolides A and B	
Lyngbya sordida	Apratoxin D	
Lyngbya confervoides	Obyanamide, largamides A–C, tiglicamides A–C	
Lyngbya sp.	Palau'amide, ulongapeptin, kempopeptins A and B	
L. bouillonii	Bouillomides A	
L. semiplena	Lyngbyastatins	
Leptolyngbya sp.	Coibamide A	
Symploca sp.	Largazole, tasipeptins A and B	
Oscillatoria margaritifera	Veraguamides	
Dichothrix utahensis	Molassamide	
Oscillatoria sp.	Largamides A–H	Cyclic peptides
L. confervoides	Pompanopeptins A and B	
Tychonema sp.	Brunsvicamides A–C	
Symploca sp.	Belamide A	Tetrapeptide
Phormidium spp.	Caylobolide B	Macrolactone
L. majuscula	Caylobolide A	
Lyngbya sp.	Lyngbyaloside B, biselyngbyaside	Glycosylated macrolide
L. majuscula	Isomalyngamide A	
L. majuscula	Antillatoxin B, carmabins A and B, hoiamides, (+)-kalkitoxin	Lipopeptide
Lyngbya sp.	Malyngamide	
Symploca sp.	Tasiamide	
L. majuscula	Dragonamide E	Linear lipopeptide

molecules (Gerwick et al., 2001). For example, *L. majuscula* yielded over 30% of all marine cyanobacterial metabolites, reflecting its notable biosynthetic capacity with regard to natural products production, and are also deemed as "super" producer of secondary metabolites (Gerwick et al., 2001). Some of the bioactive secondary metabolites of cyanobacteria are listed in Table 12.1 and Figure 12.4.

12.4   Bioactive antimicrobial compounds from marine cyanobacteria

12.4.1   Antibacterial activity

Increasing the incidence of drug-resistant bacterial pathogens, diseases may become untreatable (posed therapeutic challenges and are of great concern worldwide). Hence, there is a need to search for new antimicrobials to combat antibiotic-resistant strains

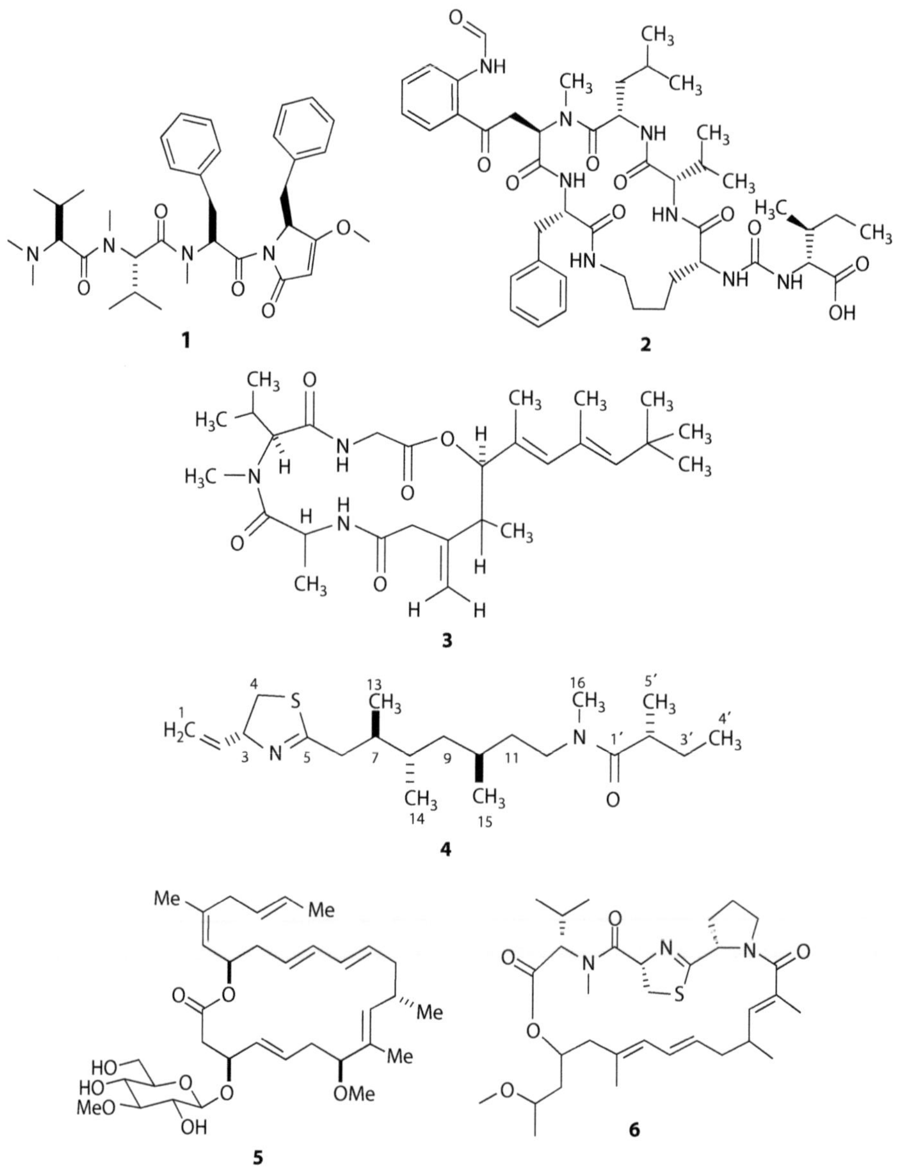

Figure 12.4 Some marine cyanobacterial secondary metabolites: (**1**) belamide A, (**2**) brunsvi-camide C, (**3**) antillatoxin A, (**4**) kalkitoxin, (**5**) biselyngbyaside 1, (**6**) alotamide A.   (*Continued*)

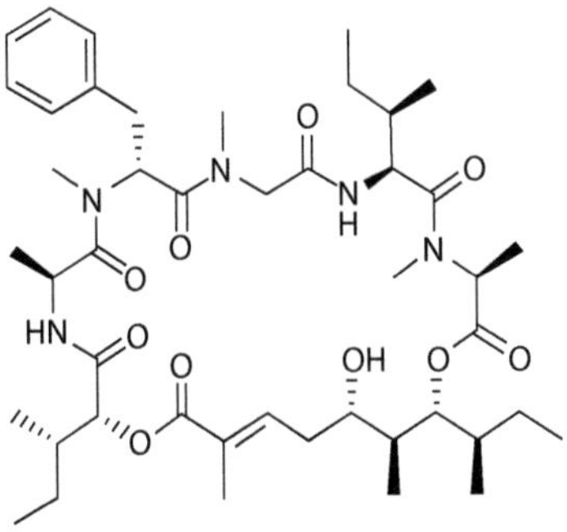

7

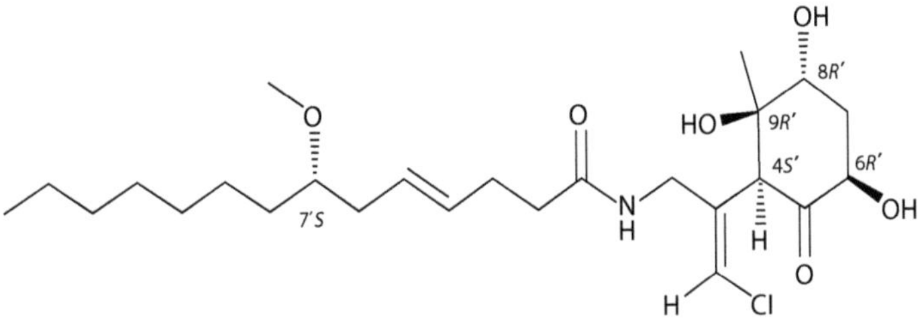

8

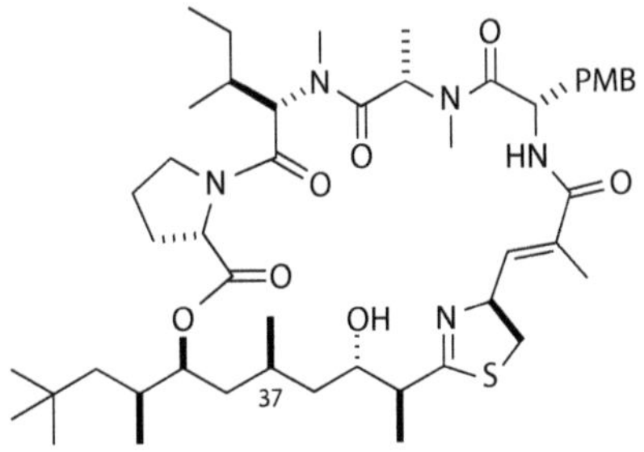

9

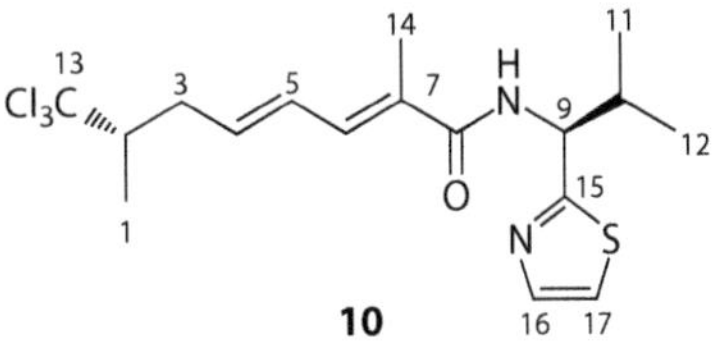

10

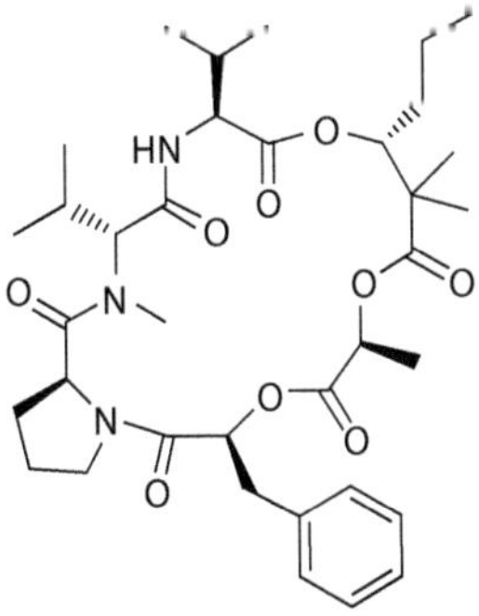

11

***Figure 12.4* (*Continued*)** Some marine cyanobacterial secondary metabolites: (**7**) lagunamide, (**8**) malyngamide 2, (**9**) apratoxin D, (**10**) herbamide B, and (**11**) palmyramide A.

of pathogenic bacteria. Marine habitat has been described recently as a "particularly promising" (Fischbach and Walsh, 2009) source of novel antibacterial marine natural products. Marine cyanobacteria are considered being one of the potential organisms useful to mankind in various ways, which contains diverse bioactive compounds responsible for antibacterial activity (Kreitlow et al., 1999; Biondi et al., 2008).

Various researchers studied the effect of cyanobacterial metabolites as antibacterial substances. Methanol extracts and culture supernatants of cyanobacteria exhibited significant antibacterial effect (Ghasemi et al., 2003; Bhateja et al., 2006). Ethyl acetate extract of *Spirulina platensis* can inhibit some gram-positive and gram-negative bacteria (Ozdemir et al., 2004). Organic extracts of *Anabaena* sp. have antibacterial properties against *Escherichia coli, Pseudomonas aeruginosa, Salmonella typhi, Klebsiella pneumoniae,* and *S. aureus* (Kaushik et al., 2009). Madhumathi et al. (2011) reported that organic extracts of *Oscillatoria laetevirens, Phormidium corium, Lyngbya martensiana, Chroococcus minor,* and *Microcystis aeruginosa* have active activities against *B. subtilis, S. aureus, Streptococcus mutans, E. coli, Micrococcus mutans,* and *K. pneumoniae*. The ethanol extracts of *Phormidium* sp. and *Microcoleus* sp. at various concentrations (0.2, 0.06, 0.03, and 0.015 g/mL) showed the antibacterial activity against *Streptococcus enteritidis* and *E. coli* (Thummajitsakul et al., 2012). Acetone extract of *Spirulina subsalsa* NTRI02 and ethanol extract of *Oscillatoria pseudogeminata* NTRI 03 showed high inhibitory activity against *P. aeruginosa* (MTCC 2453) and *S. aureus* (MTCC 96) (Reehana et al., 2012). Crude hydrophilic and lipophilic extracts of *Anabaena, Nostoc, Scytonema,* and *Microcystis* species showed activity against *Pseudomonas* sp. (Yadav et al., 2012). *Oscillatoria boryanum* has active activities against *S. mutans* (Chauhan et al., 2011). Culture liquids of *Gloeocapsa* sp. and *Synechocystis* sp. have the widest spectrum of activity with minimal inhibitory concentration (MIC) ranging from 1.56 to 12.5 mg/mL against *Streptococcus pyogenes* (Najdenski et al., 2013). *Leptolyngbya* sp. had antimicrobial activity individually and in synergy with antibiotics against gram-positive and gram-negative bacteria (Abazari et al., 2013).

Only few antibacterial compounds from cyanobacteria have been structurally characterized. Noscomin (Figure 12.5(**1**)) from *Nostoc commune* showed antibacterial activity against *B. cereus, S. epidermidis,* and *E. coli* (Mundt et al., 2003). Ambiguine isonitriles (Figure 12.5(**2**)) H and I (alkaloids) isolated from a marine cyanobacterium *Fischerella* sp. (Raveh and Carmeli, 2007) have antibacterial activity against *B. subtilis* and *Staphylococcus albus*. Carbamidocyclophanes (Figure 12.5(**4**)) are paracyclophanes isolated from *Nostoc* sp. CAVN 10 showed moderate antibacterial activity against *S. aureus* (Luke Simmons et al., 2008). Pérez Gutierrez et al. (2008a) screened two new antibacterial norbietane diterpenoids (Figure 12.5(**5**)) from cyanobacterium *Micrococcus lacustris,* showing antibacterial activity against *S. aureus, S. epidermidis, S. typhi, Vibrio cholerae, B. subtilis, B. cereus, E. coli,* and *K. pneumoniae*. Phycobiliproteins isolated from *Synechocystis* sp., *Arthrospira fusiformis, Porphyridium aerugineum,* and *Porphyridium cruentum* have antimicrobial activity (Najdenski et al., 2013). The fatty acids of *Synechocystis* sp. inhibited the growth of *B. cereus* and *E. coli*. Cyanobacterial exopolysaccharides (EPSs) are having antibacterial activity. EPSs of *Synechocystis* sp., *Gloeocapsa* sp., and *Rhodella reticulata* have high antibacterial activity (Najdenski et al., 2013). Table 12.2 shows the cyanobacterial compounds and their targeted bacteria.

12.4.2 Antifungal activity of cyanobacteria

Fungal infections caused by *Cryptococcus neoformans, Histoplasma capsulatum, Pneumocystis carinii, Aspergillus* sp., and *Candida* sp. represent an increasing threat to human health. The prevalence of these fungal infections has increased significantly during the past decades

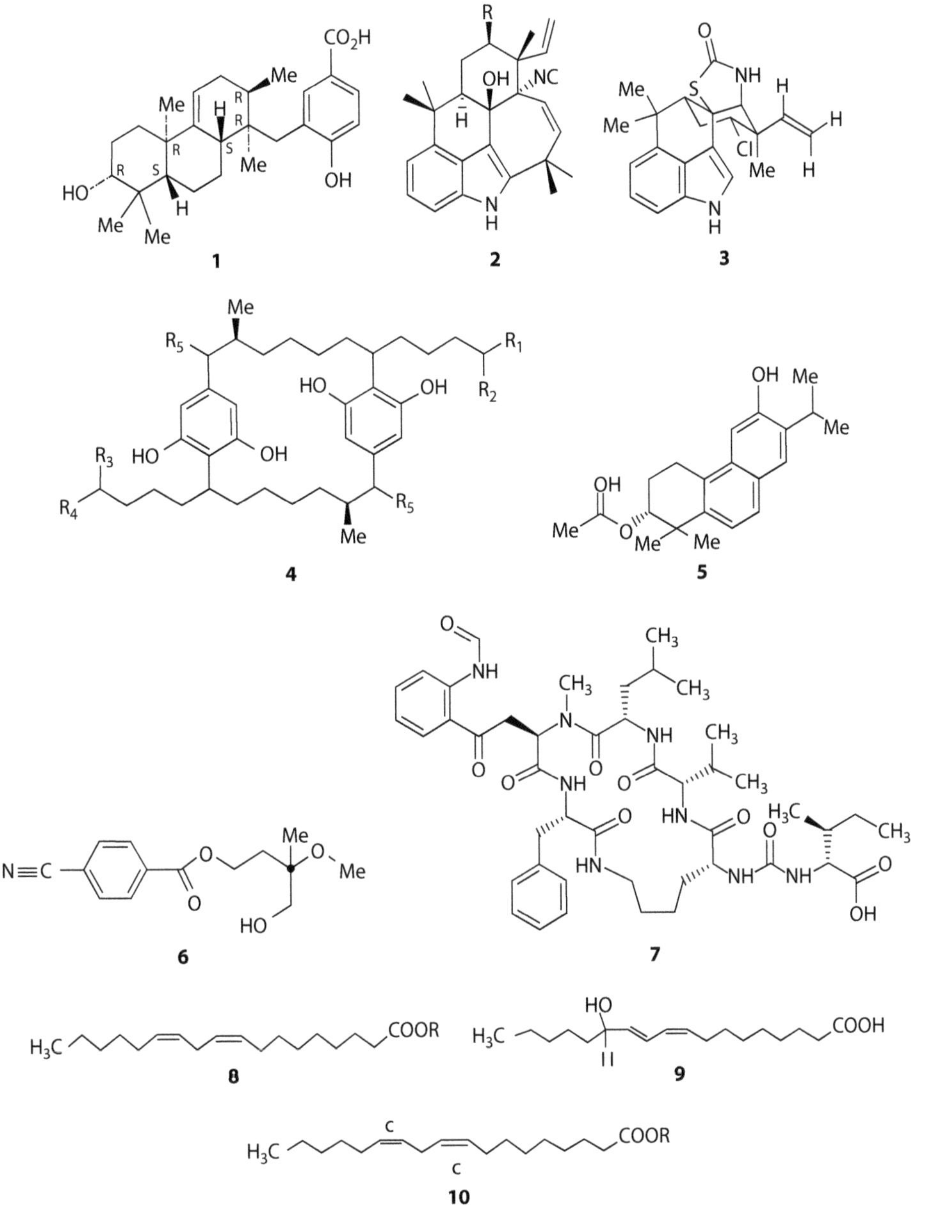

Figure 12.5 Antibacterial metabolites from cyanobacteria: (**1**) noscomin, (**2**) ambiguine 1 isonitrile, (**3**) hapalindole T, (**4**) carbamidocyclophanes, (**5**) norbietane diterpenoid, (**6**) phenolic compound, (**7**) brunsvicamide C, (**8**) α-dimorphecolic acid, (**9**) 13-hydroxy-9Z, 11E-octadecadienoic acid, and (**10**) coriolic acid.

(Vicente et al., 2003). Commercially available antifungal drugs have some side effects. Marine metabolites can be exploited in a number of different ways for the development of new antifungal drugs. The compounds extracted from the source either can be directly used as drugs as leads for the design of novel synthetic products, or as leads for the design of a novel mode of action available as new screening targets. Marine cyanobacteria have

Table 12.2 Antibacterial activity of cyanobacterial compounds

Organism	Compound	Targeted organism
Nostoc commune	Noscomin	*B. cereus, S. epidermidis, E. coli*
Nostoc muscorum	Phenolic compound	*B. subtilis, B. cereus, E. coli, S. typhi, S. aureus*
Nostoc sp. CAVN 10	Caramida cyclophanes	*S. aureus*
Fischerella sp.	Ambiguine 1 isonitrile	*E. coli, Staphylococcus albus, B. subtilis*
	Hapalindole T	*S. aureus, P. aeruginosa, S. typhi, E. coli*
Lyngbya sp.	Pahayokolide A	*Bacillus* sp.
Oscillatoria redekei HUB051, *Synechocystis* sp.	Fatty acid	*B. subtilis, Micrococcus flavus and S. aureus*
M. lacustris	Norbietane diterpenoids	*S. aureus, S. epidermidis, S. typhi, Vibrio cholerae, B. subtilis, B. cereus, E. coli,* and *K. pneumoniae*
Synechocystis sp., *Arthrospira fusiformis, Porphyridium aerugineum,* and *Porphyridium cruentum*	Phycobiliproteins	Antibacterial activity
Synechocystis sp., *Gloeocapsa* sp., and *Rhodella reticulata*	EPSs	*B. cereus, E. coli*

a number of novel metabolites having antifungal activities. However, antifungal activity is also reported with crude extract and pure compounds from cyanobacteria such as *Anabaena, Nostoc, Aphanocapsa, Synechocystis, Synechococcus, Oscillatoria, Nodularia,* and *Calothrix* (Volk and Furkert, 2006; Kim, 2006; Drobac-Čik et al., 2007; Pawar and Puranik, 2008; Ghazala et al., 2010).

Organic solvent extracts of cyanobacteria have antifungal activity. The extract from *Nostoc paludosum* shows activity against *Candida albicans* (Ramachandra Rao, 1994). Diethyl ether and acetone extract of *Spirulina platensis* had the highest antifungal activity (Ozdemir et al., 2004). Methanol extracts of *Phormidium valderianum* and *P. fragile* showed anticandidal activity (Sundararaman and Nagaraja, 2006). Likewise, acetone extract of *Phormidium corium*, methanol extract of *Lyngbya maintensiana* and diethyl ether extract of *M. aeruginosa* showed the largest inhibition zone tested against fungal pathogens (Madhumathi et al., 2011). Pyridine and *n*-butanol extracts of *Oscillatoria subbrevis, O. amphibia,* and *O. chlorina* gave maximum activity against *Aspergillus wentii* and *Candida albicans* (Prabakaran, 2011). Methanol extracts of *Oscillatoria salina* actively inhibit *Fusarium solani* and *Phormidium tenue* against *Rhizoctonia solani* (Sakthivel and Kathiresan, 2012). The cyanobacteria *Lyngbya aestuarii* and *Aphanothece bullosa* are found to be a potent source of antifungal activity (Kumar et al., 2013). *P. fragile* exhibited potential antifungal activity against *Aspergillus flavus, Candida albicans,* and *Trichoderma viride* (Senthil Kumar et al., 2013).

Rath and Priyadarshani (2013) reported that crude extracts of *Phormidium tenue* and *Phormidium* sp. had high antifungal activity.

The antimicrobial activity of the cyanobacterial extract could be due to the presence of different secondary metabolites that include polypeptides and lipopeptides

Table 12.3 Antifungal agents from cyanobacteria

Source	Compound
L. majuscula	Tanikolide, lyngbyabellin B, hectochlorin, almiramides A–C, dragonamide E
Cyanobacterial mat assemblage	Majusculoic acid
Scytonema sp.	Scytoscalarol
Tolypothrix sp.	Hassallidin A
Nodularia harveyana	Norharmane
Nostoc insulare	4'-Dihydroxy biphenyl
Fischerella muscicola	Fischerellin A
Synechocystis	Antifungal
Aphanothece bullosa	Antifungal

(Burja, et al., 2001); amides and alkaloids (Ghasemi et al., 2004); heptadecane and tetradecane (Ozdemir et al., 2004); fatty acids, tetramine, spermine, and piperazine (Prashantkumar et al., 2006; Shanab, 2007); flavonoids, triterpenoids, phenolic compounds (Figure 12.5(6)), and free hydroxyl group (Yu et al., 2009); and metabolites such as tannin, alkaloids, protein, and flavonoids (Zeeshan et al., 2010). Hassallidin A, a glycosylated lipopeptide from cyanobacterium *Hassallia* sp., has antifungal activity (Neuhof et al., 2005). Table 12.3 and Figure 12.6 show few cyanobacterial secondary metabolites used as antifungal compounds. The marine cyanobacterial genera *Lyngbya* is known to produce more than 200 compounds including antifungal such as lobocyclamides A–C and lyngbyabellin B (Milligan et al., 2000; MacMillan et al., 2002). Although the cyanobacteria *Nodularia harveyana* for norharmane (9H-pyrido[3,4-b]indole), *Nostoc insulare* for 4,4'-dihydroxy biphenyl (Hagmann and Jiittner, 1996), *Fischerella muscicola* for fischerellin A, and *Synechocystis* for partially purified AK3 as antifungal compounds (Hagmann and Jiittner, 1996; Yoon and Lee, 2009).

12.4.3   *Antiprotozoan metabolites*

Due to the limited availability of effective pharmaceutical products and serious side effects of the available therapy for protozoan infections, the search for useful antiprotozoan metabolites from natural resources is necessary (Table 12.4). Cyanobacteria are prolific producers of structurally distinct and biologically active antiprotozoan metabolites (Figure 12.7). The Panamanian International Cooperative Biodiversity Group has reported the isolation of five classes of antiprotozoal compounds from cyanobacteria (Tan et al., 2006). The crude extracts of *Nostoc commune* and *Rivularia biasolettiana* are active against *Plasmodium falciparum*, *Trypanosoma brucei rhodesiense*, and *Leishmania donovani* (Broniatowska et al., 2011). The crude extract of *Lyngbya aestuarii* (25.6 mg/mL) and *Aphanothece bullosa* (24.0 mg/mL) has antileishmanial activity (Kumar et al., 2013).

Marine cyanobacteria are extremely rich in diverse lipopeptide natural products, many of which have potent biological activities (Gerwick et al., 2001). For example, bioassay-guided fractionation of the lipophilic extract of the marine cyanobacterium *Phormidium ectocarpi* yielded hierridin B, and 2,4-dimethoxy-6-heptadecyl-phenol showed antiplasmodial activity toward *P. falciparum* (Papendorf et al., 1998). The discovery of several unique linear peptides with antiparasitic activity includes the highly modified lipopeptide gallinamide A with antimalarial activity (IC_{50} = 8.4 μM) (Linington et al., 2009), dragonamide E with antileishmanial activity (IC_{50} = 5.1 μM) (Balunas et al., 2010), and the almiramides

A–C also with promising antileishmanial activity (Sanchez et al., 2010; Uzair et al., 2012). Besides these compounds, the protease inhibitor nostocarboline (Barbaras et al., 2008), an alkaloid isolated from *Nostoc* sp., was also found to be active against *T. brucei*, *Trypanosoma cruzi*, *Leishmania donovani*, and *P. falciparum* (IC_{50} = 0.5–0.194 mM). Two new lipopeptides such as viridamides A and B are having antiprotozoal activity, isolated from *Oscillatoria nigroviridis* (Luke Simmons et al., 2008). Aerucyclamide C isolated from *M. aeruginosa* PCC 7806 has also been found to be active against *T. brucei* and the already known aerucyclamide B against *P. falciparum* (Portmann et al., 2008). In addition, new acyl praline derivatives, tumonoic acid I, from the marine cyanobacterium *Blennothrix cantharidosmum* displayed (IC_{50} 2 mM) moderate antimalarial activity (Clark et al., 2008).

12.4.4 Antiviral activity

The wide outbreaks of deadly viral diseases like dengue, human immunodeficiency virus (HIV, acquired immune deficiency syndrome), and Ebola hemorrhagic fever may have dramatic consequences. Hence, potent and safe antiviral agents are urgently needed

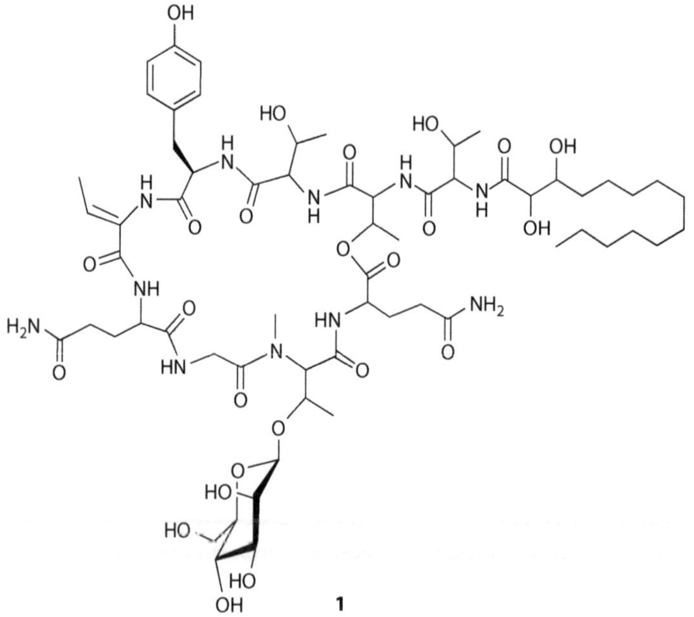

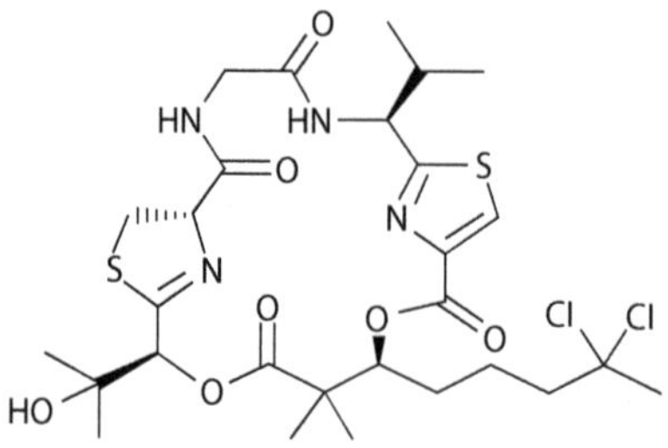

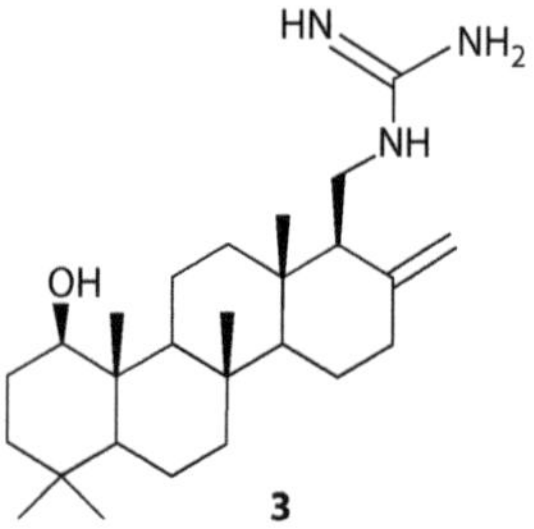

Figure 12.6 Antifungal secondary metabolites from cyanobacteria: (**1**) hassallidin A, (**2**) lyngbya-bellin B, (**3**) scytoscalarol.   *(Continued)*

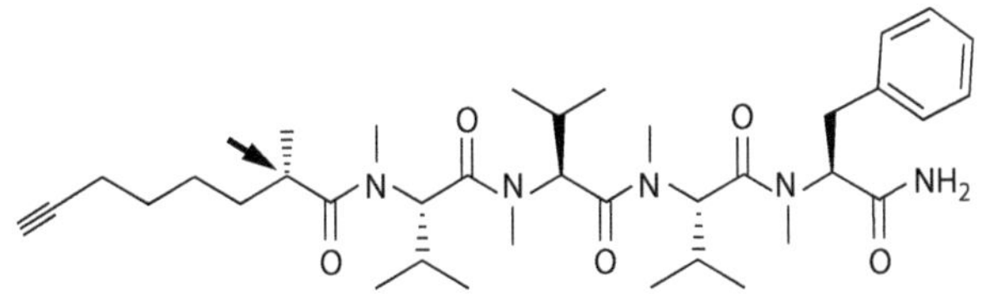

Figure 12.6 (Continued) Antifungal secondary metabolites from cyanobacteria: (**4**) dragonamide, (**5**) almiramide, (**6**) norharmane, and (**7**) majusculoic acid.

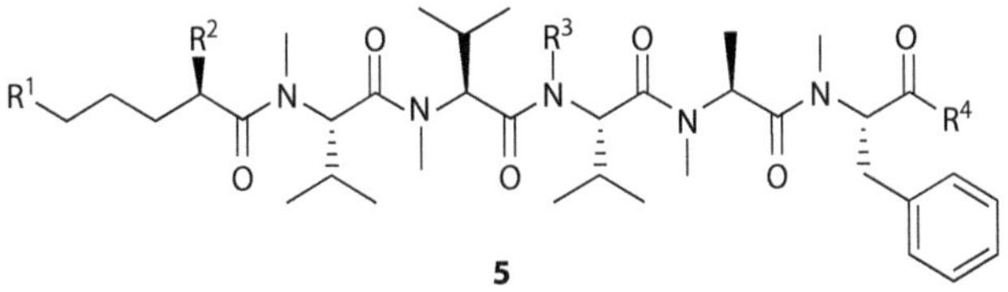

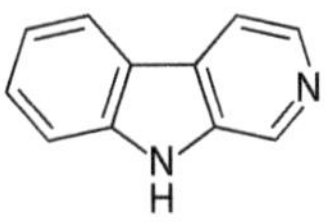

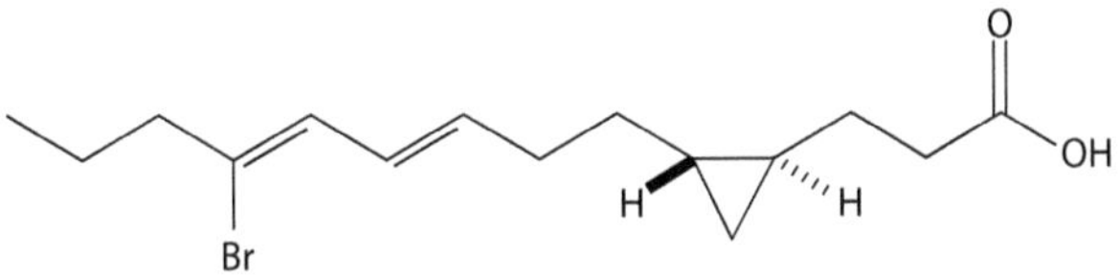

Table 12.4 Antiprotozoan compounds from cyanobacteria

Source	Compound	Activity against
Symploca sp.	Gallinamide A, symplocamide A	*P. falciparum, Trypanosoma cruzi, Leishmania donovani*
Blennothrix cantharidosmum	Tumonoic acid I	*P. falciparum*
M. aeruginosa PCC 7806	Aerucyclamide C	Antileishmanial, *T. brucei*
Oscillatoria nigroviridis	Viridamides A and B	Antiprotozoal *Trypanosoma cruzi*, Leishmania mexicana, *P. falciparum*
Nostoc sp.	Nostocarboline	*T. brucei, Trypanosoma cruzi, Leishmania donovani,* and *P. falciparum*
Phormidium ectocarpi	Hierridin B and 2,4-dimethoxy-6-heptadecyl-phenol	*P. falciparum*
Lyngbya sp.	Dragonamide E	Antileishmanial
L. majuscula	Almiramides A–C	Antileishmanial
Fischerella ambigua	Ambigol C	*Trypanosoma rhodesiense, P. falciparum*

in these situations. The scientist searching novel natural products derived from marine cyanophytes for antiviral activity has yielded a considerable number of active crude aqueous and organic solvent extracts. Marine cyanobacteria also appear to be a rich source of new antiviral compounds. Gustafson et al. (1989) used a tetrazolium-based microculture to screen extracts of marine cyanobacterial cultures, *Lyngbya* sp. (Tripathi et al., 2010), *Lyngbya lagerheimii,* and *Phormidium tenue* for inhibition of HIV-1 (Gustafson et al., 1989).

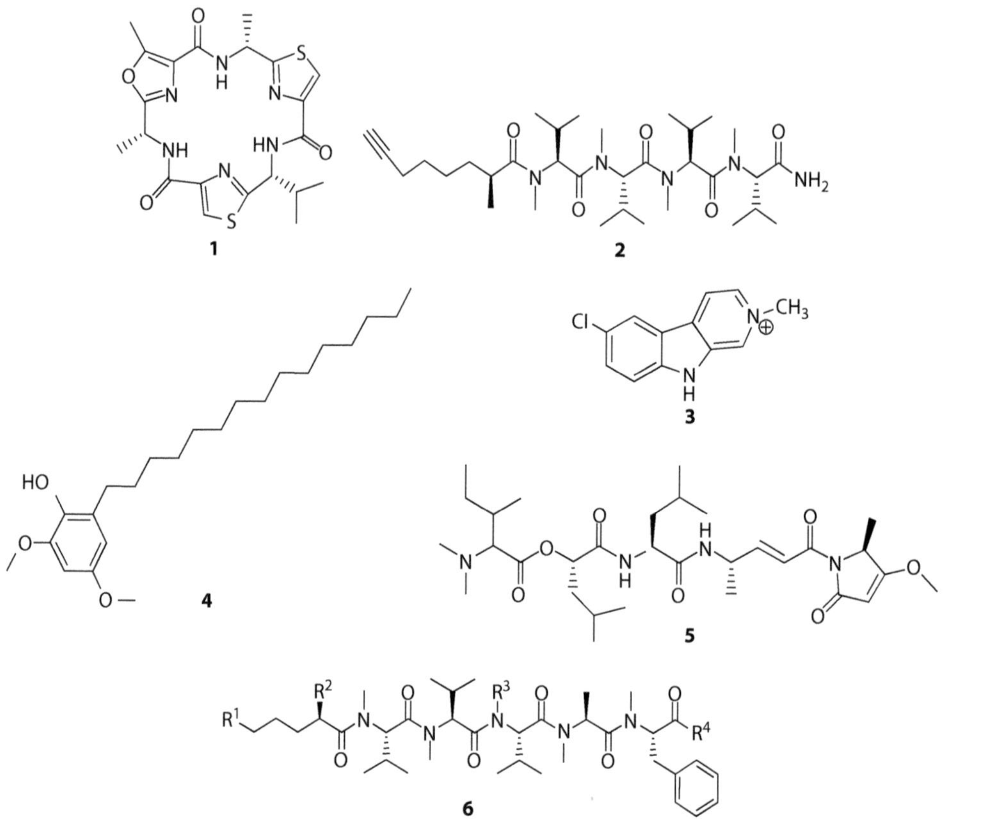

Figure 12.7 Cyanobacterial secondary metabolites showing antiprotozoan activity: (**1**) venturamide A, (**2**) dragonamide B, (**3**) nostocarboline, (**4**) hierridin B, (**5**) gallinamide A, and (**6**) almiramide.

Anti-influenza virus activity was found in extracts of *Lyngbya* prepared from early exponential growth phase cells (Armstrong et al., 1991). In the University of Hawaii, researchers screen around 10% of (n = 600) cyanobacterial extracts tested live in virus test systems for inhibition of HSV-2 and HIV type 1 (HIV-1). However, a smaller percentage of (2.5%) extracts were active against respiratory syncytial virus (Patterson et al., 1993). Lau et al. (1993) screened 2% (among 900 strains) of cultured cyanobacteria having antireverse transcriptase activity against avian myeloblastosis virus and HIV-1.

So far, few antiviral compounds were isolated from marine cyanobacteria (Table 12.5). The antiviral cancer polysaccharides such as spirulan and Ca spirulan were from *Spirulina* sp. that showed potent and broad-spectrum activity against HIV-1, HIV-2, Influenza, and a series of other enveloped viruses (Feldman et al., 1999). These sulfated polysaccharides prevent virus–cell attachment and fusion with host cells (CD4⁺) and inhibit the reverse transcriptase activity of HIV-1 (like azidothymidine). The natural sulfoglycolipids from cyanobacteria *Scytonema* sp. are also reported to inhibit HIV reverse transcriptase and DNA polymerases (Loya et al., 1998).

A polypeptide cyanovirin-N (CV-N) (11 kDa; 101 amino acid) is a carbohydrate-binding protein produced by *Nostoc ellipsosporum* showing potent *in vitro* and *in vivo* activities against HIV and other lentiviruses (Boyd et al., 1997). CV-N is a cyanobacterial protein with potent neutralizing activity against HIV-1 high affinity with gp120, and

Table 12.5 Antiviral compounds from marine cyanobacteria

Organism	Compound	Activity against
Spirulina sp.	Spirulan	HIV-1 and HIV-2 (antireverse transcriptase activity), influenza, HSV
Nostoc flagelliforme	Nostoflan	HSV-1 (HF), HSV-2, HCMV, influenza, adenovirus type 2
Nostoc ellipsosporum	CV-N	HIV-1, HIV-2, HSV-6, SIV, FIV
Scytonema varium	Scytovirin N	HIV-1
Scytonema sp.	Sulfoglycolipid	HIV-1
Trichodesmium erythraeum	Debromoaplysiatoxin, anhydrodebromoaplysiatoxin, and 3-methoxydebromoaplysiatoxin	Chikungunya virus
Microcystis ichthyoblabe	Ichthyopeptins A and B	Influenza A

it blocks the envelope glycoprotein-mediated membrane fusion reaction associated with HIV-1 entry. CV-N has blocking action at the level of gp120 interaction with coreceptor and the chemokine receptor CXCR4 as an entry receptor (Dey et al., 2000). Thus, CV-N has broad-spectrum antiviral activity, for example, (1) CV-N impairs both CD4-dependent and CD4-independent binding of sgp120 to the target cells, (2) CV-N blocks the sCD4-induced binding of sgp120 with cell-associated coreceptor CXCR4, and (3) CV-N dissociates bound sgp120 from target cells (Mori and Boyd, 2001).

Another anti-HIV protein scytovirin (cyanobacteria, *Scytonema varium*) has two carbohydrate-binding sites. This unique protein (95 residues) contains five structurally important intrachain disulfide bonds, which makes unusual clusters with aromatic residues of sites, suggesting a binding mechanism similar to other known hevein-like carbohydrate-binding proteins but differing in carbohydrate specificity. Scytovirin is used to inhibit HIV entry and prophylactic anti-HIV applications (McFeeters et al., 2007). It binds to the envelope glycoprotein of HIV (gp120, gp160, and gp41) and inactivates the virus in low nanomolar concentrations (Bokesch et al., 2003; Xiong et al., 2006, 2010). Nostoflan is an acidic polysaccharide from *Nostoc flagelliforme* that exhibits potent virucidal activity against herpes simplex virus 1 (Kenji et al., 2005).

In addition, two cyclic depsipeptides, ichthyopeptins A and B, were also isolated from *Microcystis ichthyoblabe*. They showed antiviral activity against influenza A virus with an IC_{50} value of 12.5 mg/mL (Zainuddin et al., 2007). Deepak Kumar et al. (2014) screened five different compounds from *Trichodesmium erythraeum*. The debrominated analogs such as debromoaplysiatoxin, anhydrodebromoaplysiatoxin, and 3-methoxydebromoaplysiatoxin displayed dose-dependent inhibition of Chikungunya virus (Gupta et al., 2014).

12.4.5   Antiquorum compounds

The most widely used method of antifouling activity in marine environments involves the use of highly toxic compounds, which affects the marine life. Hence, finding an alternative form of biofilm control is essential. Many potential marine-derived antifoulants have been identified from organisms that are naturally exposed to larvae of fouling organisms and produce antifouling chemicals (Tan et al., 2010).

Cyanobacteria produce a variety of potential natural bioactive metabolites that may have to prevent biofouling by colonizing organisms. Generally, metabolites with antibiotic, algicidal, cytotoxic, and enzyme-inhibiting activities may represent antiquorum activities on both micro- and macrofoulents. Molecules such as amides, alkaloids, terpenoids, fatty acids, lipopeptides, lactones, steroids, and pyrroles display antifouling activities. Phycotoxins and related products from cyanobacteria may serve as materials for anti-fouling applications (Dahms et al., 2006). Recently, more than 21 different antifouling substances from 27 strains of cyanobacteria have been reported (Dahms et al., 2006). For example, cyanobacteria *L. majuscula* produced many useful secondary metabolites that have antifeed properties, and they may inhibit invertebrate larval settlement. The metabolite dolistain 16 (EC_{50} at 0.003 μg/mL) showed high antisettlement activities against barnacle cyprids (Tan et al., 2010), honaucins A–C, potent inhibitors of bacterial quorum sensing (Choi et al., 2012), ellagic acid (Huber et al., 2003), malyngolide (Dobretsov et al., 2010), and microcolins A and B and kojic acid that inhibits the LuxR-based reporter induced by N-3-oxo-hexanoyl-L-homoserine lactone (Dobretsov et al., 2011). Honaucin A and its two natural analogs exhibit potent inhibition of both bioluminescence and a quorum sensing–dependent phenotype, in *Vibrio harveyi* BB120 (Choi et al., 2012). Malyngamides A and B, which are structurally different from malyngamide C, inhibited quorum sensing–dependent violacein production by *Chromobacterium violaceum* CV017 (Dobretsov et al., 2011) and inhibited Las-R-based bacterial luminescence (Kwan et al., 2010). Dobretsov et al. (2011) reported that kojic acid effectively inhibits the bacterial density (3.2 fold) and diatom density (4.7 fold) on glass slide at 330 μM concentration. Kojic acid has tyrosinase inhibitory action (Burdock et al., 2001), by which it inhibits fouling organisms. The mechanism of this quorum inhibition remains unknown, but it was suggested that antibiotics can change bacterial membrane permeability, thus affecting flux of quorum signals (Skindersoe et al., 2008).

The discovery of marine cyanobacterial compounds with macrofoulent antisettlement activities were revealed as a potential source of natural antifoulants controlling biofouling. A number of natural products, including hantupeptin C (Figure 12.8a), isomajusculamide (Figure 12.8e), and majusculamide A, have moderate barnacle antisettlement activities (Tan and Goh, 2009). The marine cyanobacterium *Schizothrix calcicola* produced the compound ypaoamide (Figure 12.8b), which prevents feeding of two species of rabbitfishes (Nagle and Paul, 1998). Pitipeptolide A from *L. majuscula* has chemical defense against various grazers (Cruz-Rivera and Paul, 2007) and larval recruitments of *Acropora surculosa* and *Pocillopora damicornis* (Kuffner and Paul, 2004). Lyngbyatoxin A and debromoaplysiatoxin inhibit various marine predators (Capper et al., 2006). Brown et al. (2004) demonstrated antifouling property of marine cyanobacterial (*Kyrtuthrix maculans*) compound maculalactone A that has toxicity against various species of barnacle larvae such as *Tetraclita japonica*, *Balanus amphitrite*, and *Ibla cumingii*.

12.5   Screening of bioactive compounds from cyanobacteria

Considering the great diversity of natural products from marine cyanobacteria, the appropriate methods can be rapidly used to screen different bioactive compounds from the source of great interest. To design this screening methodology, different parameters have to be considered, which include the nature of the bioactive compounds (soluble property, heat resistance, or molecular weight) and their activity (Falch et al., 1995). The screening of bioactive metabolites from the cyanobacterial source begins with suitable extraction technique, based on either expected or targeted bioactive metabolites (Figure 12.9).

Generally, solvent extraction method is used to obtain certain compounds from the source materials (Plaza et al., 2010). A large number of bioactive compounds have been

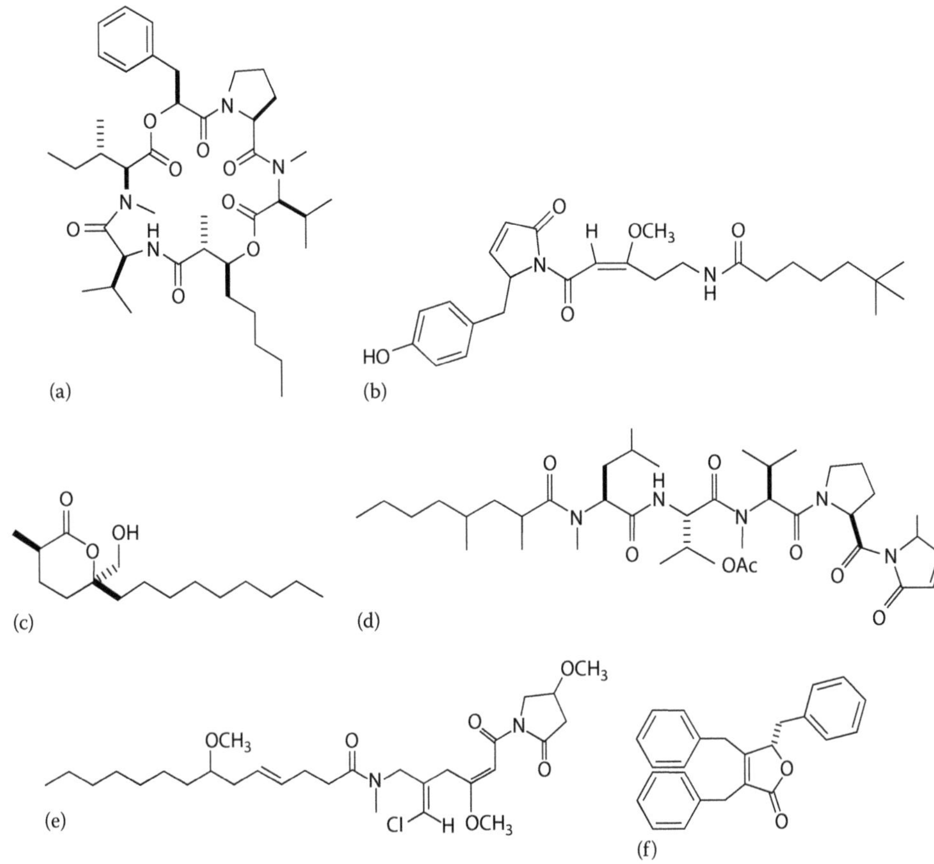

Figure 12.8 Chemical structure of cyanobacterial secondary metabolites as antifouling compounds: (a) hantupeptin C, (b) ypaoamide, (c) malyngolide, (d) microcolin B, (e) isomalyangamide A, and (f) maculalactone A.

traditionally extracted with organic solvents (hexane, ethanol, water, methanol, acetone, diethyl ether, petroleum ether), either single solvent method or fractionation methods (Starmans and Nijhuis, 1996; Plaza et al., 2010; Tokuoka et al., 2010). These organic solvents can be employed for the extraction of both polar and nonpolar organic compounds such as alkaloids, phenols, aromatic hydrocarbons, fatty acids, and oils (El Hattab et al., 2007; Plaza et al., 2010). The parameters such as influence of solvents, temperatures, and pressures might have a significant influence on the outcome of the extraction process. Invariably, solvent extraction is advantageous compared to other methods due to low processing cost and ease of operation. However, this method uses toxic solvents and requires an evaporation/concentration step for recovery (Joana Gil-Chávez et al., 2013).

After the extraction obtained using diverse conditions, these extracts must be tested for biological activities by performing the appropriate functional activity assay(s) (Figure 12.9), for example, antibacterial and antifungal screening can be done with agar spot assay, well diffusion assay, and disk diffusion assay. When an organic solvent extract shows bioactivity, the bioactive compounds may be screened and characterized subsequently.

Once the target biological activities have been confirmed, the next step involves chemical characterization of the bioactive components present in the initial extract.

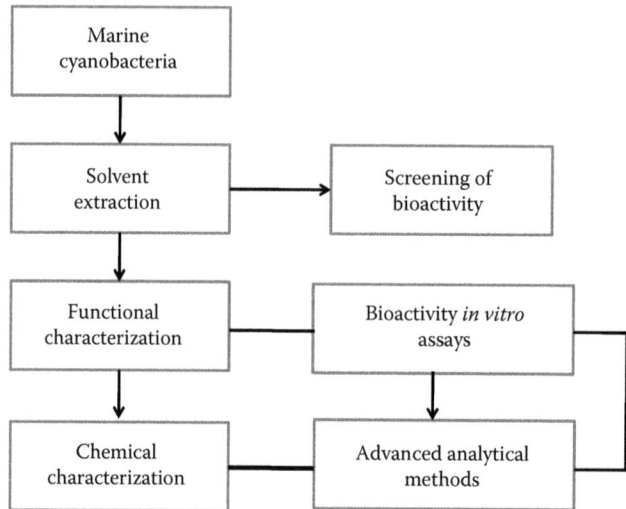

Figure 12.9 Basic scheme showing the proposed work flow for the screening of bioactive compounds from marine cyanobacteria.

Primarily, bioactive compounds should be traced by qualitative detection methods such as phenols (Harborne, 1998; Biglari et al., 2008), amino acids, sterols and saponins (Lacaille-Dubois and Wagner, 1996; Shiau et al., 2009), alkaloids (Shamsa et al., 2008), flavonoids (Mansouri et al., 2005; Wang et al., 2008), tannins (ferric chloride test), carbohydrates, proteins, and fatty acids (standard protocols). The chemistry of bioactive metabolites like toxins is known, and that has made possible the development of high-throughput analysis methods such as gas chromatography (GC), high-performance liquid chromatography (HPLC), and liquid chromatography–mass spectrometry (LC/MS). Matrix-assisted laser desorption/ionization time-of-flight mass spectrometry has been proven as a fast screening method for the detection of cyanobacterial peptide toxins and bioactive compounds (Erhard et al., 1997; Fastner et al., 2001; Welker et al., 2004). For example, to identify the individual compound, separation is needed and identification methods such as TLC, HPLC, FT-IR, and GC with mass detection (Meriluoto et al., 2004; Meriluoto and Codd, 2005) and LC/MS (Sivonen, 2000) are required.

Purified compounds' functional characterization should be assessed through the application of fast in vitro assays directed to the confirmation of the sought biological properties, for instance, antioxidant capacity assays and antimicrobial activity assays. Based on the pharmacological importance, bioactive secondary metabolites are synthesized and used for various purposes.

12.6   Biosynthetic genes for secondary metabolites

In the past two decades, biosynthesis gene clusters were assigned to an increasing number of these cyanobacterial natural products (Welker and von Döhren, 2006; Jones et al., 2009). Microbial peptides are produced by two types of biosynthetic pathways: by ribosomal synthesis and subsequent posttranslational modification and processing or by giant complex multidomain enzymes and the nonribosomal peptide synthetases (NRPSs) without the help of ribosomes and mRNA. Nonribosomal peptides are often cyclic and/or branched structures that contain nonproteinogenic amino acids including D-amino acids, with *N*-methyl and *N*-formyl groups, or glycosylated, acylated, halogenated, or hydroxylated. Nonribosomal peptides are often dimers or trimers. The biosynthetic gene of NRPS consists of modules, each being responsible for the incorporation of a single amino acid. In general, each NRPS complex contains three modules such as (a) initiation module, [F/NMT]-A-PCP-; elongation module, -(C/Cy)-[NMT]-A-PCP-[E]-; and termination module, -(TE/R). The order of these modules typically follows a colinearity rule; the succession of modules corresponds to the order of amino acids in the final product. Each module contains a number of domains; each domain plays an important role to add and modify the amino acids by their own functions.

The domains of NRPS modules are as follows: F, formylation (optional); A, adenylation (required in a module); PCP, and peptide-carrier protein with attached 4′-phosphopantetheine (required in a module); C, condensation forming the amide bond (required in a module); Cy, cyclization into thiazolines or oxazolines (optional); Ox, oxidation of thiazolines or oxazolines to thiazoles or oxazoles (optional); Red, reduction of thiazolines or oxazolines to thiazolidines or oxazolidines (optional); E, epimerization into D-amino acids (optional); NMT, *N*-methylation (optional); TE, termination by a thioesterase (only found once in a NRPS); and R, reduction to terminal aldehyde or alcohol (optional).

A minimal module is composed of an amino acid–activating adenylation (A) domain, a peptidyl carrier (PCP) domain carrying the phosphopantetheine cofactor and a condensation (C) domain (Figure 12.10) (Koglin and Walsh, 2009).

Currently, there are well over 300 proteinogenic and nonproteinogenic secondary metabolites, being reported from marine cyanobacteria (Uzair et al., 2012). These natural products represent great structural diversity, belonging to the polyketide synthase (PKS) and NRPS, as well as hybrid PKS–NRPS structural classes (Gerwick et al., 2001). To date, only few putative biosynthetic PKS–NRPS gene clusters of marine cyanobacterial molecules have been reported including barbamide, jamaicamide, and curacin A (Chang et al., 2002, 2004; Edwards et al., 2004).

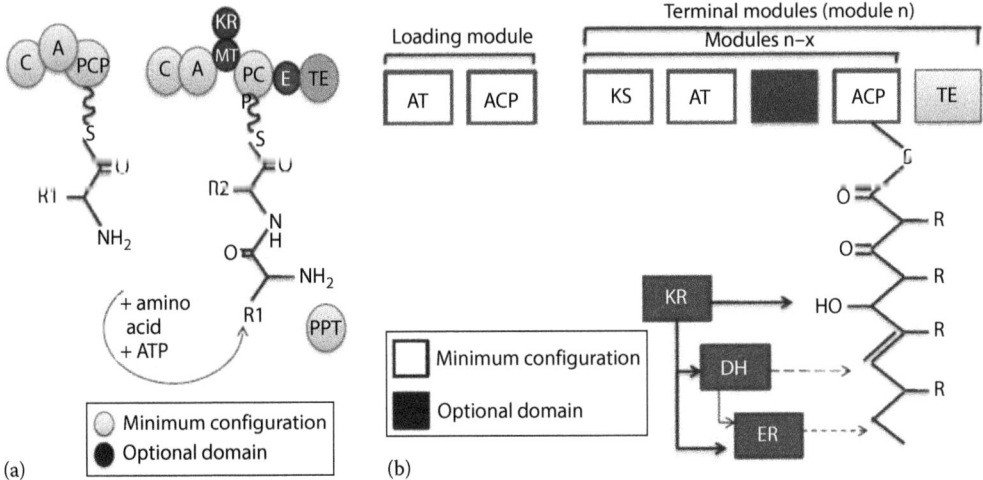

Figure 12.10 Schematic representation of enzymatic domains in (a) nonribosomal peptide synthetases and (b) polyketide synthase gene clusters. *Abbreviations:* A, adenylation domain; ACP, acyl carrier protein; AT, acyltransferase; C, condensation domain; PCP, peptidyl carrier protein; DH, dehydratase; MT, methyltransferase; E, epimerase; KS, ketosynthase; KR, ketoreductase; ER, enoyl reductase; and TE, thioesterase.

In contrast, ribosomal biosynthesis of peptides has limitations to a minimum of 20 proteinogenic amino acids. This group of peptides nevertheless displays a high diversity and a considerable biosynthetic and bioactive potential. The ribosomal prepeptides are typically composed of a leader peptide and a core peptide. Associated posttranslational modification enzymes (PTMs) catalyze different types of macrocyclizations of the core peptide and side-chain modifications of amino acids. Peptide maturation further requires cleavage of the leader peptide by processing proteases (PPs) frequently combined with transport across the plasma membrane (Oman and van der Donk, 2010).

Cyanobacterial macrolides like lyngbouillosides A, B, and C and the related compounds have attracted a lot of attention over the past decade due to their intriguing architecture, their natural scarcity, and their potential biological activities (El Marrouni et al., 2013). They are produced by modular-type PKSs resembling NRPS with respect to their modular nature. The well-studied PKS assembly is the animal fatty acid synthase (Jenke-Kodama et al., 2005) and consists of ketosynthase, acyltransferase, ketoreductase, dehydratase, enoyl reductase, and acyl carrier protein domains (Staunton and Weissman, 2001). Parts of the domains (KR, DH, ER) are optionally used leading to a different reduction state of the keto groups of polyketides.

12.7   Barbamide biosynthetic pathway

The marine cyanobacteria *L. majuscula* produces several bioactive secondary metabolites that include barbamide, curacin A, carmabin A, antillatoxin B, and malyngamide (Ramaswamy et al., 2006). Biosynthesis of these secondary metabolites may have unique structural features.

The genetic architecture and catalytic domain organization of the barbamide gene cluster (bar) is generally colinear and contains 12 open reading frames (ORFs) (Chang et al., 2002) (Figure 12.8). It is synthesized by a mixed PKS/NRPS system with unusual features that include a stand-alone PCP and unusual adenylation (A) domains, as well as a PKS module that is encoded on two separate ORFs. In addition, the biosynthetic system encodes several tailoring enzymes involved in (1) the chlorination, α-oxidation, and decarboxylation of leucine to form a trichloroisovaleric acid moiety and (2) the oxidative decarboxylation of the cysteine residue at the end of the assembly line to form the terminal thiazole ring (Figure 12.11).

1. The BarA-BarE and BarJ proteins are involved in the conversion of leucine to the tri-chloroisovaleric acid.
2. BarA contains a PCP domain.
3. BarD dwells a leucine/trichloroleucine-specific A domain, which activates leucine and is required for covalent bond formation between leucine and BarA.
4. BarC shares high sequence homology with known thioesterases (TE). BarC can mediate the release of leucine from BarA.
5. BarB1 and BarB2 are believed to be the putative halogenases (Figure 12.8). The conversion of L-leucine to the chlorinated α-keto derivative occurs by the action of BarB1 and BarB2, while the substrate is bound as a thioester intermediate to BarA.
6. BarE contains an A domain that shows high activity for both the nonchlorinated and trichlorinated α-keto derivatives of L-leucine, as well as modest activation of trichloroleucine. The chlorinated α-keto intermediate could then be released and incorporated as the loading module in BarE to initiate barbamide biosynthesis.

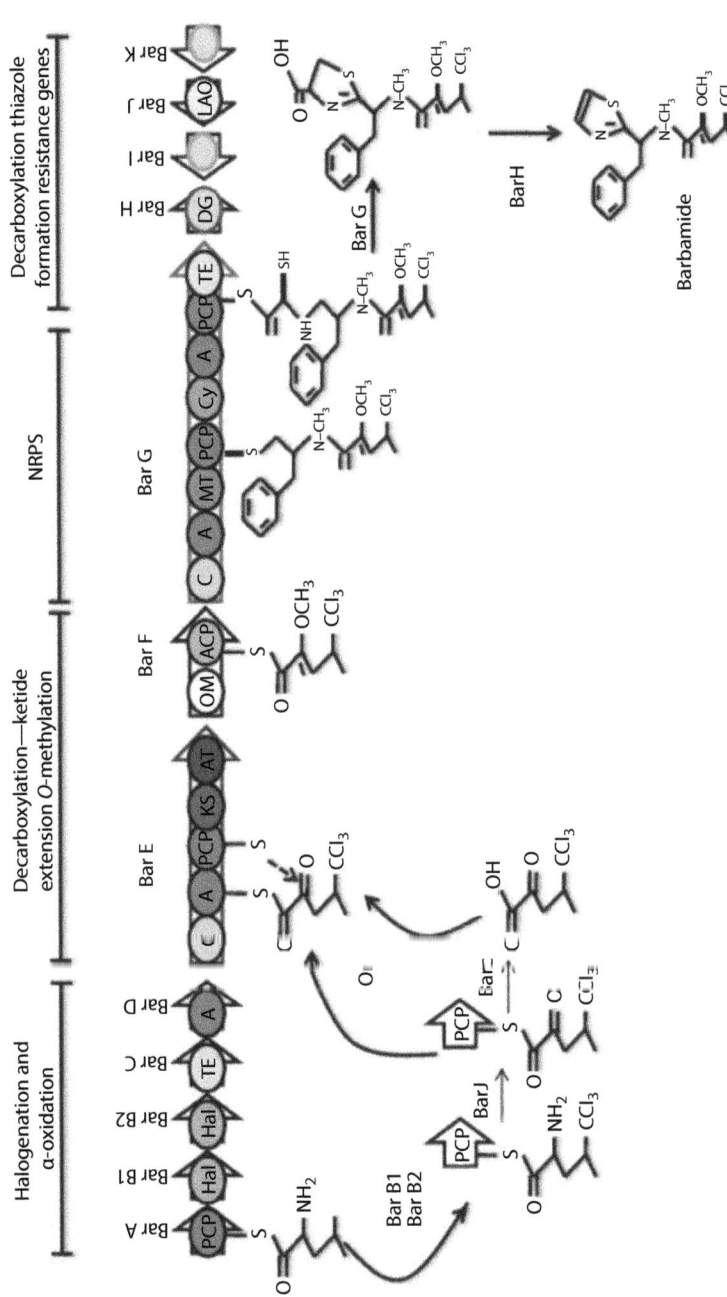

Figure 12.11 Barbamide biosynthetic gene cluster, an example for nonribosomal peptide synthetase–polyketide synthase hybrid biosynthesis.

7. In addition, BarJ, which shares homology with L-amino-acid oxidases, is proposed to function as the α-oxidase that converts trichloroleucine to the α-keto derivative.

8. BarG is predicted to catalyze the addition of *N*-methyl phenylalanine and cysteine followed by cleavage from the enzyme complex by the internal TE domain residing in BarG. The released product must then undergo oxidative decarboxylation, possibly mediated by BarH, to form the final product, barbamide.

The genetic architecture of the bar cluster is supported by several stable isotope feeding experiments that demonstrate that barbamide is derived from acetate; two *S*-adenosylmethionine-derived methyl groups; and three amino acids L-leucine, L-phenylalanine, and L-cysteine (Sitachitta and Gerwick, 1998; Sitachitta et al., 2000; Williamson et al., 2004).

12.8   Conclusion

Cyanobacteria are one of the sources of known novel bioactive compounds including toxins, polysaccharides, sterols, and vitamins with wide pharmaceutical applications. Many compounds from cyanobacteria are useful for the welfare of mankind. At present, few compounds and their analogs identified from cyanobacteria are in clinical trials, and some of them have passed different phases of clinical trials to prove their candidacy as potential drugs. Even if a single cyanobacterium, for example, *Lyngbya* sp., has produced more than 100 metabolites, we can think what kind of novelty and potentiality the cyanobacterium is having. Research is needed to unfold other hidden bioactive principles of marine cyanobacteria. One exciting feature of the biosynthetic gene clusters identified from marine cyanobacteria is the absence of self-resistance, regulatory, or transport genes. Recent chemical studies have shown that marine cyanobacteria are capable of producing a wide range of other non-PKS–NRPS molecules. Thus, there is a need for extensive research in this new emerging field for antimicrobial drug discovery.

References

Adams, D. G. 2002. Cyanobacteria in symbiosis with hornworts and liverworts. In: Rai, A. N., Bergman, B., and Rasmussen, U., eds., *Cyanobacteria in Symbiosis*. Dordrecht, the Netherlands: Kluwer Academic Publishers, pp. 117–135.

Aimi, N., Odaka, H., Sakai, S., Fujiki, H., Suganuma, M., Moore, R. E., and Patterson, G. M. L. 1990. Lyngbyatoxins B and C, two new irritants from *Lyngbya majuscula*. *J Nat Prod* 53(6): 1593–1596.

Armstrong, J. E., Janda, K. E., Alvarado, B., and Wright, A. E. 1991. Cytotoxin production by marine *Lyngbya* strain (cyanobacterium) in large scale laboratory bioreactors. *J Appl Phycol* 3(3): 277–282.

Balunas, M. J., Linington, R. G., Tidgewell, K., Fenner, A. M., Ureña, L. D., Togna, G. D., Kyle, D. E., and Gerwick, W. H. 2010. Dragonamide E, a modified linear lipopeptide from *Lyngbya majuscula* with antileishmanial activity. *J Nat Prod* 73(1): 60–66.

Barbaras, D., Kaiser, M., Brun, R., and Gademann, K. 2008. Potent and selective antiplasmodial activity of the cyanobacterial alkaloid nostocarboline and its dimers. *Bioorg Med Chem Lett* 18: 4413–4415.

Bhateja, P., Mathur, T., Pandya, M., Fatma, T., and Rattan, A. 2006. Activity of blue–green microalgae extracts against in vitro generated Staphylococcus aureus with reduced susceptibility to vancomycin. *Fitoterpia* 77: 233–235.

Biglari, F., AlKarkhi, A. F., and Easa, A. M. 2008. Antioxidant activity and phenolic content of various date palm (*Phoenix dactylifera*) fruits from Iran. *Food Chem* 107(4): 1636–1641.

Biondi, N., Tredici, M. R., Taton, A., Wilmotte, A., Hodgson, D. A., Losi, D., and Marinelli, F. 2008. Cyanobacteria from benthic mats of Antarctic lakes as a new source of bioactivities. *J Appl Microbiol* 105: 105–115.

Bokesch, H. R., O'Keefe, B. R., McKee, T. C., Pannell, L. K., Patterson, G. M., Gardella, R. S., and Boyd, M. R. 2003. A potent novel anti-HIV protein from the cultured cyanobacterium. *Scytonema varium*. *Biochemistry* 42: 2578–2584.

Boyd, M. R., Gustafson, K. R., McMahon, J. B., Shoemaker, R. H., O'Keefe, B. R., Mori, T., and Henderson, L. E. 1997. Discovery of cyanovirin-N, a novel human immune deficiency virus inactivating protein that binds viral surface envelope glycoprotein gp120: Potential application to microbicide development. *Antimicrob Agents Chemother* 41: 1521–1530.

Broniatowska, B., Allmendinger, A., Kaiser, M., Montamat-Sicotte, D., Hingley-Wilson, S., Lalvani, A., and Tasdemir, D. 2011. and cytotoxic potential of cyanobacterial (blue–green algal) extracts from Ireland. *Nat Prod Commun* 6(5): 689–694.

Browitzka, M. A. 1995. Microalgae as sources of pharmaceuticals and other biologically active compounds. *J Appl Phycol* 7: 3–15.

Brown, G. D., Wong, H. F., Hutchinson, N., Lee, S. C., Chan, B. K., and Williams, G. A. 2004. Chemistry and biology of maculalactone A from the marine cyanobacterium *Kyrtuthrix maculans*. *Phytochem Rev* 3: 381–400.

Burdock, G. A., Soni, M. G., and Carabin, I. G. 2001. Evaluation of health aspects of kojic acid in food. *Regul Toxicol Pharmacol* 33: 80–101.

Burja, A. M., Banaigs, B., Abou-Mansour, E., Burgess, J. G., and Wright, P. C. 2001. Marine cyanobacteria—A prolific source of natural products. *Tetrahedron* 57: 9347–9377.

Capper, A., Cruz-Rivera, E., Paul, V. J., and Tibbetts, I. R. 2006. Chemical deterrence of a marine cyanobacterium against sympatric and non-sympatric consumers. *Hydrobiologia* 553: 319–326.

Chang, R. L., Ghamsari, L., Manichaikul, A., Hom, E. F., Balaji, S., Fu, W., and Papin, J. A. 2011. Metabolic network reconstruction of *Chlamydomonas* offers insight into light-driven algal metabolism. *Mol Syst Biol* 7: 518.

Chang, Z., Flatt, P., Gerwick, W. H., Nguyen, V. A., Willis, C. L., and Sherman, D. H. 2002. The barbamide biosynthetic gene cluster: A novel marine cyanobacterial system of mixed polyketide synthase (PKS)-non-ribosomal peptide synthetase (NRPS) origin involving an unusual trichloroleucyl starter unit. *Gene* 296: 235–247.

Chang, Z., Sitachitta, N., Rossi, J. V., Roberts, M. A., Flatt, P. M., Jia, J., and Gerwick, W. H. 2004. Biosynthetic pathway and gene cluster analysis of curacin A, an antitubulin natural product from the tropical marine cyanobacterium *Lyngbya majuscula*. *J Nat Prod* 67: 1356–1367.

Chauhan, R., Silambarasan, S., and Abraham, J. 2011. Biodiversity of marine cyanobacteria and its antibacterial activity. *IJPI'S J Biotechnol Biother* 1(4): 15–19.

Choi, H., Mascuch, S. J., Villa, F. A., Byrum, T., Teasdale, M. E., Smith, J. E., and Gerwick, W. H. 2012. Honaucins A–C, potent inhibitors of inflammation and bacterial quorum sensing: Synthetic derivatives and structure-activity relationships. *Chem Biol* 19: 589–598.

Clark, B. R., Engene, N., Teasdale, M. E., Rowley, D. C., Matainaho, T., Valeriote, F. A., and Gerwick, W. H. 2008. Natural products chemistry and taxonomy of the marine cyanobacterium. *Blennothrix cantharidosmum*. *J Nat Prod* 71: 1530–1537.

Coates, A. R. M. and Hu, Y. 2007. Novel approaches to developing new antibiotics for bacterial infections. *Br J Pharma* 152: 1147–1154.

Cruz-Rivera, E. and Paul, V. J. 2007. Chemical deterrence of a cyanobacterial metabolite against generalized and specialized grazers. *J Chem Ecol* 33: 213–217.

Dahms, H. U., Ying, X., and Pfeiffer, C. 2006. Antifouling potential of cyanobacteria: A mini-review. *Biofouling* 22(5–6): 317–327.

Dawson, R. M. 1998. The toxicology of microcystins. *Toxicon* 36(7): 953–962.

Devillers, J., Doré, J. C., Guyot, M., Poroikov, V., Gloriozova, T., Lagunin, A., and Filimonov, D. 2007. Prediction of biological activity profiles of cyanobacterial secondary metabolites. *SAR QSAR Environ Res* 18: 629–643.

Dey, B., Lerner, D. L., Lusso, P., Boyd, M. R., Elder, J. H., and Berger, E. A. 2000. Multiple antiviral activities of cyanovirin-N: Blocking HIV type1 gp120 interaction with CD4 co receptor and inhibition of diverse enveloped viruses. *J Virol* 74: 4562–4569.

Dobretsov, S., Teplitski, M., Alagely, A., Gunasekera, S. P., and Paul, V. J. 2010. Malyngolide from the cyanobacterium *Lyngbya majuscula* interferes with quorum sensing circuitry. *Environ Microbiol Rep* 2: 739–744.

Dobretsov, S., Teplitski, M., Bayer, M., Gunasekera, S., Proksch, P., and Paul, V. J. 2011. Inhibition of marine biofouling by bacterial quorum sensing inhibitors. *Biofouling* 27(8): 893–905.

Drobac-Čik, A. V., Dulić, T. I., Stojanović, D. B., and Svirčev, Z. B. 2007. The importance of extremophile cyanobacteria in the production of biological active compounds. *Proc Nat Sci Matica Srpska Novi Sad* 112: 57–66.

Edwards, D. J., Marquez, B. L., Nogle, L. M., McPhail, K., Goeger, D. E., Roberts, M. A., and Gerwick, W. H. 2004. Structure and biosynthesis of the jamaicamides, new mixed polyketide-peptide neurotoxins from the marine cyanobacterium *Lyngbya majuscula*. *Chem Biol* 11: 817–833.

El Hattab, M., Culioli, G., Piovetti, L., Chitour, S. E., and Valls, R. 2007. Comparison of various extraction methods for identification and determination of volatile metabolites from the brown alga *Dictyopteris membranacea*. *J Chromatogr A* 1143(1–2): 1–7.

El Marrouni, A., Kolleth, A., Lebeuf, R., Gebauer, J., Prevost, S., Heras, M., and Cossy, J. 2013. Lyngbouilloside and related macrolides from marine cyanobacteria. *NPC Nat Prod Commun* 8: 1–8.

Erhard, M., von Döhren, H., and Jungblut, P. 1997. Rapid typing and elucidation of new secondary metabolites of intact cyanobacteria using MALDI-TOF mass spectrometry. *Nat Biotechnol* 15: 906–909.

Falch, B. S., König, G. M., Wright, A. D., Sticher, O., Angerhofer, C. K., Pezzuto, J. M., and Bachmann, H. 1995. Biological activities of cyanobacteria: Evaluation of extracts and pure compounds. *Planta Med* 61: 321–328.

Fastner, J., Erhard, M., and von Döhren, H. 2001. Determination of oligopeptide diversity within a natural population of *Microcystis* spp. (cyanobacteria) by typing single colonies by matrix–assisted laser desorption ionization–time of flight mass spectrometry. *Appl Environ Microbiol* 67: 5069–5076.

Feldman, S. C., Reynaldi, S., Stortz, C. A., Cerezo, A. S., and Damonte, E. B. 1999. Antiviral properties of fucoidan fractions from *Leathesia difformis*. *Phytomedicine* 6: 335–340.

Fischbach, M. A. and Walsh, C. T. 2009. Antibiotics for emerging pathogens. *Science* 325(5944): 1089–1093.

Gerwick, W. H., Tan, L., and Sitachitta, N. 2001. *The Alkaloids*, Vol. 57. San Diego, CA: Academic Press, pp. 75–184.

Ghasemi, Y., Yazdi, M. T., Shafiee, A., Amini, M., Shokravi, S., and Zarrini, G. 2004. Parsiguine, A novel antimicrobial substance from *Fischerella ambigua*. *Pharmacol Biol* 2: 318–322.

Ghasemi, Y., Yazdi, M. T., Shokravi, S., Soltani, N., and Zarrini, G. 2003. Antifungal and antibacterial activity of paddy-fields cyanobacteria from the north of Iran. *J Sci Islamic Rep Iran* 14(3): 203–209.

Ghazala, B., Naila, B., and Shameel, M. 2010. Fatty acids and biological activities of crude extracts of freshwater algae from Sindh. *Pak J Bot* 42: 1201–1212.

Gupta, D. K., Kaur, P., Leong, S. T., Tan, L. T., Prinsep, M. R., and Chu, J. J. H. 2014. Anti-chikungunya viral activities of aplysiatoxin-related compounds from the marine cyanobacterium *Trichodesmium erythraeum*. *Mar Drugs* 12(1): 115–127.

Gustafson, K. R., Cardellina, J. H., Fuller, R. W., Weislow, O. S., Kiser, R. F., Snader, K. M., and Boyd, M. R. 1989. AIDS antiviral Sulfolipids from cyanobacteria (blue-green-algae). *J Natl Cancer Inst* 81(16): 1254–1258.

Hagmann, L. and Jiittner, F. 1996. Fischerellin A, a novel photosystem-II-inhibiting allelochemical of the cyanobacterium *Fischerella muscicola* with antifungal and herbicidal activity. *Tetrahedron Lett* 37: 6539–6542.

Han, B., Goeger, D., Maier, C. S., and Gerwick, W. H. 2005. The wewakpeptins, cyclic depsipeptides from a Papua New Guinea collection of the marine cyanobacterium *Lyngbya semiplena*. *J Org Chem* 70: 3133–3139.

Harborne, J. B. 1998. *Phytochemical Methods: A Guide to Modern Techniques of Latent Analysis*. London, U.K.: Chapman and Hall Publishers, p. 278.

Huber, B., Eberl, L., Feucht, W., and Polster, J. 2003. Influence of polyphenols on bacterial biofilm formation and quorum-sensing. *Z Naturforsch C* 58: 879–884.

Jaki, B., Orjala, J., Heilmann, J., Linden, A., Vogler, B., and Sticher, O. 2000. Novel extra cellular diterpenoids with biological activity from the cyanobacterium *Nostoc commune*. *J Nat Prod* 63: 339–343.

Jaki, B., Orjala, J., and Sticher, O. 1999. A novel extracellular diterpenoid with antibacterial activity from the cyanobacterium *Nostoc commune. J Nat Prod* 62: 502–503.

Jenke-Kodama, H., Sandmann, A., Muller, R., and Dittmann, E. 2005. Evolutionary implications of bacterial polyketide synthases. *Mol Biol Evol* 22(10): 2027–2039.

Joana Gil-Chávez, G., Villa, J. A., Fernando Ayala-Zavala, J., Basilio Heredia, J., Sepulveda, D., Yahia, E. M., and González-Aguilar, G. A. 2013. Technologies for extraction and production of bioactive compounds to be used as nutraceuticals and food ingredients: An overview. *Compr Rev Food Sci Food Safety* 12(1): 5–23.

Jones, A. C., Gu, L., Sorrels, C. M., Sherman, D. H., and Gerwick, W. H. 2009. New tricks from ancient algae: Natural product biosynthesis in marine cyanobacteria. *Curr Opin Chem Biol* 13: 216–223.

Kaushik, P., Chauhan, A., and Goyal, P. 2009. Screening of *Lyngbya majuscula* for potential antibacterial activity and HPTLC analysis of active methanolic extract. *J Pure Appl Microbiol* 3: 169–174.

Kim, J. D. 2006. Screening of cyanobacteria (blue–green algae) from rice paddy soil for antifungal activity against plant pathogenic fungi. *Microbiology* 34(3): 138–142.

Koglin, A. and Walsh, C. T. 2009. Structural insights into nonribosomal peptide enzymatic assembly lines. *Nat Prod Rep* 26: 987–1000.

Kreitlow, S., Mundt, S., and Lindequist, U. 1999. Cyanobacteria—A potential source of new biologically active substance. *J Biotechnol* 70: 61–73.

Kuffner, I. B. and Paul, V. J. 2004. Effects of the benthic cyanobacterium *Lyngbya majuscula* on larval recruitment of the reef corals *Acropora surculosa* and *Pocillopora damicornis. Coral Reefs* 23: 455–458.

Kumar, M., Tripathi, M. K., Srivastava, A., Gour, J. K., Singh, R. K., Tilak, R., and Asthana, R. K. 2013. Cyanobacteria, *Lyngbya aestuarii* and *Aphanothece bullosa* as antifungal and antileishmanial drug resources. *Asian Pac J Trop Biomed* 3(6): 458–463.

Lacaille-Dubois, M. A. and Wagner, H. 1996. A review of the biological and pharmacological activities of saponins. *Phytomedicine* 2(4): 363–386.

Lau, A. F., Siedlecki, J., Anleitner, J., Patterson, G. M., Caplan, F. R., and Moore, R. E. 1993. Inhibition of reverse transcriptase activity by extracts of cultured blue–green algae (cyanophyta). *Planta Med* 59(2): 148–151.

Linington, R. G., Clark, B. R., Trimble, E. E., Almanza, A., Ureña, L. D., Kyle, D. E., and Gerwick, W. H. 2009. Antimalarial peptides from marine cyanobacteria: Isolation and structural elucidation of gallinamide A. *J Nat Prod* 72: 14–17.

Linington, R. G., Edwards, D. J., Shuman, C. F., McPhail, K. L., Matainaho, T., and Gerwick, W. H. 2008. Symplocamide A, a potent cytotoxin and chymotrypsin inhibitor from the marine Cyanobacterium *Symploca* sp. *J Nat Prod* 71(1): 22–27.

Loya, S., Reshef, V., Mizrachi, E., Silberstein, C., Rachamim, Y., Carmeli, S., and Hizi, A. 1998. The inhibition of the reverse transcriptase of HIV-1 by the natural sulfoglycolipids from cyanobacteria: Contribution of different moieties to their high potency. *J Nat Prod* 61: 891–895.

Luke Simmons, T., Engene, N., Ureña, L. D., Romero, L. I., Ortega-Barría, E., Gerwick, L., and Gerwick, W. H. 2008. Viridamides A and B, lipodepsipeptides with antiprotozoal activity from the marine cyanobacterium *Oscillatoria nigroviridis. J Nat Prod* 71: 1544–1550.

MacMillan, J. B., Ernst-Russell, M. A., de Ropp, J. S., and Molinski, T. F. 2002. Lobocyclamides A–C. Lipopeptides from a cryptic cyanobacterial mat containing *Lyngbya confervoides. J Org Chem* 67: 8210–8215.

Madhumathi, V., Deepa, P., Jeyachandran, S., Manoharan, C., and Vijayakumar, S. 2011. Antimicrobial activity of cyanobacteria isolated from freshwater lake. *Int J Microbiol Res* 2: 213–216.

Malloy, K. L., Villa, F. A., Engene, N., Matainaho, T., Gerwick, L., and Gerwick, W. H. 2011. Malyngamide 2, an oxidized lipopeptide with nitric oxide inhibiting activity from a Papua New Guinea marine cyanobacterium. *J Nat Prod* 74(1): 95–98.

Mansouri, A., Embarek, G., Kokkalou, E., and Kefalas, P. 2005. Phenolic profile and antioxidant activity of the Algerian ripe date palm fruit (*Phoenix dactylifera*). *Food Chem* 89: 411–420.

Mayer, A. M., Rodríguez, A. D., Berlinck, R. G., and Hamann, M. T. 2009. Marine pharmacology in 2005–6: Marine compounds with anthelmintic, antibacterial, anticoagulant, antifungal, anti-inflammatory, antimalarial, antiprotozoal, antituberculosis and antiviral activities; affecting the cardiovascular, immune and nervous systems and other miscellaneous mechanisms of action. *Biochim Biophys Acta* 1790(5): 283–308.

McFeeters, R. L., Xiong, C., O'Keefe, B. R., Bokesch, H. R., McMahon, J. B., Ratner, D. M., and Byrd, R. A. 2007. The novel fold of scytovirin reveals a new twist for antiviral entry inhibitors. *J Mol Biol* 369(2): 451–461.

Meriluoto, J. and Codd, G. A. 2005. *Toxic: Cyanobacterial Monitoring and Cyanotoxin Analysis*. Åbo, Finalnd: Åbo Akademi University Press (Turku), ISBN 951–765–259–3, p. 149.

Meriluoto, J., Karlsson, K., and Spoof, L. 2004. High-throughput screening of ten microcystins and nodularins, cyanobacterial peptide hepatotoxins, by reversed-phase liquid chromatography–electrospray ionisation mass spectrometry. *Chromatographia* 59: 291–298.

Milligan, K. E., Marquez, B. L., Williamson, R. T., and Gerwick, W. H. 2000. Lyngbyabellin B. A toxic and antifungal secondary metabolite from the marine cyanobacterium *Lyngbya majuscula*. *J Nat Prod* 63: 1440–1443.

Mori, T. and Boyd, M. R. 2001. Cyanovirin-N, a potent human immunodeficiency virus-inactivating protein, blocks both CD4-dependent and CD4-independent binding of soluble gp120 (sgp120) to target cells, inhibits sCD4-induced binding of sgp120 to cell-associated CXCR4 and dissociates bound sgp120 from target cells. *Antimicrob Agents Chemother* 45(3): 664–672.

Mukherjee, P., Banerjee, I., Khatoon, N., and Pal, R. 2013. Cyanobacteria as elicitor of pigment in ornamental fish *Hemigrammus caudovittatus* (Buenos Aires Tetra). *J Algal Biomass Utln* 4(3): 59–65.

Mundt, S., Kreitlow, S., and Jansen, R. 2003. Fatty acids with antibacterial activity from the cyanobacterium *Oscillatoria redekei* HUB051. *J Appl Phycol* 15: 263–267.

Nagle, D. G. and Paul, V. J. 1998. Chemical defense of a marine cyanobacterial bloom. *J Exp Mar Biol Ecol* 225: 29–38.

Najdenski, H. M., Gigova, L. G., Iliev, I. I., Pilarski, P. S., Lukavský, J., Tsvetkova, I. V., and Kussovski, V. K. 2013. Antibacterial and antifungal activities of selected microalgae and cyanobacteria. *Int J Food Sci Technol* 48(7): 1533–1540.

Neuhof, T., Schmieder, P., Preussel, K., Dieckmann, R., Pham, H., Bartl, F., and von Döhren, H. 2005. Hassallidin A, a glycosylated lipopeptide with antifungal activity from the cyanobacterium *Hassallia* sp. *J Nat Prod* 68(5): 695–700.

Newman, D. J., Cragg, G. M., and Snader, K. M. 2003. Natural products as sources of new drugs over the period 1981–2002. *J Nat Prod* 66: 1022–1037.

Nunnery, J. K., Mevers, E., and Gerwick, W. H. 2010. Biologically active secondary metabolites from marine cyanobacteria. *Curr Opin Biotechnol* 21: 787–793.

Okura, K., Matsuoka, S., and Inoue, M. 2013. The bulky side chain of antillatoxin is important for potent toxicity: Rational design of photoresponsive cytotoxins based on SAR studies. *Chem Commun (Camb)* 49(73): 8024–8026.

Oman, T. J. and van der Donk, W. A. 2010. Follow the leader: The use of leader peptides to guide natural product biosynthesis. *Nat Chem Biol* 6: 9–18.

Ozdemir, G., Ulku Karabay, N., Dalay, M. C., and Pazarbasi, B. 2004. Antibacterial activity of volatile component and various extracts of *Spirulina platensis*. *Phytother Res* 18(9): 754–757.

Palaniselvam, V. 1998. Epiphytic cyanobacteria of mangrove: Ecological, physiological and biochemical studies and their utility as biofertilizer and shrimp-feed. PhD thesis, Annamalai University, Chidambaram, Tamil Nadu, India, p. 141.

Papendorf, O., König, G. M., and Wright, A. D. 1998. Hierridin B and 2,4-dimethoxy-6-heptadecyl-phenol, secondary metabolites from the cyanobacterium *Phormidium ectocarpi* with antiplasmodial activity. *Phytochemistry* 49: 2383–2386.

Patterson, G. M. L., Baker, K. K., Baldwin, C. L., Bolis, C. M., Caplan, F. R., Larsen, L. K., Levine, I. A. et al. 1993. Antiviral activity of cultured blue-green algae (Cyanophyta). *J Phycol* 29: 125–130.

Pawar, S. T. and Puranik, P. R. 2008. Screening of terrestrial and freshwater halotolerant cyanobacteria for antifungal activities. *World J Microbiol Biotechnol* 24: 1019–1025.

Pearson, L., Mihali, T., Moffitt, M., Kellmann, R., and Neilan, B. 2010. On the chemistry, toxicology and genetics of the cyanobacterial toxins, microcystin, nodularin, saxitoxin and cylindrospermopsin. *Mar Drugs* 8: 1650–1680.

Pérez Gutiérrez, M., Suyama, T. L., Engene, N., Wingerd, J. S., Matainaho, T., and Gerwick, W. H. 2008a. Apratoxin D, a potent cytotoxic cyclodepsipeptide from papua new guinea collections of the marine cyanobacteria *Lyngbya majuscula* and *Lyngbya sordida*. *J Nat Prod* 71(6): 1099–1103.

Pérez Gutiérrez, R. M., Flores, A. M., Solís, R. V., and Jimenez, J. C. 2008b. Two new antibacterial norabietane diterpenoids from cyanobacterium. *Micrococcus lacustris. J Nat Med* 62: 328–331.

Plaza, M., Santoyo, S., Jaime, L., Reina, G. G. B., Herrero, M., Señoráns, F. J., and Ibáñez, E. 2010. Screening for bioactive compounds from algae. *J Pharm Biomed Anal* 51(2): 450–455.

Portmann, C., Blom, J. F., Kaiser, M., Brun, R., Jüttner, F., and Gademann, K. 2008. Isolation of aerucyclamides C and D and structure revision of microcyclamide 7806A, heterocyclic ribosomal peptides from *Microcystis aeruginosa* PCC 7806 and their antiparasite evaluation. *J Nat Prod* 71: 1891–1896.

Prabakaran, M. 2011. In vitro antimicrobial potentials of marine *Oscillatoria* species. *Asian J Plant Sci Res* 1: 58–64.

Prashantkumar, P., Angadi, S. B., and Vidyasagar, G. M. 2006. Antimicrobial activity of blue-green and green algae. *Indian J Pharm Sci* 68: 647–648.

Priyadarshani, I. and Rath, B. 2012. Commercial and industrial applications of micro algae—A review. *J Algal Biomass Utln* 3(4): 89–100.

Ramachandra Rao, C. S. V. 1994. Antimicrobial activity of cyanobacteria. *Indian J Mar Sci* 23: 55–56.

Ramaswamy, A. V., Flatt, P. M., Edwards, D. J., Simmons, T. L., Han, B., and Gerwick, W. H. 2006. The secondary metabolites and biosynthetic gene clusters of marine cyanobacteria. Applications in biotechnology. In: Proksch, P. and Muller, W. E. G., eds., *Frontiers in Marine Biotechnology. Horizon Biosci* 175–224.

Rath, B. and Priyadarshani, I. 2013. Antibacterial and antifungal activity of marine cyanobacteria from Odisha Coast. *Int J Curr Trends Res* 2(1): 248–251.

Raveh, A. and Carmeli, S. 2007. Antimicrobial ambiguines from the cyanobacterium *Fischerella* sp. collected in Israel. *J Nat Prod* 70: 196–201.

Reehana, N., Parveez Ahamed, A., and Thajuddin, N. 2012. *In vitro* studies on bactericidal effect of selected cyanobacteria against bacterial pathogens. *Int J Med Res* 1(7): 345–347.

Sakthivel, K. and Kathiresan, K. 2012. Antimicrobial activities of marine cyanobacteria isolated from mangrove environment of south east coast of India. *J Natl Prod* 5: 147–156.

Sanchez, L. M., Lopez, D., Vesely, B. A., Della Togna, G., Gerwick, W. H., Kyle, D. E., and Linington, R. G. 2010. Almiramides A–C: Discovery and development of a new class of leishmaniasis lead compounds. *J Med Chem* 53: 4187–4197.

Senthil Kumar, N. S., Sivasubramanian, V., and Mukund, S. 2013. Antimicrobial and Antifungal activity of extracts of *Phormidium fragile* Gomont. *J Algal Biomass Utln* 4(3): 66–71.

Shamsa, F., Monsef, H., Ghamooshi, R., and Verdian-rizi, M. 2008. Spectrophotometric determination of total alkaloids in some Iranian medicinal plants. *Thai J Pharm Sci* 32: 17–20.

Shanab, S. M. M. 2007. Bioactive Allelo-chemical compounds from *Oscillatoria* species (Egyptian Isolates). *Int J Agric Biol* 9: 617–621.

Shiau, I. L., Shih, T. L., Wang, Y. N., Chen, H. T., Lan, H. F., Lin, H. C., and Murase, Y. 2009. Quantification for saponin from a soapberry in cleaning products by a chromatographic and two colorimetric assays. *J Faculty Agri Kyushu Univ* 54: 215–221.

Singh, R. K., Tiwari, S. P., Rai, A. K., and Mohapatra, T. M. 2011. Cyanobacteria: An emerging source for drug discovery. *J Antibiot* 64: 401–412.

Sitachitta, N. and Gerwick, W. H. 1998. Grenadadiene and grenadamide, cyclopropyl-containing fatty acid metabolites from the marine cyanobacterium *Lyngbya majuscula. J Nat Prod* 61: 681–684.

Sitachitta, N., Williamson, R. T., and Gerwick, W. H. 2000. Yanucamides A and B, two new depsipeptides from an assemblage of the marine cyanobacteria *Lyngbya majuscula* and *Schizothrix* species. *J Nat Prod* 63: 197–200.

Sivonen, K. 2000. Chapter 26, Freshwater cyanobacterial neurotoxins: Ecobiology, chemistry and detection. In: Botana, L. M., ed., *Seafood and Freshwater Toxins.* New York: Marcel Dekker, Inc., pp. 567–582.

Skindersoe, M. E., Alhede, M., Phipps, R., Yang, L., Jensen, P. O., Rasmussen, T. B., and Givskov, M. 2008. Effects of antibiotics on quorum sensing in *Pseudomonas aeruginosa. Antimicrob Agents Chemother* 52: 3648–3663.

Soria-Mercado, I. E., Pereira, A., Cao, Z., Murray, T. F., and Gerwick, W. H. 2009. Alotamide A, a novel neuropharmacological agent from the marine cyanobacterium *Lyngbya bouillonii. Org Lett* 11: 4704–4707.

Starmans, D. A. and Nijhuis, H. H. 1996. Extraction of secondary metabolites from plant material: A review. *Trends Food Sci Technol* 7: 191–197.

Starr, T. J., Deig, E. F., Church, K. K., and Allen, M. B. 1962. Antibacterial and antiviral activities of algal extracts studied by acridine orange staining. *Tex Rep Biol Med* 20: 271–278.

Staunton, J. and Weissman, K. J. 2001. Polyketide biosynthesis: A millennium review. *Nat Prod Rep* 18: 380–416.

Suhail, S., Biswas, D., Farooqui, A., Arif, J. M., and Zeeshan, M. 2011. Antibacterial and free radical scavenging potential of some cyanobacterial strains and their growth characteristics. *J Chem Pharm Res* 3: 472–478.

Sundararaman, M. and Nagaraja, B. K. 2006. Investigation of marine cyanobacteria for antifungal activity against *Candida albicans*. *Seaweed Res Util* 28(1): 183–187.

Tan, G., Gyllenhaal, C., and Soejarto, D. D. 2006. Biodiversity as a source of anticancer drugs. *Curr Drug Targets* 7: 265–277.

Tan, L. T. and Goh, B. P. L. 2009. Chemical ecology of marine cyanobacterial secondary metabolites: A mini-review. *J Coast Dev* 13(1): 1–9.

Tan, L. T., Goh, B. P. L., Tripathi, A., Lim, M. G., Dickinson, G. H., Lee, S. S. C., and Teo, S. L. M. 2010. Natural antifoulants from the marine cyanobacterium *Lyngbya majuscula*. *Biofouling* 26(6): 685–695.

Tan, L. T., Okino, T., and Gerwick, W. H. 2000. Hermitamides A and B, toxic malyngamide-type natural products from the marine cyanobacterium *Lyngbya majuscula*. *Nat Prod* 63(7): 952–955.

Taton, A., Grubisic, S., Balthasart, P., Hodgson, D. A., Laybourn-Parry, J., and Wilmotte, A. 2006. Biogeographical distribution and ecological ranges of benthic cyanobacteria in East Antarctic lakes. *FEMS Microbiol Ecol* 57(2): 272–289.

Teruya, T., Sasaki, H., Fukazawa, H., and Suenaga, K. 2009. Bisebromoamide, a potent cytotoxic peptide from the marine cyanobacterium *Lyngbya* sp.: Isolation, stereostructure and biological activity. *Org Lett* 11: 5062–5065.

Thajuddin, N. and Subramanian, G. 2005. Cyanobacterial biodiversity and potential applications in biotechnology. *Curr Sci* 89: 47–57.

Thummajitsakul, S., Silprasit, K., and Sittipraneed, S. 2012. Antibacterial activity of crude extracts of cyanobacteria *Phormidium* and *Microcoleus* species. *Afr J Microbiol Res* 6(10): 2574–2579.

Tidgewell, K., Engene, N., Byrum, T., Media, J., Valeriote, F. A., and Gerwick, W. H. 2010. Evolved diversification of a modular natural product pathway: Apratoxins F and G, two cytotoxic cyclic depsipeptides from a Palmyra collection of *Lyngbya bouillonii*. *Chem Bio Chem* 11: 1458–1466.

Tokuoka, M., Sawamura, N., Kobayashi, K., and Mizuno, A. 2010. Simple metabolite extraction method for metabolic profiling of the solid-state fermentation of *Aspergillus oryzae*. *J Biosci Bioeng* 110(6): 665–669.

Tripathi, A., Puddick, J., Prinsep, M. R., Lee, P. P. F., and Tan, L. T. 2010. Hantupeptins B and C, cytotoxic cyclodepsipeptides from the marine cyanobacterium *Lyngbya majuscula*. *Phytochemistry* 71: 307–311.

Uzair, B., Tabassum, S., Rasheed, M., and Rehman, S. F. 2012. Exploring marine cyanobacteria for lead compounds of pharmaceutical importance. *Sci World J* 2012: 1–10.

Vicente, M. F., Basilio, A., Cabello, A., and Pelaez, F. 2003. Microbial natural products as a source of antifungals. *Clin Microbiol Infect* 9(1): 15–32.

Wang, Y. C., Chuang, Y. C., and Hsu, H. W. 2008. The flavonoid, carotenoid and pectin content in peels of citrus cultivated in Taiwan, *Food Chem* 106(1): 277–284.

Welker, M., Christiansen, G., and Von Döhren, H. 2004. Diversity of coexisting *Planktothrix* (cyanobacteria) chemotypes deduced by mass spectral analysis of microystins and other oligopeptides. *Arch Microbiol* 182(4): 288–298.

Welker, M. and Von Döhren, H. 2006. Cyanobacterial peptides—Nature's own combinatorial biosynthesis. *FEMS Microbiol Rev* 30(4): 530–563.

Williamson, R. T., Singh, I. P., and Gerwick, W. H. 2004. Taveuniamides: New chlorinated toxins from a mixed assemblage of marine cyanobacteria. *Tetrahedron* 60: 7025–7033.

Xiong, C., O'Keefe, B. R., Byrd, R. A., and McMahon, J. B. 2006. Potent anti-HIV activity of scytovirin domain 1 peptide. *Peptides* 27: 1668–1675.

Xiong, S., Fan, J., and Kitazato, K. 2010. The antiviral protein cyanovirin-N: The current state of its production and applications. *Appl Microbiol Biotechnol* 86(3): 805–812.

Yadav, S., Sinha, R. P., and Tyagi, M. B. 2012. Antimicrobial activity of some cyanobacteria. *Int J Pharm Pharm Sci* 4(3): 631–635.

Yoon, Y. S. and Lee, C. G. 2009. Partial purification and characterization of a novel antifungal compound against *Aspergillus* spp. from *Synechocystis* sp. PCC 6803. *Biotechnol Bioprocess Eng* 14: 383–390.

Yu, H., Jia, S., and Dai, Y. 2009. Growth characteristics of the cyanobacterium *Nostoc flagelliforme* in photoautotrophic, mixotrophic and heterotrophic cultivation. *J Appl Phycol* 21: 127–133.

Zainuddin, E. N., Mentel, R., Wray, V., Jansen, R., Nimtz, M., Lalk, M., and Mundt, S. 2007. Cyclic depsipeptides, ichthyopeptins A and B, from *Microcystis ichthyoblabe. J Nat Prod* 70: 1084–1088.

Zeeshan, M., Suhail, S., Biswas, D., Farooqui, A., and Arif, J. M. 2010. Screening of selected cyanobacterial strains for phytochemical compounds and biological activities *in vitro. Biochem Cell Arch* 10: 163–168.

chapter thirteen

Antimicrobial and natural compounds from edible mushrooms

Annamalai Panneerselvam, V. Ambikapathy, and M. Nithya

Contents

13.1   Introduction

A mushroom (or toadstool) is the fleshy, spore-bearing fruiting body of a fungus, typically produced aboveground on soil or on its food source. The standard for the name "mushroom" is the cultivated white button mushroom, *Agaricus bisporus*; hence, the word "mushroom" is most often applied to those fungi (Basidiomycota, Agaricomycetes) that have a stem (stipe), a cap (pileus), and gills (lamellae, sing, lamella) or pores on the underside of the cap. These pores or gills produce microscopic spores that help the fungus spread across the ground or its occupant surface (Figure 13.1).

    Mushroom describes a variety of gilled fungi, with or without stems, and the term is used even more generally, to describe both the fleshy fruiting bodies of some Ascomycota and the woody or leathery fruiting bodies of some Basidiomycota, depending on the context of the word.

    Forms deviating from the standard morphology usually have more specific names, such as "puffball," "stinkhorn," and "morel," and gilled mushrooms themselves are often called "agarics" in reference to their similarity to *Agaricus* or their place Agaricales. By extension, the term "mushroom" can also designate the entire fungus when in culture, the thallus (called a mycelium) of species forming the fruiting bodies called mushrooms, or the species itself (Dickinson and Lucas, 1982).

13.2   Mushroom

Mushrooms are mostly Basidiomycetes and gilled. Their spores, called basidiospores, are produced on the gills and fall in a fine rain of powder from under the caps as a result. At the microscopic level, the basidiospores are shot-off basidia and then fall between the gills in the dead-air space. As a result, for most mushrooms, if the cap is cut off and placed gill-side-down overnight, a powdery impression reflecting the shape of the gills (or pores, or spines, etc.) is formed (when the fruit body is sporulating). The color of the powdery print, called a spore print, is used to help classify mushrooms and can help to identify them. Spore print colors include white (most common), brown, black, purple-brown, pink, yellow, and creamy, but almost never blue, green, or red.

    While modern identification of mushrooms is quickly becoming molecular, the standard methods for identification are still used by most and have developed into a fine art harking back to medieval times and the Victorian era, combined with microscopic examination. The presence of juices upon breaking, bruising reactions, odors, tastes, shades of color, habitat, habit, and season are all considered by both amateur and professional mycologists. Tasting and smelling mushrooms carry their own hazards because of poisons and allergens. Chemical tests are also used for some genera.

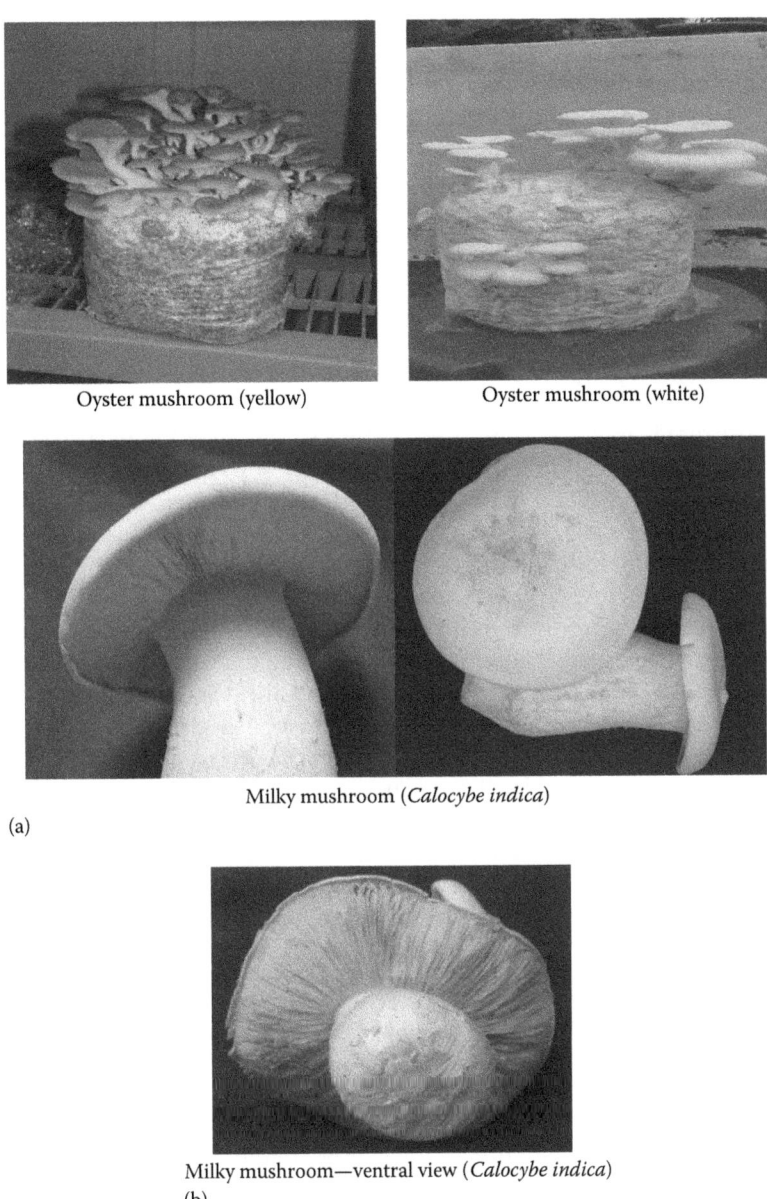

Oyster mushroom (yellow)   Oyster mushroom (white)

Milky mushroom (*Calocybe indica*)

(a)

Milky mushroom—ventral view (*Calocybe indica*)

(b)

Figure 13.1 Mushroom: close view of (a) stalk and (b) Gills plate.

In general, identification to genus can often be accomplished in the field using a local mushroom guide. Identifying species, however, requires more effort; one must remember that a mushroom develops from a button stage into a mature structure, and only the latter can provide certain characteristics needed for the identification of the species. However, overmature specimens lose features and cease producing spores. Many novices have mistaken humid water marks on paper for white spore prints or discolored paper from oozing liquids on lamella edges for colored spored prints.

13.2.1   Edible mushrooms

Edible mushrooms are the fleshy and edible fruit bodies of several species of macrofungi (fungi that bear fruiting structures that are large enough to be seen with the naked eye). They can appear either belowground (hypogeous) or aboveground (epigeous) where they may be picked by hand. Edibility may be defined by criteria that include the absence of poisonous effects on humans and desirable taste and aroma.

Edible mushrooms are consumed by humans as comestibles for their nutritional value, and they are occasionally consumed for their supposed medicinal value. Mushrooms consumed by those practicing folk medicine are known as medicinal mushrooms. While hallucinogenic mushrooms (e.g., psilocybin mushrooms) are occasionally consumed for recreational or religious purposes; they can produce severe nausea and disorientation and are therefore not commonly considered edible mushrooms.

Edible mushrooms include many fungal species that are either harvested wild or cultivated. Easily cultivatable and common wild mushrooms are often available in markets, and those that are more difficult to obtain (such as the prized truffle and matsutake) may be collected on a smaller scale by private gatherers. Some preparations may render certain poisonous mushrooms fit for consumption.

Before assuming that any wild mushroom is edible, it should be identified. Proper identification of a species is the only safe way to ensure edibility. Some mushrooms that are edible for most people can cause allergic reactions in some individuals, and old or improperly stored specimens can cause food poisoning. Deadly poisonous mushrooms that are frequently confused with edible mushrooms and responsible for many fatal poisonings include several species of the *Amanita* genus, in particular, *Amanita phalloides*, the death cap. Mushrooms growing in polluted locations can accumulate pollutants such as heavy metals.

Some species are difficult to cultivate; others (particularly mycorrhizal species) have not yet been successfully cultivated. Some of these species are harvested from the wild and can be found in markets. When in season, they can be purchased fresh, and many species are sold dried as well. The following species are commonly harvested from the wild:

- *Boletus edulis* or edible *Boletus*, native to Europe, known in Italian as *fungo porcino* (plural "porcini," pig mushroom), in German as Steinpilz (stone mushroom), in Russian as "white mushroom," in Albanian as wolf mushroom, and in French the *cèpe*. It also known as the king bolete and is renowned for its delicious flavor. It is sought after worldwide and can be found in a variety of culinary dishes.
- *Cantharellus cibarius* (the chanterelle), the yellow chanterelle is one of the best and most easily recognizable mushrooms and can be found in Asia, Europe, North America, and Australia. There are poisonous mushrooms that resemble it, though these can be confidently distinguished if one is familiar with the chanterelle's identifying features.
- *Cantharellus tubaeformis*, the tube chanterelle or yellow leg.
- *Clitocybe nuda*, Blewit (or Blewitt).
- *Cortinarius caperatus* the gypsy mushroom (recently moved from the genus *Rozites*).
- *Craterellus cornucopioides*, trompette de la mort or horn of plenty.
- *Grifola frondosa*, known in Japan as *maitake* (also "hen of the woods" or "sheep's head"); a large, hearty mushroom commonly found on or near stumps and bases of oak trees and believed to have *Macrolepiota procera* properties.
- *Gyromitra esculenta*, this "false morel" is prized by the Finns. This mushroom is deadly poisonous if eaten raw, but highly regarded when parboiled.
- *Hericium erinaceus*, a tooth fungus; also called "lion's mane mushroom."

- *Hydnum repandum*, sweet tooth fungus, hedgehog mushroom, urchin of the woods.
- *Lactarius deliciosus*, saffron milk cap, consumed around the world and prized in Russia.
- *Morchella* species (morel family), morels belong to the ascomycete grouping of fungi. They are usually found in open scrub, woodland, or open ground in late spring. When collecting this fungus, care must be taken to distinguish it from the poisonous false morels, including *Gyromitra esculenta*.
  - *Morchella conica var. deliciosa*
  - *Morchella esculenta var. rotunda*
- *Tricholoma matsutake* the matsutake, a mushroom highly prized in Japanese cuisine.
- *Tuber* species (the truffle), truffles have long eluded the modern techniques of domestication known as *trufficulture*. Although the field of trufficulture has greatly expanded since its inception in 1808, several species still remain uncultivated. Domesticated truffles include
  - *Tuber borchii*
  - *Tuber brumale*
  - *Tuber indicum*—Chinese black truffle
  - *Tuber macrosporum*—white truffle
  - *Tuber mesentericum*—Bagnoli truffle
  - *Tuber uncinatum*—Black summer truffle

13.2.1.1 Other edible wild species

Many wild species are consumed around the world. The species that can be identified *in the field* (without use of special chemistry or a microscope) and therefore safely eaten vary widely from country to country, even from region to region. This list is a sampling of lesser-known species that are reportedly edible.

Lactarius salmonicolor

- *Amanita caesarea* (Caesar's mushroom)
- *Armillaria mellea*
- *Boletus badius*
- *Chroogomphus rutilus* (pine spikes or spike caps)
- *Calvatia gigantea* (giant puffball)
- *Calocybe gambosa* (St George's mushroom)
- *Clavariaceae* species (coral fungus family)
- *Clavulinaceae* species (coral fungus family)
- *Coprinus comatus*, the shaggy mane; must be cooked as soon as possible after harvesting or the caps will first turn dark and unappetizing and then deliquesce and turn to ink, hence not being found in markets for this reason
- Corn smut
- *Cortinarius variecolor*
- *Fistulina hepatica* (beefsteak polypore or the ox tongue)
- *Hygrophorus chrysodon*

Auricularia auricula-judae

- *Lactarius salmonicolor*
- *Lactarius subdulcis* (mild milkcap)
- *Lactarius volemus*

- *Laetiporus sulphureus* (sulfur shelf), also known by names such as the "chicken mushroom" and "chicken fungus" and is a distinct bracket fungus popular among mushroom hunters
- *Leccinum aurantiacum* (red-capped scaber stalk)
- *Leccinum scabrum* (birch bolete)
- *Lepiota procera*
- *Macrolepiota procera* parasol mushroom, globally being widespread in temperate regions
- *Polyporus squamosus* (dryad's saddle and pheasant's back mushroom)
- *Polyporus mylittae*
- *Ramariaceae* species (coral fungus family)
- *Rhizopogon luteolus*
- *Russula*, some members of which are edible
- *Sparassis crispa*, also known as "cauliflower mushroom"
- *Suillus bovinus*
- *Suillus granulatus*
- *Suillus luteus*
- *Suillus tomentosus*
- *Tricholoma terreum*

13.2.1.2 Conditionally edible species

Amanita muscaria, a conditionally edible species: There are a number of fungi that are considered choice by some and toxic by others. In some cases, proper preparation can remove some or all of the toxins (Rubel and Arora, 2008):

- *Amanita muscaria* is edible if parboiled to leach out toxins. Fresh mushrooms cause vomiting, twitching, drowsiness, and hallucinations due to the presence of muscimol. Although present in *A. muscaria*, ibotenic acid is not in high enough concentration to produce any physical or psychological effects unless massive amounts are ingested.
- *Coprinopsis atramentaria* is edible without special preparation. However, consumption with alcohol is toxic due to the presence of coprine. Some other *Coprinus* spp. share this property.
- *Gyromitra esculenta* is eaten by some after it has been parboiled; however, mycologists do not recommend it. Raw *Gyromitra* are toxic due to the presence of gyromitrin, and it is not known if all of the toxin can be removed by parboiling.
- *Lactarius* spp.: Apart from *L. deliciosus* that is universally considered edible, other *Lactarius* spp. that are considered toxic elsewhere in the world are eaten in some eastern European countries and Russia after pickling or parboiling (Arora, 1986).
- *Verpa bohemica*: Considered choice by some, it even can be found for sale as a "morel," but cases of toxicity have been reported. Verpas contain toxins similar to gyromitrin and similar precautions applied.

13.2.2 Medicinal mushrooms

13.2.2.1 Ganoderma lucidum

Medicinal mushrooms are mushrooms or extracts used or studied as possible treatments for diseases. Some mushroom materials, including polysaccharides, glycoproteins, and proteoglycans, modulate immune system responses and inhibit tumor growth in preliminary research, whereas other isolates showed potential cardiovascular, antiviral, antibacterial, antiparasitic, anti-inflammatory, and antidiabetic properties. Currently, several extracts have widespread use in Japan, Korea, and China, as adjuncts to radiation treatments and chemotherapy, even though clinical evidence of efficacy in humans has not been confirmed (Smith et al., 2002).

Historically, mushrooms have long been thought to hold medicinal value, especially in traditional Chinese medicine. They have been studied in modern medical research since the 1960s, where most studies use extracts, rather than whole mushrooms. Only a few specific extracts have been tested for efficacy in laboratory research. Polysaccharide-K and lentinan are among the extracts best understood from *in vitro* research of animal models such as mice or early-stage human pilot studies (Borchers et al., 2008).

Preliminary experiments showed that glucan-containing mushroom extracts may affect the function of the innate and adaptive immune systems, functioning as bioresponse modulators. In some countries, the extracts of polysaccharide-K, schizophyllan, polysaccharide peptide, or lentinan are government-registered adjuvant cancer therapies.

13.2.2.2   Medical applications

13.2.2.2.1   Antimicrobial isolates and derivatives   Immunomodulatory protein isolated from *Ganoderma lucidum*. Antibiotics retapamulin, tiamulin, and valnemulin are derivatives of the mushroom isolate pleuromutilin. Plectasin, austrocortilutein, austrocortirubin, coprinol, oudemansin A, strobilurin, illudin, pterulone, and sparassol are antibiotics isolated from mushrooms. Researchers have isolated a number of antifungal, antiviral, and antiprotozoan isolates from mushrooms (Engler et al., 1998) (Table 13.1).

Table 13.1 Enzyme inhibitors from mushroom

Mushroom	Isolate/extract/metabolite	Enzyme inhibited
Pleurotus ostreatus	Lovastatin	HMG-CoA reductase
Hypholoma sublateritium	Clavaric acid	Farnesyl transferase
Polyozellus multiplex	Polyozellin, thelephoric acid, kynapcins	Prolyl endopeptidase
Lentinus edodes	Eritadenine	S-adenosyl-L-homocysteine hydrolase
Coprinopsis atramentaria	1-Aminocyclopropanol	Acetaldehyde dehydrogenase
Inonotus obliquus	Extract	Dipeptidyl peptidase-4
Grifola frondosa	Extract	Alpha-glucosidase
Trametes versicolor	Extract	Alpha-amylase
Pholiota squarrosa, *Daedalea*	Extract (*Pholiota squarrosa*), quercinol (*Daedalea quercina*)	Xanthine oxidase
Phellinus linteus	Phellinstatin	Enoyl-ACP reductase
Phellinus linteus	Hispidin and hypholomine B	Neuraminidase
Various	Extract	5-Alpha reductase
Various	Extract	Aromatase
Various	Peptides	Angiotensin-converting enzyme
Daedalea quercina	Quercinol	Cyclooxygenase 2
Daedalea quercina	Quercinol	Horseradish peroxidase

13.2.2.2.2   Anticancer research   Bristol-Myers Squibb manufactures paclitaxel using *Penicillium raistrickii*. In 2014, researchers reported creating a transgenic *Flammulina velutipes* that expresses the gene used to synthesize the paclitaxel precursor baccatin III (Han et al., 2014).

Mushroom isolates researched for anticancer activity include clavaric acid, a farnesyl transferase inhibitor; asparaginase, a *Flammulina velutipes* isolate; irofulven and acylfulvene, derivatives of illudin S, conjugated linoleic acid, an *Agaricus* isolate; grifolin, an *Albatrellus confluens* isolate; and clitocine, a *Leucopaxillus giganteus* isolate.

Some countries have approved mushroom extracts lentinan, polysaccharide-K, and polysaccharide peptide as immunologic adjuvants (Ina et al., 2013). There is some evidence of this use having effectiveness in prolonging and improving the quality of life for patients with certain cancers, although the Memorial Sloan Kettering Cancer Center observes that "well designed, large scale studies are needed to establish the role of lentinan as a useful adjunct to cancer treatment." According to Cancer Research UK, "there is currently no evidence that any type of mushroom or mushroom extract can prevent or cure cancer" (Gao et al., 2009).

The mushrooms credited with success against cancer belong to the genera *Phellinus, Pleurotus, Agaricus, Ganoderma, Clitocybe, Antrodia, Trametes, Cordyceps, Xerocomus, Calvatia, Schizophyllum, Flammulina, Suillus, Inonotus, Inocybe, Funlia, Lactarius, Albatrellus, Russula,* and *Fomes*. The anticancer compounds play crucial role as reactive oxygen species inducer, mitotic kinase inhibitor, antimitotic inhibitor, angiogenesis inhibitor, and topoisomerase inhibitor, leading to apoptosis and eventually checking cancer proliferation. This review updates the recent findings on the pharmacologically active compounds, their antitumor potential, and the underlying mechanism of biological action in order to raise awareness for further investigations to develop cancer therapeutics from mushrooms.

13.2.2.2.3   Antidiabetic research   Many mushroom isolates act as DPP-4 inhibitors, alpha-glucosidase inhibitors, and alpha-amylase inhibitors *in vitro*. Ternatin is a mushroom isolate that suppresses hyperglycemia (Lo and Wasser, 2011).

13.2.2.2.4   Psychotropic research   Psychotropic medicines created from ergot alkaloids include cafergot, dihydroergotamine, methysergide, methylergometrine, hydergine, nicergoline, lisuride, bromocriptine, cabergoline, and pergolide. *Polyozellus multiplex* synthesizes prolyl endopeptidase inhibitors polyozellin, thelephoric acid, and kynapcins, while *Hericium erinaceus* isolates erinacine and hericenone promote nerve growth factor synthesis and myelination *in vitro*.

Neurotrophic mushroom isolates include L-theanine, tricholomalides, scabronines, and termitomycesphins. Many mushrooms synthesize the partial, nonselective, serotonin receptor agonist/analog psilocin.

13.2.2.2.5   Nutritional research   Mushrooms are the only food source of statins like lovastatin (Atli et al., 2013). Only fungi and animals can synthesize vitamin D. Mushrooms have been verified creating D_2 (ergocalciferol), D_4 (22-dihydroergocalciferol), and vitamin D_1 (lumisterol + D_2) (Keegan et al., 2013). Mushrooms are a rare source of ergothioneine (Weigand-Heller et al., 2012) that contain ACE inhibitor peptides and are a source of prebiotic dietary fiber. Mushrooms also contain a variety of chemicals like lovastatin, cordycepin, inotilone, quercinol, antcin B, antrodioxolanone, and benzocamphorin F having preliminary research evidence for anti-inflammatory activity. Mushroom mycelia can be used to enhance the concentrations of γ-aminobutyric acid, ergothioneine, and other antioxidants in bread (Ulziijargal et al., 2013; Postemsky and Curvetto, 2014).

13.2.3   Poisonous mushrooms

Many mushroom species produce secondary metabolites that can be toxic, mind altering, antibiotic, antiviral, or bioluminescent. Although there are only a small number of deadly species, several others can cause particularly severe and unpleasant symptoms. Toxicity likely plays a role in protecting the function of the basidiocarp: the mycelium has expended considerable energy and protoplasmic material to develop a structure to efficiently distribute its spores. One defense against consumption and premature destruction is the evolution of chemicals that render the mushroom inedible, either causing the consumer to vomit the meal (emetics) or to learn to avoid consumption altogether. In addition, due to the ability of mushrooms to absorb heavy metals, including those that are radioactive, European mushrooms may, to date, include toxicity from the 1986 Chernobyl disaster and continue to be studied.

13.2.3.1   Old and new world hallucinogenic mushrooms

Most fungi are not deadly to humans, and many are perfectly edible (and are quite delicious); however, some species are poisonous and contain potent neurotoxins. Placing a silver coin in a pan of cooking mushrooms to see if it turns black is not a reliable method of testing poisonous mushrooms. Unless you understand fungal terminology and know how to use a good taxonomic key (such as *Mushrooms Demystified* by David Arora, 1986), the staff at Wayne's Word® do not encourage self-indulgence on wild mushrooms. The beautiful, red, fly agaric mushroom (*A. muscaria*) is unmistakable with its bright red cap covered with white scales. It contains the toxic alkaloid, muscimol, which is derived from ibotenic acid, which is an amino acid. In Europe, the mushrooms were reportedly left in open dishes to kill flies; however, according to some authorities, the flies are merely stunned or stupefied by the toxin and may even regain control and fly away. Although it is poisonous to humans, there are other species of *Amanita* that are much more dangerous and are potentially lethal if ingested. Some of these dangerously poisonous species are death cap (*A. phalloides*), death angel (*A. ocreata*), and panther amanita (*A. pantherina*). Fortunately, these latter deadly poisonous species are not bright red and are seldom confused with *A. muscaria*; however, they may be confused with other edible mushrooms by inexperienced gourmets. According to David Arora, species may be one of the most common causes of mushroom poisoning in the Pacific Northwest. David Isaak stated in a recent e-mail message that in the Pacific Northwest, a number of people gather *the panther* for culinary purposes, cooking it in several changes of water to remove most of the psychoactive materials (Allen and Merlin, 1992).

13.2.3.2   Amanita muscaria

When ingested by humans, *A. muscaria* may produce visions and delirium, and it is perhaps one of the oldest known hallucinogens. Recent studies suggest that this mushroom was the mysterious God-narcotic "Divine Soma" of ancient India. Thousands of years ago, Aryan conquerors who swept across India worshiped soma, drinking it in religious ceremonies. Many hymns in the Indian Rig Veda are devoted to Soma and describe the mushroom and its effects. According to the Rig Veda, Soma is without leaves, seeds, or branches, but with a head and stalk or pillar (the structure of a mushroom); its dazzling red skin is like the hide of the bull (the red cap); and its dress like that of a sheep, with woolly fragments remaining when the envelop bursts (the outer membranous envelop called the universal veil breaks as the stalk grows upward, leaving white remnants on the red cap). This is a remarkably accurate description of the fly agaric mushroom (*A. muscaria*). There are reports of Siberian tribesmen who ingested the mushroom to get intoxicated. Since the active chemical (muscimol) passes through the body relatively unaltered, others would drink the urine from these men to get high. This way a few mushrooms could inebriate many people relatively safely and efficiently. Lapland shamans eat fly agaric mushrooms for enlightenment, and some authors have postulated that this may have given rise to the flying reindeer and the red- and white-costumed Santa Claus legends.

Apparently, not everyone agrees that the Divine Soma is *A. muscaria*. According to Terrence McKenna, the active alkaloid in fly agaric mushrooms (muscimol) does not produce the psychoactive effects described in the Rig Veda and other literature. Terrence McKenna has continued to question the effects of *A. muscaria* and suggested other possible candidates. Based on firsthand experience with these hallucinogens, he has suggested that a psilocybin "magic mushroom," such as *Stropharia cubensis*, is the true Divine Soma. In fact, he also states that the use of mind-altering psilocybin mushrooms by ancient humans in Africa may have been a catalyst in the development of language and religion in primitive cultures.

A. muscaria was apparently one of the sacred hallucinogenic mushrooms of the Incas, Mayans, and Aztecs. (Other New World psychedelic genera included *Psilocybe*, *Panaeolus*, *Conocybe*, and *Stropharia*.) For the Indians of Mexico and Central and South America, partaking of these mushrooms was a deeply religious experience, enabling them to communicate with their gods. Cortez reported a mushroom (resembling *A. muscaria*) being eaten during the coronation of Montezuma, and in Guatemala, stone carvings dating back to 1000 BC depict curious figures with umbrellalike tops resembling the caps and stalks of an *Amanita* mushroom. Mushrooms are also depicted in ancient Peruvian vessels and in the Mexican codices. One drawing shows an animal-like messenger from god offering the sacred *Amanita* to a ruler seated on a throne. And, a fresco in a Roman Catholic Church in Plaincourault (Indre), France, depicts Adam and Eve on either side of a tree of knowledge that is unequivocally a branched *Amanita* mushroom. Some scholars believe that the original story of Alice's Adventures in Wonderland, where Alice speaks to a green caterpillar who is seated on a red- and white-capped mushroom, is actually the interpretation of a mushroom experience by the author Rev. C.L. Dodgson of Christ Church College in Oxford (better known by his pen name Lewis Carroll). Another hallucinogenic "high" that is commonly depicted in paintings and children's stories is the infamous, "politically incorrect" picture of a witch flying on a broom—the effects of a portion made from the deadly alkaloids of several solanaceous herbs, including jimsonweed (*Datura stramonium*).

13.3   Nutritional content of edible mushrooms

Mushrooms are a low-calorie food usually eaten cooked or raw and as garnish to a meal. Dietary mushrooms are a good source of B vitamins, such as riboflavin, niacin, and pantothenic acid, and essential minerals, such as selenium, copper, and potassium. Fat, carbohydrate, and calorie contents are low, with the absence of vitamin C and sodium. There are approximately 20 calories in an ounce of mushrooms (Table 13.2).

When exposed to ultraviolet light, natural ergosterols in mushrooms produce vitamin D_2, a process now exploited for the functional food retail market.

13.3.1   Nutrients in button mushrooms

White button mushrooms, the popular ones available in all the grocery stores, have a surprising amount of nutrients including niacin, riboflavin, folate, phosphorus, iron, pantothenic acid, zinc, potassium, copper, magnesium, vitamin B6, selenium, and thiamin.

Table 13.2 Nutritional value of edible mushroom (100 g)

Energy	113 kJ (27 kcal)
Carbohydrates	4.1 g
Fat	0.1 g
Protein	2.5 g
Thiamine (vitamin B_1)	0.1 mg (9%)
Riboflavin (vitamin B_2)	0.5 mg (42%)
Niacin (vitamin B_3)	3.8 mg (25%)
Pantothenic acid (B_5)	1.5 mg (30%)
Vitamin C	0 mg (0%)
Calcium	18 mg (2%)
Phosphorus	120 mg (17%)
Potassium	448 mg (10%)
Sodium	6 mg (0%)
Zinc	1.1 mg (12%)

Source: USDA Nutrient Database.
Percentages are roughly approximated using U.S. recommenda-
tions for adults.

In addition, white button mushroom extract has been found to reduce the size of some cancer tumors and slow down the production of some cancer cells. It is most prominently linked to reducing the risk of breast and prostate cancers (Koyyalamudi et al., 2009).

13.3.2   Great weight loss food

For those who are always looking for nutritious weight loss foods to pack into our diets, mushrooms are a less well-known option. Mushrooms are low in calories, carbohydrates, fats, and sodium. However, like watermelon, they are very high in water content (80%–90% water) and fiber that makes them a great diet food.

13.3.3   Excellent source of potassium

Most people think that bananas are the food of high potassium, but it may surprise you to learn that mushrooms out rank bananas on the potassium chart. Potassium helps the body process sodium and lower blood pressure, so people with hypertension or high risk of stroke can enjoy tremendous health benefits from a regular dose of mushrooms in their diet.

13.3.4   Disease-fighting properties

All mushrooms are an excellent source of the antioxidant, and the selenium works with vitamin E to protect cells from damaging free radicals. Some studies also indicate that antioxidants are some of the best nutrients for preventing and fighting cancers. Like almonds, mushrooms are becoming more popular for their cancer-fighting and disease-protecting properties.

Shitake mushrooms in particular are also high in the beta-glucan lentinan. Lentinan has been linked with strengthening the immune system and helping combat illnesses that attack the immune system like AIDS. In addition, mushroom extract has been linked to some treatments for both migraines and mental disorders.

13.3.5   Metabolism support nutrient

The human metabolism relies on a healthy dose of protein, fiber, and vitamin B to keep it functional and robust. Mushrooms rank high in all three of these metabolism-supporting nutrients.

13.3.6   Great source of heart-healthy copper

Copper is one of the less talked about minerals that is essential to the body, but that the body cannot make on its own. Copper has properties that help protect our cardiovascular system, and just one small serving of mushrooms contains more than 20% of the copper we need daily.

 With our fast-paced lifestyles and the highly processed foods, we may frequently find ourselves eating in haste; the magnesium, potassium, phosphorus, and selenium nutrients found in a single dish of mushrooms can really make up for some of the deficiencies we struggle to combat in our diets.

13.4   Bioactive compounds of edible mushrooms

Both the fruit bodies and the mycelia of *M. esculenta* contain an uncommon amino acid, *cis*-3-amino-L-proline; this amino acid does not appear to be protein bound. In addition to *M. esculenta*, the amino acid is known to exist only in *M. conica* and *M. crassipes* (Moriguchi et al., 1979).

 Laboratory experiments using rodent models suggest that the polysaccharides from *M. esculenta* fruit bodies have several medicinal properties, including antitumor effects, immunoregulatory properties, fatigue resistance, and antiviral effects. Extracts from the fruit bodies have antioxidant properties. It also has been shown that the polysaccharides from *M. esculenta* mycelia have antioxidant activity. The fungus is listed in the IUCN National Register of medicinal plants in Nepal. *M. esculenta* is also used in traditional Chinese medicine to treat indigestion, excessive phlegm, and shortness of breath (Rotzoll et al., 2005; Nitha and Janardhanan, 2008).

 Secondary metabolites were either absent or present in very different ratios and in general showed significantly less potency in cultivated Chaga. Cultivated Chaga furthermore results in a reduced diversity of phytosterols, particularly lanosterol, an intermediate in the synthesis of ergosterol and lanostane-type triterpenes. This effect was partially reversed by the addition of silver ion, an inhibitor of ergosterol biosynthesis.

 The major active compounds found are unsaturated fatty acids such as linoleic acid and conjugated linoleic acid. The interaction of linoleic acid and conjugated linoleic acid with aromatase mutants expressed in Chinese hamster ovary cells showed that these fatty acids inhibit aromatase with similar potency and that mutations at the active site regions affect its interaction with these two fatty acids.

 Due to their natural polyisoprene content (1.1%–7.7% by dry weight of fruit bodies), *L. volemus* fruit bodies can also be used to produce rubber. The chemical structure of rubber from the mushroom consists of a high-molecular-mass homologue of polyprenol, arranged as a dimethylallyl group, two *trans* isoprene units, and a long sequence of *cis* isoprenes (between 260 and 300 units), terminated by a hydroxyl or fatty acid ester. Biosynthetically, the creation of the polyisoprene begins with the compound *trans, trans*-farnesyl pyrophosphate, and is thought to terminate by esterification of polyisoprenyl pyrophosphate. The enzyme isopentenyl-diphosphate delta-isomerase has been identified as required for the initiation of rubber synthesis in *L. volemus* and several other *Lactarius* species.

13.4.1   Alkaloids

G. lucidum contains other compounds often found in fungal materials, including polysaccharides (such as beta-glucan), coumarin, mannitol, and alkaloids (Paterson, 2006). Ergotamine is the secondary metabolite and the principal alkaloid produced by the ergot fungus *Claviceps purpurea* and related family in the fungi Clavicipitaceae. It possesses structural similarity to several neurotransmitters and has biological activity as vasoconstrictor. It is used medicinally for the treatment of acute migraine attacks. Its medicinal usage began in the sixteenth century to induce childbirth. It has been used to prevent postpartum hemorrhage (bleeding after birth). The mechanism of action of ergotamine is complex. The molecules share structural similarity with neurotransmitters such as serotonin, dopamine, and epinephrine and can thus bind to several receptors (Schardl, 2000).

13.4.1.1   Psilocybin mushrooms
Psilocybin (also known as psilocybine) is a psychedelic alkaloid of the tryptamine family, found in psilocybin mushrooms. It is present in hundreds of species of fungi, including those of the genus *Psilocybe*, such as *Psilocybe cubensis* and *Psilocybe semilanceata*. But it is also reportedly isolated from a dozen or so other genera. Psilocybin mushrooms are commonly called "magic mushrooms" or more simply "shrooms." The psilocybin content of psychoactive mushrooms varies and depends on species, growth, drying conditions, and mushroom sizes. The intensity and duration of recreational and entheogenic use of psilocybin mushrooms vary depending on species of mushrooms, dosage, individual physiology, set, and setting.

13.4.1.2   Psilocybin biochemistry
Psilocybin is a product that is converted into the pharmacologically active compound psilocin in the body by dephosphorylation. This chemical reaction takes place under strongly acidic conditions or enzymatically by phosphatases in the body. Psilocybin is a zwitterionic alkaloid that is soluble in water, moderately soluble in methanol and ethanol, and insoluble in most organic solvents. Mature mycelium contains some amount of psilocybin, which can be extracted with an acidic solution, usually of citric acid or ascorbic acid (vitamin C). Young mycelium (recently germinated from spores) does not contain substantial amounts of alkaloids. It is also known to mimic the effects of serotonin.

13.4.1.3   Psilocybin and medicine
In a current study of psilocybin led by Charles Grob, 12 subjects are being administered with either the hallucinogen or a placebo in two separate sessions. Grob hopes to reduce the psychological distress (e.g., obsessive–compulsive behavior) that is associated with death by treating patients with psilocybin.

13.4.1.4   Psilocybin effects
The effects of psilocybin are often pleasant, even ecstatic, including a deep sense of connection to others, confusion, hilarity, and a general feeling of connection to nature and the universe. Difficult trips may occur when psychedelic compounds are taken in a nonsupportive or inadequate environment, by an inexperienced person, in an unexpectedly high dose, or when the substance triggers difficult areas of one's psyche.

At low doses, hallucinatory effects occur, including walls that seem to breathe, a vivid enhancement of colors, and the animation of organic shapes. At higher doses, experiences tend to be less social and more entheogenic, often catalyzing intense spiritual experiences.

A very small number of people are unusually sensitive to psilocybin's effects, where doses as little as 0.25 g of dried *Psilocybe cubensis* mushrooms (normally a threshold dose of around 2 mg psilocybin) can result in effects usually associated with medium and high doses. Likewise, there are some people who require relatively high doses of psilocybin to gain low-dose effects. Individual brain chemistry and metabolism play a major role in determining a person's response to psilocybin. Psilocybin is metabolized mostly in the liver where it becomes psilocin. It is broken down by the enzyme monoamine oxidase (MAO). MAO inhibitors have been known to sustain the effects of psilocybin for longer periods of time; people who are taking an MAOI for a medical condition (or are seeking to potentiate the mushroom experience) should be careful.

Mental and physical tolerances to psilocybin build and dissipate quickly. Taking psilocybin more than three or four times in a week (especially 2 days in a row) can result in diminished effects. Tolerance dissipates after a few days, so frequent users often keep doses spaced 5–7 days apart to avoid the effect.

13.4.1.5 *Adverse effects to psilocybin*

Individuals that have relatives with schizophrenia should be very careful about consuming psilocybin or any hallucinogenic drug at all due to the risk of triggering a psychosis. Because of the ease of cultivating psilocybin mushrooms or gathering wild species, purified psilocybin is practically nonexistent on the illegal drug market.

The psychoactive alkaloid in the teonanacatl mushrooms is psilocybin, a potent indole alkaloid. Psilocin, a dephosphorylated version of psilocybin, is about 10 times stronger. After ingestion by humans, psilocybin is automatically converted into psilocin. Most psilocybin-containing mushrooms have only a trace of psilocin. According to David Arora (1986), the common psilocybin mushroom of the Pacific coast of North America, *Psilocybe cyanescens*, has a higher concentration of natural psilocin and is appropriately named "potent psilocybe." Although two of the most famous species of psilocybin mushrooms are *Psilocybe mexicana* and *Stropharia cubensis*, there are literally dozens of other species in the four aforementioned genera with similar hallucinogenic properties. In fact, Paul Stamets (*Psilocybin Mushrooms of the World*, 1996) describes all of the species and includes color photographs. Like so many little brown mushrooms (LBMs), they are difficult to identify unless you are familiar with mushroom structure and spore taxonomy and have a good compound microscope at your disposal. In fact, two deadly look-alike LBMs (*Galerina autumnalis* and *Pholiotina filaris*) resemble certain species of *Psilocybe*. The small ring on their stems (called an annulus) and rusty brown spores (rather than black spores) are *dead* giveaways to avoid these potentially lethal mushrooms (Figure 13.2).

Indole alkaloids contain the indole carbon–nitrogen ring that is also found in the fungal alkaloids ergine and psilocybin, the neurotransmitter serotonin, and the mind-altering drug LSD. These alkaloids may interfere or compete with the action of serotonin in the brain.

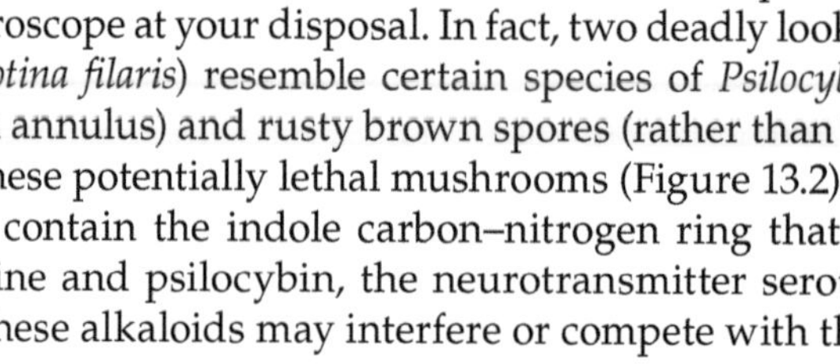

Figure 13.2 Structure of serotonin and psilocybin.

13.4.2   Flavonoids

L. deliciosus, Sarcodon imbricatus, and *Tricholoma portentosum* contain more flavonoids. The total phenols and flavonoids were the major components found in the mushroom extracts; ascorbic acid was found in small amounts (0.18–0.52 mg/g), and β-carotene and lycopene were only found in vestigial amounts (<91 μg/g). *L. deliciosus* contains 8.14 ± 0.81 (mg/g) of flavonoids, *S. imbricatus* 2.82 ± 0.09 (mg/g), and *T. portentosum* 0.40 ± 0.02 (mg/g).

The extracts with the entire mushroom showed higher phenolic and flavonoid contents than either the cap or the stipe. Also, the amount of phenolic and flavonoid compounds in the cap methanolic extracts was higher than the amount found in stipe extracts. The antimicrobial screening of phenolic compounds is extracted from *L. deliciosus, S. imbricatus,* and *T. portentosum* against *Bacillus cereus, Bacillus subtilis* (gram-positive bacteria), *E. coli,* and *Pseudomonas aeruginosa* (gram-negative bacteria) and *Candida albicans* and *Cryptococcus neoformans* (fungi). The minimal inhibitory concentrations (MICs) for bacteria and fungi were determined as an evaluation of the antimicrobial activity of the tested mushrooms. The diameters of the inhibition zones corresponding to the MICs are also presented. All the mushrooms revealed antimicrobial activity showing different selectivities and MICs for each microorganism. *L. deliciosus* showed better results than *T. portentosum* and *S. imbricatus* (lower MICs), which is in agreement with the higher content of phenols and flavonoids found in the first species.

The entire and the cap mushroom extracts inhibited *B. cereus, B. subtilis, P. aeruginosa, C. albicans,* and *C. neoformans,* while the stipe mushroom extract only inhibited *B. cereus, P. aeruginosa,* and *C. neoformans.* Only this mushroom revealed activity against *P. aeruginosa* (gram-negative bacteria) and *C. albicans* (fungi) and for the gram-positive bacteria showed lower MICs. The *T. portentosum* extract was effective only against gram-positive bacteria (*B. cereus, B. subtilis*) and *C. neoformans.* When the cap was used, the same selectivity was obtained, but with higher MICs; the stipe extract was only effective against *B. cereus.* In fact, the content in total phenols and flavonoids for the stipe extracts was always lower than in the other extracts. As expected due to its lower content in bioactive compounds, *S. imbricatus* was the less effective (higher MICs) mushroom, showing activity only against *B. cereus* and *C. neoformans.* The cap extract was selective for *B. cereus,* while the stripe extract was not effective against the tested microorganisms (Turkoglu et al., 2007).

13.4.3   Phenols

Phenolic acids can be found in mushroom basidiomycete species. For example, protocatechuic acid and pyrocatechol are found in *Agaricus bisporus* as well as other phenylated substances like phenylacetic and phenylpyruvic acids. Other compounds like atromentin and thelephoric acid can also be isolated from fungi in the Agaricomycetes class. Orobol, an isoflavone, can be isolated from *Aspergillus niger* (Table 13.3).

Omphalotus nidiformis is not edible. Although reputedly mild tasting, eating it will result in vomiting, which generally occurs 30 min to 2 h after consumption and lasts for several hours. There is no diarrhea, and the patients recover without lasting ill effects. Its toxicity was first mentioned by Anthony M. Youngin in his 1982 guidebook *Common Australian Fungi.* The toxic ingredient of many species of *Omphalotus* is a sesquiterpene compound known as illudin S. This, along with illudin M and a cometabolite illudosin, has been identified in *O. nidiformis.* The two illudins are common to the genus *Omphalotus* and not found in any other basidiomycete mushroom. Additional three compounds unique to *O. nidiformis* have been identified and named illudins F, G, and H.

Table 13.3 Mushroom metabolites

Mushroom	Metabolite
Pleurotus ostreatus	Lovastatin
Hypholoma sublateritium	Clavaric acid
Polyozellus multiplex	Polyozellin, thelephoric acid, kynapcins
Lentinus edodes	Eritadenine
Coprinopsis atramentaria	1-Aminocyclopropanol
Pholiota squarrosa, Daedalea quercina	Extract (*Pholiota squarrosa*), quercinol (*Daedalea quercina*)
Phellinus linteus	Phellinstatin
Phellinus linteus	Hispidin and hypholomine B
Daedalea quercina	Quercinol

Extracts of several species of Australian mushrooms have been investigated for cytotoxicity to cancer cells; material from *O. nidiformis* showed marked toxicity to gastric (AGS), colon (HT-29), and estrogen-independent breast cancer (MDA-MB-231) cell lines. Irofulven, a compound derived from illudin S, is undergoing phase II clinical trials as a possible therapy for various types of cancers. Fruit body extracts have antioxidant and free radical scavenging properties, which may be attributed to the presence of phenolic compounds (Ribéreau-Gayon, 1972).

13.4.4 Sterols

G. lucidum produces a group of triterpenes, called ganoderic acids, which have a molecular structure similar to that of steroid hormones. Sterols isolated from the mushroom include ganoderol, ganoderic acid, ganoderiol, ganodermanontriol, lucidadiol, and ganodermadiol.

Lactarius volemus fruit bodies contain a unique sterol molecule called volemolide, a derivative of the common fungal sterol ergosterol that may have application in fungal chemotaxonomy. A 2001 study identified further nine sterols, three of which were previously unknown to science. According to the authors, these types of highly oxygenated compounds—similar to sterols found in marine soft corals and sponges—are rare in fungi. The mushroom also contains volemitol (D-glycero-D-mannoheptitol), a seven-carbon sugar alcohol first isolated from the species by the French scientist Émile Bourquelot in 1889. Volemitol occurs as a free sugar in many plant and brown algal species.

13.4.5 Terpenes

G. lucidum contains terpenoids that have been on the less volatile triterpenoid (triterpene) and sterol-type compounds. The triterpene chemical structure is based on the ground structure of lanosterol, which is an important intermediate in the biosynthetic pathway for steroids and triterpenes in microorganisms and animals. Sterols, compounds closely related to triterpenoids, are also found in *Ganoderma*. They have been isolated from the fruiting body and mycelium and have also been shown to exhibit potent cytotoxic activity. A specific sterol, ergosterol peroxide, was isolated from *G. lucidum* and has been shown to enhance the inhibitory effect of linoleic acid on the inhibition of mammalian DNA polymerase-β (Hseu et al., 1996).

13.4.6 Triterpenes

G. lucidum produces a group of triterpenes, called ganoderic acids, which have a molecular structure similar to that of steroid hormones. It also contains other compounds often found in fungal materials, including polysaccharides (such as beta-glucan), coumarin, mannitol, and alkaloids. Sterols isolated from the mushroom include ganoderol, ganoderenic acid, ganoderiol, ganodermanontriol, lucidadiol, and ganodermadiol (Shiao et al., 1988).

13.4.7 Saponins

Saponins are toxic to some microorganisms and to animals, which includes the growth of mycelium of *Pleurotus sapidus*, *G. lucidum*, *Cantharellus cibarius*, *Laccaria amethystina*, *Clitocybe odora*, *Lepista nuda*, *Lepista saeva*, *L. deliciosus*, *Laccaria laccata*, *Pleurotus ostreatus*, and *Hericium erinaceus* having the saponins with antioxidant activities.

13.4.8 Tannins

Wild edible Nigerian mushrooms including *Cantharellus cibarius*, *L. amethystina*, *Clitocybe odora*, *L. nuda*, *Macrolepiota procera*, *Lepista saeva*, *L. deliciosus*, *Laccaria laccata*, *Pleurotus ostreatus*, and *Hericium erinaceus* were investigated. The mushrooms were harvested fresh, sun dried, pulverized, and analyzed according to standard procedures. Proximate analysis showed high level of proteins (14.03%–60.38%), crude fibers (3.94%–20.36%), carbohydrates (4.17%–32.50%), ashes (17.44%–33.60%), fats (1.29%–14.29%), and folic acids (4.75–5.51 g/g) in all species. Mineral analysis of all species indicated the presence of potassium, sodium, magnesium, manganese, calcium, copper, and iron. Potassium is of the highest amount in all species of plant (1370–5710 g/100 g). High antioxidant activity was also observed in these mushrooms with the species *L. amethystina* and *L. nuda* exhibiting the strongest antioxidant activity with values as high as 53.64 and 53.65 nm, respectively. Phytochemical screening revealed that above 10 species have the presence of varying quantities of alkaloids, flavonoids, saponins, and tannins (Lattif et al., 1996).

13.4.9 Anthraquinones

13.4.9.1 Red from mushrooms

The Dermocybe family of mushroom produces oranges and reds. The addition of an iron mordant gives darker shades and almost black ones. With older mushrooms, longer cooking times or the addition of ammonia can give lilac shades. Low heat or the addition of acid or vinegar gives warm reds, for example, *Dictyophora cinnabarina* and *Cortinarius semisanguineus*.

13.4.9.2 Blue from mushrooms

Blue was the most used color in many of the fabrics from the grave sites. Blue may have come from dyers.

Woad, *Isatis tinctoria*: Blue from the mushrooms *Thelephora*, with iron or tin mordants, that yields greens and blues.
Hydnellum suaveolens, S. imbricatus: Yields blues if it is old and its top has darkened.
Hapalopilus rutilans: With ammonia, yields strong, colorfast violet-blue shades.
Cortinarius violaceus: Produces violet-blue shades, with an iron mordant and dark greys.

13.5   Antimicrobial compounds and their applications

Most studies on mushrooms with antibacterial activity describe the action of its extracts without identifying the compounds responsible for this activity. However, some compounds have been described as active against gram-positive bacteria. Five of these compounds are terpenes. Confluentin (1A), grifolin (1B), and neogrifolin (1C) from *Albatrellus fletti* showed activity against *B. cereus* and *Enterococcus faecalis*. The best result was for *Enterococcus faecalis* (MIC 0.5–1.0 mg/mL) (Liu et al., 2010). Ganomycins A and B isolated from *Ganoderma pfeifferi* showed activity against *B. subtilis*, *Micrococcus flavus*, and *Staphylococcus aureus* (15–25 mm zones of inhibition at a concentration of 250 µg/mL) (Mothana et al., 2000).

A steroid, 3,11-dioxolanosta-8,24(Z)-diene-26-oic acid (2), was isolated from the *Jahnoporus hirtus* mushroom and revealed activity against *B. cereus* and *Enterococcus faecalis* (Liu et al., 2010).

Four sesquiterpenes with antimicrobial activity were described. The enokipodins A, B, C, and D, isolated from the mycelium of *Flammulina velutipes*, with activity against *B. subtilis*, but only enokipodins A and C, showed activity against *S. aureus* (Ishikawa et al., 2001).

Oxalic acid (3), an organic acid isolated from the mycelium of *Lentinus edodes*, showed activity against *B. cereus*, *S. aureus*, and *Streptococcus faecalis* (Bender et al., 2003).

Coloratin A, a benzoic acid derivative isolated from *Xylaria intracolorata*, inhibited *S. aureus* (Quang et al., 2006).

Eight compounds of anthraquinone derivatives were also reported due to their antibacterial activities. 6-Methylxanthopurpurin-3-*O*-methyl ether, austrocortilutein, austrocortilutein, austrocortirubin, and torosachrysone, isolated from the mushroom *Cortinarius basirubencens*, and physcion, erythroglaucin, and emodin isolated from other species of *Cortinarius*, were all effective against *S. aureus* (Beattie et al., 2010).

CSAP (*Cordyceps sinensis* antibacterial protein-*N*-terminal sequence ALATQHGAP), isolated from *Cordyceps sinensis*, showed strong activity against *S. aureus* and poor activity against *B. subtilis*. However, the antibacterial action of this protein was bacteriostatic (Zheng et al., 2006).

The ribonuclease, isolated from *Pleurotus sajor-caju*, showed activity against *S. aureus*, acting on RNA (Ngai and Ng, 2004).

The peptide plectasin, isolated from *Pseudoplectania nigrella*, is a macromolecule belonging to the class of defensins, present in animals and plants, which act at the cell wall, more specifically in the synthesis of peptidoglycan. This peptide showed activity against *B. cereus*, *Bacillus thuringiensis*, *Corynebacterium diphtheriae*, *Enterococcus faecalis*, *E. faecium*, (VREF), *S. aureus*, (MRSA), *Staphylococcus epidermidis* (MRSE), *Streptococcus pneumoniae*, (PRSP), and *Streptoccus pyogenes*.

The *in vitro* action of plectasin against *Streptococcus pneumoniae* is comparable to the action of penicillin and vancomycin (Mygind et al., 2005).

The peptides peptaibol boletusin, peptaibol chrysospermin 3, and peptaibol chrysospermin 5 (isolated from Boletus spp.) allow for the opening of pores for ion transport and showed activity against *B. subtilis*, *Corynebacterium lilium*, and *S. aureus*. The peptaibol chrysospermin 3 also showed activity against *Streptococcus* sp. (Lee et al., 1999).

Fraction B from *Pycnoporus sanguineus*, whose main constituent is a phenoxazin-3-one-type pigment, showed activity against *S. aureus* and *Streptococcus* A, B, C, and G. Lower values of MIC were obtained against *Streptococcus* strains (Smania et al., 1995).

13.6 Conclusion

Antibiotic resistance among microbes urgently necessitates the development of novel antimicrobial agents such as alternate therapies using natural products. Many pharmaceutical substances with potent and unique health enhancing properties have been isolated from medicinal mushrooms and distributed worldwide. Mushroom-based products either from the mycelia or fruiting bodies are consumed in the form of capsules, tablets, or extracts. It has been reported by many workers that fruit bodies of different mushrooms like *Lactarius* sp., *Fomitopsis* sp., *Boletus* sp., *Pleurotus tuber-regium*, *L. deliciosus*, *S. imbricatus* and *T. portentosum*, *Russula delica*, *Pleurotus eryngii* var. *ferulae*, *Infundibulicybe geotropa*, *L. controversus*, *L. delicious* and *Phellinus hartigii*, *Lactarius indigo*, and *Stereum ostrea* contain a wide range of antimicrobial activity. With an increasing number of bacteria developing resistance to commercial antibiotics, such as MSRA (methicillin-resistant *S. aureus* and *Pseudomonas*), extracts and derivatives from mushrooms hold great promise for novel medicines in modern times.

Acknowledgment

The authors are grateful to the secretary and correspondent, the principal, the dean faculty of sciences, and the staff members of the Department of Botany and Microbiology, AVVM Sri Pushpam College, Poondi, Thanjavur, Tamil Nadu, India.

References

Allen, J. W. and Merlin, M. D. 1992. Psychoactive fungi use in Koh Samui and Koh Pha-Ngan, Thailand. *J Ethnopharmacol* 35(3): 205–228.

Arora, D. 1986. *Mushrooms Demystified*, 2nd ed. Ten Speed Press, Berkeley, CA.

Atli, B., Yamac, M., and Yildiz, Z. 2013. Optimization of submerged fermentation conditions for lovastatin production by the culinary-medicinal oyster mushroom, *Pleurotus ostreatus* (Higher Basidiomycetes). *Int J Med Mushrooms* 15(5): 487–495.

Beattie, K. D., Rouf, R., Gander, L., May, T. W., Ratkowsky, D., Donner, C. D., Gill, M., Grice, I. D., and Tiralongo, E. 2010. Antibacterial metabolites from Australian macrofungi from the genus *Cortinarius*. *Phytochemistry* 71: 948–955.

Bender, S., Dumitrache, C. N., Backhaus, J., Christie, G., Cross, R. F., Lonergan, G. T., and Baker, W. L. 2003. A case for caution in assessing the antibiotic activity of extracts of culinary-medicinal Shiitake mushroom [*Lentinus edodes* (Berk.) Singer] (Agaricomycetideae). *Int J Med Mushrooms* 5: 31–35.

Borchers, A. T., Krishnamurthy, A., Keen, C. L., Meyers, F. J., and Gershwin, M. E. 2008. The immunobiology of mushrooms. *Exp Biol Med* 233(3): 259–276.

Dickinson, C. and Lucas, J. 1982. *VNR Color Dictionary of Mushrooms*. Van Nostrand Reinhold, New York, pp. 9–11.

Engler, M., Anke, T., and Sterner, O. 1998. Production of antibiotics by *Collybia nivalis*, *Omphalotus olearius*, a Favolaschia and a *Pterula* species on natural substrates. *Z Naturforsch C* 53(5–6): 318–324.

Gao, Y., Xu, H., Lu, Z., and Xu, Z. 2009. Quantitative determination of steroids in the fruiting bodies and submerged-cultured mycelia of *Inonotus obliquus*. *Se Pu* 27(6): 745–749.

Han, F., Kang, L. Z., Zeng, X. L., Ye, Z. W., Guo, L. Q., and Lin, J. F. 2014. Bioproduction of baccatin III, an advanced precursor of paclitaxol with transgenic *Flammulina velutipes* expressing 10-Deacetylbaccatin III-10-*O*-acetyl transferase gene. *J Sci Food Agric* 94(12): 2376–2383.

Hseu, R. S., Wang, H. H., Wang, H. F., and Moncalvo, J. M. 1996. Differentiation and grouping of isolates of the *Ganoderma lucidum* complex by random amplified polymorphic DNA-PCR compared with grouping on the basis of internal transcribed spacer sequences. *Appl Environ Microbiol* 62(4): 1354–1363.

Ina, K., Kataoka, T., and Ando, T. 2013. The use of lentinan for treating gastric cancer. *Anticancer Agents Med Chem* 13(5): 681–688.

Ishikawa, N. K., Fukushi, Y., Yamaji, K., Tahara, S., and Takahashi, K. 2001. Antimicrobial cuparene-type sesquiterpenes, enokipodins C and D, from a mycelial culture of *Flammulina velutipes*. *J Nat Prod* 64: 932–934.

Keegan, R. J., Lu, Z., Bogusz, J. M., Williams, J. E., and Holick, M. F. 2013. Photobiology of vitamin D in mushrooms and its bioavailability in humans. *Dermatoendocrinol* 5(1): 165–176.

Koyyalamudi, S. R., Jeong, S. C., Song, C. H., Cho, K. Y., and Pang, G. 2009. Vitamin D2 formation and bioavailability from *Agaricus bisporus* button mushrooms treated with ultraviolet irradiation. *J Agric Food Chem* 57(8): 3351–3355.

Lattif, L. A., Daran, A. B. M., and Mohammed, A. B. 1996. Relative distribution of minerals in the pileus and stalk of some selected edible mushroom. *Food Chem* 56: 155–121.

Lee, S. J., Yeo, W. H., Yun, S., and Yoo, D. 1999. Isolation and sequence analysis of new peptaibol, boletusin, from *Boletus* spp. *J Pept Sci* 5: 374–378.

Liu, X. T., Winkler, A. L., Schwan, W. R., Volk, T. J., Rott, M. A., and Monte, A. 2010. Antibacterial compounds from mushrooms I: A lanostane-type triterpene and prenylphenol derivates from *Jahnoporus hiritus* and *Albatrellus fletti* and their activities against *Bacillus cereus* and *Enterococcus faecalis*. *Planta Med* 76: 182–185.

Lo, H. C. and Wasser, S. P. 2011. Medicinal mushrooms for glycemic control in diabetes mellitus: History, current status, future perspectives, and unsolved problems (review). *Int J Med Mushrooms* 13(5): 401–426.

Moriguchi, M., Sada, S.-I., and Hatanaka, S.-I. 1979. Isolation of cis-3-amino-L-proline from cultured mycelia or *Morchella esculenta*. *Appl Environ Microbiol* 38(5): 1018–1019.

Mothana, R. A. A., Jansen, R., Julich, W. D., and Lindequist, U. 2000. Ganomycins A and B, new antimicrobial Farnesyl hydroquinones from the Basidiomycete *Ganoderma pfeifferi*. *J Nat Prod* 63: 416–418.

Mygind, P. H., Fischer, R. L., Schnorr, K. M., Hansen, M. T., Sonksen, C. P., Ludvigsen, S., Raventos, D. et al. 2005. Plectasin is a peptide antibiotic with therapeutic potential from a saprophytic fungus. *Nature* 437: 975–980.

Ngai, P. H. K. and Ng, T. B. 2004. A ribonuclease with antimicrobial, antimitogenic and antiproliferative activities from the edible mushroom *Pleurotus sajor-caju*. *Peptides* 25: 11–17.

Nitha, B. and Janardhanan, K. K. 2008. Aqueous-ethanolic extract of morel mushroom mycelium *Morchella esculenta*, protects cisplatin and gentamicin induced nephrotoxicity in mice. *Food Chem Toxicol* 46(9): 3193–3199.

Paterson, R. R. 2006. *Ganoderma*—A therapeutic fungal biofactory. *Phytochemistry* 67(18): 1985–2001.

Postemsky, P. and Curvetto, N. 2014. Enhancement of wheat grain antioxidant activity by solid state fermentation with *Grifola* spp. *J Med Food* 17(5): 543–549.

Quang, D. N., Bach, D. D., Hashimoto, T., and Asakawa, Y. 2006. Chemical constituents of the Vietnamese inedible mushroom *Xylaria intracolorata*. *Nat Prod Res* 20: 317–321.

Ribéreau-Gayon, P. 1972. *Plant Phenolics*. Oliver and Boyd, Edinburgh.

Rotzoll, N., Dunkel, A., and Hofmann, T. 2005. Activity-guided identification of (S)-malic acid 1-O-D-glucopyranoside (morelid) and gamma-aminobutyric acid as contributors to umami taste and mouth-drying oral sensation of morel mushrooms (*Morchella deliciosa* Fr.). *J Agric Food Chem* 53(10): 4149–4156.

Rubel, W. and Arora, D. 2008. A study of cultural bias in field guide determinations of mushroom edibility using the iconic mushroom, *Amanita muscaria*, as an example. *Econ Bot* 62(3): 223–243.

Schardl, C. L. 2000. Symbiotic parasites and mutualistic pathogens: Clavicipitaceous symbionts of grasses. In: *Fungal Pathology*, Kronstad, J. W. (ed.). Dordrecht, the Netherlands: Kluwer Academic Publishers, pp. 307–345.

Shiao, M. S., Lin, L. J., and Yeh, S. F. 1988. Triterpenes in *Ganoderma lucidum*. *Phytochemistry* 27: 873–875.

Smania, A., Monache, F. D., Smania, E. F. A., Gil, M. L., Benchetrit, L. C., and Cruz, F. S. 1995. Antibacterial activity of a substance produced by the fungus *Pycnoporus sanguineus* (Fr.) Murr. *J Ethnopharmacol* 45: 177–181.

Smith, J. E., Rowan, N. J., and Sullivan, R. 2002. *Medicinal Mushrooms for Cancer.* Cancer Research UK, London, U.K.

Turkoglu, A., Duru, M. E., Mercan, N., Kivrak, I., and Gezer, K. 2007. Antioxidant and antimicrobial activities of *Laetiporus sulphureus* (Bull.) Murrill. *Food Chem* 101: 267–273.

Ulziijargal, E., Yang, J. H., Lin, L. Y., Chen, C. P., and Mau, J. L. 2013. Quality of bread supplemented with mushroom mycelia. *Food Chem* 138(1): 70–76.

Weigand-Heller, A. J., Kris-Etherton, P. M., and Beelman, R. B. 2012. The bioavailability of ergothioneine from mushrooms (*Agaricus bisporus*) and the acute effects on antioxidant capacity and biomarkers of inflammation. *Prev Med* 54(Suppl): S75–S78.

Zheng, H., Maoqing, Y., Liqiu, X., Wenjuan, T., Lian, L., and Guolin, Z. 2006. Purification and characterization of an antibacterial protein from the cultured mycelia of *Cordyceps sinensis*. *Wuhan Univ J Nat Sci* 11: 709–714.

chapter fourteen

Aspergillosis and its resistance
Marine natural products as future treatment

Kumar Saurav, Subhasish Saha, Manoj Singh, Soumik Sarkar,
Dharumadurai Dhanasekaran, and K. Kannabiran

Contents

14.1 Introduction

Over the past two decades, fungal infections have increased significantly and have been associated with increased morbidity and mortality. As advances in medical care have improved the survival of patients with severe and life-threatening illnesses, the more aggressive nature of such care has led to a rapid increase in the number of immunosuppressed populations. These changes have been correlated with a substantial increase in the rate of invasive fungal infections, mainly resulting from the rapid increase in the number of at-risk patients. Despite the remarkable progress in antifungal drug research in the past decade, difficulty in prompt treatment along with the complexity of the clinical characteristics of at-risk patients continues to make it a great challenge for the healthcare workers and researchers. The most commonly recognized causes of opportunistic invasive fungal infections traditionally are *Candida albicans, Cryptococcus neoformans,* and *Aspergillus fumigatus*. Along with the widespread use of antifungal prophylaxis, the epidemiology of infection has shifted toward non*albicans Candida*, non*fumigatus Aspergillus*, opportunistic yeastlike fungi (e.g., *Trichosporon* and *Rhodotorula* spp.), Zygomycetes, and hyaline molds (e.g., *Fusarium* and *Scedosporium* spp.). These new and emerging fungi are characterized, and they exhibit greater resistance to standard antifungal drugs. Invasive fungal infections due to these previously rare fungi are also more difficult to diagnose and are associated with even higher mortality rates.

Aspergillus is a ubiquitous soil-dwelling organism found in organic debris, dust, compost, foods, spices, and rotted plants. The genus *Aspergillus* includes over 185 species. Around 20 species have so far been reported as causative agents of opportunistic infections in humans. Among these, *Aspergillus fumigatus* is the most commonly isolated species, followed by *Aspergillus flavus* and *Aspergillus niger*. *Aspergillus oryzae*, *Aspergillus terreus*, *Aspergillus ustus*, and *Aspergillus versicolor* are among the other species less commonly isolated as opportunistic pathogens (Rinaldi, 1983). *Aspergillus* spp. are well known to play a role in three different clinical settings in humans: (1) opportunistic infections, (2) allergic states, and (3) toxicoses. Immunosuppression is the major factor predisposing to development of opportunistic infections (Ho and Yuen, 2000). *Aspergillus*, like other filamentous fungi, is primarily acquired from an inanimate reservoir, usually by the inhalation of airborne spores. The organism grows best at 37°C, and the small spores (2–3 μm) are easily inhaled and deposited deep in the lungs, leading to a variety of clinical syndromes (Figure 14.1). These infections may present in a wide spectrum, varying from local involvement to dissemination and as a whole called aspergillosis. Among all filamentous fungi, *Aspergillus* is in general the most commonly isolated one in invasive infections. It is the second most commonly recovered fungus in opportunistic mycoses following *Candida*. Although these are distinct pulmonary entities, on rare occasions, one condition may change to another; for example, an aspergilloma may change to invasive pulmonary aspergillosis (IPA) (Tomee et al., 1995).

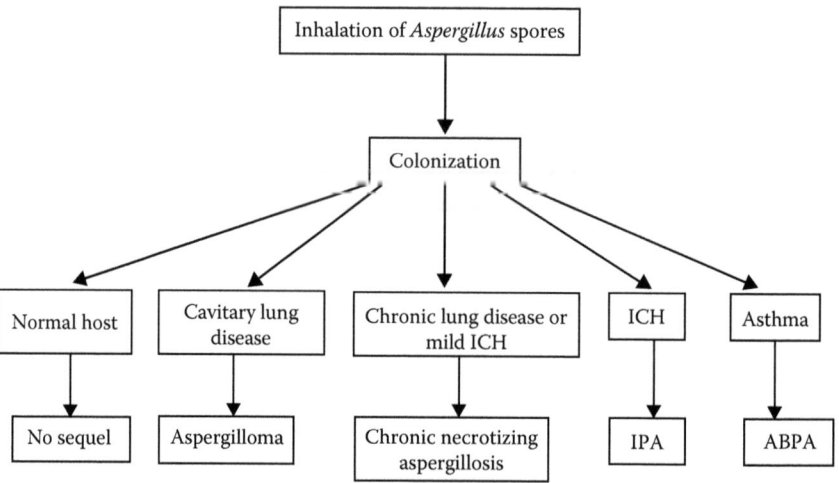

Figure 14.1 The clinical spectrum of conditions associated with inhalation of *Aspergillus* spores. ICH, immunocompromised host; IPA, invasive pulmonary aspergillosis; ABPA, allergic bronchopulmonary aspergillosis.

14.1.1 Aspergilloma

This is the most common and best-recognized form of pulmonary involvement due to *Aspergillus*. The aspergilloma (fungal ball) consists of masses of fungal mycelia, inflammatory cells, fibrin, mucus, and tissue debris, usually developing in a preformed lung cavity. Although other fungi such as *Zygomycetes* and *Fusarium* may cause the formation of a fungal ball, *Aspergillus* sp., specifically *A. fumigatus*, are by far the most common etiologic agents.

14.1.2 Chronic necrotizing aspergillosis (CNA)

Also called semi-invasive aspergillosis, this entity was first described in two reports in 1981 and 1982 (Gefter et al., 1981; Binder et al., 1982). Chronic necrotizing aspergillosis (CNA) is an indolent, destructive process of the lung due to the invasion by *Aspergillus* species (usually *A. fumigatus*). This entity is different from aspergilloma in that there is local invasion of the lung tissue, and a preexisting cavity is not needed, although a cavity with a fungal ball may develop in the lung as a secondary phenomenon due to destruction by the fungus.

14.1.3 Invasive pulmonary aspergillosis (IPA)

The vast majority of IPA cases are seen in immunocompromised patients (Hibberd and Rubin, 1994). Neutropenia is the most important risk factor, and it is estimated that IPA accounts for 7.5% of all infections in neutropenic patients following induction therapy for acute myelogenous leukemia. The risk of IPA increases with the duration of neutropenia (i.e., neutrophil count, <500 cells/μL) and is estimated to be 1% per day for the first 3 weeks, after which time it increases to 4% per day. Transplantation, especially lung and bone marrow transplantation (BMT), is another increasingly significant risk factor for IPA (Soubani et al., 1996). It is estimated that 5% of BMT recipients develop IPA with mortality rates ranging between 30% and 80% (Wald et al., 1997). In BMT recipients, IPA may be seen in the first few weeks after the procedure with delayed engraftment or graft failure and more commonly in the setting of treatment with corticosteroids for graft-versus-host disease.

14.1.4 Allergic bronchopulmonary aspergillosis

Allergic bronchopulmonary aspergillosis (ABPA) is a hypersensitivity reaction to *Aspergillus* antigens, mostly due to *A. fumigatus*. It is typically seen in patients with long-standing asthma or cystic fibrosis, and it is estimated that 7%–14% of corticosteroid-dependent asthma patients and 6% of patients with cystic fibrosis meet the diagnostic criteria for ABPA (Basich et al., 1981; Mroueh and Spock, 1994). The factors leading to ABPA are not clearly understood. It is believed that *Aspergillus*-specific, IgE-mediated type I hypersensitivity reactions and *Aspergillus*-specific, IgG-mediated type III hypersensitivity reactions play a central role in the pathogenesis of ABPA.

14.2   Drug resistance

Empirical antifungal therapy using parenteral amphotericin B deoxycholate (AMB), first-generation azoles (e.g., ketoconazole), and 5-fluocytosine (5-FC) were the only available therapeutic options for fungal infections. However, AMB has significant dose-limiting nephrotoxicity and infusion-related toxicity, while the first-generation azoles and 5-FC have a relatively limited therapeutic spectrum and potency. Thus, the introduction of potent, broad-spectrum oral and parenteral second-generation triazoles such as fluconazole and itraconazole since the early 1990s has been a welcomed addition to our antifungal armamentarium. Fluconazole, a fungistatic triazole, has been shown to be clinically effective against yeasts, dermatophytes, and dimorphic fungi (Galgiani, 1990; Grant and Clissold, 1990), while itraconazole, which has a broader spectrum of antifungal activity, is effective against moulds, including *Aspergillus* species and various pheohyphomycetes as well as endemic fungi (Boogaerts and Maertens, 2001). Both have limitations, such as a relatively narrow antifungal spectrum, propensity for selection of resistant strains (fluconazole) (Kontoyiannis and Bodey, 2002), erratic bioavailability of the capsule form, adverse gastrointestinal side effects to the oral solution form (itraconazole), and many other clinically important drug–drug interactions (itraconazole).

The development of lipid formulations of AMB that lack much of the parent compound's nephrotoxicity, new extended spectrum third-generation triazoles and the echinocandins, and a new class of cell wall acting parenteral antifungals are some of the major recent advancements in antifungal therapy (Kontoyiannis, 2001). Three lipid formulations of AMB are commercially available in the United States and most of the western European countries: liposomal AMB (L-AMB), AMB lipid complex (ABLC), and AMB colloidal dispersion (ABCD). Although these lipid products represent attractive alternatives to delivery of AMB from a toxicity standpoint, their use should be limited to patients who cannot tolerate, or whose infection does not respond to, AMB, because they are all substantially more expensive than the parent drug, with L-AMB being the most expensive.

14.2.1   Resistance to amphotericin B

The mode of action and the known mechanisms of resistance of *A. fumigatus* against current antifungal agents are summarized in Table 14.1. It is suggested that the mechanisms of action of amphotericin are mainly by promotion of oxidative damage of cell membranes through the generation of reactive oxygen species (Sterling and Merz, 1998; Moore et al., 2000). Clinical resistance of IA to AMB-based therapy is observed frequently in clinical practice (Kontoyiannis and Bodey, 2002; Patterson, 2002). Typically, such infections with unfavorable outcome occur in severely immunocompromised patients with multiple underlying host factors. Several preclinical studies have documented the resistance of *A. terreus* to AMB (Walsh et al., 2003; Lionakis et al., 2005). Although the mechanisms of resistance of *A. terreus* to AMB have not been elucidated, it appears that this fungus has much less ergosterol (the target molecule of AMB action) in its fungal membrane than the more AMB-susceptible *A. fumigates* (Walsh et al., 2003). It is true that both primary and secondary *in vitro* resistance of *A. fumigatus* to AMB have been reported (Verweij et al., 1998; Manavathu et al., 2000), but such resistance is generally considered a rare phenomenon (Moosa et al., 2002; Paterson et al., 2003; Dannaoui et al., 2004).

Table 14.1 Mode of action and the known mechanisms of resistance of *A. fumigatus* against current antifungal agents

Antifungal agent	Mechanisms of action	Mechanisms of resistance
Polyenes 　Amphotericin B 　Lipid formulations 　Liposomal nystatin Azoles 　Itraconazole 　Voriconazole 　Posaconazole Echinocandins 　Caspofungin 　Micafungin 　Anidulafungin Allylamines 　Terbinafine	Fungicidal Interaction with ergosterol, intercalation of fungal membrane that leads to increased permeability to univalent and divalent cations and cell death Alternative mechanism of action through oxidation of fungal membrane Inhibition of P-450 14-a-DM (ERG 11), accumulation of lanosterol leading to perturbation of fungal cell membrane Inhibition of cell wall glucan synthesis, leading to susceptibility of the fungal cell to osmotic lysis Fungistatic Inhibition of squalene epoxidase (Erg 1), with subsequent ergosterol depletion and accumulation of toxic sterol intermediates	Decreased access of AMB to drug target in fungal membrane (a) Altered membrane ergosterol content and reduced intercalation (b) Increased cell wall rigidity (c) Sequestration of fungi to lysosomes Decreased oxidative damage (a) Overexpression of catalases and superoxide dismutases of *A. fumigatus* (b) Anaerobic environment Increased drug efflux Overexpression of target enzyme Decreased affinity to the binding site Upregulation of homeostatic stress response pathways (HSP 90; calcineurin) Altered drug uptake Exogenous cholesterol import (rescue ergosterol depletion) Upregulation of homeostatic stress response pathways (HSP 90; calcineurin) Overexpression of target site Upregulation of genes encoding for β-glucan synthetase Overexpression of genes related to transport of cell wall components Increased drug efflux Overexpression of target site (ERG 1) Overexpression of salicylate monooxygenase (drug degradation)

Source: Chamilos, G. and Kontoyiannis, D.P., *Drug Resist. Update*, *8*(6), 344, 2005.

　　Manavathu et al. (2005) have reported recently on *in vivo/in vitro* correlation for AMB in a mouse model of *A. fumigatus* infection, but their results were from tests on a laboratory-generated AMB-resistant *A. fumigatus* isolate. Other investigators were unable to correlate the *in vitro* susceptibility of *A. fumigatus* to AMB with the outcome in animal model studies (Johnson et al., 2000).

14.2.2 *Resistance to azoles*

The triazoles are antifungal agents that act by blocking the ergosterol biosynthetic pathway at the C-14 demethylation stage. Triazoles bind to lanosterol 14-α demethylase (14α-DM, or Cyp51), a cytochromes P450 enzyme that is encoded by the ERG 11 gene

(Lamb et al., 1999; Odds et al., 2003). This leads to ergosterol depletion and accumulation of lanosterol and other toxic 14-α methylated sterols.

Cowen and Lindquist (2005) showed a novel molecular mechanism of resistance of *Aspergillus* and other opportunistic fungi that involves HSP90, a molecular chaperone with an essential role in folding, transport, and maturation of key regulatory proteins under stress-induced conditions. Using *S. cerevisiae* mutants that expressed high or low levels of HSP90, they demonstrated that HSP90 is required both for the emergence of drug resistance to azoles and echinocandins and for the continued drug resistance once it has occurred. Furthermore, the investigators found that HSP90 function in drug resistance is mediated by the calcineurin pathway, which is known to be implicated in virulence of several pathogenic fungi and also enables fungal cells to tolerate drugs that block ergosterol biosynthesis.

Liu et al. (2003) have recently reported that cross-resistance or tolerance mechanisms between different classes of azoles in *A. fumigatus* strains sequentially exposed to azoles. Serial passages of 10 *A. fumigatus* spp. in plates containing FLU, an azole with minimal activity against *Aspergillus* spp., resulted in attenuation of *in vivo* fungicidal activity of ITC and VRC against all isolates tested. The degree of azole cross-resistance is azole specific, and its pattern might reflect the similarities in molecular structures among different triazoles (Xiao et al., 2004) (Figure 14.2). Similarly, in a large surveillance study in Sweden that included 400 clinical and 150 environmental *A. fumigatus* isolates, there was no evidence of cross-resistance between ITC and VRC in 10 (2.5%) clinical and 36 (24%) environmental isolates that displayed high ITC MICs (≥2 µg/mL). In a recent study of 596 clinical and environmental isolates of *A. fumigatus*, although there was no evidence of azole resistance, there was a trend toward extended loss of susceptibility across VRC and ITC, POS and ITC, and VRC and POS (Guinea et al., 2005). Broad-spectrum cross-resistance among all the azoles has been shown in *A. fumigatus* causing IA in a patient who was on prolonged ITC secondary prophylaxis (Warris et al., 2002).

Balajee et al. (2004) have identified 10 variants of multidrug-resistant *A. fumigatus* clinical isolates, all of which had an unusual sporulation pattern and a unique mitochondrial cytochrome *b* sequence. These isolates exhibited increased MICs not only against all the triazoles tested but also against AMB.

14.2.3 Resistance to echinocandins

Echinocandins are a unique class of antifungals that target the fungal cell wall by noncompetitive inhibition of 1,3-β-D-glucan synthetase, a fungus-specific enzyme target. Cross-resistance between echinocandins and the other classes of antifungals has not been described thus far. Warn et al. (2003) have shown a murine model of disseminated

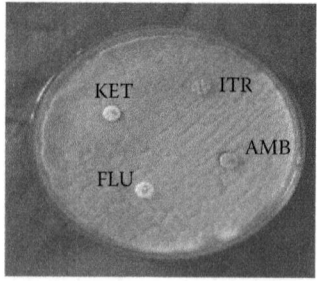

Figure 14.2 Drug-resistant pattern of *Aspergillus* isolate to various standard drugs by disc diffusion assay. (From Kumar, S. and Kannabiran, K., *J. Med. Mycol.*, 20(2), 101, 2010.)

aspergillosis; the echinocandin micafungin has significant activity against both an ITC-resistant *A. fumigatus* isolate and an AMB-resistant *A. terreus* isolate.

14.2.4   Resistance to allylamine

Terbinafine is an oral allylamine that blocks ergosterol biosynthesis by inhibiting a membrane-bound squalene epoxidase encoded by the *ERG1* gene (Ryder, 1992). Inhibition of squalene epoxidase leads to ergosterol deficiency along with accumulation of the toxic squalene, which probably accounts for the *in vivo* reported fungicidal activity of TRB. Although TRB is indicated exclusively for skin and nail infections, it is suggested that it might have a role in the treatment of IA, alone or in combination with triazoles (Schiraldi et al., 1996).

The decreased sensitivity of *Aspergillus* sp. against available drugs and lack of its effectiveness to combat such pathogens continued to be the major challenges for health-care-providers. Hence, there is an urgent need to search for novel bioactive compounds from natural sources with having a broad spectrum of activity with less side effects against drug-resistant *Aspergillus* strains for effective control and management of aspergillosis.

14.3   Antifungal therapy

Aspergillus sp. are commonly found in soil, water, and decaying material all over the world. Unlike invasive candidiasis, invasive aspergillosis (IA) occurs predominantly in highly immunocompromised patients (Baddley et al., 2001; Marr et al., 2002). *Aspergillus fumigatus* is the most common causative species of IA, but nonfumigatus *Aspergillus* such as *Aspergillus niger*, *Aspergillus flavus*, and *Aspergillus terreus* have been increasingly reported (Baddley et al., 2001; Marr et al., 2002). IA is an emerging condition among patients with other causes of immunosuppression, such as organ transplantation, in patients with advanced acquired immunodeficiency syndrome (AIDS), and patients treated with newer immunosuppressive agents such as infliximab (Patterson, 2005; Pfaller et al., 2006). The underlying disease condition greatly determines the level of risk for patients to become infected by *Aspergillus* species (Table 14.2). However, the frequency of cases observed by individual physicians may vary greatly for different patient groups, depending on the risk period. For some patients, this is lifelong, such as patients with acquired immunodeficiency syndrome (AIDS) or chronic granulomatous disease, whereas in other patient groups, such as those treated for acute myeloid leukemia, the risk period is the highest.

Since 1970, a number of autopsy studies have been published that indicate that the epidemiology of invasive fungal infections in immunocompromised patients is changing. Two decades ago, autopsy studies found *Candida* to be the predominant pathogen. However, studies of autopsy cases from Europe, the United States, and Japan show that the number of patients dying from invasive aspergillosis has increased significantly over the past two decades.

14.3.1   Current antifungal agents

The antifungal agents currently available for the treatment of systemic fungal infections are amphotericin B, lipid formulations of amphotericin B, 5-fluorocytosine, and the azoles, miconazole, ketoconazole, fluconazole, and itraconazole. Currently, the criteria used for the selection of optimal new drug candidates include inhibitors of fungal cell wall biosynthesis, potency comparable to amphotericin B, safety comparable to fluconazole, fungicidal activity *in vitro*, and fungicidal activity *in vivo*.

Table 14.2 Spectrum of patient groups or conditions with increased risk of invasive
Aspergillus infection categorized by infection risk

High (>10%)	Moderate (1%–10%)	Low (<1%)	Negligible
Chronic granulomatous disease	AIDS patients	Systemic lupus erythematosus on prednisolone	Normal healthy persons
Lung transplant recipients	Liver, heart, or pancreas transplant recipients	Diabetes mellitus	Hospitalized patients without neutropenia or treatment with corticosteroids or other risk conditions
Acute myeloid leukemia	Acute myeloid leukemia	Alcoholism	
Allogeneic BMT recipients with GVHD > grade II	Allogeneic BMT recipients without GVHD of grade I	Corticosteroid treatment	
	Autologous BMT recipients, intensive care patients on steroids, severe combined immunodeficiency syndrome	Patients treated for solid tumors	
		Intensive care patients	
		Agammaglobulinemia	
		Kidney transplant recipients	
	Lymphoma	Influenza	
	Major burn (e.g., >30%)	Surgery in contaminated operating room air	
		Major trauma	

Source: Shao, P.L. et al., *Int. J. Antimicrob. Ag.*, 30(6), 487, 2007.
BMT, bone marrow transplantation; GVHD, graft-versus-host disease.

Amphotericin B (AMB) is a polyene antifungal agent first isolated from *Streptomyces nodosus* in 1955. AMB is an elongated cyclic molecule consisting of hydrophilic polyhydroxyl and hydrophobic polyene domains (Figure 14.3).

Amphotericin B complexed with ergosterol disrupts the fungal plasma membrane, which leads to increased membrane permeability, leakage of the cytoplasmic contents, and finally death to the fungal cell. Liposomal preparations of amphotericin B have been used to reduce the nephrotoxicity of conventional amphotericin B. The polyenes are fungicidal and have the broadest spectrum of antifungal activity among others (Hiemenz and Walsh, 1996; Andriole, 1999). They have excellent *in vitro* activity against *Blastomyces dermatitidis, Coccidioides immitis, Cryptococcus neoformans, Histoplasma capsulatum, Paracoccidioides brasiliensis, Sporotrichum* species, and *Torulopsis (Candida) glabrata*. It also has excellent activity against *Candida albicans* and most other *Candida* species

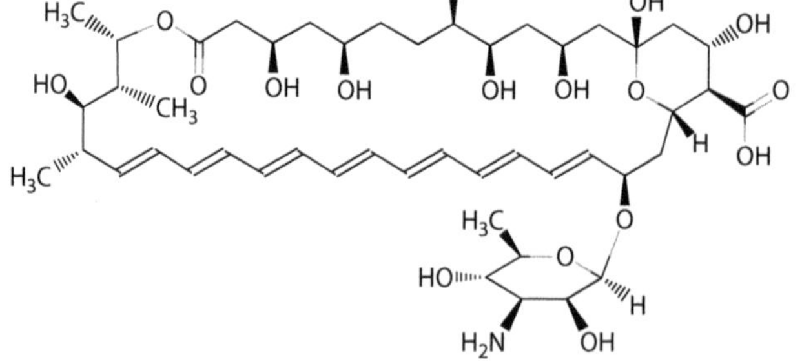

Figure 14.3 Chemical structure of amphotericin B. (From National Library of Medicine ChemIDPlus.)

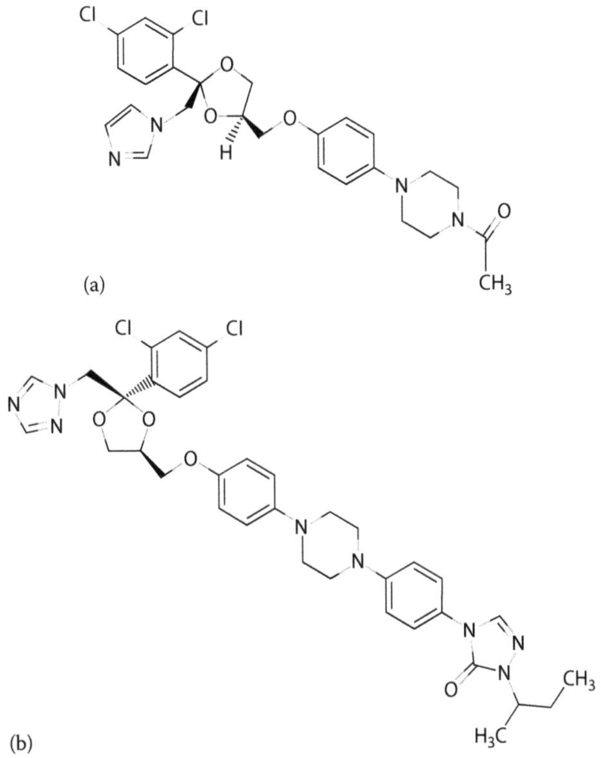

(a)

(b)

Figure 14.4 Chemical structure of (a) ketoconazole and (b) itraconazole.

except for *Candida lusitaniae*. It has variable activity against *Aspergillus* species and *Zygomycetes* (*Mucor*) species, whereas *Fusarium, Trichosporon* species, and *Pseudallescheria boydii* are often resistant (Andriole, 1999).

Initially different azole compounds, imidazoles, clotrimazole, miconazole, and keto-conazole (Figure 14.4), followed by the triazoles, fluconazole, and itraconazole inhibit fungal cytochrome P450 3A-dependent C 14-demethylase, which is responsible for the conversion of lanosterol to ergosterol, thereby depleting ergosterol in the fungal cell mem-brane (Andriole, 1999). Azole antifungal activity varies with each compound, and clini-cal efficacy may not coincide with *in vitro* activity. Azoles are active against *C. albicans, C. neoformans, C. immitis, H. capsulatum, B. dermatitidis, P. brasiliensis*, and *C. glabrata*, and some *Aspergillus* spp., *Fusarium* spp., and *Zygomycetes* are resistant to currently available azoles (Groll et al., 1998).

14.3.2 *Antifungal agents under clinical trial*

Isavuconazole is a new triazole currently undergoing phase III clinical trials. This compound has shown *in vitro* activity against a large number of clinically important yeasts and moulds including *Aspergillus* spp., *Fusarium* spp., *Scedosporium* spp., *Candida* spp., the *Zygomycetes*, and *Cryptococcus* spp. Similar to voriconazole, reduced *in vitro* activity is seen against *Histoplasma capsulatum*. *In vivo* efficacy has been demonstrated in murine models of invasive aspergillosis and candidiasis. Additionally, there are several potential pharmacokinetic and drug–drug interaction advantages of this compound over existing antifungal agents.

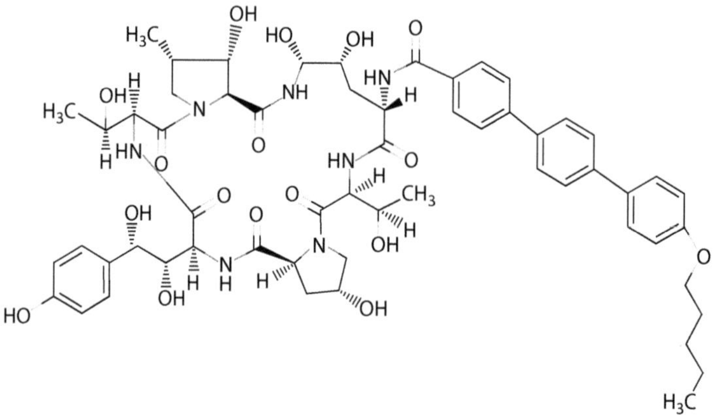

Figure 14.5 Chemical structure of anidulafungin. (From National Library of Medicine ChemIDPlus.)

Posaconazole works by disrupting the close packing of acyl chains of phospholipids, impairing the functions of certain membrane-bound enzyme systems such as ATPase and enzymes of the electron transport system and thus inhibiting growth of the fungi. It does this by blocking the synthesis of ergosterol by inhibiting the enzyme lanosterol 14-α-demethylase and accumulation of methylated sterol precursors. Posaconazole is significantly more potent at inhibiting 14-α demethylase than itraconazole. Recent advancement in the clinical effectiveness of posaconazole prophylaxis was made in a trial for patients with acute myelogenous leukemia (AML), where no significant differences were observed for persistent fever, pneumonia, lung infiltrates indicative of invasive pulmonary aspergillosis, or attributable and overall mortality (Vehreschild et al., 2010).

There are two echinocandins, anidulafungin (Figure 14.5) and aminocandin, currently undergoing clinical evaluation. Anidulafungin was developed by Eli Lilly and licensed to Vicuron Pharmaceuticals in May 1999 and is a semisynthetic derivative of echinocandin B, a fungal metabolite originally isolated from *Aspergillus rugulovalvus* (formerly *Aspergillus rugulosus*). Vicuron has completed a phase III trial of anidulafungin for the treatment of esophageal candidiasis, and phase III trials are in progress for the treatment of invasive aspergillosis and candidiasis/candidemia. Recently, phase IV trial of the drug to evaluate a short course of anidulafungin, followed optionally by oral voriconazole, for the treatment of candidemia and invasive candidiasis has completed in October 2009 (USA clinicaltrials. gov, National Institute of Health).

Indevus licensed the echinocandin, aminocandin (HMR- 3270), from Novexel, France (Originally Sanofi-Aventis antiinfective group) in April 2003 and initiated phase I clinical trials against systemic fungal infections in February 2004. Although the structure of aminocandin is not yet in the public domain, it is known to be a semisynthetic derivative of deoxymulundocandin, a natural product originally isolated from the fungus *Aspergillus sydowii* by Hoechst India in 1992.

14.3.3 Natural products as antifungal agents

Pharmacological screening and usage of natural products for the treatment of human diseases have had a long history from Ayurvedic medicine to modern drugs (Swerdlow, 2000). The majority of modern drugs are reported to be mostly from natural products

(Newman et al., 2003). In the past few decades, a worldwide increase in the incidence of fungal infections has been observed as well as a rise in the resistance of some species of fungus to different fungicidals used in medicinal practice. The majority of clinically used antifungals have various drawbacks in terms of toxicity, efficacy, and cost, and their frequent use has led to the emergence of resistant strains. Hence, there is a great demand for novel antifungals belonging to a wide range of structural classes, selectively acting on new targets with fewer side effects. One approach might be the testing of natural products for their antifungal activities as potential sources for drug development. As reported by Butler (2005), about 70 natural products or natural product derivatives are currently undergoing clinical trials in different parts of the world including United States, Europe, Japan, and Korea, out of which 30 were derived from microorganism and 12 were marine derived.

14.3.4 *Microbial natural products as a source of antifungals*

Microbial natural products are the important source of both existing and new drugs. The exploration of microorganisms as a source of therapeutically useful compounds is one of the recent developments and less well-known history than the use of plants and plant extracts in human medicine. Secondary metabolites are defined as naturally produced substances that do not play an explicit role in the organisms that produce them. These microorganisms may have evolved the ability to produce such compounds because of the selection advantages conferred upon them as a result of the interactions of the compounds with specific receptors in other organisms (Demain, 1983; Omura 1986). One of the targets for novel antifungals under active investigation is the fungal cell wall. Antifungal agents acting on this target are inherently selective. Fungal cell wall composition varies among species, but it generally has three polymeric components: glucan, chitin, and mannoproteins. Inhibitors of glucan synthesis have been shown to possess antifungal activity *in vitro* as well as *in vivo* in many different animal models. Some of the microbial natural products that are inhibitors of glucan synthesis are listed in Table 14.3.

The classical inhibitors of chitin synthesis are nikkomycins and polyoxins. The substrate analogues of UDP-N-acetylglucosamine (building block for chitin biosynthesis) were isolated from two different *Streptomyces* species: *S. tendae* (nikkomycin) and *S. cacaoi var. anoensis* (polyoxin) (Cabib, 1991). Nikkomycins exhibit activity against dimorphic fungi but low activity against yeast and filamentous fungi. Nikkomycins and polyoxins are currently used exclusively as agricultural fungicides, due to their modest activity against human pathogens (Cohen, 1993). Recently, two novel antifungal compounds were found: phellinsin A and arthrichtin. Phellinsin A is a phenolic compound exhibiting antifungal activity against human pathogens such as *Trichophyton mentagrophytes* and *A. fumigatus* and very weak activity against other human pathogens such as *C. neoformans* and *Coccidioides immitis*. Arthrichitin is a cyclic depsipeptide isolated from *Arthrinium phaeospermum* and from the marine fungus *Hypoxylon oceanicum* (Vijayakumar et al., 1996) with moderate activity against *Candida* spp., *Trichophyton* spp., and several other phytopathogens.

Mannoproteins are the third main component of the fungal cell wall. They form the outer layer of the cell wall and contain as much as 50% carbohydrate. The majority of the cell wall mannoproteins are anchored by β-(1,6)- and β-(1,3)-glucan and play several important functions of fungal membrane. Inhibitors of the mannoproteins function are the

Table 14.3 Microbial natural product inhibitors of glucan synthesis

Compounds	Producing species
Lipopeptides	
Echinocandin B	*Aspergillus nidulans*
Aculeacin	*Aspergillus rugulosus*
Mulundocandin	*Aspergillus aculeatus*
Sporiofungins	*Aspergillus sydowii*
Pneumocandins	*Penicillium arenicola*
Cryptocandin	*Cryptosporiopsis* sp.
WFI 1899 and related sulfate derivatives	*Glarea lozoyensis*
FR901469	*Pezicula* sp.
Arborcandins	*Cryptosporiopsis* sp.
Clavariopsins	*Cryptosporiopsis quercina*
Glycolipids	*Coleophoma empetri*
Papulacandins	*Coleophoma crateriformis*
Corynecandin	*Tolypocladium parasiticum*
Mer-WF 3010	*Chalaria* sp.
Fusacandin	Unidentified fungus
BU-4794F	Unidentified fungus
L-687781	*Clavariopsis aquatic*
Acidic terpenoids	*Papularia sphaerosperma*
Enfumafungin	*Coryneum modonium*
Arundifungin	*Phialophora cyclaminis*
Ascoteroside	*Fusarium sambucinum*
Ergokonin A	*Gilmaniella* sp.
	Dictyochaeta simplex
	Hormonema sp.
	Arthrinium arundinis
	A. phaeospermum
	Leotiales anamorphs
	Coelomycete undetermined
	Ascotricha amphitricha
	Mycoleptodiscus atromaculans
	Trichoderma longibrachiatum
	T. koningii
	T. viride

Source: Vicente, M.F. et al., *Clin. Microbiol. Infect.*, 9(1), 15, 2003.

pradimicin/benanomycin family, whose chemical structure possesses a benzo [a] naph-thacenequinone skeleton (Fromtling, 1998). The free carboxyl group of these compounds interacts with the saccharide portion of cell surface mannoprotein, which is followed by disruption of the plasma membrane and leakage of intracellular potassium. These anti-fungal (and antiviral) agents were produced by *Actinomadura* sp. These antibiotics exhib-ited remarkable *in vivo* activity against systemic fungal infections caused by *C. albicans*, *A. fumigatus*, and *C. neoformans* in mice.

Table 14.4 Microbial natural products as inhibitors of sphingolipid biosynthesis
and protein synthesis

Compounds	Producing species
Sphingolipid biosynthesis	*Aspergillus fumigatus Paecilomyces variotii Streptomyces* sp.
Sphingofungins	*Trichoderma viride*
Lipoxamycin	*Isaria sinclairii*
Viridiofungins	*Fusarium moniliforme*
Myriocin	*Sporormiella australis*
Fumonisin B1	*Aureobasidium pullulans*
Australifungin	Unidentified sterile fungus
Aureobasidin A	*Micromonospora chalcea*
Khafrefungin	*Streptomyces galbus Micromonospora* sp.
Rustmicin	*Micromonospora* sp.
Galbonolide B	*Sporomiella minimoides*
Minimoidin	*Sordaria araneosa*
Protein synthesis	*Zopfiella marina*
Sordarin	*Penicillium minioluteum*
Zofimarin	Unidentified sterile fungus
BE31405	*Xylaria* sp.
SCH57404	*Hypoxylon croceum*
Xylarin	*Graphium putredinis*
Hypoxysordarin	
GR135402	

Source: Adapted from Vicente, M.F. et al., *Clin. Microbiol. Infect.*, 9(1), 15, 2003.

Sphingolipids, although present in relatively small proportion in the fungal cytoplasmic membrane, are essential for cellular functions (Wells and Lester, 1983), and inhibition of sphingolipid synthesis results in growth inhibition and cell death (Zweerink et al., 1992; Mandala et al., 1995). A list of few compounds, which are inhibitors of sphingolipid biosynthesis and protein synthesis, is provided in Table 14.4.

14.3.5 Marine microbial natural products as a source of antifungals

Microorganisms from extreme environments have gained considerable attention in recent years because of their diversity and biological activities, mainly due to their ability to produce novel chemical compounds of high commercial value. Microbial sources serve as a template for the isolation of many bioactive anticancer compounds. Several earlier studies reported on the antifungal activity of several novel marine natural products isolated from marine algae, fungi, bacteria, sponges, and sea stars (Mayer and Lehmann, 2000; Mayer and Hamann, 2002, 2005; Mayer et al., 2007). A list of antifungal compounds isolated from marine microorganisms is provided in Table 14.5.

Six new bengazole derivatives and a new bengamide L were reported from the sponge *Pachastrissa* sp. (Fernández et al., 1999). Although no studies on mechanism of action of the compound, the bengazole derivatives were observed to be active against *Candida albicans* with a minimum inhibitory concentration (MIC) of 0.8–1.5 μg/mL. Oceanapiside, a

Table 14.5 List of antifungal compounds isolated from marine microorganisms

Compounds	Organism	Pharmacological activity
Bengazole, bengamide	Sponges	*C. albicans* inhibition
Oceanapiside	Sponges	*C. glabrata* inhibition
Spongistatin I	Sponges	*A. nidulans* inhibition
Tanikolide	Bacterium	*C. albicans* inhibition
Theopederins F–J	Sponges	*S. cerevisiae* inhibition
Basiliskamides A and B	Bacterium	*C. albicans* and *A. fumigatus* inhibition
Corticatic acids A and E	Sponges	*C. albicans* and *A. fumigatus* inhibition
Swinhoeiamide A	Sponges	*C. albicans* and *A. fumigatus* inhibition
Patagonicoside A	Sea cucumber	*Cladosporium cucumerinum* inhibition
Oxybis methyl phenol	Fungus	*C. albicans*, *T. rubrum*, and *A. niger* inhibition
Polyester 15G256h	Fungus	Cell wall biosynthesis inhibition
Astroscleridae sterol	Sponges	*S. cerevisiae* inhibition
Dysidea arenaria sterol	Sponges	Fluconazole resistance reversal in *C. albicans*
Massadine	Sponges	Geranylgeranyltransferase I inhibition
Naamine G	Sponges	*C. herbarum* inhibition
Utenospongin B	Sponges	*C. tropicales* and *F. oxysporum* inhibition
(2S,3R)-2-Aminododecan-3-ol	Ascidian	*C. albicans* inhibition
Capisterones A and B	Algae	Enhancement of fluconazole activity
Dysidea herbacea phenol	Sponges	*C. albicans* and *A. niger* inhibition
Spongistatin	Sponges	Broad panel of yeasts and filamentous fungi
Halocidin	Ascidian	*C. albicans* inhibition
Hassallidin A	Bacterium	*C. albicans* and *A. fumigates* inhibition
Latrunculins	Sponges	*C. albicans* inhibition comparable to clotrimazole
Majusculoic acid	Bacterium	*C. albicans* inhibition, less potent than fluconazole

Sources: Mayer, A.M. and Lehmann, V.K., *Pharmacologist*, 42, 62, 2000; Mayer, A.M. and Hamann, M.T., *Comp. Biochem. Physiol.*, 132(3), 315, 2002.

new glycosidicyamino alcohol lipid from the sponge *Oceanapia philippensis*, demonstrated antifungal activity against the fluconazole-resistant yeast *Candida glabrata* (MIC = 10 μg/mL) (Nicholas et al., 1999). This is the most noteworthy study with a marine-derived antifungal compound having broad-spectrum activity against pathogens. Ovechkina et al. (1999) demonstrated potent microtubule-severing activity with spongistatin 1, a macrocyclic lactone isolated from the sponge *Hyrtios erecta*. Tsukamoto et al. (1999) isolated five new bioactive metabolites, theopederins F–J, from the sponge *Theonella swinhoei*.

Laboratory cultures of the marine bacterium *Bacillus laterosporus* produced the novel polyketides basiliskamides A and B (Figure 14.6) (Barsby et al., 2002). Both compounds showed potent activity against *Candida albicans* (MIC = 1.0 and 3.1 μg/mL, respectively) and *Aspergillus fumigatus* (MIC = 2.5 and 5.0 μg/mL, respectively), which was comparable to amphotericin B.

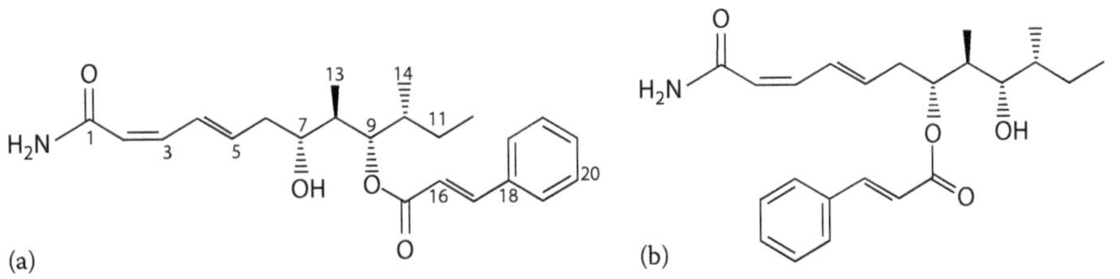

Figure 14.6 Chemical structure of novel polyketides basiliskamides A (a) and basiliskamides B (b) isolated from marine bacterium *Bacillus laterosporus*. (From National Library of Medicine ChemIDPlus.)

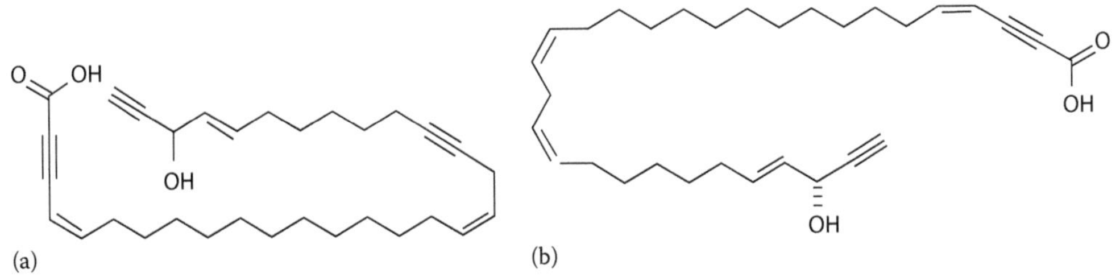

Figure 14.7 Two novel polyacetylenic acids, corticatic acids A (a) and corticatic acids E (b) isolated from the marine sponge *Petrosia corticata*. (Source: National Library of Medicine ChemIDPlus.)

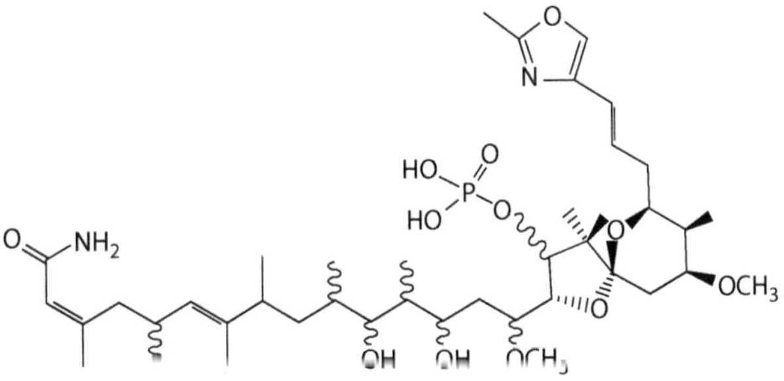

Figure 14.8 Chemical structure of a novel antifungal calyculin derivative swinhoeiamide A, isolated from the marine sponge *Theonella swinhoei*. (Source: National Library of Medicine ChemIDPlus.)

Two novel polyacetylenic acids, corticatic acids A and E (Figure 14.7) isolated from the marine sponge *Petrosia corticata* (Nishimura et al., 2002), were shown to inhibit geranyl-geranyltransferase type I (GGTase I), an enzyme involved in fungal cell wall biosynthesis. Interestingly, while corticatic acids A and E inhibited *C. albicans* with IC_{50} values of 3.3 and 7.3 μM, the fact that there is little sequence identity between human and *Candida* GGTase I suggested that these marine compounds may become leads for novel and "selective anti-fungal agents."

Swinhoeiamide A (Figure 14.8), a novel antifungal calyculin derivative, was isolated from the marine sponge *Theonella swinhoei* (Edrada et al., 2002). Swinhoeiamide A showed strong antifungal activity toward *C. albicans* and *A. fumigatus* (MIC = 1.2 and 1.0 μg/mL, respectively).

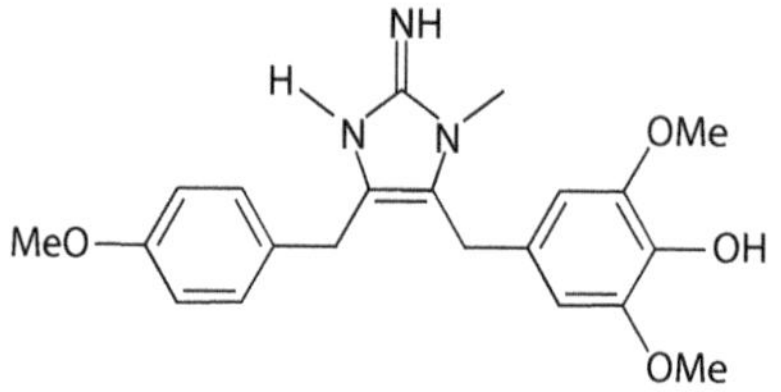

Figure 14.9 Chemical structure of novel imidazole alkaloid, naamine G, isolated from marine sponge *Leucetta chagosensis*. (From National Library of Medicine ChemIDPlus.)

Yang et al. (2003) reported a new sterol sulfate isolated from a deep water marine sponge of the family Astroscleridae, which exhibited antifungal activity against "super-sensitive" *Saccharomyces cerevisiae* (MIC = 15 μg/mL). Jacob et al. (2003) investigated the antifungal properties of a previously described sterol isolated from the marine sponge *Dysidea arenaria*. Interestingly, they observed a reversal of fluconazole resistance from 300 to 8.5 μM when combined with 3.8 μM of the *Dysidea arenaria* sterol, putatively as a result of inhibition of the MDR1-type efflux pump in multidrug-resistant *C. albicans*.

A novel imidazole alkaloid, naamine G (Figure 14.9), was reported from the Indonesian marine sponge *Leucetta chagosensis* that exhibited strong antifungal activity against the phytopathogenic fungus *Cladosporium herbarum* (Hassan et al., 2004).

It remains to be determined if this compound will also be effective against fungi that infect mammalian hosts. Rifai et al. (2004) reported that untenospongin B (Figure 14.10a), isolated from the Moroccan marine sponge *Hippospongia communis*, was more potent than amphotericin B, a clinically used antifungal agent, against *Candida tropicalis* (MIC = 4–8 μg/mL) and *Fusarium oxysporum* (MIC = 2–4 μg/mL). Kossuga et al. (2004) reported a new anti-fungal agent polyketide, (2S,3R)-2-aminododecan-3-ol (Figure 14.10b), isolated from the Brazilian ascidian *Clavelina oblonga*, which was very active against *C. albicans* (MIC = 0.7 ± 0.05 μg/mL). Although the mechanism of action of this compound remains undetermined, its bioactivity was comparable to the clinically used antifungal agents nystatin (MIC = 1–4 μg/mL) and ketoconazole (MIC = 1–4 μg/mL).

Sionov et al. (2005) observed that a phenol compound (Figure 14.11) from the marine sponge *Dysidea herbacea* had significant activity against the human fungal pathogens *C. albicans* and *Aspergillus fumigatus* (MIC = 1.95–7.8 μg/mL), and the activity is comparable with the clinically used antifungal amphotericin B (MIC = 1–2 μg/mL).

Pettit et al. (2005) extended the *in vitro* and *in vivo* pharmacology of the marine spon-gistatin 1 (Figure 14.12) isolated from the marine sponge *Hyrtios erecta*, a previously described anticancer agent. The macrocyclic lactone polyether was shown to be fungi-cidal to 74 reference strains and clinical isolates (MIC = 1–32 μg/mL), including several fungal strains resistant to the clinically used drugs flucytosine, ketoconazole, and fluco-nazole. Furthermore, mechanism of action studies revealed that spongistatin disrupted

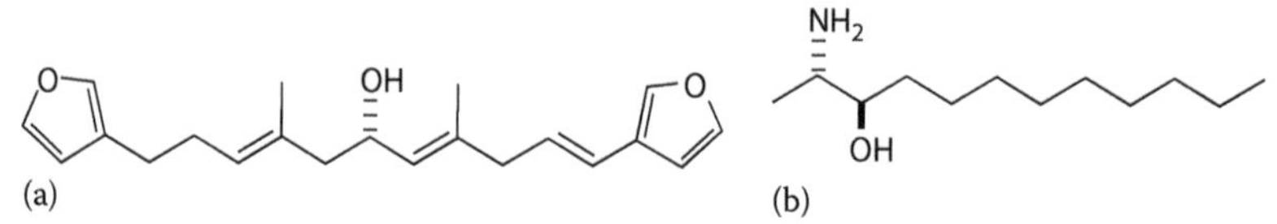

Figure 14.10 Chemical structure of (a) untenospongin B, isolated from the Moroccan marine sponge *Hippospongia communis*. (b) (2S,3R)-2-aminododecan-3-ol, isolated from the Brazilian ascid-ian *Clavelina oblonga*. (From National Library of Medicine ChemIDPlus.)

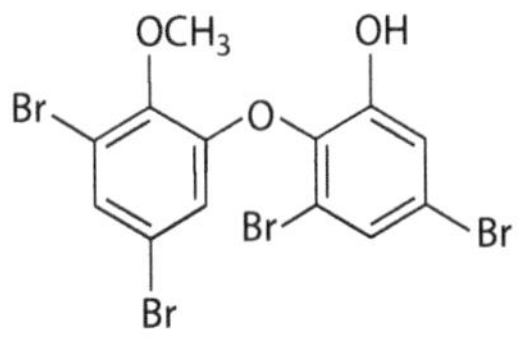

Figure 14.11 Chemical structure of a phenol compound extracted from the marine sponge *Dysidea herbacea*. (From National Library of Medicine ChemIDPlus.)

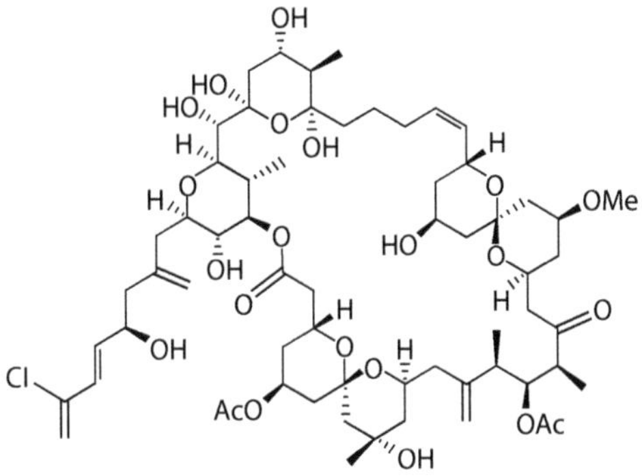

Figure 14.12 Chemical structure of spongistatin 1 isolated from the marine sponge *Hyrtios erecta*. (From National Library of Medicine ChemIDPlus.)

cytoplasmatic and spindle microtubules in *Cryptococcus neoformans* in a time- and concentration-dependent manner, preventing nuclear migration and both nuclear and cellular cell division.

14.4 Antifungal compounds from marine actinomycetes

Actinomycetes are a group of prokaryotic organisms, are gram-positive bacteria that grew extensively in soils rich with organic matter, and are capable of producing several secondary metabolites (Demain and Sanchez, 2009). Majority of the studies on extremophilic organisms, however, have been confined to the isolation and characterization of extremophilic bacteria, but in general, marine actinomycetes are relatively less explored for novel bioactive secondary metabolites till date. The marine environment remains as a virtually untapped source for novel actinomycetes. The distribution and abundance of actinomycetes generally depend on various ecological habitats, which include beach sand and seawater.

An important reason for discovering novel secondary metabolites is to circumvent the problem of resistant pathogens, which are no longer susceptible to the currently used drugs. The number of deaths due to these clever pathogenic organisms is on the rise, which needs to be controlled. Secondary metabolites from marine actinomycetes may form the basis for the synthesis of novel therapeutic drugs, which may be efficient to combat a range of resistant microbes. The exploitation of marine actinomycetes as a source for discovery of novel secondary metabolites yielded numerous novel antifungal metabolites in the past decade.

Bonactin (Figure 14.13), an antimicrobial ester, was isolated from the liquid culture of a *Streptomyces* sp. BD21–2 obtained from a shallow-water sediment sample collected at Kailua Beach, Oahu, Hawaii, (USA) (Schumacher et al., 2003). Bonactin displays antimicrobial activity against gram-positive and gram-negative bacteria as well as against several fungi.

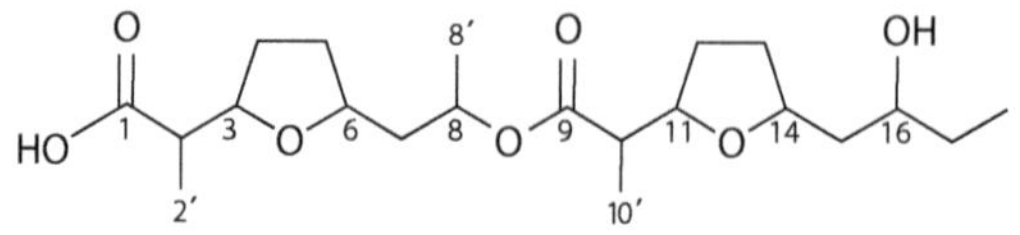

Figure 14.13 Chemical structure of bonactin—an ester compound isolated from *Streptomyces* sp. BD21–2. (From National Library of Medicine ChemIDPlus.)

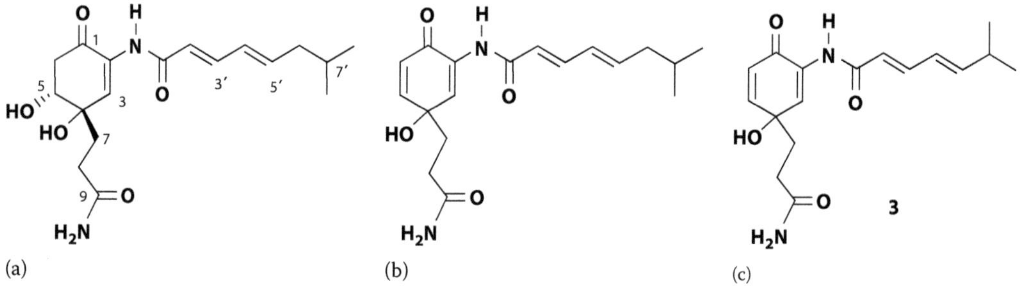

(a)   (b)   (c)

Figure 14.14 Daryamide A (a), daryamide B (b), and daryamide C (c), a polyketides isolated from *Streptomyces* sp. CNQ-085.

Daryamides (Figure 14.14) are cytotoxic and antifungal polyketides isolated from culture broth of a *Streptomyces* strain, CNQ-085. These bioactive compounds have been shown to exhibit moderate cytotoxicity against the human colon carcinoma cell line HCT-116 and moderate antifungal activities against *Candida albicans* (Asolkar et al., 2006). Similarly, chandrananimycins, isolated from marine *Actinomadura* sp. MO48, have been shown to exhibit antibacterial, anticancer, and antifungal activities (Maskey et al., 2003).

References

Andriole, V. T. 1999. Current and future antifungal therapy: New targets for antifungal agents. *J Antimicrob Chemother*, 44(2), 151–162.

Asolkar, R. N., Jensen, P. R., Kauffman, C. A., and Fenical, W. 2006. Daryamides AC, weakly cytotoxic polyketides from a marine-derived actinomycete of the genus *Streptomyces* strain CNQ-085. *J Nat Prod*, 69(12), 1756–1759.

Baddley, J. W., Stroud, T. P., Salzman, D., and Pappas, P. G. 2001. Invasive mold infections in allogeneic bone marrow transplant recipients. *Clin Infect Dis*, 32(9), 1319–1324.

Balajee, S. A., Weaver, M., Imhof, A., Gribskov, J., and Marr, K. A. 2004. *Aspergillus fumigatus* variant with decreased susceptibility to multiple antifungals. *Antimicrob Agents Chemother*, 48(4), 1197–1203.

Barsby, T., Kelly, M. T., and Andersen, R. J. 2002. Tupuseleiamides and basiliskamides, new acyldipeptides and antifungal polyketides produced in culture by a *Bacillus laterosporus* isolate obtained from a tropical marine habitat. *J Nat Prod*, 65(10), 1447–1451.

Basich, J. E., Graves, T. S., Baz, M. N., Scanlon, G., Hoffmann, R. G., Patterson, R., and Fink, J. N. 1981. Allergic bronchopulmonary aspergillosis in corticosteroid-dependent asthmatics. *J Allergy Clin Immunol*, 68(2), 98–102.

Binder, R. E., Faling, L. J., Pugatch, R. D., Mahasaen, C., and Snider, G. L. 1982. Chronic necrotizing pulmonary aspergillosis: A discrete clinical entity. *Medicine*, 61(2), 109.

Boogaerts, M. and Maertens, J. 2001. Clinical experience with itraconazole in systemic fungal infections. *Drugs*, 61(1), 39–47.

Butler, M. S. 2005. Natural products to drugs: Natural product derived compounds in clinical trials. *Nat Prod Rep*, 22(2), 162–195.

Cabib, E. 1991. Differential inhibition of chitin synthetases 1 and 2 from *Saccharomyces cerevisiae* by polyoxin D and nikkomycins. *Antimicrob Agents Chemother, 35*(1), 170–173.

Chamilos, G. and Kontoyiannis, D. P. 2005. Update on antifungal drug resistance mechanisms of *Aspergillus fumigatus. Drug Resist Update, 8*(6), 344–358.

Cohen, E. 1993. Chitin synthesis and degradation as targets for pesticide action. *Arch Insect Biochem Physiol, 22*(1–2), 245–261.

Cowen, L. E. and Lindquist, S. 2005. Hsp90 potentiates the rapid evolution of new traits: Drug resistance in diverse fungi. *Science, 309*(5744), 2185–2189.

Dannaoui, E., Meletiadis, J., Tortorano, A. M., Symoens, F., Nolard, N., Viviani, M. A., and Grillot, R. 2004. Susceptibility testing of sequential isolates of *Aspergillus fumigatus* recovered from treated patients. *J Med Microbiol, 53*(2), 129–134.

Demain, A. L. 1983. New applications of microbial products. *Science, 219*(4585), 709–714.

Demain, A. L. and Sanchez, S. 2009. Microbial drug discovery: 80 years of progress. *J Antibiot, 62*(1), 5–16.

Edrada, R. A., Ebel, R., Supriyono, A., Wray, V., Schupp, P., Steube, K., van Soest, R., and Proksch, P. 2002. Swinhoeiamide A, a new highly active calyculin derivative from the marine sponge *Theonella swinhoei. J Nat Prod, 65*(8), 1168–1172.

Fernández, R., Dherbomez, M., Letourneux, Y., Nabil, M., Verbist, J. F., and Biard, J. F. 1999. Antifungal metabolites from the marine sponge Pachastrissa sp.: New bengamide and bengazole derivatives. *J Nat Pro, 62*(5), 678–680.

Fromtling, R. A. 1998. Human mycoses and current antifungal therapy. *Drug News Perspect, 11*(3), 185.

Galgiani, J. N. 1990. Fluconazole, a new antifungal agent. *Ann Intern Med, 113*(3), 177–179.

Gefter, W. B., Weingrad, T. R., Epstein, D. M., Ochs, R. H., and Miller, W. T. 1981. "Semi-invasive" pulmonary aspergillosis: A new look at the spectrum of *Aspergillus* infections of the lung. *Radiology, 140*(2), 313–321.

Grant, S. M. and Clissold, S. P. 1990. Fluconazole. A review of its pharmacodynamic and pharmacokinetic properties, and therapeutic potential in superficial and systemic mycoses. *Drugs, 39*, 877–916.

Groll, A. H., Piscitelli, S. C., and Walsh, T. J. 1998. Clinical pharmacology of systemic antifungal agents: A comprehensive review of agents in clinical use, current investigational compounds, and putative targets for antifungal drug development. *Adv Pharmacol, 44*, 343–500.

Guinea, J., Recio, S., Peláez, T., Torres-Narbona, M., and Bouza, E. 2008. Clinical isolates of *Aspergillus* species remain fully susceptible to voriconazole in the post-voriconazole era. *Antimicrob Agents Chemother, 52*, 3444–3446.

Hassan, W., Edrada, R., Ebel, R., Wray, V., Berg, A., van Soest, R., Wiryowidagdo, S., and Proksch, P. 2004. New imidazole alkaloids from the Indonesian sponge *Leucetta chagosensis. J Nat Prod, 67*(5), 817–822.

Hibberd, P. L. and Rubin, R. H. 1994. Clinical aspects of fungal infection in organ transplant recipients. *Clin Infect Dis, 19*(Supplement 1), S33–S40.

Hiemenz, J. W. and Walsh, T. J. 1996. Lipid formulations of amphotericin B: Recent progress and future directions. *Clin Infect Dis, 22*(Supplement 2), S133–S144.

Ho, P. L. and Yuen, K. Y. 2000. Aspergillosis in bone marrow transplant recipients. *Crit Rev Oncol Hematol, 34*(1), 55–69.

Jacob, M. R., Hossain, C. F., Mohammed, K. A., Smillie, T. J., Clark, A. M., Walker, L. A., and Nagle, D. G. 2003. Reversal of fluconazole resistance in multidrug efflux-resistant fungi by the dysidea a renaria sponge sterol 9α, 11α-epoxycholest-7-ene-3β, 5α, 6α, 19-tetrol 6-acetate. *J Nat Prod, 66*(12), 1618–1622.

Johnson, E. M., Oakley, K. L., Radford, S. A., Moore, C. B., Warn, P., Warnock, D. W., and Denning, D. W. 2000. Lack of correlation of in vitro amphotericin B susceptibility testing with outcome in a murine model of *Aspergillus* infection. *J Antimicrob Chemother, 45*(1), 85–93.

Kontoyiannis, D. P. 2001. A clinical perspective for the management of invasive fungal infections: Focus on IDSA guidelines. *Pharmacotherapy, 21*(8P2), 175S–187S.

Kontoyiannis, D. P. and Bodey, G. 2002. Invasive aspergillosis in 2002: An update. *Eur J Clin Microbiol Infect Dis, 21*(3), 161–172.

Kossuga, M. H., MacMillan, J. B., Rogers, E. W., Molinski, T. F., Nascimento, G. G., Rocha, R. M., and Berlinck, R. G. 2004. (2S,3R)-2-aminododecan-3-ol, a new antifungal agent from the ascidian *Clavelina oblonga*. *J Nat Prod, 67*(11), 1879–1881.

Kumar, S. and Kannabiran, K. 2010. Antifungal activity of *Streptomyces* VITSVK5 spp. against drug resistant *Aspergillus* clinical isolates from pulmonary tuberculosis patients. *J Med Mycol, 20*(2), 101–107.

Lamb, D., Kelly, D., and Kelly, S. 1999. Molecular aspects of azole antifungal action and resistance. *Drug Resist Update, 2*(6), 390–402.

Lionakis, M. S., Lewis, R. E., Torres, H. A., Albert, N. D., Raad, I. I., and Kontoyiannis, D. P. 2005. Increased frequency of non-fumigatus *Aspergillus* species in amphotericin B- or triazole–pre-exposed cancer patients with positive cultures for *Aspergilli*. *Diagn Microbiol Infect Dis, 52*(1), 15–20.

Liu, W., Lionakis, M. S., Lewis, R. E., Wiederhold, N., May, G. S., and Kontoyiannis, D. P. 2003. Attenuation of itraconazole fungicidal activity following preexposure of *Aspergillus fumigatus* to fluconazole. *Antimicrob Agents Chemother, 47*(11), 3592–3597.

Manavathu, E. K., Cutright, J. L., Loebenberg, D., and Chandrasekar, P. H. 2000. A comparative study of the in vitro susceptibilities of clinical and laboratory-selected resistant isolates of Aspergillus spp. to amphotericin B, itraconazole, voriconazole and posaconazole (SCH 56592). *J Antimicrob Chemother, 46*(2), 229–234.

Manavathu, E. K., Cutright, J. L., and Chandrasekar, P. H. 2003. In vivo resistance of a laboratory-selected *Aspergillus fumigatus* isolate to Amphotericin B. *Antimicrob Agents Chemother, 49*(1), 428–430.

Mandala, S. M., Thornton, R. A., Frommer, B. R., Curotto, J. E., Rozdilsky, W., Kurtz, M. B., Giacobbe, R. A. et al. 1995. The discovery of australifungin, a novel inhibitor of sphinganine N-acyltransferase from *Sporormiella australis*. Producing organism, fermentation, isolation, and biological activity. *J Antibiotics, 48*(5), 349–356.

Marr, K. A., Carter, R. A., Crippa, F., Wald, A., and Corey, L. 2002. Epidemiology and outcome of mould infections in hematopoietic stem cell transplant recipients. *Clin Infect Dis, 34*(7), 909–917.

Maskey, R. P., Li, F. C. S., Qin, H. H., Fiebig, and Laatsch, H. 2003. Chandrananimycins A–C: Production of novel anticancer antibiotics from a marine *Actinomadura* sp. isolate M048 by variation of medium composition and growth conditions. *J Antibiot, 56*, 622–629.

Mayer, A. M. and Hamann, M. T. 2002. Marine pharmacology in 1999: Compounds with antibacterial, anticoagulant, antifungal, anthelmintic, anti-inflammatory, antiplatelet, antiprotozoal and antiviral activities affecting the cardiovascular, endocrine, immune and nervous systems, and other miscellaneous mechanisms of action. *Comp Biochem Physiol, 132*(3), 315–339.

Mayer, A. M. and Hamann, M. T. 2005. Marine pharmacology in 2001–2002: Marine compounds with anthelmintic, antibacterial, anticoagulant, antidiabetic, antifungal, anti-inflammatory, antimalarial, antiplatelet, antiprotozoal, antituberculosis, and antiviral activities; affecting the cardiovascular, immune and nervous systems and other miscellaneous mechanisms of action. *Comp Biochem Physiol C Pharmacol 140*(3), 265–286.

Mayer, A. M. and Lehmann, V. K. 2000. Marine pharmacology in 1998: Marine compounds with antibacterial, anticoagulant, antifungal, anti-inflammatory, anthelmintic, antiplatelet, antiprotozoal, and antiviral activities; with actions on the cardiovascular, endocrine, immune, and nervous systems; and other miscellaneous mechanisms of action. *Pharmacologist, 42*, 62–69.

Mayer, A. M., Rodríguez, A. D., Berlinck, R. G., and Hamann, M. T. 2007. Marine pharmacology in 2003–4: Marine compounds with anthelmintic antibacterial, anticoagulant, antifungal, anti-inflammatory, antimalarial, antiplatelet, antiprotozoal, antituberculosis, and antiviral activities; affecting the cardiovascular, immune and nervous systems, and other miscellaneous mechanisms of action. *Comp Biochem Physiol C Pharmacol, 145*(4), 553–581.

Moore, C. B., Sayers, N., Mosquera, J., Slaven, J., and Denning, D. W. 2000. Antifungal drug resistance in *Aspergillus*. *J Infect, 41*(3), 203–220.

Moosa, M. Y. S., Alangaden, G. J., Manavathu, E., and Chandrasekar, P. H. 2002. Resistance to amphotericin B does not emerge during treatment for invasive aspergillosis. *J Antimicrob Chemother, 49*(1), 209–213.

Mroueh, S. and Spock, A. 1994. Allergic bronchopulmonary aspergillosis in patients with cystic fibrosis. *Chest, 105*(1), 32–36.

Newman, D. J., Cragg, G. M., and Snader, K. M. 2003. Natural products as sources of new drugs over the period 1981–2002. *J Nat Prod, 66*(7), 1022–1037.

Nicholas, G. M., Hong, T. W., Molinski, T. F., Lerch, M. L., Cancilla, M. T., and Lebrilla, C. B. 1999. Oceanapiside, an antifungal bis-α,ω-amino alcohol glycoside from the marine sponge *Oceanapia phillipensis. J Nat Prod, 62*(12), 1678–1681.

Nishimura, S., Matsunaga, S., Shibazaki, M., Suzuki, K., Harada, N., Naoki, H., and Fusetani, N. 2002. Corticatic acids D and E, polyacetylenic geranylgeranyltransferase type I inhibitors, from the marine sponge *Petrosia corticata. J Nat Prod, 65*(9), 1353–1356.

Odds, F. C., Brown, A. J., and Gow, N. A. 2003. Antifungal agents: Mechanisms of action. *Trends Microbiol, 11*(6), 272–279.

Omura, S. 1986. Philosophy of new drug discovery. *Microbiol Rev, 50*(3), 259.

Ovechkina, Y. Y., Pettit, R. K., Cichacz, Z. A., Pettit, G. R., and Oakley, B. R. 1999. Unusual antimicrotubule activity of the antifungal agent spongistatin 1. *Antimicrob Agents Chemother, 43*(8), 1993–1999.

Paterson, P. J., Seaton, S., Prentice, H. G., and Kibbler, C. C. 2003. Treatment failure in invasive aspergillosis: Susceptibility of deep tissue isolates following treatment with amphotericin B. *J Antimicrob Chemother, 52*(5), 873–876.

Patterson, T. F. 2002. New agents for treatment of invasive aspergillosis. *Clin Infect Dis, 35*(4), 367–369.

Patterson, T. F. 2005. Advances and challenges in management of invasive mycoses. *Lancet, 366*(9490), 1013–1025.

Pettit, R. K., Woyke, T., Pon, S., Cichacz, Z. A., Pettit, G. R., and Herald, C. L. 2005. *In vitro* and *in vivo* antifungal activities of the marine sponge constituent spongistatin. *Med Mycol, 43*(5), 453–463.

Pfaller, M. A., Pappas, P. G., and Wingard, J. R. 2006. Invasive fungal pathogens: Current epidemiological trends. *Clin Infect Dis, 43*(Supplement 1), S3–S14.

Rifai, S., Fassouane, A. F., Kijjoa, A., and Van Soest, R. 2004. Antimicrobial activity of untenospongin B, a metabolite from the marine sponge *Hippospongia communis* collected from the Atlantic coast of Morocco. *Mar Drugs, 2*(3), 147–153.

Rinaldi, M. 1983. Invasive aspergillosis. *Rev Infect Dis, 5,* 1061–1077.

Ryder, N. S. 1992. Terbinafine: Mode of action and properties of the squalene epoxidase inhibition. *Br J Derm, 126,* 2–7.

Schiraldi, G. F., Lo Cicero, S., Colombo, M. D., Rossato, D., Ferrarese, M., and Soresi, E. 1996. Refractory pulmonary aspergillosis: Compassionate trial with terbinafine. *Br J Derm, 134*(s46), 25–29.

Schumacher, R. W., Talmage, S. C., Miller, S. A., Sarris, K. E., Davidson, B. S., and Goldberg, A. 2003. Isolation and structure determination of an antimicrobial ester from a marine sediment-derived bacterium. *J Nat Prod, 66*(9), 1291–1293.

Shao, P. L., Huang, L. M., and Hsueh, P. R. 2007. Recent advances and challenges in the treatment of invasive fungal infections. *Int J Antimicrob Agents, 30*(6), 487–495.

Sionov, E., Roth, D., Sandovsky-Losica, H., Kashman, Y., Rudi, A., Chill, L., Berdicevsky, I., and Segal, E. 2005. Antifungal effect and possible mode of activity of a compound from the marine sponge *Dysidea herbacea. J Infect, 50*(5), 453–460.

Soubani, A. O., Miller, K. B., and Hassoun, P. M. 1996. Pulmonary complications of bone marrow transplantation. *Chest, 109*(4), 1066–1077.

Sterling, T. R. and Merz, W. G. 1998. Resistance to amphotericin B: Emerging clinical and microbiological patterns. *Drug Resist Update, 1*(3), 161–165.

Swerdlow, J. L. 2000. *Nature's Medicine—Plants that Heal.* Random House, Inc., New York, p. 56.

Tomee, J. F., van der Werf, T. S., Latge, J. P., Koeter, G. H., Dubois, A. E., and Kauffman, H. F. 1995. Serologic monitoring of disease and treatment in a patient with pulmonary aspergilloma. *Am J Respir Crit Care Med, 151*(1), 199–204.

Tsukamoto, S., Matsunaga, S., Fusetani, N., and Toh-e, A. 1999. Theopederins FJ: Five new antifungal and cytotoxic metabolites from the marine sponge, *Theonella swinhoei. Tetrahedron, 55*(48), 13697–13702.

Vehreschild, J. J., Rüping, M. J. G. T., Wisplinghoff, H., Farowski, F., Steinbach, A., Sims, R., and Cornely, O. A. 2010. Clinical effectiveness of posaconazole prophylaxis in patients with acute myelogenous leukaemia (AML): A 6 year experience of the Cologne AML cohort. *J Antimicrob Chemother*, 65(7), 1466–1471.

Verweij, P. E., Oakley, K. L., Morrissey, J., Morrissey, G., and Denning, D. W. 1998. Efficacy of LY303366 against amphotericin B-susceptible and-resistant *Aspergillus fumigatus* in a murine model of invasive aspergillosis. *Antimicrob Agents Chemother*, 42(4), 873–878.

Vicente, M. F., Basilio, A., Cabello, A., and Pelaez, F. 2003. Microbial natural products as a source of antifungals. *Clin Microbiol Infect*, 9(1), 15–32.

Vijayakumar, E. K. S., Roy, K., Chatterjee, S., Deshmukh, S. K., Ganguli, B. N., Fehlhaber, H. W., and Kogler, H. 1996. Arthrichitin. A new cell wall active metabolite from *Arthrinium phaeospermum*. *J Org Chem*, 61(19), 6591–6593.

Wald, A., Leisenring, W., van Burik, J. A., and Bowden, R. A. 1997. Epidemiology of *Aspergillus* infections in a large cohort of patients undergoing bone marrow transplantation. *J Infect Dis*, 175(6), 1459–1466.

Walsh, T. J., Petraitis, V., Petraitiene, R., Field-Ridley, A., Sutton, D., Ghannoum, M. et al. 2003. Experimental pulmonary aspergillosis due to *Aspergillus terreus*: Pathogenesis and treatment of an emerging fungal pathogen resistant to amphotericin B. *J Infect Dis*, 188(2), 305–319.

Warn, P. A., Morrissey, G., Morrissey, J., and Denning, D. W. 2003. Activity of micafungin (FK463) against an itraconazole-resistant strain of *Aspergillus fumigatus* and a strain of *Aspergillus terreus* demonstrating *in vivo* resistance to amphotericin B. *J Antimicrob Chemother*, 51(4), 913–919.

Warris, A., Weemaes, C. M., and Verweij, P. E. 2002. Multidrug resistance in *Aspergillus fumigatus*. *N Engl J Med*, 347(26), 2173–2174.

Wells, G. B. and Lester, R. L. 1983. The isolation and characterization of a mutant strain of *Saccharomyces cerevisiae* that requires a long chain base for growth and for synthesis of phosphosphingolipids. *J Biol Chem*, 258(17), 10200–10203.

Xiao, L., Madison, V., Chau, A. S., Loebenberg, D., Palermo, R. E., and McNicholas, P. M. 2004. Three-dimensional models of wild-type and mutated forms of cytochrome P450 14α-sterol demethylases from *Aspergillus fumigatus* and *Candida albicans* provide insights into posaconazole binding. *Antimicrob Agents Chemother*, 48(2), 568–574.

Yang, S. W., Chan, T. M., Pomponi, S. A., Chen, G., Loebenberg, D., Wright, A., Patel, M., Gullo, V., Pramanik, B., and Chu, M. 2003. Structure elucidation of a new antifungal sterol sulfate, Sch 575867, from a deep-water marine sponge (Family: Astroscleridae). *J Antibiot (Tokyo)* 56(2), 186–189.

Zweerink, M. M., Edison, A. M., Wells, G. B., Pinto, W., and Lester, R. L. 1992. Characterization of a novel, potent, and specific inhibitor of serine palmitoyltransferase. *J Biol Chem*, 267(35), 25032–25038.

section two

Broad spectrum antimicrobial compounds from animals

chapter fifteen

Secondary metabolites from microorganisms isolated from marine sponges from 2000 to 2012

Mohammad F. Mehbub, Christopher M.M. Franco, and Wei Zhang

Contents

15.1   Introduction

The continuously developing resistance of pathogenic bacteria, the reemergence of viral diseases, and cancers that are still incurable make us redouble our efforts to find cures to alleviate human vulnerabilities. The introduction of new drugs is crucial, as older antibiotics and drugs begin to lose their efficacy. Rationally designed drugs have made inroads into the pharmaceutical industry, but natural products continue to introduce the chemical diversity required to maintain a contribution of around 60% of the drugs, directly or after chemical modification, which are available in the market (Newman and Cragg, 2007). Microbial natural products contribute more than 40% of new chemical entities reported between 1981 and 2010 (Newman et al., 2003; Baltz et al., 2005; Koehn and Carter, 2005; Fisher, 2014).

Whereas the majority of microorganisms that produce valuable products have been obtained predominantly from terrestrial sources, there has been a recent trend to collect microorganisms that are associated with other life forms such as endophytes of plants (Govindasamy et al., 2014). There has also been an evolving realization that the oceans, which cover a larger proportion of the earth's surface, have a different microbial diversity, and here too there are associations with marine life forms such as sponges. It was noted that due to the aqueous milieu, their metabolites have more potent activities and are usually structurally very different from those found from terrestrial-based samples. Recent studies have borne out this hypothesis, and the marine environment is proving to be an eclectic source of novel chemical diversity that is contributing to drug discovery. Many bioactive substances have been isolated from a variety of marine organisms, including phytoplankton, bryozoans, algae, sponges, tunicates, and molluscs (Faulkner, 2002; Proksch et al., 2002; Zhang et al., 2005; Mehbub et al., 2014). Microorganisms associated with these animals have shown the ability to adapt, and this adaptation capacity may be the key reason for their secondary metabolite production capacity (Piel, 2004, 2009; König et al., 2006; Brady et al., 2009; Valliappan et al., 2014).

Marine sponges (phylum Porifera) are of particular interest because they are remarkable filter feeders; some can filter 24 m^3 kg^{-1} sponge day^{-1} (Vogel, 1977). During the filtration process, they concentrate bacterial cells that are otherwise diluted in seawater. They harbor dense and diverse microbial consortia, which comprise as much as 40% of sponge tissue volume and span all three domains of life (Taylor et al., 2007). This makes sponges excellent models for the study of marine host-associated bacteria as they represent a substantial reservoir of novel microbial diversity (Taylor et al., 2004). Therefore, in recent years, the search has intensified for microorganisms from sponges with the expectation that novel compounds will result from their screening.

There are more than 8500 sponge species (Van Soest et al., 2012) that contain very diverse microbial consortia (Taylor et al., 2004); and, it has been reported that individual species from sponges during the past decade contains at least a few different types of bioactive natural products (Mehbub et al., 2014). Therefore, sponges could be termed the "drugstore of the sea" (Blunt et al., 2009). Many of these compounds likely serve as agents of defense that protect the immobile animals from being overgrown or ingested (Pawlik, 1992; Paul and Ritson-Williams, 2008), but for most substances, an ecological function has not been experimentally demonstrated. Likewise, the often-stated question whether sponge-derived natural products are biosynthesized by sponges or by associated microorganisms remains largely unanswered (Faulkner et al., 1993; Piel et al., 2004). Insights into this issue could have a significant impact on marine pharmacology. For most compounds, drug development is currently not possible due to a limited access to the biological material. If the actual source is a bacterium, supply could be ensured by cultivating the producer or by isolating the biosynthetic genes and expressing the pathway in culturable bacteria (Piel, 2006; Schmitt et al., 2008). The advantage of the latter approach is that it should be generally applicable to a wide range of compounds independent of cultivation. Although the genetic tools to express bacterial pathways are in principle available (Fujii, 2009), the application of this strategy to sponge symbionts is currently highly challenging for several reasons.

This review focuses on microorganisms from sponges that have been reported to produce secondary metabolites or bioactive compounds from 2000 to 2012. The microorganisms reported here are the bacteria, with actinobacteria looked at separately due to their recognized ability to produce a wide range of secondary metabolites, and fungi, including yeast. Data have been compiled from the published literature and data reviewed by Faulkner (2002) and Blunt et al. (2003, 2004, 2005, 2006, 2007, 2008, 2009, 2010, 2011, 2012, 2013, 2014) from Natural Product Reports.

Species isolated previously from marine sediments but subsequently reported from sponges, for example, *Salinispora* strains isolated from the Great Barrier Reef sponge *Pseudoceratina clavata* (Kim et al., 2005), will not be included.

15.2 Microbial sources

The microbial populations include archaea and bacteria (Webster et al., 2001; Hentschel et al., 2003), fungi (Höller et al., 2000), cyanobacteria (Thacker and Starnes, 2003), unicellular algae (Vacelet, 1981), dinoflagellates (Garson et al., 1998), and actinobacteria (Maldonado et al., 2005b), which make up at least half the tissue volume in some sponge species (Vacelet and Donadey, 1977; Hentschel et al., 2003).

Members of the phylum Actinobacteria and specifically the order Actinomycetales have been identified as abundant members of sponge-associated microbial communities (Hentschel et al., 2002; Zhang et al., 2006). However, in this survey, it is evident that there has been a shift from actinobacteria to fungi as the main microorganisms. The number of fungi from sponge samples reported as producers of new compounds is now at least three times higher compared to the reports for actinobacteria from the same source.

Over the 2000–2012 period, a total of 269 new compounds were isolated from sponge-associated microbes (Figure 15.1). Of these, 186 new compounds were isolated from 27 genera of fungi including two compounds from one yeast genus, compared to just 56 new compounds from seven actinobacterial genera and 27 new compounds from seven bacterial genera (Table 15.1). Confirming the switch to fungal sources are the 69 publications on fungal metabolites compared to 20 for actinobacteria and 15 for bacteria during this review period. However, our understanding of the sponge-associated fungal, actinobacterial, and bacterial communities and their structures is still inadequate.

15.2.1 Marine fungi

Fungi from marine sponges are now the primary source of novel metabolites from microbial sources that have industrial as well as medicinal values. By 1998, more than 100 metabolites from marine-derived fungi had been described by different researchers (Biabani and Laatsch, 1998). However, the majority of the reports focused on natural product

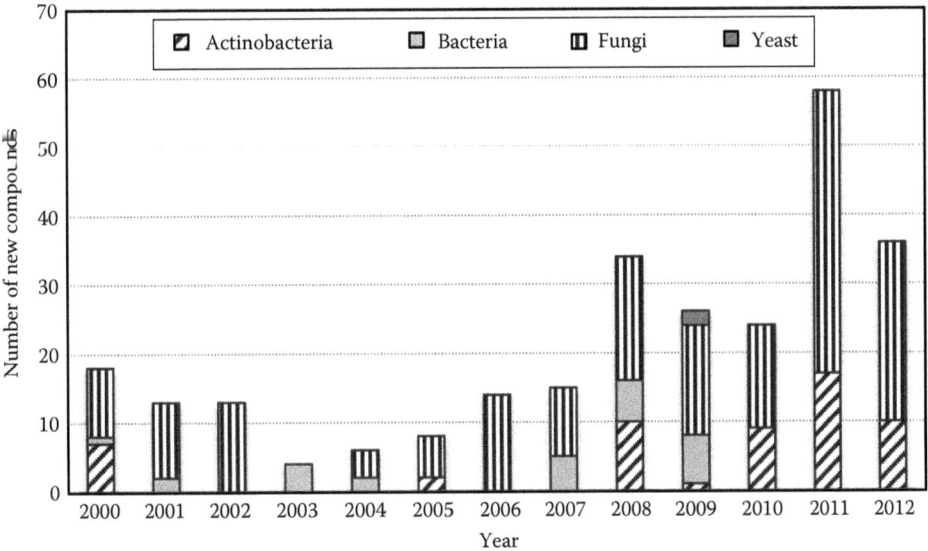

Figure 15.1 The total number of new compounds isolated from sponge-associated microorganisms from 2000 to 2012.

Table 15.1 New compounds isolated from microorganisms from marine sponges from 2000 to 2012: their chemical class, country of collection, and reported activity

Compound name	Chemical class	Country of collection	Name of microorganism	Type of microorganism	Name of sponge	Reported activity	References
GGL.1 1,2-O-diacyl-3-[α-glucopyranosyl-(1–6)-α-glucopyranosyl)]glycerol	Glycoglycerolipid	Croatia	*Microbacterium* sp.	Actinobacterium	*Halichondria panicea*	Anticancer	Wicke et al. (2000)
GGL.2 1-O-acyl-3-[R-glucopyranosyl-(1–3)-(6-O-acyl-R-mannopyranosyl)] glycerol	Glycoglycerolipid	Croatia	*Microbacterium* sp.	Actinobacterium	*Halichondria panicea*	Anticancer	Wicke et al. (2000)
GGL.3 1-O-acyl-3-[6-O-acetyl-R-glucopyranosyl-(1–3)-(6-O-acyl-R-mannopyranosyl)] glycerol	Glycoglycerolipid	Croatia	*Microbacterium* sp.	Actinobacterium	*Halichondria panicea*	Anticancer	Wicke et al. (2000)
GGL.4 1,2-O-diacyl-3-[α-galactofuranosyl)]glycerol	Glycoglycerolipid	Croatia	*Microbacterium* sp.	Actinobacterium	*Halichondria panicea*	Anticancer	Wicke et al. (2000)
DPG tetraacyldiphosphatidyl glycerol	Diphosphatidylglycerol	Croatia	*Microbacterium* sp.	Actinobacterium	*Halichondria panicea*	Anticancer	Wicke et al. (2000)
Compound 1 ($C_{25}H_{31}N_3O_6$)/β-aminopimelic acid	Tripeptide	Bulgaria	*Pseudomonas/ Alteromonas*	Bacterium	*Dysidea fragilis*	Antiviral	De Rosa et al. (2000)
4′-N-Methyl-5′-hydroxystaurosporine	Indolocarbazole alkaloid	Spain	*Micromonospora* sp.	Actinobacterium	*Clathrina coriacea*	Cytotoxic	Hernández et al. (2000)
5′-Hydroxystaurosporine	Indolocarbazole alkaloid	Spain	*Micromonospora* sp.	Actinobacterium	*Clathrina coriacea*	Cytotoxic	Hernández et al. (2000)
Iso-cladospolide B	Hexaketide	Indonesia	Not identified	Fungus	Not identified	NR	Smith et al. (2000)
Seco-patulolide C	Hexaketide	Indonesia	Not identified	Fungus	Not identified	NR	Smith et al. (2000)
Pandangolide 1	Polyketide	Indonesia	Not identified	Fungus	Not identified	NR	Smith et al. (2000)
Pandangolide 2	Polyketide	Indonesia	Not identified	Fungus	Not identified	NR	Smith et al. (2000)
Spiciferone A	Spiciferone derivative	Indonesia	*Drechslera hawaiiensis*	Fungus	*Callyspongia aerizusa*	NR	Edrada et al. (2000)
Spiciferone B	Spiciferone derivative	Indonesia	*Drechslera hawaiiensis*	Fungus	*Callyspongia aerizusa*	NR	Edrada et al. (2000)

(Continued)

Table 15.1 (Continued) New compounds isolated from microorganisms from marine sponges from 2000 to 2012: their chemical class, country of collection, and reported activity

Compound name	Chemical class	Country of collection	Name of microorganism	Type of microorganism	Name of sponge	Reported activity	References
Spiciferol A	Spiciferone derivative	Indonesia	*Drechslera hawaiiensis*	Fungus	*Callyspongia aerizusa*	NR	Edrada et al. (2000)
Spiciferone A (1)	Spiciferone derivative	Indonesia	*Drechslera hawaiiensis*	Fungus	*Callyspongia aerizusa*	NR	Edrada et al. (2000)
Butoxyl-spiciferin	Spiciferone derivative	Indonesia	*Drechslera hawaiiensis*	Fungus	*Callyspongia aerizusa*	NR	Edrada et al. (2000)
Asperic acid	Pyran derivative	United States	*Aspergillus niger*	Fungus	*Hyrtios proteus*	NR	Varoglu and Crews (2000)
Nafuredin	Lactone	Palau	*Aspergillus niger*	Fungus	Not identified	Cytotoxic	Takano et al. (2001)
2-Acetamido-2-deoxy-D-galacturonic acid	Amino sugar	Russia	*Pseudoalteromonas distincta*	Bacterium	Not identified	NR	Muldoon et al. (2001)
5-Acetamido-3,5,7,9-tetradeoxy-7-formamido-L-glycero-L-manno-nonulosonic acid	Amino sugar	Russia	*Pseudoalteromonas distincta*	Bacterium	Not identified	NR	Muldoon et al. (2001)
Aspergione A	Chromone	Indonesia	*Aspergillus versicolor*	Fungus	*Xestospongia exigua*	NR	Lin et al. (2001a)
Aspergione B	Chromone	Indonesia	*Aspergillus versicolor*	Fungus	*Xestospongia exigua*	NR	Lin et al. (2001a)
Aspergione C	Chromone	Indonesia	*Aspergillus versicolor*	Fungus	*Xestospongia exigua*	NR	Lin et al. (2001a)
Aspergione D	Chromone	Indonesia	*Aspergillus versicolor*	Fungus	*Xestospongia exigua*	NR	Lin et al. (2001a)
Aspergione E	Chromone	Indonesia	*Aspergillus versicolor*	Fungus	*Xestospongia exigua*	NR	Lin et al. (2001a)
Aspergione F	Chromone	Indonesia	*Aspergillus versicolor*	Fungus	*Xestospongia exigua*	NR	Lin et al. (2001a)
Aspergillone	Chromone	Indonesia	*Aspergillus versicolor*	Fungus	*Xestospongia exigua*	NR	Lin et al. (2001b)

(Continued)

Table 15.1 (Continued) New compounds isolated from microorganisms from marine sponges from 2000 to 2012: Their chemical class, country of collection, and reported activity

Compound name	Chemical class	Country of collection	Name of microorganism	Type of microorganism	Name of sponge	Reported activity	References
Aspergillodiol	Chromone	Indonesia	*Aspergillus versicolor*	Fungus	*Xestospongia exigua*	NR	Lin et al. (2001b)
Aspergillol	Chromone	Indonesia	*Aspergillus versicolor*	Fungus	*Xestospongia exigua*	NR	Lin et al. (2001b)
12-Acetyl-aspergillol	Chromone	Indonesia	*Aspergillus versicolor*	Fungus	*Xestospongia exigua*	NR	Lin et al. (2001b)
Lunatin	Anthraquinone	Indonesia	*Curvularia lunata*	Fungus	*Niphates olmeda*	Antibacterial	Jadulco et al. (2002)
Herbarin A	α-Pyrone	France	*Cladosporium herbarum*	Fungus	*Aplysina aerophoba*	Miscellaneous	Jadulco et al. (2002)
Herbarin B	α-Pyrone	France	*Cladosporium herbarum*	Fungus	*Aplysina aerophoba*	Miscellaneous	Jadulco et al. (2002)
Herbaric acid	Phthalide	Indonesia	*Cladosporium herbarum*	Fungus	*Callyspongia aerizusa*	NR	Jadulco et al. (2002)
Varitriol	Macrotetrolide	Venezuela	*Emericella variecolor*	Fungus	Not identified	Anticancer	Malmstrom et al. (2002)
Varioxirane	NR	Venezuela	*Emericella variecolor*	Fungus	Not identified	NR	Malmstrom et al. (2002)
Dihydroterrein	NR	Venezuela	*Emericella variecolor*	Fungus	Not identified	NR	Malmstrom et al. (2002)
Varixanthone	NR	Venezuela	*Emericella variecolor*	Fungus	Not identified	Antimicrobial	Malmstrom et al. (2002)
Xestodecalactone B	Decalactone	Indonesia	*Penicillium* cf. *montanense*	Fungus	*Xestospongia exigua*	Antiyeast	Edrada et al. (2002)
Microsphaerone A	γ-Pyrone	France	*Microsphaeropsis* sp.	Fungus	*Aplysina aerophoba*	Anticancer	Wang et al. (2002)
Microsphaerone B	γ-Pyrone	France	*Microsphaeropsis* sp.	Fungus	*Aplysina aerophoba*	Anticancer	Wang et al. (2002)
Xestodecalactone A	Decalactone	Indonesia	*Penicillium* cf. *montanense*	Fungus	*Xestospongia exigua*	NR	Edrada et al. (2002)

(Continued)

Table 15.1 (Continued) New compounds isolated from microorganisms from marine sponges from 2000 to 2012: Their chemical class, country of collection, and reported activity

Compound name	Chemical class	Country of collection	Name of microorganism	Type of microorganism	Name of sponge	Reported activity	References
Xestodecalactone C	Decalactone	Indonesia	*Penicillium* cf. *montanense*	Fungus	*Xestospongia exigua*	NR	Edrada et al. (2002)
YM-266183	Thiopeptide	Japan	*Bacillus cereus*	Bacterium	*Halichondria japonica*	Antibacterial	Nagai et al. (2003)
YM-266184	Thiopeptide	Japan	*Bacillus cereus*	Bacterium	*Halichondria japonica*	Antibacterial	Nagai et al. (2003)
Pseudoalterobactin A	Siderophore	Palau	*Pseudoalteromonas* sp.	Bacterium	*Cinachyrella australiensis*	NR	Kanoh et al. (2003)
Pseudoalterobactin B	Siderophore	Palau	*Pseudoalteromonas* sp.	Bacterium	*Cinachyrella australiensis*	NR	Kanoh et al. (2003)
Cyclo-(glycyl-L-seryl-L-prolyl-L-glutamyl)	Cyclic peptide	Italy	*Ruegeria* sp.	Bacterium	*Suberites domuncula*	Antibacterial	Mitova et al. (2004)
Cyclo-(glycyl-L-prolyl-L-glutamyl)	Cyclic peptide	Italy	*Ruegeria* sp.	Bacterium	*Suberites domuncula*	Antibacterial	Mitova et al. (2004)
Communesin C	Indole alkaloid (communesin derivative)	Italy	*Penicillium* sp.	Fungus	*Axinella verrucosa*	Anticancer	Jadulco et al. (2004)
Communesin D	Indole alkaloid (communesin derivative)	Italy	*Penicillium* sp.	Fungus	*Axinella verrucosa*	Anticancer	Jadulco et al. (2004)
Petrosifungins A	Cyclic peptide	Italy	*Penicillium brevicompactum*	Fungus	*Petrosia ficiformis*	NR	Bringmann et al. (2004)
Petrosifungins B	Cyclic peptide	Italy	*Penicillium brevicompactum*	Fungus	*Petrosia ficiformis*	NR	Bringmann et al. (2004)
Sorbicillactone A	Sorbicillin alkaloid	Italy	*Penicillium chrysogenum*	Fungus	*Ircinia fasciculata*	Anti-HIV	Bringmann et al. (2005)
Sorbicillactone B	Sorbicillin alkaloid	Italy	*Penicillium chrysogenum*	Fungus	*Ircinia fasciculata*	Anti-HIV	Bringmann et al. (2005)
Sorbivinetone	Sorbicillin alkaloid	Italy	*Penicillium chrysogenum*	Fungus	*Ircinia fasciculata*	NR	Bringmann et al. (2005)

(Continued)

Table 15.1 (Continued) New compounds isolated from microorganisms from marine sponges from 2000 to 2012: Their chemical class, country of collection, and reported activity

Compound name	Chemical class	Country of collection	Name of microorganism	Type of microorganism	Name of sponge	Reported activity	References
(S)-2,4-Dihydroxy-1-butyl(4-hydroxy)benzoate	Benzoate	China	*Penicillium aurantiogriseum*	Fungus	*Mycale plumose*	Cytotoxic	Xin et al. (2005)
Dehydroxynocardamine	Cyclic peptide	Korea	*Streptomyces* sp.	Actinobacterium	Not identified	Miscellaneous	Lee et al. (2005)
Desmethylenylnocardamine	Cyclic peptide	Korea	*Streptomyces* sp.	Actinobacterium	Not identified	Miscellaneous	Lee et al. (2005)
Clonostachysin A	Cyclic peptide	Japan	*Clonostachys rogersoniana*	Fungus	*Halichondria japonica*	Miscellaneous	Adachi et al. (2005)
Clonostachysin B	Cyclic peptide	Japan	*Clonostachys rogersoniana*	Fungus	*Halichondria japonica*	Miscellaneous	Adachi et al. (2005)
Guangomide A	Cyclic depsipeptide	Papua New Guinea	Not identified fungus	Fungus	*Ianthella* sp.	Antibacterial	Amagata et al. (2006)
Guangomide B	Cyclic depsipeptide	Papua New Guinea	Not identified fungus	Fungus	*Ianthella* sp.	Antibacterial	Amagata et al. (2006)
Homodestcardin	Cyclic depsipeptide	Papua New Guinea	Not identified fungus	Fungus	*Ianthella* sp.	NR	Amagata et al. (2006)
RHM1	Peptide	Papua New Guinea	*Acremonium* sp.	Fungus	*Teichaxinella* sp.	Cytotoxic	Boot et al. (2006)
RHM2	Octapeptide	Papua New Guinea	*Acremonium* sp.	Fungus	*Teichaxinella* sp.	Cytotoxic	Boot et al. (2006)
Tropolactone A	Meroterpene	United States	*Aspergillus* sp.	Fungus	Not identified	Cytotoxic	Cueto et al. (2006)
Tropolactone B	Meroterpene	United States	*Aspergillus* sp.	Fungus	Not identified	Cytotoxic	Cueto et al. (2006)
Tropolactone C	Meroterpene	United States	*Aspergillus* sp.	Fungus	Not identified	Cytotoxic	Cueto et al. (2006)

(Continued)

Table 15.1 (Continued) New compounds isolated from microorganisms from marine sponges from 2000 to 2012: Their chemical class, country of collection, and reported activity

Compound name	Chemical class	Country of collection	Name of microorganism	Type of microorganism	Name of sponge	Reported activity	References
Tropolactone D	Meroterpene	United States	*Aspergillus* sp.	Fungus	Not identified	Cytotoxic	Cueto et al. (2006)
IB-01212	Cyclic depsipeptide	Japan	*Clonostachys* sp.	Fungus	Not identified	Cytotoxic	Cruz et al. (2006)
Roridin R	Macrocyclic trichothecene	Indonesia	*Myrothecium* sp.	Fungus	Not identified	Cytotoxic	Xu et al. (2006)
12-Hydroxyroridin E	Macrocyclic trichothecene	Indonesia	*Myrothecium* sp.	Fungus	Not identified	Cytotoxic	Xu et al. (2006)
Roridin Q	Macrocyclic trichothecene	Indonesia	*Myrothecium* sp.	Fungus	Not identified	Cytotoxic	Xu et al. (2006)
2′,3′-Deoxyroritoxin D	Macrocyclic trichothecene	Indonesia	*Myrothecium roridum*	Fungus	Not identified	Cytotoxic, antiyeast	Xu et al. (2006)
5-*cis*-3-Oxo-C12-HSL (compound 1)	N-Acyl-L-homoserine lactone	Norway	*Mesorhizobium* sp.	Bacterium	*Phakellia ventilabrum*	Antibacterial, cytotoxic	Krick et al. (2007)
5-*cis*-3-Oxo-C12-homoserine lactone	N-Acyl-L-homoserine lactone	Norway	*Mesorhizobium* sp.	Bacterium	*Phakellia ventilabrum*	NR	Krick et al. (2007)
Bromoalterochromide A	Brominated chromopeptide	Australia	*Pseudoalteromonas maricaloris*	Bacterium	*Fascaplysinopsis reticulata*	Cytotoxic	Speitling et al. (2007)
Bromoalterochromide A′	Brominated chromopeptide	Australia	*Pseudoalteromonas maricaloris*	Bacterium	*Fascaplysinopsis reticulata*	Cytotoxic	Speitling et al. (2007)
Aurantiomide B	Quinazoline alkaloid	China	*Penicillium aurantiogriseum*	Fungus	*Mycale plumose*	Cytotoxic	Xin et al. (2007)
Aurantiomide C	Quinazoline alkaloid	China	*Penicillium aurantiogriseum*	Fungus	*Mycale plumose*	Cytotoxic	Xin et al. (2007)
Glyco-C$_{30}$-carotenoic acid	Diapolycopenedioic acid	NR	*Rubritalea squalenifaciens*	Bacterium	*Halichondria okadai*	Miscellaneous	Shindo et al. (2007)
Trichodermanone A	Sorbicillin–polyketide	Dominica	*Trichoderma* sp.	Fungus	*Agelas dispar*	Miscellaneous	Neumann et al. (2007)
Trichodermanone B	Sorbicillin–polyketide	Dominica	*Trichoderma* sp.	Fungus	*Agelas dispar*	Miscellaneous	Neumann et al. (2007)

(Continued)

Table 15.1 (Continued) New compounds isolated from microorganisms from marine sponges from 2000 to 2012: Their chemical class, country of collection, and reported activity

Compound name	Chemical class	Country of collection	Name of microorganism	Type of microorganism	Name of sponge	Reported activity	References
Trichodermanone C	Sorbicillin–polyketide	Dominica	*Trichoderma* sp.	Fungus	*Agelas dispar*	Miscellaneous	Neumann et al. (2007)
Trichodermanone D	Sorbicillin–polyketide	Dominica	*Trichoderma* sp.	Fungus	*Agelas dispar*	Miscellaneous	Neumann et al. (2007)
Aurantiomide A	Quinazoline alkaloid	China	*Penicillium aurantiogriseum*	Fungus	*Mycale plumose*	NR	Xin et al. (2007)
Aspinotriol A	Pentaketide	Micronesia	*Aspergillus ostianus*	Fungus	Not identified sponge	NR	Kito et al. (2007)
Aspinotriol B	Pentaketide	Micronesia	*Aspergillus ostianus*	Fungus	Not identified sponge	NR	Kito et al. (2007)
Aspinonediol	Pentaketide	Micronesia	*Aspergillus ostianus*	Fungus	Not identified sponge	NR	Kito et al. (2007)
Streptophenazine A	Phenazine	Germany	*Streptomyces* sp.	Actinobacterium	*Halichondria panicea*	Antibacterial	Mitova et al. (2008)
Streptophenazine B	Phenazine	Germany	*Streptomyces* sp.	Actinobacterium	*Halichondria panicea*	Antibacterial	Mitova et al. (2008)
Streptophenazine C	Phenazine	Germany	*Streptomyces* sp.	Actinobacterium	*Halichondria panicea*	Antibacterial	Mitova et al. (2008)
Streptophenazine D	Phenazine	Germany	*Streptomyces* sp.	Actinobacterium	*Halichondria panicea*	Antibacterial	Mitova et al. (2008)
Streptophenazine E	Phenazine	Germany	*Streptomyces* sp.	Actinobacterium	*Halichondria panicea*	Antibacterial	Mitova et al. (2008)
Streptophenazine F	Phenazine	Germany	*Streptomyces* sp.	Actinobacterium	*Halichondria panicea*	Antibacterial	Mitova et al. (2008)
Streptophenazine G	Phenazine	Germany	*Streptomyces* sp.	Actinobacterium	*Halichondria panicea*	Antibacterial	Mitova et al. (2008)
Streptophenazine H	Phenazine	Germany	*Streptomyces* sp.	Actinobacterium	*Halichondria panicea*	Antibacterial	Mitova et al. (2008)
Chlorohydroaspyrone A	Aspyrone derivative (polyketide)	Korea	*Exophiala* sp.	Fungus	*Halichondria panicea*	Antibacterial	Zhang et al. (2008a)

(Continued)

Table 15.1 (Continued) New compounds isolated from microorganisms from marine sponges from 2000 to 2012: Their chemical class, country of collection, and reported activity

Compound name	Chemical class	Country of collection	Name of microorganism	Type of microorganism	Name of sponge	Reported activity	References
Chlorohydroaspyrone B	Aspyrone derivative (polyketide)	Korea	*Exophiala* sp.	Fungus	*Halichondria panicea*	Antibacterial	Zhang et al. (2008a)
Scopularide A	Cyclic depsipeptide	Croatia	*Scopulariopsis brevicaulis*	Fungus	*Tethya aurantium*	Antibacterial, cytotoxic	Yu et al. (2008)
Scopularide B	Cyclic depsipeptide	Croatia	*Scopulariopsis brevicaulis*	Fungus	*Tethya aurantium*	Antibacterial, cytotoxic	Yu et al. (2008)
Gymnastatin Q	NR	Japan	*Gymnascella dankaliensis*	Fungus	*Halichondria japonica*	Anticancer	Amagata et al. (2008)
Dihydroinfectopyrone	Pyrone	Thailand	Order pleosporales	Fungus	Not identified	Anticancer	Proksch et al. (2008)
Aspergillide A	Polyketide	Micronesia	*Aspergillus ostianus*	Fungus	Not identified	Cytotoxic	Kito et al. (2008)
Aspergillide B	Polyketide	Micronesia	*Aspergillus ostianus*	Fungus	Not identified	Cytotoxic	Kito et al. (2008)
Aspergillide C	Polyketide	Micronesia	*Aspergillus ostianus*	Fungus	Not identified	Cytotoxic	Kito et al. (2008)
Gymnastatin R	NR	Japan	*Gymnascella dankaliensis*	Fungus	*Halichondria japonica*	Cytotoxic	Amagata et al. (2008)
Dankastatin A	NR	Japan	*Gymnascella dankaliensis*	Fungus	*Halichondria japonica*	Cytotoxic	Amagata et al. (2008)
Dankastatin B	NR	Japan	*Gymnascella dankaliensis*	Fungus	*Halichondria japonica*	Cytotoxic	Amagata et al. (2008)
(Z)-6-Benzylidene-3-hydroxymethyl-1,4-dimethyl-3-methylsulfanylpiperazine-2,5-dione	Diketopiperazine	Thailand	*Strain CRIF2 (order Pleosporales)*	Fungus	Not identified	Cytotoxic	Prachyawarakorn et al. (2008)
(3S,3′R)-3-(3′-Hydroxybutyl)-7-methoxyphthalide	Diketopiperazine	Thailand	*Strain CRIF2 (order Pleosporales)*	Fungus	Not identified	Cytotoxic	Prachyawarakorn et al. (2008)
Diapolycopenedioic acid xylosyl ester A	Glyco-C_{30}-carotenoic acid	Japan	*Rubritalea squalenifaciens*	Bacterium	*Halichondria okadai*	Miscellaneous	Shindo et al. (2008)

(Continued)

Table 15.1 (Continued) New compounds isolated from microorganisms from marine sponges from 2000 to 2012: Their chemical class, country of collection, and reported activity

Compound name	Chemical class	Country of collection	Name of microorganism	Type of microorganism	Name of sponge	Reported activity	References
Diapolycopenedioic acid xylosyl ester B	Glyco-C$_{30}$-carotenoic acid	Japan	*Rubritalea squalenifaciens*	Bacterium	*Halichondria okadai*	Miscellaneous	Shindo et al. (2008)
Diapolycopenedioic acid xylosyl ester C	Glyco-C$_{30}$-carotenoic acid	Japan	*Rubritalea squalenifaciens*	Bacterium	*Halichondria okadai*	Miscellaneous	Shindo et al. (2008)
Compound 3 (C$_{10}$H$_{11}$NO$_4$)	Siderophore	Indonesia	*Pseudoalteromonas* sp.	Bacterium	*Halisarca ectofibrosa*	Miscellaneous	You et al. (2008)
Cebulactam A1	Macrolactam	Philippines	*Saccharopolyspora cebuensis*	Actinobacterium	*Haliclona* sp.	NR	Pimentel-Elardo et al. (2008)
Cebulactam B1	Macrolactam	Philippines	*Saccharopolyspora cebuensis*	Actinobacterium	*Haliclona* sp.	NR	Pimentel-Elardo et al. (2008)
Cyclo-[phenylalanyl-leucyl]2	Peptide	Thailand	*Pseudoalteromonas* sp.	Bacterium	*Halisarca ectofibrosa*	NR	Rungprom et al. (2008)
Cyclo-[leucyl-isoleucyl]2	Peptide	Thailand	*Pseudoalteromonas* sp.	Bacterium	*Halisarca ectofibrosa*	NR	Rungprom et al. (2008)
1-(2,8-Dihydroxy-1,2,6-trimethyl-1,2,6,7,8,8ahexahydro-naphthalen-1-yl)-3-methoxy-propan-1-one	Polyketide	NR	*Mycelia sterilia*	Fungus	Not identified	NR	Hao et al. (2008)
4,8-Dihydroxy-7-(2-hydroxy-ethyl)-6-methoxy-3,4-dihydro-2-naphthalen-1-one	Polyketide	NR	*Mycelia sterilia*	Fungus	Not identified	NR	Hao et al. (2008)
1-Methyl-naphthalene-2,6-dicarboxylic acid	Polyketide	NR	*Mycelia sterilia*	Fungus	Not identified	NR	Hao et al. (2008)
Circumdatin J	Alkaloid	Micronesia	*Aspergillus ostianus*	Fungus	Not identified	NR	Ookura et al. (2008)
6-Hydroxymethyl-1-phenazine-carboxamide	Phenazine	Korea	*Breibacterium* sp. KMD 003	Bacterium	*Callyspongia* sp.	Antibacterial	Choi et al. (2009)
1,6-Phenazinedimethanol	Phenazine	Korea	*Breibacterium* sp. KMD 003	Bacterium	*Callyspongia* sp.	Antibacterial	Choi et al. (2009)

(Continued)

Table 15.1 (Continued) New compounds isolated from microorganisms from marine sponges from 2000 to 2012: Their chemical class, country of collection, and reported activity

Compound name	Chemical class	Country of collection	Name of microorganism	Type of microorganism	Name of sponge	Reported activity	References
Anguyclinone (PM070747)	Benz[a]anthraquinone	Tanzania	*Saccharopolyspora taberi* (PEM-06-F23–019B)	Actinobacterium	Unidentified	Anticancer	Pérez et al. (2009)
Hydroxydecylparaben, 4-hydroxybenzoic acid 3-hydroxy-decyl ester	Paraben	France	*Microbulbifer* sp.	Bacterium	*Leuconia nivea*	Antimicrobial	Quévrain et al. (2009)
Methyldecylparaben, 4-hydroxybenzoic acid methyl-decyl ester	Paraben	France	*Microbulbifer* sp.	Bacterium	*Leuconia nivea*	Antimicrobial	Quévrain et al. (2009)
Hydroxymethyldecylparaben, 4-hydroxybenzoic acid 3-hydroxy-methyl-decyl ester	Paraben	France	*Microbulbifer* sp.	Bacterium	*Leuconia nivea*	Antimicrobial	Quévrain et al. (2009)
Dodec-5-enylparaben, 4-hydroxybenzoic acid dodec-5-enyl ester	Paraben	France	*Microbulbifer* sp.	Bacterium	*Leuconia nivea*	Antimicrobial	Quévrain et al. (2009)
Aspergillusol A	Tyrosine	Thailand	*Aspergillus aculeatus*	Fungus	*Xestospongia testudinaria*	Antiyeast, cytotoxic	Ingavat et al. (2009)
Beauversetin	Tetramic acid	Germany	*Beauveria bassiana*	Fungus	*Myxilla incrustans*	Cytotoxic	Neumann et al. (2009)
Epoxyphomalin A	Prenylated polyketide	Dominica	*Phoma* sp.	Fungus	*Ectyplasia perox*	Cytotoxic	Mohamed et al. (2009)
Epoxyphomalin B	Prenylated polyketide	Dominica	*Phoma* sp.	Fungus	*Ectyplasia perox*	Cytotoxic	Mohamed et al. (2009)
2-(1H-Indol-3-yl)ethyl 2-hydroxypropanoate	Indole	Japan	*Pichia membranifaciens*	Yeast	*Halichondria okadai*	Miscellaneous	Sugiyama et al. (2009)
2-(1H-Indol-3-yl)ethyl 5-hydroxypentanoate	Indole	Japan	*Pichia membranifaciens*	Yeast	*Halichondria okadai*	Miscellaneous	Sugiyama et al. (2009)
JBIR-37	Glycosyl benzenediol	Japan	*Acremonium* sp.	Fungus	Not identified	NR	Izumikawa et al. (2009)

(Continued)

Table 15.1 (Continued) New compounds isolated from microorganisms from marine sponges from 2000 to 2012: Their chemical class, country of collection, and reported activity

Compound name	Chemical class	Country of collection	Name of microorganism	Type of microorganism	Name of sponge	Reported activity	References
JBIR-38	Glycosyl benzenediol	Japan	*Acremonium* sp.	Fungus	Not identified	NR	Izumikawa et al. (2009)
Paecilopyrone A	α-Pyrone	Korea	*Paecilomyces lilacinus*	Fungus	*Petrosia* sp.	NR	Elbandy et al. (2009)
Paecilopyrone B	α-Pyrone	Korea	*Paecilomyces lilacinus*	Fungus	*Petrosia* sp.	NR	Elbandy et al. (2009)
Phomaligol B	Cyclohexenone	Korea	*Paecilomyces lilacinus*	Fungus	*Petrosia* sp.	NR	Elbandy et al. (2009)
Phomaligol C	Cyclohexenone	Korea	*Paecilomyces lilacinus*	Fungus	*Petrosia* sp.	NR	Elbandy et al. (2009)
Neobacillamide A	Alkaloid	China	*Bacillus vallismortis* C89	Bacterium	*Dysidea avara*	NR	Yu et al. (2009)
Chlorocylindrocarpol	Sesquiterpene	Korea	*Acremonium* sp.	Fungus	*Stelletta* sp.	NR	Zhang et al. (2009)
Acremofuranone A	Sesquiterpene	Korea	*Acremonium* sp.	Fungus	*Stelletta* sp.	NR	Zhang et al. (2009)
Acremofuranone B	Sesquiterpene	Korea	*Acremonium* sp.	Fungus	*Stelletta* sp.	NR	Zhang et al. (2009)
Dihydroxybergamotene	Sesquiterpene	Korea	*Acremonium* sp.	Fungus	*Stelletta* sp.	NR	Zhang et al. (2009)
Trichopyrone [6-(4-hydroxy-1-pentenyl)-4-meth oxy-3-methyl-2H-pyran-2-one]	Pyranone	Dominica	*Trichoderma viride*	Fungus	*Agelas dispar*	NR	Abdel-Lateffa et al. (2009)
JBIR-15	Aspochracin derivative	Japan	*Aspergillus sclerotiorum*	Fungus	*Mycale* sp.	NR	Motohashi et al. (2009)
Mayamycin	Polyketide	Germany	*Nocardiopsis* sp.	Actinobacterium	*Halichondria panicea*	Antibacterial, cytotoxic	Schneemann et al. (2010a)
JBIR-58	Salicylamide	Japan	*Streptomyces* sp. SpD081030ME-02	Actinobacterium	Not identified	Cytotoxic	Ueda et al. (2010a)

(Continued)

Table 15.1 (Continued) New compounds isolated from microorganisms from marine sponges from 2000 to 2012: Their chemical class, country of collection, and reported activity

Compound name	Chemical class	Country of collection	Name of microorganism	Type of microorganism	Name of sponge	Reported activity	References
Fellutamide C	Lipopeptide	Japan	*Aspergillus versicolor*	Fungus	*Petrosia* sp.	Cytotoxic	Lee et al. (2010)
Epoxyphomalin D	Prenylated polyketide	Dominica	*Paraconiothyrium sporulosum*	Fungus	*Ectyplasia perox*	Cytotoxic	Mohamed et al. (2010)
JBIR-97	NR	Japan	*Tritirachium* sp.	Fungus	*Pseudoceratina purpurea*	Cytotoxic	Ueda et al. (2010b)
JBIR-98	NR	Japan	*Tritirachium* sp.	Fungus	*Pseudoceratina purpurea*	Cytotoxic	Ueda et al. (2010b)
JBIR-99	NR	Japan	*Tritirachium* sp.	Fungus	*Pseudoceratina purpurea*	Cytotoxic	Ueda et al. (2010b)
Trichoderin A	Aminolipopeptide	NR	*Trichoderma* sp.	Fungus	Not identified	Antituberculosis	Pruksakorn et al. (2010)
Trichoderin A1	Aminolipopeptide	NR	*Trichoderma* sp.	Fungus	Not identified	Antituberculosis	Pruksakorn et al. (2010)
Trichoderin B	Aminolipopeptide	NR	*Trichoderma* sp.	Fungus	Not identified	Antituberculosis	Pruksakorn et al. (2010)
JBIR-65	Diterpene	Japan	*Actinomadura* sp. SpB081030SC-15	Actinobacterium	Not identified	Miscellaneous	Takagi et al. (2010a)
JBIR-34	Indole-containing peptide	Japan	*Streptomyces* sp. Sp080513GE-23	Actinobacterium	*Haliclona* sp.	Miscellaneous	Motohashi et al. (2010)
JBIR-35	Indole-containing peptide	Japan	*Streptomyces* sp. Sp080513GE-23	Actinobacterium	*Haliclona* sp.	Miscellaneous	Motohashi et al. (2010)
Nocapyrone A	γ-Pyrone	Germany	*Nocardiopsis* sp.	Actinobacterium	*Halichondria panicea*	NR	Schneemann et al. (2010b)
Nocapyrone B	γ-Pyrone	Germany	*Nocardiopsis* sp.	Actinobacterium	*Halichondria panicea*	NR	Schneemann et al. (2010b)
Nocapyrone C	γ-Pyrone	Germany	*Nocardiopsis* sp.	Actinobacterium	*Halichondria panicea*	NR	Schneemann et al. (2010b)
Nocapyrone D	γ-Pyrone	Germany	*Nocardiopsis* sp.	Actinobacterium	*Halichondria panicea*	NR	Schneemann et al. (2010b)

(Continued)

Table 15.1 (Continued) New compounds isolated from microorganisms from marine sponges from 2000 to 2012: Their chemical class, country of collection, and reported activity

Compound name	Chemical class	Country of collection	Name of microorganism	Type of microorganism	Name of sponge	Reported activity	References
JBIR-74	Roquefortine C analog (mycotoxin)	Japan	*Aspergillus* sp.	Fungus	Not identified	NR	Takagi et al. (2010b)
JBIR-75	Roquefortine C analog (mycotoxin)	Japan	*Aspergillus* sp.	Fungus	Not identified	NR	Takagi et al. (2010b)
Epoxyphomalin C	Prenylated polyketide	Dominica	*Paraconiothyrium sporulosum*	Fungus	*Ectyplasia perox*	NR	Mohamed et al. (2010)
Epoxyphomalin E	Prenylated polyketide	Dominica	*Paraconiothyrium sporulosum*	Fungus	*Ectyplasia perox*	NR	Mohamed et al. (2010)
Sorbicillinone A	Sorbicillin	Italy	*Penicillium chrysogenum*	Fungus	*Ircinia fasciculata*	NR	Bringmann et al. (2010)
Sorbifuranone B	Sorbicillin	Italy	*Penicillium chrysogenum*	Fungus	*Ircinia fasciculata*	NR	Bringmann et al. (2010)
Sorbifuranone A	Sorbicillin	Italy	*Penicillium chrysogenum*	Fungus	*Ircinia fasciculata*	NR	Bringmann et al. (2010)
Marilone B	Phthalide (polyketide)	Australia	*Stachylidium* sp.	Fungus	*Callyspongia* sp. cf. *C. flammea*	Antagonistic of serotonin receptor	Almeida et al. (2011)
Butyrlolactone-VI	Dibenzylbutyrolactone	Chile	*Aspergillus* sp. (2P-22)	Fungus	*Cliona chilensis*	Antibacterial	San-Martin et al. (2011)
Cillifuranone	Intermediate in sorbifuranone	Croatia	*Penicillium chrysogenum* strain LF066	Fungus	*Tethya aurantium*	Antibiotic	Wiese et al. (2011)
Insuetolide A	Meroterpene	Israel	*Aspergillus aculeatus*	Fungus	*Psammocinia* sp.	Antifungal	Cohen et al. (2011)
Bendigole D	3-Keto sterol	NR	*Actinomadura* sp.	Actinobacterium	*Suberites japonicus*	Anti-inflammatory, cytotoxic	Simmons et al. (2011)
Bendigole F	3-Keto sterol	NR	*Actinomadura* sp.	Actinobacterium	*Suberites japonicus*	Anti-inflammatory	Simmons et al. (2011)

(Continued)

Table 15.1 (Continued) New compounds isolated from microorganisms from marine sponges from 2000 to 2012: Their chemical class, country of collection, and reported activity

Compound name	Chemical class	Country of collection	Name of microorganism	Type of microorganism	Name of sponge	Reported activity	References
(3S,8R)-Methyl 8-hydroxy-3-methoxycarbonyl-2-methylenenonanoate	Hexylitaconic acid	Korea	*Penicillium* sp.	Fungus	*Stelletta* sp.	Anti-inflammatory	Li et al. (2011b)
(3S)-Methyl-9-hydroxy-3-methoxycarbonyl-2-methylenenonanoate	Hexylitaconic acid	Korea	*Penicillium* sp.	Fungus	*Stelletta* sp.	Anti-inflammatory	Li et al. (2011b)
Marilone A	Phthalide (polyketide)	Australia	*Stachylidium* sp.	Fungus	*Callyspongia* sp. cf. C. *flammea*	Antimalarial and anticancer	Almeida et al. (2011)
Marilone C	Phthalide (polyketide)	Australia	*Stachylidium* sp.	Fungus	*Callyspongia* sp. cf. C. *flammea*	Antimalarial and Anticancer	Almeida et al. (2011)
Myrocin D	Diterpene	Italy	*Arthrinium* sp.	Fungus	*Geodia cydonium*	Anticancer	Ebada et al. (2011)
22-Deoxythiocoraline	Thiocoraline analog	United States	*Verrucosispora* sp.	Actinobacterium	*Chondrilla caribensis* f. *caribensis*	Cytotoxic	Wyche et al. (2011)
Thiochondrilline C	Thiocoraline analog	United States	*Verrucosispora* sp.	Actinobacterium	*Chondrilla caribensis* f. *caribensis*	Cytotoxic	Wyche et al. (2011)
12-Sulfoxythiocoraline	Thiocoraline analog	United States	*Verrucosispora* sp.	Actinobacterium	*Chondrilla caribensis* f. *caribensis*	Cytotoxic	Wyche et al. (2011)
Acremostrictin	Tricyclic lactone	Korea	*Acremonium strictum*	Fungus	Not identified	Antibacterial, miscellaneous	Julianti et al. (2011)
Insuetolide C	Meroterpene	Israel	*Aspergillus aculeatus*	Fungus	*Psammocinia* sp.	Cytotoxic	Cohen et al. (2011)
(E)-6-(4'-Hydroxy-2'-butenoyl)-strobilactone A	Sesquiterpene	Israel	*Aspergillus aculeatus*	Fungus	*Psammocinia* sp.	Cytotoxic	Cohen et al. (2011)

(Continued)

Table 15.1 (Continued) New compounds isolated from microorganisms from marine sponges from 2000 to 2012: Their chemical class, country of collection, and reported activity

Compound name	Chemical class	Country of collection	Name of microorganism	Type of microorganism	Name of sponge	Reported activity	References
Fellutamide F	Lipopeptide	Korea	*Aspergillus versicolor*	Fungus	*Petrosia* sp.	Cytotoxic	Lee et al. (2011)
Dihydrotrichodermolide	Polyketide (sorbicillin dimer)	NR (East Pacific)	*Phialocephala* sp.	Fungus	*Stelletta* sp.	Cytotoxic	Li et al. (2011a)
Dihydrodemethylsorbicillin	Polyketide (sorbicillin monomer)	NR (East Pacific)	*Phialocephala* sp.	Fungus	*Stelletta* sp.	Cytotoxic	Li et al. (2011a)
Phialofurone	Benzofuranone	NR (East Pacific)	*Phialocephala* sp.	Fungus	*Stelletta* sp.	Cytotoxic	Li et al. (2011a)
Bendigole E	3-Keto sterol	NR	*Actinomadura* sp.	Actinobacterium	*Suberites japonicus*	NR	Simmons et al. (2011)
JBIR-56	Peptide	Japan	*Streptomyces* sp.	Actinobacterium	Not identified	NR	Motohashi et al. (2011)
JBIR-57	Peptide	Japan	*Streptomyces* sp.	Actinobacterium	Not identified	NR	Motohashi et al. (2011)
Streptomycindole	Indole alkaloid	China	*Streptomyces* sp.	Actinobacterium	*Craniella australiensis*	NR	Huang et al. (2011)
Thiochondrilline A	Thiocoraline analog	United States	*Verrucosispora* sp.	Actinobacterium	*Chondrilla caribensis f. caribensis*	NR	Wyche et al. (2011)
Thiochondrilline B	Thiocoraline analog	United States	*Verrucosispora* sp.	Actinobacterium	*Chondrilla caribensis f. caribensis*	NR	Wyche et al. (2011)
Arthrinin A	Diterpene	Italy	*Arthrinium* sp.	Fungus	*G. cydonium*	NR	Ebada et al. (2011)
Arthrinin B	Diterpene	Italy	*Arthrinium* sp.	Fungus	*G. cydonium*	NR	Ebada et al. (2011)
Arthrinin C	Diterpene	Italy	*Arthrinium* sp.	Fungus	*G. cydonium*	NR	Ebada et al. (2011)
Arthrinin D	Diterpene	Italy	*Arthrinium* sp.	Fungus	*G. cydonium*	NR	Ebada et al. (2011)
Asperaculin A	Sesquiterpene	Thailand	*Aspergillus aculeatus*	Fungus	*Xestospongia testudinaria*	NR	Ingavat et al. (2011)
Pre-aurantiamine	Diketopiperazine	Thailand	*Aspergillus aculeatus*	Fungus	*Stylissa flabelliformis*	NR	Antia et al. (2011)

(*Continued*)

Table 15.1 (Continued) New compounds isolated from microorganisms from marine sponges from 2000 to 2012: Their chemical class, country of collection, and reported activity

Compound name	Chemical class	Country of collection	Name of microorganism	Type of microorganism	Name of sponge	Reported activity	References
(−)-9-Hydroxyhexylitaconic acid	Diketopiperazine	Thailand	*Aspergillus aculeatus*	Fungus	*S. flabelliformis*	NR	Antia et al. (2011)
(−)-9-Hydroxyhexylitaconic acid-4-methyl ester	Diketopiperazine	Thailand	*Aspergillus aculeatus*	Fungus	*S. flabelliformis*	NR	Antia et al. (2011)
Insuetolide B	Meroterpene	Israel	*Aspergillus aculeatus*	Fungus	*Psammocinia* sp.	NR	Cohen et al. (2011)
Austalide M	Meroterpene	Italy	*Aspergillus* sp.	Fungus	*Tethya aurantium*	NR	Zhou et al. (2011)
Austalide N	Meroterpene	Italy	*Aspergillus* sp.	Fungus	*Tethya aurantium*	NR	Zhou et al. (2011)
Austalide O	Meroterpene	Italy	*Aspergillus* sp.	Fungus	*Tethya aurantium*	Cytotoxic	Zhou et al. (2011)
Austalide P	Meroterpene	Italy	*Aspergillus* sp.	Fungus	*Tethya aurantium*	Cytotoxic	Zhou et al. (2011)
Austalide Q	Meroterpene	Italy	*Aspergillus* sp.	Fungus	*Tethya aurantium*	Cytotoxic	Zhou et al. (2011)
(3S,8R)-8-Hydroxy-3-carboxy-2-methylenenonanoic acid	Hexylitaconic acid	Korea	*Penicillium* sp.	Fungus	*Stelletta* sp.	NR	Li et al. (2011b)
(3S)-9-Hydroxy-3-carboxy-2-methylenenonanoic acid	Hexylitaconic acid	Korea	*Penicillium* sp.	Fungus	*Stelletta* sp.	NR	Li et al. (2011b)
Stachyline A	Tyrosine-derived metabolite	Australia	*Stachylidium* sp.	Fungus	*Callyspongia* sp. cf. *C. flammea*	NR	Almeida et al. (2010)
Stachyline B	Tyrosine-derived metabolite	Australia	*Stachylidium* sp.	Fungus	*Callyspongia* sp. cf. *C. flammea*	NR	Almeida et al. (2010)
Stachyline C	Tyrosine-derived metabolite	Australia	*Stachylidium* sp.	Fungus	*Callyspongia* sp. cf. *C. f/C. fla*	NR	Almeida et al. (2010)

(Continued)

Table 15.1 (Continued) New compounds isolated from microorganisms from marine sponges from 2000 to 2012: Their chemical class, country of collection, and reported activity

Compound name	Chemical class	Country of collection	Name of microorganism	Type of microorganism	Name of sponge	Reported activity	References
Stachyline D	Tyrosine-derived metabolite	Australia	*Stachylidium* sp.	Fungus	*Callyspongia* sp. cf. *C. flammea*	NR	Almeida et al. (2010)
Compound 1, new (5a,6a)-isomer of ophiobolin H	Sesterterpene	NR (Adriatic Sea)	*Aspergillus ustus*	Fungus	*Suberites domuncula*	No significant cytotoxicity found	Liu et al. (2011)
Compound 2, (5a,6a)-5-O-methylophiobolin H	Sesterterpene	NR (Adriatic Sea)	*Aspergillus ustus*	Fungus	*Suberites domuncula*	No significant cytotoxicity found	Liu et al. (2011)
Compound 3, 5-O-methylophiobolin H	Sesterterpene	NR (Adriatic Sea)	*Aspergillus ustus*	Fungus	*Suberites domuncula*	No significant cytotoxicity found	Liu et al. (2011)
Compound 4, s(6a)-21,21-O-dihydroophiobolin G	Sesterterpene	NR (Adriatic Sea)	*Aspergillus ustus*	Fungus	*Suberites domuncula*	No significant cytotoxicity found	Liu et al. (2011)
Compound 5, (6a)-18,19,21,21-Otetrahydro-18,19-dihydroxyophiobolin G	Sesterterpene	NR (Adriatic Sea)	*Aspergillus ustus*	Fungus	*Suberites domuncula*	No significant cytotoxicity found	Liu et al. (2011)
Tetromycin 1	Tetromycin	France	*Streptomyces axinellae*	Actinobacterium	*Axinella polypoides*	Antiparasitic, inhibition of cathepsin L-like proteases	Pimentel-Elardo et al. (2011)
Tetromycin 2	Tetromycin	France	*Streptomyces axinellae*	Actinobacterium	*Axinella polypoides*	Antiparasitic, inhibition of cathepsin L-like proteases	Pimentel-Elardo et al. (2011)
Tetromycin 3	Tetromycin	France	*Streptomyces axinellae*	Actinobacterium	*Axinella polypoides*	Antiparasitic, inhibition of cathepsin L-like proteases	Pimentel-Elardo et al. (2011)

(Continued)

Table 15.1 (Continued) New compounds isolated from microorganisms from marine sponges from 2000 to 2012: Their chemical class, country of collection, and reported activity

Compound name	Chemical class	Country of collection	Name of microorganism	Type of microorganism	Name of sponge	Reported activity	References
Tetromycin 4	Tetromycin	France	*Streptomyces axinellae*	Actinobacterium	*Axinella polypoides*	Antiparasitic, inhibition of cathepsin L-like proteases	Pimentel-Elardo et al. (2011)
Lobophorin C	Kijanimicin	China	*Streptomyces microflavus*	Actinobacterium	*Hymeniacidon perlevis*	Cytotxic	Wei et al. (2011)
Lobophorin D	Kijanimicin	China	*Streptomyces microflavus*	Actinobacterium	*Hymeniacidon perlevis*	Anticancer	Wei et al. (2011)
JBIR-109	Trichostatin analog	Japan	*Streptomyces sp. strain RM72*	Actinobacterium	Not identified	Anticancer	Hosoya et al. (2012)
JBIR-110	Trichostatin analog	Japan	*Streptomyces sp. strain RM72*	Actinobacterium	Not identified	Anticancer	Hosoya et al. (2012)
JBIR-111	Trichostatin analog	Japan	*Streptomyces sp. strain RM72*	Actinobacterium	Not identified	Anticancer	Hosoya et al. (2012)
Urdamycinone E	Glycosylated benz[α]anthraquinone	Thailand	*Streptomycessp. BCC45596*	Actinobacterium	*Xestospongia* sp.	Antimalarial, antituberculosis	Supong et al. (2012)
Urdamycinone G	Glycosylated benz[α]anthraquinone	Thailand	*Streptomyces sp. BCC45596*	Actinobacterium	*Xestospongia* sp.	Antimalarial, antituberculosis	Supong et al. (2012)
Dehydroxyaquayamycin	Glycosylated benz[α]anthraquinone	Thailand	*Streptomyces sp. BCC45596*	Actinobacterium	*Xestospongia* sp.	Antimalarial, antituberculosis	Supong et al. (2012)
Acremolin	Methyl guanine	Korea	*Acremonium strictum*	Fungus	Not identified Choristida sponge	Cytotoxic	Julianti et al. (2012)
Disydonol A	Sesquiterpene	China	*Aspergillus* sp.	Fungus	*Xestospongia testudinaria*	Cytotoxic	Sun et al. (2012)
Disydonol C	Sesquiterpene	China	*Aspergillus* sp.	Fungus	*Xestospongia testudinaria*	Cytotoxic	Sun et al. (2012)
Aspergilusidone A	Depsidone	Thailand	*Aspergillus unguis CRI282–03*	Fungus	Not identified	Cytotoxic	Sureram et al. (2012)
Aspergilusidone B	Depsidone	Thailand	*Aspergillus unguis CRI282–03*	Fungus	Not identified	Cytotoxic	Sureram et al. (2012)

(Continued)

Table 15.1 (Continued) New compounds isolated from microorganisms from marine sponges from 2000 to 2012: Their chemical class, country of collection, and reported activity

Compound name	Chemical class	Country of collection	Name of microorganism	Type of microorganism	Name of sponge	Reported activity	References
Aspergilusidone C	Depsidone	Thailand	*Aspergillus unguis* CRI282–03	Fungus	Not identified	Cytotoxic	Sureram et al. (2012)
New compound	Diaryl ether	Thailand	*Aspergillus unguis* CRI282–03	Fungus	Not identified	Cytotoxic	Sureram et al. (2012)
Aspergiterpenoid A	Sesquiterpene	China	*Aspergillus* sp.	Fungus	*Xestospongia testudinaria*	Antibacterial, cytotoxic	Li et al. (2012)
(–)-Sydonol	Sesquiterpene	China	*Aspergillus* sp.	Fungus	*Xestospongia testudinaria*	Antibacterial, cytotoxic	Li et al. (2012)
(–)-Sydonic acid	Sesquiterpene	China	*Aspergillus* sp.	Fungus	*Xestospongia testudinaria*	Antibacterial, cytotoxic	Li et al. (2012)
(–)-5-(hydroxymethyl)-2-(2′,6′,6′-trimethyltetrahydro-2H-pyran-2-yl)phenol	Sesquiterpene	China	*Aspergillus* sp.	Fungus	*Xestospongia testudinaria*	Antibacterial, cytotoxic	Li et al. (2012)
JBIR 124	Sorbicillin	Japan	*Penicillium citrinum* sp. I080624C1f01	Fungus	Not identified	Miscellaneous	Kawahara et al. (2012)
Cyclodysidin A	Cyclic lipopeptide	Croatia	*Streptomyces strain* RV15	Actinobacterium	*Dysidea tupha*	NR	Abdelmohsen et al. (2012)
Cyclodysidin B	Cyclic lipopeptide	Croatia	*Streptomyces strain* RV15	Actinobacterium	*Dysidea tupha*	NR	Abdelmohsen et al. (2012)
Cyclodysidin C	Cyclic lipopeptide	Croatia	*Streptomyces strain* RV15	Actinobacterium	*Dysidea tupha*	NR	Abdelmohsen et al. (2012)
Cyclodysidin D	Cyclic lipopeptide	Thailand	*Streptomyces strain* RV15	Actinobacterium	*Dysidea tupha*	NR	Abdelmohsen et al. (2012)
Disydonol B	Sesquiterpene	China	*Aspergillus* sp.	Fungus	*Xestospongia testudinaria*	None	Sun et al. (2012)
1-Hydroxy-10-methoxy-dibenz[b,e]oxepin-6,11-dione	Dibenz[b,e]oxepine	Japan	*Beauveria bassiana* TPU942	Fungus	Not identified	None	Yamazaki et al. (2012)
(5E)-2-methyl-5-[(1′R*, 5′R*)-2-methylidene-7-oxobicyclo[3.2.1]oct-6-ylidene]-4-oxopentanoic acid	Sesquiterpene	Thailand	*Emericellopsis minima*	Fungus	*Hyrtios erecta*	None	Pinheiro et al. (2012)

(Continued)

Table 15.1 (Continued) New compounds isolated from microorganisms from marine sponges from 2000 to 2012: Their chemical class, country of collection, and reported activity

Compound name	Chemical class	Country of collection	Name of microorganism	Type of microorganism	Name of sponge	Reported activity	References
Eurocristatine	Diketopiperazine	Thailand	*Eurotium cristatum*	Fungus	Not identified	None	Gomes et al. (2012)
JBIR-113	Depsipeptide	Japan	*Penicillium* sp.	Fungus	Not identified	NR	Kawahara et al. (2012)
JBIR-114	Depsipeptide	Japan	*Penicillium* sp.	Fungus	Not identified	NR	Kawahara et al. (2012)
JBIR-115	Depsipeptide	Japan	*Penicillium* sp.	Fungus	Not identified	NR	Kawahara et al. (2012)
Cyclomarinone	Phthalide (polyketide)	Australia	*Stachylidium* sp.	Fungus	*Callyspongia* sp. cf. *C. flammea*	NR	Almeida et al. (2012)
Maristachone A	Phthalide (polyketide)	Australia	*Stachylidium* sp.	Fungus	*Callyspongia* sp. cf. *C. flammea*	NR	Almeida et al. (2012)
Maristachone B	Phthalide (polyketide)	Australia	*Stachylidium* sp.	Fungus	*Callyspongia* sp. cf. *C. flammea*	NR	Almeida et al. (2012)
Maristachone C	Phthalide (polyketide)	Australia	*Stachylidium* sp.	Fungus	*Callyspongia* sp. cf. *C. flammea*	NR	Almeida et al. (2012)
Maristachone D	Phthalide (polyketide)	Australia	*Stachylidium* sp.	Fungus	*Callyspongia* sp. cf. *C. flammea*	NR	Almeida et al. (2012)
Maristachone E	Phthalide (polyketide)	Australia	*Stachylidium* sp.	Fungus	*Callyspongia* sp. cf. *C. flammea*	NR	Almeida et al. (2012)
Marilactone	Phthalide (polyketide)	Australia	*Stachylidium* sp.	Fungus	*Callyspongia* sp. cf. *C. flammea*	NR	Almeida et al. (2012)

NR, not reported.

chemistry rather than describing the biological aspects of fungi associated with sponges (Höller et al., 2000). Fungal associations with sponges have been well established (Yarden, 2014) though not well characterized (Suryanarayanan, 2012) and require further in-depth study (Webster and Taylor, 2012). In this report, the fungi isolated from sponges between 2000 and 2012 that were reported to produce novel compounds belong to the following genera: *Acremonium, Arthrinium, Aspergillus, Beauveria, Cladosporium, Clonostachys, Curvularia, Drechslera, Emericella, Emericellopsis, Eurotium, Exophiala, Gymnascella, Microsphaeropsis, Myrothecium, Paecilomyces, Paraconiothyrium, Penicillium, Phialocephala, Phoma, Scopulariopsis, Stachylidium, Trichoderma,* and the yeast *Pichia* (Table 15.1). *Aspergillus* spp. produced 59 new compounds followed by *Penicillium* spp. that produced 26. Of a total of 272 natural products that were isolated from fungi during the period of 1970–2002, 28% came from sponges (Bugni and Ireland, 2004). During the current review period, 69% of the 269 new compounds were produced by fungi from sponges.

As the ecology of fungi in the marine environment is revealed through molecular ecology methods (Gao et al., 2008), the relationships between fungi and their hosts in the marine environment will help our understanding of the marine ecosystem and could lead to improved collection and isolation methods and the identification of chemically unexplored species (Bugni and Ireland, 2004; Abdelmohsen et al., 2014).

15.2.2   Actinobacteria

Actinobacterial associations have been found in reef and deepwater sponges, and evidence for sponge-specific symbioses exists (Hentschel et al., 2006). Moreover, marine actinobacteria in particular have yielded a higher percentage of novel secondary metabolites as compared to fungi (Lam, 2006) in the past. In this review, we report on only seven genera of actinobacteria isolated from sponges that have been reported to produce 57 new compounds. The genera are *Actinomadura, Microbacterium, Micromonospora, Nocardiopsis, Saccharopolyspora,* and *Verrucosispora,* with *Streptomyces* spp. producing the highest number of compounds (32).

In at least one case, actinobacterial symbionts such as a *Micromonospora* sp. have been shown to produce bioactive compounds (manzamines) that have no terrestrial equivalents (Montalvo et al., 2005) and are of considerable interest. Sponges in the South China Sea harbor a large diversity of actinobacteria and show evidence of host specificity (Li et al., 2006).

15.2.3   Bacteria

Bacteria associated with sponges that were reported to produce new metabolites were from four classes and seven genera only—Alphaproteobacteria (*Mesorhizobium, Ruegeria*), Firmicutes (*Bacillus*), Gammaproteobacteria (*Microbulbifer, Pseudoalteromonas, Pseudomonas*), and Verrucomicrobia (e.g., *Rubritalea*). However, no new compound has been reported from sponge-associated cyanobacteria since 2000, which revealed that these photosynthetic microorganisms were given less attention.

15.3   Sponge sources

Over the 13-year period (2000–2012), from a total of 104 reported studies, which would have used at least one sample per study, 35 sponge genera and 30 unidentified sponges were obtained from 23 countries or regions around the world. The sponge genera with the highest number of reported compounds are *Halichondria* with 34 compounds from three species from 12 separate studies. All three groups of microorganisms were isolated, but from different studies. This was a common feature of the studies, where each publication reported on compounds from only one type of microbe. *Xestospongia* with 25 compounds from fungi and actinobacteria were reported in eight separate studies, *Callyspongia* spp. with 22 compounds from fungi and bacteria in six separate studies, *Stelletta* spp. with 11 compounds from fungi only in four separate studies, and *Suberites* spp. with 10 compounds from all three types of microorganism in three studies. The rest of the sponge samples produced between 1 and 8 compounds were reported in 1–3 studies for each genus.

15.4   Chemical diversity among the microbial metabolites and their activities

A diverse array of compounds has been discovered from sponges such as terpenoids, steroids, phenolic compounds, alkaloids, polysaccharides, peptides, polyketides, and fatty acids, which all showed potential biological activity (Mehbub et al., 2014). Among the 269 new compounds isolated from sponge-associated microbes, only 159 compounds showed bioactivity that includes antibacterial, anticancer/cytotoxicity, antifungal, anti-HIV, anti-inflammatory, antiparasitic, antimalarial, antituberculosis, antiviral, and antiyeast (Figure 15.2). It was noted that the fungi associated with sponges in particular produce similar classes of compounds as sponges. Of particular note are the multiple congeners, with about two-thirds of the compounds produced as a set of three or more congeners, which is common in secondary metabolism. At least 40% of compounds had no activity reported when the publication of the compounds first appeared in the literature. Subsequent testing is likely to have taken place as it is unlikely that these new resources would not be assessed for their bioactive potential. However, it was noted that there were few screening assays reported that were based on receptor-binding activity or mode of action studies.

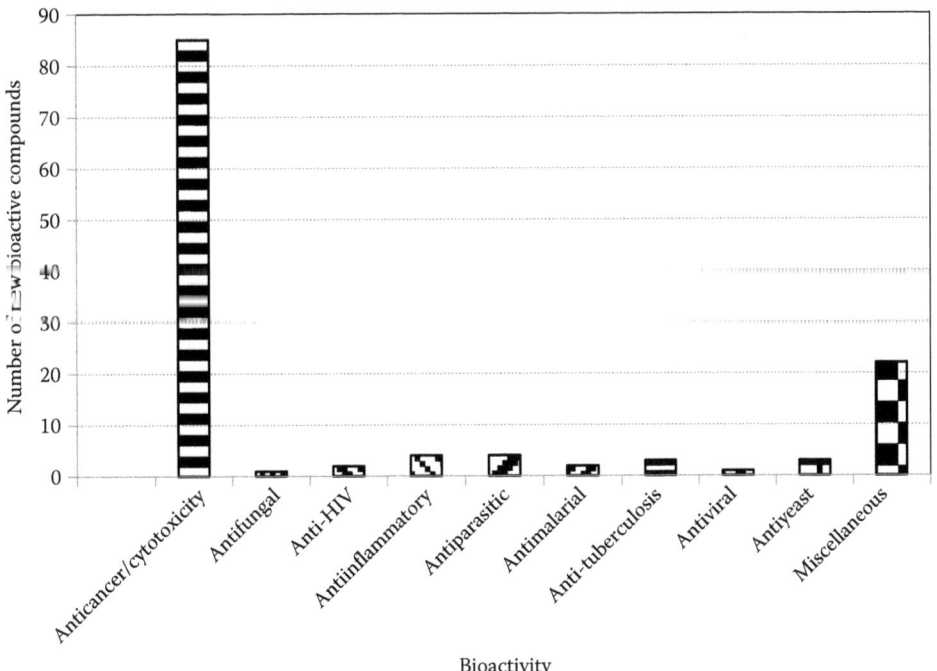

Figure 15.2 The reported bioactivity of compounds isolated from microorganisms associated with marine sponges from 2000 to 2012.

15.5   Identification of microbial diversity

Identification of microorganisms is important because biotechnology relies on defining their activities (Hugenholtz and Pace, 1996). Microbial species identified rapidly by sequencing their signature ribosomal genes; 16S rRNA for bacteria and actinobacteria and 28S rRNA for fungi are the most common. Full characterization is achieved using polyphasic methodology that employs both phenotypic and genotypic characteristics (Vandamme et al., 1996; Achtman and Wagner, 2008). Phenotypic methods include classical taxonomy, colony characteristics, biochemical and physiological studies, numerical taxonomy, cell wall composition, fatty acid analyses, isoprenoid quinones and whole-cell protein analyses, polyamines, cytochromes, and in some cases advanced spectroscopic methods (FTIR, pyrolysis mass spectrometry, UV resonance Raman spectroscopy). The genotypic methods include 16S and 28S rRNA gene sequencing and, increasingly, multilocus gene sequencing, DNA base content, and DNA–DNA hybridization (Sarethy et al., 2014).

15.6   Isolation techniques for bacteria, actinobacteria, and fungi from sponges

The use of improved cultivation approaches for the discovery of marine natural products from marine microorganisms is of paramount importance for the development of new pharmaceuticals (Bull et al., 2005). Conventional isolation methods are used to cultivate the most abundant microorganisms, but more selective methods are required to culture the uncommon and new microbial species and genera that have not been isolated previously (Ferguson et al., 1984; Eilers et al., 2000). It has been shown recently that previously uncultivable microorganisms could be grown in pure culture if provided with the chemical components of their natural environment (Kaeberlein et al., 2002; Rappe et al., 2002).

The main parameters that have influenced the isolation of novel strains include enrichment, pretreatment of samples, choice of nutrients in isolation, application of appropriate antibiotics, use of seawater and salt concentrations, inclusion of sponge extracts, and above all cultivation techniques that mimic nature (Hameş-kocabaş and Uzel, 2012). Moreover, temperature, pH, culture conditions, incubation time, and removal of emerging isolates are also very important (Kaewkla and Franco, 2013).

Using specific enrichment techniques, primarily by varying the media, hundreds of different organisms from the family of Streptosporangiacae were isolated, and many were hypothesized to produce novel antibiotics on the basis of observations of other members of this family (Lazzarini et al., 2001). In a separate example, by changing the media in combination with using selective agents for motile microorganisms, two new antibiotic-producing actinobacterial species were discovered, demonstrating the usefulness of a guided-culturing approach for the isolation of rare strains (Otoguro et al., 2001). Finally, the culturing of rare organisms has been improved by the development of new techniques. This is illustrated by the use of microcapsules, derived from a single encapsulated cell, for high-throughput screening (Zengler et al., 2005). These culturing techniques will undoubtedly improve access to natural products as new microorganisms are isolated in large numbers.

15.6.1   Sponge sampling and isolation of microorganisms

Through the use of culture-independent molecular techniques, new insights into the structure of microbial communities have been gained (Vanwonterghem et al., 2014). Molecular tools have a great potential to assist in developing strategies for identifying previously uncultured bacteria. Identification of organisms via molecular tools can also help in determining appropriate culture media, most effective for isolation and purification of specific microorganisms. These tools allow us to further investigate or exploit microorganisms (Stewart, 2012).

After deciding on the sponges to be sampled and their location, during the sponge collection, careful attention is required to minimize contamination. Sterile ziplock bags or sterile tubes are used accompanied by the rapid transfer of samples to avoid any possible contamination from runoff (Fenical and Jensen, 2006).

Sterilized seawater is used to remove loosely attached microorganisms, and then the sponge sample is cut into small pieces (~1 cm^3). These parts are then homogenized in a prechilled sterilized mortar and pestle using sterilized seawater or phosphate-buffered saline (Zhang et al., 2008b; Abdelmohsen et al., 2010). The process of isolation of actinobacteria or fungi is to dilute the crushed sponge samples with sterile seawater or one-quarter strength Ringer's solution (Pathom-Aree et al., 2006; Bredholdt et al., 2007). After mixing by either vortexing or shaking, they are transferred to the isolation agar media in petri dishes (Jensen et al., 1991; Pathom-Aree et al., 2006).

Isolation media that were used for actinobacteria and for fungi from sponges are presented in Table 15.2 and Table 15.3, respectively.

Table 15.2 Media used for the successful isolation of actinobacteria from sponges

Medium ingredients	References
GA: 6 mL 100% glycerol, 1 g arginine, 1 g K$_2$HPO$_4$, 0.5 g MgSO$_4$ · 7H$_2$O, 18 g agar, and 1 L natural seawater.	Mincer et al. (2002)
Noble: 8 g noble (purified) agar, 500 mg mannitol, 100 mg peptone, 1 L natural seawater, rifampicin (5 µg mL^{-1}), and cycloheximide (100 µg mL^{-1}).	Jensen et al. (2005)
Seawater: Eighteen grams of agar, 1 L natural seawater, rifampicin (5 µg mL^{-1}), and cycloheximide (100 µg mL^{-1}).	Jensen et al. (2005)
PA: 2 g peptone, 0.1 g asparagine, 4 g sodium propionate, 0.5 g K$_2$HPO$_4$, 0.1 g MgSO$_4$, 0.01 g FeSO$_4$ · 7H$_2$O, 5 g glycerol, 20 g NaCl, 18 g Difco Bacto agar, 1 l distilled water, K$_2$Cr$_2$O$_7$ (50 µg mL^{-1}), and 15 µg nalidixic acid (15 µg mL^{-1}).	Zhang et al. (2008b)
Artificial seawater medium (per liter): 23.0 g NaCl, 0.75 g KCl, 1.47 g CaCl$_2$ · 2H$_2$O, 5.08 g MgCl$_2$ · 6H$_2$O, 6.16 g MgSO$_4$ · 7H$_2$O, 5 g NH$_4$Cl, 3.5 g yeast extract, 3.5 g peptone, 0.89 g Na$_2$HPO$_4$ · 2H$_2$O, and 20 g glucose.	Wicke et al. (2000)
Bennett agar (per liter): 1 g yeast extract, 1 g beef extract, 2 g tryptone, 10 g dextrose, and 15 g agar.	Lee et al. (2005)
Bennett agar (as previous) supplemented with nalidixic acid (0.02%) and cycloheximide (0.02%).	Pérez et al. (2009)
Gause's starch medium: 20 g soluble starch, 1 g KNO$_3$, 0.5 g K$_2$HPO$_4$, 0.5 g MgSO$_4$ · 7H$_2$O, 0.5 g NaCl, 0.01 g 7H$_2$O, and 18 g agar 1 l water containing 50% natural seawater, and 0.1 mg mL^{-1} K$_2$Cr$_2$O$_7$. (The pH of the solution was set to 7.2 before sterilization.)	Li et al. (2005)
RH: 10 g raffinose, 1 g L-histidine, 1 g K$_2$HPO$_4$, 0.5 g MgSO$_4$ · 5H$_2$O, 0.01 g FeSO$_4$, and 15 g agar; pH 7.2.	Maldonado et al. (2009)
Starch casein: 10 g soluble starch, 0.3 g casein, 2 g K$_2$HPO$_4$, 2 g KNO$_3$, 2 g NaCl, 0.05 g MgSO$_4$ · 7H$_2$O, 0.02 g CaCO$_3$, 0.01 g FeSO$_4$ · 7H$_2$O, 15 g agar, and 1 L distilled water.	Maldonado et al. (2005b)
Salts: solution A (750 mL artificial seawater containing 1 g K$_2$HPO$_4$ and 10 g Bacto Agar) and solution B (250 mL artificial seawater containing 1 g KNO$_3$, 1 g MgSO$_4$ · 7H$_2$O, 1 g CaCl$_2$ · 2H$_2$O, 0.2 g FeCl$_3$, 0.1 g MnSO$_4$ · 7H$_2$O). (Solutions A and B are autoclaved separately and mixed and supplemented with 1 mL trace element solution)	Magarvey et al. (2004)

Table 15.3 Media used for isolation of fungi from sponges

Formula	References
YMP (per liter): 3 g yeast extract, 3 g malt extract, 5 g peptone, 10 g glucose, and 24.4 g sea salt. (The pH of the medium was adjusted to 7.2–7.4 using 0.1 N NaOH or HCl prior to inoculation.)	Jadulco et al. (2004)
Oatmeal agar: supplemented with 100% seawater.	Cruz et al. (2006)
MEA medium (prepared with 75% seawater, obtained locally): 20 g glucose, 20 g malt extract, 20 g agar, 1 g peptone, penicillin (10,000 units mL^{-1}), and streptomycin (5 mg mL^{-1}).	Lee et al. (2010)
GPY agar: based on natural seawater of 30 PSU, containing per liter 1 g glucose, 0.5 g peptone, 01 g yeast extract, and 15 g agar.	Wiese et al. (2011)
PDA: potato dextrose agar (Difco) with 250 mg L^{-1}. (Chloramphenicol was also mended with different fungicides.)	Paz et al. (2010)
Potato carrot agar (KM) per liter: 20 g cooked and mashed potatoes, 20 g cooked and mashed carrots, 20 g agar with and without artificial sea salts, and cyclosporine A (0.5 mg L^{-1}).	Proksch et al. (2008), Höller et al. (2000)
Modified ISP-4	Huang et al. (2012)

15.6.2   Pretreatment of sponges for selective isolation of actinobacteria

Pretreatment methods have been found successful to isolate rare or selective actinobacteria. Physical methods such as drying the sponges in laminar airflow and dilution have been described (Jensen et al., 2005; Mehbub and Amin, 2012). In addition, dilution and incubation in water bath at 50°C for 6 min and 40°C for 60 min have been described (Kim et al., 2005; Selvin et al., 2009).

However, many pretreatment methods that have been described for isolation of actinobacteria from marine sediments can also be applied for sponge samples: mechanical disruption of sponge tissue using glass beads (Maldonado et al., 2009); physical treatment such as drying, stamping, and dilution (Mincer et al., 2002; Jensen et al., 2005; Gontang et al., 2007); heat treatment such as using dry heat and incubation in different temperatures (Jensen et al., 1991; Mincer et al., 2002, 2005; Bredholt et al., 2008); freezing (Jensen et al., 2005; Bredholdt et al., 2007); radiation (Bredholdt et al., 2007); heat and radiation (Eccleston et al., 2008); centrifugation (Maldonado et al., 2005b); and chemical treatment such as using 1.5% phenol (Bredholt et al., 2008). It has been established that for obligate marine actinobacteria, sodium or salt is a prerequisite (Maldonado et al., 2005a). And, methods could be used to isolate actinobacteria according to the desired result. Even careful use of antibiotics is important for isolating targeted actinobacteria, such as the *Salinispora* strain for which cycloheximide and rifampicin are commonly used (Mincer et al., 2005). Besides, it has been revealed that UV irradiation is another effective method for getting selective isolates of different actinobacterial genera. For instance, high frequency irradiation helped to isolate *Streptosporangium* and *Rhodococcus*, and extremely high frequency irradiation was favorable for isolating *Nocardiopsis*, *Nocardia*, and *Streptosporangium* spp., and UV radiation was effective for isolating *Nocardiopsis*, *Nocardia*, and *Pseudonocardia* spp. (Bredholdt et al., 2007).

Careful selection of different nutrient sources is very important to isolate novel bacteria or fungi because the microbes residing in the sponges are likely to grow under environmental conditions that mimic their source. Moreover, the use of sponge extracts, sediment extracts, different salts, and natural seawater has been practiced in the past. In addition, soluble starch, glycerol, glucose, raffinose, and mannitol were popular as carbon sources, while peptone, yeast, casein, nitrate, histidine, and L-asparagine were popular as nitrogen sources (Jensen et al., 2005; Maldonado et al., 2005b, 2009; Mincer et al., 2005; Gontang et al., 2007; Kennedy et al., 2009).

15.7   Concluding remarks

The discovery of new compounds from the oceans is increasing with over 1000 new compounds per year reported from marine sponges alone. In light of this abundance, the numbers of new compounds reported from microorganisms associated with sponges may seem less significant at around 35 compounds per year in recent years. However, the currently insurmountable problem of getting a continuous supply of a sponge metabolite for drug development if chemical synthesis is not economic will lead to a more intensive search for microbial sources, while efforts are made to culture sponge cells with the capability to produce the desired compounds on a large scale.

It is not surprising that fungi have overtaken the prokaryotes in the search for new secondary metabolites due to their ability to synthesize compounds belonging to the same chemical classes as their sponge hosts. To date, there have been no reports of fungal symbionts producing a sponge-derived metabolite. While fungi from sponge continue to be the dominant type of microbial producers of new compounds, it is important that studies on their ecology and interactions be undertaken to improve our understanding of their role in this ecosystem. One of the benefits will be to realize new isolation methods and techniques to reveal a broader diversity than is currently known from sponge samples. It was noted that a high proportion of the new metabolites are reported without any reported bioactivity associated with them; when there are reports, the bioactivity is from a nonspecific evaluation such as antimicrobial activity, rather than a specific mode of action assay based on a new target site that might reveal a new compound early. Most reports of anticancer activity are also based on nonspecific cytotoxic activity against one or more mammalian cell lines without further detailed assessment as to their mechanisms of action.

Further analysis of the reporting of bioactivity also shows that most compounds are reported to be tested in one assay. Therefore, the challenge is not only to find new compounds but also to develop mechanisms to evaluate their activity in a range of assays. This will allow the full potential of these varied molecules to be realized in the search for new and effective therapies for the range of diseases that are still awaiting a cure. Microbial secondary metabolites are produced by organisms that can be readily scaled up so that the supply of the compound can meet future demand.

References

Abdel-Lateffa, A., Fischa, K., and Wrightd, A. D. 2009. Trichopyrone and other constituents from the marine sponge-derived fungus *Trichoderma* sp. *Z. Naturforsch. B.* 64:186–192.

Abdelmohsen, U. R., Bayer, K., and Hentschel, U. 2014. Diversity, abundance and natural products of marine sponge-associated actinomycetes. *Nat. Prod. Rep.* 31:381–399.

Abdelmohsen, U. R., Pimentel-Elardo, S. M., Hanora, A., Radwan, M., Abou-El-Ela, S. H., Ahmed, S., and Hentschel, U. 2010. Isolation, phylogenetic analysis and anti-infective activity screening of marine sponge-associated actinomycetes. *Mar. Drugs* 8:399–412.

Abdelmohsen, U. R., Zhang, G., Philippe, A., Schmitz, W., Pimentel-Elardo, S. M., Hertlein-Amslinger, B., and Bringmann, G. 2012. Cyclodysidins A–D, cyclic lipopeptides from the marine sponge-derived *Streptomyces* strain RV15. *Tetrahedron Lett.* 53:23–29.

Achtman, M. and Wagner, M. 2008. Microbial diversity and the genetic nature of microbial species. *Nat. Rev. Microbiol.* 6:431–440.

Adachi, K., Kanoh, K., Wisespongp, P., Nishijima, M., and Shizuri, Y. 2005. Clonostachysins A and B, new anti-dinoflagellate cyclic peptides from a marine-derived fungus. *J. Antibiot.* 58:145–150.

Almeida, C., Eguereva, E., Kehraus, S., and Konig, G. M. 2012. Unprecedented Polyketides from a marine sponge-associated *Stachylidium* sp. *J. Nat. Prod.* 76:322–326.

Almeida, C., Kehraus, S., Prudêncio, M., and König, G. M. 2011. Marilones A–C, phthalides from the sponge-derived fungus *Stachylidium* sp. *Beilstein J. Org. Chem.* 7:1636–1642.

Almeida, C., Part, N., Bouhired, S., Kehraus, S., and König, G. M. 2010. Stachylines A–D from the sponge-derived fungus *Stachylidium* sp. *J. Nat. Prod.* 74:21–25.

Amagata, T., Morinaka, B. I., Amagata, A., Tenney, K., Valeriote, F. A., Lobkovsky, E., and Crews, P. 2006. A chemical study of cyclic depsipeptides produced by a sponge-derived fungus. *J. Nat. Prod.* 69:1560–1565.

Amagata, T., Tanaka, M., Yamada, T., Minoura, K., and Numata, A. 2008. Gymnastatins and dankastatins, growth inhibitory metabolites of a *Gymnascella* species from a *Halichondria* sponge. *J. Nat. Prod.* 71:340–345.

Antia, B. S., Aree, T., Kasettrathat, C., Wiyakrutta, S., Ekpa, O. D., Ekpe, U. J., and Kittakoop, P. 2011. Itaconic acid derivatives and diketopiperazine from the marine-derived fungus *Aspergillus aculeatus* CRI322–03. *Phytochemistry.* 72:816–820.

Baltz, R. H., Miao, V., and Wrigley, S. K. 2005. Natural products to drugs: Daptomycin and related lipopeptide antibiotics. *Nat. Prod. Rep.* 22:717–741.

Biabani, M. A. F. and Laatsch, H. 1998. Advances in chemical studies on low-molecular weight metabolites of marine fungi. *J. Prakt. Chem.* 340:589–607.

Blunt, J. W., Copp, B. R., Hu, W.-P., Munro, M. H. G., Northcote, P. T., and Prinsep, M. R. 2007. Marine natural products. *Nat. Prod. Rep.* 24:31–86.

Blunt, J. W., Copp, B. R., Hu, W.-P., Munro, M. H. G., Northcote, P. T., and Prinsep, M. R. 2008. Marine natural products. *Nat. Prod. Rep.* 25:35–94.

Blunt, J. W., Copp, B. R., Hu, W.-P., Munro, M. H. G., Northcote, P. T., and Prinsep, M. R. 2009. Marine natural products. *Nat. Prod. Rep.* 26:170–244.

Blunt, J. W., Copp, B. R., Keyzers, R. A., Munro, M. H. G., and Prinsep, M. R. 2012. Marine natural products. *Nat. Prod. Rep.* 29:144–222.

Blunt, J. W., Copp, B. R., Keyzers, R. A., Munro, M. H. G., and Prinsep, M. R. 2013. Marine natural products. *Nat. Prod. Rep.* 30:237–323.

Blunt, J. W., Copp, B. R., Keyzers, R. A., Munro, M. H. G., and Prinsep, M. R. 2014. Marine natural products. *Nat. Prod. Rep.* 31:160–258.

Blunt, J. W., Copp, B. R., Munro, M. H. G., Northcote, P. T., and Prinsep, M. R. 2003. Marine natural products. *Nat. Prod. Rep.* 20:1–48.

Blunt, J. W., Copp, B. R., Munro, M. H. G., Northcote, P. T., and Prinsep, M. R. 2004. Marine natural products. *Nat. Prod. Rep.* 21:1–49.

Blunt, J. W., Copp, B. R., Munro, M. H. G., Northcote, P. T., and Prinsep, M. R. 2005. Marine natural products. *Nat. Prod. Rep.* 22:15–61.

Blunt, J. W., Copp, B. R., Munro, M. H. G., Northcote, P. T., and Prinsep, M. R. 2006. Marine natural products. *Nat. Prod. Rep.* 23:26–78.

Blunt, J. W., Copp, B. R., Munro, M. H. G., Northcote, P. T., and Prinsep, M. R. 2010. Marine natural products. *Nat. Prod. Rep.* 27:165–237.

Blunt, J. W., Copp, B. R., Munro, M. H. G., Northcote, P. T., and Prinsep, M. R. 2011. Marine natural products. *Nat. Prod. Rep.* 28:196–268.

Boot, C. M., Tenney, K., Valeriote, F. A., and Crews, P. 2006. Highly N-methylated linear peptides produced by an atypical sponge-derived *Acremonium* sp. *J. Nat. Prod.* 69:83–92.

Brady, S. F., Simmons, L., Kim, J. H., and Schmidt, E. W. 2009. Metagenomic approaches to natural products from free-living and symbiotic organisms. *Nat. Prod. Rep.* 26:1488–1503.

Bredholdt, H., Galatenko, O. A., Engelhardt, K., Fjaervik, E., Terekhova, L. P., and Zotchev, S. B. 2007. Rare actinomycete bacteria from the shallow water sediments of the Trondheim fjord, Norway: Isolation, diversity and biological activity. *Environ. Microbiol.* 9:2756–2764.

Bredholt, H., Fjaervik, E., Johnsen, G., and Zotchev, S. B. 2008. *Actinomycetes* from sediments in the Trondheim fjord, Norway: Diversity and biological activity. *Mar. Drugs* 6:12–24.

Bringmann, G., Lang, G., Bruhn, T., Schäffler, K., Steffens, S., Schmaljohann, R., and Imhoff, J. F. 2010. Sorbifuranones A–C, sorbicillinoid metabolites from *Penicillium* strains isolated from Mediterranean sponges. *Tetrahedron* 66:9894–9901.

Bringmann, G., Lang, G., Gulder, T. A., Tsuruta, H., Mühlbacher, J., Maksimenka, K., and Müller, W. E. 2005. The first sorbicillinoid alkaloids, the antileukemic sorbicillactones A and B, from a sponge-derived *Penicillium chrysogenum. Tetrahedron* 61:7252–7265.

Bringmann, G., Lang, G., Steffens, S., and Schaumann, K. 2004. Petrosifungins A and B, novel cyclodepsipeptides from a sponge-derived strain of *Penicillium brevicompactum. J. Nat. Prod.* 67:311–315.

Bugni, T. S. and Ireland, C. M. 2004. Marine-derived fungi: A chemically and biologically diverse group of microorganisms. *Nat. Prod. Rep.* 21:143–163.

Bull, A. T., Stach, J. E., Ward, A. C., and Goodfellow, M. 2005. Marine actinobacteria: Perspectives, challenges, future directions. *Antonie Van Leeuwenhoek* 87:65–79.

Choi, E. J., Kwon, H. C., Ham, J., and Yang, H. O. 2009. 6-Hydroxymethyl-1-phenazine-carboxamide and 1,6-phenazinedimethanol from a marine bacterium, *Brevibacterium* sp. KMD 003, associated with marine purple vase sponge. *J. Antibiot.* 62:621–624.

Cohen, E., Koch, L., Thu, K. M., Rahamim, Y., Aluma, Y., Ilan, M., and Carmeli, S. 2011. Novel terpenoids of the fungus *Aspergillus insuetus* isolated from the Mediterranean sponge *Psammocinia* sp. collected along the coast of Israel. *Biorg. Med. Chem.* 19:6587–6593.

Cruz, L. J., Insua, M. M., Baz, J. P. et al. 2006. IB-01212, a new cytotoxic cyclodepsipeptide isolated from the marine fungus *Clonostachys* sp. ESNA-A009. *J. Org.* 71:3335–3338.

Cueto, M., MacMillan, J. B., Jensen, P. R., and Fenical, W. 2006. Tropolactones A–D, four meroterpenoids from a marine-derived fungus of the genus *Aspergillus. Phytochemistry* 67:1826–1831.

De Rosa, S., De Giulio, A., Tommonaro, G., Popov, S., and Kujumgiev, A. 2000. A β-Amino acid containing tripeptide from a *Pseudomonas–Alteromonas* bacterium associated with a Black Sea sponge. *J. Nat. Prod.* 63:1454–1455.

Ebada, S. S., Schulz, B., Wray, V., Totzke, F., Kubbutat, M. H., Müller, W. E., and Proksch, P. 2011. Arthrinins A–D: Novel diterpenoids and further constituents from the sponge derived fungus *Arthrinium* sp. *Biorg. Med. Chem.* 19:4644–4651.

Eccleston, G. P., Brooks, P. R., and Kurtböke, D. I. 2008. The occurrence of bioactive micromonosporae in aquatic habitats of the Sunshine Coast in Australia. *Mar. Drugs* 6:243–261.

Edrada, R. A., Ebel, R., Supriyono, A., Wray, V., Schupp, P., Steube, K., and Proksch, P. 2002. Swinhoeiamide A, a new highly active Calyculin derivative from the marine sponge *Theonella swinhoei. J. Nat. Prod.* 65:1168–1172.

Edrada, R. A., Wray, V., Berg, A., Grafe, U., Sudarsono, U., Brauers, G., and Proksch, P. 2000. Original communications-novel spiciferone derivatives from the fungus *Drechslera hawaiiensis* isolated from the marine sponge *Callyspongia aerizusa. Z. Naturforsch.* 55:218–221.

Eilers, H., Pernthaler, J., and Amann, R. 2000. Succession of pelagic marine bacteria during enrichment: A close look at cultivation-induced shifts. *Appl. Environ. Microbiol.* 66:4634–4640.

Elbandy, M., Shinde, P. B., Hong, J., Bae, K. S., Kim, M. A., Lee, S. M., and Jung, J. H. 2009. α-Pyrones and yellow pigments from the sponge-derived fungus *Paecilomyces lilacinus. Bull. Korean Chem. Soc.* 30:189.

Faulkner, D., Hai-yin, H., Unson, M., Bewley, C., and Garson, M. 1993. New metabolites from marine sponges: Are symbionts important? *Gazz. Chim. Ital.* 123:301–307.

Faulkner, D. J. 2002. Marine natural products. *Nat. Prod. Rep.* 19:1–48.

Fenical, W. and Jensen, P. R. 2006. Developing a new resource for drug discovery: Marine actinomycete bacteria. *Nat. Chem. Biol.* 2:666–673.

Ferguson, R. L., Buckley, E., and Palumbo, A. 1984. Response of marine bacterioplankton to differential filtration and confinement. *Appl. Environ. Microbiol.* 47:49–55.

Fisher, J. 2014. *Modern NMR Techniques for Synthetic Chemistry.* CRC Press.

Fujii, I. 2009. Heterologous expression systems for polyketide synthases. *Nat. Prod. Rep.* 26:155–169.

Gao, Z., Li, B., Zheng, C., and Wang, G. 2008. Molecular detection of fungal communities in the Hawaiian marine sponges *Suberites zeteki* and *Mycale armata. Appl. Environ. Microbiol.* 74:6091–6101.

Garson, M., Flowers, A. E., Webb, R. I., Charan, R. D., and McCaffrey, E. J. 1998. A sponge/dinoflagellate association in the haplosclerid sponge *Haliclona* sp.: Cellular origin of cytotoxic alkaloids by Percoll density gradient fractionation. *Cell Tissue Res.* 293:365–373.

Gomes, N. M., Dethoup, T., Singburaudom, N., Gales, L., Silva, A., and Kijjoa, A. 2012. Eurocristatine, a new diketopiperazine dimer from the marine sponge-associated fungus *Eurotium cristatum*. *Phytochem Lett.*

Gontang, E. A., Fenical, W., and Jensen, P. R. 2007. Phylogenetic diversity of gram-positive bacteria cultured from marine sediments. *Appl. Environ. Microbiol.* 73:3272–3282.

Govindasamy, V., Franco, C. M., and Gupta, V. V. 2014. Endophytic actinobacteria: Diversity and ecology. *Advances in Endophytic Research*. Springer, pp. 27–59.

Hameş-kocabaş, E. and Uzel, A. 2012. Isolation strategies of marine-derived actinomycetes from sponge and sediment samples. *J. Microbiol. Methods.* 88:342–347.

Hao, G., Qing-Hua, Z., Miao-Miao, J., Jin-Shan, T., Cheng-Du, M., Kui, H., and Xin-Sheng, Y. 2008. Polyketides from a marine sponge-derived fungus *Mycelia sterilia* and proton–proton long-range coupling. *Magn. Reson. Chem.* 46:1148–1152.

Hentschel, U., Fieseler, L., Wehrl, M. et al. 2003. Microbial diversity of marine sponges. *Prog. Mol. Subcell. Biol.* 37:59–88.

Hentschel, U., Hopke, J., Horn, M., Friedrich, A. B., Wagner, M., Hacker, J., and Moore, B. S. 2002. Molecular evidence for a uniform microbial community in sponges from different oceans. *Appl. Environ. Microbiol.* 68:4431–4440.

Hentschel, U., Usher, K. M., and Taylor, M. W. 2006. Marine sponges as microbial fermenters. *FEMS Microbiol. Ecol.* 55:167–177.

Hernández, L. M., Blanco, J. A., Baz, J. P., Puentes, J. L., Millán, F. R., Vázquez, F. E., Fernández-Chimeno, R. I., and Grávalos, D. G. 2000. 4'-N-methyl-5'-hydroxystaurosporine and 5'-hydroxystaurosporine, new indolocarbazole alkaloids from a marine *Micromonospora* sp. strain. *J. Antibiot.* 53:895–902.

Höller, U., Wright, A. D., Matthee, G. F., Konig, G. M., Draeger, S., Aust, H. J., and Schulz, B. 2000. Fungi from marine sponges: Diversity, biological activity and secondary metabolites. *Mycol. Res.* 104:1354–1365.

Hosoya, T., Hirokawa, T., Takagi, M., and Shin-ya, K. 2012. Trichostatin analogues JBIR-109, JBIR-110, and JBIR-111 from the marine sponge-derived *Streptomyces* sp. RM72. *J. Nat. Prod.* 75:285–289.

Huang, H., Yang, T., Ren, X., Liu, J., Song, Y., Sun, A., and Ju, J. 2012. Cytotoxic angucycline class glycosides from the deep sea actinomycete *Streptomyces lusitanus* SCSIO LR32. *J. Nat. Prod.* 75:202–208.

Huang, X. L., Gao, Y., Xue, D. Q., Liu, H. L., Peng, C. S., Zhang, F. L., and Guo, Y. W. 2011. Streptomycindole, an Indole alkaloid from a marine *Streptomyces* sp. DA22 associated with South China sea sponge *Craniella australiensis*. *Helv. Chim. Acta.* 94:1838–1842.

Hugenholtz, P. and Pace, N. R. 1996. Identifying microbial diversity in the natural environment: A molecular phylogenetic approach. *Trends Biotechnol.* 14:190–197.

Ingavat, N., Dobereiner, J., Wiyakrutta, S., Mahidol, C., Ruchirawat, S., and Kittakoop, P. 2009. Aspergillusol A, an α-glucosidase inhibitor from the marine-derived fungus *Aspergillus aculeatus*. *J. Nat. Prod.* 72:2049–2052.

Ingavat, N., Mahidol, C., Ruchirawat, S., and Kittakoop, P. 2011. Asperaculin A, a sesquiterpenoid from a marine-derived fungus, *Aspergillus aculeatus*. *J. Nat. Prod.* 74:1650–1652.

Izumikawa, M., Khan, S. T., and Komaki, H. 2009. JBIR-37 and -38, novel glycosyl benzenediols, isolated from the sponge-derived fungus, *Acremonium* sp. SpF080624G1f01. *Biosci. Biotechnol. Biochem.* 73:2138–2140.

Jadulco, R., Edrada, R. A., Ebel, R., Berg, A., Schaumann, K., Wray, V., and Proksch, P. 2004. New communesin derivatives from the fungus *Penicillium* sp. derived from the Mediterranean sponge *Axinella verrucosa*. *J. Nat. Prod.* 67:78–81.

Jensen, P. R., Dwight, R., and Fenical, W. 1991. Distribution of actinomycetes in near-shore tropical marine sediments. *Appl. Environ. Microbiol.* 57:1102–1108.

Jensen, P. R., Gontang, E., Mafnas, C., Mincer, T. J., and Fenical, W. 2005. Culturable marine actinomycete diversity from tropical Pacific Ocean sediments. *Environ. Microbiol.* 7:1039–1048.

Julianti, E., Oh, H., Jang, K. H., Lee, J. K., Lee, S. K., Oh, D. C., and Shin, J. 2011. Acremostrictin, a highly oxygenated metabolite from the marine fungus *Acremonium strictum*. *J. Nat. Prod.* 74:2592–2594.

Julianti, E., Oh, H., Lee, H.-S., Oh, D.-C., Oh, K.-B., and Shin, J. 2012. Acremolin, a new 1H-azirine metabolite from the marine-derived fungus *Acremonium strictum*. *Tetrahedron Lett.* 53:2885–2886.

Kaeberlein, T., Lewis, K., and Epstein, S. S. 2002. Isolating "uncultivable" microorganisms in pure culture in a simulated natural environment. *Science* 296:1127–1129.

Kaewkla, O. and Franco, C. M. 2013. *Kribbella endophytica* sp. *nov.*, an endophytic actinobacterium isolated from the surface-sterilized leaf of a native apricot tree. *Int. J. Syst. Evol. Microbiol.* 63:1249–1253.

Kanoh, K., Kamino, K., Leleo, G., Adachi, K., and Shizuri, Y. 2003. Pseudoalterobactin A and B, new siderophores excreted by marine bacterium *Pseudoalteromonas* sp. KP20–4. *J. Antibiot.* 56:871–875.

Kawahara, T., Takagi, M., and Shin-ya, K. 2012. Three new depsipeptides, JBIR-113, JBIR-114 and JBIR-115, isolated from a marine sponge-derived *Penicillium* sp. fS36. *J. Antibiot.* 65:147–150.

Kennedy, J., Baker, P., Piper, C., Cotter, P. D., Walsh, M., Mooij, M. J., and Dobson, A. D. 2009. Isolation and analysis of bacteria with antimicrobial activities from the marine sponge *Haliclona simulans* collected from Irish waters. *Mar. Biotechnol.* 11:384–396.

Kim, T. K., Garson, M. J., and Fuerst, J. A. 2005. Marine actinomycetes related to the 'Salinospora' group from the Great Barrier Reef sponge *Pseudoceratina clavata*. *Environ. Microbiol.* 7:509–518.

Kito, K., Ookura, R., Yoshida, S., Namikoshi, M., Ooi, T., and Kusumi, T. 2007. Pentaketides relating to aspinonene and dihydroaspyrone from a marine-derived fungus, *Aspergillus ostianus*. *J. Nat. Prod.* 70:2022–2025.

Kito, K., Ookura, R., Yoshida, S., Namikoshi, M., Ooi, T., and Kusumi, T. 2008. New cytotoxic 14-membered macrolides from marine-derived fungus *Aspergillus ostianus*. *Org. Lett.* 10:225–228.

Koehn, F. E. and Carter, G. T. 2005. The evolving role of natural products in drug discovery. *Nat. Rev. Drug Discovery.* 4:206–220.

König, G. M., Kehraus, S., Seibert, S. F., Abdel-Lateff, A., and Müller, D. 2006. Natural products from marine organisms and their associated microbes. *ChemBioChem* 7:229–238.

Krick, A., Kehraus, S., Eberl, L., Riedel, K., Anke, H., Kaesler, I., and König, G. M. 2007. A marine *Mesorhizobium* sp. produces structurally novel long-chain N-acyl-L-homoserine lactones. *Appl. Environ. Microbiol.* 73:3587–3594.

Lam, K. S. 2006. Discovery of novel metabolites from marine actinomycetes. *Curr. Opin. Microbiol.* 9:245–251.

Lazzarini, A., Cavaletti, L., Toppo, G., and Marinelli, F. 2001. Rare genera of actinomycetes as potential producers of new antibiotics. *Antonie Van Leeuwenhoek* 79:399–405.

Lee, H.-S., Shin, H. J., Jang, K. H., Kim, T. S., Oh, K.-B., and Shin, J. 2005. Cyclic peptides of the nocardamine class from a marine-derived bacterium of the genus *Streptomyces*. *J. Nat. Prod.* 68:623–625.

Lee, Y. M., Dang, H. T., Li, J., Zhang, P., Hong, J., Lee, C. O., and Jung, J. H. 2011. A cytotoxic fellutamide analogue from the sponge-derived fungus *Aspergillus versicolor*. *Bull. Korean Chem. Soc.* 32:3817–3820.

Lee, Y. M., Hong, J., Lee, C.-O., Bae, K. S., Kim, D.-K., and Jung, J. H. 2010. A cytotoxic lipopeptide from the sponge-derived fungus *Aspergillus versicolor*. *Bull. Korean Chem. Soc.* 31:205–208.

Li, C. S., An, C. Y., Li, X. M., Gao, S. S., Cui, C. M., Sun, H. F., and Wang, B. G. 2011a. Triazole and dihydroimidazole alkaloids from the marine sediment-derived fungus *Penicillium paneum* SD-44. *J. Nat. Prod.* 74:1331–1334.

Li, D., Xu, Y., Shao, C. L., Yang, R. Y., Zheng, C. J., Chen, Y. Y., and Wang, C. Y. 2012. Antibacterial Bisabolene-type sesquiterpenoids from the sponge-derived fungus *Aspergillus* sp. *Mar. Drugs* 10:234–241.

Li, F., Maskey, R. P., Qin, S., Sattler, I., Fiebig, H. H., Maier, A., and Laatsch, H. 2005. Chinikomycins A and B: Isolation, structure elucidation, and biological activity of novel antibiotics from a marine *Streptomyces* sp. isolate M045#, 1. *J. Nat. Prod.* 68:349–353.

Li, J. L., Zhang, P., Lee, Y. M., Hong, J., Yoo, E. S., Bae, K. S., and Jung, J. H. 2011b. Oxygenated hexylitaconates from a marine sponge-derived fungus *Penicillium* sp. *Chem. Pharm. Bull.* 59:120–123.

Li, Z.-Y., He, L.-M., Wu, J., and Jiang, Q. 2006. Bacterial community diversity associated with four marine sponges from the South China Sea based on 16S rDNA-DGGE fingerprinting. *J. Exp. Mar. Biol. Ecol.* 329:75–85.

Lin, W. H., Fu, H. Z., Li, J., and Proksch, P. 2001a. Novel chromone derivatives from marine fungus *Aspergillus versicolor* isolated from the sponge *Xestospongia exigua*. *Chin. Chem. Lett.* 12:235–238.

Lin, W. H., Li, J., Fu, H. Z., and Proksch, P. 2001b. Four novel hydropyranoindeno-derivatives from marine fungus *Aspergillus versicolor*. *Chin. Chem. Lett.* 12:435–438.

Liu, H. B., Edrada-Ebel, R., Ebel, R., Wang, Y., Schulz, B., Draeger, S., and Proksch, P. 2011. Ophiobolin sesterterpenoids and pyrrolidine alkaloids from the sponge-derived fungus *Aspergillus ustus*. *Helv. Chim. Acta.* 94:623–631.

Magarvey, N. A., Keller, J. M., Bernan, V., Dworkin, M., and Sherman, D. H. 2004. Isolation and characterization of novel marine-derived actinomycete taxa rich in bioactive metabolites. *Appl. Environ. Microbiol.* 70:7520–7529.

Maldonado, L., Fragoso-Yáñez, D., Pérez-García, A., Rosellón-Druker, J., and Quintana, E. 2009. Actinobacterial diversity from marine sediments collected in Mexico. *Antonie Van Leeuwenhoek* 95:111–120.

Maldonado, L. A., Fenical, W., Jensen, P. R., Kauffman, C. A., Mincer, T. J., Ward, A. C., and Goodfellow, M. 2005a. *Salinispora arenicola* gen. nov., sp. nov. and *Salinispora tropica* sp. nov., obligate marine actinomycetes belonging to the family Micromonosporaceae. *Int. J. Syst. Evol. Microbiol.* 55:1759–1766.

Maldonado, L. A., Stach, J. E. M., Pathom-Aree, W., Ward, A. C., Bull, A. T., and Goodfellow, M. 2005b. Diversity of cultivable actinobacteria in geographically widespread marine sediments. *Antonie Van Leeuwenhoek* 87:11–18.

Malmstrøm, J., Christophersen, C., Barrero, A. F., Oltra, J. E., Justicia, J., and Rosales, A. 2002. Bioactive metabolites from a marine-derived strain of the fungus *Emericella variecolor*. *J. Nat. Prod.* 65:364–367.

Mehbub, M., Lei, J., Franco, C., and Zhang, W. 2014. Marine sponge derived natural products between 2001 and 2010: Trends and opportunities for discovery of bioactives. *Mar. Drugs* 12:4539–4577.

Mehbub, M. F. and Amin, A. K. M. R. 2012. Isolation and identification of actinobacteria from two South Australian marine sponges *Aplysilla rosea* and *Aplysina* sp. *BRP.* 7:345–354.

Mincer, T. J., Fenical, W., and Jensen, P. R. 2005. Culture-dependent and culture-independent diversity within the obligate marine actinomycete genus *Salinispora*. *Appl. Environ. Microbiol.* 71:7019–7028.

Mincer, T. J., Jensen, P. R., Kauffman, C. A., and Fenical, W. 2002. Widespread and persistent populations of a major new marine actinomycete taxon in ocean sediments. *Appl. Environ. Microbiol.* 68:5005–5011.

Mitova, M., Popov, S., and De Rosa, S. 2004. Cyclic peptides from a *Ruegeria* strain of bacteria associated with the sponge *Suberites domuncula*. *J. Nat. Prod.* 67:1178–1181.

Mitova, M. I., Lang, G., Wiese, J., and Imhoff, J. F. 2008. Subinhibitory concentrations of antibiotics induce phenazine production in a marine *Streptomyces* sp. *J. Nat. Prod.* 71:824–827.

Mohamed, I. E., Gross, H., Pontius, A., Kehraus, S., Krick, A., Kelter, G., and König, G. M. 2009. Epoxyphomalin A and B, prenylated polyketides with potent cytotoxicity from the marine-derived fungus *Phoma* sp. *Org. Lett.* 11:5014–5017.

Mohamed, I. E., Kehraus, S., Krick, A., König, G. M., Kelter, G., Maier, A., and Gross, H. 2010. Mode of action of epoxyphomalins A and B and characterization of related metabolites from the marine-derived fungus *Paraconiothyrium* sp. *J. Nat. Prod.* 73:2053–2056.

Montalvo, N. F., Mohamed, N. M., Enticknap, J. J., and Hill, R. T. 2005. Novel actinobacteria from marine sponges. *Antonie Van Leeuwenhoek* 87:29–36.

Motohashi, K., Inaba, K., Fuse, S., Doi, T., Izumikawa, M., Khan, S. T., and Shin-ya, K. 2011. JBIR-56 and JBIR-57, 2(1H)-pyrazinones from a marine sponge-derived *Streptomyces* sp. SpD081030SC-03. *J. Nat. Prod.* 74:1630–1635.

Motohashi, K., Inaba, S., Takagi, M., and Shin-ya, K. 2009. JBIR-15, a new aspochracin derivative, isolated from a sponge-derived fungus, *Aspergillus sclerotiorum* Huber Sp080903f04. *Biosci. Biotechnol. Biochem.* 73:1898–1900.

Motohashi, K., Takagi, M., and Shin-ya, K. 2010. Tetrapeptides possessing a unique skeleton, JBIR-34 and JBIR-35, isolated from a sponge-derived actinomycete, *Streptomyces* sp. Sp080513GE-23. *J. Nat. Prod.* 73:226–228.

Muldoon, J., Shashkov, A. S., Sof'ya, N. S., Tomshich, S. V., Komandrova, N. A., Romanenko, L. A., and Savage, A. V. 2001. Structure of an acidic polysaccharide from a marine bacterium *Pseudoalteromonas distincta* KMM 638 containing 5-acetamido-3,5,7,9-tetradeoxy-7-formamido-L-glycero-L-manno-nonulosonic acid. *Carbohydr. Res.* 330:231–239.

Nagai, K., Kamigiri, K., Arao, N., Suzumura, K. I., Kawano, Y., Yamaoka, M., and Suzuki, K. 2003. YM-266183 and YM-266184, novel thiopeptide antibiotics produced by *Bacillus cereus* isolated from a marine sponge. I. Taxonomy, fermentation, isolation, physico-chemical properties and biological properties. *J. Antibiot.* 56:123–128.

Newman, D. J. and Cragg, G. M. 2007. Natural products as sources of new drugs over the last 25 years. *J. Nat. Prod.* 70:461–477.

Newman, D. J., Cragg, G. M., and Snader, K. M. 2003. Natural products as sources of new drugs over the period 1981–2002. *J. Nat. Prod.* 66:1022–1037.

Neumann, K., Abdel-Lateff, A., Wright, A. D., Kehraus, S., Krick, A., and König, G. M. 2007. Novel Sorbicillin derivatives with an unprecedented carbon skeleton from the sponge-derived fungus *Trichoderma* species. *Eur. J. Org. Chem.* 2007:2268–2275.

Neumann, K., Kehraus, S., Gütschow, M., and König, G. 2009. Cytotoxic and HLE-inhibitory tetramic acid derivatives from marine-derived fungi. *Nat. Prod. Commun.* 4:347–354.

Ookura, R., Kito, K., Ooi, T., Namikoshi, M., and Kusumi, T. 2008. Structure revision of circumdatins A and B, benzodiazepine alkaloids produced by marine fungus *Aspergillus ostianus*, by x-ray crystallography. *J. Org.* 73:4245–4247.

Otoguro, M., Hayakawa, M., Yamazaki, T., and Iimura, Y. 2001. An integrated method for the enrichment and selective isolation of *Actinokineospora* spp. in soil and plant litter. *J. Appl. Microbiol.* 91:118–130.

Pathom-Aree, W., Stach, J. E., Ward, A. C., Horikoshi, K., Bull, A. T., and Goodfellow, M. 2006. Diversity of actinomycetes isolated from Challenger Deep sediment (10,898 m) from the Mariana Trench. *Extremophiles* 10:181–189.

Paul, V. J. and Ritson-Williams, R. 2008. Marine chemical ecology. *Nat. Prod. Rep.* 25:662–695.

Pawlik, J. R. 1992. Chemical ecology of the settlement of benthic marine invertebrates. *Oceanogr. Mar. Biol. Annu. Rev.* 30:273–335.

Paz, Z., Komon-Zelazowska, M., Druzhinina, I. S., Aveskamp, M. M., Shnaiderman, A., Aluma, Y., and Yarden, O. 2010. Diversity and potential antifungal properties of fungi associated with a Mediterranean sponge. *Fungal Divers.* 42:17–26.

Pérez, M., Schleissner, C., Rodríguez, P., Zúñiga, P., Benedit, G., Sánchez-Sancho, F., and de la Calle, F. 2009. PM070747, a new cytotoxic angucyclinone from the marine-derived *Saccharopolyspora taberi* PEM-06-F23–019B. *J. Antibiot.* 62:167–169.

Piel, J. 2004. Metabolites from symbiotic bacteria. *Nat. Prod. Rep.* 21:519–538.

Piel, J. 2006. Bacterial symbionts: Prospects for the sustainable production of invertebrate-derived pharmaceuticals. *Curr. Med. Chem.* 13:39–50.

Piel, J. 2009. Deep sequencing reveals exceptional diversity and modes of transmission for bacterial sponge symbionts. *Nat. Prod. Rep.* 26:338–362.

Piel, J., Hui, D., Wen, G., Butzke, D., Platzer, M., Fusetani, N., and Matsunaga, S. 2004. Antitumor polyketide biosynthesis by an uncultivated bacterial symbiont of the marine sponge *Theonella swinhoei*. *Proc. Natl. Acad. Sci. U S A* 101:16222–16227.

Pimentel-Elardo, S. M., Buback, V., Gulder, T. A., Bugni, T. S., Reppart, J., Bringmann, G., and Hentschel, U. 2011. New tetromycin derivatives with anti-trypanosomal and protease inhibitory activities. *Mar. Drugs* 9:1682–1697.

Pimentel-Elardo, S. M., Gulder, T. A. M., Hentschel, U., and Bringmann, G. 2008. Cebulactams A1 and A2, new macrolactams isolated from *Saccharopolyspora cebuensis*, the first obligate marine strain of the genus *Saccharopolyspora*. *Tetrahedron Lett.* 49:6889–6892.

Pinheiro, Â., Dethoup, T., Bessa, J., Silva, A. M. S., and Kijjoa, A. 2012. A new bicyclic sesquiterpene from the marine sponge associated fungus *Emericellopsis minima*. *Phytochem. Lett.* 5:68–70.

Prachyawarakorn, V., Mahidol, C., Sureram, S., Sangpetsiripan, S., Wiyakrutta, S., Ruchirawat, S., and Kittakoop, P. 2008. Diketopiperazines and phthalides from a marine derived fungus of the order Pleosporales. *Planta Med.* 74:69–72.

Proksch, P., Ebel, R., Edrada, R., Riebe, F., Liu, H., Diesel, A., and Schulz, B. 2008. Sponge-associated fungi and their bioactive compounds: The *Suberites case. Bot. Mar.* 51:209–218.

Proksch, P., Edrada, R., and Ebel, R. 2002. Drugs from the seas—Current status and microbiological implications. *Appl. Microbiol. Biotechnol.* 59:125–134.

Pruksakorn, P., Arai, M., Kotoku, N., Vilchèze, C., Baughn, A. D., Moodley, P., and Kobayashi, M. 2010. Trichoderins, novel aminolipopeptides from a marine sponge-derived *Trichoderma* sp., are active against dormant mycobacteria. *Bioorg. Med. Chem. Lett.* 20:3658–3663.

Quévrain, E., Domart-Coulon, I., Pernice, M., and Bourguet-Kondracki, M. L. 2009. Novel natural parabens produced by a *Microbulbifer bacterium* in its calcareous sponge host *Leuconia nivea. Environ. Microbiol.* 11:1527–1539.

Rappe, M. S., Connon, S. A., Vergin, K. L., and Giovannoni, S. J. 2002. Cultivation of the ubiquitous SAR11 marine bacterioplankton clade. *Nature* 418:630–633.

Rungprom, W., Siwu, E. R., Lambert, L. K., Dechsakulwatana, C., Barden, M. C., Kokpol, U., and Garson, M. J. 2008. Cyclic tetrapeptides from marine bacteria associated with the seaweed *Diginea* sp. and the sponge *Halisarca ectofibrosa. Tetrahedron* 64:3147–3152.

San-Martin, A., Rovirosa, J., Vaca, I., Vergara, K., Acevedo, L., Vina, D., and Chamy, M. C. 2011. New butyrolactone from a marine-derived fungus *Aspergillus* sp. *J. Chil. Chem. Soc.* 56:625–627.

Sarethy, I. P., Pan, S., and Danquah, M. K. 2014. Modern taxonomy for microbial diversity, biodiversity. In *The Dynamic Balance of the Planet*, Grillo, O. (Ed.), pp. 51–68.

Schmitt, S., Angermeier, H., Schiller, R., Lindquist, N., and Hentschel, U. 2008. Molecular microbial diversity survey of sponge reproductive stages and mechanistic insights into vertical transmission of microbial symbionts. *Appl. Environ. Microbiol.* 74:7694–7708.

Schneemann, I., Kajahn, I., Ohlendorf, B., Zinecker, H., Erhard, A., Nagel, K., and Imhoff, J. F. 2010a. Mayamycin, a cytotoxic polyketide from a *Streptomyces* strain isolated from the marine sponge *Halichondria panicea. J. Nat. Prod.* 73:1309–1312.

Schneemann, I., Ohlendorf, B., Zinecker, H., Nagel, K., Wiese, J., and Imhoff, J. F. 2010b. Nocapyrones A–D, γ-pyrones from a *Nocardiopsis* strain isolated from the marine sponge *Halichondria panicea. J. Nat. Prod.* 73:1444–1447.

Selvin, J., Shanmughapriya, S., Gandhimathi, R., Kiran, G. S., Ravji, T. R., Natarajaseenivasan, K., and Hema, T. A. 2009. Optimization and production of novel antimicrobial agents from sponge associated marine actinomycetes *Nocardiopsis dassonvillei* MAD08. *Appl. Microbiol. Biotechnol.* 83:435–445.

Shindo, K., Asagi, E., Sano, A., Hotta, E., Minemura, N., Mikami, K., and Maoka, T. 2007. Diapolycopenedioic acid xylosyl ester, a novel glyco-C-carotenoic acid produced by a new marine bacterium *Rubritalea squalenifaciens. Tetrahedron Lett.* 48:2725–2727.

Shindo, K., Asagi, E., Sano, A., Hotta, E., Minemura, N., Mikami, K., Tamesada, E., Misawa, N., and Maoka, T. 2008. Diapolycopenedioic acid xylosyl esters A, B, and C, novel antioxidative glyco-C30-carotenoic acids produced by a new marine bacterium *Rubritalea squalenifaciens. J. Antibiot.* 61:185–191.

Simmons, L., Kaufmann, K., Garcia, R., Schwar, G., Huch, V., and Muller, R. 2011. Bendigoles D–F, bioactive sterols from the marine sponge-derived *Actinomadura* sp. SBMs009. *Bioorg. Med. Chem.* 19:6570–6575.

Smith, C. J., Abbanat, D., Bernan, V. S., Maiese, W. M., Greenstein, M., Jompa, J., and Ireland, C. M. 2000. Novel polyketide metabolites from a species of marine fungi. *J. Nat. Prod.* 63:142–145.

Speitling, M., Smetanina, O. F., Kuznetsova, T. A., and Laatsch, H. 2007. Bromoalterochromides A and A′, unprecedented chromopeptides from a marine *Pseudoalteromonas maricaloris* strain KMM 636T. *J. Antibiot.* 60:36–42.

Stewart, E. J. 2012. Growing unculturable bacteria. *J. Bacteriol.* 194:4151–4160.

Sugiyama, Y., Ito, Y., Suzuki, M., and Hirota, A. 2009. Indole derivatives from a marine sponge-derived yeast as DPPH radical scavengers. *J. Nat. Prod.* 72:2069–2071.

Sun, L. L., Shao, C. L., Chen, J. F., Guo, Z. Y., Fu, X. M., Chen, M., and Wang, C. Y. 2012. New bisabolane sesquiterpenoids from a marine-derived fungus *Aspergillus* sp. isolated from the sponge *Xestospongia testudinaria. Bioorg. Med. Chem. Lett.* 22:1326–1329.

Supong, K., Thawai, C., Suwanborirux, K., Choowong, W., Supothina, S., and Pittayakhajonwut, P. 2012. Antimalarial and antitubercular C-glycosylated benz[α]anthraquinones from the marine-derived *Streptomyces* sp. BCC45596. *Phytochem. Lett.* 5:651–656.

Sureram, S., Wiyakrutta, S., Ngamrojanavanich, N., Mahidol, C., Ruchirawat, S., and Kittakoop, P. 2012. Depsidones, aromatase inhibitors and radical scavenging agents from the marine-derived fungus *Aspergillus unguis* CRI282–03. *Planta Med.* 78:582–588.

Suryanarayanan, T. S. 2012. The diversity and importance of fungi associated with marine sponges. *Bot. Mar.* 55:553–564.

Takagi, M., Motohashi, K., Khan, S. T., Hashimoto, J., and Shin-ya, K. 2010a. JBIR-65, a new diterpene, isolated from a sponge-derived *Actinomadura* sp. SpB081030SC-15. *J. Antibiot.* 63:401–403.

Takagi, M., Motohashi, K., and Shin-ya, K. 2010b. Isolation of 2 new metabolites, JBIR-74 and JBIR-75, from the sponge-derived *Aspergillus* sp. fS14. *J. Antibiot.* 63:393–395.

Takano, D., Nagamitsu, T., Ui, H., Shiomi, K., Yamaguchi, Y., Masuma, R., and Ōmura, S. 2001. Absolute configuration of nafuredin, a new specific NADH-fumarate reductase inhibitor. *Tetrahedron Lett.* 42:3017–3020.

Taylor, M. W., Radax, R., Steger, D., and Wagner, M. 2007. Sponge-associated microorganisms: Evolution, ecology, and biotechnological potential. *Microbiol. Mol. Biol. Rev.* 71:295–347.

Taylor, M. W., Schupp, P. J., Dahllöf, I., Kjelleberg, S., and Steinberg, P. D. 2004. Host specificity in marine sponge-associated bacteria, and potential implications for marine microbial diversity. *Environ. Microbiol.* 6:121–130.

Thacker, R. and Starnes, S. 2003. Host specificity of the symbiotic cyanobacterium *Oscillatoria spongeliae* in marine sponges, *Dysidea* spp. *Mar. Biol.* 142:643–648.

Ueda, J.-Y., Khan, S. T., Takagi, M., and Shin-ya, K. 2010a. JBIR-58, a new salicylamide derivative, isolated from a marine sponge-derived *Streptomyces* sp. SpD081030ME-02. *J. Antibiot.* 63:267–269.

Ueda, J.-Y., Takagi, M., and Shin-ya, K. 2010b. New xanthoquinodin-like compounds, JBIR-97, -98 and-99, obtained from marine sponge-derived fungus *Tritirachium* sp. SpB081112MEf2. *J. Antibiot.* 63:615–618.

Vacelet, J. 1981. Algal-sponge symbioses in the coral reefs of New Caledonia: A morphological study. *Proceedings of the 4th International Coral Reef Symposium*, Manila, May 18–22, 1981, Vol 2. pp. 713–719.

Vacelet, J. and Donadey, C. 1977. Electron microscope study of the association between some sponges and bacteria. *J. Exp. Mar. Biol. Ecol.* 30:301–314.

Valliappan, K., Sun, W., and Li, Z. 2014. Marine actinobacteria associated with marine organisms and their potentials in producing pharmaceutical natural products. *Appl. Microbiol. Biotechnol.* 98:7365–7377.

Van Soest, R. W., Boury-Esnault, N., Vacelet, J., Dohrmann, M., Erpenbeck, D., De Voogd, N. J., and Hooper, J. N. 2012. Global diversity of sponges (Porifera). *PLoS One* 7:e35105.

Vandamme, P., Pot, B., Gillis, M., De Vos, P., Kersters, K., and Swings, J. 1996. Polyphasic taxonomy, a consensus approach to bacterial systematics. *Microbiol. Rev.* 60:407–438.

Vanwonterghem, I., Jensen, P. D., Ho, D. P., Batstone, D. J., and Tyson, G. W. 2014. Linking microbial community structure, interactions and function in anaerobic digesters using new molecular techniques. *Curr. Opin. Biotechnol.* 27:55–64.

Varoglu, M. and Crews, P. 2000. Biosynthetically diverse compounds from a saltwater culture of sponge-derived *Aspergillus niger*. *J. Nat. Prod.* 63:41–43.

Vogel, S. 1977. Current-induced flow through living sponges in nature. *Proc. Natl. Acad Sci. U S A* 74:2069–2071.

Wang, C.-Y., Wang, B.-G., Brauers, G., Guan, H.-S., Proksch, P., and Ebel, R. 2002. Microsphaerones A and B, two novel γ-pyrone derivatives from the sponge-derived fungus *Microsphaeropsis* sp. *J. Nat. Prod.* 65:772–775.

Webster, N. S. and Taylor, M. W. 2012. Marine sponges and their microbial symbionts: Love and other relationships. *Environ. Microbiol.* 14:335–346.

Webster, N. S., Wilson, K. J., Blackall, L. L., and Hill, R. T. 2001. Phylogenetic diversity of bacteria associated with the marine sponge *Rhopaloeides odorabile*. *Appl. Environ. Microbiol.* 67:434–444.

Wei, R. B., Xi, T., Li, J., Wang, P., Li, F. C., Lin, Y. C., and Qin, S. 2011. Lobophorin C and D, new kijanimicin derivatives from a marine sponge-associated actinomycetal strain AZS17. *Mar. Drugs* 9:359–368.

Wicke, C., Hüners, M., Wray, V., Nimtz, M., Bilitewski, U., and Lang, S. 2000. Production and structure elucidation of glycoglycerolipids from a marine sponge-associated *Microbacterium* species. *J. Nat. Prod.* 63:621–626.

Wiese, J., Ohlendorf, B., Blümel, M., Schmaljohann, R., and Imhoff, J. F. 2011. Phylogenetic identification of fungi isolated from the marine sponge *Tethya aurantium* and identification of their secondary metabolites. *Mar. Drugs* 9:561–585.

Wyche, T. P., Hou, Y., Braun, D., Cohen, H. C., Xiong, M. P., and Bugni, T. S. 2011. First natural analogs of the cytotoxic thiodepsipeptide thiocoraline A from a marine *Verrucosispora* sp. *J. Org.* 76:6542–6547.

Xin, Z.-H., Zhu, W.-M., Gu, Q.-Q., Fang, Y.-C., Duan, L., and Cui, C.-B. 2005. A new cytotoxic compound from *Penicillium aurantiogriseum*, symbiotic or epiphytic fungus of sponge *Mycale plumose*. *Chin. Chem. Lett.* 16:1227–1229.

Xin, Z. H., Fang, Y., Du, L., Zhu, T., Duan, L., Chen, J., and Zhu, W. M. 2007. Aurantiomides A–C, quinazoline alkaloids from the sponge-derived fungus *Penicillium aurantiogriseum* SPO–19. *J. Nat. Prod.* 70:853–855.

Xu, J., Takasaki, A., Kobayashi, H., Oda, T., Yamada, J., Mangindaan, R. E., and Namikoshi, M. 2006. Four new macrocyclic trichothecenes from two strains of marine-derived fungi of the genus *Myrothecium*. *J. Antibiot.* 59:451–455.

Yamazaki, H., Rotinsulu, H., Kaneko, T., Murakami, K., Fujiwara, H., Ukai, K., and Namikoshi, M. 2012. A new dibenz [b,e] oxepine derivative, 1-hydroxy-10-methoxy-dibenz[b,e]oxepin-6,11-dione, from a marine-derived fungus, *Beauveria bassiana* TPU942. *Mar. Drugs* 10:2691–2697.

Yarden, O. 2014. Fungal association with sessile marine invertebrates. *Front. Microbiol.* 5:228.

You, M.-X., Zhang, H.-P., and Hu, C.-Q. 2008. Isolation and characterization of three siderophores from marine bacteria. *Chin. J. Chem.* 26:1332–1334.

Yu, L.-L., Li, Z.-Y., Peng, C.-S., Li, Z.-Y., and Guo, Y.-W. 2009. Neobacillamide A, a novel thiazole-containing alkaloid from the marine bacterium *Bacillus vallismortis* C89, associated with South China Sea sponge *Dysidea avara*. *Helv. Chim. Acta.* 92:607–612.

Yu, Z., Lang, G., Kajahn, I., Schmaljohann, R., and Imhoff, J. F. 2008. Scopularides A and B, cyclodepsipeptides from a marine sponge-derived fungus, *Scopulariopsis brevicaulis*. *J. Nat. Prod.* 71:1052–1054.

Zengler, K., Walcher, M., Clark, G., Haller, I., Toledo, G., Holland, T., and Keller, M. 2005. High-throughput cultivation of microorganisms using microcapsules. *Methods Enzymol.* 397:124–130.

Zhang, D., Yang, X., Kang, J. S., Choi, H. D., and Son, B. W. 2008a. Chlorohydroaspyrones A and B, antibacterial aspyrone derivatives from the marine-derived fungus *Exophiala* sp. *J. Nat. Prod.* 71:1458–1460.

Zhang, P., Bao, B., Dang, H. T., Hong, J., Lee, H. J., Yoo, E. S., and Jung, J. H. 2009. Anti-inflammatory sesquiterpenoids from a sponge-derived fungus *Acremonium* sp. *J. Nat. Prod.* 72:270–275.

Zhang, H., Lee, Y. K., Zhang, W., and Lee, H. K. 2006. Culturable actinobacteria from the marine sponge *Hymeniacidon perleve*: Isolation and phylogenetic diversity by 16S rRNA gene-RFLP analysis. *Antonie Van Leeuwenhoek* 90:159–169.

Zhang, H., Zhang, W., Jin, Y., Jin, M., and Yu, X. 2008b. A comparative study on the phylogenetic diversity of culturable actinobacteria isolated from five marine sponge species. *Antonie Van Leeuwenhoek* 93:241–248.

Zhang, L., An, R., Wang, J., Sun, N., Zhang, S., Hu, J., and Kuai, J. 2005. Exploring novel bioactive compounds from marine microbes. *Curr. Opin. Microbiol.* 8:276–281.

Zhou, Y., Mándi, A., Debbab, A., Wray, V., Schulz, B., Müller, W. E., and Aly, A. H. 2011. New austalides from the sponge-associated fungus *Aspergillus* sp. *Eur. J. Org. Chem.* 2011:6009–6019.

Broad spectrum antimicrobial compounds from plants and rhizosphere microorganisms

chapter sixteen

Antimicrobial compounds and their chemical entities on therapeutic herbals for agricultural and medical applications

P. Rajesh, K. Natarajan, and V. Rajesh Kannan

Contents

16.1 Introduction

The major challenge facing us today is how to control diseases and their consequent effects on humans, animals, and plants in our society as well as on modern medicines. Following the inevitable inadequate use of modern medicines or antibiotics, microorganisms have developed resistance, and the world has turned to traditional antibiotic therapy for therapeutical application to control diseases in humans and plants.

However, most antibiotics kill the microorganisms by cell wall damage as well as inhibition of cellular processes essential for survival, which leads to a strong selective pressure to develop resistance against antibiotics. In other words, the long-term use of large quantities of different antibiotics has contributed to antibiotic resistance in both agriculturally and medically important bacteria, including in *Pseudomonas aeruginosa* and *Staphylococcus aureus*. Therefore, society urgently needs an alternative approach to develop effective therapies against microbial pathogens without affecting their growth in order to avoid creating antibiotic-resistant bacteria in any species that belongs to animals or plants (Keyser et al., 2008).

While dealing with antimicrobial compounds, the plant kingdom's ability to inter-convert and transform organic compounds in evidence-based biological activities varies widely. Plants are effective in synthesizing organic compounds through the process of photosynthesis. Other than plants, microorganisms and animals must obtain the raw materials containing organic compounds for their diet. In this way, metabolic pathways interact using the catabolic degradation of food, while anabolic pathways biosynthesize the specialized molecules. These two processes in combination define metabolism, and it can be further divided into processes called primary and secondary metabolisms. Primary metabolism involves processes that are basically the same in most organisms, such as pathways for modifying and synthesizing carbohydrates, proteins, fats, and nucleic acids. However, there are several organisms that also posses the ability to synthesize special-ized compounds that are often unique to that species, and they are defined as secondary metabolism and the processes are not essential for life.

Secondary metabolites are not produced under all conditions and provisions; in many cases, their actual function is unknown and cannot be defined in a simple manner. However, due to their higher investment in energy and carbon, the secondary metabolite compounds probably have important ecological roles as protection against either biotic factors such as herbivory, predation, and competition or abiotic factors such as UV light. Other potential roles are included in a special character of plants that may be used as volatile attractants. Whatever their specific functions, they probably play some role that benefits the producing organism. It is within this area of secondary metabolites that most biologically active compounds are found, although the actual mechanism in the boundary of the primary and secondary metabolites is not known.

Secondary metabolites are produced in plants besides in primary biosynthetic and metabolic routes of compounds aimed at plant growth and development as discussed earlier. They can be regarded as products of biochemical side tracks in plant cells and not needed for daily functioning of the plant. Phylogenetically, the secondary bioactive compounds in plants appear to be randomly synthesized, but they are not useless junk. Several of them are found to hold important roles and functions in living plants. For example, the flavonoids can protect against free radicals generated during photosyn-thesis and reduce aging as well as cell and tissue damage. The terpenoids may attract pollinators or seed dispersers and inhibit competing plants. Alkaloids usually ward off herbivore animals or insect attacks (phytoalexins), and other secondary metabolites function as cellular signaling molecules or have other functions in the plants with their specific actions. The plants producing bioactive secondary metabolites seem to be the rule rather than the exception. Thus, most plants, even common food and feed plants, are capable of producing such compounds. However, the typical poisonous or medici-nal plants contain higher concentrations of more potent bioactive compounds than food and feed plants.

This chapter is on the significance, role, biosynthetic pathway, and classification of the secondary metabolites, perpetual usage of antimicrobial compounds from various plant-derived products to control agriculture and medically important pathogens, and also deals with the resistance power of modern antibiotics.

16.2   Mode of action of synthetic antimicrobial agents

Antimicrobials and specifically antibiotics are defined as low-molecular-weight organic natural products made by plants that are active against microorganisms at lower concentration.

This activity develops through a limited number of mechanisms; antimicrobials interfere with cell wall synthesis, cell membrane integrity, protein synthesis, DNA replication and repair, transcription, and intermediate metabolism (Figure 16.1).

The compounds blocking the cell wall biosynthesis inhibit enzymes involved in the synthesis of different components of the cell wall, while the secondary metabolite compounds interfering with cell membrane integrity disorganize the structure or inhibit the function of bacterial membranes. Antimicrobials affecting protein synthesis act in impairing the ribosomal subunits, binding to prevent the translation and binding to cause wrong translation, and producing toxic and altered proteins. Compounds having an effect on DNA replication and repair inhibit enzymes such as gyrase, topoisomerase, and N-methyltransferase. Similarly, compounds affecting transcription inhibit the subunits of the bacterial RNA polymerase blocking the entry of the first nucleotide necessary to activate the polymerase. Finally, some antimicrobials interfere with intermediate metabolism by inhibiting enzymes involved in the biosynthesis of different substances (Demain, 1999) such that the genetic modification with antimicrobial resistance was intended to discover new antimicrobial agents from the plant products (Figure 16.2).

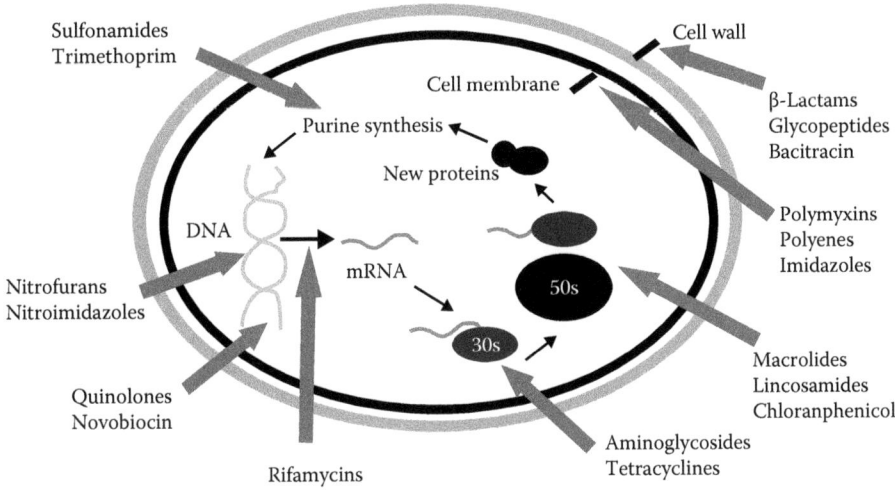

Figure 16.1 Mode of action in synthetic antimicrobial agents.

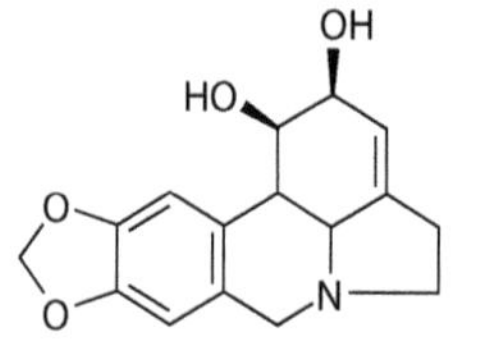

Lycroine (**1**)

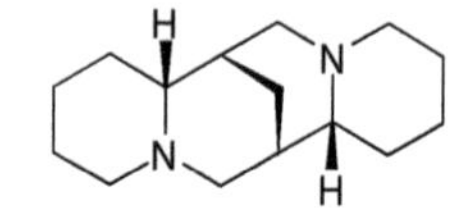

Sparteine (**2**)

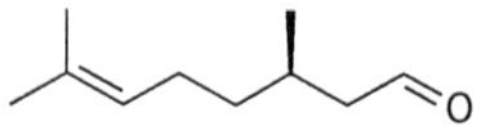

Citronellal (**3**)

Cineole (**4**)

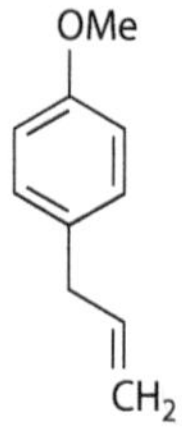

Essential oils (**5**)

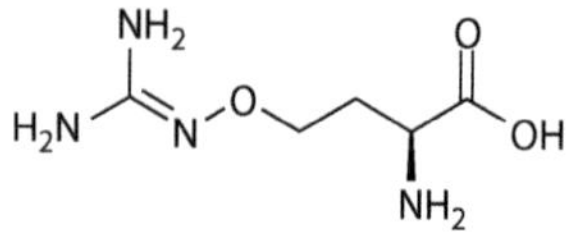

Canavanine (**6**)

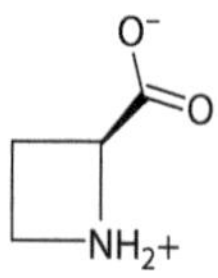

Azetidine-2-carboxylic acid (**7**)

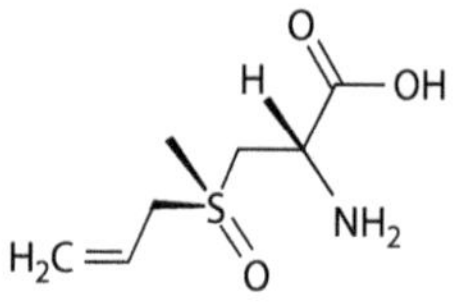

Alliin (**8**)

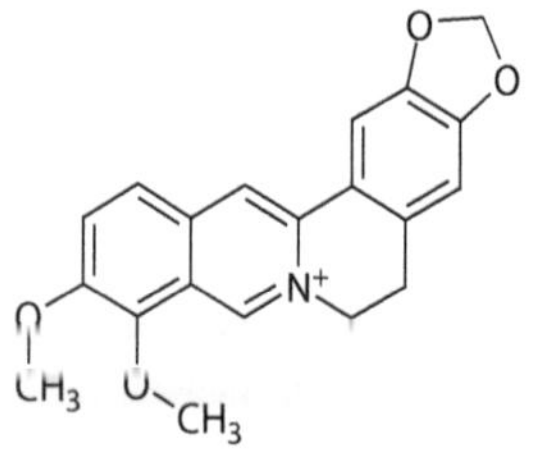

Berberine (**9**)

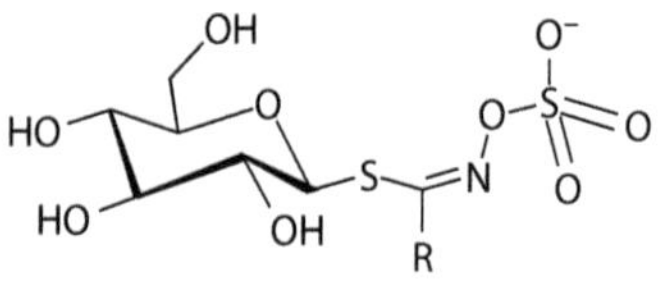

Glucosinolates (**10**)

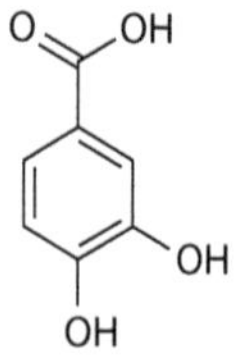

Protocatechuic acid (**11**)

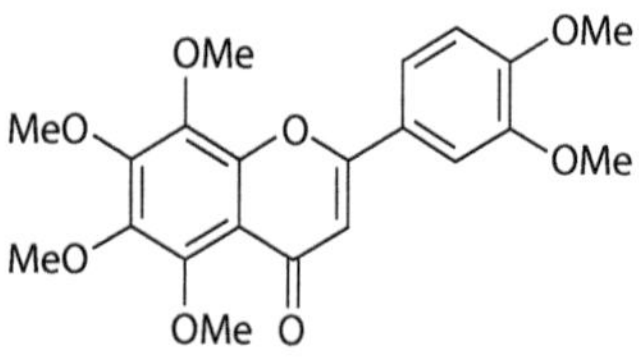

Nobiletin (**12**)

Figure 16.2 Secondary metabolic chemical compounds from plant metabolisms (drawn using ISIS/ChemDraw).   (*Continued*)

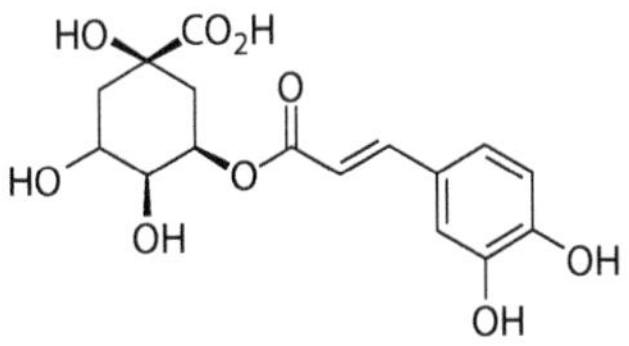

Chlorogenic acid (**13**)

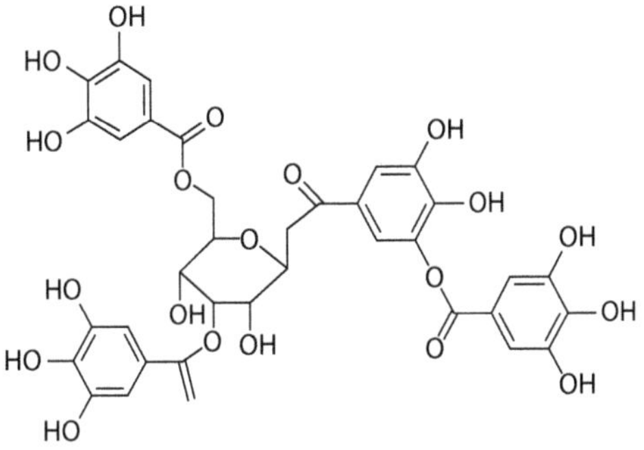

Tannins (**14**)

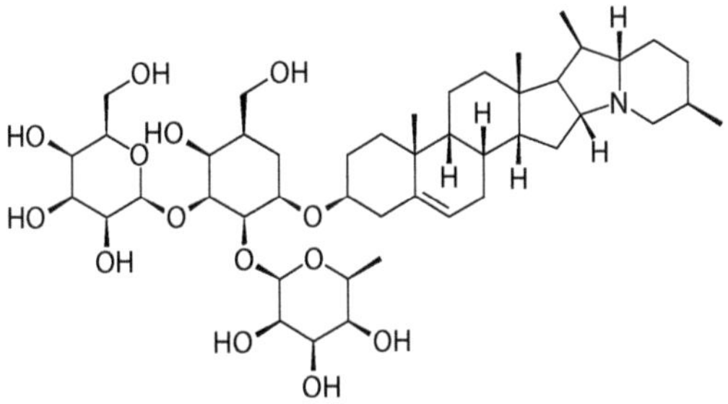

Solanine (**15**)

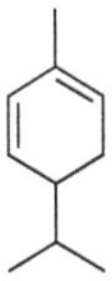

Phellandrene (**16**)

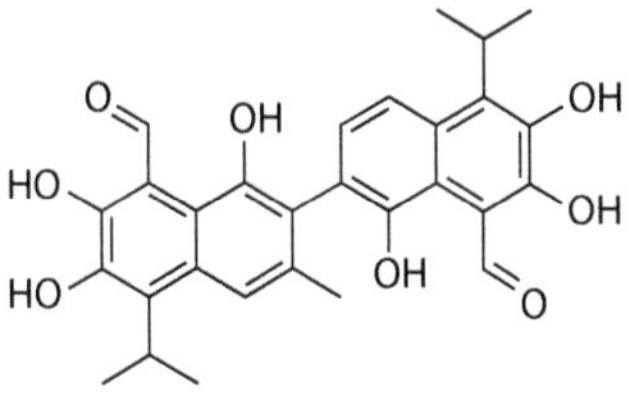

Gossypol (**17**)

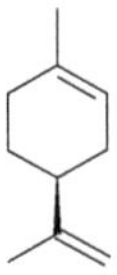

Limonene (**18**)

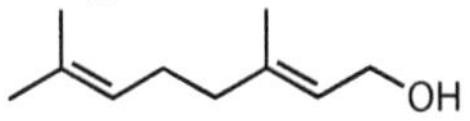

Geraniol (**19**)

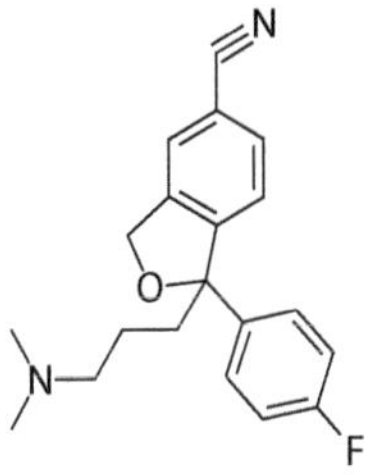

Citrol (**20**)

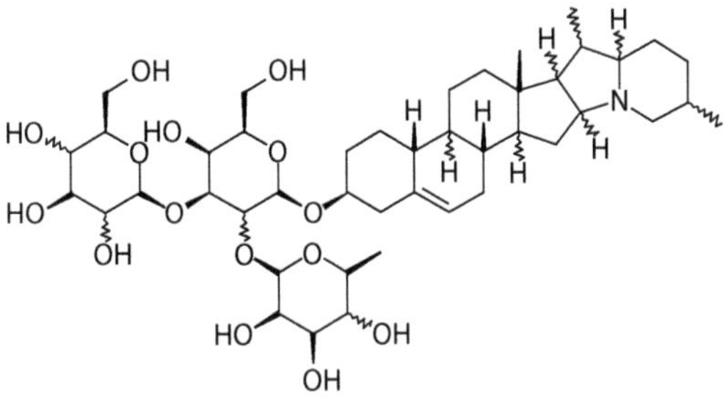

Saponins (**21**)

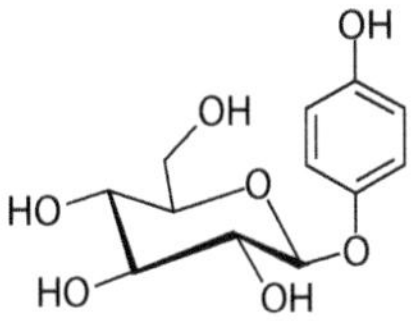

Arbutine (**22**)

Figure 16.2 (Continued) Secondary metabolic chemical compounds from plant metabolisms (drawn using ISIS/ChemDraw). *(Continued)*

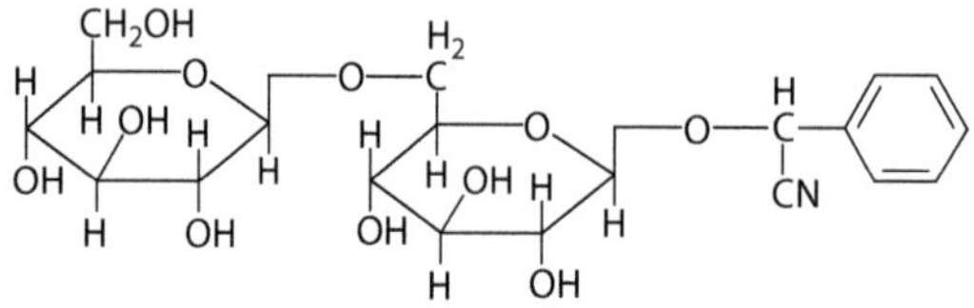

Cyanogenic glycosides (**23**)

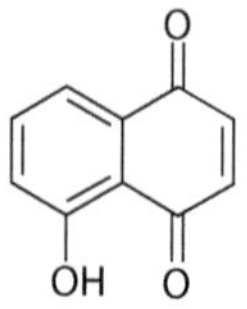

Juglone (**24**)

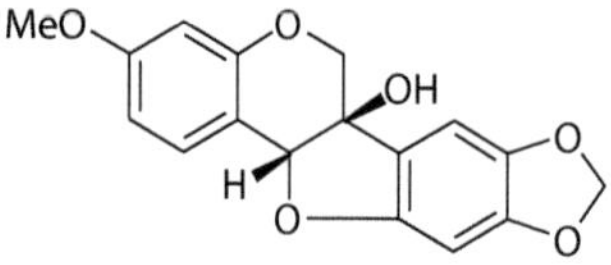

Pisatin (**25**)

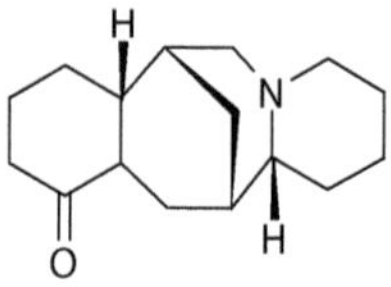

Lupanine (**26**)

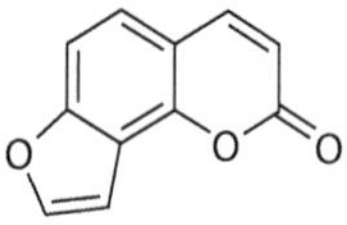

Furanocoumarin (**27**)

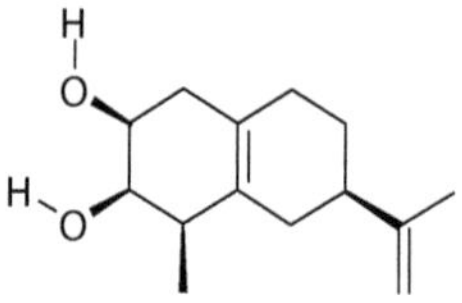

Rishitin (**28**)

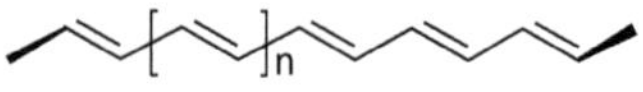

Polyacetylenes (**29**)

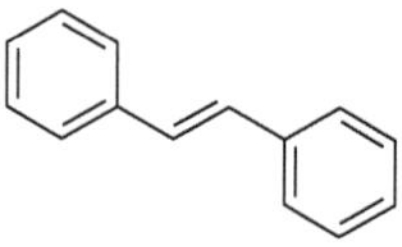

Stilbenes (**30**)

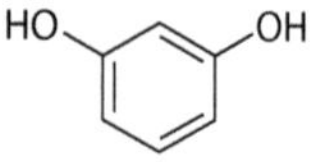

Simple phenols (**31**)

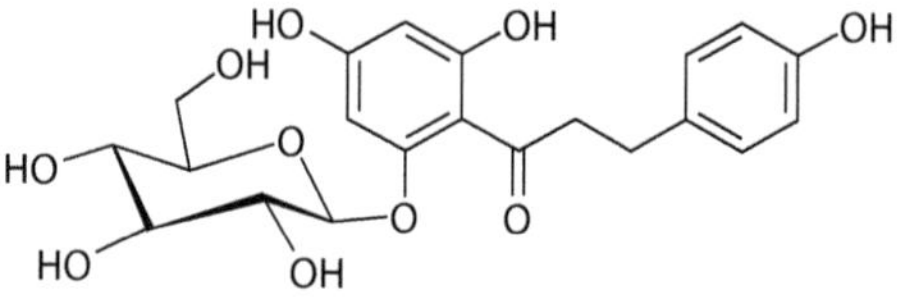

Phloridzin (**32**)

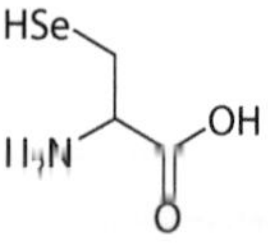

Nonprotein amino acid (**33**)

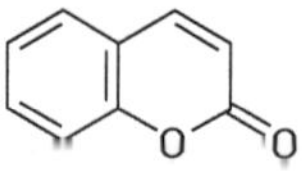

Coumarins (**34**)

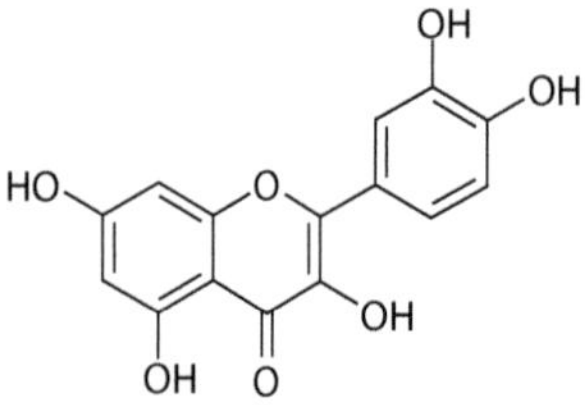

Quercetin (**35**)

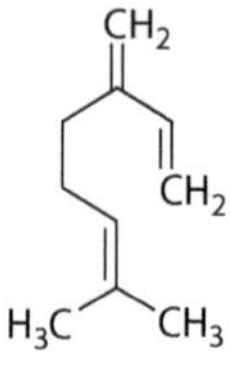

Monoterpenes (**36**)

Figure 16.2 (Continued) Secondary metabolic chemical compounds from plant metabolisms (drawn using ISIS/ChemDraw).   *(Continued)*

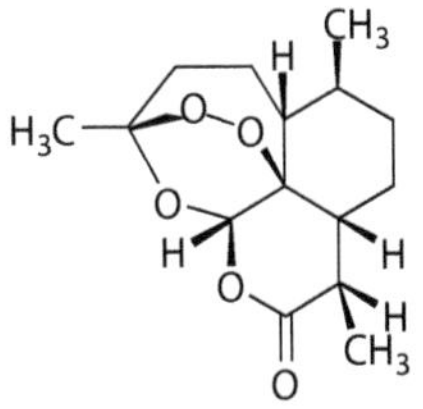

Sesquiterpene lactones (**37**)

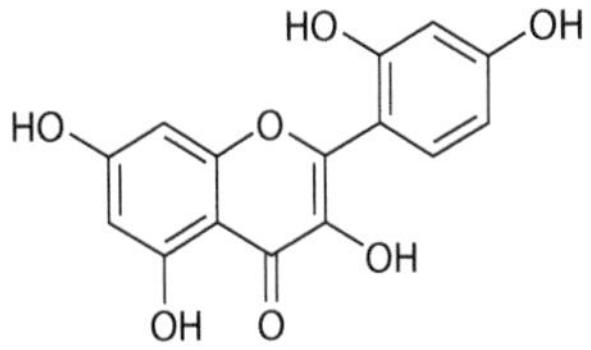

Morin (**38**)

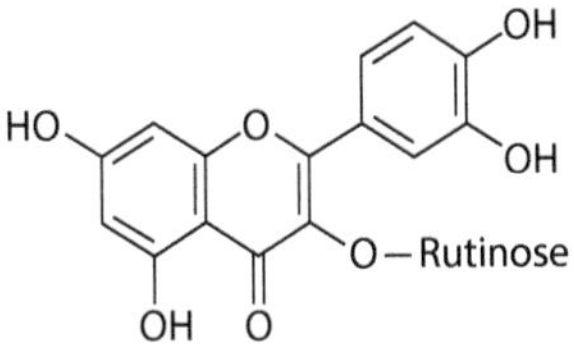

Rutin (**39**)

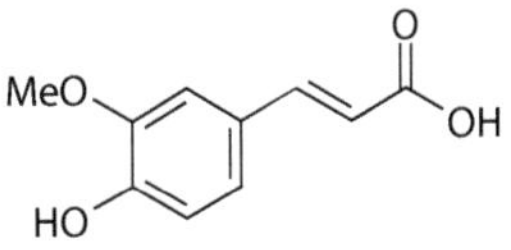

Ferulic acid (**40**)

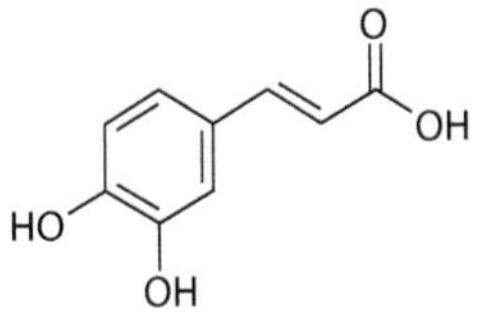

Caffeic acid (**41**)

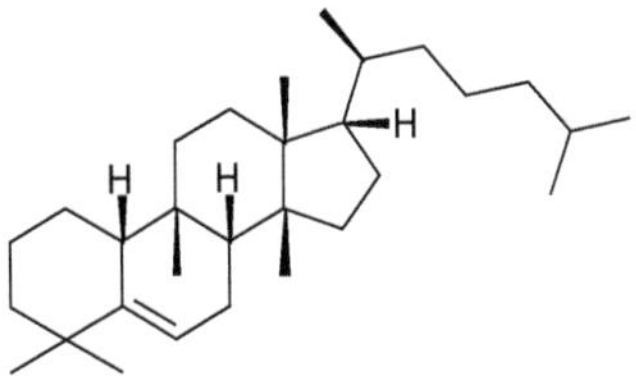

Cucurbitacin (**42**)

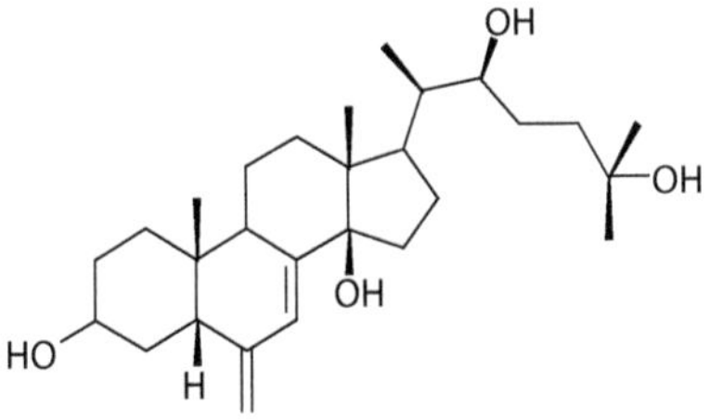

Phytoecdysone (**43**)

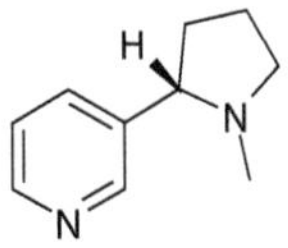

Nicotine (**44**)

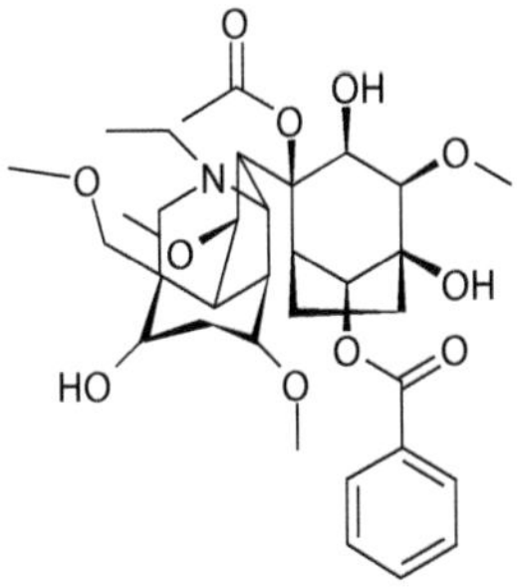

Aconitine (**45**)

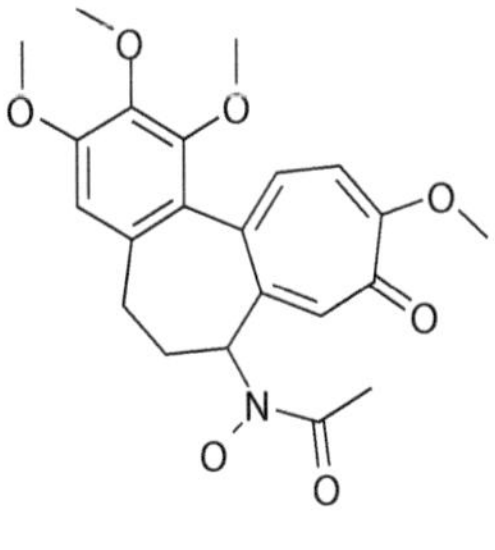

Colchicines (**46**)

*Figure 16.2 (**Continued**)* Secondary metabolic chemical compounds from plant metabolisms (drawn using ISIS/ChemDraw). *(Continued)*

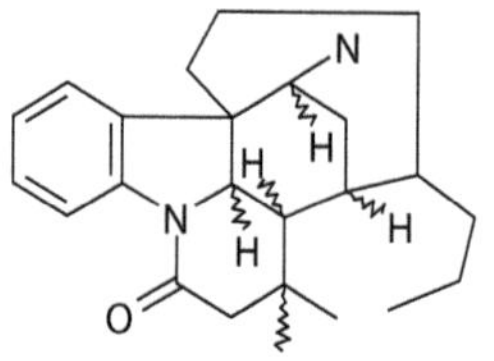

Strychnine (**47**)

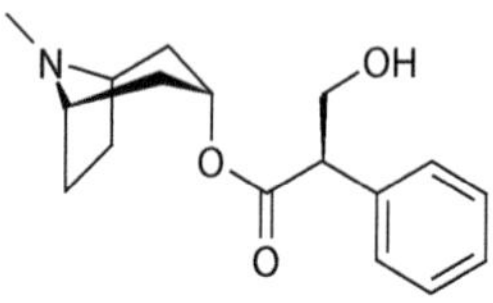

Atropine (**48**)

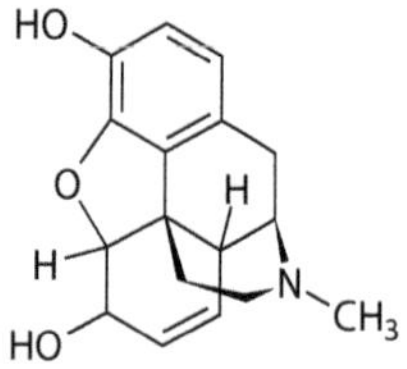

Morphine (**49**)

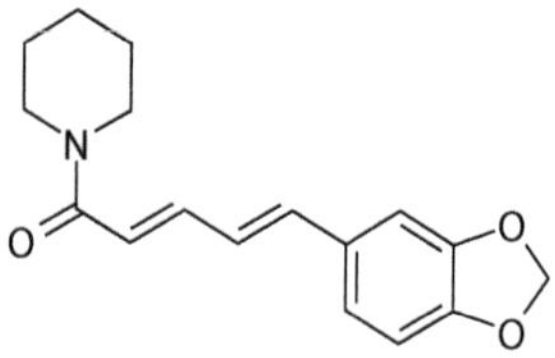

Piperine (**50**)

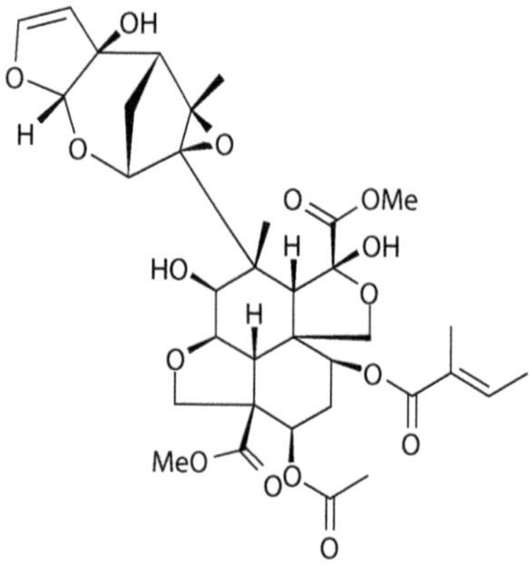

Azadirachtin (**51**)

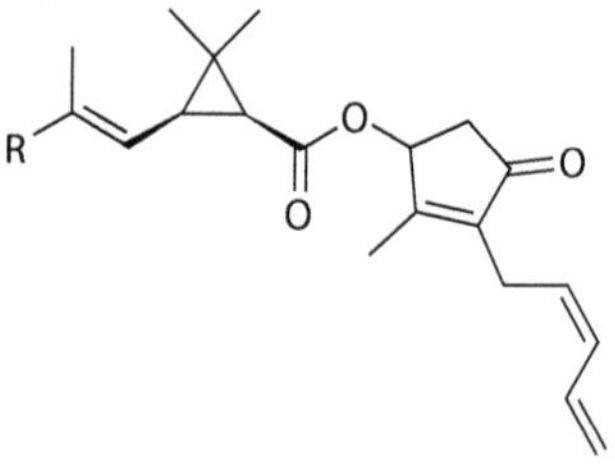

Pyrethrins (**52**)

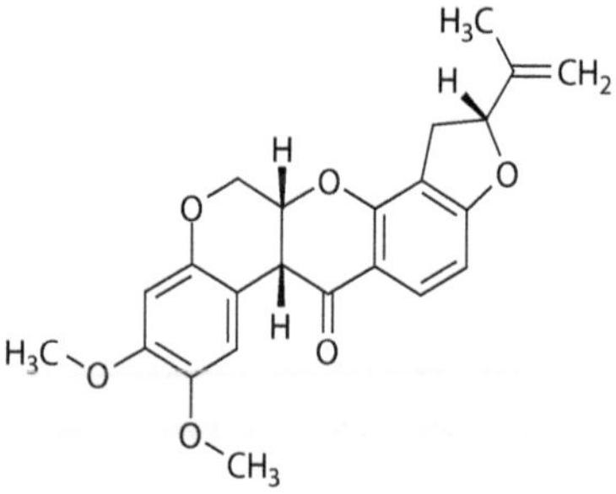

Rotenone (**53**)

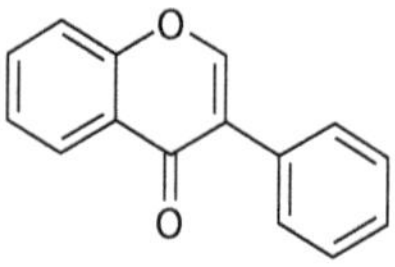

Isoflavones (**54**)

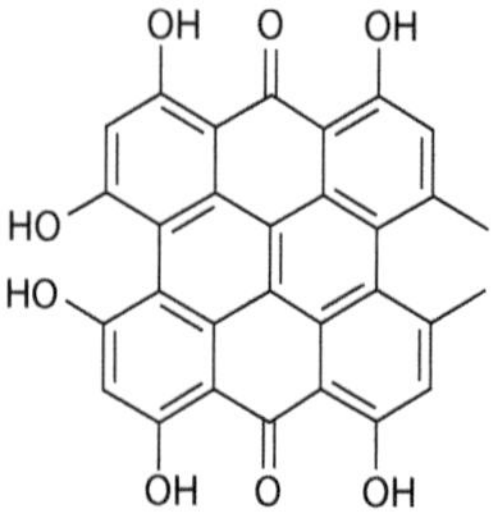

Hypericin (**55**)

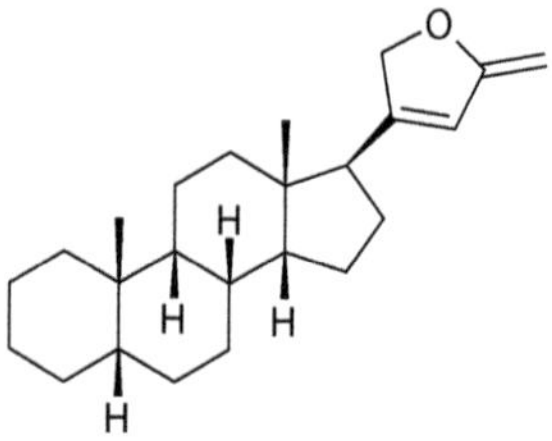

Cardenolides (**56**)

Figure 16.2 (Continued) Secondary metabolic chemical compounds from plant metabolisms (drawn using ISIS/ChemDraw).

16.3   Natural plant products as antimicrobial agents

The multidrug-resistant strains of many microorganisms have revealed exploration of alternative antimicrobial agents. Medicinal plants have become the focus of intense study in terms of validation of their traditional uses through the determination of their actual antimicrobial effects. Synthetic drugs are not only expensive and inadequate for the treatment of diseases but also often have adulterations and side effects; other than the aforementioned panic condition, various microorganisms have made them resistant following continuous usage of the antibiotics. Therefore, a search for new infection fighting strategies to control microbial infections is imperative (Sieradzki et al., 1999).

Plants have a long history of antibiotic usage for curing disease used as antimicrobial, including antibacterial, antiviral, and antifungal agents in plants as well as for medical purposes. Attention was refocused on plant origin, antimicrobial and antidermatophytic agents after the discovery of penicillin, which is considered to be one of the most important lifesaving phytodrugs and virtually known as the secondary metabolites. Since then, many extracts from different plants have been tested for their antimicrobial activities, but there are so many unopened pages of plant life that require more careful investigation to reveal their hidden characteristics of secondary metabolites. Natural products are generally harmless or have minimum side effects when compared with synthetic drugs. In addition to the antimycotic drugs, the effect of certain indigenous plant extracts on the spore germination of dermatophytes *in vitro* has also been studied, but the work is fragmentary. Antifungal activities of some plants have been reported by various researchers throughout the world (Sharma, 1988; Tewari et al., 1990; Caceres et al., 1993; Rajendheran et al., 1998; Farombi, 2003; Nair et al., 2005; Mahesh and Satish, 2008; Prusti et al., 2008).

In India, references to the curative properties of some herbs (as available in the *Rigveda*) seem to be the earliest record of the use of secondary metabolites for medicinal purposes. Singh et al. (1988) reported the exhibiting activity of the essential oil of *Eucalyptus rostrata* leaves against four human pathogens, namely, *Trichophyton mentagrophytes*, *Epidermophyton floccosum*, and *Microsporum gypseum* and *M. canis*. Iyer and Williamson (1991) investigated the efficacy of some plant extracts and their secondary metabolites against *Trichophyton* species. *Rhododendron arboreum*, *Rhussemialata*, *Terminalia bellerica*, and *Woodfordia fruticosa* showed antibacterial activity preferably to gram-negative organisms. Generally, gram-negative bacteria are more resistant than gram-positive bacteria (Kelmanson et al., 2000; Parekh et al., 2005).

16.4   Antimicrobial compounds

Plants produce and accumulate an enormous variety of secondary metabolites (often referred to as natural products). Many of these products have well-defined ecological roles in plant defense and in mediating interactions of plants with other organisms or microorganisms, while others have yet unknown biological roles. More than 45,000 different chemical structures of natural products have been identified. The largest group of natural products is the terpenes with more than 25,000 structure elucidations. Additionally, more than 2000 alkaloids (**57**) and about 8000 phenolic derivatives are known (Croteau et al., 2000). Despite the large number of different proven chemical structures, there are an amazingly low number of biochemical pathways by which

Table 16.1 Biological activities of some secondary metabolites as plant-derived phytochemicals

Biological activity	Secondary metabolic compounds
Antiviral	Lycroine (A) (**1**) and sparteine (A) (**2**)
Antibacterial	Citronellal (T) (**3**), cineole (T) (**4**), many essential oils (T) (**5**), canavanine (NP) (**6**), azetidine-2-carboxylic acid (NP) (**7**), alliin (NP) (**8**), berberine (A) (**9**), and glucosinolates (**10**)
Antifungal	Protocatechuic acid (P) (**11**), nobiletin (F) (**12**), chlorogenic acid (P) (**13**), tannin (P) (**14**), solanine (A) (**15**), phellandrene (T) (**16**), gossypol (T) (**17**), limonene (T) (**18**), geraniol (T) (**19**), citrol (T) (**20**), saponins (S) (**21**), canavanine (NP) (**6**), arbutine (Q) (**22**), cyanogenic glycosides (CG) (**23**), glucosinolates (**10**), juglone (Q) (**24**), pisatin (F) (**25**), lupanine (A) (**26**), furanocoumarin (**27**), rishitin (T) (**28**), polyacetylenes (**29**), and stilbenes (**30**)
Allelopathy	Simple phenols (**31**); juglone (Q) (**24**); phloridzin (P) (**32**); nonprotein amino acids (**33**); coumarins (**34**); quercetin (F) (**35**); polyacetylenes (**29**); monoterpenes (T) (**36**) and sesquiterpene lactones (**37**)
Toxicity/repellency against insects	Quercetin (Q) (**35**); morin (F) (**38**); rutin (F) (**39**); ferulic acid (P) (**40**); caffeic acid (P) (**41**); juglone (Q) (**24**); tannins (P) (**14**); limonene (T) (**18**); gossypol (T) (**17**); cucurbitacin (T) (**42**); saponins (T) (**21**); phytoecdysone (T) (**43**); glucosinolates (**10**); monoterpenes (**36**); steroidal and quinolizidine alkaloids, nicotine (A) (**44**); aconitine (A) (**45**); colchicines (A) (**46**); strychnine (A) (**47**); atropine (A) (**48**); morphine (A) (**49**); berberine (A) (**9**); piperine (A) (**50**); nonprotein amino acids (**33**); cyanogenic glycosides (**23**); sesquiterpene lactones (T) (**37**); azadirachtin (T) (**51**); protease inhibitors, lectins; pyrethrins (**52**); rotenone (**53**); and coumarins (**34**)
Vertebrates	Isoflavones (F) (**54**), coumarins (**34**), juglone (Q) (**24**), tannins (**14**), hypericin (F) (**55**), gossypol (T) (**17**), essential oils (T) (**5**), saponins (T) (**21**), sesquiterpene lactones (T) (**37**), cardenolides (T) (**56**), glucosinolates (**10**), alkaloids (**57**), nonprotein amino acids (**33**), and cyanogenic glycosides (**23**)
Attractants	Flavonoids (**58**), anthocyanins (**59**), monoterpenes (**36**), sesquiterpenes (**60**), carotenoids (T) (**61**), betalains (A) (**62**), quinines (**63**), phenols (**31**), amino sugars (**64**), amino acids (**65**), and lipids (**66**)

Sources: Schlee, D., *Okologische Biochemie*, Springer, Berlin, Germany, 1986; Swain, T., *Annu. Rev. Plant Physiol.*, 28, 479, 1977; Wink, M., Chemical ecology of quinolizidine alkaloids, in: Waller, G.R., ed., *Allelochemicals: Role in Agriculture, Forestry and Ecology*, Am. Chem. Soc., 330, 524, 1987.

A, alkaloid; F, flavonoid; P, simple phenol; S, saponins; T, terpene; NP, nonprotein amino acids; Q, quinone.

these compounds are biosynthesized. Some of the biological activities in secondary metabolites of plants have been derived in Table 16.1.

16.5   *History of antimicrobial compounds*

The use of secondary metabolites can be traced back to 2600 BC, and the first records were written on hundreds of clay tablets in cuneiform from Mesopotamia (Newman et al., 2000).

Then more than 4000 years ago, plant secondary metabolites were used as mixtures or plant extracts mainly in food, medicine, and to make poison. Although, plant secondary metabolites have been benefiting humans for thousands of years, their mysteries were disclosed in late two centuries. The isolation of the first pure natural product, morphine (**49**) from opium poppy (*Papaver somniferum*), by German pharmacist Friedrich Wilhelm Serturner in 1806 opened the door to a new era of scientific plant secondary metabolite research (Croteau et al., 2000).

He demonstrated that the active principle of plant extract could be attributed to a single organic compound, which can be isolated. This finding initiated natural product chemistry and speeded up developments in synthetic, analytical, and pharmaceutical chemistry. Even today, plant secondary metabolites are closely involved in human life; more than 30% of drugs are derived directly or indirectly from natural products and sources.

Plant secondary metabolites were thought to be simply waste products. In the 1970s, however, new research showed that secondary plant metabolites play an indispensable role in the survival of a plant in its environment. One of the most ground-breaking ideas of this time argued that environmental conditions affected the evolution of plant secondary metabolites, and this indicated the high gene plasticity of secondary metabolites, but this theory was ignored for about half a century before gaining acceptance. Recently, the research around secondary plant metabolites is focused on the gene level and the genetic diversity of plant metabolites. Biologists are trying to trace back genes to their origin and reconstruct evolutionary pathways (Hartmann, 2007).

Plant secondary metabolism primarily has been delivered and discussed in the latter half of the nineteenth century; however, there was still much confusion over what the exact function and use these secondary metabolites had. The well-known secondary plant metabolites were by-products of the primary metabolism and were not crucial to the plant's survival. Early research only succeeded as far as categorizing the secondary plant metabolites but did not give real insight into the actual function of the secondary plant metabolites. In the early half of the 1900s, the main research around plant secondary metabolism was dedicated to the formation of secondary metabolites in plants, and this research was compounded by the use of tracer techniques that made deducing metabolic pathways much easier. However, there was still not much research being conducted into the functions of secondary plant metabolites until around the twenty-first century (Hartmann, 2007).

The secondary metabolites with antimicrobial compounds have been used in the past decades in a wide manner. It was not until 1870s that Tyndall, Pasteur, and Roberts reported on the antagonistic effects of various virulent organisms, and the antibiotics era began in 1929 with the discovery of penicillin by Fleming. The products of penicillin effectively started and augmented in the 1940s. In spite of this, there were only two classes of naturally occurring β-lactam antibiotics known as penicillin and cephalosporin until 1970. About three decades, there was a consequence on the antibiotic products from plants and microorganisms, but there were still revisions. The plant- and microorganism-based natural products mentioned before are widely divergent in their chemical structures and their antimicrobial properties. Unique peptides are having antimicrobial properties over 400 numbers that have been proven from diverse sources including plants, bacteria, fungi, and vertebrates (Lehrer and Ganz, 1996; Steinstraesser et al., 2002).

16.6   Significance and importance of secondary metabolites

Crudely, there were huge peptides inducing the epithelial surfaces and concurrent responses to the invading microorganisms. In addition, some of the peptides from the plant-derived secondary metabolites are involved in the chemotaxis and promote wound healing and contribute to the adaptive immunity by mobilizing memory T cells and immature dendritic cells. In parallel to the screening of new antibiotics, they have been the spotlight in finding low-molecular-weight secondary metabolites with other biological activities. The secondary metabolites have been found to play a significant role in activities such as enzyme inhibition and plant growth stimulation and been also used in the production of herbicides, insecticides, and anthelmintics. Examples of two major roles the secondary metabolites have contributed were, first, the screening of secondary metabolites, which has been found useful for antibiotics and, second, the development of unknown compounds using new technologies for detecting inhibition of enzymes and biological activities and other targets. In addition, they play a significant role in agriculture as well as in clinical medicines due to their function as antimicrobials used in the treatment of microbial infections in plants and humans, respectively (Kieslich, 1986).

From the past research, there were extensive studies reported about their origin and functionality. Being ineffective secondary metabolites have been accepted as waste products, but under the pressure of natural selection, they have evolved as messenger molecules that must have endured long enough to shuttle between the various components of the microbial community.

This fact would explain why secondary metabolites have the tendency to become small molecules as a natural consequence of their functions (Jarvis, 1995).

16.7   Role of secondary metabolism in plants

Secondary metabolites play a major role in the adaptation of plants to the changing environment and in overcoming stress constraints. This flows from the large complexity of chemical types and interactions underlying various functions: structure stabilization, determined by polymerization, condensation of phenols (31) and quinones, or electrostatic interactions of polyamines with negatively charged loci in cell components (Edreva et al., 2008).

Primary metabolism in plants constitutes all metabolic pathways that are essential to the plant's survivability. Primary metabolites are compounds directly involved in the growth and development of a plant, whereas secondary metabolites are compounds produced in other metabolic pathways, however important, and they are not essential to the functioning of the plant. However, the secondary plant metabolites are useful in the long term, often for defense purposes, and give plants characteristics such as color that is known as pigments like chlorophyll and carotenoids. The plant secondary metabolites are also used in signaling and regulation of primary metabolic pathways. The plant hormones are secondary metabolites, often used to regulate the metabolic activity within cells and overall development of the plant. As mentioned before, the plant's secondary metabolites help it maintain an intricate balance with the environment, often adapting to match the environmental needs. Plant metabolites that color the plant are a good example of this as the coloring of a plant can attract pollinators and also defend against attack by animals.

The hypothesis of the secondary metabolites went over a year, and they are as follows:

- Secondary metabolites act as an alternative defense mechanism, because only the organisms lacking an immune system are prolific producers of these compounds.
- They have sophisticated structures, mechanisms of action, and complex and energetically expensive pathways.
- Secondary metabolites act in the competition between microorganisms, plants, and animals.
- They are produced by biosynthetic genes clusters, which would only be selected if the product conferred a selective advantage. Some particularities of these gene clusters are the absence of nonfunctional genes and the presence of resistance and regulatory genes.
- The production of secondary metabolites with antibiotic activities is temporarily related with sporulation when the cells are particularly sensitive to competitors and require special protection when a nutrient runs out.

Furthermore, the wide diversity of secondary metabolites suggests a broad range of functions. Nevertheless, these functions could depend on the conditions, optimal or not, surrounding the producer microorganism. Finally, due to their crucial importance, the study and exploitation of secondary metabolites continue to progress despite the lack of agreement regarding why microbes produce such chemical diversity of antimicrobial compounds (Demain and Fang, 2000).

16.8   Biosynthetic pathway of secondary metabolites

The biosynthesis of secondary metabolites involves a series of reactions that are enzymatically catalyzed using several common mechanisms such as electrophilic additions in alkylation, aldol, transamination, and decarboxylation, glycosylation, and redox reactions (Figure 16.3). These secondary metabolites are synthesized at higher cost, requiring a steady flow of precursors from primary metabolism together with energy-rich cofactors like ATP and NAD(P)H. But the plants can compensate for the production of secondary metabolites higher in carbon through photosynthesis. However, for those metabolites

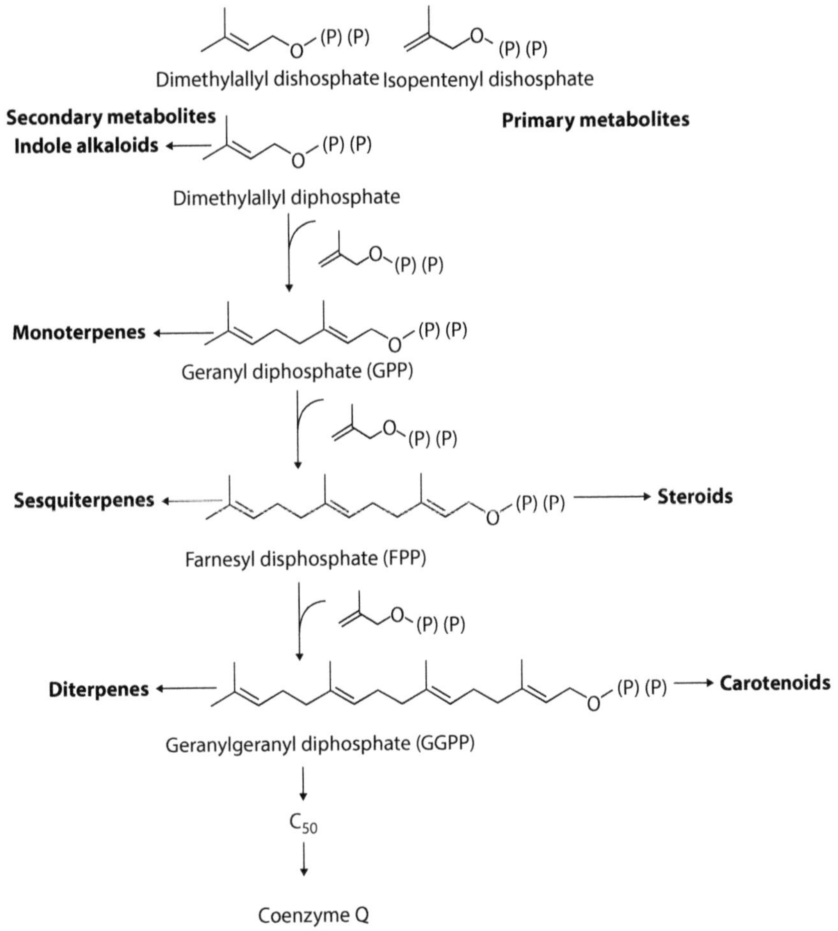

Figure 16.3 Biosynthetic pathway (mevalonate pathway). (From Keller, N.P., *Nat. Rev.*, 3, 937, 2005.)

higher in limiting compounds, like nitrogen in alkaloids (**57**), secondary metabolism can compete with primary pathways, such as in protein biosynthesis.

The ability to synthesize secondary metabolites has been selected through the course of evolution in different plant lineages when such compounds address specific needs such as the following:

Florescent volatiles and pigments have evolved to attract insect pollinators and thus enhance fertilization.

Toxic chemicals have evolved to ward off pathogens and herbivores or to suppress the growth of neighboring plants through synthesis.

Chemicals found in fruits prevent spoilage and act as signals (in the form of color, aroma, and flavor) in the presence of sugars, vitamins, and flavors for animals that eat the fruit and therefore help to disperse the seeds.

Other chemicals serve cellular functions that are unique to the particular plant in which they occur (e.g., resistance to salt or drought).

Terpenes are biosynthesized by the isopentenoid pathway that includes two main metabolic branches, the mevalonic acid pathway (Figure 16.3), by which sesqui- and triterpenes are biosynthesized, and the deoxyxylulose diphosphate pathway, by which mono-, di-, and tetraterpenes are formed (Croteau et al., 2000). Most phenolic derivatives are biosynthesized either by the shikimic acid pathway or through the malonate–acetate pathway, and most alkaloids (**57**) are biosynthesized from amino acids. The monoterpenoids such as linalool, menthol, thymol, and limonene (**18**) are derived from geranyl diphosphate. The different conversions from GPP are catalyzed by a group of enzymes termed monoterpene synthases. These enzymes share many properties such as cofactor requirements, molecular size, and protein sequence similarity, but still they are different enough to allow for the catalysis of the different products according to their specificity (McGarvey and Croteau, 1995). Sesqui- and triterpenes, as well as many triterpene-derived metabolites such as the saponins (**21**) and sterols, are synthesized from the 15-carbon intermediate farnesyl diphosphate (FPP).

Similar to monoterpene synthases, minor changes in the sesquiterpene synthase enzymes (and their genes) are responsible for catalyzing the conversion of FPP to the different sesquiterpenes, respectively. Still, at least in angiosperms, sesquiterpene synthase genes are significantly similar to each other and also similar to angiosperm monoterpene synthases (Bohlmann et al., 2000). In an analogous way, it seems that the ubiquitous pathway to lignin has been specifically adapted in many plants for the production of unique natural products and modifications of existing genes and enzymes to allow for the specific conversions. Thus, phenylpropanoids are mostly derived from L phenylalanine, not only a precursor of lignin, anthocyanins, and other flavonoids (**58**) but also a precursor of t-anethole in anise (*Pimpinella anisum*) (Manitto et al., 1974a) and infennel (*Foeniculum vulgare*) (Kaneko, 1960). The L-phenylalanine is also a precursor of cinnamaldehyde in cinnamon in *Cinnamomum zeylanicum* (Senanayake et al., 1977) and of estragole in sweet basil in *Ocimum basilicum* (Manitto et al., 1974b).

16.9   Classes of secondary metabolites

There is no rigid scheme for classifying natural products and their immense diversity in structure, function, and biosynthesis that allows them to fit neatly into a few simple categories. The plants produce thousands of organic compounds that are traditionally divided into two large classes, that is, primary and secondary metabolites. The term natural product is generally taken to mean a secondary metabolite—a small molecule that is not essential to the growth and development of the producing organism and is not classified by structure. Well, over 3,000,000 secondary metabolites probably exist, which are generally classified into five categories: terpenoids (**73**), steroids, fatty acid–derived substances and polyketides, alkaloids (**57**), and nonribosomal polypeptides and enzyme cofactors.

The plant has been derived secondary metabolites more than 1,000,000 chemical compounds in various agricultural and medicinal purposes (Hadacek, 2002). The most characteristic feature of secondary metabolites is the largest diversity of chemical types, embracing representatives of all main classes of organic compounds: aliphatic, including polyamines (**67**), ethylene (**68**), and isoprene (**69**); aromatic, such as phenolic alcohols (**70**), phenolic acids (**71**), and unsaturated aromatic carbonic acids (**72**); hydroaromatic terpenoids (**73**) and jasmonic acid (**74**); and heterocyclic compounds, such as flavonoids (**58**) and indole derivatives (**75**). Unique carbon skeletons occur along with multiplicity of functional groups that were tabulated in Table 16.2 (Edreva et al., 2008).

Table 16.2 Diverse chemical types of secondary metabolites

Chemical types	Chemical formula	Representative
Aliphatic	$H_2N-CH_2-CH_2-CH_2-CH_2-NH_2$	Polyamines (**67**)
	$H_2C=CH_2$	Ethylene (**68**)
	$H_2C=C(CH_3)-CH=CH_2$	Isoprene (**69**)
Aromatic	4-hydroxycinnamyl alcohol structure (HO–C₆H₄–CH=CH–CH₂OH)	Phenolic alcohols (**70**)
	salicylic acid structure (C₆H₄ with COOH and OH)	Phenolic acids (**71**)
	ferulic acid structure (HO, OMe substituted C₆H₃–CH=CH–COOH)	Unsaturated aromatic carbonic acids (**72**)
Hydroaromatic	terpenoid structure	Terpenoids (**73**)
	jasmonic acid structure (cyclopentanone with COOH and pentenyl chain)	Jasmonic acid (**74**)
Heterocyclic	flavonoid (quercetin-type) structure with OH groups	Flavonoids (**58**)
	indole structure	Indole derivatives (**75**)

The majority of secondary metabolites belong to one of the families, each of which has particular structural characteristics arising from the way in which they are built up in nature (biosynthesis). The following are some other vast classifications of secondary metabolites:

- Terpenoids (**73**) are defined by about 29,000 compounds, and terpenes constitute the largest class of secondary metabolites and are united by their common biosynthetic origin from acetyl-coA or glycolytic intermediates. A vast majority of the different terpene structures produced by plants as secondary metabolites are presumed to be involved in defense as toxins and feeding deterrents to a large number of plant-feeding insects and mammals.
- Monoterpenes are expanded by about 1000 compounds, and their numerous derivatives are important agents of insect toxicity. They show strong insecticidal responses (neurotoxin) to insects like beetles, wasps, moths, and bees and are a popular ingredient in commercial insecticides because of their low persistence in the environment and low mammalian toxicity.
- Sesquiterpenes were defined by around 3000 compounds with their role in plant defense characterized by a five-membered lactone ring as a cyclic ester and their strong feeding repellence to many herbivorous insects and mammals. Sesquiterpenes also play primarily regulatory roles in the initiation and maintenance of seed and bud dormancy and plants' response to water stress by modifying the membrane properties and acting as transcriptional activators. In addition, this increases the cytosolic calcium concentration and causes alkalinization of the cytosol.
- Diterpenes are established by around 1000 compounds and are used as skin irritants and internal toxins to mammals. It plays various detrimental roles in numerous plant developmental processes such as seed germination, leaf expansion, flower and fruit set, dry mass and biomass production, stomatal conductance, and CO_2 fixation and also the exertion of their numerous physiological effects via specific enzymes, the synthesis of which they induce by influencing the basic process of translocation and transcription.
- Triterpenes, steroids, and saponins (**21**) were expanded around 4000 compounds, and several steroid alcohols (sterols) are important component of plant cell membranes, especially in the plasma membrane as regulatory channels, and maintain permeability to small molecules by decreasing the motion of fatty acid chains.
- Phenolics are in a wide range of about 8000 discovered compounds, and the plants produce a large variety of secondary products that contain a phenol group, a hydroxyl functional group in an aromatic ring called phenol, and a chemically heterogeneous group.
- They could play an important role in the plant defense system against pests and diseases including root parasitic nematodes.
- Flavonoids (**58**) are explained in account of around 2000 compounds. This type of secondary metabolites is one of the largest classes of plant phenolics, which perform very different functions in plant system including pigmentation and defense. Two other major groups of flavonoids (**58**) found in flowers are flavones and flavonols that function to protect cells from UV radiation because they accumulate in epidermal layers of leaves as well as stems and absorb light strongly in the UV-B region while letting visible wavelengths throughout uninterrupted.
- Polyacetylenes are detected at around 1000 compounds, and they are elaborated.

- Polyketides were described to about 750 chemical compounds, and they are not defined in a detailed manner.
- Phenylpropanoids and its derived compounds were found about 500 numbers without definitions.
- Nitrogen-containing compounds are biosynthesized from common amino acids, and all are of considerable interest because of their role in the antiherbivore defense and toxicity to humans.
- Alkaloids (**57**) were coined around the numbers 12,000, and generally, most of them are toxic to some degree and appear to serve primarily in defense against microbial infection and herbivore attack. They are usually synthesized from one of the few common amino acids, particularly from the aspartic acid, lysine, tyrosine, and tryptophan.
- Nonprotein amino acids are defined and found to be present at around 600 compounds, and many plants also contain unusual amino acids called nonprotein amino acids (**33**) that incorporated into proteins but are present as free forms and act as protective defensive substances. They exert their toxicity in various ways, and some of them block the synthesis of or uptake of protein amino acid, while others can be mistakenly incorporated into proteins.
- Cyanogenic glycoside compounds are found to be about 100, and they constitute a group of N-containing protective compounds other than alkaloids (**57**) and release the poisonous hydrogen cyanide (HCN). They are not in themselves toxic but are readily broken down to give off volatile poisonous substances like HCN and hydrogen sulfide (H_2S) when the plant is crushed.
- Glucosinolates are found to be of 100 compounds in nature.
- Amines are detected in and around 100 compounds.

16.10   Genetics of secondary metabolites

The genes regulating and ensuring synthesis of secondary metabolites and their expression can be grouped into five classes:

1. Structural genes that code for enzymes involved in the biosynthesis
2. Regulatory genes that determine the induction or repression of the structural genes
3. Genes that determine the resistance of the producing organism
4. Genes that control the permeability of the compound
5. Genes that control primary pathways

The genetic regulation of all aforementioned genes is highly complicated because many environmental and microbial factors affect the production of these compounds (Romero-Tabarez, 2004). The genes coding for biosynthesis of secondary metabolites has some features that can be summarized as follows:

a. They are arranged in clusters on the chromosome.
b. These genes are organized in several transcription units, not in a single operon.
c. Within each cluster, at least some transcription units are controlled by the products of pathway-specific regulatory genes.
d. Specific regulatory genes are situated in the closest proximity of biosynthetic genes.
e. Gene expression is primarily controlled at the level of transcription but also at the level of translation.

f. Genes coding for resistance to a given metabolite are closely linked with the cluster of biosynthetic genes. Usually, genes coding for at least two different types of resistance to a produced secondary metabolite can be found within the gene cluster.

g. Expression of genes coding for a secondary metabolite, including genes coding for resistance to it, is controlled by central overruling regulatory circuits often exhibiting pleiotropic regulatory effects (Romero-Tabarez, 2004).

16.11   Exploitation of secondary metabolites for human welfare

16.11.1   Agricultural benefits of secondary metabolites from medicinal plants

In the field of agriculture, the plant breeders select the varieties that provide maximal yields in combination with optimal quality and resistance against pathogens, herbivores, and other environmental stressors. Behind these panic conditions, feeding on one part of the plant can induce systemic production of these chemicals in undamaged plant tissues, and once released, these chemicals can act as signals to neighboring plants to begin producing similar compounds. Production of these chemicals exacts a high metabolic cost on the host plant, so many of these compounds are not produced in large quantities until insects have begun to feed. The primary metabolites are substances produced by all plant cells that are directly involved in the growth, development, or reproduction processes; examples include sugars, proteins, amino acids, and nucleic acids. The secondary metabolites are not directly involved in the growth or reproduction, but they are often involved in the plant defense mechanism. These compounds usually belong to one of these three large chemical classes: terpenoids (**73**), phenolics, and alkaloids (**57**) as discussed earlier (Hassan Adeyemi, 2011).

The chemical protection plays a decisive role in the resistance of plants against pathogens and herbivores. The so-called secondary metabolites, which are a characteristic feature of plants, are especially important and can protect plants against a wide variety of microorganisms (bacteria, viruses, and fungi) and herbivores (arthropods, vertebrates). With the condition that all pathogens evolved in the plant systems, the chemical barrier or bioactive secondary metabolites will be used to treat the disease or simply used as the defense mechanism to decrease the pathogenic involvement. For example in chemical defense compounds, lupins that are of alkaloid-free varieties (sweet lupins) have been selected by plant breeders due to their high susceptibility to a wide range of herbivores to which the alkaloid-rich wild types were resistant (Wink, 1988).

Plant surfaces are usually covered by a hydrophobic layer consisting of antibiotic and repellent cuticular waxes, which contains other secondary metabolites such as alkaloids (**57**), flavonoids (**58**), terpenoids (**73**), and steroids. The presence of carbohydrates and lignins imparts in their constitution in the cell wall as well as proliferation in the infection and wound sight of the plant. The synthesis of inhibitory proteins or enzymes could degrade the microbial cell walls and other microbial constituents, such as peroxidase and phenoloxidase that could help to inactivate phytotoxins of microbial origin. The synthesis of low-molecular-weight compounds produces secondary metabolites that possessed repellent or toxic properties against microorganisms and herbivores (Levin, 1976; Swain, 1977; Harborne, 1982). Color and scent performance by secondary metabolites such as flavonoids (**58**), terpenoids (**73**), and other volatiles can underlie attraction or repelling of insects and herbivores, while toxins can be involved in plant–plant allelopathic interactions (Hadacek, 2002).

16.11.2   Agricultural benefits of bactericidal compounds from medicinal plants

The *Panax ginseng* root extracts have been used for antibacterial compounds in bacteri-cidal activities and have been reported to have ginsenosides as their corresponding sec-ondary metabolites (**76**) (Osbourn, 1996; Sparg et al., 2004). Sixteen coumarins (**34**), that is, 7,8-dihydroxycoumarin (**77**), umbelliferone (**78**), scoparone (6,7-dimethoxycoumarin) (**79**), esculetin (**80**), 6-hydroxy-7-methoxycoumarin (**81**), herniarin (7-methoxycouma-rin) (**82**), scopoletin (**83**), 6,7-diacetoxy coumarin (**84**), 6-methoxy-7-acetylcoumarin (**85**) and 6-acetoxy-7-methoxycoumarin (**86**), 8-methoxypsoralen (**87**), 8-acetyl-7-hydroxycou-marin (**88**), 7,8-dihydroxy-6-methoxycoumarin (**89**), 6,7-dimethoxy-4-methylcoumarin (**90**), 5,7-dihydroxy-4-methylcoumarin (**91**), 4-hydroxycoumarin (**92**), and 4-hydroxy-6,7-dimethylcoumarin (**93**), were tested against a panel of bacteria. Their antibacterial activ-ity was determined depending on the microorganism such as *Bacillus subtilis, Escherichia coli, Proteus mirabilis, Klebsiella pneumoniae, Salmonella typhi, Salmonella* sp., *Shigella boy-dii, Shigella* sp., *Enterobacter aerogenes, Enterobacter agglomerans, Sarcinalutea, Staphylococcus epidermidis, Staphylococcus aureus, Yersinia enterocolitica,* and *Vibrio cholerae.* In addition, 7,8-dihydroxy-6-methoxycoumarin (**89**), 8-acetyl-7-hydroxycoumarin (**90**), 7,8-dihydroxycou-marin (**77**), 6,7-dimethoxy-4-methylcoumarin (**90**), scoparone (**79**), and esculetin (**80**) have been found and reported an activity against *V. cholera,* and 8-acetyl-7-hydroxycoumarin (**88**), 7,8-dihydroxy-6-methoxycoumarin (**89**), and 7,8-dihydroxycoumarin (**77**) were the most active phytoconstituents. The dimethoxy compounds 6,7-dimethoxy-4-methylcoumarin (**90**) and scoparone (**79**) showed and reported a significant activity against fungal strains, especially *T. mentagrophytes* and *Rhizoctonia solani* (Cespedes et al., 2006), which are effective repellants against viruses.

16.11.3   Agricultural benefits of fungicide compounds from medicinal plants

The *Panax ginseng* root extracts have also been used for antifungal compounds in fungi-cide activities, and ginsenosides as their secondary metabolites (**76**) have been reported by Osbourn (1996) and Sparg et al. (2004). The wyerone acid (**94**) is produced by the broad bean (*Vicia faba*) when it is attacked by microorganisms and thus used as fun-gicides. The evaluated fungi were *Aspergillus niger, Penicillium notatum, Fusarium monili-forme, Fusarium sporotrichum, R. solani,* and *T. mentagrophytes.* The most active compounds against gram-positive and gram-negative bacteria were the dihydroxylated coumarins (**34**): 7,8-dihydroxycoumarin (**77**) and 7,8-dihydroxy-6-methoxycoumarin (**89**) (Cespedes et al., 2006).

16.11.4   Agricultural benefits of insecticidal compounds from medicinal plants

The toxicity of secondary metabolites to insects and numerous feeding experiments has been carried out, in which secondary compounds accumulated in plants are shown and reported as toxic to insect herbivores that do not normally encounter them, and thus some of these compounds have been used as insecticides. The three secondary compounds used as insecticides are as follows: (1) pyrethroid insecticides, which are used widely in agri-culture and horticulture and as active constituents of derris powder; (2) popular domes-tic insecticide; and (3) nicotine (**44**) (tobacco, *Nicotiana tabacum*) used in smoke bombs for greenhouses.

Eight numbers of isolated compounds were tested for insecticidal activity against female adults of *Aedes aegypti*: β-sitosterol-3-*O*-β-ᴅ-glucoside (**95**), palmitic acid (**96**), linaro-side (**97**), homoplantagenin (**98**), 5,7,3'-trihydroxy-6,4'-dimethoxy flavone-7-*O*-glucoside and shikimic acid (**99**), protochatecuic acid (**100**), blepharin (**101**), and acetoside (**102**) for adulticidal activity, respectively (Amin et al., 2012). The larvicidal activity of palmitic acid against *Culex quinquefasciatus*, *Anopheles stephensi*, and *A. aegypti* was previously reported by Abdul-Rahman et al. (2000).

Narbon bean (*Vicia narbonensis*), which is a grain legume, emits a foul smell to deter animals from feeding on its seeds. The grass pea (*Lathyrus sativus*), which is also a grain legume, has an enormous potential to be a food source since not only it has a nutritious seed but it also has a high drought resistance. Unfortunately, the seed contains a number of toxic nonprotein amino acids (**33**) including β-oxalyl diaminopropionic acid (**103**), which was reported to cause slow progressive paralysis.

Dioclea megacarpa seeds, from a tree in the Costa Rica rain forests, are quite remarkable in that they contain up to 13% dry weight of the toxic amino acid canavanine (**6**). Canavanine (**6**) is similar in structure to arginine, and insects mistakenly incorporate canavanine (**6**) during protein synthesis instead of arginine. The results are that proteins fail to have the desired properties just like nonfunctioning enzymes and the insects die. However, one species of beetle (*Caryedes brasiliensis*) has adapted to this toxin and can discriminate between arginine and canavanine (**6**). This beetle can provide rich food supply for itself without competition from other insects.

Lupanine (**26**), nicotine (**44**), tomatine (**104**), solanine (**105**), ergotamine (**106**), and lysergic acids (**107**) were also natural secondary metabolites under the category of alkaloids (**57**), which are susceptible to insect attacks at any use. In terpenes, isoprene (**108**), citronellal (**3**) (oil of lemon), menthol (**109**) (peppermint oil), zingiberene (**110**) (ginger oil), azadirachtin (**51**), and pyrethrins (**52**) were also reported for insecticidal activities.

Populus trees are deploying various types of defenses, one of which is the production of a myriad of secondary compounds that are produced from the shikimate-phenylpropanoid pathway. The plant secondary metabolites are used in activities against herbivores as insecticides such as salicin (**111**), salicortin (**112**), tremuloidin (**113**), tremulacin (**114**), populoside (**115**), trichocarposide (**116**), p-coumarate (**117**), caffeate (**118**), ferulate (**119**), sinapate (**120**), coumaroyl quinate (**121**), caffeoyl quinate (**122**), feruloyl quinate (**123**), pinocembrin chalcone (**124**), naringenin chalcone (**125**), eriodictyol chalcone (**126**), homoeriodictyol chalcone (**127**), pinobanksin (**128**), aromadendrin (**129**), taxifolin (**130**), dihydromyricetin (**131**), galangin (**132**), kaempferol (**133**), quercetin (**35**), myricetin (**134**), pinocembrin (**135**), naringenin (**136**), eriodictyol (**137**), homoeriodictyol (**138**), chrysin (**139**), apigenin (**140**), luteolin (**141**), condensed tannins (**142**), isoprene (**143**), (E)-β-ocimene (**144**), linalool (**145**), (E,E) α farnesene (**146**), germacrene D (**147**), (Z) 3 hexenal (**148**), (Z)-3-hexen-1-ol (**149**), and (Z)-3-hexenyl acetate (**150**) (Chen et al., 2009).

The *Ginkgo biloba* leaf extracts have been used for insecticidal and antifeedant activities, and their secondary metabolites are reported to be ginkgolides A (**151**), B (**152**), C (**153**), and J (**154**) (Ahn et al., 1997), and the *Panax ginseng* root extracts and their secondary metabolites ginsenosides (**76**) are also reported useful in insecticidal activities (Osbourn, 1996; Sparg et al., 2004).

The use of pyrethrum in agriculture was recommended in controlling the attacks of tulip bulb mites, iris root aphids, black citrus aphids, pea aphids, potato aphids, grape leafhoppers, cotton leafhoppers, grape trips, bean trips, cabbage worms, tomato fruit worms, parsley caterpillars, red-humped caterpillars, and many other insects. Pyrethrin (**52**) alone is a naturally occurring insecticide, and industries and researchers found that synthetic insecticides are derived from them. However, the already prepared and marketed synthetic pyrethroid-derived compounds were proven to be very harmful for human beings. They are known as DDT, permethrin, fenvalerate, deltamethrin, etofenprox, acrinathrin, and bifenthrin. Resin acids include pimaric acid (**155**) (from *Pinus* species) and podocarpic acid from *Podocarpus cupressinum*; manoyl oxides (**156**) were also used as insecticides.

16.11.5   Medical benefits of secondary metabolites from medicinal plants

The discovery of new bioactive natural products is still a fascinating field in organic chemistry as demonstrated by the recent paradigms of the anticancer drug epothilone, the immunosuppressant rapamycin, or the proteasome inhibitor salinosporamide, to name a few, but there are hundreds of possible examples. Finding new secondary metabolites is a prerequisite for the development of novel pharmaceuticals, and this is an especially urgent task in the case of antibiotics due to the rapid spreading of bacterial resistances and the emergence of multiresistant pathogenic strains, which poses severe clinical problems in the treatment of infectious diseases. This Thematic Series on the biosynthesis and function of secondary metabolites deals with the discovery of new biologically active compounds from all kinds of sources, including plants, bacteria, and fungi, and also with their biogenesis. Biosynthetic aspects are closely related to functional investigations, because a deep understanding of metabolic pathways to natural products not only on a chemical but also on a genetic and enzymatic level allows for the expression of whole biosynthetic gene clusters in heterologous hosts. This technique can make interesting, new secondary metabolites available from noncultivable microorganisms or may be used to optimize their availability by fermentation, for further research and also for production in the pharmaceutical industry.

Especially fascinating is the intrinsic logic of the polyketide and nonribosomal peptide biosynthetic machineries, which is strongly correlated with the logic of fatty acid biosynthesis as part of the primary metabolism. Insights into the mechanisms of modular polyketide and nonribosomal peptide assembly lines open up the possibility for direct modifications, for example, the oxidation states of the natural product's carbon backbone by simple domain knockouts within the responsible megasynthases or the introduction of a variety of alternative biosynthetic starters by mutasynthesis approaches, thus leading to new variants of known metabolites, which may have improved the properties for therapeutic use. Another interesting aspect is the usage of enzymes in chemical transformations, which can provide synthetic organic chemistry with efficient access route to typically chiral building blocks that may otherwise be difficult to obtain (Jeroen, 2011). Many of these secondary metabolic compounds are valued for their pharmacological activities and industrial or agricultural properties that increase the commercial value of crops (Bingham et al., 1998; Paganga et al., 1999).

16.11.6   Antibacterial agents from synthesized secondary metabolites of plants

The secondary metabolites of many plants, such as essential oils (**5**), bisabolol (**157**), ngaione (**158**), hymenoxin (**159**), and santonin (**160**), were reported to possess antimicrobial properties in antibacterial and antifungal activities (Bruneton, 1999; Heinrich et al., 2004) used for medicinal purposes. Phaseolin (**161**) (bean) and isopimpinellin (**162**) (parsley) were reported as the most significant plant secondary metabolites for the antibacterial activity against virulence of medical pathogens.

The common herbs tarragon and thyme both contain caffeic acid (**41**), which are effective agents against bacteria. The catechol (**163**) and pyrogallol (**164**) both are hydroxylated phenols, shown to be toxic to microorganisms. For that reason, the potential range of quinone and coumarin (**34**) antimicrobial effect is great. There are reports of antimicrobial properties associated with polyamines (in particular, spermidine (**165**)), isothiocyanates (**166**), thiosulfinates (**167**), and glucosides (**168**) (Ciocan and Bara, 2007).

16.11.7   Antifungal agents from synthesized secondary metabolites of plants

A study has shown that the antifungal agents derived from the plant heartwood of *Pinus* species have used secondary metabolites such as resveratrol (**169**), pinosylvin (**170**), and stilbenoids (**171**) (Harborne, 1998; Heinrich et al., 2004). The major components of this mixture are a group of structurally related compounds, that is, apigeninidin (**172**) and luteolinidin (**173**) (Brinker and Seigler, 1991); pisatin (**25**), maackiain (**174**), daidzein (**175**), and medicarpin (**176**) (Graham, 1995); glyceollin (**177**) and keivitone (**178**) (Shanker et al., 2003); phaseolin (**161**) (Durango et al., 2002); emodin-a (**179**) (Izhaki, 2002), malvone A (**180**) (Veshkurova et al., 2006), umbelliferone (**78**), and scopoletin (**83**) (Uritani and Hoshiya, 1953; Minamikawa et al., 1964; Tanaka et al., 1983); scoparone (**79**) (Afek et al., 1986; Kim et al., 1992), ayapin (**181**), and resveratrol (**169**) (Langcake and Pryce, 1976); pterostilbene (**182**) (Bala et al., 2000; Jeandet et al., 2002; Commun et al., 2003) and ε-viniferin (**183**) (Dercks and Creasy, 1989); pinosylvins (**170**), pinosylvinmonomethyl ether (**184**), and pinosylvin dimethyl ether (**185**); carvacrol (**186**) and thymol (**187**) (Muller-Riebau et al., 1995); rishi-tin (**28**), lubimin (**188**), and solavetivone (**189**) (Kuc 1995); niveusin-ß (**190**) and costunolide (**191**) (Grayer and Harborne, 1994); cichoralexin (**192**), lactucin (**193**), momilactones A (**194**) (Kabuto et al., 1973) and B (**195**), sanguinarine (**196**), brassinin (**197**), ciclobrassinin (**198**), brassilexin (**199**), wasalexin A (**200**), and arvelexin (**201**) (Pedras et al., 2003); caulexins A (**202**), B (**203**), and C (**204**) (Pedras and Jha, 2006); and camalexins (**205**) and wyerone acid (**94**) (Buzi et al., 2003), which were reported to have antifungal properties derived from various plant genus and species.

The aforementioned common herbs, tarragon and thyme, are also effective agents against fungi. Lawsone (2-hydroxy-1,4-naphthoquinone) (**206**) and Me-lawsone (2-methoxy-1,4-naphthoquinone) (**207**) are the two main antifungal naphtha quinones found naturally in *Impatiens balsamina* (Wanchai, 1998). The metabolites of many plants, such as essential oils (**5**), bisabolol (**157**), ngaione (**158**), hymenoxin (**159**), and santo-nin (**160**), were reported to possess antimicrobial properties in antifungal activities (Bruneton, 1999; Heinrich et al., 2004) used for medicinal purposes.

16.11.8   Antiviral agents from synthesized secondary metabolites of plants

The phytoalexin resveratrol (3,4,5-trihydroxystilbene) is produced within a range of edible plants in response to tissue damage and environmental stressors such as fungal and viral attacks and moreover the antiviral effects against the *herpes* simplex virus (Docherty et al., 1999; Faith et al., 2006).

16.12   Summary

Plants produce thousands of secondary metabolites of diverse chemical nature.

These compounds and chemical entities play major and important roles in protecting plants under adverse conditions. In such way, there was a huge preparation developed under agricultural and medical purposes, which was known as pharmaceutical preparation. It created a revolution in the world day by day due to the panic conditions caused by the plant and animal diseases. Furthermore, nowadays, the beneficiary effects have been elevated by the application secondary metabolites, which leads to the protection of plant and animal health. Moreover, the secondary metabolites from several plants are used in the production of medicines, dyes, insecticides, flavors, and fragrances.

Unraveling plant secondary metabolism is the way to successful applications in molecular farming, health food, functional food, and plant resistance. Various pathways have been altered using genes encoding biosynthetic enzymes or regulatory proteins and show enormous potential for the genetic engineering of plant secondary metabolism. Recent achievements have been made in the metabolic engineering of plant secondary metabolism.

In another way, biosynthetic aspects are closely related to functional investigations, because of the deep understanding of metabolic pathways to natural products, not only on a chemical but also on a genetic and enzymatic level, that allows for the expression of whole biosynthetic gene clusters in heterologous hosts. Secondary metabolites play a major role in the adaptation of plants to the changing environment and in overcoming stress constraints. This results in a duly large complexity of chemical entities and interactions underlying various functions: structure stabilization, determined by polymerization, condensation of phenols and quinones, and electrostatic interactions of polyamines with negatively charged loci in cell components; photoprotection, which is related to absorbance of visible light and UV radiation due to the presence of double bonds; antioxidant and antiradical activities, governed by the availability of OH, NH_2, and SH groupings, as well as aromatic nuclei and unsaturated aliphatic chains; and signal transduction.

In discussions about secondary metabolites, there have been reviews made that contribute to the various classifications and methods used for these metabolites in the past 10 decades. In such way, the biosynthetic pathway of secondary metabolites, their role in the plant's metabolism, their significance and importance in genetics, the history of antimicrobial compounds, the natural plant products as antimicrobial agents, the mode of action of synthetic antimicrobial agents, and the classifications of secondary metabolites in the field of agricultural and medical applications are discussed in this chapter in a critical manner.

Their major classifications were defined and found as aliphatic, aromatic, hydroaromatic, and heterocyclic in their chemical aspects. Other subclassifications and chemical entities were often based on their major chemical components. About 207 chemical units were reviewed here and used for the purpose of agricultural and medical applications. In a more detailed discussion, the chemical entities of compounds were used in the treatment of plant diseases such as red rot disease and tikka disease and all insecticidal attacks and used as antibacterial, antifungal, and insecticidal agents in the field of agriculture. On the other hand, the chemical units were used for the purpose of treating animal and human diseases like arthritis, cancers, diabetes, inflammations, ulcers, and wounds.

16.13   Conclusion

This chapter is concluded with the major classes of phytochemical compounds that are found in medicinal plants and perpetually used for various antibacterial, antifungal, and insecticidal activities in the field of agriculture, which employs secondary metabolites in the treatment of plant disorders such as sheath blight, leaf blast, red rot disease, tikka disease, and insect attacks. In addition, these secondary metabolites are used in the composition of antibacterial, antifungal, and antiviral compounds in the medical field to treat various debilitating human diseases including arthritis, cancer, diabetes, and immunodeficiency diseases. About 207 compounds from plants with secondary metabolites, which are most useful in agricultural and medical applications, were reviewed here. This review about secondary metabolites as bioactive constituents will be helpful for future research and development (Figures 16.4 and 16.5).

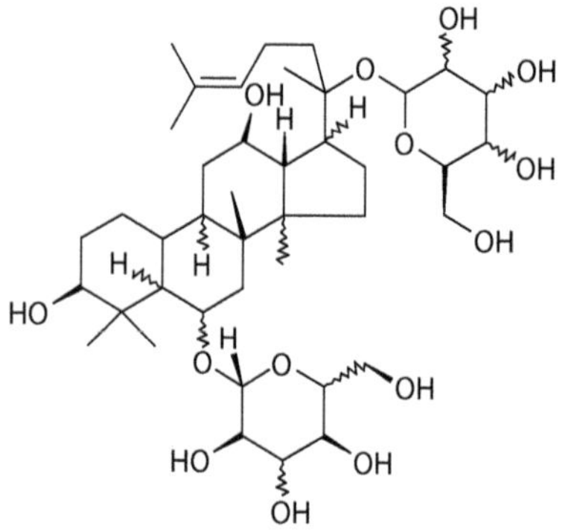

Ginsenoside (**76**)

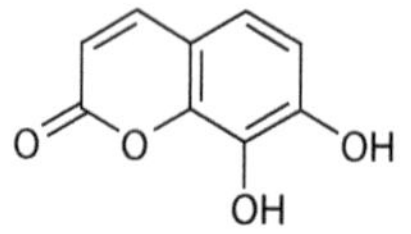

7,8-dihydroxycoumarin (**77**)

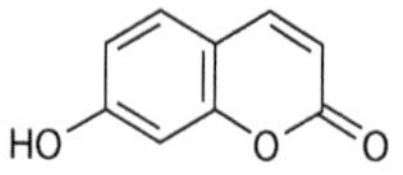

Umbelliferone (**78**)

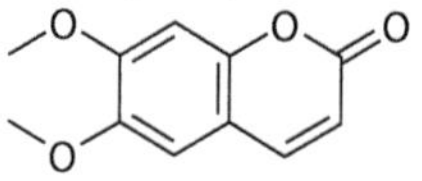

Scoparone (6,7-dimethoxycoumarin) (**79**)

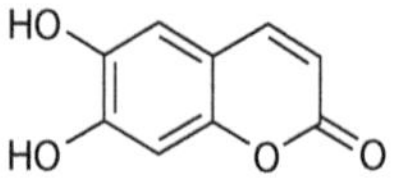

Esculetin (**80**)

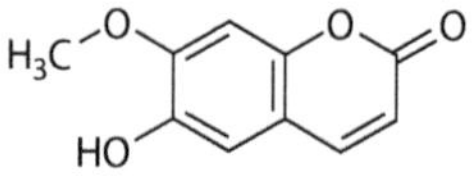

6-hydroxy-7-methoxycoumarin (**81**)

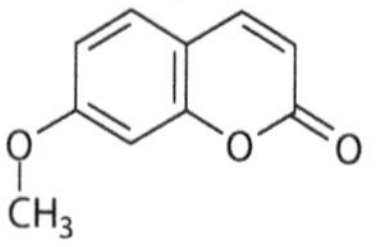

Herniarin (7-methoxycoumarin) (**82**)

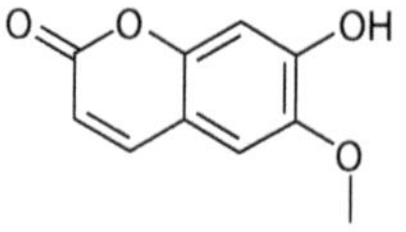

Scopoletin (**83**)

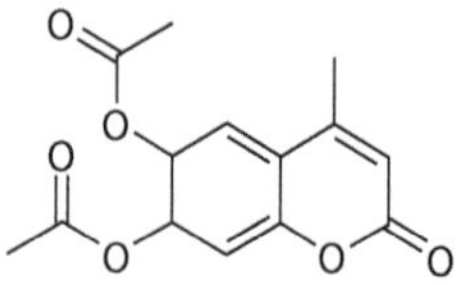

6,7-diacetoxy coumarin (**84**)

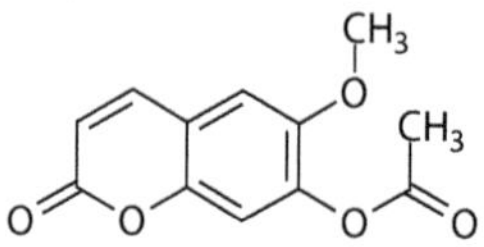

6-methoxy-7-acetylcoumarin (**85**)

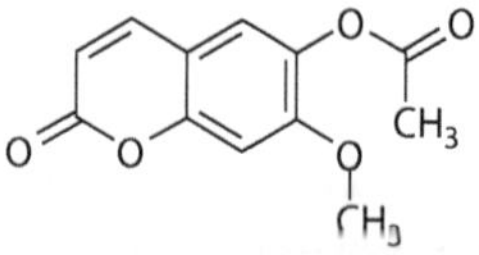

6-acetoxy-7-methoxycoumarin (**86**)

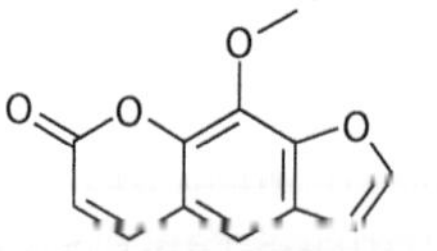

8-methoxypsoralen (**87**)

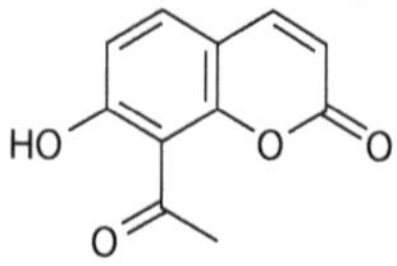

8-acetyl-7-hydroxycoumarin (**88**)

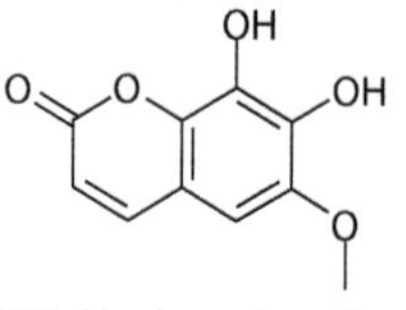

7,8-dihydroxy-6-methoxycoumarin (**89**)

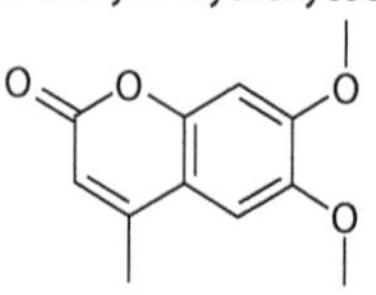

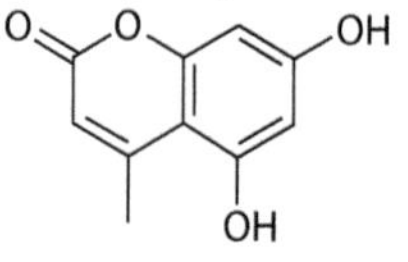

Figure 16.4 Secondary metabolic chemical compounds from plants for antibacterial, antifungal, and insecticidal activities (drawn using ISIS/ChemDraw). (*Continued*)

6,7-dimethoxy-4-methylcoumarin (**90**)

5,7-dihydroxy-4-methylcoumarin (**91**)

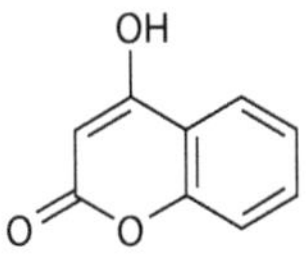

4-hydroxycoumarin (**92**)

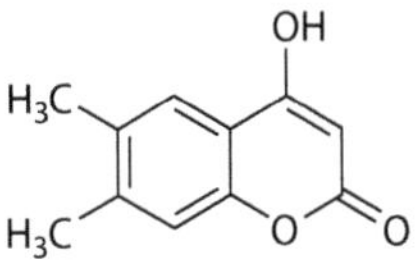

4-hydroxy-6,7-dimethylcoumarin (**93**)

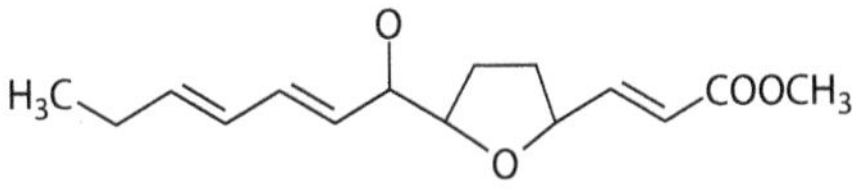

Wyerone acid (**94**)

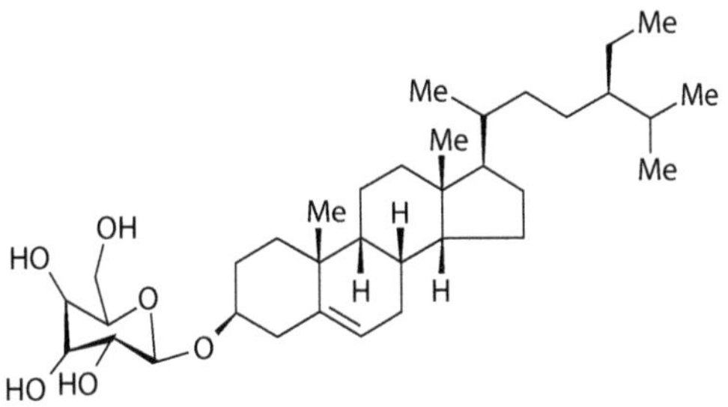

β-Sitosterol-3-*O*-β-ᴅ-glucoside (**95**)

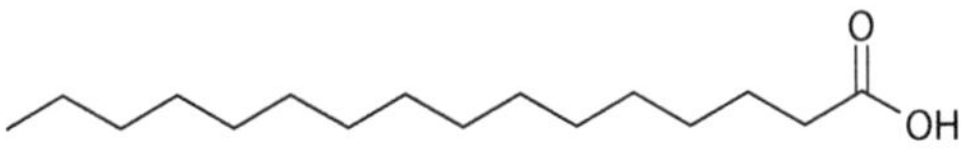

Palmitic acid (**96**)

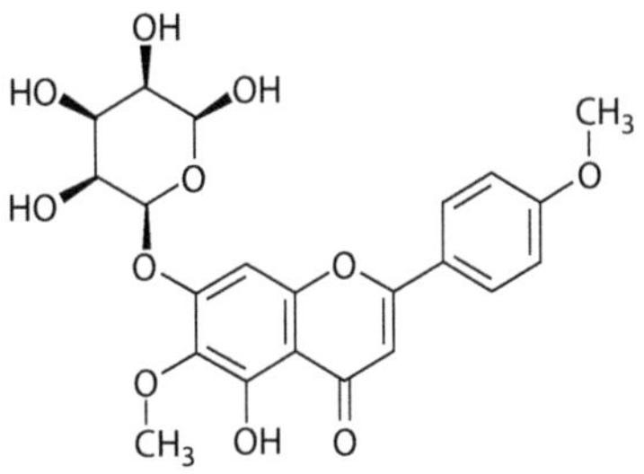

Linaroside (**97**)

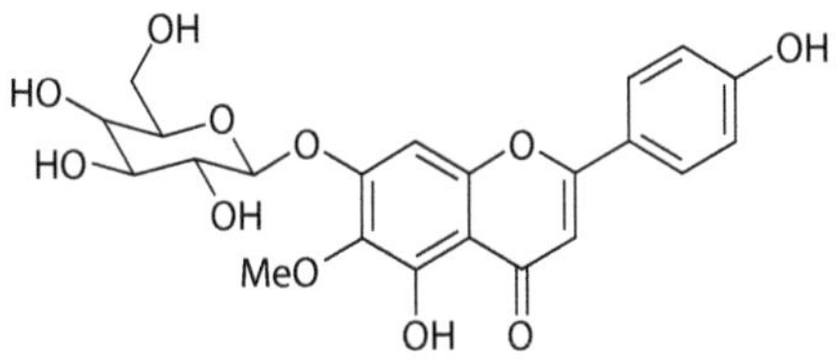

Homoplantagenin (**98**)

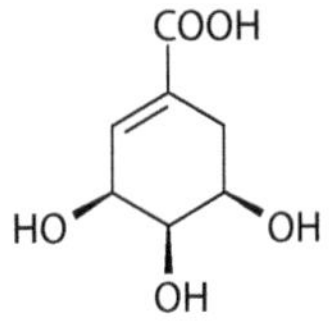

Tuliroside acid (**99**)

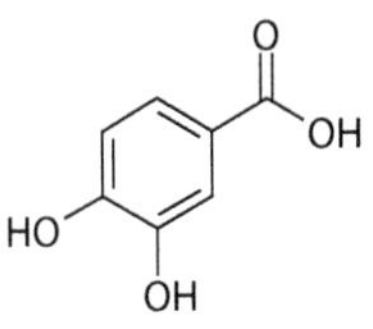

Protochatecuic acid (**100**)

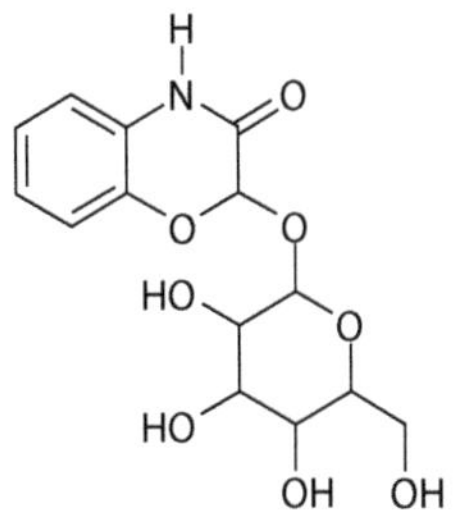

Blepharin (**101**)

Figure 16.4 (Continued) Secondary metabolic chemical compounds from plants for antibacterial, antifungal, and insecticidal activities (drawn using ISIS/ChemDraw).   (*Continued*)

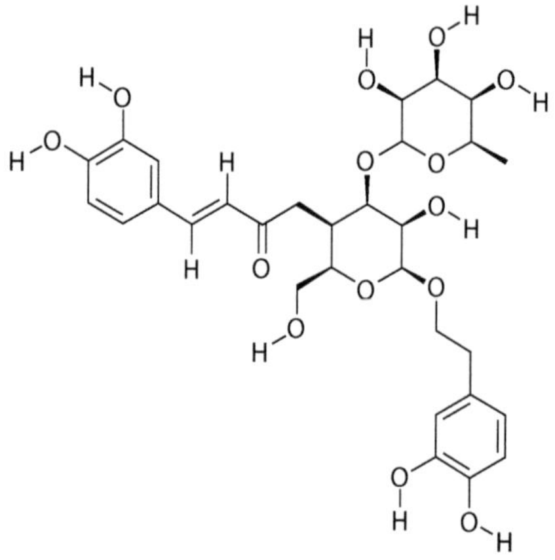

Acetoside (**102**)

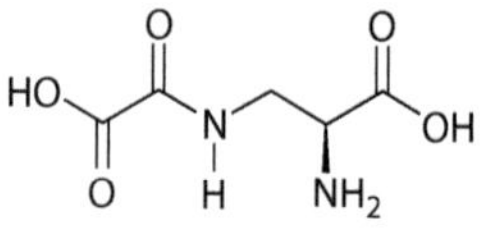

β-Oxalyl diaminopropionic acid (**103**)

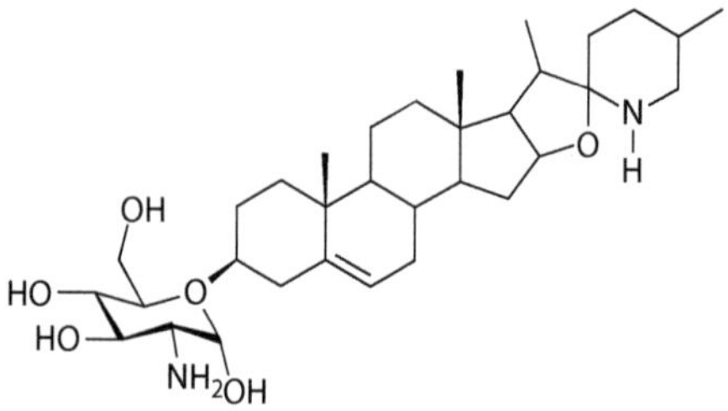

Tomatine (**104**)

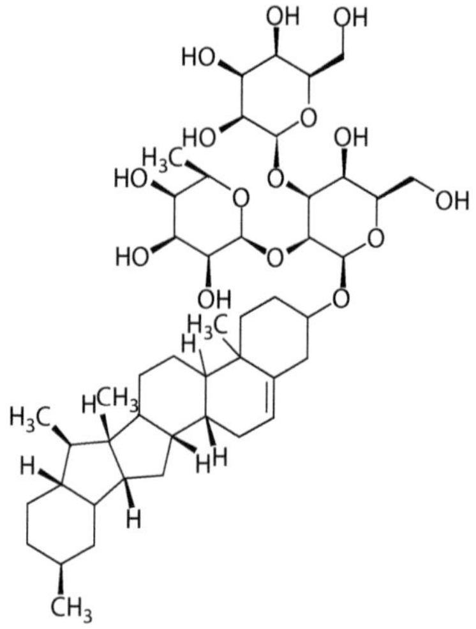

Solanine (**105**)

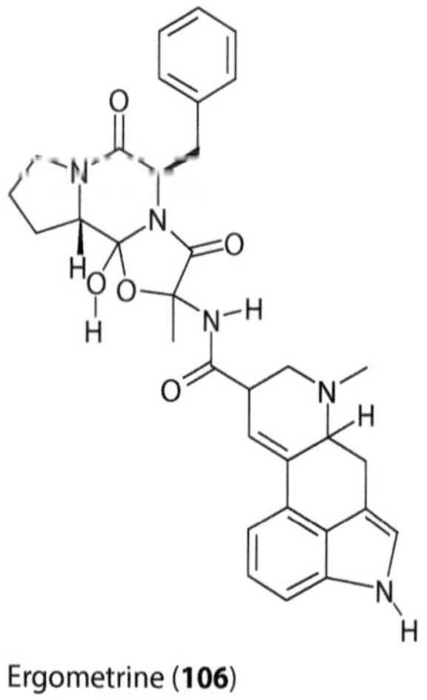

Ergometrine (**106**)

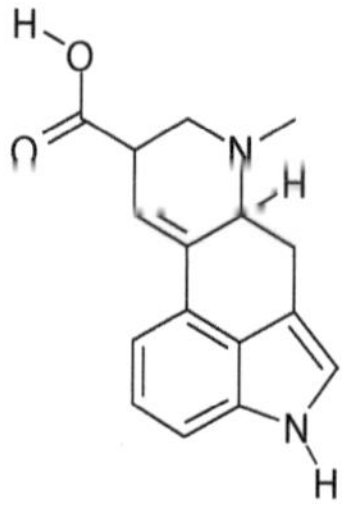

Lysergic acid (**107**)

Figure 16.4 (Continued) Secondary metabolic chemical compounds from plants for antibacterial, antifungal, and insecticidal activities (drawn using ISIS/ChemDraw).   (*Continued*)

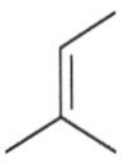

Isoprene (**108**)

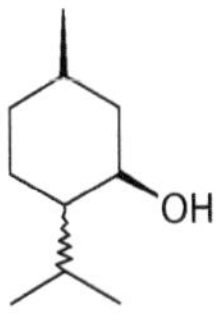

Menthol (**109**)

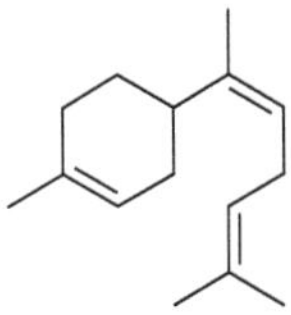

Zingiberene (**110**)

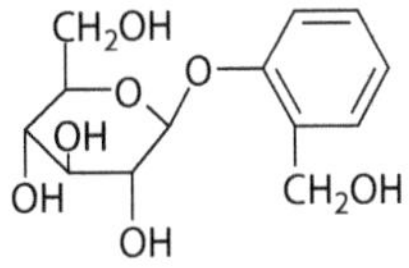

Salicin (**111**)

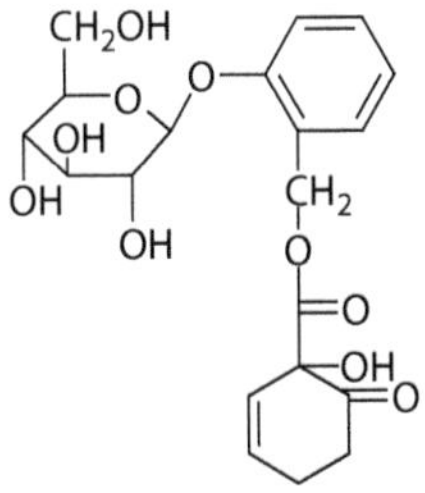

Salicortin (**112**)

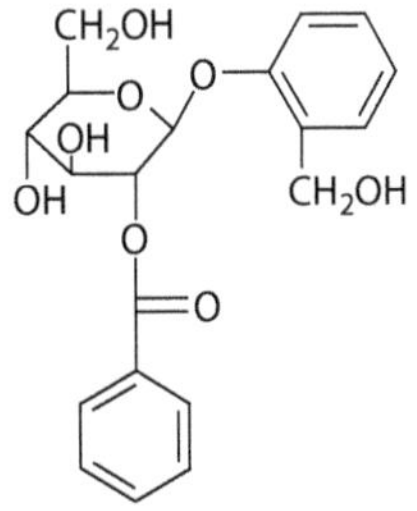

Tremuoloidin (**113**)

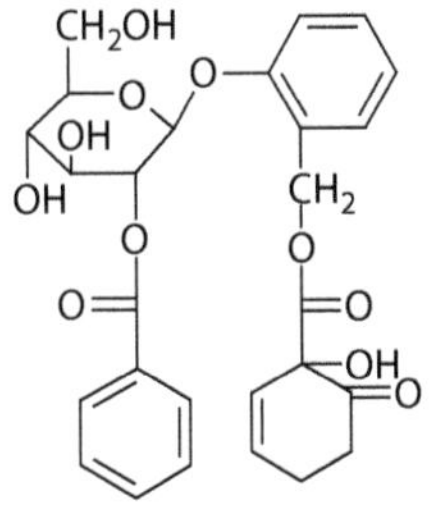

Tremulacin (**114**)

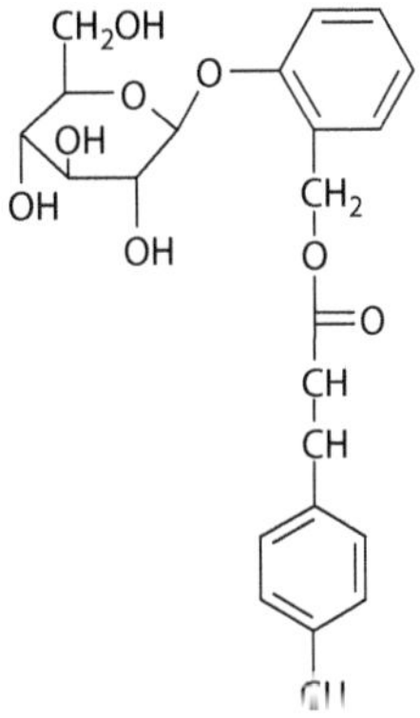

Populoside (**115**)

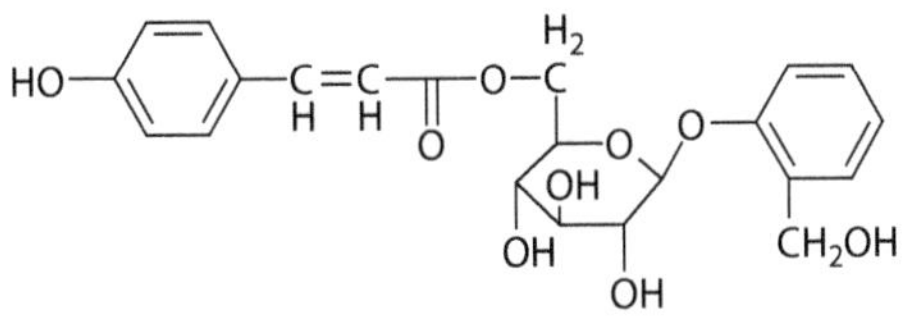

Trichocarposide (**116**)

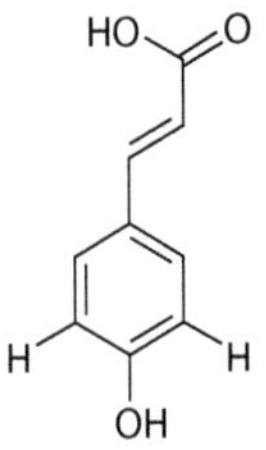

p-Coumarate (**117**)

Figure 16.4 (Continued) Secondary metabolic chemical compounds from plants for antibacterial, antifungal, and insecticidal activities (drawn using ISIS/ChemDraw). *(Continued)*

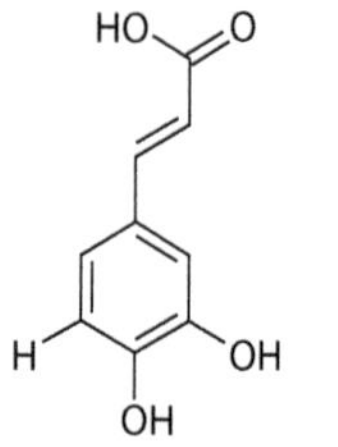

Caffeate (**118**)

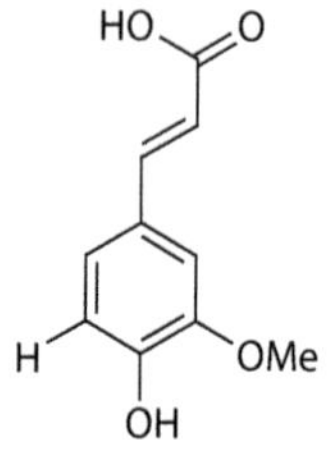

Ferulate (**119**)

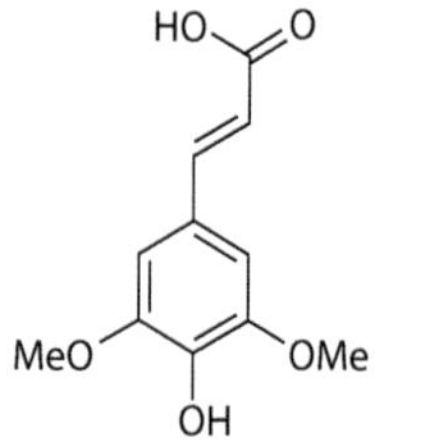

Sinapate (**120**)

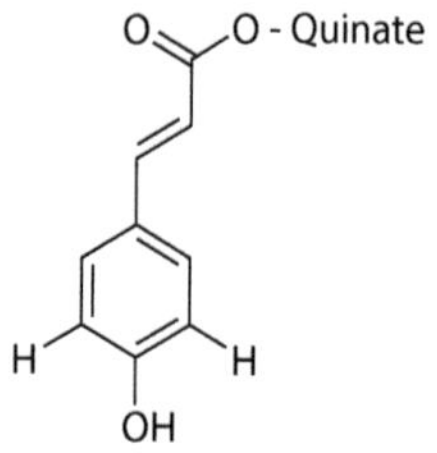

Coumaroyl quinate (**121**)

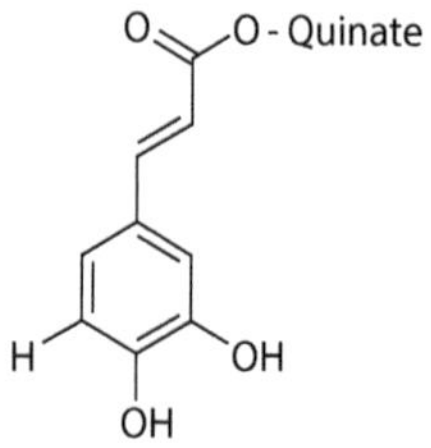

Caffeoyl quinate (**122**)

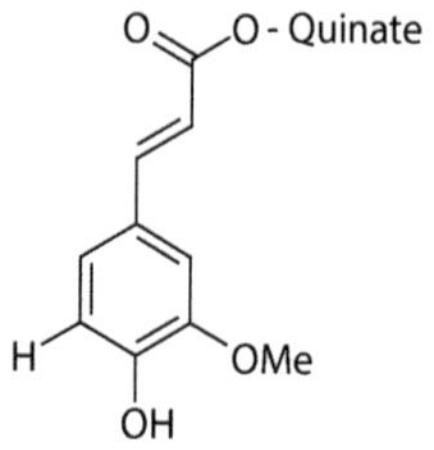

Feruloyl quinate (**123**)

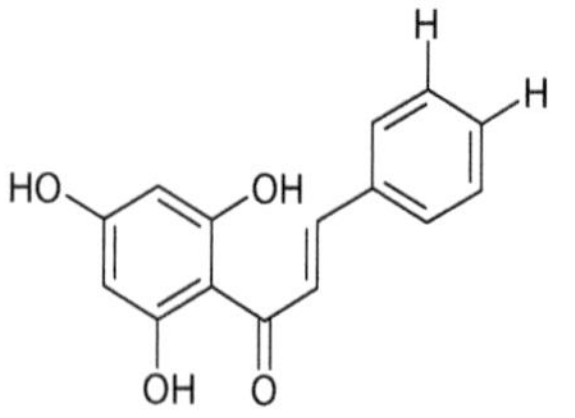

Pinocembrin chalcone (**124**)

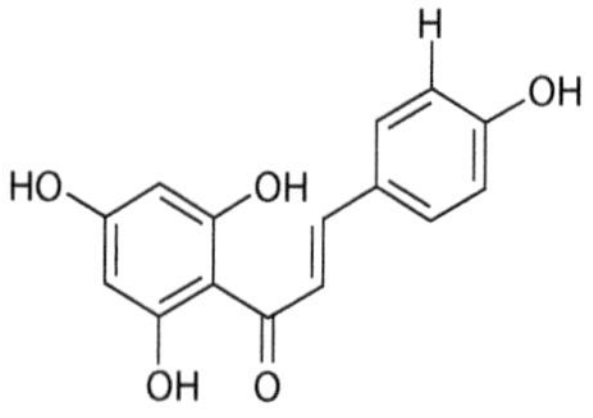

Naringenin chalcone (**125**)

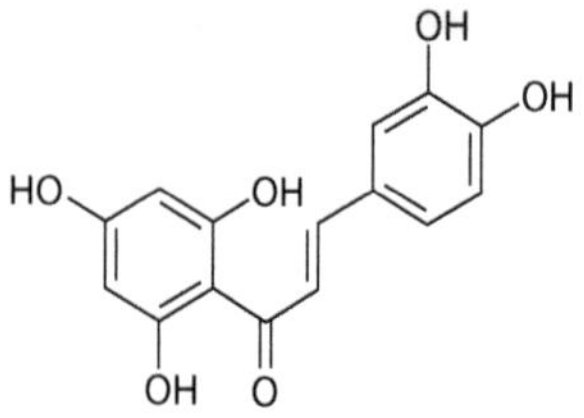

Eriodictyol chalcone (**126**)

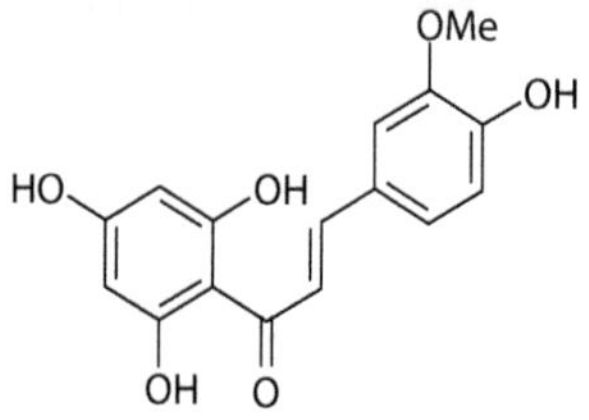

Homoeriodictyol chalcone (**127**)

Figure 16.4 (Continued) Secondary metabolic chemical compounds from plants for antibacterial, antifungal, and insecticidal activities (drawn using ISIS/ChemDraw).   *(Continued)*

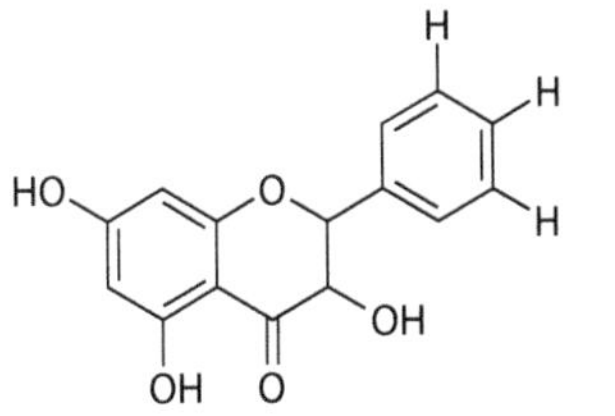

Pinobanksin (**128**)

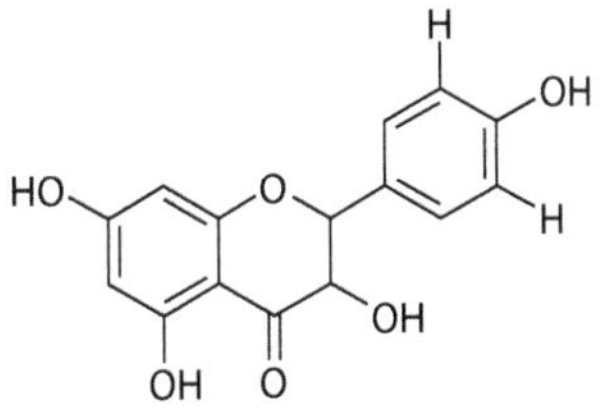

Aromadendrin (**129**)

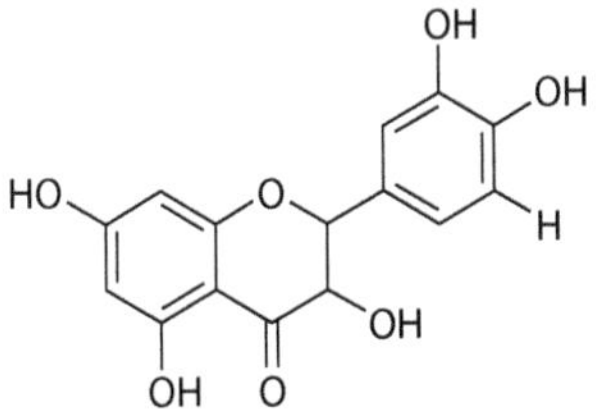

Taxifolin (**130**)

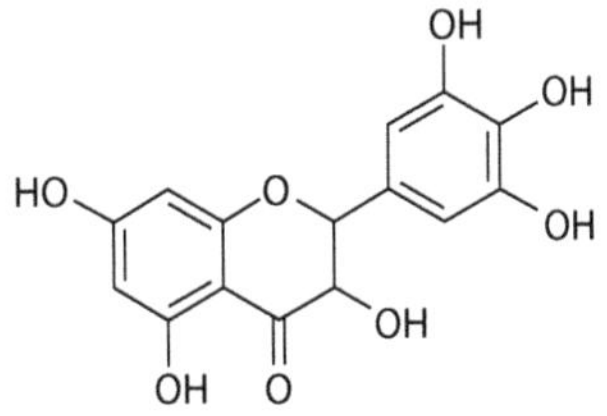

Dihydromyricetin (**131**)

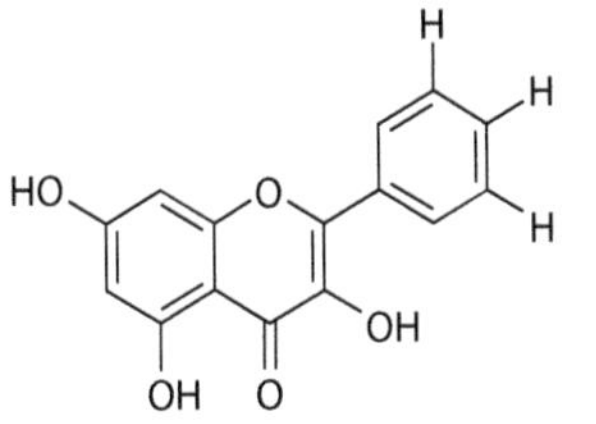

Galangin (**132**)

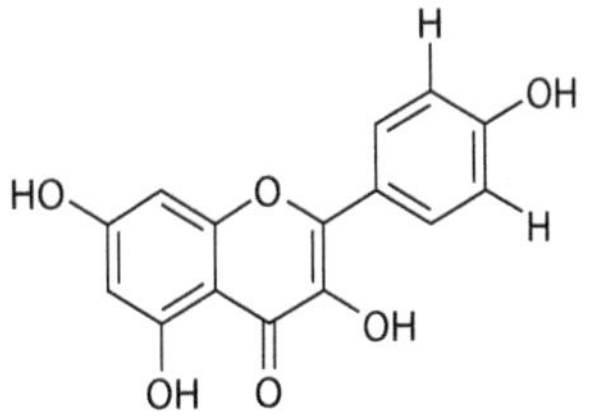

Kaempferol (**133**)

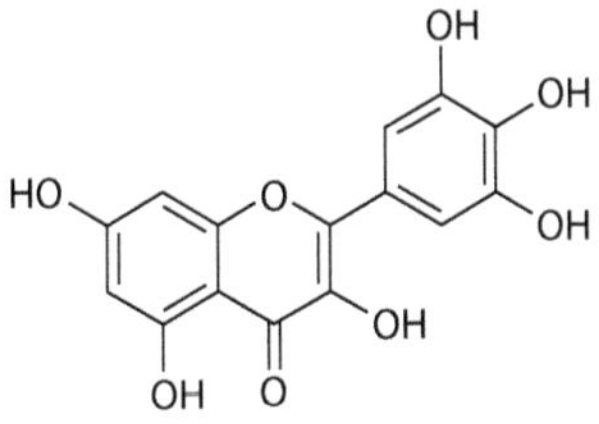

Myricetin (**134**)

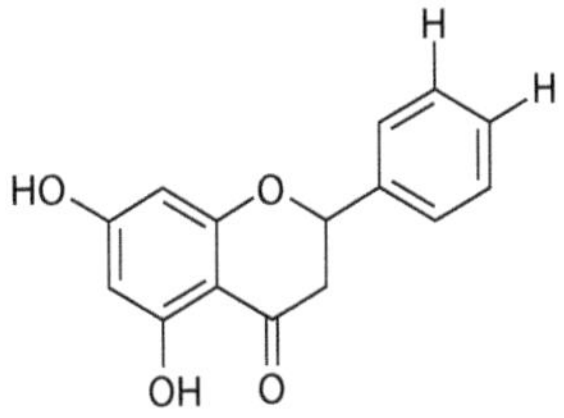

Pinocembrin (**135**)

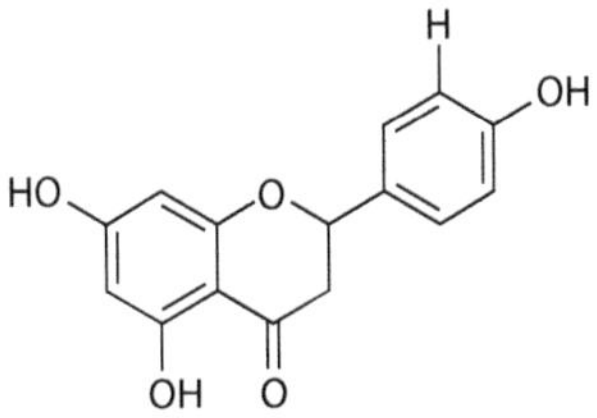

Naringenin (**136**)

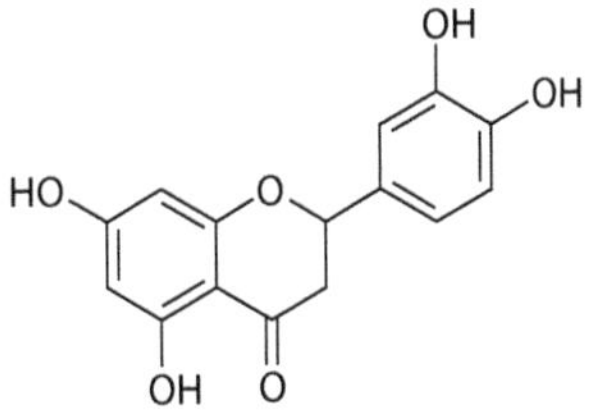

Eriodictyol (**137**)

Figure 16.4 (Continued) Secondary metabolic chemical compounds from plants for antibacterial, antifungal, and insecticidal activities (drawn using ISIS/ChemDraw).   *(Continued)*

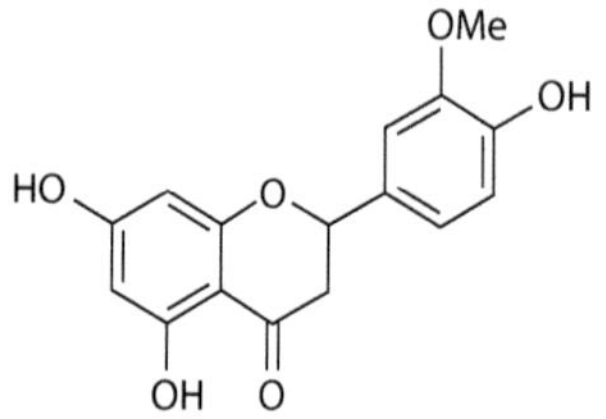

Homoeriodictyol (**138**)

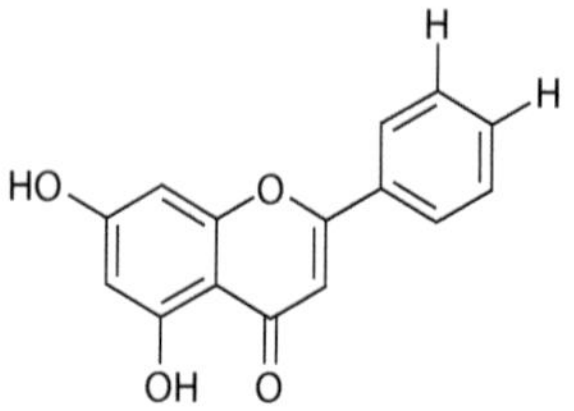

Chrysin (**139**)

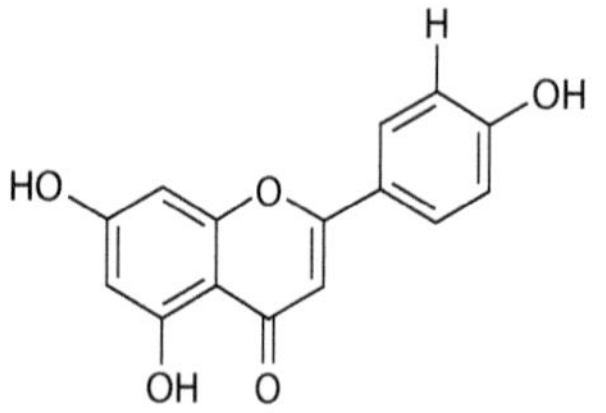

Apigenin (**140**)

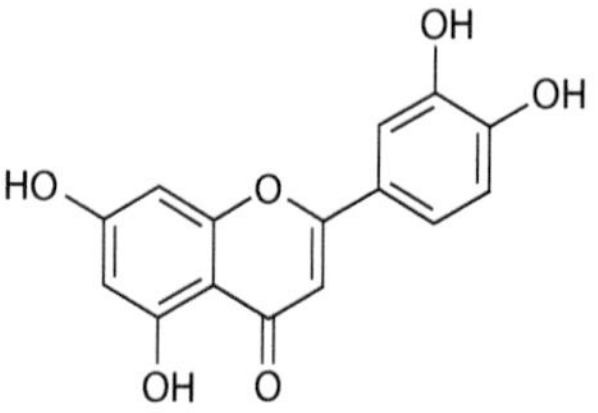

Luteolin (**141**)

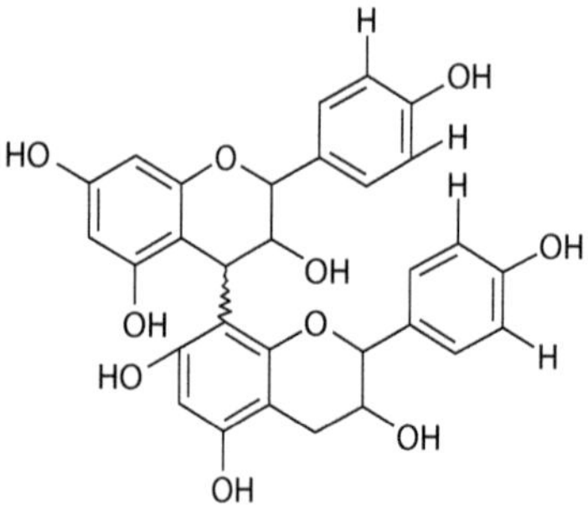

Condensed tannins (**142**)

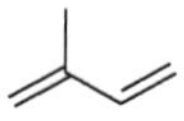

Isoprene (**143**)

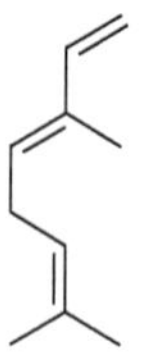

(E)-β-ocimene (**144**)

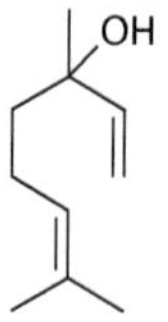

Linalool (**145**)

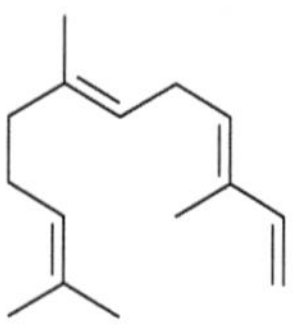

(E,E)-α-farnesene (**146**)

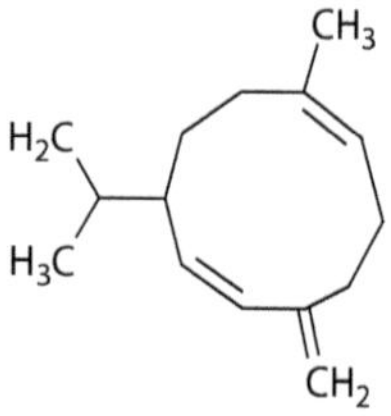

Germacrene D (**147**)

Figure 16.4 (Continued) Secondary metabolic chemical compounds from plants for antibacterial, antifungal, and insecticidal activities (drawn using ISIS/ChemDraw).   (*Continued*)

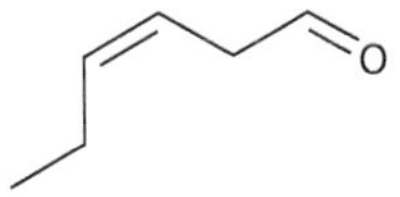

(Z)-3-hexenal (**148**)

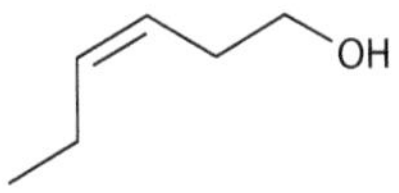

(Z)-3-hexen-1-ol (**149**)

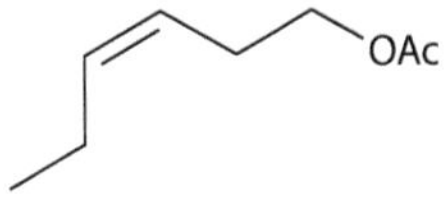

(Z)-3-hexenyl acetate (**150**)

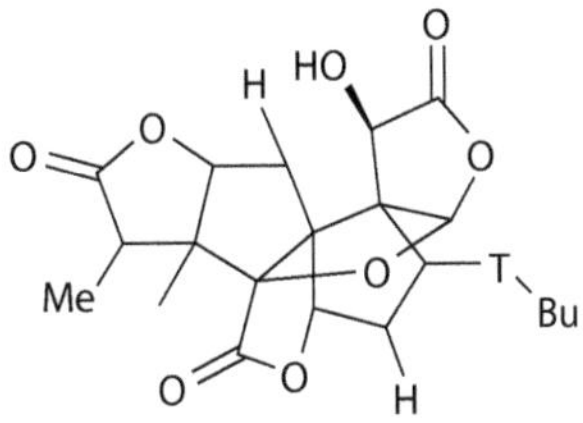

Ginkgolides A (**151**)

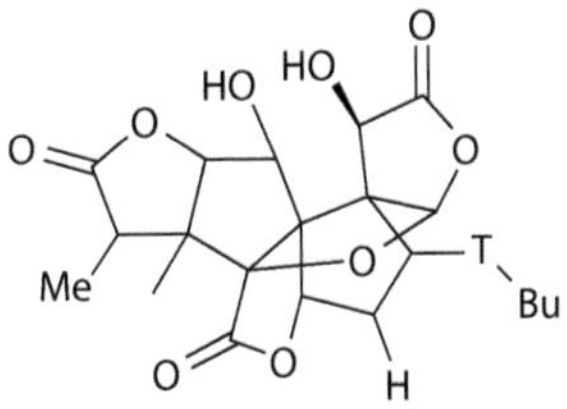

Ginkgolides B (**152**)

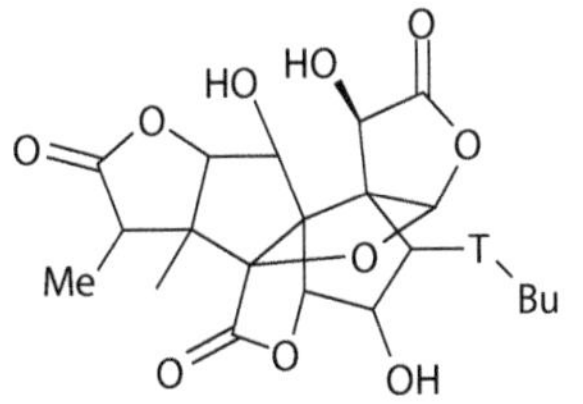

Ginkgolides C (**153**)

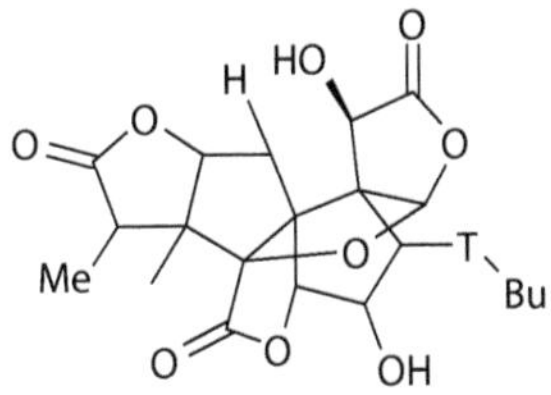

Ginkgolides J (**154**)

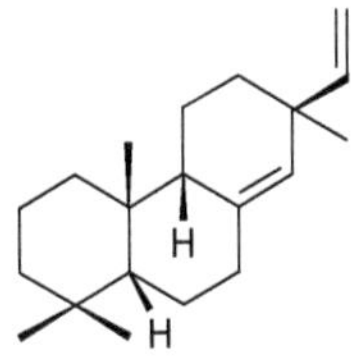

Pimaric acid (**155**)

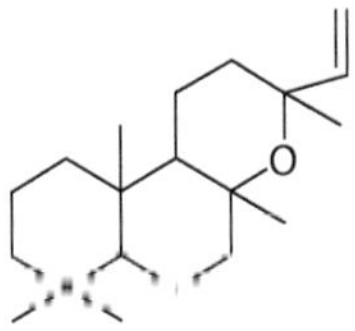

Manoyl oxide (**156**)

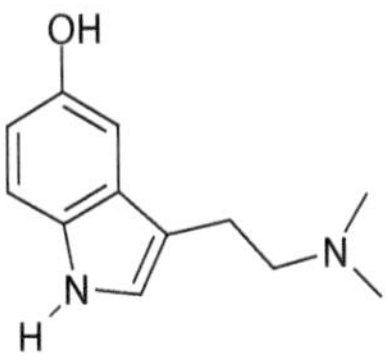

Alkaloids (**57**)

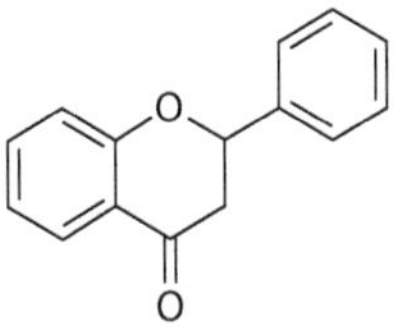

Flavonoids (**58**)

Figure 16.4 (Continued) Secondary metabolic chemical compounds from plants for antibacterial, antifungal, and insecticidal activities (drawn using ISIS/ChemDraw). *(Continued)*

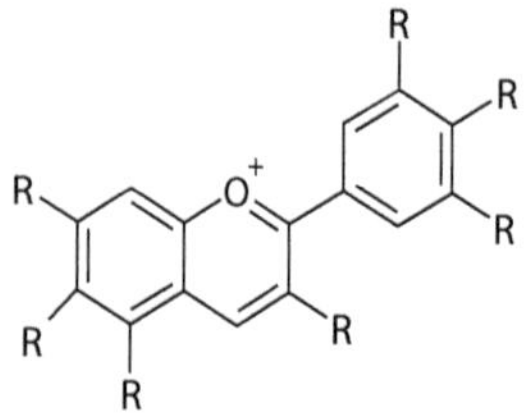

Anthocyanins (**59**)

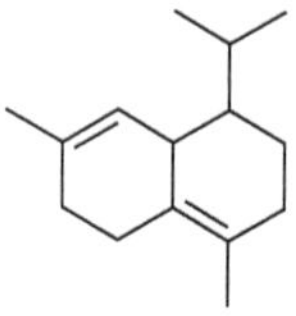

Sesquiterpenes (**60**)

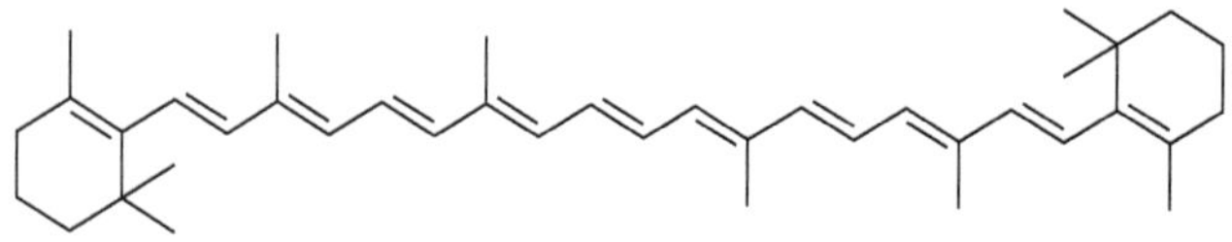

Carotenoids (**61**)

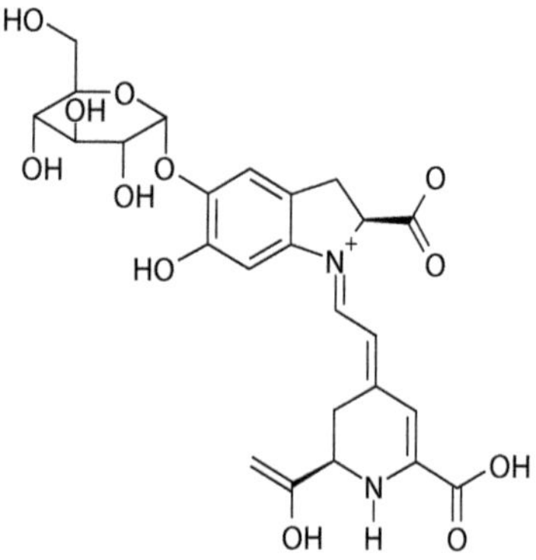

Betalains (**62**)

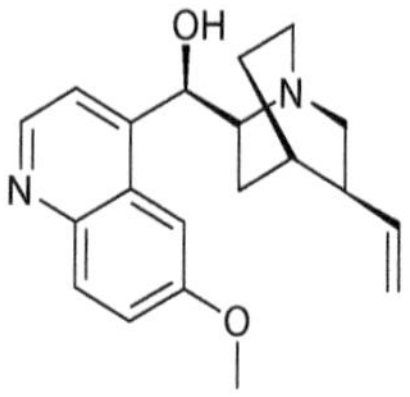

Quinine (**63**)

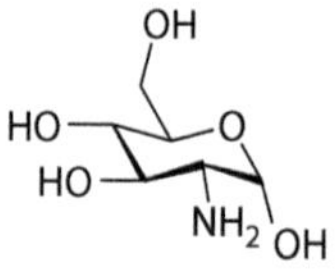

Amino sugars (**64**)

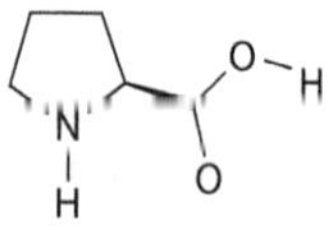

Amino acids (**65**)

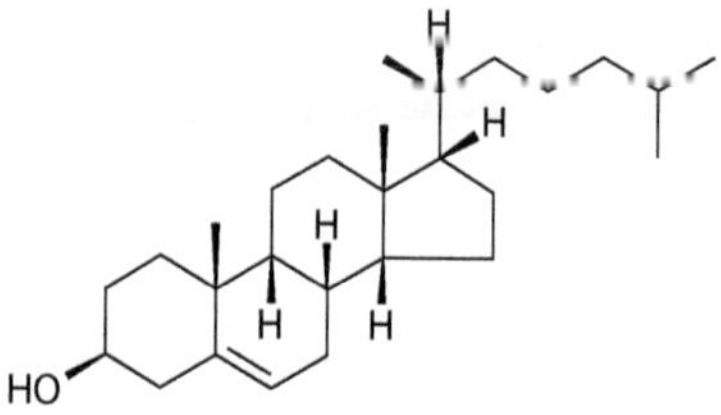

Lipids (**66**)

Figure 16.4 (Continued) Secondary metabolic chemical compounds from plants for antibacterial, antifungal, and insecticidal activities (drawn using ISIS/ChemDraw).

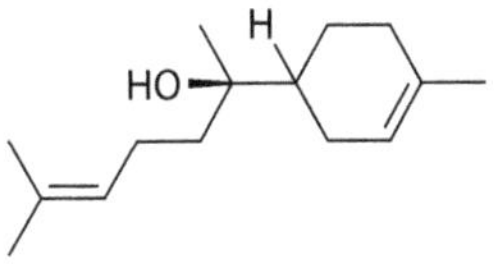

Bisabolol (**157**)

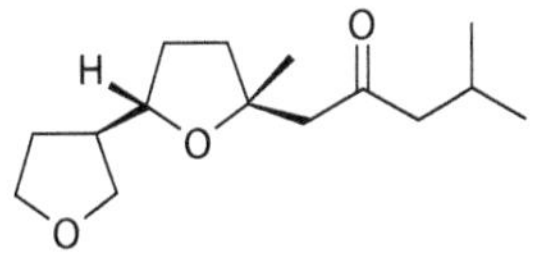

Ngaione (**158**)

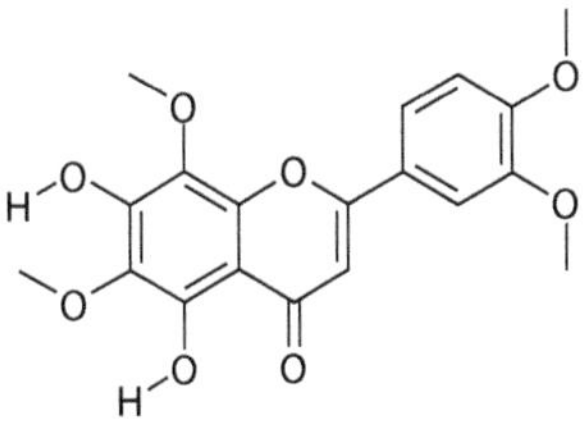

Hymenoxin (**159**)

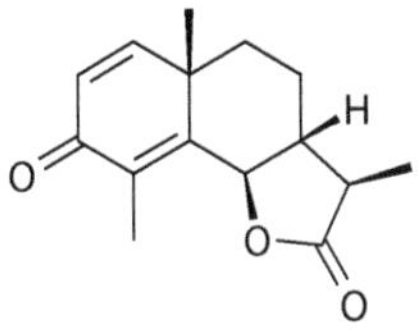

Santonin (**160**)

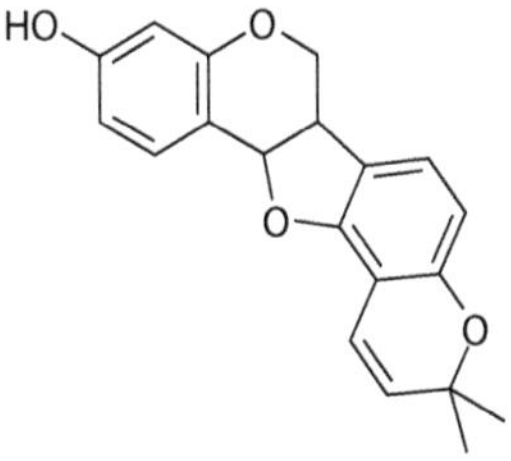

Phaseolin (**161**)

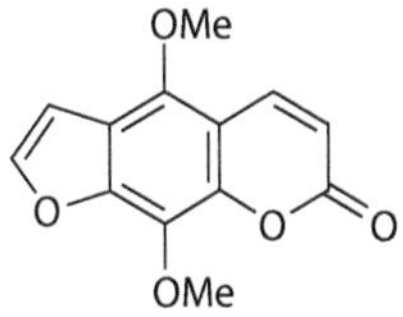

Isopimpinellin (**162**)

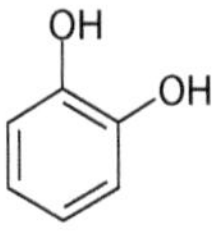

Catechol (**163**)

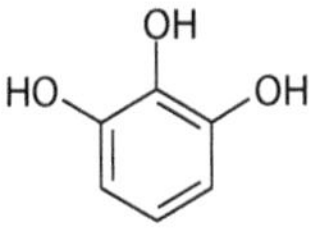

Pyrogallol (**164**)

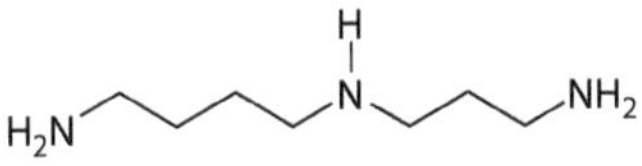

Spermidine (**165**)

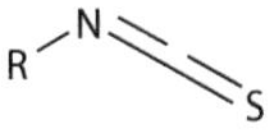

Isothiocyanates (**166**)

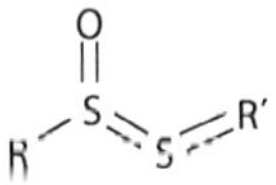

Thiosulfinate (**167**)

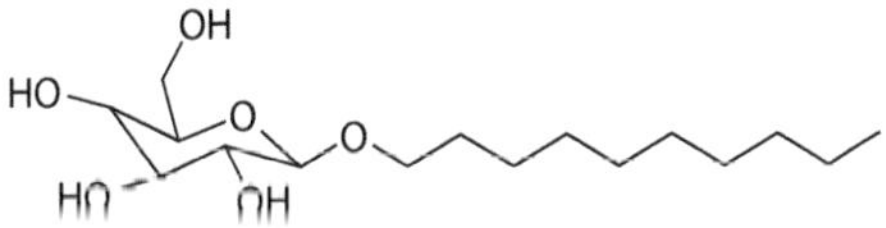

Glucosides (**168**)

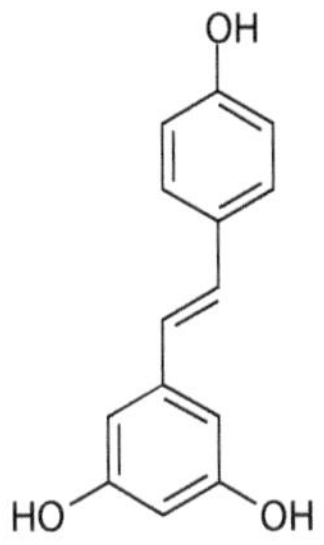

Resveratrol (**169**)

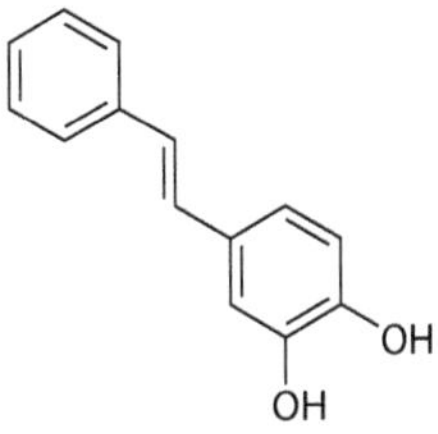

Pinosylvins (**170**)

Figure 16.5 Secondary metabolic chemical compounds from plants for antibacterial, antifungal, and antiviral activities (drawn using ISIS/ChemDraw). *(Continued)*

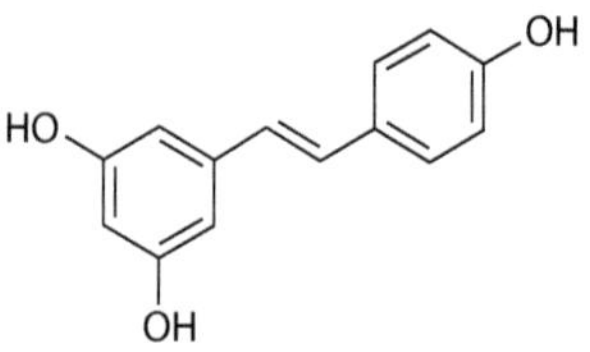

Stilbenoids **(171)**

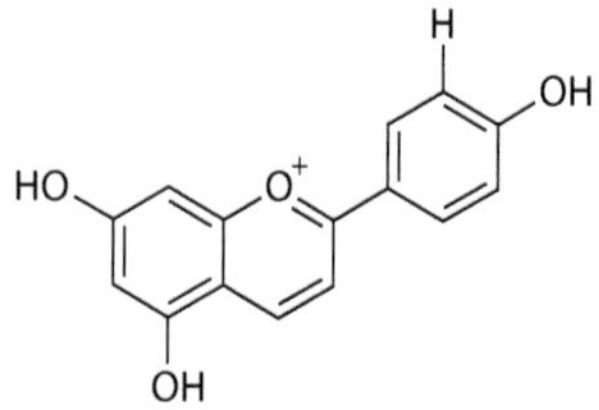

Apigeninidin **(172)**

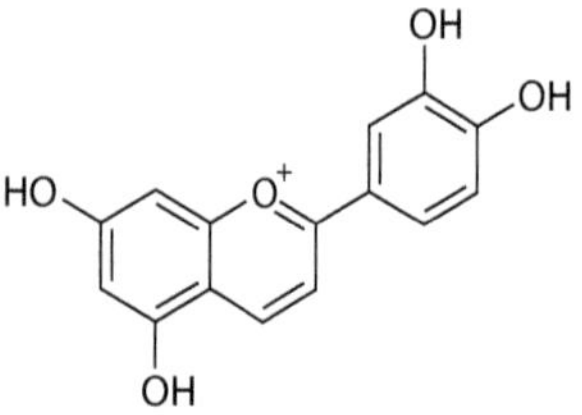

Luteolinidin **(173)**

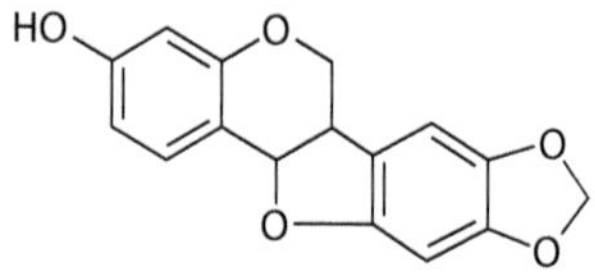

Maackiain **(174)**

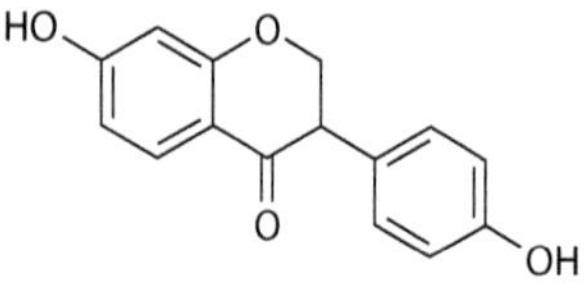

Daidzein **(175)**

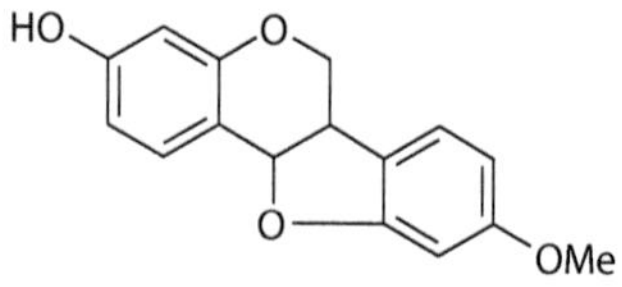

Medicarpin **(176)**

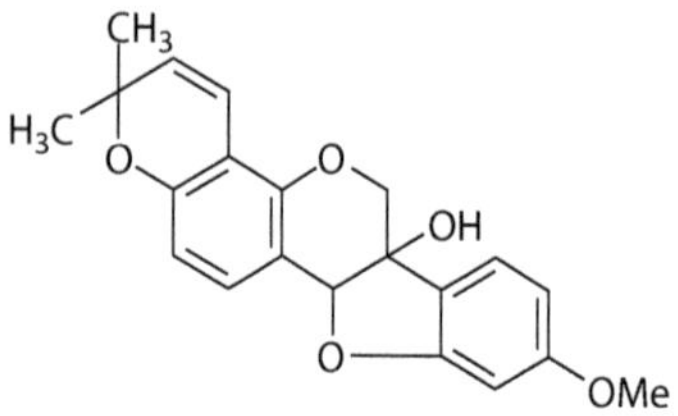

Glyceollin **(177)**

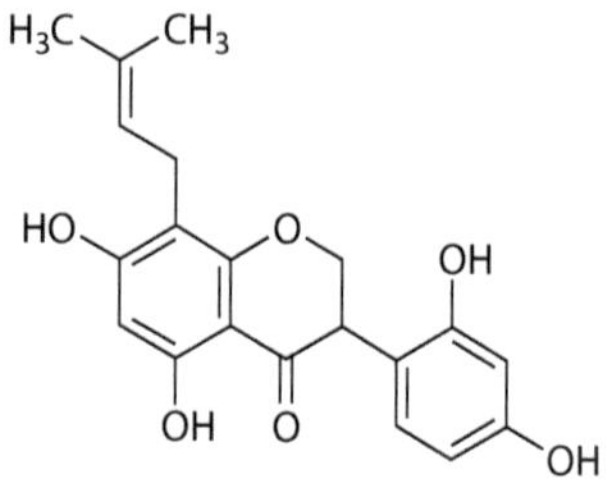

Keivitone **(178)**

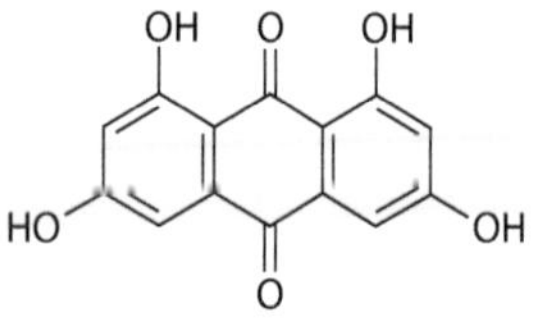

Emodin-A **(179)**

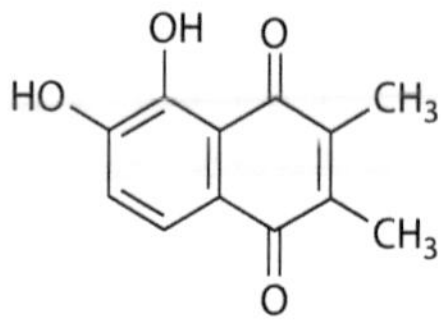

Malvone A **(180)**

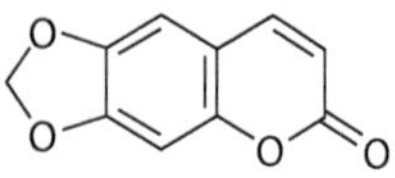

Ayapin **(181)**

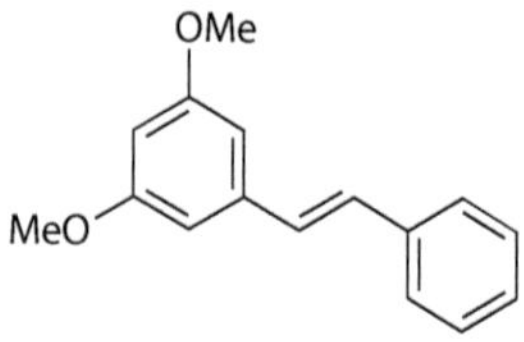

Pterostilbene **(182)**

Figure 16.5 **(*Continued*)** Secondary metabolic chemical compounds from plants for antibacterial, antifungal, and antiviral activities (drawn using ISIS/ChemDraw).   (*Continued*)

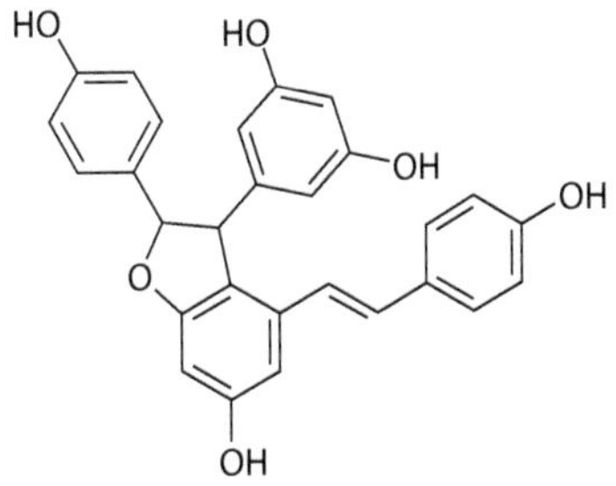

ε-Viniferin (**183**)

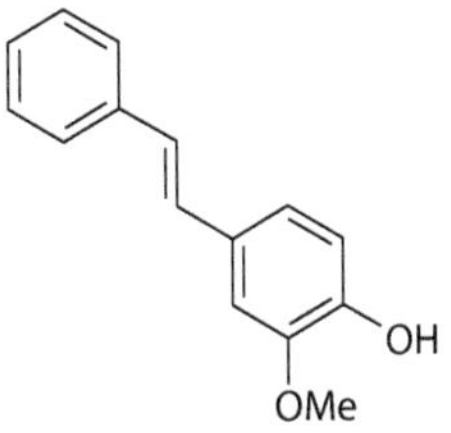

Pinosylvinmonomethyl ether (**184**)

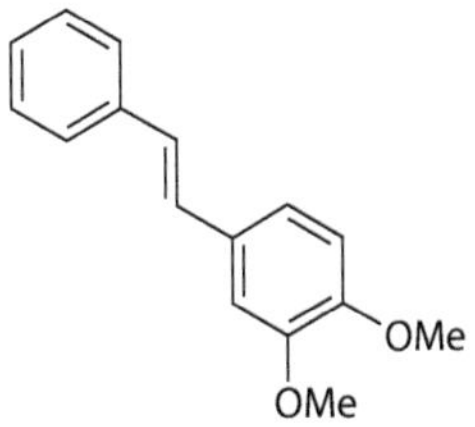

Pinosylvin dimethyl ether (**185**)

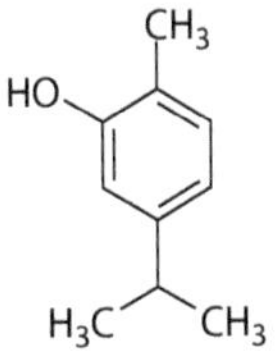

Carvacrol (**186**)

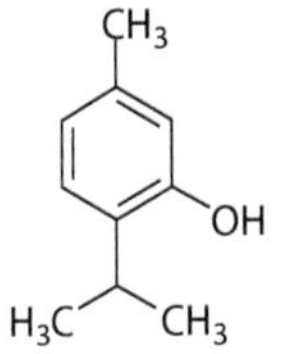

Thymol (**187**)

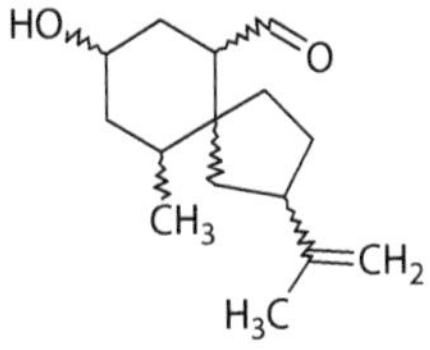

Lubimin (**188**)

Solavetivone (**189**)

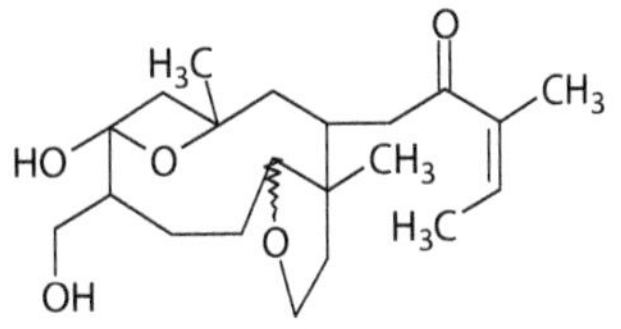

Niveusin-ß (**190**)

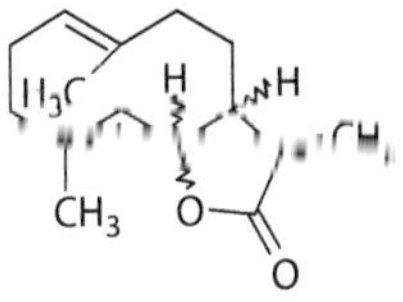

Costunolide (**191**)

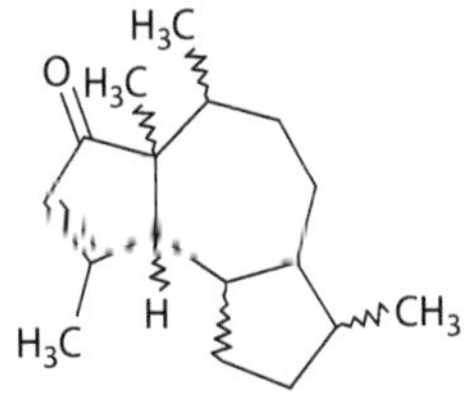

Cichoralexin (**192**)

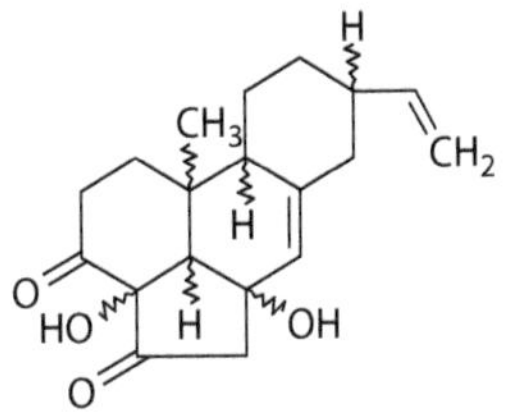

Lactucin (**193**)

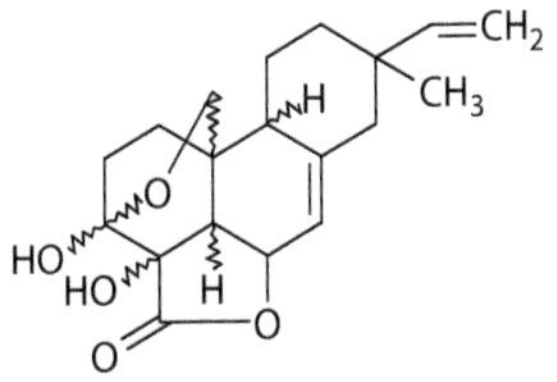

Momilactone A (**194**)

Figure 16.5 (Continued) Secondary metabolic chemical compounds from plants for antibacterial, antifungal, and antiviral activities (drawn using ISIS/ChemDraw). *(Continued)*

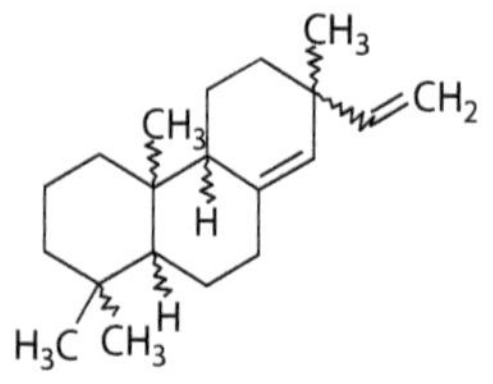

Momilactone B (**195**)

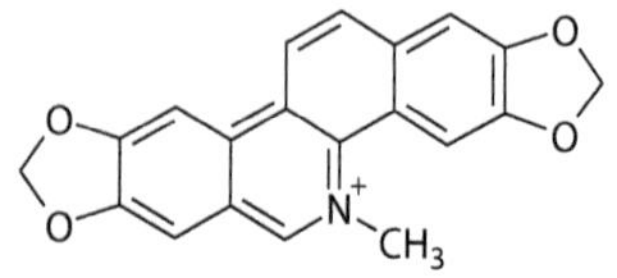

Sanguinarine (**196**)

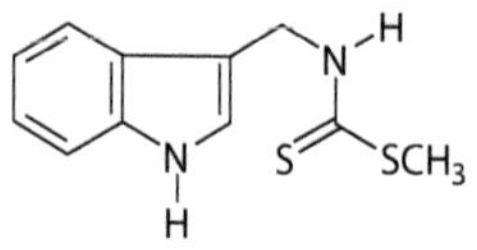

Brassinin (**197**)

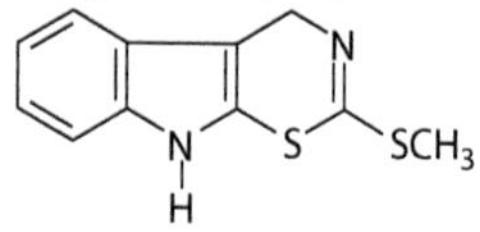

Ciclobrassinin (**198**)

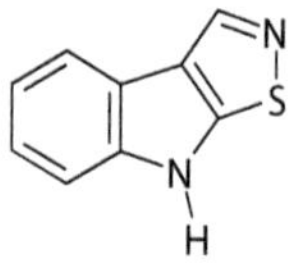

Brassilexin (**199**)

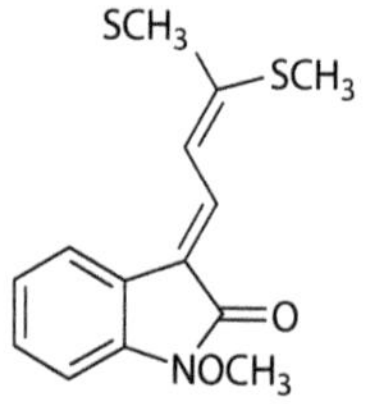

Wasalexin A (**200**)

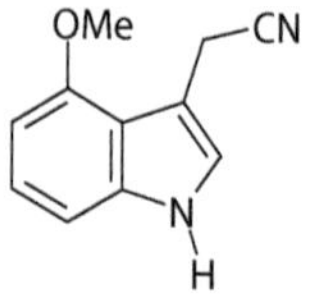

Arvelexin (**201**)

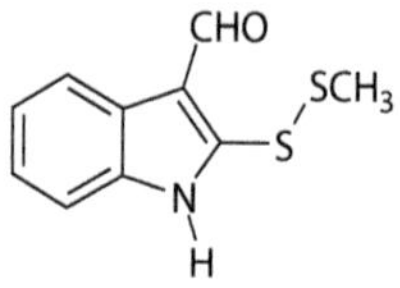

Caulexin A (**202**)

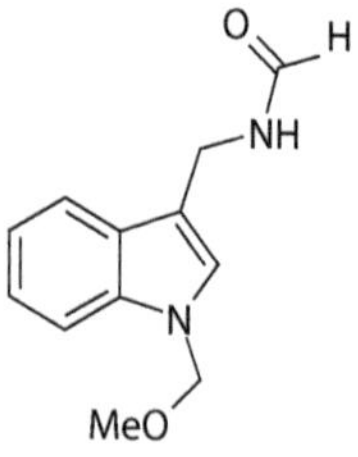

Caulexin B (**203**)

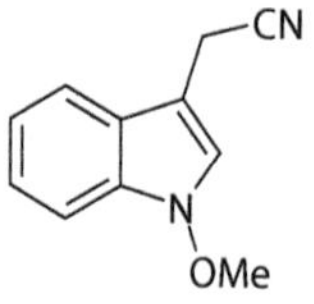

Caulexin C (**204**)

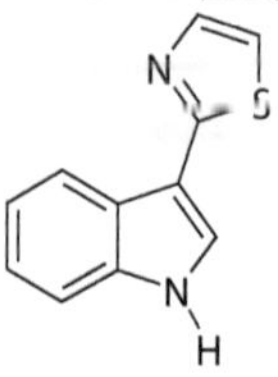

Camalexins (**205**)

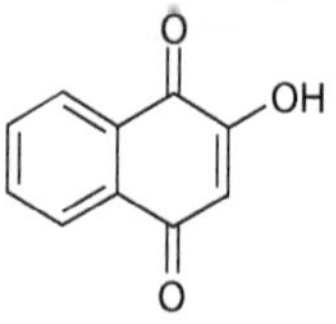

Lawsone (**206**)

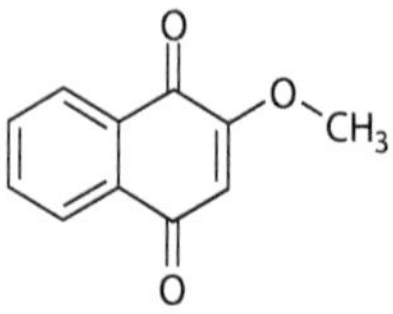

Me-lawsone (**207**)

Figure 16.5 (Continued) Secondary metabolic chemical compounds from plants for antibacterial, antifungal, and antiviral activities (drawn using ISIS/ChemDraw).

References

Abdul-Rahman, A., Gopalakrishnan, G., Ghouse, B.S., Arumugam, S., and Himalayan, B. 2000. Effect of *Feronia limonia* on mosquito larvae. *Fitoterapia* 71: 553.

Afek, U., Sztejnberg, A., and Carmely, S. 1986. 6,7–Dimethoxycoumarin, a citrus phytoalexin conferring resistance against *Phytophthora gummosis*. *Phytochemistry* 25: 1855–1856.

Ahn, Y., Kwon, M., Park, H., and Han, C. 1997. Potent insecticidal activity of *Ginkgo biloba* derived trilactone terpenes against *Nilaparvata lugens*. *Phytochem. Pest Control* 658: 90–105.

Amin, E., Radwan, M.M., El-Hawary, S.S., Fathy, M.M., Mohammed, R., Becne, J.J., and Khan, I. 2012. Potent insecticidal secondary metabolites from the medicinal plant *Acanthus montanus*. *Rec. Nat. Prod.* 6(3): 301–305.

Bala, A., Kollmann, A., Ducrot, P., Majira, A., Kerhoas, L., Leroux, P., Delorme, P., and Einhorn, J. 2000. Cis-ε-viniferin, a new antifungal resveratrol dehydrodimer from *Cyphostemma crotalarioides* roots. *J. Phytopathol.* 148: 29–32.

Bingham, S.A., Atkinson, C., Liggins, J., Bluck, L., and Coward, A. 1998. Phytoestrogens: Where are we now. *J. Nutr.* 79: 393–406.

Bohlmann, J., Gershenzon, J., and S. Aubourg. 2000. Biochemical, molecular, genetic and evolutionary aspects of defense-related terpenoid metabolism in conifers. In: Romeo, J.T., Ibrahim, R., Varin, L. eds., *Evolution of Metabolic Pathways. Recent Advances in Phytochemistry*, vol. 34. Elsevier Science, Oxford, U.K., pp. 109–150.

Brinker, A.M. and Seigler, D.S. 1991. Isolation and identification of piceatannol as a phytoalexin from sugarcane. *Phytochemistry* 30: 3229–3232.

Bruneton, J. 1999. *Pharmacognosy, Phytochemistry and Medicinal Plants*. Intercept Ltd., Paris, France.

Buzi, A., Chilosi1, G., Timperio, A., Zolla, L., Rossall, S., and Magro1, P. 2003. Polygalacturonase produced by *Botrytis fabae* as elicitor of two furanoacetylenic phytoalexins in Vicia faba pods. *J. Plant Pathol.* 85(2): 111–116.

Caceres, A., Lopez, B., Juarez, X., Aguila, J., Garcia, S., and Del Aguila, J. 1993. Plants used in Guatemala for the treatment of dermatophytic infections. 2. Evaluation of antifungal activity of seven American plants. *J. Ethanopharm.* 40: 207–213.

Cespedes, C.L., Avila, J.G., Martinez, A., Serrato, B., Calderon-Mugica, J.C., and Salgado Garciglia, R. 2006. Antifungal and antibacterial activities of Mexican tarragon (*Tagetes lucida*). *J. Agric. Food Chem.* 54: 3521–3527.

Chen, F., Liu, C.-J., Tschaplinski, T.J., and Zhao, N. 2009. Genomics of secondary metabolism in populus: Interactions with biotic and abiotic environments. *Crit. Rev. Plant Sci.* 28: 375–392.

Ciocan, I.D. and Bara, I.I. 2007. Plant products as antimicrobial agents. Analele Stiintifice ale Universitatii, Alexandru Ioan Cuza, Sectiunea Genetica Si Biologie Moleculara, TOM VIII, North Carolina.

Commun, C., Mauro, M., Chupeau, Y., Boulay, M., Burrus, M., and Jeandet, P. 2003. Phytoalexin production in grapevine protoplasm during isolation and culture. *Plant Physiol. Biochem.* 41: 317–323.

Croteau, R., Kutchan, T.M., and Lewis, N.G. 2000. Natural products (secondary metabolites). In: Buchanan, B., Gruissem, W., and Jones, R. (eds.), *Biochemistry and Molecular Biology of Plants*. American Society of Plant Physiologists, Rockville, MD, pp. 1250–1318.

Demain, A.L. 1999. Pharmaceutically active secondary metabolites of microorganisms. *Appl. Microbiol. Biotechnol.* 52(4): 455–463.

Demain, A.L. and Fang, A. 2000. The natural functions of secondary metabolites. *Adv. Biochem. Eng. Biotechnol.* 69: 1–39.

Dercks, W. and Creasy, L. 1989. The significance of stilbene phytoalexins in the *Plasmopara viticola* grape vine interaction. *Physiol. Mol. Plant Pathol.* 34: 189–202.

Docherty, J., Fu, M., Stiffler, B., Limperos, R., Pokabla, C., and DeLucia, A. 1999. Resveratrol inhibition of herpes simplex virus replication. *Antiviral Res.* 43: 135–145.

Durango, D., Quiñones, W., Torres, F., Rosero, Y., Gil, J., and Echeverri, F. 2002. Phytoalexin accumulation in Colombian bean varieties and aminosugars as elicitors. *Molecules* 7: 817–832.

Edreva, A., Velikova, V., Tsonev, T., Dagnon, S., Gurel, A., Aktas, L., and Gesheva, E. 2008. Stress-protective role of secondary metabolites: Diversity of functions and mechanisms. *Gen. Appl. Plant Physiol.* 34(1–2): 67–78.

Faith, S.A., Sweet, T.J., Bailey, E., Booth, T., and Docherty, J.J. 2006. Resveratrol suppresses nuclear factor-[kappa] B in herpes simplex virus infected cells. *Antiviral Res.* 72: 242–251.

Farombi, E.O. 2003. African indigenous plants with chemotherapeutic potential and biotechnological approach to the production of bioactive prophylactic agents. *Afr. J. Biotechnol.* 2: 662–671.

Graham, T.L. 1995. Cellular biochemistry of phenyl propanoids responses of soybean to infection by *Phytophthora sojae*. In: Daniel, M. and Purkayastha, R.P. (eds.), *Handbook of Phytoalexin Metabolism and Action*. Marcel Dekker, New York, pp. 85–116.

Grayer, R.J. and Harborne, J.J. 1994. A survey of antifungal compounds from higher plants. *Phytochemistry* 37: 19–42.

Hadacek, F. 2002. Secondary metabolites as plant traits: Current assessment and future perspectives. *Crit. Rev. Plant Sci.* 21: 273–322.

Harborne, J.B. 1982. Recent advances in chemical ecology. *Nat. Prod. Rep.* 3: 323–344.

Hartmann, T. 2007. From waste products to ecochemicals: Fifty years research of plant secondary metabolism. *Phytochemistry* 68(22–24): 2831–2846.

Hassan Adeyemi, M.M. 2011. A review of secondary metabolites from plant materials for post harvest storage. *Int. J. Pure Appl. Sci. Technol.* 6(2): 94–102.

Heinrich, M., Barnes, J., Gibbons, S., and Williamson, E.M. 2004. *Fundamentals of Pharmacognosy and Phytotherapy*. Churchill Livingstone, Elsevier Science Ltd., London, U.K.

Iyer, S.R. and Williamson, D. 1991. Studies on the occurrence of keratinophilic fungi in the soils of Rajasthan- II. *Indian J. Phytopathol.* 44: 487–490.

Izhaki, I. 2002. Emodin-α secondary metabolites with multiple ecological functions in higher plants—Research review. *New Physiol.* 155: 205–217.

Jarvis, B.B. 1995. Secondary metabolites and their role in evolution. *An. Acad. Bras. Cienc.* 67(3): 329–345.

Jeandet, P., Douillet, A., Bessis, R., Debord, S., Sbaghi, M., and Adrian, M. 2002. Phytoalexins from the Vitaceae, biosynthesis, phytoalexin gene expression in transgenic plants, antifungal activity and metabolism. *J. Agric. Food Chem.* 50: 2731–2741.

Jeroen, S.D. 2011. Biosynthesis and function of secondary metabolites. *Beilstein J. Org. Chem.* 7: 1620–1621.

Kabuto, K.C., Sasaki, N., Tsunagawa, M., Aizawa, H., Fujita, K., Kato, Y., Kitahara, Y., and Takahashi, N. 1973. Momilactones, growth inhibitors from rice, *Oryza sativa* L. *Tetrahedron Lett.* 14: 3861–3864.

Kaneko, K. 1960. Biogenetic studies of natural products. V. Biosynthesis of anethole by *Foeniculum vulgare. Chem. Pharm. Bull. (Japan)* 8: 875–879.

Keller, N.P., Turner, G., and Bennett, J.W. 2005. Fungal secondary metabolism from biochemistry to genomics. *Nat. Rev.* 3: 937–947.

Kelmanson, J.E., Jager, A.K., and Van, S.J. 2000. Zulu medicinal plants with antibacterial activity. *J. Ethnopharmacol.* 69: 241–246.

Keyser, P., Elofsson, M., Rosell, S., and Wolf Watz, H. 2008. Virulence blockers as alternatives to antibiotics: Type III secretion inhibitors against gram-negative bacteria. *J. Intern. Med.* 264: 17–29.

Kieslich, K. 1986. Production of drugs by microbial biosynthesis and biotransformation. Possibilities, limits and future developments (1st communication). *Arzneimittelforschung* 36(4): 774–778.

Kim, S.H., Kang, S.S., and Kim, C.M. 1992. Coumarin glycosides from the roots of *Angelica dahurica. Arch. Pharm. Res.* 15: 73–77.

Kuc, J. 1995. Phytoalexins stress metabolism and disease resistance in plants. *Annu. Rev. Phytopathol.* 33: 275–297.

Langcake, P. and Pryce, R.J. 1976. The production of resveratrol by *Vitis vinifera* and other members of the Vitaceae as a response to infection or injury. *Physiol. Plant Pathol.* 9: 77–86.

Lehrer, R.I. and Ganz, T. 1996. Endogenous vertebrate antibiotics. Defensins, protegrins and other cysteine-rich antimicrobial peptides. *Ann. N. Y. Acad. Sci.* 797: 228–239.

Levin, D.A. 1976. The chemical defenses of plants to pathogens and herbivores. *Ann. Rev. Ecol. Syst.* 7: 121–159.

Mahesh, B. and Satish, S. 2008. Antimicrobial activity of some medicinal plants against plant and human pathogen. *World J. Agric. Sci.* 4: 839–843.

Manitto, P., Monti, D., and Gramatica, P. 1974a. Biosynthesis of anethole in *Pimpinella anisum L.* *Tetrahedron Lett.* 17: 1467–1568.

Manitto, P., Monti, D., and Gramatica, P. 1974b. Biosynthesis of phenylpropanoid compounds. Part I. Biosynthesis of eugenol in *Ocimum basilicum.* JCS *Perkin* I: 1727–1731.

McGarvey, D.J. and Croteau, R. 1995. Terpenoid metabolism. *Plant Cell.* 7: 1015–1026.

Minamikawa, T., Akazawa, T., and Uritani, I. 1964. Two glucosides of coumarin derivatives in sweet potato roots infected by *Ceratocystis fimbriata. Agric. Biol. Chem.* 28: 230–233.

Muller-Riebau, F., Beger, B., and Yegen, O. 1995. Chemical composition and fungitoxic properties to phytopathogenic fungi of essential oils selected aromatic plants growing wild in Turkey. *J. Agric. Food Chem.* 43: 2262–2266.

Nair, R., Kalariya, T., and Chanda, S. 2005. Antibacterial activity of some selected Indian medicinal flora. *Turk J. Biol.* 29: 41–47.

Newman, D.J., Cragg, G.M., and Snader, K.M. 2000. The influence of natural products upon drug discovery. *Nat. Prod. Rep.* 17: 215–234.

Osbourn, A. 1996. Saponins and plant defence: A soap story. *Trends Plant Sci.* 1: 4–9.

Paganga, G., Miller, N., and Rice-Evans, C.A. 1999. The polyphenolic content of fruit and vegetables and their antioxidant activities. What does a serving constitute? *Free Radic. Res.* 30: 153–162.

Parekh, P., McDaniel, M.C., Ashen, M.D., Miller, J.I., Sorrentino, M., Chan, V., Roger, S., Blumenthal, R.S., and Sperling, L.S. 2005. Diets and cardiovascular disease: An evidence-based assessment. *J. Am. Coll Cardiol.* 45: 1379–1387.

Pedras, M.S. and Jha, M. 2006. Toward the control of *Leptosphaeria maculans*: Design, syntheses, biological activity, and metabolism of potential detoxification inhibitors of the crucifer phytoalexin brassinin. *Bioorg. Med. Chem.* 14: 4958–4979.

Pedras, M.S., Chumala, P.B., and Suchy, M. 2003. Phytoalexins from Thlaspiarvense, a wild crucifer resistant to virulent *Leptosphaeria maculans*, structures, synthesis and antifungal activity. *Phytochemistry* 64: 949–956.

Prusti, A., Mishra, S.R., Sahoo, S., and Mishra, S.K. 2008. Antibacterial activity of some Indian medicinal plants. *Ethnobot. Leaflets* 12: 227–230.

Rajendheran, J., Mani, A.M., and Navaneethakannan, K. 1998. Antibacterial activity of some selected medicinal plants. *Geobios* 25: 280–282.

Romero-Tabarez, M. 2004. Discovery of the new antimicrobial compound 7-O-malonyl macrolactin A. Aus Puerto Berrio, Kolumbien. PhD dissertation, Puerto Berrio, Colombia.

Schlee, D. 1986. *Okologische Biochemie.* Springer, Berlin, Germany.

Senanayake, U.M., Wills, R.B.H., and Lee, T.H. 1977. Biosynthesis of eugenol and cinnamic aldehyde in *Cinnamomum zeylanicum. Phytochemistry* 16: 2032–2033.

Shanker, V., Bhalla, K., and Trillium, R. 2003. An overview of the non-mevalonic pathway for terpenoid biosynthesis in plants. *J. Biosci.* 28: 637–646.

Sharma, A. 1988. Secondary metabolites from tissue cultures of some medicinally important plants. PhD thesis, University of Rajasthan, Jaipur, India.

Sieradzki, K., Wu, S.W., and Tomasz, A. 1999. Inactivation of the methicillin resistance gene mecA in vancomycin-resistant *Staphylococcus aureus. Micro. Drug Resist.* 5(4): 253–257.

Singh, S., Singh, S.K., and Tripathi, S.C. 1988. Fungitoxic properties of essential oil of *Eucalyptus rostrata. Indian Perfum.* 32: 190–193.

Sparg, S.G., Light, M.E., and Van Staden, J. 2004. Biological activities and distribution of plant saponins. *J. Ethnopharmacol.* 94: 219–43.

Steinstraesser, L., Tack, B.F., Waring, A.J., Hong, T., Boo, L.M., Fan, M.H., Remick, D.I., Su, G.L., Lehrer, R.I., and Wang, S.C. 2002. Activity of novispirin G10 against *Pseudomonas aeruginosa* in vitro and in infected burns. *Antimicrob. Agents Chemother.* 46(6): 1837–1844.

Swain, T. 1977. Secondary compounds as protective agents. *Annu. Rev. Plant Physiol.* 28: 479–501.

Tanaka, Y., Data, E.S., Hirose, S., Taniguchi, T., and Uritani, I. 1983. Biochemical changes in secondary metabolites in wounded and deteriorated cassava roots. *Agric. Bio. Chem.* 47: 693–700.

Tewari, L.C., Agarwal, R.G., Pandey, M.J., Uniyal, M.R., and Pandey, G. 1990. Some traditional folk medicine from the Himalayas (U. P. Region). *Aryavaidyan* 4: 49–57.

Uritani, I. and Hoshiya, I. 1953. Phytopathological chemistry of the black-rotted sweet potato. Part 6. Isolation of coumarin substances from potato and their physiology. *J. Agric. Chem. Soc.* 27: 161–164.

Veshkurova, O., Golubenko, Z., Pshenichnov, E., Arzanova, I., Uzbekov, V., Sultanova, E., Salikhov, S., Williams, H.J., Reibenspies, J.H., and Puckhaber, L.S. 2006. Malvone A phytoalexin found in *Malva sylvestris* (family Malvaceae). *Phytochemistry* 67(21): 2376–2379.

Wanchai, D.E. 1998. Chasing the key enzymes of secondary metabolite-biosynthesis from thai medicinal plants. Invited lecture presented at the *International Conference on Biodiversity and Bioresources: Conservation and Utilization*, Phuket, Thailand. Other presentations are published in *Pure Appl. Chem.* 70(11): 1998.

Wink, M. 1987. Chemical ecology of quinolizidine alkaloids. In: Waller, G.R. (ed.), *Allelochemicals: Role in Agriculture, Forestry and Ecology. Am. Chem. Soc.* 330: 524–533.

Wink, M. 1988. Plant breeding: Importance of plant secondary metabolites for protection against pathogens and herbivores. *Theor. Appl. Genet.* 75: 225–233.

chapter seventeen

Role of antimicrobial compounds from Trichoderma *spp. in* plant disease management

Someshwar Bhagat, O.P. Sharma, Ajanta Birah, Natarajan Amaresan, Israr Ahmad, Nasim Ahmad, and C. Chattopadhyay

Contents

17.1   Introduction

Antimicrobial compounds such as secondary metabolites and cell wall–degrading enzymes produced by *Trichoderma* spp. play a pivotal role in plant disease management. *Trichoderma* spp. are well-known biocontrol agents that are successfully used as biopesticides worldwide, and many among them are producers of secondary metabolites with antimicrobial properties. Antimicrobial compounds produced by *Trichoderma* spp. are often coupled with other mechanisms of biocontrol such as mycoparasitism and the production of cell wall–degrading enzymes, competition for nutrients/space, and induced resistance in the plant. The production of antimicrobial compounds by *Trichoderma* spp. is strain specific, which includes volatile and nonvolatile antimicrobial compounds, such as 6-pentyl-α-pyran-2-one, gliotoxin, viridin, harzianopyridone, harziandione, and peptaibols. Synergistic effects between cell wall–degrading enzymes and plethora of secondary metabolites/antibiotics of *Trichoderma* spp. on fungal pathogen growth have been studied and well documented (Vinale et al., 2008). Ghisalberti and Sivasithamparam (1991) have classified antimicrobial compounds into three categories: (1) volatile antibiotics, that is, 6-pentyl-α-pyrone and most of the isocyanide derivates; (2) water-soluble compounds, that is, heptelidic acid or koningic acid; and (3) peptaibols, which are linear oligopeptides of amino acids. The involvement of toxic metabolites in plant disease development and the interactions between beneficial and pathogenic fungi have been conclusively demonstrated (Howell, 1998).

Trichoderma spp. are the most frequently isolated soil fungi and present in plant root systems. These fungi are opportunistic, avirulent plant symbionts (Harman et al., 2004) and function as parasites and antagonists of many plant pathogenic fungi. *Trichoderma* spp. are among the most-studied fungal biocontrol agents and commercially marketed as biopesticides, biofertilizers, and soil amendments (Harman et al., 2004; Lorito et al., 2004). *Trichoderma*, a filamentous soil-inhabiting mycoparasite, has been used in commercial preparation for biological control of many fungal-induced plant diseases (Bhagat and Pan, 2007, 2010; Bhagat et al., 2013). *Trichoderma* has multifaceted function in agriculture, and it can work in several ways: (1) colonization by establishing themselves within diverse microbial communities in the rhizosphere, (2) control of pathogenic and competitive/deleterious microflora by using a variety of mechanisms (antimicrobial compounds), (3) improvement of the crop health, and (4) stimulation of root growth (Harman et al., 2004; Vinale et al., 2008; Bhagat and Pan, 2010; Bhagat et al., 2013).

The first record of antibiotics exudation in the growth medium of *Trichoderma lignorum* was made by Weindling (1934) followed by isolation of crystalline compound from there (Weindling and Emerson, 1936) and isolation of the same compound from *Gliocladium fimbriatum* also with high fungicidal activity by Weindling (1937), which received much attention among the researchers and opened new area of research for plant disease management.

Several cell wall–degrading enzymes from different strains of *Trichoderma* have been purified and characterized. Interestingly, when tested alone or in combinations, the purified proteins showed inhibitory activity toward a broad spectrum of fungal pathogens (*Rhizoctonia solani*, *Fusarium* sp., *Alternaria* sp., *Ustilago* sp., *Venturia* sp., and *Colletotrichum* sp., as well as the Oomycetes *Pythium* sp. and *Phytophthora* sp., which lack chitin in their cell walls).

Direct application of antimicrobial compounds produced by *Trichoderma* spp. and several other biocontrol agents, instead of the whole live organisms, has numerous advantages in plant disease management and may be more acceptable for the major stakeholders. They have very short shelf life and are difficult to store even for a shorter period. The selective production of active compounds may be performed by modifying the growth conditions of microbes (including *Trichoderma*), that is, type and composition of culture medium, temperature for incubation, and pH.

Trichoderma spp. produce various kinds of secondary metabolites that are chemically different natural compounds of relatively low molecular weight, which are mainly produced by microorganisms and typically associated to individual genera, species, or strains. Secondary metabolites are biosynthesized from primary metabolites in specialized pathways (i.e., polyketides or mevalonate pathways derived from acetyl coenzyme A or amino acids). These compounds show several biological activities possibly related to survival functions of the organism, such as competition against other micro- and macroorganisms, antibiosis, symbiosis, and metal transport, and play an important role in plant disease management. However, biological activities are not necessarily confined to one specific group or any particular single metabolite (Hanson, 2003, 2005). Some of fungal antimicrobial compounds can modify the growth and the metabolism of plants, whereas others seem to target specific fungal activities such as sporulation and hyphal elongation (Keller et al., 2005). Some of the secondary metabolites produced by *Trichoderma* and *Gliocladium* species are as follows.

17.2   *Antimicrobial compounds in plant disease management*

The secondary metabolites are the major constituent of antimicrobial compounds of *Trichoderma*, which are a heterogeneous group of natural compounds that are considered to help the producing organism in survival and basic functions, namely, competition, symbiosis, metal transport, and differentiation. *Trichoderma* spp. are well-known producers of antimicrobial compounds that are toxic for phytopathogenic fungi (Ghisalberti, 2002). Some of antimicrobial compounds that directly affect plant disease management are as follows.

Pyrones are one of the first volatile antifungal compounds isolated from *Trichoderma* species. This compound was identified by Collins and Halim (1972) in *Trichoderma viride*. Later on, it has been isolated from several *Trichoderma* species and strains. The pyrone 6-pentyl-2*H*-pyran-2-one (6-pentyl-α-pyrone) is a metabolite commonly purified in the culture filtrate of different *Trichoderma* species (*T. viride, T. atroviride, T. harzianum, T. koningii*) and is directly linked with the coconut aroma released by axenically developed colonies. 6-Pentyl-α-pyrone has shown both *in vivo* and *in vitro* antifungal activities toward several plant pathogenic fungi, and a strong relationship has been found between the biosynthesis of this metabolite and the biocontrol ability of the producing microbe (Bhagat et al., 2013; Vinale et al., 2014). Takahiro et al. (2013) have isolated another pyrone named cytosporone S from a *Trichoderma* sp., which has been reported to have *in vitro* antimicrobial activity against several bacteria and fungi (Table 17.1).

Koninginins are complex pyrones isolated from *T. koningii, T. harzianum*, and *T. aureoviride*. Koninginins A, B, D, E, and G showed *in vitro* antibiotic activity toward the take-all pathogen *Gaeumannomyces graminis* var. *tritici*. Koninginin D can also inhibit the growth of other important soilborne plant pathogens such as *Pythium middletonii, R. solani, Phytophthora cinnamomi, Fusarium oxysporum*, and *Bipolaris sorokiniana*.

Another antifungal compound viridin, isolated from *T. koningii, T. viride*, and *T. virens* (Singh et al., 2005; Vinale et al., 2014) prevented spore germination of *Botrytis allii, Colletotrichum lini, Fusarium caeruleum, Penicillium expansum, Aspergillus niger*, and *Stachybotrys atra* (Vinale et al., 2014). *T. viride, T. hamatum*, and certain *Gliocladium* species produce viridiol, another member of viridins, with antifungal and phytotoxic metabolite. *T. viride, T. hamatum*, and certain *Gliocladium* species produce viridiol, a similar antifungal and phytotoxic metabolite for which the *in vivo* activity has been demonstrated (Howel and Stipanovic, 1994; Vinale et al., 2014).

Harzianopyridone, a nitrogen heterocyclic compound, obtained from *T. harzianum* metabolite is a potent antibiotic effective against *B. cinerea, R. solani* (Vinale et al., 2014), *G. graminis* var. *tritici*, and *P. ultimum* (Vinale et al., 2006). A new compound named harzianic acid, characterized by the presence of a pyrrolidindione ring system, has been isolated from *T. harzianum*. This tetramic acid derivative exhibited *in vitro* antibiotic activity against *Pythium irregulare, Sclerotinia sclerotiorum*, and *R. solani* (Vinale et al., 2009). Recently, Vinale et al. (2013) have also reported harzianic acid from *Trichoderma* with antimicrobial and plant growth promotion activities.

The azaphilones isolated from *T. harzianum* T22 are natural products containing a highly oxygenated bicyclic core and a chiral quaternary center. A new azaphilone, named T22 azaphilone, showed a marked growth inhibition of several plant pathogens (*R. solani, P. ultimum*, and *G. graminis* var. *tritici*) *in vitro* (Vinale et al., 2006).

Harzianolide and its derivatives, butenolide and deydroharzianolide, were isolated from different strains of *T. harzianum* (Vinale et al., 2006, 2014). The *in vitro* antifungal activities of these antimicrobial compounds were established against several phytopathogenic agents (Vinale et al., 2006, 2014). A novel hydroxy-lactone derivative, named cerinolactone, has been isolated from culture filtrates of *Trichoderma cerinum*. The isolated compound showed *in vitro* antifungal activity against *R. solani, P. ultimum*, and *B. cinerea*.

Trichoderma spp. also produce isocyano metabolites. But isolation and separation of these compounds are very difficult due to their instability (Reino et al., 2008). Dermadin

Table 17.1 Antimicrobial activities of secondary metabolites isolated from *Trichoderma* sp.

Metabolites derived from	*Trichoderma* spp.	Antimicrobial activity
Acetate derivative		
Ferulic acid	*Gliocladium virens*	Antiviral, bactericide, and fungicide
Tricarboxylic acid derivatives		
Viridiofungins A, B, C	*T. viride*	Antifungal
Polyketides		
Benzoquinones and quinhydrones	*G. roseum*	Antibiotic
Oxygen heterocyclic compounds		
Nectariapyron	*G. vermoesenii*	Antibiotic
Harzianopyridone	*T. harzianum*	Antifungal
Butenolide harzianolide	*T. harzianum*	Antifungal
Harzianic acid	*T. harzianum*	Antimicrobial
Pyrones		
6-pentyl-*a*-pyrone	*Trichoderma* species	Antifungal Antimicrobial
Dehydroderivative of 6-pentyl-*a*-pyrone	*T. viride* and *T. koningii*	Antifungal
Massoilactone and *d*-decanolactone	*Trichoderma* species	Antifungal
	T. koningii and *T. harzianum*	Antifungal
Octaketides	*G. virens* and *T. viride*	Antifungal
Sesquiterpenes		
Daucane sesquiterpenes		
Trichothecenes	*Trichoderma* species	Antifungal
Triterpenes and sterol		
Helvolic acid	*Gliocladium* species	Antibiotic
Ergokonin A and B	*T. koningii*	Antifungal
Viridin	*Gliocladium virens*	Antibiotic
Isocyano derivatives		
Dermadine	*Trichoderma* spp.	Antibiotic and antifungal
Diketopiperazines		
Gliotoxin	*Gliocladium virens*	Antibiotic, antiviral
Epitrisulfide	*Gliocladium virens*	Antibiotic
Diketopiperazines		
Gliovirin	*Gliocladium virens*	Antibiotic, antiviral
Polypeptides		
Trichopolyns	*T. polysporum*	Antibiotic

Source: Sivasithamparam, K. and Ghisalberti, E.L., Secondary metabolism in *Trichoderma* and *Gliocladium*, In: *Trichoderma* and *Gliocladium*, Vol. 1, Harman, G.E., Kubicek, C.P., Eds., London, U.K.,Taylor & Francis Ltd., pp. 139–191, 1998.

from *T. viride, T. koningii,* and *T. hamatum* is an antibiotic metabolite. The isonitrile tricho-viridin isolated from *T. koningii* and *T. viride* showed *in vitro* antimicrobial properties. Several dermadin and trichoviridin analogues have also been isolated.

Gliotoxin and gliovirin are the two most important *Trichoderma* antimicrobial compounds belonging to diketopiperazines. P group strains of *Trichoderma* (*Gliocladium*) *virens* produced the gliovirin, which was very active against *P. ultimum,* but inactive against *R. solani.* Similarly, strains of the Q group produced gliotoxin, which was very active against *R. solani,* but less active against *P. ultimum* (Howell, 1999; Vinale et al., 2014). Seedling bioassay tests revealed that the strains of the P group were able to effectively control *Pythium* damping off on cotton, while Q group strains gave better results toward the same disease caused by *R. solani* (Howell, 1991; Howell et al., 1993; Vinale et al., 2014). These studies clearly indicated the potential role of antibiotic production in the biocontrol mechanism of the gliotoxin/gliovirin producers. The *T. virens veA* ortholog *vel1* (VELVET protein Vel1) is involved in regulation of gliotoxin biosynthesis, biocontrol activity, and many other secondary metabolism-related genes (Mukherjee et al., 2010, 2013).

Peptaibols are linear peptides rich in nonproteinogenic amino acids (α-aminoisobutyric acid and isovaline); acetylated at the N-terminal group and the C-terminus is an amino alcohol (phenylalaninol, valinol, leucinol, isoleucinol, or tryptophanol). Lorito et al. (1996) demonstrated that peptaibols inhibited β-glucan synthase activity in the host fungus, but acting synergistically with *T. harzianum* β-glucanases. The inhibition of glucan synthase prevented the reconstruction of the pathogen cell walls, thus facilitating the disruptive action of β-glucanases. The most important peptaibol is the *T. viride* alamethicin. The terms peptaibiome and peptaibiomics (peptide antibiotics or peptaibiotics) have been suggested to describe the analysis and study of all peptaibols expressed in an organism or tissue using spectrometric methods, like LC/ESI-MSn or intact-cell MALDI-TOF (Neuhof et al., 2007). Mukherjee et al. (2012b) reviewed the genes that determine the range of secondary metabolites produced by *Trichoderma,* including both useful and toxic compounds. Polyketide synthases and nonribosomal peptide synthases (NRPSs) define two major classes of secondary metabolites (Mukherjee et al., 2012a; Baker et al., 2012).

17.3 *Inhibition of bioactive molecules produced by other fungi*

Some of antimicrobial compounds (secondary metabolites) produced by *Trichoderma* spp. inhibit certain bioactive molecules formed by other fungi. They may be produced by certain antagonistic and endophytic fungi also in pure culture. Detoxification of fungal toxins by beneficial microorganisms also represents interesting possibilities for disease management. Pure cultures of fungi able to detoxify mycotoxins have been obtained from complex microbial populations by using enrichment culture techniques (Karlovsky, 1999). *T. viride* and other fungal species were able to degrade aflatoxin B1. Moreover, *T. harzianum* hydrolases were able to degrade aflatoxin B1 (AFB1) and ochratoxin A (OTA) *in vitro*.

In substrates with high carbon/nitrogen ratios, *T. virens* produced a metabolite similar to the antibiotic viridin, called viridiol that acts as a plant growth inhibitor. This compound, isolated also from *T. hamatum,* inhibited the 5′-hydroxyaverantin dehydrogenase, involved in aflatoxin biosynthesis, thus reducing or completely blocking the production of this mycotoxin during the fungal interactions (Sakuno et al., 2000; Wipf and Kerekes, 2003). Elad (1996) found that mycoparasitism and antibiosis were not the main biocontrol mechanisms of *T. harzianum* strain T39 against *B. cinerea.* These authors indicated that the antagonist interferes with the infection process by affecting the pathogen conidia in the early stages of the interaction. They suggested that *T. harzianum* (T39) acts directly by inhibiting *B. cinerea* hydrolytic enzymes or indirectly by blocking plant responses that induce enzymatic activity in *B. cinerea*. Subsequently, Elad and Kapat (1999) demonstrated that a *Trichoderma* protease is involved in the biocontrol of *B. cinerea* by degrading the hydrolytic enzymes required for infection by the pathogen.

17.4   Compounds favor competition for nutrients

Competition for carbon, nitrogen, and iron plays an important role during the interactions between beneficial and pathogenic fungi and is often associated with the biocontrol mechanisms of nonpathogenic *Fusarium* and *Trichoderma* species (Vinale et al., 2008). *Trichoderma* has a strong ability to mobilize and take up soil nutrients, which makes it more efficient and competitive than many other soil microbes. This process could be related also to the production of organic acids, namely, fumaric, gluconic, and citric acids, which decrease soil pH and allow the solubilization of micronutrients, phosphates, and mineral cations such as iron, manganese, and magnesium (Vinale et al., 2008). Iron is a mineral essential nutrient for numerous microorganisms, both bacteria and fungi. In the aerobic environment (with oxygen and neutral pH), iron exists mainly as Fe^{3+} in immobilized form rather than hydroxides and oxyhydroxides forms, making it unavailable for microbial growth (Miethke, 2013). Microorganisms that excrete siderophores are able to grow in natural environments poor in iron using residual immobilized iron. Most fungi produce various siderophores, which help the microbes to overcome adverse conditions (Winkelmann, 2007). The production of microbial siderophores can be beneficial to plants for two reasons: (1) siderophores can solubilize iron unavailable for the plant, and (2) siderophore production by nonpathogenic microorganisms can also suppress the growth of plant pathogens by depriving them of iron sources. The majority of the fungal siderophores isolated so far belong to hydroxamate class and can be divided into three structural families: fusarinines, coprogens, and ferrichromes. Fungi typically produce more than one siderophore, even if restricted to a particular family.

Segarra et al. (2010) analyzed the effect of iron during the interaction of *Trichoderma* (T34) with *F. oxysporum* f.sp. *lycopersici* on tomato plants, to establish the importance of iron concentration for the activity of a *Trichoderma asperellum* strain (T34). These results clearly indicated a role of siderophores during the interaction with the pathogen and the plants, although no siderophores have been isolated and characterized thus far from the culture filtrate of this beneficial microbe (De Santiago et al., 2009). But Vinale et al. (2013) demonstrated the ability of tetramic acid from *Trichoderma* sp. to bind with a good affinity to Fe^{3+}, which explains the mechanism of iron solubilization that significantly changes nutrient availability in the soil environment for other microorganisms and the host plant. In addition to iron transport, siderophores produced by microorganisms (including *Trichoderma* sp.) have other functions and effects, including enhancement of virulence of pathogens, storage of intracellular iron, and suppression of microbial growth during the competition with other microbes (Miethke, 2013).

Anke et al. (1991) isolated siderophores from all different structural families simultaneously from a culture filtrate of *Trichoderma* spp. The culture filtrate of this fungus obtained in iron-deficient condition contained coprogen, coprogen B, fusarinine C, and ferricrocin. *T. harzianum* produced the highest number of siderophores and did not have any unique compound, while *Trichoderma reesei* biosynthesized one cisfusarinine as the major siderophore and three others that were present only in *T. harzianum*. The variation of siderophores produced by *Trichoderma* spp. is expected due to further modifications of the NRPS products rather than diverse NRPS-encoding genes (Lehner et al., 2013).

17.5 Compounds favor systemic-induced resistance

Several antimicrobial compounds (metabolites) produced by *Trichoderma* are involved in the induction of plant resistance, such as (1) proteins with enzymatic activity, that is, xylanase; (2) avirulence-like gene products able to induce defense reactions in plants; and (3) low-molecular-weight compounds released from either fungal or plant cell walls, by specific enzyme activities. Some of the low-molecular-weight degradation products released from fungal cell walls have been purified and characterized that consist of short oligosaccharides with two types of monomers, with and without an amino acid residue (Woo et al., 2006; Woo and Lorito, 2007). These compounds elicited a reaction in the plant when applied to leaves or when injected into root or leaf tissues. They also stimulated the biocontrol ability of fungi such as *Trichoderma* by activating the mycoparasitic gene expression cascade. Some *Trichoderma* secondary metabolites may act as elicitors of plant defense mechanisms against pathogens. A reduction of disease symptoms on tomato and canola seedlings treated with 6PP and inoculated with the pathogens *B. cinerea* or *Leptosphaeria maculans* has been reported. Moreover, soil drench applications of 6PP 4 days before inoculation with *Fusarium moniliforme* showed considerable suppression of seedling blight and significant plant growth promotion, as compared to untreated control (El-Hasan and Buchenauer, 2009). Application of 6PP on maize seedlings distinctly enhanced the activities of β-1,3-glucanase, peroxidase, and polyphenoloxidase in both shoot and root tissues indicating an induction of defense responses in maize plants (El-Hasan and Buchenauer, 2009).

Peptaibols are another class of plant defense elicitors produced by *Trichoderma*. Application of alamethicin, a long sequence peptaibol with 20 residues produced by *T. viride*, induced defense responses in *Phaseolus lunatus* (lima bean) (Engelberth et al., 2000) and *Arabidopsis thaliana* (Chen et al., 2003). Mukherjee et al (2012b) demonstrated that mutation in one of the polyketide synthase/nonribosomal peptide synthases (PKS/NRPS) hybrid genes caused reduction in induction of the defense response gene phenylalanine ammonia lyase of *T. virens*, suggesting a putative role for the associated metabolite (Mukherjee et al., 2012b). These results provide evidence that a PKS/NRPS hybrid enzyme responsible for the metabolite production is involved in *Trichoderma*–plant interactions resulting in induction of defense response in maize.

17.6 Compounds that enhance plant growth, development, and yield

Trichoderma species can improve the plant growth, development, and yield (Harman et al., 2004; Bhagat et al., 2013; Mukherjee et al., 2013). Several strains belonging to the *Trichoderma* genus have been found to be able to stimulate plant development (Kumar et al., 2011), especially at the root level (i.e., formation of more lateral roots), by activating an auxin-dependent mechanism (Contreras-Cornejo et al., 2009; Vinale et al., 2014) and/or producing indole-3-acetic acid (IAA) or auxin analogues (Hoyos-Carvajal et al., 2009; Vinale et al., 2014). Growth promotion of plant by antimicrobial compounds of *Trichoderma* has been demonstrated (Vinale et al., 2012b). Koninginins, 6PP, trichocaranes A–D, harzianopyridone, cyclonerodiol, harzianolide, and harzianic acid are examples of isolated compounds that promote plant growth in a concentration-dependent manner (Vinale et al., 2014). A novel metabolite, named cerinolactone, has been isolated and characterized from *T. cerinum* and was able to positively alter the growth of tomato seedlings 3 days after treatment (Vinale et al., 2012a, 2014). The dose–effect response of plant growth to secondary metabolites produced by *Trichoderma* spp. clearly indicated need for further systematic research. These metabolites seem to act as auxin-like molecules, which have a positive effect at low concentrations with inhibitory effect at higher doses. A hormone activity was detected on etiolated pea stems treated with harzianolide and 6PP. These compounds also affected the growth of tomato and canola seedlings in a manner depending on the concentration and/or the application method used (Vinale et al., 2008, 2014).

17.7   Conclusion

A plethora of antimicrobial compounds produced by *Trichoderma* spp. have been isolated and characterized. The fungal secondary metabolites with a direct antimicrobial activity against plant pathogens have been mainly isolated from biocontrol strains of the genus *Trichoderma* (Sharma et al., 2004). Even though many secondary metabolites are known, elite strains usually produce only a few main secondary metabolites (Bhagat, 2008; Mukherjee et al., 2012a; Bhagat et al., 2013). The quality and quantity of secondary metabolites synthesized depend on (1) the compound considered, (2) the species and the strain, (3) the occurrence of other microorganisms, (4) the equilibrium among elicited biosynthesis and biotransformation rate, and (5) the growth conditions. Furthermore, in some cases, the biocontrol agent was able to modulate the production of toxic secondary metabolites according to the presence or the absence of the target pathogen (Vinale et al., 2009; Bhagat and Pan, 2010).

Interestingly, antimicrobial compounds produced by *Trichoderma* spp. may also be involved in biocontrol mechanism through the inhibition of bioactive products. It is well recognized that biocontrol fungi, such as selected agents of *Trichoderma* spp., are able to produce compounds with multiple activities, including direct/indirect toxic effects against plant pathogens, plant defense induction, or growth promotion (Vinale et al., 2008; Sharma et al., 2004). A hormone-like effect has been proposed for some *Trichoderma* secondary metabolites, and specific antimicrobial compounds having this characteristic have been detected in plant–fungus cultures. In fact, treatment with *Trichoderma* metabolites produces extensive changes of the plant expressome, proteome, and metabolome, by acting on specific pathways involved in the synthesis of major growth hormones and induction of resistance to biotic/abiotic stresses and nutrient uptake (Vinale et al., 2012; Mukherjee et al., 2012a; Bhagat et al., 2013). These recent findings have suggested new strategies for the development of novel bioformulations based on antimicrobial compounds alone or in combination with live organisms, in order to maximize the desired beneficial effects and reduce the risks associated with the release of microorganisms into the diverse crop ecosystem.

References

Almassi, F., Ghisalberti, E.L., Narbey, M.J., and Sivasithamparam, K. 1991. New antibiotics from strains of *Trichoderma harzianum*. *J Nat Prod* 54: 396–402.

Anke, H., Kinn, J., Bergquist, K.E., and Sterner, O. 1991. Production of siderophores by strains of the genus *Trichoderma*—Isolation and characterization of the new lipophilic coprogen derivative, palmitoylcoprogen. *Biometals* 4: 176–80.

Baker, S.E., Perrone, G., Richardson, N.M., Gallo, A., and Kubicek, C.P. 2012. Phylogenetic analysis and evolution of polyketide synthase-encoding genes in *Trichoderma*. *Microbiology* 158: 147–154.

Bhagat, S., Bambawale, O.M., Tripathi, A.K., Ahmad, I., and Srivastava, R.C. 2013. Biological management of fusarial wilt of tomato by *Trichoderma* spp. in Andamans. *Ind J Hort* 70: 397–403.

Bhagat, S. and Pan, S. 2007. Mass multiplication of *T. harzianum* on agricultural byproducts and their evaluation in management of seedling blight (*R. solani*) of mung bean and collar rot (*S. rolfsii*) of groundnut. *Ind J Agric Sci* 77: 583–588.

Bhagat, S. and Pan, S. 2010. Biological management of root and collar rot of French bean. *Ind J Agric Sci* 42: 42–50.

Bhagat, S. 2008. Biocontrol potential of some isolates of *Trichoderma* spp. from Andaman and Nicobar Islands. A Ph.D. thesis submitted to Bidhan Chandra Krishi Viswavidyalaya, Mohanpur, West Bengal, India.

Brian, P.W. and McGowan, J.C. 1945. Viridin: A highly fungistatic substance produced by *Trichoderma viride. Nature* 156: 144–145.

Chen, F., D'Auria, J.C., and Tholl, D. 2003. An *Arabidopsis thaliana* gene for methylsalicylate biosynthesis, identified by a biochemical genomics approach, has a role in defence. *Plant J* 36: 577–588.

Claydon, N., Hanson, J.R., Truneh, A., and Avent, A.G. 1991. Harzianolide, a butenolide metabolite from cultures of *Trichoderma harzianum. Phytochemistry* 30: 3802–3803.

Collins, R.P. and Halim, A.F. 1972. Characterisation of the major aroma constituent of the fungus *Trichoderma viride (Pers.). J Agric Food Chem* 20: 437–438.

Contreras-Cornejo, H.A., Macìas-Rodrìguez, L., Cortés-Penagos, C., and López-Bucio, J. 2009. *Trichoderma virens*, a plant beneficial fungus, enhances biomass production and promotes lateral root growth through an auxin-dependent mechanism in *Arabidopsis. Plant Physiol* 149: 1579–1592.

De Santiago, A., Quintero, J.M., Avilés, M., and Delgado, A. 2009. Effect of *Trichoderma asperellum* strain T34 on iron nutrition in white lupin. *Soil Biol Biochem* 41: 2453–2459.

Dickinson, J.M., Hanson, J.R., Hitchcock, P.B., and Claydon, N. 1989. Structure and biosynthesis of harzianopyridone, an antifungal metabolite of *Trichoderma harzianum. J Chem Soc Perkin Trans* 1: 1885–1887.

Elad, Y. 1996. Mechanisms involved in the biological control of *Botrytis cinerea* incited diseases. *Eur J Plant Pathol* 102: 719–732.

Elad, Y. and Kapat, A. 1999. The role of *Trichoderma harzianum* protease in the biocontrol of *Botrytis cinerea. Eur J Plant Pathol* 105: 177–189.

El-Hasan, A. and Buchenauer, H. 2009. Actions of 6-pentyl-alpha-pyrone in controlling seedling blight incited by *Fusarium moniliforme* and inducing defence responses in maize. *J Phytopathol* 157: 697–707.

Engelberth, J., Koch, T., Schuler, G., Bachmann, N., Rechtenbach, J., and Boland, W. 2000. Ion channel-forming alamethicin is a potent elicitor of volatile biosynthesis and tendril coiling. Cross talk between jasmonate and salicylate signaling in lima bean. *Plant Physiol* 125: 369–377.

Ghisalberti, E.L. 2002. Anti-infective agents produced by the hyphomycetes genera *Trichoderma* and *Glioclaudium. Curr Med Chem* 1: 343–374.

Ghisalberti, E.L. and Sivasithamparam, K. 1991. Antifungal antibiotics produced by *Trichoderma* spp. *Soil Biol Biochem* 23: 1011–1020.

Golder, W.S. and Watson, T.R. 1980. Lanosterol derivatives as precursors in the biosynthesis of viridin. Part 1. *J Chem Soc Perkin Trans* 1: 422–425.

Hanson, J.R. 2003. *Natural Products: The Secondary Metabolites.* Vol. 17 Ed. Cambridge, U.K.: Royal Society of Chemistry.

Hanson, J.R. 2008. *The Chemistry of Fungi.* Ed. Cambridge, U.K.: Royal Society of Chemistry.

Harman, G.E., Howell, C.R., Viterbo, A., Chet, I., and Lorito, M. 2004. *Trichoderma* species opportunistic, avirulent plant symbionts. *Nat Rev Microbiol* 2: 43–56.

Hoyos-Carvajal, L., Orduz, S., and Bissett, J. 2009. Growth stimulation in bean (*Phaseolus vulgaris* L.) by *Trichoderma. Biol Control* 51: 409–416.

Howell, C.R. 1991. Biological control of *Pythium* damping off of cotton with seed coating preparations of *Gliocladium virens. Phytopathology* 81: 738–741.

Howell, C.R., Stipanovic, R.D., and Lumsden, R.D. 1993. Antibiotic production by strains of *Gliocladium virens* and its relation to the biocontrol of cotton seedling diseases. *Biocontrol Sci Technol* 3: 435–441.

Howell, C.R. 1998. The role of antibiosis in biocontrol. In: Kubiecek, C.P., Harman, G.E., editors. *Trichoderma and Gliocladium*, Vol 1. London, U.K.: Taylor & Francis; 173–184.

Howell, C.R. 1999. Selective isolation from soil and separation *in vitro* of P and Q strains of *Trichoderma virens* with differential media. *Mycologia* 91: 930–934.

Howell, C.R. 2003. Mechanisms employed by *Trichoderma* species in the biological control of plant diseases: The history and evolution of current concepts. *Plant Dis* 87: 4–10.

Karlovsky, P. 1999. Biological detoxification of fungal toxins and its use in plant breeding, feed and food production. *Nat Toxins* 7: 1–23.

Keller, N.P., Turner, G., and Bennett, J.W. 2005. Fungal secondary metabolism—From biochemistry to genomics. *Nat Rev Microbiol* 3: 937–947.

Lehner, S.M., Atanasova, L., and Neumann, N.K. 2013. Isotope-assisted screening for iron-containing metabolites reveals high diversity among known and unknown siderophores produced by *Trichoderma* spp. *Appl Environ Microbiol* 79: 18–31.

Leong, J. 1986. Siderophores: Their biochemistry and possible role in the biocontrol of plant pathogens. *Ann Rev Phytopathol* 24: 187–209.

Lorito, M. 2008. A novel role for *Trichoderma* secondary metabolites in the interactions with plants. *Physiol Mol Pl Pathol* 72: 80–86.

Lorito, M., Farkas, V., Rebuffat, S., Bodo, B., and Kubicek, C.P. 1996. Cell wall synthesis is a major target of mycoparasitic antagonism by *Trichoderma harzianum*. *J Bacteriol* 178: 6382–6385.

Miethke, M. 2013. Molecular strategies of microbial iron assimilation: From high-affinity complexes to cofactor assembly systems. *Metallomics* 5: 15–28.

Moffatt, J.S., Bu'Lock, J.D., and Yuen, T.H. 1969. Viridiol, a steroid-like product from *Trichoderma viride*. *J Chem Soc Chem Commun* 14: 849.

Mukherjee, P.K. and Kenerley, C.M. 2010. Regulation of morphogenesis and biocontrol properties in *Trichoderma virens* by a VELVET protein, Vel1. *Appl Environ Microbiol* 76: 2345–2352.

Mukherjee, P.K., Horwitz, B.A., and Kenerley, C.M. 2012a. Secondary metabolism in *Trichoderma*—A genomic perspective. *Microbiology* 158: 35–45.

Mukherjee, P.K., Buensanteai, N., Moran-Diez, M.E., Druzhinina, I.S., and Kenerley, C.M. 2012b. Functional analysis of non-ribosomal peptide synthetases (NRPSs) in *Trichoderma virens* reveals a polyketide synthase (PKS)/NRPS hybrid enzyme involved in the induced systemic resistance response in maize. *Microbiology* 158: 155–165.

Mukherjee, P.K., Horwitz, B.A., Herrera-Estrella, A., Schmoll, M., and Kenerley, C.M. 2013. *Trichoderma* research in the genome era. *Ann Rev Phytopathol* 51:105–129.

Neuhof, T., Dieckmann, R., Druzhinina, I.S., Kubicek, C.P., and von Dohren, H. 2007. Intact-cell MALDI-TOF mass spectrometry analysis of peptaibol formation by the genus *Trichoderma/Hypocrea*: Can molecular phylogeny of species predict peptaibol structures? *Microbiology* 153: 3417–3437.

Ordentlich, A., Wiesman, Z., Gottlieb, H.E., Cojocaru, M., and Chet, I. 1992. Inhibitory furanone produced by the biocontrol agent *Trichoderma harzianum*. *Phytochemistry* 31: 485–486.

Reino, J.L., Guerrero, R.F., Hernández-Galán, R., and Collado, I.G. 2008. Secondary metabolites from species of the biocontrol agent *Trichoderma*. *Phytochem Rev* 7: 89–123.

Sakuno, E., Yabe, K., Hamasaki, T., and Nakajima, H.A. 2000. New inhibitor of 5'-hydroxyaverantin dehydrogenase, an enzyme involved in aflatoxin biosynthesis, from *Trichoderma hamatum*. *J Nat Prod* 63: 1677–1678.

Scarselletti, R. and Faull, J.L. 1994. *In Vitro* activity of 6-pentyl-a-pyrone, a metabolite of *Trichoderma harzianum*, in the inhibition of *Rhizoctonia solani* and *Fusarium oxysporum* f sp *lycopersici*. *Mycol Res* 98:1207–1209.

Segarra, G., Casanova, E., Avilès, M., and Trillas, I. 2010. *Trichoderma asperellum* strain T34 controls Fusarium wilt disease in tomato plants in soilless culture through competition for iron. *Microb Ecol* 59: 141–149.

Sharma, O.P., Jeswani, M.D., Bambawale, O.M., and Murthy, K.S. 2004. *Trichoderma* as a potential bio-agent for the management of diseases: Taxonomy, biology and mode of action. In: *Microbial Biotechnology*, Trivedi, P.C., Eds. Jaipur, India: Avishankar Publishers & Distributors.

Sivasithamparam, K. and Ghisalberti, E.L. 1998. Secondary metabolism in *Trichoderma* and *Gliocladium*. In: *Trichoderma and Gliocladium*, Vol. 1, Harman, G.E., Kubicek, C.P., Eds. London, U.K.: Taylor & Francis Ltd. pp. 139–191.

Singh, S., Dureja, P., Tanwar, R.S., & Singh, A. 2005. Production and antifungal activity of secondary metabolites of *Trichoderma virens*. *Pestic Res J* 17: 26–29.

Takahiro, I., Nonaka, K., Suga, T., Masuma, R., Omura, S., and Shiomi, K. 2013. Cytosporone S with antimicrobial activity, isolated from the fungus *Trichoderma* sp. FKI-6626. *Bioorg Med Chem Lett* 23: 679–681.

Vinale, F., Marra, R., Scala, F., Ghisalberti, E.L., Lorito, M., and Sivasithamparam, K. 2006. Major secondary metabolites produced by two commercial *Trichoderma* strains active against different phytopathogens. *Lett App Microbiol* 43: 143–148.

Vinale, F., Sivasithamparam, K., Ghisalberti, E.L., Marra, R., Woo, S.L., and Lorito, M. 2008. *Trichoderma*-plant-pathogen interactions. *Soil Biol Biochem* 40: 1–10.

Vinale, F., Ghisalberti, E.L., and Sivasithamparam, K. 2009. Factors affecting the production of *Trichoderma harzianum* secondary metabolites during the interaction with different plant pathogens. *Lett Appl Microbiol* 48: 705–711.

Vinale, F., Arjona, G.I., and Nigro, M. 2012a. Cerinolactone, a hydroxylactone derivative from *Trichoderma cerinum*. *J Nat Prod* 75: 103–106.

Vinale, F., Sivasithamparam, K., Ghisalberti, E.L., Ruocco, M., Woo, S., and Lorito, M. 2012b. *Trichoderma* secondary metabolites that affect plant metabolism. *Nat Prod Commun* 7: 1545–1550.

Vinale, F., Sivasithamparam, K., Ghisalberti, E.L., Woo, S.L., Nigro, M., Marra, R., Lombardi, N., Pascale, A., Ruocco, M., Lanzuise, S., Manganiello, G., and Lorito, M. 2014. *Trichoderma* secondary metabolites active on plants and fungal pathogens. *The Open Mycol J* 8: 127–139.

Vinale, F., Nigro, N., Sivasithamparam, K., Flematti, G., Ghisalberti, E.L., Ruocco, M., Varlese, R. et al. 2013. Harzianic acid: A novel siderophore from *Trichoderma harzianum*. *FEMS Microbiol Lett* 347: 123–129.

Weindling, R. 1934. Studies on a lethal principle effective in the parasitic action of *Trichoderma lignorum* on *Rhizoctonia solani* and other soil fungi. *Phytopathology* 24: 1153–1179.

Weindling, R. 1937. The isolation of toxin substances from the culture filtrates of *Trichoderma* and *Gliocladium*. *Phytopathology* 27: 1175–1177.

Weindling, R. and Emersion, O.H. 1936. The isolation of a toxic substance from the culture filtrate of a *Trichoderma*. *Phytopathology* 26: 1068–1070.

Winkelmann, G. 2007. Ecology of siderophores with special reference to the fungi. *Biometals* 20: 379–392.

Wipf, P. and Kerekes, A.D. 2003. Structure reassignment of the fungal metabolite TAEMC161 as the phytotoxin viridiol. *J Nat Prod* 66: 716–871.

Woo, S.L. and Lorito, M. 2007. Exploiting the interactions between fungal antagonists, pathogens and the plant for biocontrol. In: Vurro, M. and Gressel, J., Eds., *Novel Biotechnologies for Biocontrol Agent Enhancement and Management*. Amsterdam, the Netherlands: IOS, Springer Press, pp. 107–130.

Woo, S.L., Scala, F., Ruocco, M., and Lorito, M. 2006. The molecular biology of the interactions between *Trichoderma* spp., phytopathogenic fungi, and plants. *Phytopathology* 96: 181–185.

Worasatit, N., Sivasithamparam, K., Ghisalberti, E.L., and Rowland, C. 1994. Variation in pyrone production, pectic enzymes and control of rhizoctonia root rot of wheat among single-spore isolates of *Trichoderma koningii*. *Mycol Res* 98: 1357–1363.

chapter eighteen

Antimicrobial compounds from rhizosphere bacteria and their role in plant disease management

Natarajan Amaresan, Nallanchakravarthula Srivathsa, Velusamy Jayakumar, Someshwar Bhagat, and Nooruddin Thajuddin

Contents

18.1 Introduction

Plants are in intimate relationship with microorganisms, and many of those associations are permanent. Microbial association starts when a seed falls onto the soil and continues with its germination, and its growth ceases at one point and it dies. The populations of microorganisms surrounding/on the surface of plants exceeds 10^6 bacteria/g of leaves and 10^9 bacteria/g of roots. Many of these organisms are possibly saprophytic in nature (no deleterious or beneficial effects can be conferred by their presence), nourishing themselves from the nutrients produced by the plants. However, the presence of some microorganisms can be detrimental in nature (e.g., pathogens), while some may be beneficial in nature (e.g., plant growth-promoting rhizobacteria (PGPR), biocontrol agents (BCA), nitrogen-fixing bacteria, and other symbionts).

Plants allocate carbon below ground in the form of root exudates, thereby influencing the microbial communities (Morgan et al., 2005; Drigo et al., 2010). The term "rhizosphere" was coined by Lorenz Hiltner (Hiltner, 1904) to define the volume of soil in close proximity to roots that are characterized by elevated microbial populations. The rhizosphere is under the continuous influence of living roots and the rich nutrient supply (rhizodeposition), which enables microorganisms to have direct influence on plant growth. The root exudates consist of simple and complex sugars, growth regulators, primary and secondary compounds such as amino acids, organic acids, phenolic acids, flavonoids, fatty acids, enzymes, steroids, alkaloids, and vitamins (Uren, 2000; Philippot et al., 2013). Some of these root exudates are known to play a role in shaping the microbial communities in the rhizosphere (Haichar et al., 2008; Bressen et al., 2009; Drigo et al., 2010).

PGPR are one of the most commonly studied rhizosphere components in terms of direct plant growth promotion and biological control and are isolated from different environments (Lugtenberg and Kamilova, 2009; Amaresan et al., 2014a, b). Bacteria that are intimately associated with plant roots, for example, endophytes, are also known for their plant growth promotion and biological control (Kumar et al., 2011; Amaresan et al., 2012a). The rhizobacteria are the dominant driving forces in recycling of soil nutrients, and subsequently, they are crucial for soil fertility (Glick, 2012). Rhizobacteria or the bacteria in the rhizosphere help the plant directly or indirectly through various mechanisms and associations. Nitrogen fixation, synthesis of various phytohormones, and solubilization of various minerals are some of the direct effects (Basak and Biswas, 2010; Panhwar et al., 2012), and production of antagonistic substances specifically against plant pathogens is known to be an indirect effect (Hao et al., 2011).

The antagonistic potential of these rhizobacteria as well as root-associated bacteria has been shown to control various root, foliage, and postharvest diseases of agricultural crops (Glick and Bashan, 1997; Goel et al., 2002; Weller, 2007; Amaresan et al., 2012b). Not only bacteria but also fungi are known to have antagonistic effects against plant pathogens by producing an array of antagonistic compounds (Kumar et al., 2012; Philippot et al., 2013). When these organisms are used in a controlled manner, they can enhance overall soil fertility and plant health (Berg, 2009). Rhizobacteria are being used as biocontrol agents, as they are in the vicinity of plant roots and can act as frontline of defense against soilborne plant pathogens (Dowling and O'Gara, 1994). Biocontrol can be explained as the "harnessing of disease-suppressive microorganisms for plant protection."

The synergistic plant growth–promoting effects of fluorescent *Pseudomonas* sp. and *Mesorhizobium* sp. *cicer* strain in both sterile and "wilt sick" soil conditions of chickpea crop were reported by Sindhu et al. (2002). Their coinoculation resulted in the enhanced nodulation by *Mesorhizobium* sp. as well as increased shoot dry weight in comparison with uninoculated controls. The disease suppression and/or the plant growth–promoting effects could be the result of various multifarious interactions with the other members of the rhizosphere community.

18.2   Plant–microbe interaction in rhizosphere

Most microorganisms are uncultivable in nature (Torsvik and Ovreas, 2002), and it is important for us to unravel their interactions with plants. With the advent of cultivation-independent methods and molecular methods, the untangling of the relationships involved was made easier to study. It is of fundamental importance to understand the various mechanisms and processes that regulate soil ecosystem functioning. A recent review (Philippot et al., 2013) entitled "Going back to the roots: The microbial ecology of the rhizosphere" discusses the various microbial interactions and processes that are occurring in the rhizosphere and their importance to sustainable agriculture, among others. For sustainable agriculture, it is crucial to understand the importance of biofertilizers and biopesticides in enhancing nutrient acquisition and sustainable plant protection (Huang et al., 2014).

Plants pump carbon below ground in the form of root exudates. The plant roots select from a wide pool of soil microorganisms (Nallanchakravarthula et al., 2014); the selected microorganisms play a role in plant protection and nutrient acquisition by releasing various antimicrobial compounds or increasing nutrient uptake. The rhizosphere has been described as both a playground and battlefield for soilborne pathogens and beneficial microorganisms (Raaijmakers et al., 2008). Plant carbon is also known to attract plant pathogens, thereby competition occurs in between the pathogens and other microorganisms, wherein the rhizosphere microbial community dynamics change. There are two types of plant-originated organic compounds that are released into their surrounding environment. One of which is exuded by the plants at the root surface, termed as root exudates. A rough estimate by Uren (2000) on root exudates indicates about 10% of the photosynthate is being allocated to the root. The second one originates from the decaying plant cells as a result of cell death. As a result, the rhizosphere is considered as nutrient rich (heterotrophic medium) when compared with bare soils, which remain nonconducive or poor medium for plant growth. The quality and quantity of root exudates depend on the plant species, cultivar, age, and the environmental conditions (Rovira, 1956; Haichar et al., 2008). There is accumulating evidence that different plant species select different microorganisms (Costa et al., 2006; Haichar et al., 2008).

18.3   Rhizosphere bacteria in plant disease management

Many rhizosphere bacteria are known to produce antagonistic compounds that play a role in encouraging or inhibiting soilborne plant pathogens. The populations of antagonistic microorganisms can be increased artificially by fertilizer application, organic amendments, foliar spraying of chemicals, etc. To be a successful antagonist, rhizobacteria should also compete for nutrients. The root exudate composition is known to influence rhizosphere microbial communities (Bressen et al., 2009; Drigo et al., 2010). In a study by Wu et al. (2008) susceptible cotton cultivar to *Verticillium dahliae* is known to produce more amino acids and sugars in comparison with resistant cultivar. In another recent study (Li et al., 2013) of peanut cultivars, the susceptible cultivar root exudate composition significantly differed in sugars, alanine, and total amino acids, whereas *p*-hydroxybenzoic acid, benzoic acid, *p*-coumaric acid, and total phenolic acids were higher in midresistant cultivar. It was also shown that the plant can exploit the microbial consortia from soil in order to counteract the plant pathogen (Mendes et al., 2011). Rudrappa et al. (2008) demonstrated the stimulation of growth of *Bacillus subtilis* in response to signals released by *Arabidopsis* plants challenged by *Pseudomonas syringae* pv. tomato. In *in vitro* studies, *Pseudomonas* isolates have been shown to inhibit *Phytophthora capsici* by the production of biosurfactant (Ozyilmaz and Benlioglu, 2013). Soil amendments are also shown to increase the suppressiveness to plant pathogens and modulate bacterial members (Giotis et al., 2012; Cretoiu et al., 2013). It has newly been reported that many pathogens are attacking the new crop plants, implying the need to identify and formulate new compounds or biocontrol agents or methods to counteract the pathogens (Jayakumar et al., 2009; Singh et al., 2011; Sharma et al., 2013).

18.4   Antimicrobial compounds of rhizosphere bacteria

Common soil bacterial genera such as *Pseudomonas*, *Burkholderia*, *Enterobacter*, and *Bacillus* are also known to be successful endophytes (Lodewyckx et al., 2002; Amaresan et al., 2012a, b, 2014b). Many secondary metabolites are being produced by these genera such as antibiotics, anticancer, volatile organic compounds, antifungal, antiviral, insecticidal, and immunosuppressant compounds. In spite of various compounds isolated from these endophytes and rhizosphere bacteria, they are still an untapped array of various secondary metabolites that remain to be exploited.

Antibiotics and its production are known to be a major mechanism by which rhizobacteria play a key role in plant protection. Their production has also been reported from various extreme sources. Using purified antibiotics, they were shown to suppress diseases by mutant analysis and biochemical studies. They subdue or, in some cases, kill the plant pathogenic fungi by inhibition of spore germination, fungistasis, and lysis of fungal mycelia and/or by exerting fungicidal effects. DeCoste et al. (2010) showed an increase in the populations of *Pseudomonas* sp. LBUM300 associated with strawberry plants exposed to *V. dahliae*. *V. dahliae* presence increased the expression of "hcnC" gene, but the presence of *Pseudomonas* spp. did not alter the colonization by the pathogen. It was reported that antibiotic compounds such as 2,4-diacetylphloroglucinol (2,4-DAPG), oomycin A, phenazines, pyrrolnitrin, pyrroles, and pyocyanin are produced by rhizobacteria, which are known to inhibit the plant pathogen growth (Bender et al., 1999).

18.4.1   2,4-Diacetylphloroglucinol

2,4-DAPG is a natural phenolic compound found in certain strains of gram-negative bacteria such as *Pseudomonas fluorescens*. It is active against organisms ranging from viruses, bacteria, plants, to nematodes; its enhanced activity has always been found against plant pathogens (Keel et al., 1992; Maurhofer et al., 1992; Mazzola et al., 1995; Cronin et al., 1997; Delany et al., 2001; Dwivedi and Johri, 2003; Siddiqui and Shaukat, 2003a; Islam and Fukushi, 2010). 2,4-DAPG from *Pseudomonas fluorescens* CHA0 has been shown to induce plant resistance against pathogens (Iavicoli et al., 2003; Siddiqui and Shaukat, 2003b, 2004). It has also been shown that application of 2,4-DAPG-producing pseudomonads increased crop soybeans yields (McSpadden Gardener et al., 2006). *Pseudomonas protegens* is also known to produce 2,4-DAPG and showed similar effect on plant pathogens such as *P. fluorescens* (Ramette et al., 2011). Its activity was observed both *in vitro* and *in vivo*, when tested with both wild type and nonproducing mutants (Vincent et al., 1991; Fenton et al., 1992; Keel et al., 1992). In a comparison study, the 2,4-DAPG-producing strains protected plants better than the nonproducing strains (Cronin et al., 1997; Rezzonico et al., 2007). The polyketide 2,4-DAPG has shown to inhibit the growth of plant pathogenic bacteria such as *Erwinia carotovora* (Cronin et al., 1997), *Pythium* spp., and other pathogenic fungi such as *Rhizoctonia solani*, *Thielaviopsis basicola*, and *Gaeumannomyces graminis* var. *tritici* (Howell and Stipanovic 1979; Keel et al., 1992; Shanahan et al., 1992; Cronin et al., 1997), including nematodes (Cronin et al., 1997). The best-known example of disease suppression by the 2,4-DAPG-producing strains is take all of wheat caused by *G. graminis* var. *tritici* (Mendes et al., 2011; Raaijmakers and Mazzola, 2012; Kwak and Weller, 2013).

The molecular mechanism of 2,4-DAPG action against a soilborne phytopathogenic peronosporomycete *Pythium ultimum* var. *sporangiferum* has been described by de Souza et al. (2003) wherein it alters the plasma membrane and vacuolizes, thereby disintegrating the hyphal cell contents. The motility of zoospores and zoo sporogenesis in downy mildew pathogen, *Plasmopara viticola*, and a damping-off pathogen, *Aphanomyces cochlioides*, is known to be effected by its derivatives (Islam and von Tiedemann, 2011).

18.4.2 Phenazines

Phenazine-1-carboxylic acid is an antifungal metabolite (Haynes et al., 1956; Thomashow and Weller, 1988; Hass and Defago, 2005) that is produced by *Pseudomonas*. It is a heterocyclic nitrogen compound showing colored pigmentation and is produced exclusively by bacteria belonging to genera *Pseudomonas, Streptomyces, Nocardia, Sorangium, Brevibacterium, Burkholderia* (Turner and Messenger, 1986), and *Bacillus* (Kim, 2000). Other derivatives of phenazines include pyocyanin (King et al., 1954), phenazine-1-carboxamide (Birkofer, 1947), idoinin (Gerber, 1969), aeruginosin A (Holliman, 1969), and aeruginosin B (Herbert and Holliman, 1969).

The biocontrol capacity of *P. fluorescens* 2–79 (Thomashow and Weller, 1988), *Pseudomonas chlororaphis* PCL1391, and *Pseudomonas aeruginosa* PNA1 (Tambong and Hofte, 2001) is known to be caused by production of phenazines. It was also suggested in controlling soilborne plant pathogen *R. solani* (Rosales et al., 1995; Huang et al., 2004). Bull et al. (1991) showed that *P. fluorescens* 2–79 controls *G. graminis* var. *tritici*, which causes take all of wheat by production of phenazine-1-carboxylic acid. They showed that the population of phenazine-producing strains is inversely proportional to the number of lesions caused by the pathogen and reported during primary infection of roots; phenazine-1-carboxylic acid is a major factor in suppression of *G. graminis* var. *tritici*. These heterocyclic antibiotics are reported to show wide-spectrum antimicrobial activity, particularly by *Pseudomonas* spp. (Hu et al., 2005; Sunish Kumar et al., 2005; Ravindra Naik and Sakthivel, 2006). Attempts are made in order to understand the scale and quantitative aspects of phenazine production in natural settings (Mavrodi et al., 2012).

18.4.3 Pyrrolnitrin

Very few gram-negative bacteria such as *Enterobacter agglomerans, Serratia* spp., and *Pseudomonas* spp. are known to produce this compound. It is an organohalogenic compound derived from tryptophan and has an antifungal activity (Hammer et al., 1999; Haas and Defago, 2005; Costa et al., 2009). Its production has been linked with certain bacteria in controlling plant diseases, especially fungal pathogens (Haas and Defago, 2005; Costa et al., 2009). *Burkholderia cepacia* strain 5.5B was identified with the production of pyrrolnitrin, in suppression of stem rot of poinsettia caused by *R. solani*. This compound has also shown to be an antagonistic toward *Botrytis cinerea, R. solani*, and *Sclerotinia sclerotiorum* (Hammer and Evensen, 1993; Fernando et al., 2005).

18.4.4 Bacteriocins

Bacteriocins are another group of antibiotic compounds produced by bacteria. They are produced by many gram-negative and gram-positive bacteria and are known to inhibit other related strains of same species due to their high specificity. Application of such bacteria for controlling soilborne and phyllosphere-inhabiting bacterial plant pathogens seems to be promising. An avirulent strain of *Agrobacterium* is known to produce a peptide bacteriocin trifolitoxin that enhances the biological control of *Agrobacterium vitis* crown gall (Herlache and Triplett, 2002). In a field study, the bean nodulation of *Rhizobium etli* CE3 increases with the expression of trifolitoxin genes from *Rhizobium leguminosarum* bv.

trifolii T24 (Robleto et al., 1998). *Serratia plymithicum* produces "serracin P," a phage tail–like bacteriocin, which was employed in controlling the fire blight caused by *Erwinia amylovora* (Jabrane et al., 2002). *Xanthomonas campestris* pv. *glycines* shows antibacterial activity against phytopathogenic *Xanthomonas* spp. by secretion of "glycinecin A" (Heu et al., 2001). *Pseudomonas syringae* subsp. *savastanoi*, the causal agent of olive knot disease, is known to be inhibited by bacteriocin produced from *Pseudomonas syringae* pv. *ciccaronei*.

18.5   Mechanism of action

Rhizobacteria impedes growth of various phytopathogenic microorganisms using diverse mechanisms that include (1) antibiosis, (2) lytic enzyme productions, (3) degradation of pathogen virulence, (4) production of siderophores, (5) reduction in ethylene production, and (6) induction of systemic resistance.

18.5.1   Antibiosis

In general, rhizobacteria show antibiosis activity in inhibiting a wide variety of microorganisms. PGPR are known to exhibit antibiosis to native microflora (Burr et al., 1978), but it was not known whether it was a result of antagonism in soil resulted from antibiosis or competition or both. In *in vitro* studies, it is unclear that the antibiosis effect by PGPR is linked to production of inhibitory substances on root surfaces (Kloepper and Schroth, 1981). Antibiotic production by microorganisms was demonstrated in soil organic matter (Wright, 1956). In spite of their importance, determining their role in the rhizosphere remains elusive. Rationally, the antibiotic production in the rhizosphere would equip the producers to be better colonizers than their counterparts in competing against nutrients.

Frequently, antibiosis is used by many biocontrol agents such as fluorescent *Pseudomonas* spp., *Bacillus* spp., *Streptomyces* spp., and *Trichoderma* spp. A wide array of chemicals have been identified, and their roles in suppression of many plant pathogens have been documented (Fravel, 1988; Loper and Lindow, 1993; Weller and Thomashow, 1993, Raaijmakers and Mazzola, 2012). Not only antibiotics but also bacteriocins, enzymes such as cell wall–degrading enzymes, and volatile compounds with antifungal activity show antibiosis. *Pseudomonas fluorescens* CHAO is known to produce siderophores, phenazines, 2,4-diacetylphloroglucinol, and cyanide, and various combinations of these metabolites are responsible for its antagonism against *G. graminis* var. *tritici* and *Chalara elegans* (Defago and Haas, 1990).

18.5.2   Lytic enzyme production

Microorganisms are known to excrete hydrolytic enzymes that are lytic in nature. These enzymes play an important role in mycoparasitism. Fungal cell walls are made up of chitin and β-1,3-glucans (Lam and Gaffney, 1993). The action of chitinases or β-1,3-glucanase (laminarinase) in the lysis of fungal cell walls, in combination as well as alone, is well documented (Harman et al., 1993; Lam and Gaffney, 1993; Lorito et al., 1993, 1994a, b).

Pseudomonas stutzeri has been shown to produce chitinase and laminarinase, which caused the lysis of *Fusarium solani* germ tube and its mycelia. Markedly, this effect is more on mycelial growth inhibition rather than spore germination (Lim et al., 1991). The prevalence of diseases caused by *R. solani*, *Sclerotium rolfsii*, and *Pythium ultimum* was declined due to the enzymatic action of β-1,3-glucanase produced by *Pseudomonas cepacia*. In a greenhouse study, a chitinase (*ChiA*)-deficient *Serratia marcescens* mutant was shown to increase the fungal pathogen germ tube elongation and *Fusarium* wilt of pea seedling (Lam and

Gaffney, 1993). In a separate study, a non-biocontrol agent *Escherichia coli* is transformed with ChiA from *S. marcescens*; the resultant transgenic bacterium showed a reduced disease incidence of southern blight of bean caused by *S. rolfsii* (Shapira et al., 1989).

18.5.3 Degradation of pathogen virulence

Degradation of pathogen virulence is another way of biological control. For example, albicidin toxin produced by *Xanthomonas albilineans* is detoxified by *Pantoea dispersa* (Zhang and Birch, 1996). Binding of proteins to toxins is another way of mediation of their detoxification effect. The toxins produced by *Klebsiella oxytoca* and *Alcaligenes denitrificans* are known to be inhibited reversibly by unknown protein (Walker et al., 1988; Basnayake and Birch, 1995). Enzymes also play a role in the detoxification. *X. albilineans* produces albicidin, a polyketide–peptide compound that is known to inhibit the supercoiling activity of DNA gyrase both in plants and bacteria; *P. dispersa* esterase is known to detoxify irreversibly this toxin (Zhang and Birch, 1997). Certain phytotoxins (fusaric acid) produced by *Fusarium* species are known to be detoxicated by *B. cepacia* and *Ralstonia solanacearum* by hydrolyzing fusaric acid (Toyoda et al., 1988). Pathogen toxins show broad-spectrum activity against other microorganisms, especially its competitors, or even detoxify the antibiotics produced by their antagonists as a self-defense mechanism (Schouten et al., 2004).

18.5.4 Competition

Nutrient limitation occurs in poor soils. Most often, essential elements for microbial or plant growth are not in free forms, and they bound to soil particles or organic matter or form chemical complexes, which need to be mineralized for their availability. Of these nutrients, iron is very important, and its bioavailability is limited by the solubility of Fe^{3+}. Microorganisms produce siderophores to chelate iron from minerals, especially under nutrient-limiting conditions. The studies on siderophores started decades ago on the discovery of fungal ferrichrome (Neilands, 1952). Siderophores transport not only iron but also other elements such as Al, Cd, Cu, Ga, In, Pb, and Zn, as well as with radionuclides including U and Np (Kiss and Farkas, 1998; Neubauer et al., 2000a, b). Siderophores also play a role in antagonism against plant pathogens (Buyer and Leong, 1986; Solanki et al., 2014; Suman and Veena, 2014). Many soil bacteria belonging to *Erwinia*, *Pseudomonas*, *Nocardia*, *Streptomyces*, *Arthrobacter*, and *Chromobacterium* are known to produce siderophores (Meyer and Abdallah, 1980; Muller and Raymond, 1984; Buyer and Leong, 1986; Dorrier et al., 1990; Berner and Winkelmann, 1990; Gunter et al., 1993; Wei et al., 2007). The presence of these siderophores has been suggested to depend on the soil physiochemical and biological properties (Bossier et al., 1988; Nelson et al., 1988). It was shown that there was an interspecies utilization of the siderophores in fluorescent pseudomonad, and it was suggested that a specific outer membrane receptor protein might play a role for this interspecies siderophore utilization (Buyer and Leong 1986).

18.5.5 Induced resistance

Some of the biocontrol agents are known to induce certain physiological changes to the plant, thereby increasing its tolerance/resistance toward a pathogen by induced systemic resistance (ISR). This resistance can be localized or systemic in nature. The genes that are responsible for ISR are also known to induce systemic acquired resistance (SAR), but not always. When a localized infection occurs or an attenuated pathogen attacks the plant, SAR will be expressed as a resistance to wide pathogens (Uknes et al., 1992). Jones and Dangl (2006) describe how a plant immune system responds in general to microbes including nonpathogens. There are two pathways by which host resistance is mediated. First, SAR is mediated by salicylic acid that initiates the expression of pathogenesis-related proteins including a variety of enzymes. Second pathway involves the ISR, which is mediated by jasmonic acid and/or ethylene. BCA are known to induce disease resistance in many ways such as *Bacillus mycoides*; a biocontrol agent is able to stimulate sugar beet to produce peroxidase, chitinase, and β-1,3-glucanase (Bargabus et al., 2003). *Bacillus subtilis* strains GB03 and IN97 and *Pseudomonas putida* shown to produce 2,3-butanediol and lipopolysaccharide in *Arabidopsis* (Ryu et al., 2004; Meziane et al., 2005), siderophore production by *S. marcescens* in cucumber (Press et al., 2001). Root colonizers such as *Pseudomonas* spp. and *Trichoderma* spp. are found to be potential elicitors of plant host defenses. It was also reported that PGPR strains upon inoculation elicit SAR and ISR by salicylic acid, siderophore, lipopolysaccharides, and 2,3-butanediol and other volatile substances (van Loon et al., 1998; Ryu et al., 2004; Ongena et al., 2004). Quorum sensing is also known to play a role in the ISR. *S. marcescens* 90–166 has been reported to elicit ISR in tobacco plants in a pathogen-dependent manner (Ryu et al., 2013).

18.6   *Formulations and field application*

The biocontrol agents are used for the control of soilborne, seed-borne, and airborne fungal, bacterial, and viral plant pathogens. Apart from controlling the plant pathogens, they are also reported to enhance crop growth. There are a number of carrier substrates that are available to formulate biocontrol organisms for field applications. Some of the carrier molecules include diatomaceous earth granules, wheat bran, wheat bran–saw dust mixture, wheat bran–peat mixture, vermiculite–wheat bran acid formulation, alginate pellets, talc, granules, wettable powder, pellets, sticks, and powder.

The preparatory method of talc-based formulation of *Pseudomonas* is given in the succeeding text; proteose peptone, 20 g; dipotassium hydrogen phosphate, 2.5 g; glycerol, 15 mL; magnesium sulphate, 6 g; and distilled water, 1000 mL.

King's B broth medium is prepared in conical flasks and sterilized in autoclave for 20 min. A loop of antagonistic bacteria from 24 h old culture is inoculated into the broth medium and incubated in a rotary shaker at 150 rpm for 2 days at room temperature. The grown-up culture will be used directly as inoculum or used as mother culture for mass production in the fermenter.

For mass production by fermentation, method is being employed, using King's B broth medium. The media would be prepared and sterilized; after sterilization, the mother culture is added to the fermenter at 100 mL/10 L of medium and maintained in optimum condition for 2–3 days. One kilogram of talc powder is taken in a metal tray, and the pH was adjusted to neutral by adding $CaCO_3$ at 150 g/kg. The bacterial growth from the conical flask or fermenter is mixed with talc powder at 1:2.5 ratio (i.e., 400 mL culture in 1 kg of talc powder). The mixture is air dried and mixed with 5 g of carboxymethyl cellulose per 1 kg of talc powder. It is then packed in polythene bag, sealed, and can be used within 4 months.

Application to crop

1. Seed treatment
   - 4 g/kg of seed
   - 600 g in 65 L of water—soak 50 kg seed for 12 h
2. Seedling dip—1%
3. Soil application—2.5 kg/ha
4. Spray—0.1%

18.7 Future perspectives

Many rhizobacteria are known to exert their beneficial effects under laboratory conditions; but when tested under field or greenhouse conditions, they fail to show their beneficial effects. There could be many reasons for their failure. To be a better BCA or PGPR, they should show successful root colonization, be able to produce secondary metabolites, and be able to withstand both biotic and abiotic stresses. Understanding their mechanisms that are pivotal for their successful expression would lead us to isolate improved bacteria. Commercially available biocontrol agents are not genetically modified, and the employment of wild-type or isolated strains as a biocontrol agent is likely to continue. Engaging rational screening methods that stimulate field conditions are expected to generate new isolates. Employing the use of multiple organisms for biocontrol is being increasingly used for effective plant disease management and growth promotion. This could be attributed to the various arrays of secondary metabolites produced by the different microorganisms present in the mixed culture. It might be possible to formulate a microbial consortium that has improved biocontrol properties, especially under natural conditions such as siderophore producers, phloroglucinols, and phenazines. But much is needed to be done on the impact of biocontrol agents and PGPRs and their associations including their limitations. The advent of the advanced molecular methods and high-throughput sequencing may reveal the intricacies involved in their associations and mechanisms, thereby providing an empirical basis for sustainable agriculture.

References

Amaresan N., Jayakumar V., Kumar K., and Thajuddin N. 2012a. Isolation and characterization of plant growth promoting endophytic bacteria and their effect on tomato (*Lycopersicon esculentum*) and chilli (*Capsicum annuum*) seedling growth. *Ann Microbiol.* 62:805–810.

Amaresan N., Jayakumar V., Kumar K., and Thajuddin N. 2012b. Endophytic bacteria from tomato and chilli, their diversity and antagonistic potential against *Ralstonia solanacearum*. *Arch Phytopathol Pl Protect.* 45:344–355.

Amaresan N., Kumar K., Sureshbabu K., and Madhuri K. 2014a. Plant growth-promoting potential of bacteria isolated from active volcano sites of Barren Island, India. *Lett Appl Microbiol.* 58:130–137.

Amaresan N., Jayakumar V., and Thajuddin N. 2014b. Isolation and characterisation of endophytic bacteria associated with chilli (*Capsicum annuum*) promising in commercial agricultural ecosystem. *Ind J Biotechnol.* 13:247–255.

Bargabus, R. L., Zidack, N. K., Sherwood, J. E., and Jacobsen, B. J. 2003. Oxidative burst elicited by *Bacillus mycoides* isolate Bac J, a biological control agent, occurs independently of hypersensitive cell death in sugar beet. *Mol Plant Microbe Interact.* 16:1145–1153.

Basak, B. B., and Biswas, D. R. 2010. Co-inoculation of potassium solubilizing and nitrogen fixing bacteria on solubilization of waste mica and their effect on growth promotion and nutrient acquisition by a forage crop. *Biol Fertil Soils.* 46:641–648.

Basnayake, W. V. S., and Birch, R. G. 1995. A gene from *Alcaligenes denitrificans* that confers albicidin resistance by reversible antibiotic binding. *Microbiology.* 141:3551–560.

Bender, C. L., Rangaswamy, V., and Loper, J. E. 1999. Polyketide production by plant-associated pseudomonads. *Ann Rev Phytopathol.* 37:175–196.

Berg, G. 2009. Plant–microbe interactions promoting plant growth and health: Perspectives for controlled use of microorganisms in agriculture. *Appl Microbiol Biotechnol.* 84:11–18.

Berner, I., and Winkelmann, G. 1990. Ferrioxamine transport mutants and the identification of the ferrioxamine receptor protein FoxA in *Erwinia herbicola* (*Enterobacter agglomerans*). *Biol Metals.* 2:197–202.

Berner, I., Konetschny Rapp, S., Jung, G., and Winkelmann, G. 1988. Characterization of ferrioxamine E 449 as the principal siderophore of *Erwinia herbicola* (*Enterobacter agglomerans*). *Biol Metals*. 1:51–56.

Birkofer, L. 1947. Chlororaphin, ein weiteres farbiges stoffwechselproduckt des *Bacillus pyocyaneus*. *Chem Ber*. 80:212–214.

Bossier, P., Hofte, M., and Verstraete, W. 1988. Ecological significance of siderophores in soil. *Adv Microb Ecol*. 10:385–414.

Bull, C. T., Weller, D. M., and Thomashow, L. S. 1991. Relationship between root colonization and suppression of *Gaeumannomyces graminis* var. *tritici* by *Pseudomonas fluorescens* strain 2–79. *Phytopathology*. 81:954–959.

Burr, T. J., Schroth, M. N., and Suslow, T. 1978. Increased potato yields by treatment of seed pieces with specific strains of *Pseudomonas fluorescens* and *P. putida*. *Phytopathology*. 68:1377–1383.

Buyer, S. J. and Leong, J. 1986. Iron transport-mediated antagonism between plant growth-promoting and plant deleterious *Pseudomonas* strains. *J Biol Chem*. 261:791–794.

Chin-A-Woeng, T. F. C., Bloemberg, G. V., van der Bij, A. J., et al., 1998. Biocontrol by phenazine-1-carboxamide-producing *Pseudomonas chlororaphis* PCL1391 of tomato root rot caused by *Fusarium oxysporum* f. sp. *radicis-lycopersici*. *Mol Plant-Microbe Interact*. 11:1069–1077.

Costa, R., Götz, M., Mrotzek, N., Lottmann, J., Berg, G., and Smalla, K. 2006. Effects of site and plant species on rhizosphere community structure as revealed by molecular analysis of microbial guilds. *FEMS Microbiol Ecol*. 56:236–249.

Costa, R., van Aarle, I. M., Mendes, R., and van Elsas, J. D. 2009. Genomics of pyrrolnitrin biosynthetic loci: Evidence for conservation and whole-operon mobility within gram-negative bacteria. *Environ Microbiol*. 11:159–175.

Cretoiu, M.S., Korthals, G.W., Visser, J.H.M., and Elsas, J.D. van. 2013. Chitin amendment increases soil suppressiveness toward plant pathogens and modulates the actinobacterial and oxalobacteraceal communities in an experimental agricultural field. *Appl Environ Microbiol*. 79:5291–5301.

Cronin, D., Moenne-Loccoz, Y., Fenton, A., et al., 1997. Role of 2,4-diacetylphloroglucinol in the interactions of the biocontrol *Pseudomonas* strain F113 with the potato cyst nematode *Globodera rostochiensis*. *Appl Environ Microbiol*. 63:1357–1361.

DeCoste, N.J., Gadkar, V.J., and Filion, M. 2010. *Verticillium dahliae* alters *Pseudomonas* spp. populations and HCN gene expression in the rhizosphere of strawberry. *Can J Microbiol*. 56:906–915.

Defago, G. and Haas, D. 1990. *Pseudomonas* as antagonists of soil borne plant pathogens: Mode of action and genetic analysis. In: *Soil Biochemistry*, ed. J. M. Bollag and G. Stotzky, 249–291. New York: Marcel Dekker.

Delany, I. R., Walsh, U. F., Ross, I. et al., 2001. Enhancing the biocontrol efficacy of *Pseudomonas fluorescens* F113 by altering the regulation and production of 2,4-diacetylphloroglucinol. *Plant Soil*. 232:195–205.

de Souza, J. T., Arnould, C., Deulvot, C., et al., 2003. Effect of 2,4-diacetylphloroglucinol on *Pythium*: Cellular responses and variation in sensitivity among propagules and species. *Phytopathol*. 93:966–975.

Dowling, D. N. and O'Gara, F. 1994. Metabolites of *Pseudomonas* involved in the biocontrol of plant disease. *Trends in Biotechnol*. 12:133–141.

Drigo, B., Pijl, A. S., Duyts, H., Kielak, et al., 2010. Shifting carbon flow from roots into associated microbial communities in response to elevated atmospheric CO2. *Proc Natl Acad Sci*. 107:10938–10942.

Dwivedi, D. and Johri, B. N. 2003. Antifungals from fluorescent pseudomonads: Biosysnthesis and regulation. *Curr Sci*. 85:1693–1703.

Fenton, A. M., Stephens, P. M., Crowley, J., et al., 1992. Exploitation of gene(s) involved in 2,4-diacetylphloroglucinol biosynthesis to confer a new biocontrol capability to a *Pseudomonas* strain. *Appl Environ Microbiol*. 58:3873–3878.

Fernando, W. D. G., Nakkeeran, S., and Zhang, Y. 2005. Biosynthesis of antibiotics by PGPR and its relation in biocontrol of plant diseases. In *PGPR: Biocontrol and Biofertilizer*, ed. Z. A. Siddiqui, 67–109. the Netherlands: Springer Science.

Fravel, D. 1988. Role of antibiosis in biocontrol of plant disease. *Annu Rev Phytopathol*. 26:75–91.

Gerber, N. N. 1969. New microbial phenazines. *J Heterocyclic Chem.* 6:297–300.

Giotis, C., Theodoropoulou, A., Cooper, J., Hodgson, R., Shotton, P., Shiel, R., et al., 2012. Effect of variety choice, resistant rootstocks and chitin soil amendments on soil-borne diseases in soil-based, protected tomato production systems. *Eur J Plant Pathol.* 134:605–617.

Glick, B. R. 2012. Plant growth-promoting bacteria: Mechanisms and applications. *Scientifica.* 2012:1–15.

Glick, B. R. and Bashan, Y. 1997. Genetic manipulation of plant growth-promoting bacteria to enhance biocontrol of phytopathogens. *Biotechnol Adv.* 15:353–378.

Goel, A. K., Sindhu, S. S., and Dadarwal, K. R. 2002. Stimulation of nodulation and plant growth of chickpea (*Cicer arietinum* L.) by *Pseudomonas* spp. antagonisctic to fungal pathogens. *Biol Fertil Soils.* 36:391–396.

Gunter, K., Toupet, C., and Schupp, T. 1993. Characterization of an iron-regulated promoter involved in desferrioxamine B synthesis in *Streptomyces pilosus*: Repressor-binding site and homology to the diphtheria toxin gene promoter. *J Bacteriol.* 175:3295–3302.

Haas, D. and Keel, C. 2003. Regulation of antibiotic production in root-colonizing *Pseudomonas* spp. and relevance for biological control of plant disease. *Annu Rev Phytopathol.* 41:117–153.

Haichar, F. el. Z., Marol, C., Berge, O., et al., 2008. Plant host habitat and root exudates shape soil bacterial community structure. *ISME J* 2:1221–1230.

Hammer, P. E. and Evensen, K. B. 1993. Post harvest control of *Botrytis cinerea* on cut flowers with pyrrolnitrin. *Plant Dis.* 77:283–286.

Hammer, P. E., Burd, W., Hill, D. S., Ligon, J. M., and van Pee, K. 1999. Conservation of the pyrrolnitrin biosynthetic gene cluster among six pyrrolnitrin-producing strains. *FEMS Microbiol Lett.* 180:39–44.

Hao, D., Gao, P., Liu, P., Zhao, J. et al. 2011. AC3–33, a novel secretory protein, inhibits Elk1 transcriptional activity via ERK pathway. *Mol Biol Rep.* 38:1375–1382.

Harman, G. E., Hayes, C. K., Lorito, M. et al., 1993. Chitinolytic enzymes of *Trichoderma harzianum* purification of chitobiosidase and endochitinase. *Phytopathology.* 83:313–318.

Hass, D., and Defago, G. 2005. Biological control of soil borne pathogens by fluorescent pseudomonads. *Nat Rev Microbiol.* 3:307–319.

Haynes, W. C., Stodola, F. H., Locke, J. M. et al., 1956. *Pseudomonas aureofaciens* Kluyver and Phenazine α-carboxylic acid, its characteristic pigment. *J Bacteriol.* 72:412–417.

Herbert, R. B. and Holliman, F. G. 1969. Pigments of *Pseudomonas aeruginosa. J Biol Chem.* 159:725–750.

Herlache, T. C. and Triplett, E. W. 2002. Expression of a crown gall biological control phenotype in an avirulent strain of *Agrobacterium vitis* by addition of the trifolitoxin production and resistance genes. *BMC Biotechnol.* 2:2.

Heu, S., Oh, J., Kang, Y., Ryu, S., Cho, S. K., Cho, Y., and Cho, M. 2001. gly gene cloning and expression and purification of glycinecin A, a bacteriocin produced by *Xanthomonas campestris* pv. *glycines* 8ra. *Appl Environ Microbiol.* 67:4105–4110.

Hiltner, L. 1904. Uber neuere erfahrungen und probleme auf dem gebiet der bodenbakteriologie unter besonderer berucksichtigung der grundungung und brache. *Arb Dtsch Landwirtsch Ges.* 98:59–78.

Holliman, F. G. 1969. Pigments of *Pseudomonas* sp. part I. Structure and synthesis of aeruginosin A. *J Chem Soc C.* 18:2514–2516.

Howell, C. R., and Stipanovic, R. D. 1979. Control of *Rhizoctonia solani* on cotton seedlings with *Pseudomonas fluorescens* and with an antibiotic produced by the bacterium. *Phytopathol.* 69:480–482.

Hu, H. B., Xu, Y. Q., Chen, F., Zhang, X. H., and Hur, B. K. 2005. Isolation and characterization of a new fluorescent *Pseudomonas* strain that produces both phenazine-1-carboxylic acid and pyoluteorin. *J Microbiol Biotechnol.*15:86–90.

Huang, Z. Y., Bonsall, R. F., Mavrodi, D. V., Weller, D. M., and Thomashow, L. S. 2004. Transformation of *Pseudomonas fluorescens* with genes for biosynthesis of phenazine-1-carboxylic acid improves biocontrol of *Rhizoctonia* root rot and in situ antibiotic production. *FEMS Microbiol Ecol.* 49:243–251.

Huang, X. F., Chaparro, J. M., Reardon, K. F., Zhang, R., Shen, Q., and Vivanco, J. M. 2014. Rhizosphere interactions: Root exudates, microbes, and microbial communities. *Botany.* 92:267–275.

Iavicoli, A., Boutet, E., Buchal, A., and Métraux, J. P. 2003. Induced systemic resistance in *Arabidopsis thaliana* in response to root inoculation with *Pseudomonas fluorescens* CHA0. *Mol Plant-Microbe Interact.* 16: 851–858.

Islam, M. T. and Fukushi, Y. 2010. Growth inhibition and excessive branching in *Aphanomyces cochlioides* induced by 2,4-diacetylphloroglucinol is linked to disruption of filamentous actin cytoskeleton in the hyphae. *World J Microbiol Biotechnol.* 26:1163–1170.

Islam, M. T. and von Tiedemann, A. 2011. 2,4-Diacetylphloroglucinol suppresses zoosporogenesis and impairs motility of Peronosporomycete zoospores. *World J Microbiol Biotechnol.* 27:2071–2079.

Jabrane, A., Sabri, A., Compère, P., et al., 2002. Characterization of serracin P, a phage-tail-like bacteriocin, and its activity against *Erwinia amylovora*, the fire blight pathogen. *Appl Environ Microbiol.* 68:5704–5710.

Jayakumar V., Usha Rani G.K., Amaresan N., and Rajalakshmi S. 2009. First report of anthracnose disease of black pepper (*Piper nigrum*) caused by an unknown species of *Colletotrichum*. *Plant Disease.* 93:199–199.

Jones, J. D. and Dangl, J. L. 2006. The plant immune system. *Nature.* 444:323–329.

Keel, C., Schnider, U., Maurhofer, M., et al., 1992. Suppression of root diseases by *Pseudomonas fluorescens* CHA0: Importance of the bacterial secondary metabolite, 2,4-diacetylphloroglucinol. *Mol Plant Microbe Interact.* 5:4–13.

Kim, K. 2000. Phenazine-1-carboxylic acid resistance in phenazine-1-carboxylic producing *Bacillus* sp B-6. *J Biochem Mol Biol.* 33:332–336.

King, E. O., Ward, M. K., and Raney, D. E. 1954. Two simple media for the demonstration of pyocyanin and fluorescein. *J Lab Clin Med.* 44:301–307.

Kiss, T. and Farkas, E. 1998. Metal-binding ability of desferrioxamine B. *J Inclusion Phenomena Mol Recog Chem.* 32:385–403.

Kloepper, J. W. and Schroth, M. N. 1981. Relationship of *in vitro* antibiosis of plant growth-promoting rhizobacteria to plant growth and displacement of root microflora. *Phytopathology.* 71:1020–1024.

Kumar, K., Amaresan, N., Bhagat, S., Madhuri, K., and Srivastava, R. C. 2011. Isolation and characterization of rhizobacteria associated with coastal agricultural ecosystem of rhizosphere soils of cultivated vegetable crops. *World J Microbiol Biotechnol.* 27:1625–1632.

Kumar, K., Amaresan, N., Bhagat, S., Madhuri, K., and Srivastava, R. C. 2012. Isolation and characterization of *Trichoderma* spp. for antagonistic activity against root rot and foliar pathogens. *Ind J Microbiol.* 52:137–144.

Kwak, Y. S. and Weller, D. M. 2013. Take-all of wheat and natural disease suppression: A review. *Plant Pathol. J.* 29: 125–135.

Lam, S. T., and Gaffney, T. D. 1993. Biological activities of bacteria used in plant pathogen control. In: *Biotechnology in Plant Disease Control*, ed. I. Chet, 291–320. New York: John Wiley & Sons.

Li, X., Zhang, T., Wang, X., Hua, K., Zhao, L., and Han, Z. 2013. The composition of root exudates from two different resistant peanut cultivars and their effects on the growth of soil-borne pathogen. *Int J Biol Sci.* 9: 164–173.

Lim, H. S., Kim, Y. S., and Kim, S. D. 1991. *Pseudomonas stutzeri* YPL-1 genetic transformation and antifungal mechanism against *Fusarium solani*, an agent of plant root rot. *Appl Environ Microbiol.* 57:510–516.

Lodewyckx, C., Vangronsveld, J., Porteous, F., et al., 2002. Endophytic bacteria and their potential applications. *Crit Rev Plant Sci.* 21:583–606.

Loper, J. E. and Lindow, S. E. 1993. Roles of competition and antibiosis in suppression of plant diseases by bacterial control agents. In: *Pest Management: Biologically Based Technologies*, ed. R. D. Lumsden, and J. L. Vaughn, 144–155. Washington, DC: American Chemical Society.

Lorito, M., Harman, G. E., Hayes, C. K., et al., 1993. Chitinolytic enzymes produced by *Trichoderma harzianum*: Antifungal activity of purified endochitinase and chitobiosidase. *Phytopathology.* 83:302–307.

Lorito, M., Hayes, C. K., Di Pietro, A., Woo, S. L., and Harman, G. E. 1994a. Purification, characterization, and synergistic activity of a glucan 1,3-glucosidase and an N-acetyl-beta-glucosaminidase from *Trichoderma harzianum*. *Phytopathology.* 84:398–405.

Lorito, M., Hayes, C. K., Zoina, A., Scala, F., Del-Sorbo, G., Woo, S. L., and Harman, G. E. 1994b. Potential of genes and gene products from *Trichoderma* spp. and *Gliocladium* spp. for the development of biological pesticides. *Mol Biotechnol.* 2:209–217.

Lugtenberg, B. and Kamilova, F. 2009. Plant growth-promoting rhizobacteria. *Annu Rev Microbiol.* 63:541–556.

Maurhofer, M., Keel, C., Schnider, U., Viosard, C., Haas, D., and Defago, G. 1992. Influence of enhanced antibiotic production in *Pseudomonas fluorescens* strain CHA0 on its disease suppressive capacity. *Phytopathology.* 82:190–195.

Mavrodi, D.V., Mavrodi, O.V., Parejko, J.A., Bonsall, R.F., Kwak, Y.-S., Paulitz, T.C., et al., 2012. Accumulation of the antibiotic phenazine-1-carboxylic acid in the rhizosphere of dryland cereals. *Appl Environ Microbiol.* 78:804–812.

Mazzola, M., Fujimoto, D. K., Thomashow, L. S., and Cook, R. J. 1995. Variation in sensitivity of *Gaeumannomyces graminis* to antibiotics produced by fluorescent pseudomonads spp. and effect on biological control of take-all disease. *Appl Environ Microbiol.* 61:2554–2559.

McSpadden Gardener, B., Benitez, M. S., Camp, A., and Zumpetta, C. 2006. Evaluation of a seed treatment containing a *phlD*+ strain of *Pseudomonas fluorescens* on organic soybeans, 2005. Biological and cultural tests for control of plant diseases report. 21: FC046.

Mendes, R., Kruijt, M., de Bruijn, I., Dekkers, E., van der Voort, M., Schneider, J.H.M., et al., (2011) Deciphering the rhizosphere microbiome for disease-suppressive bacteria. *Science.* 332:1097–1100.

Meyer, J. M. and Abdallah, M. A. 1980. The siderochromes of non fluorescent pseudomonads: Production of nocardamine by *Pseudomonas stutzeri. J Gen Microbiol.* 118:125–129.

Meziane, H., van der Sluis, I., van Loon, L. C., Höfte, M., and Bakker, P. A. 2005. Determinants of *Pseudomonas putida* WCS358 involved in inducing systemic resistance in plants. *Mol Plant Pathol.* 6:177–185.

Morgan, J. A. W., Bending, G., and White, P. 2005. Biological costs and benefits to plant-microbe interactions in the rhizosphere. *J Exp Bot.* 56:1729–1739.

Muller, G. and Raymond, K. N. 1984. Specificity and mechanism of ferrioxamine-mediated iron transport in *Streptomyces pilosus. J Bacteriol.* 160:304–312.

Nallanchakravarthula, S., Mahmood, S., Alström, S., and Finlay, R. D. 2014. Influence of soil type, cultivar and *Verticillium dahliae* on the structure of the root and rhizosphere soil fungal microbiome of Strawberry. *PLoS ONE.* 9:e111455.

Neilands, J. B. 1952. A crystalline organo-iron pigment from a rust fungus (*Ustilago sphaerogena*). *J Am Chem Soc.* 74:4846.

Nelson, M., Cooper, C. R., Crowley, D. E., Reid, C. P. P., and Szaniszlo, P. J. 1988. An *Escherichia coli* bioassay of individual siderophores in soil. *J Plant Nutr.* 11:915–924.

Neubauer, U., Furrer, G., Kayser, A., and Schulin, R. 2000a. Siderophores, NTA, and citrate: Potential soil amendments to enhance heavy metal mobility in phytoremediation. *Int J Phytoremediation.* 2:353–368.

Neubauer, U., Nowak, B., Furrer, G., and Schulin, R. 2000b. Heavy metal sorption on clay minerals affected by the siderophore desferrixamine B. *Environ Sci Technol.* 34:2749–2755.

Ongena, M., Duby, F., Rossignol F., et al., 2004. Stimulation of the lipoxygenase pathway is associated with systemic resistance induced in bean by a nonpathogenic *Pseudomonas* strain. *Mol Plant Microbe Interact.* 17:1009–1018.

Ozyilmaz, U. and Benlioglu, K. 2013. Enhanced biological control of Phytophthora blight of pepper by biosurfactant-producing *Pseudomonas. Plant Pathol J.* 29(4):418–426.

Panhwar, Q. A., Othman, R., Rahman, Z. A., Meon, S., and Ismail, M. R. 2012. Isolation and characterization of phosphate-solubilizing bacteria from aerobic rice. *Afr J Biotechnol.* 11:2711–2719.

Philippot, L., Raaijmakers, J. M., Lemanceau, P., and van der Putten, W. H. 2013. Going back to the roots: The microbial ecology of the rhizosphere. *Nat Rev Microbiol.* 11:789–799.

Press, C. M., Loper, J. E., and Kloepper, J. W. 2001. Role of iron in rhizobacteria-mediated induced systemic resistance of cucumber. *Phytopathology.* 91:593–598.

Raaijmakers, J. M., and Mazzola, M. 2012. Diversity and natural functions of antibiotics produced by beneficial and plant pathogenic bacteria. *Annu Rev Phytopathol.* 50:403–424.

Raaijmakers, J. M., Paulitz, T. C., Steinberg, C., et al., 2008. The rhizosphere: A playground and battle-field for soilborne pathogens and beneficial microorganisms. *Plant Soil.* 321:341–361.

Ramette, A., Frapolli, M., Saux, M. F. L., et al., 2011. *Pseudomonas protegens* sp. *nov.*, widespread plant-protecting bacteria producing the biocontrol compounds 2,4-diacetylphloroglucinol and pyo-luteorin. *Syst Appl Microbiol.* 34:180–188.

Ravindra Naik, P. and Sakthivel, N. 2006. Functional characterization of a novel hydrocarbonoclas-tic *Pseudomonas* sp. strain PUP6 with plant-growth-promoting traits and antifungal potential. *Res Microbiol.* 157:538–546.

Rezzonico, F., Zala, M., Keel, C., et al., 2007. Is the ability of biocontrol fluorescent pseudomonads to produce the antifungal metabolite 2,4-diacetylphloroglucinol really synonymous with higher plant protection? *New Phytol.* 173:861–872.

Robleto, E. A., Kmiecik, K., Oplinger, E. S., Nienhuis, J., and Triplett, E. W. 1998. Trifolitoxin produc-tion increases nodulation competitiveness of *Rhizobium etli* CE3 under agricultural conditions. *Appl Environ Microbiol.* 64:2630–2633.

Rosales, A. M., Thomashow, L., Cook, R. J., and Mew, T. W. 1995. Isolation and identification of antifungal metabolites produced by rice-associated antagonistic *Pseudomonas* spp. *Phytopathology* 85:1028–1032.

Rovira, A. D. 1956. Plant root excretions in relation to the rhizosphere effect. *Plant Soil.* 7:178–194.

Rudrappa, T., Czymmek, K. J., Paré, P. W., and Bais, H. P. 2008. Root-secreted malic acid recruits beneficial soil bacteria. *Plant Physiol.* 148:1547–1556.

Ryu, C. M, Farag, M. A., Hu, C. H., et al., 2004. Bacterial volatiles induce systemic resistance in *Arabidopsis. Plant Physiol.* 134:1017–1026.

Ryu, C.-M., Choi, H. K., Lee, C.-H., Murphy, J. F., Lee, J.-K., and Kloepper, J. W. 2013. Modulation of quorum sensing in acylhomoserine lactone-producing or degrading tobacco plants leads to alteration of induced systemic resistance elicited by the Rhizobacterium *Serratia marcescens* 90–166. *Plant Pathol. J.* 29: 182–192.

Schouten, A., van Den Berg, G., Edel-Hermann, V., et al., 2004. Defense responses of *Fusarium oxy-sporum* to 2,4-diacetylphloroglucinol, a broad-spectrum antibiotic produced by *Pseudomonas fluorescens. Mol Plant-Microbe Interact.* 17:1201–1211.

Shanahan, P., O'sullivan, D. J., Simpson, P., Glennon, J. D., and O'Gara, F. 1992. Isolation of 2,4-diacetylphloroglucinol from a fluorescent pseudomonad and investigation of physiologi-cal parameters influencing its production. *Appl Environ Microbiol.* 58:353–358.

Shapira, R., Ordentlich, A., Chet, I., and Oppenheim, A. B. 1989. Control of plant diseases by chitin-ase expressed from cloned DNA in *Escherichia coli. Phytopathology.* 79:1246–1249.

Sharma P., Meena P. D., and Singh Y. P. 2013. New record of twig blight on *Catharanthus roseus* in India. *Afr J Microbiol Res.* 7(38):4680–4682.

Siddiqui, I. A. and Shaukat, S. S. 2003a. Plant species, host age and host genotype effects on *Meloidogyne incognita* biocontrol by *Pseudomonas fluorescens* strain CHA0 and its genetically-modified derivatives. *J Phytopathol.* 151:231–238.

Siddiqui, I. A. and Shaukat, S. S. 2003b. Suppression of root-knot disease by *Pseudomonas fluorescens* CHA0 in tomato: Importance of bacterial secondary metabolite, 2,4-diacetylphloroglucinol. *Soil Biol Biochem.* 35:1615–1623.

Siddiqui, I. A. and Shaukat, S. S. 2004. Suppression of *Meloidogyne incognita* by *Pseudomonas fluroscens* strain CHA0 and its genetically modified derivatives: II. The influence of sodium chloride. *Nematol Mediterranea.* 32:127–130.

Sindhu, S. S., Suneja, S., Goel, A. K., Parmar, N., and Dadarwal, K. R. 2002. Plant growth promoting effects of *Pseudomonas* sp. on coinoculation with *Mesorhizobium* sp. *cicer* strain under sterile and wilt sick soil conditions. *Appl Soil Ecol.* 19:57–64.

Singh N., Sharma P., and Verma O.P. 2011. First report of *Choanophora* sp. causing twig blight of *Boerhavia diffusa* in India. *New Dis Rep.* 23:29. doi:10.5197/j.2044–0588.2011.023.029

Solanki, M. K., Singh, R. K., Srivastava, S., et al., 2014. Isolation and characterization of siderophore producing antagonistic rhizobacteria against *Rhizoctonia solani. J Basic Microbiol* 54:585–597.

Suman, K. and Veena, K. 2014. Effect of antagonistic rhizobacteria coinoculated with *Mesorhizobium ciceris* on control of fusarium wilt in chickpea (*Cicer arietinum* L.). *Afr J Microbiol Re.* 8:1255–1265.

Sunish Kumar, R., Ayyadurai, N., Pandiaraja, P., et al., 2005. Characterization of antifungal metabolite produced by a new strain *Pseudomonas aeruginosa* PUPa3 that exhibits broad spectrum antifungal activity and biofertilizing traits. *J Appl Microbiol.* 98:145–154.

Tambong, J. T. and Hofte, M. 2001. Phenazines are involved inbiocontrol of *Pythium myriotylum* on cocoyam by *Pseudomonas aeruginosa* PNA1. *Eur J Plant Pathol.* 107:511–521.

Thomashow, L. S. and Weller, D. M. 1988. Role of a phenazine antibiotic from *Pseudomonas fluorescens* in biological control of *Gaeumannomyces graminis* var. *tritici. J Bacteriol.* 170:3499–3508.

Torsvik, V. and Ovreas, L. 2002. Microbial diversity and function in soil: From genes to ecosystems. *Curr Opin Microbiol.* 5:240–245.

Toyoda, H., Hashimoto, H., Utsumi, R., Kobayashi, H., and Ouchi, S. 1988. Detoxification of fusaric acid by a fusaric acid-resistant mutant of *Pseudomonas solanacearum* and its application to biological control of fusarium wilt of tomato. *Phytopathology.* 78:1307–1311.

Turner, J. M. and Messenger, A. J. 1986. Occurrence, biochemistry and physiology of phenazine pigment production. *Adves Microb Physiol.* 27:211–275.

Uknes, S., Mauch-Mani, B., Moyer, M., et al., 1992. Acquired resistance in *Arabidopsis. Plant Cell.* 4:645–656.

Uren, N. C. 2000. Types, amounts, and possible functions of compounds released into the rhizosphere by soil-grown plants. In: *The Rhizosphere: Biochemistry and Organic Substances at the Soil-Plant Interface,* ed. R. Pinton, Z. Varanini, and P. Nannipiero, 19–40. New York: Marcel Dekker.

Van Loon, L. C., Bakker, P. A., and Pieterse, C. M. 1998. Systemic resistance induced by rhizosphere bacteria. *Annu Rev Phytopathol.* 36:453–483.

Vincent, M. N., Harrison, L. A., Brackin, J. M., et al., 1991. Genetic analysis of the antifungal activity of a soilborne *Pseudomonas aureofaciens* strain. *Appl Environ Microbiol.* 57:2928–2934.

Walker, M. J., Birch, R. G., and Pemberton, J. M. 1988. Cloning and characterization of an albicidin resistance gene from *Klebsiella oxytoca. Mol Microbiol.* 2:443–454.

Wei, X., Sayavedra-Soto, L. A., and Arp, D. J. 2007. Characterization of the ferrioxamine uptake system of *Nitrosomonas europaea. Microbiology.* 153:3963–3972.

Weller, D. M. 2007. *Pseudomonas* biocontrol agents of soil borne pathogens: Looking back over 30 years. *Phytopathology.* 97:250–256.

Weller, D. M. and Thomashow, L. S. 1993. Use of rhizobacteria for biocontrol. *Curr Opin Biotechnol.* 4:306–311.

Wright, J. M. 1956. The production of antibiotics in soil. IV. Production of antibiotics in coats of seeds grown in soil. *Ann Appl Biol.* 44:561–566.

Wu, Y., Fang, W., Zhu, S., Jin, K., and Ji, D. 2008. The effects of cotton root exudates on the growth and development of *Verticillium dahliae. Front Agric China.* 2:435–440.

Zhang, L. and Birch, R. G. 1996. Biocontrol of sugar cane leaf scald disease by an isolate of *Pantoea dispersa* which detoxifies albicidin phytotoxins. *Lett Appl Microbiol.* 22:132–136.

Zhang, L. and Birch, R. G. 1997. The gene for albicidin detoxification from *Pantoea dispersa* encodes an esterase and attenuates pathogenicity of *Xanthomonas albilineans* to sugarcane. *Proc Natl Acad Sci USA.* 94:9984–9989.

Synthetic chemical compounds as broad spectrum antimicrobials

chapter nineteen

Microbe-mediated synthesis of silver nanoparticles

A new drug of choice against pathogenic microorganisms

Deene Manikprabhu and Wen-Jun Li

Contents

19.1   Introduction

In writing this chapter, the first question that came to our mind was "what are these nanoparticles?" What is so interesting about them that researchers around the globe are behind it? Well the answer lies in the unique properties possessed by these particles and, besides, due to the fact that they can be revolutionized the way we want. The term *nano* is adapted from the Greek word meaning "dwarf" and acts as a bridge between bulk materials and atomic or molecular structures (Thakkar et al., 2010).

Although the concept of nanoparticles was first presented by Richard Feynman through his famous lecture entitled "There's a plenty of room at the bottom" at the American Institute of Technology (Hulkoti and Taranath, 2014), nanoparticles have been used since antiquity. For example, the Chinese used gold nanoparticles as an inorganic dye to introduce red color into their ceramic porcelains more than a thousand years ago.

A Roman period glass called the Lycurgus Cup contained metal nanoparticles, which provided beautiful colors. In medieval times, nanoparticles were used for decoration of cathedral windows. Siddhars, the great Indian ancient scientists, practiced the use of Rajat Bhasma (Rajat is silver and Bhasma means fine powder) in medicine (Hansen et al., 2008; Manikprabhu and Lingappa, 2014).

Among various nanoparticles, silver nanoparticles are of great importance. The word "silver" is of Gothic origin, meaning shiny white; the Latin name "argentines" originates from an Aryan root, which means white and shining. Silver has been popular for domestic use since the ancient times; historically, silver was equated with the moon due to white brightness of silver. Silver antimicrobial properties were known from antiquity, having historical associations with humans dating back to 4000 BC. Silver vessels were used to preserve water and wine. Hippocrates, the father of medicine, promoted the use of silver for healing injuries. Alexander the Great was advised by Aristotle to store water in silver vessels and boil prior to use. Evidence of the use of silver nitrate as an antibacterial agent in the Roman pharmacopeia also exists. During the late eighteenth century, Crede, a German obstetrician, popularized the use of prophylactic 1% silver nitrate eye solution for the prevention of ophthalmia neonatorum. During the mid-nineteenth century, Joseph Lister and Marion Sims promoted the use of silver wire sutures in order to reduce the incidences of septic complications (Pradeep and Anshup, 2009).

At present, silver nanoparticles are used as antimicrobial agents in most public places such as elevators and railway stations in China. The mutation resistant antimicrobial activities of silver are being used in different pharmaceutical formulations such as antibacterial clothing, burn ointments, and coating for medical devices (Prabhu and Poulose, 2012; Manikprabhu and Lingappa, 2013a). Various physical and chemical methods were reported for the synthesis of silver nanoparticles, but most of these methods cause potential environmental and biological hazards. Compared to physical and chemical methods, biological synthesis using microbes and plants was regarded as a safe and eco-friendly process (Manikprabhu and Lingappa, 2013b). In this chapter, we focus on microbe-mediated synthesis of silver nanoparticles and antimicrobial activity against various pathogenic microorganisms.

19.2   Microbe-mediated synthesis of silver nanoparticles

Microbe-mediated synthesis of silver nanoparticles (Figure 19.1) requires a special ability, that is, "resistance of the organism to withstand silver ions." It is noted that those microorganisms that synthesize silver nanoparticles are vulnerable to higher concentrations of silver ions. Even though the organism has the resistance to silver ions, it becomes useless at the higher concentration. That is why silver can be called a "moiety with two functions." One is inducing the organism to synthesize nanoparticles at lower concentration; another is the induction of cell death at higher concentration (Deepak et al., 2011). The synthesis of nanoparticles was preliminary confirmed by a color change from colorless to brown. Further, synthesis was confirmed by UV–visible absorption spectroscopy; the maximum absorption between 400 and 450 nm due to surface plasmon resonance in the visible region indicates the synthesis of silver nanoparticles. In x-ray diffraction pattern, the intense peaks corresponding to (111), (200), (220), and (311) correspond to crystalline nature of silver nanoparticles (Manikprabhu and Lingappa, 2013a).

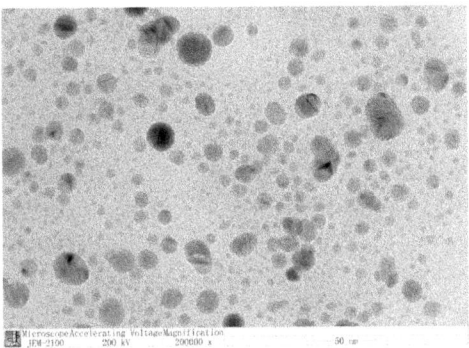

Figure 19.1 Biological synthesis of silver nanoparticles.

19.2.1 Synthesis of silver nanoparticles using bacteria

One reason why bacteria are preferred for nanoparticle synthesis is because they are easy to manipulate. The first evidence of bacteria-synthesizing nanoparticles was established using *Pseudomonas stutzeri* AG259 strain, which was isolated from a silver mine (Slawson et al., 1992).

The most widely accepted mechanism of silver biosynthesis is the presence of the nitrate reductase enzyme. During the reduction, nitrate is converted into nitrite and the electron is transferred to the silver ion; hence, the silver ion is reduced to silver (Ag^+ to Ag^0) (Prabhu and Poulose, 2012). Further, in different microorganisms, various enzymes are believed to take part in the bioreduction process involving the transport of electrons from certain electron donors to metal electron acceptors. Some studies of nonenzymatic reduction mechanism suggested that some organic functional groups of microbial cell walls could be responsible for the bioreduction process (Baker et al., 2013), for example, cells of *Lactobacillus* sp. A09 can reduce silver ions by the interaction of the silver ions with the groups on the microbial cell wall (Fu et al., 2000). Apart from these, proteins and microbial pigment were also used as reducing and stabilizing agent for the synthesis of nanoparticles (Jain et al., 2011; Manikprabhu and Lingappa, 2013a). All the mentioned mechanisms could result in the intracellular or extracellular complexion and the deposition of metal nanoparticles.

The mechanisms involved in the intracellular synthesis of nanoparticles are as follows. First, the cell wall of the microorganism is being negatively charged, which interacts electrostatically with the positively charged metal ions. Then, the enzymes present in the cell wall bioreduce the metal ions. Finally, aggregation of particles and synthesis of nanoparticles take place. The advantages of intracellular nanoparticles synthesis are that, the nanoparticles are small in size and nearly monodispersed, but in order to release intracellular-synthesized nanoparticles, additional processing steps such as ultrasound treatment or reaction with suitable detergents are required (Mukherjee et al., 2001; Sharma et al., 2007). Various bacteria have been reported for intracellular synthesis of nanoparticles like recently, an airborne *Bacillus* sp. isolated from the atmosphere was found to reduce Ag^+ ions to Ag^0. This bacterium accumulated metallic silver of 5–15 nm in size in the periplasmic space of the cell. *Lactobacillus* sp. in buttermilk produced silver crystals of well-defined morphology within the cell with no disturbance in its viability (Nair and Pradeep, 2002; Di Gregorio et al., 2005). Spherical silver nanoparticles of size 10–20 nm

Table 19.1 Synthesis of silver nanoparticles by different bacteria

Synthesized from	Location	References
Staphylococcus aureus	Extracellular	Nanda and Saravanan (2009)
Bacillus subtilis EWP-46	Extracellular	Velmurugan et al. (2014)
Pseudomonas aeruginosa	Extracellular	Kumar and Mamidyala (2011)
Escherichia coli	Extracellular	Gurunathan et al. (2009)
Bacillus licheniformis	Extracellular	Kalishwaralal et al. (2008)
Klebsiella pneumoniae	Extracellular	Kalpana and Lee (2013)
Enterobacter cloacae	Extracellular	Shahverdi et al. (2007)
Brevibacterium casei	Extracellular	Kalishwaralal et al. (2010)
Salmonella typhimurium	Extracellular	Ghorbani (2013)
Bacillus megaterium	Extracellular	Karkaj et al. (2013)

were synthesized using *Proteus mirabilis* PTCC 1710 (Samadi et al., 2009). Synthesis at extreme conditions was also reported, for example, *Corynebacterium* sp. SH09 produced silver nanoparticles at 60°C on the cell wall in the size range of 10–15 nm (Zhang et al., 2005). Further synthesis at high concentration of silver nitrate was also reported by the metal-tolerant bacteria *Idiomarina* sp. PR58–8 (Seshadri et al., 2012). Although several bacteria for intracellular synthesis of silver nanoparticles were reported, most of them are difficult to implement for industrial use due to the tedious recovery process. In this regard, extracellular synthesis of silver nanoparticles is the current focus of research. Extracellular synthesis using bacteria isolated from different environments was reported. *Bacillus* strain CS 11 isolated from the industrial area produce spherical extracellular silver nanoparticles of 42–92 nm size range (Das et al., 2014). Gram-negative marine bacteria *Pseudomonas aeruginosa* (Shivakrishna et al., 2013) and *Stenotrophomonas* synthesized silver nanoparticles in a range of 35–46 and 40–60 nm, respectively (Malhotra et al., 2013). Synthesis using the gram-positive marine bacteria *Bacillus* sp. was also reported (Maruthamuthu, 2012). The first thermophilic bacterium reported for silver nanoparticles was *Geobacillus stearothermophilus*, which synthesized nanoparticles extracellularly (Fayaz et al., 2011). Similarly, the extremophilic bacteria *Ureibacillus thermosphaericus* strain reported synthesis of silver nanoparticles at 80°C with the particle size range of 10–100 nm (Juibari et al., 2011). The endophytic bacterium *Bacillus cereus* (Sunkar and Nachiyar, 2012), the psychrophilic bacteria *Pseudomonas antarctica*, *Pseudomonas proteolytica*, *Pseudomonas meridiana*, *Arthrobacter kerguelensis*, and *Arthrobacter gangotriensis*, and the mesophilic bacteria *Bacillus indicus* and *Bacillus cecembensis* were also reported for green synthesis of silver nanoparticles (Shivaji et al., 2011). Further, detailed information regarding extracellular synthesis of nanoparticles by bacteria is mentioned in Table 19.1.

19.2.2 *Synthesis of silver nanoparticles using actinomycetes*

Actinomycetes are gram-positive bacteria but share some important characteristics of fungi and at present are the center of attraction due to their ability to produce a large number of secondary metabolites. The first actinomycete that was reported to synthesize nanoparticles was *Thermomonospora* sp. mainly due to its ability to survive in a wide range of environmental conditions (Rautaray et al., 2004). Since then, many other actinomycetes were explored for the synthesis of nanoparticles. *Streptomyces* sp. 09 PBT 005 was able to synthesize silver nanoparticles extracellularly, which had antibacterial activity (Kumar et al., 2015). Similarly, *Streptomyces aureofaciens*, *Streptomyces coelicolor* klmp33, *Streptomyces rochei, Streptomyces* sp. ERI-3, and *Streptomyces hygroscopicus* were also reported to synthesize extracellular silver nanoparticles (Prabhu et al., 2011; Manikprabhu and Lingappa, 2013a; Golinska et al., 2014; Zonooz and Salouti, 2011; Sadhasivam et al., 2010). Further reports on marine *Streptomyces parvulus* SSNP11 and *Streptomyces albidoflavus* CNP10 reduced silver ions by reducing nitrate to nitrite and ammonium were also reported (Shetty and Kumar, 2012; Baker et al., 2013). *Streptomyces* sp. BDUKAS10 isolated from mangrove reported to synthesize nanoparticles (Sivalingam et al., 2012). Nanoparticles synthesized by *Streptomyces* sp. VITPK1 showed anticandidal activity (Sanjenbam et al., 2014). Similarly, silver nanoparticles synthesized using *Streptomyces* sp. JF714876 (Vidyasagar et al., 2012), *Streptomyces* sp. JAR1 (Chauhan et al., 2013), and *Streptomyces* sp. VITSTK7 (Thenmozhi et al., 2013) showed good antimicrobial activity. Not only the genus *Streptomyces* was the center of attraction, but other genera such as *Rhodococcus* sp., *Actinomycetes* sp., and *Nocardiopsis* sp. (Golinska et al., 2014) were also reported to synthesize silver nanoparticles.

19.2.3   Synthesis of silver nanoparticles using fungi

Fungi are eukaryotic, sporeforming, and filamentous branched organisms. Fungi can accumulate metals by physicochemical and biological mechanisms, including extracellular binding by metabolites and polymers, binding to specific polypeptides, and metabolism-dependent accumulation (Gade et al., 2010). Many fungi such as *Cladosporium cladosporioides* (Balaji et al., 2009), *Neurospora crassa* (Castro-Longoria et al., 2011), and *Penicillium brevicompactum* (Shaligram et al., 2009) were exploited for the synthesis of metal nanoparticles. Intracellular synthesis of silver nanoparticles by *Verticillium* of size range 2–25 nm deposited to the surface of the cytoplasmic membrane was reported by Sastry et al. (2003). "Green synthesis" of highly stabilized nanocrystalline silver particles by a nonpathogenic and agriculturally important fungus, *Trichoderma asperellum*, was reported by Mukherjee et al. (2008). The marine-derived fungus *Aspergillus flavus* synthesized intracellular silver nanoparticles in acidic pH, while the same fungus in alkaline pH range supported rapid extracellular synthesis of silver nanoparticles (Vala et al., 2014). This indicates that the synthesis on nanoparticles is pH dependent. Not only terrestrial fungi were explored but marine fungi too. The marine fungi *Penicillium fellutanum* isolated from coastal mangrove sediment synthesized nanoparticles using proteins as a reducing agent (Kathiresan et al., 2009). The mushroom *Agaricus bisporus* was reported to synthesize nanoparticles, which showed remarkable antimicrobial activity (Dhanasekaran et al., 2013). Similarly, the edible mushroom *Pleurotus florida* synthesized polydispersed nanoparticles of size 20 ± 5nm (Bhat et al., 2011). Further detailed information regarding extracellular synthesis of nanoparticles is mentioned in Table 19.2.

The significant drawback of using these bio-entities in nanoparticles synthesis is that the genetic manipulation of eukaryotic organisms as a means of over-expressing specific enzymes is difficult when compared with bacteria.

19.3   Antimicrobial mechanism of silver nanoparticles

Silver nanoparticles are ideal bacterial agents due to their slow oxidation and liberation of Ag^+ ions to the surroundings. Moreover, the small size of these particles facilitates their penetration through cell membranes to affect intracellular processes from inside. Additionally, the excellent antibacterial properties exhibited by nanoparticles are due to their well-developed surface that provides maximum contact with the environment (Krutyakov et al., 2008).

Table 19.2 Synthesis of silver nanoparticles by different fungi

Synthesized from	Location	Size of nanoparticles (nm)	References
Aspergillus fumigatus	Extracellular	5–25	Bhainsa and D'Souza (2006)
Phanerochaete chrysosporium	Extracellular	50–200	Vigneshwaran et al. (2006)
Fusarium solani	Extracellular	5–35	Ingle et al. (2009)
Amylomyces rouxii strain KSU-09	Extracellular	5–27	Musarrat et al. (2010)
Alternaria alternate	Extracellular	20–60	Gajbhiye et al. (2009)
Aspergillus niger	Extracellular	15–20	Gade et al. (2008)
Fusarium acuminatum	Extracellular	4–50	Ingle et al. (2008)

The exact antimicrobial mechanism of silver nanoparticles is still not clear. However, various theories of the action of silver nanoparticles on microbes to cause the antimicrobial effect were proposed. One is that silver nanoparticles have the ability to anchor to the bacterial cell wall and subsequently penetrate inside the cell, thereby causing structural changes in the cell membrane like the permeability of the cell membrane leading to death of the cell (Manikprabhu and Lingappa, 2013a). Another theory is that the formation of free radicals by the silver nanoparticles may be considered to be another mechanism by which the cells die. The electron spin resonance spectroscopy studies suggested that there is formation of free radicals by the silver nanoparticles when in contact with the bacteria, and these free radicals have the ability to damage the cell membrane and make it porous, which can ultimately lead to cell death (Danilcauk et al., 2006). Yet another theory has also been proposed that there can be a release of silver ions by the nanoparticles, which can interact with the thiol groups of many vital enzymes and inactivate them and finally lead to the death of the cell (Matsumura et al., 2003). Another fact is that DNA has sulfur and phosphorus as its major components; the nanoparticles can act on these soft bases and destroy the DNA, which would definitely lead to cell death. The interaction of the silver nanoparticles with the sulfur and phosphorus of the DNA can lead to problems in the DNA replication of the bacteria and thus terminate the microbes (Manikprabhu and Lingappa, 2013a). It has also been found that the nanoparticles can modulate the signal transduction in bacteria. It is a well-established fact that phosphorylation of protein substrates in bacteria influences bacterial signal transduction. Nanoparticles dephosphorylate the peptide substrates on tyrosine residues, which lead to signal transduction inhibition and thus the stoppage of growth (Prabhu and Poulose, 2012).

19.3.1 *Antibacterial activity of silver nanoparticles*

Silver nanoparticles have been demonstrated as an effective antibacterial agent against a broad bacterial spectrum, including both gram-negative and gram-positive bacteria (Marambio-Jones and Hoek, 2010). Silver nanoparticles showed antibacterial activity against various pathogenic bacteria like *Escherichia coli, Bacillus subtilis, Enterococcus faecalis, Salmonella typhimurium, Staphylococcus epidermidis*, and *Staphylococcus aureus* (Schrofel et al., 2014; Gopinath et al., 2015). The investigation of antimicrobial activity of silver nanoparticles on gram-negative and gram-positive bacteria showed that *E. coli* was inhibited at a low concentration of Ag-NPs (3.3 nM), which was 10 times less than the minimum inhibitory concentration on *S. aureus* (33 nM), that may be due to difference in cell membrane composition and permeability of gram-negative and gram-positive bacteria (Tran and Le, 2013). This conclusion was supported by Jung et al. (2008), which suggested that the thickness of the peptidoglycan layer of gram-positive bacteria may prevent the action of the silver ions to some extent.

The antibacterial activity of silver nanoparticles against different drug-resistant pathogens like ampicillin-resistant *E. coli* and erythromycin-resistant *Streptococcus pyogenes* was evaluated, and the minimum inhibitory concentrations and minimum bactericidal concentrations of silver nanoparticles ranged between 30 and 100 mM (Lara et al., 2010). Recently, our study of methicillin-resistant *S. aureus* showed encouraging results; silver nanoparticles of average size 50 nm showed good antimicrobial activity; further, the synergistic activity of silver nanoparticles and antibiotics increases antimicrobial activity (Manikprabhu and Lingappa, 2013a). Similarly, antibacterial activity of silver nanoparticles against extended-spectrum beta-lactamase *E. coli* was also reported (Manikprabhu and Lingappa, 2014).

Apart from pH, the size of nanoparticles varied the antimicrobial activity. The antimicrobial activity of silver nanoparticles against *Streptococcus mutans* varied according to size; the minimum inhibitory concentration increased with increase in size (Espinosa-Cristobal et al., 2009). Similarly, antimicrobial activity of silver nanoparticles is dose dependent. Further, different shapes of silver nanoparticles have different antimicrobial activity. Pal et al. (2007) demonstrated that truncated triangular silver nanoplates displayed the strongest antimicrobial action against *E. coli*, when compared with spherical and rod shaped silver nanoparticles.

19.3.2   Antifungal activity of silver nanoparticles

Fungi are becoming a major concern of the world as their mycotoxins not only affect plants and animals, but also humans, causing many diseases and economic damage (Sanchez-Hervas et al., 2008). In this regard, silver nanoparticles were used as an antifungal agent against various pathogens like *Bipolaris sorokiniana, Magnaporthe grisea* (Jo et al., 2009), and *Trichophyton mentagrophytes* (Kim et al., 2008). The antifungal activity of silver nanoparticles on dermatophyte like *Candida albicans* was reported and suggested that silver nanoparticles exert an antifungal activity by disrupting the structure of the cell membrane and inhibiting the normal budding process due to the destruction of the membrane integrity (Kim et al., 2009). Fungus mediated synthesis of silver nanoparticles (*Alternaria alternate*) showed antifungal activity against *Phoma glomerata, Phoma herbarum, Fusarium semitectum, Trichoderma* sp., and *Candida albicans*. Further, the synergetic of silver nanoparticles with fluconazole increased the antifungal activity (Gajbhiye et al., 2009). This indicates that silver nanoparticles are good antimicrobial agents.

19.4   Antimicrobial activity of silver nanoparticles in various applications

19.4.1   Silver nanoparticles as a dressing material

Wound dressing materials play a major part in wound management. Dr. Robert Burrell is said to develop the world's first nanosilver-based wound dressing in 1995 (Ahmad et al., 2011). He developed Acticoat that speeds up the healing process and removes scars if any.

In recent times, the development of resistant strains of pathogens has become a major problem; to overcome this problem, newly designed wound dressing has provided a major breakthrough for the treatment of infection and wounds. Silver dressings make use of delivery systems that release silver in different concentrations. But different factors like the distribution of silver in the dressing, its chemical and physical forms, and affinity of dressing to moisture also influence the killing of microorganisms (Rai et al., 2009). Many advancements in dressing material were made; recently, one dressing material was designed that has the potential to change color when the antibiotic is released and hence alerting that there is an infection in the wound. Experts believe that this dressing has great potential in treating burn victims who are susceptible to toxic shock syndrome. With the advent of such a system, there can be a reduction in antibiotic resistance (Tian et al., 2007).

12.4.2   Silver nanoparticles as air disinfectant

Bioaerosols are airborne particles of biological origins, including viruses, bacteria, and fungi, which are capable of causing infectious, allergenic, or toxigenic diseases. Several silver nanoparticle–based air disinfectant were formulated. The antimicrobial effect of silver nanoparticles on bacterial contamination of activated carbon filters was studied. The results showed that silver nanoparticle–activated carbon filters were effective for the removal of bioaerosols. The antibacterial activity analysis of silver nanoparticle–activated carbon filters indicated that two bacteria, *B. subtilis* and *E. coli*, were completely inhibited within 10 and 60 min, respectively (Jung et al., 2008).

12.4.3   Silver nanoparticles in wound healing

Wound healing is a complex process and has been the subject of intense research for a long time. The recent emergence of nanotechnology has provided a new therapeutic modality in silver nanoparticles for use in burn wounds. The wound-healing properties of silver nanoparticles in an animal model showed rapid healing in a dose-dependent manner (Rickman et al., 2003).

12.4.4   Silver nanoparticles for agriculture application

Nanopesticides and nanoherbicides were being extensively used in agriculture. Several industries were making formulations that contain 100–250 nm nanoparticles that are more soluble in water thus increasing their activity. The water- or oil-based nanoemulsions contained uniform suspensions of pesticide or herbicide nanoparticles of 200–400 nm, which are used to control many pests (Yakub and Soboyejo, 2012).

12.4.5   Silver nanoparticles for disinfecting drinking water

Water is one of the most important substances on Earth and is essential to all living things.

Contamination of drinking water and the subsequent outbreak of waterborne diseases are the leading cause of death in many developing nations. Significant interest has arisen in the use of silver nanoparticles for water disinfection. Silver nanoparticles decorated onto porous ceramic materials are used as an antibacterial water filter when tested against *E. coli*. It was found that at a flow rate of 0.01 L min^{-1}, the output count of *E. coli* was zero (Ahmad et al., 2011). Similarly, silver nanoparticles binding with polyurethane foams resulted due to interaction of nitrogen atom with the polyurethane foams, which showed disinfection ability. At a flow rate of 0.5 L min^{-1}, the output count of *E. coli* was found nil when the input water had a bacterial load of 10^5 CFU mL^{-1} (Jung et al., 2008).

Acknowledgments

This work was supported by the Key Project of International Cooperation of Ministry of Science and Technology (MOST) (No. 2013DFA31980) and Yunnan Provincial Natural Science Foundation (2013FA004). W.-J. Li was also supported by Guangdong Province Higher Vocational Colleges and Schools Pearl River Scholar Funded Scheme (2014).

References

Ahmad, N., Sharma, S., Singh, V. N., Shamsi, S. F., Fatma, A., and Mehta, B. R. 2011. Biosynthesis of silver nanoparticles from *Desmodium triflorum*: A novel approach towards weed utilization. *Biotechnology Research International*, 2011, 1–8.

Baker, S., Harini, B. P., Rakshith, D., and Satish, S. 2013. Marine microbes: Invisible nanofactories. *Journal of Pharmaceutical Research*, 6, 383–388.

Balaji, D. S., Basavaraja, S., Deshpande, R., Bedre Mahesh, D., Prabhakar, B. K., and Venkataraman, A. 2009. Extracellular biosynthesis of functionalized silver nanoparticles by strains of *Cladosporium cladosporioides* fungus. *Colloids Surfaces B*, 68, 88–92.

Bhainsa, K. C. and D'Souza, S. F. 2006. Extracellular biosynthesis of silver nanoparticle using the fungus *Aspergillus fumigatus*. *Colloids Surfaces B*, 47, 160–164,

Bhat, R., Deshpande, R., Ganachari, S. V., Huh, D. S., and Venkataraman, A. 2011. Photo-irradiated biosynthesis of silver nanoparticles using edible mushroom *Pleurotus florida* and their antibacterial activity studies. *Bioinorganic Chemistry and Applications*, 2011, 1–7.

Castro-Longoria, E., Vilchis-Nestor, A. R., and Avalos-Borja, M. 2011. Biosynthesis of silver, gold and bimetallic nanoparticles using the filamentous fungus *Neurospora crassa*. *Colloids Surfaces B*, 83, 42–48.

Chauhan, R., Kumar, A., and Abraham, J. 2013. A biological approach to the synthesis of silver nanoparticles with *Streptomyces* sp JAR1 and its antimicrobial activity. *Scientia Pharmaceutica*, 81, 607–621.

Danilcauk, M., Lund, A., Saldo, J., Yamada, H., and Michalik, J. 2006. Conduction electron spin resonance of small silver particles. *Spectrochimica Acta A*, 63, 189–191.

Das, V. L., Thomas, R., Varghese, R. T., Soniya, E. V., Mathew, J., and Radhakrishnan, E. K. 2014. Extracellular synthesis of silver nanoparticles by the *Bacillus* strain CS 11 isolated from industrialized area. *3 Biotech*, 4, 121–126.

Deepak, V., Kalishwaralal, K., Pandian, S. R. K., and Gurunathan, S. 2011. An insight into the bacterial biogenesis of silver nanoparticles, industrial production and scale-up. M. Rai and N. Duran (eds.), in *Metal Nanoparticles in Microbiology*, pp. 17–35. Springer: Berlin/Heidelberg.

Dhanasekaran, D., Latha, S., Saha, S., Thajuddin, N., and Panneerselvam, A. 2013. Extracellular biosynthesis, characterisation and in-vitro antibacterial potential of silver nanoparticles using *Agaricus bisporus*. *Journal of Experimental Nanoscience*, 8, 579–588.

Di Gregorio, S., Lampis, S., and Vallini, G. 2005. Selenite precipitation by a rhizospheric strain of *Stenotrophomonas* sp. isolated from the root system of *Astragalus bisulcatus*: A biotechnological perspective. *Environment International*, 31, 233–241.

Espinosa-Cristobal, L. F., Martinez-Castanon, G. A., Martinez-Martinez, R. E., Loyola-Rodriguez, J. P., Patino-Marin, N., Reyes-Macias, J. F., and Ruiz, F. 2009. Antibacterial effect of silver nanoparticles against *Streptococcus mutans*. *Materials Letters*, 63, 2603–2606.

Fayaz, A. M., Girilal, M., Rahman, M., Venkatesan, R., and Kalaichelvan, P. T. 2011. Biosynthesis of silver and gold nanoparticles using thermophilic bacterium *Geobacillus stearothermophilus*. *Process Biochemistry*, 46, 1958–1962.

Fu, J. K., Liu, Y. Y., Gu, P. Y., Tang, D. L., Lin, Z. Y., Yao, B. X., and Weng, S. Z. 2000. Spectroscopic characterization on the biosorption and bioreduction of Ag(I) by *Lactobacillus* sp. A09. *Acta Physico-Chimica Sinica*, 16, 770–782.

Gade, A. K., Bonde, P., Ingle, A. P., Marcato, P. D., Duran, N., and Rai, M. K. 2008. Exploitation of *Aspergillus niger* for synthesis of silver nanoparticles. *Journal of Biobased Materials and Bioenergy*, 2, 243–247.

Gade, A., Ingle, A., Whiteley, C., and Rai, M. 2010. Mycogenic metal nanoparticles: Progress and applications. *Biotechnology Letters*, 32, 593–600.

Gajbhiye, M., Kesharwani, J., Ingle, A., Gade, A., and Rai, M. 2009. Fungus-mediated synthesis of silver nanoparticles and their activity against pathogenic fungi in combination with fluconazole. *Nanomedicine: Nanotechnology, Biology and Medicine*, 5, 382–386.

Ghorbani, H. R. 2013. Biosynthesis of silver nanoparticles using *Salmonella typhimurium*. *Journal of Nanostructure in Chemistry*, 29, 1–4.

Golinska, P., Wypij, M., Ingle, A. P., Gupta, I., Dahm, H., and Rai, M. 2014. Biogenic synthesis of metal nanoparticles from actinomycetes: Biomedical applications and cytotoxicity. *Applied Microbiology and Biotechnology*, 98, 8083–8097.

Gopinath, P. M., Narchonai, G., Dhanasekaran, D., Ranjani, A., and Thajuddin, N. 2015. Mycosynthesis, characterization and antibacterial properties of AgNPs against multidrug resistant (MDR) bacterial pathogens of female infertility cases. *Asian Journal of Pharmaceutical Sciences*, 10(2), 138–145.

Gurunathan, S., Kalishwaralal, K., Vaidyanathan, R., Venkataraman, D., Pandian, S. R. K., Muniyandi, J., and Eom, S. H. 2009. Biosynthesis, purification and characterization of silver nanoparticles using *Escherichia coli*. *Colloids Surfaces B*, 74, 328–335.

Hansen, S. F., Maynard, A., Baun, A., and Tickner, J. A. 2008. Late lessons from early warnings for nanotechnology. *Nature Nanotechnology*, 3, 444–447.

Hulkoti, N. I. and Taranath, T. C. 2014. Biosynthesis of nanoparticles using microbes—A review. *Colloids Surfaces B*, 121, 474–483.

Ingle, A., Gade, A., Pierrat, S., Sonnichsen, C., and Rai, M. 2008. Mycosynthesis of silver nanoparticles using the fungus *Fusarium acuminatum* and its activity against some human pathogenic bacteria. *Current Nanoscience*, 4, 141–144.

Ingle, A., Rai, M., Gade, A., and Bawaskar, M. 2009. *Fusarium solani*: A novel biological agent for the extracellular synthesis of silver nanoparticles. *Journal of Nanoparticle Research*, 11, 2079–2085.

Jain, N., Bhargava, A., Majumdar, S., Tarafdar, J. C., and Panwar, J. 2011. Extracellular biosynthesis and characterization of silver nanoparticles using *Aspergillus flavus* NJP08: A mechanism perspective. *Nanoscale*, 3, 635–641.

Jo, Y. K., Kim, B. H., and Jung, G. 2009. Antifungal activity of silver ions and nanoparticles on phytopathogenic fungi. *Plant Disease*, 93, 1037–1043.

Juibari, M. M., Abbasalizadeh, S., Jouzani, G. S., and Noruzi, M. 2011. Intensified biosynthesis of silver nanoparticles using a native extremophilic *Ureibacillus thermosphaericus* strain. *Materials Letters*, 65, 1014–1017.

Jung, W. K., Koo, H. C., Kim, K. W., Shin, S., Kim, S. H., and Park, Y. H. 2008. Antibacterial activity and mechanism of action of the silver ion in *Staphylococcus aureus* and *Escherichia coli*. *Applied and Environmental Microbiology*, 74, 2171–2178.

Kalishwaralal, K., Deepak V., Ramkumarpandian S., Nellaiah H., and Sangiliyandi, G. 2008. Extracellular biosynthesis of silver nanoparticles by the culture supernatant of *Bacillus licheniformis*. *Materials Letters*, 62, 4411–4413.

Kalishwaralal, K., Deepak, V., Pandian, S. R. K., Kottaisamy, M., BarathManiKanth, S., Kartikeyan, B., and Gurunathan, S. 2010. Biosynthesis of silver and gold nanoparticles using *Brevibacterium casei*. *Colloids Surfaces B*, 77, 257–262.

Kalpana, D. and Lee, Y. S. 2013. Synthesis and characterization of bactericidal silver nanoparticles using cultural filtrate of simulated microgravity grown *Klebsiella pneumoniae*. *Enzyme and Microbial Technology*, 52, 151–156.

Karkaj, O. S., Salouti, M., Zanjani, R. S., and Derakhshan, F. K. 2013. Extracellular deposition of silver nanoparticles by *Bacillus megaterium*. *Synthesis and Reactivity in Inorganic, Metal-Organic, and Nano-Metal Chemistry*, 43, 903–906.

Kathiresan, K., Manivannan, S., Nabeel, M. A., and Dhivya, B. 2009. Studies on silver nanoparticles synthesized by a marine fungus, *Penicillium fellutanum* isolated from coastal mangrove sediment. *Colloids Surfaces B*, 71, 133–137.

Kim, K. J., Sung, W. S., Suh, B. K., Moon, S. K., Choi, J. S., Kim, J. G., and Lee, D. G. 2008. Antifungal effect of silver nanoparticles on dermatophytes. *Journal of Microbiology and Biotechnology*, 18, 1482–1484.

Kim, K. J., Sung, W. S., Suh, B. K., Moon, S. K., Choi, J. S., Kim, J. G., and Lee, D. G. 2009. Antifungal activity and mode of action of silver nano-particles on *Candida albicans*. *Biometals*, 22, 235–242.

Krutyakov, Y. A., Kudrinskiy, A. A., Olenin, A. Y., and Lisichkin, G. V. 2008. Synthesis and properties of silver nanoparticles: Advances and prospects. *Russian Chemical Reviews*, 77, 233–257.

Kumar, C. G. and Mamidyala, S. K. 2011. Extracellular synthesis of silver nanoparticles using culture supernatant of *Pseudomonas aeruginosa*. *Colloids Surfaces B*, 84, 462–466.

Kumar, P. S., Balachandran, C., Duraipandiyan, V., Ramasamy, D., Ignacimuthu, S., and Al-Dhabi, N. A. 2015. Extracellular biosynthesis of silver nanoparticle using *Streptomyces* sp. 09 PBT 005 and its antibacterial and cytotoxic properties. *Applied Nanoscience*, 5(2), 169–180,

Lara, H. H., Ayala-Núñez, N. V., Turrent, L. D. C. I., and Padilla, C. R. 2010. Bactericidal effect of silver nanoparticles against multidrug-resistant bacteria. *World Journal of Microbiology and Biotechnology*, 26, 615–621.

Malhotra, A., Dolma, K., Kaur, N., Rathore, Y. S., Mayilraj, S., and Choudhury, A. R. 2013. Biosynthesis of gold and silver nanoparticles using a novel marine strain of *Stenotrophomonas*. *Bioresource Technology*, 142, 727–731.

Manikprabhu, D. and Lingappa, K. 2013a. Antibacterial activity of silver nanoparticles against methicillin-resistant *Staphylococcus aureus* synthesized using model *Streptomyces* sp. pigment by photo-irradiation method. *Journal of Pharmaceutical Research*, 6, 255–260.

Manikprabhu, D. and Lingappa, K. 2013b. Microwave assisted rapid and green synthesis of silver nanoparticles using a pigment produced by *Streptomyces coelicolor* KLMP33. *Bioinorganic Chemistry and Applications*, 2013, 1–5.

Manikprabhu, D. and Lingappa, K. 2014. Synthesis of silver nanoparticles using the Streptomyces coelicolor klmp33 2 pigment: An antimicrobial agent against extended-spectrum beta-lactamase (ESBL) producing *Escherichia coli*. *Materials Science and Engineering: C*, 45, 434–437.

Marambio-Jones, C. and Hoek, E. M. 2010. A review of the antibacterial effects of silver nanomaterials and potential implications for human health and the environment. *Journal of Nanoparticle Research*, 12, 1531–1551.

Maruthamuthu, S. 2012. Extracellular synthesis of silver nanoparticles by marine thermophilic bacteria. *International Journal of Pharmaceutical and Biological Archives*, 3, 1418–1423.

Matsumura, Y., Yoshikata, K., Kunisaki, S. I., and Tsuchido, T. 2003. Mode of bacterial action of silver zeolite and its comparison with that of silver nitrate. *Applied Environmental Microbiology*, 69, 4278–4281.

Mukherjee, P., Ahmad, A., Mandal, D., Senapati, S., Sainkar, S. R., Khan, M. I., Ramani, R., Parischa, R., Ajayakumar, P. V., Alam, M., Sastry, M., and Kumar, R. 2001. Bioreduction of AuCl₄ ions by the fungus *Verticillium* sp. and surface trapping of the gold nanoparticles formed. *Angewandte Chemie International Edition*, 40, 3585–3588.

Mukherjee, P., Roy, M., Mandal, B. P., Dey, G. K., Mukherjee, P. K., Ghatak, J., Tyagi, A. K., and Kale, SP. 2008. Green synthesis of highly stabilized nanocrystalline silver particles by a non-pathogenic and agriculturally important fungus *T. asperellum*. *Nanotechnology*, 19(7), 075103.

Musarrat, J., Dwivedi, S., Singh, B. R., Al-Khedhairy, A. A., Azam, A., and Naqvi, A. 2010. Production of antimicrobial silver nanoparticles in water extracts of the fungus *Amylomyces rouxii* strain KSU-09. *Bioresource Technology*, 101, 8772–8776.

Nair, B. and Pradeep, T. 2002. Coalescence of nanoclusters and formation of submicron crystallites assisted by *Lactobacillus* strains. *Crystal Growth & Design*, 2, 293–298.

Nanda, A. and Saravanan, M. 2009. Biosynthesis of silver nanoparticles from *Staphylococcus aureus* and its antimicrobial activity against MRSA and MRSE. *Nanomedicine: Nanotechnology, Biology and Medicine*, 5, 452–456.

Pal, S., Tak, Y. K., and Song, J. M. 2007. Does the antibacterial activity of silver nanoparticles depend on the shape of the nanoparticle? A study of the gram-negative bacterium *Escherichia coli*. *Applied and Environmental Microbiology*, 73, 1712–1720.

Prabhu, S. and Poulose, E. K. 2012. Silver nanoparticles: Mechanism of antimicrobial action, synthesis, medical applications, and toxicity effects. *International Nano Letters*, 32, 1–10.

Prabhu, K. V., Sundaramoorthi, C., and Devarasu, S. 2011. Biosynthesis of silver nanoparticles from *Streptomyces aureofaciens. Journal of Pharmaceutical Research*, 4, 820–822.

Pradeep, T. and Anshup 2009. Noble metal nanoparticles for water purification: A critical review, *Thin Solid Films*, 517, 6441–6478.

Rai, M., Yadav, A., and Gade, A. 2009. Silver nanoparticles as a new generation of antimicrobials. *Biotechnology Advances*, 27, 76–83.

Rautaray, D., Ahmad, A., and Sastry, M. 2004. Biological synthesis of metal carbonate minerals using fungi and actinomycetes. *Journal of Material Chemistry*, 14, 2333–2340.

Rickman, D., Luvall, J. C., Shaw, J., Mask, P., Kissel, D., and Sullivan, D. 2003. Precision agriculture: changing the face of farming. *Geotimes*, November (feature article).

Sadhasivam, S., Shanmugam, P., and Yun, K. 2010. Biosynthesis of silver nanoparticles by *Streptomyces hygroscopicus* and antimicrobial activity against medically important pathogenic microorganisms. *Colloids Surfaces B*, 81, 358–362.

Samadi, N., Golkaran, D., Eslamifar, A., Jamalifar, H., Fazeli, M. R., and Mohseni, F. A. 2009. Intra/extracellular biosynthesis of silver nanoparticles by an autochthonous strain of *Proteus mirabilis* isolated from photographic waste. *Journal of Biomedical Nanotechnology*, 5, 247–253.

Sanchez-Hervás, M., Gil, J. V., Bisbal, F., Ramón, D., and Martínez-Culebras, P. V. 2008. Mycobiota and mycotoxin producing fungi from cocoa beans. *International Journal of Food Microbiology*, 125, 336–340.

Sanjenbam, P., Gopal, J. V., and Kannabiran, K. 2014. Anticandidal activity of silver nanoparticles synthesized using *Streptomyces* sp. VITPK1. *Journal de Mycologie Médicale*, 24, 211–219.

Sastry, M., Ahmad, A., Islam Khan, M., and Kumar, R. 2003. Biosynthesis of metal nanoparticles using fungi and actinomycete. *Current Science*, 85, 162–170.

Schrofel, A., Kratošová, G., Safarik, I., Safarikova, M., Raska, I., and Shor, L. M. 2014. Applications of biosynthesized metallic nanoparticles—A review. *Acta Biomaterials*, 10, 4023–4042.

Seshadri, S., Prakash, A., and Kowshik, M. 2012. Biosynthesis of silver nanoparticles by marine bacterium, *Idiomarina* sp. PR58–8. *Bulletin of Materials Science*, 35, 1201–1205.

Shahverdi, A. R., Minaeian, S., Shahverdi, H. R., Jamalifar, H., and Nohi, A. A. 2007. Rapid synthesis of silver nanoparticles using culture supernatants of *Enterobacteria*: A novel biological approach. *Process Biochemistry*, 42, 919–923.

Shaligram, N. S., Bule, M., Bhambure, R., Singhal, R. S., Singh, S. K., Szakacs, G., and Pandey, A. 2009. Biosynthesis of silver nanoparticles using aqueous extract from the compactin producing fungal strain. *Process Biochemistry*, 44, 939–943.

Sharma, N. C., Sahi, S. V., Nath, S., Parsons, J. G., Gardea-Torresde, J. L., and Pal, T. 2007. Synthesis of plant-mediated gold nanoparticles and catalytic role of biomatrix-embedded. *Environmental Science and Technology*, 41, 5137–5142.

Shetty, P. R. and Kumar, Y. S. 2012. Characterization of silver nanoparticles synthesized by using marine isolate *Streptomyces albidoflavus. Journal of Microbiology and Biotechnology*, 22, 614–621.

Shivaji, S., Madhu, S., and Singh, S. 2011. Extracellular synthesis of antibacterial silver nanoparticles using psychrophilic bacteria. *Process Biochemistry*, 46(9), 1800–1807.

Shivakrishna, P., Krishna, M. R. P. G., and Charya, M. S. 2013. Synthesis of silver nano particles from marine bacteria *Pseudomonas aerogenosa. Octa Journal of Biosciences*, 1, 108–114.

Sivalingam, P., Antony, J. J., Siva, D., Achiraman, S., and Anbarasu, K. 2012. Mangrove *Streptomyces* sp. BDUKAS10 as nanofactory for fabrication of bactericidal silver nanoparticles. *Colloids Surfaces B*, 98, 12–17.

Slawson, R. M., Van Dyke, M. I., Lee, H., and Trevors, J. T. 1992. Germanium and silver resistance, accumulation and toxicity in microorganisms. *Plasmid*, 27, 73–79.

Sunkar, S. and Nachiyar, C. V. 2012. Biogenesis of antibacterial silver nanoparticles using the endophytic bacterium *Bacillus cereus* isolated from *Garcinia xanthochymus. Asian Pacific Journal of Tropical Biomedicine*, 2, 953–959.

Thakkar, K. N., Mhatre, S. S., and Parikh, R. Y. 2010. Biological synthesis of metallic nanoparticles. *Nanomedicine: Nanotechnology, Biology and Medicine*, 6, 257–262.

Thenmozhi, M., Kannabiran, K., Kumar, R., and Khanna, V. G. 2013. Antifungal activity of *Streptomyces* sp. VITSTK7 and its synthesized Ag$_2$O/Ag nanoparticles against medically important *Aspergillus* pathogens. *Journal de Mycologie Médicale*, 23, 97–103.

Tian, J., Wong, K. K., Ho, C. M., Lok, C. N., Yu, W. Y., Che, C. M., and Tam, P. K. 2007. Topical delivery of silver nanoparticles promotes wound healing. *ChemMedChem*, 2, 129–136.

Tran, Q. H. and Le, A. T. 2013. Silver nanoparticles: Synthesis, properties, toxicology,applications and perspectives. *Advances in Natural Sciences: Nanoscience and Nanotechnology*, 4(3), 033001.

Vala, A. K., Shah, S., and Patel, R. 2014. Biogenesis of silver nanoparticles by marine-derived fungus *Aspergillus flavus* from Bhavnagar Coast, Gulf of Khambhat, India. *Journal of Marine Biology & Oceanography*, 3(1), 2.

Velmurugan, P., Iydroose, M., Mohideen, M. H. A. K., Mohan, T. S., Cho, M., and Oh, B. T. 2014. Biosynthesis of silver nanoparticles using *Bacillus subtilis* EWP-46 cell-free extract and evaluation of its antibacterial activity. *Bioprocess and Biosystems Engineering*, 37, 1527–1534.

Vidyasagar, G. M., Shankaravva, B., Begum, R., and Imrose, R. R. 2012. Antimicrobial activity of silver nanoparticles synthesized by *Streptomyces* species JF714876. *International Journal of Pharmaceutical Sciences and Nanotechnology*, 5, 1638–1642.

Vigneshwaran, N., Kathe, A. A., Varadarajan, P. V., Nachane, R. P., and Balasubramanya, R. H. 2006. Biomimetics of silver nanoparticles by white rot fungus, *Phaenerochaete chrysosporium*. *Colloids Surfaces B*, 53, 55–59.

Yakub, I. and Soboyejo, W. O. 2012. Adhesion of *E. coli* to silver or copper coated porous clay ceramic surfaces. *Journal of Applied Physics*, 111(12), 124324.

Zhang, H., Li, Q., Lu, Y., Sun, D., Lin, X., Deng, X., and Zheng, S. 2005. Biosorption and bioreduction of diamine silver complex by *Corynebacterium*. *Journal of Chemical Technology and Biotechnology*, 8, 285–290.

Zonooz, N. F. and Salouti, M. 2011. Extracellular biosynthesis of silver nanoparticles using cell filtrate of *Streptomyces* sp. ERI-3. *Scientia Iranica*, 18, 1631–1635.

chapter twenty

Nanomaterials

Source of antimicrobial products

*Atanu Bhattacharyya, P.M. Gopinath, A. Ranjani,
and Dharumadurai Dhanasekaran*

Contents

20.1   Introduction

Nanoparticles (NPs) are molecular aggregates having a dimension between 1 and 100 nm. They possess different physicochemical (strength, electrical, and optical) properties due to the variation in surface area (Bhattacharyya et al., 2011, 2014). NPs occur in nature, as volcanic dust, lunar dust, and mineral composites. Natural or engineered NPs, also defined as waste or anthropogenic particles, may be formed as a result of industrial processes, like diesel exhaust, coal combustion, and welding fumes. These nanomaterials are mostly carbon-based (fullerene, single- and multiwalled carbon nanotube) and metal-based (quantum dots, nanogold, nanozinc, and nanoaluminum) materials (Bhattacharyya et al., 2011, 2014). These materials are used in several biological processes. A few studies have focused on the effects and mechanisms of nanomaterials on bacteria (Bhattacharyya, 2009; Xu et al., 2009). The functional aspects of nanomaterials are unique, and these studies have been reported with the aim to provide further insight of bacteria and nanomaterials (Bhattacharyya et al., 2007, 2014). Nanoscale metal oxides like TiO_2, ZnO, and Al_2O_3 dendrimers (nanosized polymers built from branched units) are performed with atom–atom interaction (Bhattacharyya et al., 2014).

Currently, scientists are interested in materials that are effective at the nanoscale such as gold and silver because of their natural and chemical characteristics.

The introduction of inorganic nanomaterials in this century is a unique phenomenon, and their novel characteristic features are mentioned earlier. The inorganic nanomaterials exhibit well-adopted physical, chemical, and biological properties, though all the mentioned properties may vary in size based on their character as well as on specific pH and temperature. In the biological field, nanomaterials are very much adopted due to their specific selectivity toward the targeted materials in the biological system. Moreover, recent studies clearly denote that the prepared metal oxide nanomaterials have increased antibacterial activity (Hamouda et al., 1999; Pal et al., 2007). Several experimental data are coming through that can predict in an effective way that inorganic metal silver is an antimicrobial agent (Table 20.1). These mentioned inorganic metals exhibit a very robust antibacterial activity (Novak and Feldheim, 2000; Wiley et al., 2005; Yamanaka et al., 2005). Now, it is well known that the compound and simple nature of silver nanomaterials are the reason for the unique bactericidal activity (Pal et al., 2007; Gopinath et al., 2015). Moreover, it should be remembered that random use of silver nanoparticles (AgNPs) in various fields has some possibility of toxicity in human beings as well as in the environment or in the ecosystem. The toxicity and cellular uptake experiments of several NPs clearly denote that the carbohydrate coating on silver nanoparticle modulates both oxidative stress and cellular uptake, and it can expose some toxicity in the physiological system. In this context, it has been observed that the bioactivity of AgNPs and other nanomaterials can change by using a carbohydrate coat (Kennedy et al., 2014). Different types of bacterial cells are multidrug tolerant and, therefore, able to survive antibiotic treatment. However, recent technological advances in microfluidics and reporter genes have improved this scenario. Here, we summarize recent progress in the field, revealing the ubiquitous bacterial stress alarm one ppGpp as an emerging central regulator of multidrug tolerance and persistence, both in stochastically and environmentally induced persistence. In several different organisms, toxin–antitoxin modules function as effectors of ppGpp-induced persistence (Maisoneuve and Gerdes, 2014). Therefore, the effects of different NPs on several biofilms and also on several bacteria have been dealt with in this chapter.

20.2   Utility and beautility of nanomaterials

Dwarf nanomaterial is an advanced scientific technique in the twenty-first century. Nanotechnology is becoming a revolutionary feild in its integration with the green chemistry approach. Several strategies of nanomaterials involved in strain selection, cultivation modes, recombinant gene expression, metabolic engineering, and protein redesign and reengineering and its predictive modeling will allow creating some utility in nanobioreactor process. A new nanobiotechnology arena with a high potential impact in many fields is gradually developing. Nanotechnology is becoming a new field of increasing research and industrial interest since 1980s. Nanotechnology can be defined as the manipulation of atom-by-atom interactions in nature. It can take part in the processes of chemical and biological functions. In general, the nano-object properties depend on chemical composition, but also on size, shape, composition, and their environment (like pH and temperature) including their spatial distribution. It is clear that synthesis techniques can affect considerably the properties of the nano-objects. The synthesis techniques can be categorized into top-down and bottom-up phenomena. The top-down techniques work with the material in its bulk form, and size reduction of nanoscale is made via specialized ablations (e.g., lithography, thermal decomposition, laser ablation). In this context, engineered nanomaterials have received a particular attention for their positive impact in improving, among others,

Table 20.1 Antimicrobial activity of metal nanoparticles

Nanoparticle type	Size (Average)	Organism tested	References
ZnO	13 nm	*Staphylococcus aureus*	Reddy et al. (2007)
ZnO	60 nm	*S. aureus*	Jones et al. (2008)
ZnO	40 nm	*S. aureus, Escherichia coli*	Nair et al. (2009)
ZnO	12 nm	*E. coli*	Padmavathy and Vijayaraghavan (2008)
ZnO ions	N/A	*Pseudomonas aeruginosa, S. aureus, Candida albicans*	McCarthy et al. (1992)
Silver	21 nm	*E. coli, Vibrio cholerae, Salmonella typhi, P. aeruginosa*	Morones et al. (2005)
Silver	Triangles (50 nm)	*E. coli*	Pal et al. (2007)
Silver	12 nm	*E. coli*	Sondi and Salopek-Sondi (2004)
Silver	13.5 nm	*S. aureus, E. coli*	Kim et al. (2007)
Cu	100 nm	*E. coli, Bacillus subtilis*	Yoon et al. (2007)
Fe_3O_4	9 nm	*S. aureus*	Tran et al. (2010)
Fe_3O_4	8 nm	*Staphylococcus epidermidis*	Taylor and Webster (2009)
Al_2O_3	11 nm	*E. coli*	Simon-Deckers et al. (2009)
Al_2O_3	60 nm	*E. coli, B. subtilis, Pseudomonas fluorescens*	Jiang et al. (2009)
TiO_2	17 nm	*E. coli*	Simon-Deckers et al. (2009)
SiO_2	20 nm	*E. coli, B. subtilis, P. fluorescens*	Jiang et al. (2009)
Chitosan	40 nm	*E. coli, S. aureus*	Qi et al. (2004)
Carboxyl-grafted SPIONs	10–20 nm	*S. aureus*	Subbiahdoss et al. (2012)
APTES-grafted SPIONs	10–20 nm		
PEGylated SPIONs	10–20 nm		
Ag-coated SPIONs	15–20 nm		Khan (2012)
Ag–Au-coated SPIONs	20–30 nm		
Au-coated SPIONs	25–40 nm		
ZnO	<100 nm	Halophilic bacterium sp. *EMB4*	Sinha et al. (2011)
Ag	<100 nm		
Ag caron complex-L-tyrosine polyphosphate NP(SCC23-LTP NPs)	700–800 nm	Vancomycin-resistant *Enterococcus*	Leid et al. (2012)
ZnO	<100 nm	*B. subtilis*	Sinha et al. (2011)
Ag	<100 nm		
Ag	2–4 nm		Ruparelia et al. (2008)
CuO	8–10 nm		

(Continued)

Table 20.1 (Continued) Antimicrobial activity of metal nanoparticles

Nanoparticle type	Size (Average)	Organism tested	References
Al$_2$O$_3$	40–70 nm		Jiang et al. (2009)
TiO$_2$	40–60 nm		
Cu-doped TiO$_2$ nanoparticles	20 nm	*Mycobacterium smegmatis*	Wu et al. (2010)
Ag Caron complex-L-tyrosine polyphosphate NP (SCC23-LTP NPs)	800 nm	*Klebsiella pneumoniae*	Juan et al. (2010)
Ag	43 nm		Khan et al. (2011)
NO	10–15 nm		Friedman et al. (2011)
NO	8–15 nm	*P. aeruginosa*	Friedman et al. (2011)
NO-releasing MAP3 (*N*-methyl amino propyltrimethoxysilane) Si NPs	80–100 nm		Hetrick et al. (2009)
TiO$_2$	10–25 nm		Tsuang et al. (2008)
Ag	1–10 nm		Morones et al. (2005)
ZnO	10–20 nm		Feris et al. (2009)
ZnO	25–40 nm	*Salmonella typhimurium*	Kumar et al. (2011)
TiO$_2$	40–60 nm		
Cu-doped TiO$_2$ NPs	20 nm	*Shewanella oneidensis MR-1*	Wu et al. (2010)
Ag	10 nm	*Pseudomonas putida KT2442*	Gajjar et al. (2009)
CuO	25–40 nm		
ZnO	50–70 nm		
TiO$_2$	<25 nm	*Cupriavidus metallidurans*	Simon-Deckers et al. (2009)
Al$_2$O$_3$	<25 nm		
Multiwalled carbon nanotubes	<25 nm		
Ag	16–70 nm	*Enterobacter* sp. ANT 02, *P. aeruginosa* ANT 04, *K. pneumoniae* ANT 03, and *E. coli* ANT 01	Gopinath et al. (2015)
Ag	—	*S. typhi, E. coli, Klebsiella* sp., *Pseudomonas* sp., *Enterobacter* sp., *Proteus* sp., *S. aureus, S. paratyphi*	Dhanasekaran et al. (2013)

consumer products, pharmaceutics, cosmetics, transportation, energy, and also agriculture. It is the beauty of nanomaterials (Cauerhff and Castro, 2013).

Moreover, after Richard Feynman delivered his famous lecture, "There's plenty of room at the bottom Nature Nanotechnology looks at its influence on subsequent developments in nanoscience and technology" Feynman would like to consider the atom-by-atom reactions.

Now, let us consider some examples; suppose that many molecules are in a specific tank, in order to double the density; having the same speed and exhibiting the same temperature where they belong. Then, to a close approximation, the number of collisions will be doubled, and since each will be just as "energetic" as before, the pressure is proportional to the density. If we consider the true nature of the forces between the atoms, naturally, we would expect a slight decrease in pressure because of the attraction between the atoms and the pressure is proportional to the density. In such a way, complex molecules are produced. Thus the complex materials are broken down with the specific instruments for observing whether the said materials are nanomaterials or not. The nanomaterials possess three important characters: increases in electrical potential and chemical reaction and development of magnetic power. These can all be observed through SEM (Figure 20.1), TEM, and also atomic force microscope (AFM) (Figure 20.2). The scanning tunneling microscope and its offspring, the AFM, are

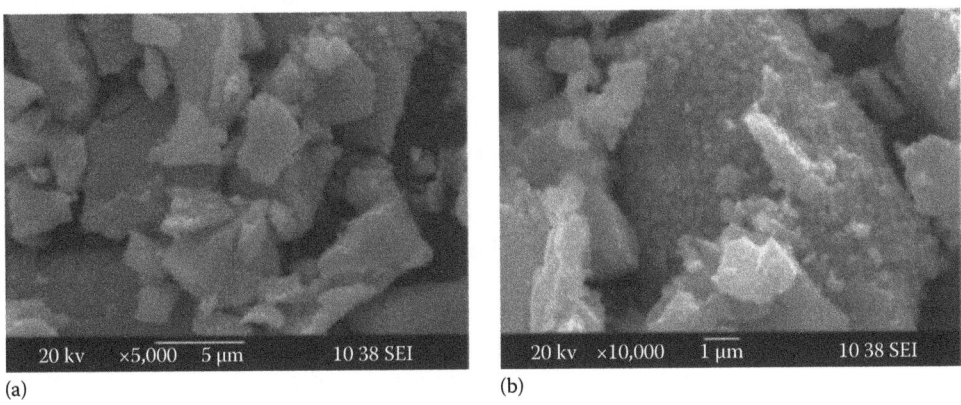

Figure 20.1 Scanning electron microscope image of silver nanoparticles synthesized using fungal extract. It reveals that the particles were roughly spherical to oval in nature with a little aggregation. The aggregation of AgNPs occurred during drying process: (a) 5,000x magnification and (b) 10,000x magnification. (Adapted from Gopinath, P.M. et al., *Asian J. Pharm. Sci.*, 10(2), 138, 2015.).

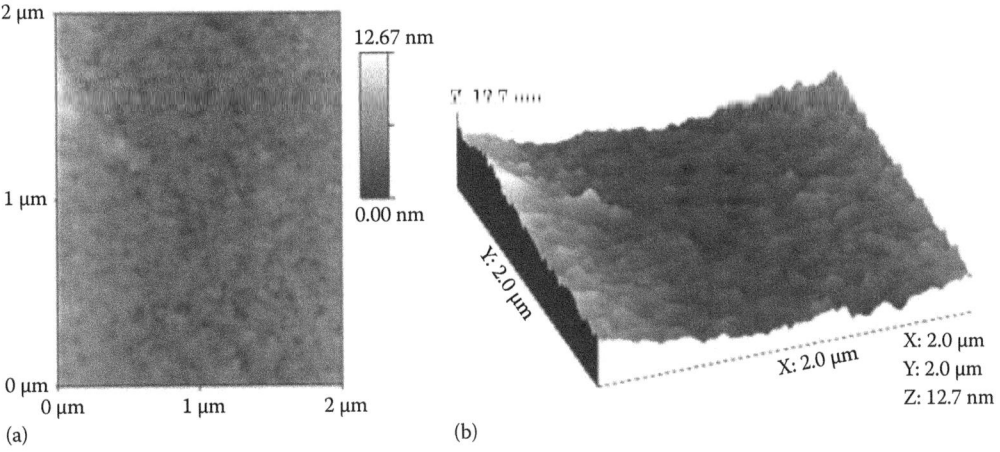

Figure 20.2 Atomic force microscope (AFM) images of AgNPs: (a) 2D view and (b) 3D view. The depth image of AFM shows the spherical arrangement of silver nanoparticles within the diameter range of 6.3–12.67 nm (Adapted from Gopinath, P.M. et al., *Asian J. Pharm. Sci.*, 10(2), 138, 2015.)

synonymous with nanotechnology, and one might assume that it was inevitable that nanotechnology became possible because of these three instruments (Toumey, 2010; Mody, 2011).

After the discovery of nanomaterials, nanotechnology gradually developed. The proposed technology was adopted in several applications in different areas including biological sciences and also medical sciences, which may be considered as the beautility of nanoscience (Yin et al., 2013; Ezzat et al., 2014).

20.3   Different forms of nanomaterials and their roles

Nanotechnology is rapidly growing with NPs produced and utilized in a wide range of commercial products throughout the world. For example, silver nanoparticles (AgNPs) are used in electronics, biosensing, clothing, food industry, paints, sunscreens, cosmetics, and medical devices. In nature, though there are several nanomaterials, we would like to examine some important nanomaterials and their roles. One of the important NPs is the silver nanoparticle. A large number of *in vitro* studies indicate that AgNPs are toxic to mammalian cells derived from the skin, liver, lung, brain, vascular system, and reproductive organs. Interestingly, some studies have shown that this particle has the potential to induce genes associated with cell cycle progression, DNA damage, and apoptosis in human cells at noncytotoxic doses. Cytotoxicity occurs due to random use of silver nanomaterials (Ahamed et al., 2010). Nanoscience has been established as a new interdisciplinary science that can be defined as a whole knowledge on fundamental properties of nanosized materials and thus observed as a tremendous effect on biological and chemical applications (Makwana et al., 2014). Inorganic NPs find applications in therapeutic agents, electrochemical biosensors, and as catalysts for removal of water pollutants. Moreover, silver nanoparticles have gained significant interest over the years due to their remarkable optical properties and have been used as sensors for various metal ions (Makwana et al., 2014). Silver nanoparticles showed higher antimicrobial efficiency compared to silver salts because of their extremely large surface area, with diameters generally smaller than 100 nm containing 20–15,000 silver atoms, providing better contact with microorganisms (Makwana et al., 2014).

Nanosized particles/fillers, both inorganic and organic materials, have unique chemical, physical, and biological functions and have been extensively studied as biomaterials or biofunctional materials. Nanocomposite hydrogels, which combine the advantages of both nanofillers and hydrogel matrices, may result in improved mechanical and biological properties and find potential biomedical applications. This chapter reviews recent developments in the synthesis, preparation, and characterization of nanocomposite hydrogels and their biomedical applications, such as drug delivery matrices and tissue engineering scaffolding (Song et al., 2015). It is a known fact that silver nanoparticles possess one positive charge. Naturally, this positive charge may act on negatively charged particles. Thus, this property of silver nanoparticles can exhibit one antibacterial property (Liu et al., 2010). Silver nanoparticles always perform their size- and surface-area-dependent functions on bacterial biofilms. These processes can save fruits, vegetables, herbs, breads, cheeses, soups, sauces, meats, etc., from bacterial contamination (Patil et al., 2011). The growth of microbes on textiles during use and storage negatively affects the wearer as well as the textile itself. The detrimental effects can be controlled by durable antimicrobial finishing of the textile using broad-spectrum biocides or by incorporating the biocide into synthetic fibers during extrusion. The application methods of antimicrobial agents and some of the most recent developments in antimicrobial treatments of textiles use various active agents such as silver, quaternary ammonium salts, polyhexamethylene biguanide, triclosan, chitosan, dyes and regenerable N-halamine compounds, and peroxy acids (Gao and Cranston, 2008). Surface-enhanced Raman scattering schemes have been employed to detect and identify small molecules like nucleic acids, lipids, peptides, and proteins for *in vivo* cellular sensing (Bantz et al., 2011).

The therapeutic use of gold can be traced back to Chinese medical practices in 2500 BC. Red colloidal gold is still used in Indian Ayurveda medicine for rejuvenation and revitalization during old age under the name of Swarna Bhasma ("Swarna" meaning gold, "Bhasma" meaning ash). Gold also has a long history of use in the Western world as nerving, a substance that could revitalize people suffering from nervous conditions. In the sixteenth century, gold was recommended for the treatment of epilepsy. In the beginning of the nineteenth century, gold was used in the treatment of syphilis. Following the discovery of the bacteriostatic effect of gold cyanide toward the tubercle bacillus by Robert Koch, gold-based therapy for tuberculosis was introduced in the 1920s (Daniel and Astruc, 2004). The major clinical uses of gold compounds are in the treatment of rheumatic diseases including psoriasis, juvenile arthritis, palindromic rheumatism, and discoid lupus erythematosus. Au particles are particularly and extensively exploited in organisms because of their biocompatibility (Li et al., 2014). Gold nanoparticles (AuNPs) generally are considered to be biologically inert but can be engineered to possess chemical or photothermal functionality. In near-infrared (NIR) irradiation, Au-based nanomaterials, Au nanospheres, Au nano-cages, and Au nanorods with characteristic NIR absorption can destroy cancer cells and bacteria via photothermal heating (Thirumurugan and Kaur, 2013). Au-based NPs can be combined with photosensitizers for photodynamic antimicrobial chemotherapy. Au nanorods conjugated with photosensitizers can kill methicillin-resistant *Staphylococcus aureus* (MRSA) by photodynamic antimicrobial chemotherapy and NIR photothermal radiation (Daniel and Astruc, 2004; Li et al., 2014). A hydrophilic photosensitizer, toluidine blue O, was conjugated on the surface of Au nanorods for photodynamic antimicrobial chemotherapy. Au nanorods served as both photodynamic and photothermal agents, and they inactivated MRSA. The combined effect of photodynamic antibacterial chemotherapy (PACT) and hyperthermia has enhanced the antimicrobial effect of AuNP. The study clearly showed that gold nanorods conjugated with a hydrophilic photosensitizer such as toluidine blue O act as dual-function agents in photodynamic inactivation and hyperthermia against MRSA. Light-absorbing AuNPs conjugated with specific antibodies have also been exploited to photothermally kill *S. aureus*. AuNPs have attracted the interest of scientists for over a century but research in this field has considerably accelerated since 2000 with the synthesis of numerous 1D, 2D, and 3D shapes as well as hollow AuNP structures. Recent studies have focused on functionalizing AuNPs as photothermal agents for hyperthermically killing pathogens (Li et al., 2014). The cancer biomarker can be detected by gold plasmatic nanodevices (Perozziello et al., 2014).

The efficacy of the antibacterial activity of AuNPs can be increased by adding anti-biotics. The antimicrobial activity of the antibiotic vancomycin was enhanced by coating with AuNP against vancomycin-resistant *Enterococci* (Li et al., 2014). The coating of amino-glycoside antibiotics with AuNPs has an antibacterial effect on a range of gram-positive and gram-negative bacteria (Table 20.1). Cofactor (a second-generation β-lactam antibiotic)–reduced AuNPs have potent antimicrobial activity on both gram-positive (*S. aureus*) and gram-negative bacteria (*Escherichia coli*) compared to cofactor and AuNPs alone. Further, the AuNPs generate holes in the cell wall, resulting in the leakage of cell contents and cell death. It is also possible that AuNPs bind to the DNA of bacteria and inhibit the uncoiling and transcription of DNA (Li et al., 2014). Recently, bimetallic NPs have received consider-able attention for their unique optical, magnetic, and catalytic properties, which are very different from those of their monometallic NP components. Among the bimetallic NPs, gold–palladium (Au–Pd) is one of the most attractive systems because of its promising use as a catalyst in CO oxidation, vinyl acetate monomer synthesis, hydrodechlorination of CClF2, hydrogenation of hydrocarbon, cyclotrimerization of acetylene, and others for biological application (Ding et al., 2010). Moreover, nanopores can help in the DNA analysis of gold nanodrills.

AuNPs can be used to coat a wide variety of surfaces, for instance, implants and fabrics for treatment of wounds and glass surface maintenance. Thus, they help to maintain the glass hygienic conditions in homes, in hospitals, and also in other places. As they possess a high surface area, they accelerate biological reactions, and thus their possibility to disinfect has been observed (Mironava et al., 2013). It has been observed that the conjugant of Fe_3O_4 and AuNPs can inhibit the multiplication of *E. coli* (Chatterjee et al., 2011).

20.4   Titanium dioxide nanoparticles

TiO_2 exhibits its photocatalytic activity with strong oxidizing power when illuminated with UV light at wavelength of less than 385 nm. TiO_2 particles catalyze the killing of bacteria on illumination by near-UV light (Stepanov et al., 2013; Sagadevan et al., 2014; Vasantharaja et al., 2015). There are also studies on bactericidal activity of nitrogen-doped metal oxide nanocatalysts on *E. coli* biofilms and on the photocatalytic oxidation of biofilm components on TiO_2-coated surfaces (Table 20.1). The TiO_2 photocatalyst acts as an alternative means of self-disinfecting materials. Though toxic effects are there, of course, that depends on the dose and the surface area of the said NP (Vasantharaja et al., 2015).

20.5   Zinc oxide nanoparticles

Nanotechnology has played an extremely important role in the design, synthesis, and characterization of various new and novel energy materials and catalysts for processing fuels from fossil fuel resources such as coal, petroleum, and natural gas. Today, fossil fuels still account for 90% of the world's energy consumption, and their use is expected to peak around the year 2050 (Liu et al., 2010). Thus, the use of nanotechnology to develop a suite of sustainable energy production schemes is one of the most important scientific challenges of the twenty-first century. The challenge is to design, synthesize, and characterize new functional nanomaterials with controllable sizes, shapes, and/or structures. Moreover, the goal of nanotechnology is to build nanodevices that are intelligent, multifunctional, exceptionally small, and extremely sensitive and have low power consumption. When a nanodevice is required for applications such as *in vivo* biomedical sensors, a nanoscale power source is required (Liu et al., 2010; Pan et al., 2010). Nanosized zinc oxide particles are usually used in the cotton and wool fabrics industries (Becheri et al., 2008). ZnO NPs cross-interact with a critical tumor-suppressive process and can protect pathways like apoptosis, senescence, and cell cycle progression. ZnO NPs also induce the endogenous genetic, transcriptomic, and proteomic landscape of the target cells (Ng et al., 2011). Moreover, the nano–ZnO-coated materials are being used and exhibit higher activity than fabrics and were significantly higher than the bulk ZnO (Yadav et al., 2006). ZnO nanowires (NWs) and nanobelts have been widely studied as a key 1D oxide nanomaterial for numerous applications. Due to the unique piezoelectric and semiconducting coupled properties, a range of novel nanodevices of ZnO have been developed, such as nanogenerators and 6,9-piezoelectric field effect transistors (Hu et al., 2010). ZnO NWs have attracted a great deal of interest because of their unique semiconducting, piezoelectric, biocompatible, and optoelectronic properties, which are fundamental for their application in electronics, optoelectronics, biology, environmental science, and energy (Yang et al., 2014). ZnO is a piezoelectric material with a wide direct bandgap of 3.37 eV and a large exciting binding energy of 60 meV (Hu et al., 2010). It has been demonstrated to have enormous applications in building up electronic, optoelectronic, electrochemical, and electromechanical nanodevices, such as UV lasers, light-emitting diodes (Hu et al., 2010), field emission devices, solar cells, high-performance nanosensors, piezoelectric nanogenerators, and nanopiezotronics (Xu et al., 2010). Thus, ZnO is used in fabricating flexible fiber nanogenerators that can be used for smart shirts, flexible electronics, and medical applications (Li and Wang, 2010). It has been reported that the gel-combustion method helps to prepare the zinc oxide NPs, which exhibit the antimicrobial activity on *Klebsiella* sp., *E. coli*, *Pseudomonas aeruginosa*, and *S. aureus* (Ramasami et al., 2015) (Table 20.1).

20.6   Magnesium oxide and other nanoparticles

Magnesium hydroxide, $Fe_2O_3^-$ (Kumari et al., 2012), and other NPs like magnesium oxide (mesoporous), sulfur dioxide (SO_2), nitrogen oxide (NO), and calcium oxide and magnetic γ-Fe_2O_3 NPs coated with poly-L-cysteine for chelation of As(III), Cu(II), Cd(II), Ni(II), Pb (II), and Zn(II) all are metal-binding proteins and can detect biomarker proteins in the physiological system (White et al., 2009).

20.7   Copper oxide nanoparticles

Among all the metal oxides, copper oxide nanomaterials have attracted more attention due to their unique properties. Cuprous oxide (Cu_2O) is a p-type semiconductor with a direct bandgap of 2.17 eV (Akimoto et al., 2006). In recent years, there is a growing interest to synthesize Cu_2O nanostructures not only for the development of synthetic strategies but also for the examination of their sensing, catalytic, electrical, and surface properties. It has been suggested that CuO may find potential application as an antimicrobial agent (Table 20.1) as it can be prepared with an extremely high surface area (Gopalakrishnan et al., 2012).

20.8   Iron nanoparticles and aluminum nanoparticles

The effects of iron oxide and tetroxide NPs over biofilm formation on different biomaterial surfaces and pluronic-coated surfaces were examined. There was significant reduction in bacterial adhesion on pluronic-coated surfaces compared to other surfaces. Subsequently, bacteria were allowed to grow for 24 h in the presence of different concentrations of iron oxide NPs. A significant reduction in biofilm growth was observed in the presence of the highest concentration of iron oxide NPs on pluronic-coated surfaces compared to other surfaces. Therefore, a combination of polymer brush coating and iron oxide NPs could show a significant reduction in biofilm formation (Thukkaram et al., 2014). Moreover, aluminum oxide NPs have a wide range of applications in industrial and personal care products. The growth inhibitory effect of alumina NPs over a wide concentration range (10–1000 µg/mL) was observed on *E. coli* (Thukkaram et al., 2014) (Table 20.1).

20.9   Magnetosomes nanoparticles

Magnetosomes are membranous prokaryotic structures present in magnetotactic bacteria (MTB). They contain 15–20 magnetite crystals (about 20–110 nm) that together act like a compass needle to orient MTB in geomagnetic fields. Magnetoaerotaxis of MTB is based on intracellular membrane-enclosed magnetite (Fe_3O_4) or greigite (Fe_3S_4) crystals, which are the magnetosomes. The chain-like arrangement of magnetosomes allows adding up of all the individual magnetic moments, which results in a dipole strong enough to align the cell within Earth's weak magnetic field. In the alphaproteobacterium *Magnetospirillum gryphiswaldense*, magnetosome biomineralization and chain assembly are under genetic control. Each magnetite crystal within a magnetosome is surrounded by a lipid bilayer, and specific soluble and transmembrane proteins are sorted to the membrane. Recent research has shown that magnetosomes are invaginations of the inner membrane and not freestanding vesicles (Katzmann et al., 2013). Magnetite-bearing magnetosomes have also been found in eukaryotic magnetotactic algae, with each cell containing several thousand crystals. Overall, magnetosome crystals have high chemical purity, narrow size ranges, and species-specific crystal morphologies and exhibit specific arrangements within the cell. These features indicate that the formation of magnetosomes is under precise biological control and is mediated through biomineralization. MTB usually mineralize either iron oxide magnetosomes, which contain crystals of magnetite (Fe_3O_4), or iron sulfide magnetosomes, which contain crystals of greigite (Fe_3S_4). Magnetosome crystals are typically 35–120 nm long, which makes them single domain. Single-domain crystals have the maximum possible magnetic moment per unit volume for a given composition (Komeili et al., 2006).

In the course of routine investigations of seawater mud samples, Blakemore discovered bacteria (1975), which surprisingly showed a magnetotactic behavior, that is, they were moving within a magnetic field toward a magnetic pole. From the freshwater, he isolated a type of *Spirillum*, which he called *Aquaspirillum magneto tacticum*. The magnetotaxis is enabled by special organelles, the so-called magnetosomes, which turned out to be crystals consisting of magnetite (Fe_3O_4) with a parallelepiped or an orthohexagonal prismatic structure storing high amounts of iron. The highest amount of iron ever found was 0.27% in *Desulfotomaculum orientis*, while *E. coli* merely contains 0.014% iron (Oberhack et al., 1987).

Magnetosome bacteria (MB) are generally found in the oxic–anoxic transition zone and exhibit two distinct types of responses to applied magnetic fields. The first is axial magnetoaerotaxis whereby the bacteria use the magnetic field as an axis of swimming with no preference for either pole. Most MB, by contrast, is polar magnetoaerotactic bacteria, meaning that they persistently swim toward only one pole of the magnet. The vast majority of MB in the Northern Hemisphere are north seeking, and those in the Southern Hemisphere are south seeking. This default swimming pattern of polar MB points them toward the bottom of their aquatic habitats where the oxygen levels are low (Araujo et al., 2015). MTB produce intracellular organelles called magnetosomes that are magnetic NPs composed of magnetite (Fe_3O_4) or greigite (Fe_3S_4) enveloped by a lipid bilayer. The synthesis of a magnetosome is through a genetically controlled process in which the bacterium has control over the composition, direction of crystal growth, and the size and shape of the mineral crystal. As a result of this control, magnetosomes have narrow and uniform size ranges, relatively specific magnetic and crystalline properties, and an enveloping biological membrane. These features are not observed in magnetic particles produced abiotically, and thus magnetosomes are of great interest in biotechnology. Most currently described MTB have been isolated from saline or brackish environments, and the availability of their genomes has contributed to a better understanding and culturing of these fastidious microorganisms (Araujo et al., 2015).

Nanometer-sized magnetic particles are of great interest in biotechnology since they have a large surface area that can be used for anchoring relatively large amounts of specific molecules and can be easily manipulated using an external magnetic field.

These magnetic particles including magnetosomes can be bound to proteins, cells, viruses, or genes of interest that can then be subsequently separated using magnetic techniques. The particles most often used for these types of studies consist of iron oxides, especially magnetite and maghemite (γ-Fe_2O_3), that are more stable than iron sulfides such as greigite. They have been used in various biomedical applications such as immunoassays, cell separation, hyperthermia protocols (treatment of cancer by localized heating), drug delivery, and nuclear magnetic resonance (Ana-Carolina et al., 2015).

20.10 Use of nanomaterials and their risk management

As we have discussed previously, the use of nanomaterials in different sectors vary properly. It is very amazing that quantum phenomena occur at the scale of single atoms and small molecules and nanotechnology will bring up new materials with improved and novel physical, chemical, and biological properties (Duncan, 2005). New architectures from individual biomolecules and biomacromolecules will possess novel functions. Here again, we will discuss in short form the uses of important nanomaterials. The silver nanoparticle is the best one, as its surface-to-volume ratio is high. Moreover, we know nanoscale materials have been used for decades in applications ranging from window glass and sunglasses to car bumpers and paints. Now, nanotechnology is part of scientific disciplines like chemistry, biology, electronics, physics, and engineering (Duncan, 2005). This technology is also being used in manufacturing, computer chips, medical diagnosis, health care, energy, biotechnology, space exploration and security, and so on. Thus, nanotechnology is expected to have a significant impact on our society. Day by day, the potential of nanotechnology is gradually applied in medical sciences. We can expect to see an increase of environmental risk due to the use of engineered nanoscale materials in nature (Dreher, 2004; Gao et al., 2015). Toxicity aspects of nanotechnology are not sufficiently focused on broader issues of interest to humanity such as resources (water, energy, and food) and the environment (Renn and Roco, 2006; Xiao et al., 2015). Naturally, the knowledge of potential human and environmental exposures combined with dose factor response to toxicity information is more necessary to determine the real or perceived risk factors of nanomaterials (Renn and Roco, 2006; Gao et al., 2015; Xiao et al., 2015). Risk analysis is broadly defined here to include risk assessment, risk characterization, risk communication, risk management, and policy relating to risk. Recent interests include risks to human health and the environment, both built and natural. We consider in this chapter the threats from physical, chemical, and biological agents and from a variety of human activities as well as natural events (Gao et al., 2015; Xiao et al., 2015). We may assume that nanomaterials may create some toxicity due to inhalation through oral or dermal routes (Renn and Roco, 2006; Xiao et al., 2015). This key concept of nanotoxicology, including the significance of dose, dose rate, dose metric, and biokinetics, is very essential in the coming decade. This will help us to know the characterization of critical physicochemical properties of the used NPs, specifically surface properties that influence their biological/toxicological properties of the cell and biokinetics. If we know the problems with the use of NPs in the environment, it will help us to solve them in future. Applications of nanotechnology are already having an impact on a number of products in the present decade. The nanometer range (generally below 100 nm) particles are specifically exploited with their functional properties. Thus, with new electronic, magnetic, optical properties, catalytic activities, solubility, and transport properties, nanosubstances execute a great role in nature in the third decade. Therefore, clear knowledge-based legislation will establish concise guidelines for the use of nanomaterials.

20.11   Conclusion

Currently, with the advance of nanotechnology, there is a possibility of using nanosensors to track the cold chain and thus make the storage system more efficient. The insertion points for nanotechnology in sensing applications are many. Nanotechnology has the potential to enable the vision of future sensor technology and sensing systems. The high surface-to-volume ratio of NWs and other nonmaterials will lead to increased sensitivity of the transducer in the sensor. Market potentials of nanotechnology in the energy conversion sector will mainly arise in the fields of thin-layer solar cells and fuel cell technology. Apart from potentially low production costs and a more flexible scalability, thin-layer solar cells bear the advantage of more consistent performance—even at fluctuating temperatures and suboptimum radiation conditions (angle of incidence, clouds). This opens up new applications, such as flat roofs or extensive solar plants.

The surface biocides are agents intended to help maintain the hygienic condition of the food contact surface by preventing or reducing microbial growth and helping "cleanability." There should be no preservative effect on the food. Surface biocides may have a useful function in food-processing equipment. Since nanomaterials have a very high surface area to mass ratio, materials such as nano–silver zinc oxide or magnesium oxide may have an effective action as a surface biocide in food contamination. The effect of polymerization or processing on the size or shape or surface chemistry of NPs has to be developed for our survival. Now, it has been established that nanomaterials could elevate the migration of nonnano-ingredients or could cause an undesirable reaction of the product during the processing and fabrication of food-packaging materials, such as in human physiology. Still, we do not clearly understand the impact of nanomaterials in waste disposal streams. More research is needed to understand risks for the sake of sustainable use of nanotechnology in different fields.

Acknowledgments

We thank Dr. H. Raja Naika, Department of Studies and Research in Environmental Science, Bharat Ratna; Prof. C.N.R. Rao Block, Tumkur University, India; and Dr. K. Palanichelvam, Department of Biotechnology, Kalasalingam University, Krishnankoil, Tamil Nadu, India, for their encouragement in writing this article.

References

Ahamed, M., AlSalhi, M.S., and Siddiqui, M.K.J. 2010. Silver nanoparticle applications and human health. *Clinica Chimica Acta*, 411(23), 1841–1848.

Akimoto, K., Ishizuka, S., Yanagita, M., Nawa, N., Paul, G.K. and Sakurai, T. 2006. Thin film deposition of Cu_2O and application for solar cells. *Solar Energy*, 80, 715–722.

Araujo, A.C.V., Abreu, F., Silva, K.T., Bazylinski, D.A., and Lins, U. 2015. Magnetotactic bacteria as potential sources of bioproducts. *Marine Drugs*, 13(1), 389–430.

Bantz, K.C., Meyer, A.F., Wittenberg, N.J., Im, H., Kurtuluş, Ö., Lee, S.H., and Haynes, C.L. 2011. Recent progress in SERS biosensing. *Physical Chemistry Chemical Physics*, 13(24), 11551–11567.

Becheri, A., Dürr, M., Nostro, P.L., and Baglioni, P. 2008. Synthesis and characterization of zinc oxide nanoparticles: Application to textiles as UV-absorbers. *Journal of Nanoparticle Research*, 10(4), 679–689.

Bhattacharyya, A. 2009. Nanoparticles-from drug delivery to insect pest control. *Akshar*, 1(1), 1–7.

Bhattacharyya, A., Barik, B., Kundu, P., Mandal, D.N., Das, A., Sen, P., Rao, C.V., and Mandal, S. 2007. Bioactivity of nanoparticles and allelochemicals on stored grain pest-*Sitophilus oryzae* (L.) (Coleoptera: Curculionidae). *27th Annual Session of the Academy of Environmental Biology and National Symposium of Biomarkers of Environmental Problems*, October 26–28, 2007, Jointly organized by Department of Zoology and Department of Environmental Sciences, Charan Singh University, Meerut, Uttar Pradesh, India, 28pp.

Bhattacharyya, A., Chandrasekhar, R., Chandra, A.K., Epidi, T.T., and Prakasham, R.S. 2014. Application of nanoparticles in sustainable agriculture: Its current status. *International Book Mission*, Academic Publisher, Manhattan, KS, pp. 429–448.

Bhattacharyya, A., Datta, P.S., Chaudhuri, P., and Barik, B.R. 2011. Nanotechnology—A new frontier for food security in socio economic development. *Disaster, Risk and Vulnerability Conference 2011 School of Environmental Sciences*, Mahatma Gandhi University, Kerala, India in association with the *Applied Geoinformatics for Society and Environment*, Stuttgart, Germany, March 12–14, 2011, pp. 116–120.

Blakemore, R. 1975. Magnetotactic bacteria. *Science*, 190(4212), 377–379.

Cauerhff, A. and Castro, G.R. 2013. Bionanoparticles, a green nanochemistry approach. *Electronic Journal of Biotechnology*, 16(3), 11–11.

Chatterjee, S., Bandyopadhyay, A., and Sarkar, K. 2011. Effect of iron oxide and gold nanoparticles on bacterial growth leading towards biological application. *Journal of Nanobiotechnology*, 9(34), 1.

Daniel, M.C. and Astruc, D. 2004. Gold nanoparticles: Assembly, supramolecular chemistry, quantum-size-related properties, and applications toward biology, catalysis, and nanotechnology. *Chemical Reviews*, 104(1), 293–346.

Dhanasekaran, D., Latha, S., Saha, S., Thajuddin, N., and Panneerselvam, A. 2013. Extracellular biosynthesis, characterisation and in-vitro antibacterial potential of silver nanoparticles using *Agaricus bisporus*. *Journal of Experimental Nanoscience*, 8(4), 579–588.

Ding, Y., Fan, F., Tian, Z., and Wang, Z.L. 2010. Atomic structure of Au–Pd bimetallic alloyed nanoparticles. *Journal of the American Chemical Society*, 132(35), 12480–12486.

Dreher, K.L. 2004. Health and environmental impact of nanotechnology: Toxicological assessment of manufactured nanoparticles. *Toxicological Sciences*, 77, 3–5.

Duncan, R. 2005. The dawning era of polymer therapeutics. *Nature Reviews Drug Discovery*, 2(5), 347–360.

Ezzat, A., Abdelhamid, A.O., Amin, A.I., El Azeem, A., Amal, S., Fouda, M.F.R., and Mohammed, D.M. 2013. Biochemical studies on the effect of nano particles of some nutrients on apoptosis modulation of breast cancer cells in experimental animals. *Journal of Applied Sciences Research*, 9(1), 658–665.

Feris, K., Otto, C., Tinker, J., Wingett, D., Punnoose, A., Thurber, A., and Pink, D. 2009. Electrostatic interactions affect nanoparticle-mediated toxicity to gram-negative bacterium *Pseudomonas aeruginosa* PAO1. *Langmuir*, 26(6), 4429–4436.

Friedman, A., Blecher, K., Sanchez, D., Tuckman-Vernon, C., Chouchulla, D., Friedman, J.M., Martinez, L.R., and Nosanchuk, J.D. 2011. Susceptibility of gram-positive and -negative bacteria to novel nitric oxide-releasing nanoparticle technology. *Virulence*, 2, 217–221.

Gajjar, P., Pettee, B., Britt, D.W., Huang, W., Johnson, W.P., and Anderson, A.J. 2009. Antimicrobial activities of commercial nanoparticles against an environmental soil microbe, *Pseudomonas putida* KT2440. *Journal of Biological Engineering*, 3(9), 1–13.

Gao, Y. and Cranston, R. 2008. Recent advances in antimicrobial treatments of textiles. *Textile Research Journal*, 78(1), 60–72.

Gao, Y., Sarfraz, M.K., Clas, S.D., Roa, W., and Löbenberg, R. 2015. Hyaluronic acid-tocopherol succinate-based self-assembling micelles for targeted delivery of rifampicin to alveolar macrophages. *Journal of Biomedical Nanotechnology*, 11(8), 1312–1329.

Gopalakrishnan, K., Ramesh, C., Ragunathan, V., and Thamilselvan, M. 2012. Antibacterial activity of Cu_2O nanoparticles on *E. coli* synthesized from tridax procumbens leaf extract and surface coating with polyaniline. *Digest Journal of Nanomaterials and Biostructures*, 7, 833–839.

Gopinath, P.M., Narchonai, G., Dhanasekaran, D., Ranjani, A., and Thajuddin, N. 2015. Mycosynthesis, characterization and antibacterial properties of AgNPs against multidrug resistant (MDR) bacterial pathogens of female infertility cases. *Asian Journal of Pharmaceutical Sciences*, 10(2), 138–145.

Hamouda, T., Hayes, M., Cao, Z., Tonda, R., Johnson, K., Craig, W., Brisker, J., and Baker. J. 1999. A novel surfactant nanoemulsion with broad-spectrum sporicidal activity against *Bacillus* species. *The Journal of Infectious Diseases*, 180, 1939–1949.

Hetrick, E.M., Shin, J.H., Paul, H.S., and Schoenfisch, M.H. 2009. Anti-biofilm efficacy of nitric oxide-releasing silica nanoparticles. *Biomaterials*, 30(14), 2782–2789.

Hu, Y., Chang, Y., Fei, P., Snyder, R.L., and Wang, Z.L. 2010. Designing the electric transport characteristics of ZnO micro/nanowire devices by coupling piezoelectric and photoexcitation effects. *ACS Nano*, 4(2), 1234–1240.

Jiang, W., Mashayekhi, H., and Xing, B. 2009. Bacterial toxicity comparison between nano-and micro-scaled oxide particles. *Environmental Pollution*, 157(5), 1619–1625.

Jones, N., Ray, B., Ranjit, K.T., and Manna, A.C. 2008. Antibacterial activity of ZnO nanoparticle suspensions on a broad spectrum of microorganisms. *FEMS Microbiology Letters*, 279(1), 71–76.

Juan, L., Zhimin, Z., Anchun, M., Lei, L., and Jingchao, Z. 2010. Deposition of silver nanoparticles on titanium surface for antibacterial effect. *International Journal of Nanomedicine*, 5, 261.

Katzmann, E., Eibauer, M., Lin, W., Pan, Y., Plitzko, J.M., and Schüler, D. 2013. Analysis of magnetosome chains in magnetotactic bacteria by magnetic measurements and automated image analysis of electron micrographs. *Applied and Environmental Microbiology*, 79(24), 7755–7762.

Kennedy, D.C., Orts-Gil, G., Lai, C.H., Müller, L., Haase, A., Luch, A., and Seeberger., P.H. 2014. Carbohydrate functionalization of silver nanoparticles modulates cytotoxicity and cellular uptake. *Journal of Nanobiotechnology*, 12, 59.

Khan, A.U. 2012. Medicine at nanoscale: A new horizon. *International Journal of Nanomedicine*, 7, 2997.

Khan, Z., Hussain, J.I., Kumar, S., Hashmi, A.A., and Malik, M.A. 2011. Silver nanoparticles: Green route, stability and effect of additives. *Journal of Biomaterials and Nanobiotechnology*, 2(04), 390.

Kim, J.S., Kuk, E., Yu, K.N., Kim, J.H., Park, S.J., Lee, H.J., and Cho, M.H. 2007. Antimicrobial effects of silver nanoparticles. *Nanomedicine: Nanotechnology, Biology and Medicine*, 3(1), 95–101.

Komeili, A., Li, Z., Newman, D.K., and Jensen, G.J. 2006. Magnetosomes are cell membrane invaginations organized by the actin-like protein MamK. *Science*, 311(5758), 242–245.

Kumar, N., Shah, V., and Walker, V.K. 2011. Perturbation of an arctic soil microbial community by metal nanoparticles. *Journal of Hazardous Materials*, 190(1), 816–822.

Kumari, M., Rajak, S., Singh, S.P., Kumari, S.I., Kumar, P.U., Murty, U.S., and Rahman, M.F. 2012. Repeated oral dose toxicity of iron oxide nanoparticles: Biochemical and histopathological alterations in different tissues of rats. *Journal of Nanoscience and Nanotechnology*, 12(3), 2149–2159.

Leid, J.G., Ditto, A.J., Knapp, A., Shah, P.N., Wright, B.D., Blust, R., and Cope, E.K. 2012. In vitro antimicrobial studies of silver carbene complexes: Activity of free and nanoparticle carbene formulations against clinical isolates of pathogenic bacteria. *Journal of Antimicrobial Chemotherapy*, 67(1), 138–148.

Li, N., Zhao, P., and Astruc, D. 2014. Anisotropic gold nanoparticles: Synthesis, properties, applications, and toxicity. *Angewandte Chemie International Edition*, 53(7), 1756–1789.

Li, Z. and Wang, Z.L. 2011. Air/liquid-pressure and heartbeat-driven flexible fiber nanogenerators as a micro/nano-power source or diagnostic sensor. *Advanced Materials*, 23(1), 84–89.

Liu, C.J., Burghaus, U., Besenbacher, F., and Wang, Z.L. 2010. Preparation and characterization of nanomaterials for sustainable energy production. *ACS Nano*, 4(10), 5517–5526.

Maisoneuve, E. and Gerdes, K. 2014. Molecular mechanisms underlying bacterial persisters. *Cell*, 157(3):539–48.

Makwana, B.A., Vyas, D.J., Bhatt, K.D., Jain, V.K., and Agrawal, Y.K. 2014. Highly stable antibacterial silver nanoparticles as selective fluorescent sensor for Fe^{3+} ions. *Spectrochimica Acta Part A: Molecular and Biomolecular Spectroscopy*, 134, 73–80.

McCarthy, T.J., Zeelie, J.J., and Krause, D.J. 1992. The antimicrobial action of zinc ion/antioxidant combinations. *Journal of Clinical Pharmacy and Therapeutics*, 17(1), 51–54.

Mironava, T., Hadjiargyrou, M., Simon, M., and Rafailovich, H. 2013. Gold nanoparticles cellular toxicity and recovery: Adipose derived stromal cells. *Nanotoxicology*, 8, 1–13.

Mody, C. 2011. *Instrumental Community: Probe Microscopy and the Path to Nanotechnology.* MIT Press, Cambridge, MA.

Morones, J.R., Elechiguerra, J.L., Camacho, A., Holt, K., Kouri, J.B., Ramírez, J.T., and Yacaman, M.J. 2005. The bactericidal effect of silver nanoparticles. *Nanotechnology*, 16(10), 2346.

Nair, S., Sasidharan, A., Rani, V.D., Menon, D., Nair, S., Manzoor, K., and Raina, S. 2009. Role of size scale of ZnO nanoparticles and microparticles on toxicity toward bacteria and osteoblast cancer cells. *Journal of Materials Science: Materials in Medicine*, 20(1), 235–241.

Ng, K.W., Khoo, S.P., Heng, B.C., Setyawati, M.I., Tan, E.C., Zhao, X., and Loo, J.S. 2011. The role of the tumor suppressor p53 pathway in the cellular DNA damage response to zinc oxide nanoparticles. *Biomaterials*, 32(32), 8218–8225.

Novak, J.P. and Feldheim, D.L. 2000. Assembly of phenylacetylene-bridged silver and gold nanoparticle arrays. *Journal of the American Chemical Society*, 122, 3979–3980.

Oberhack, M., Süssmuth, R., and Frank, H. 1987. Magnetotactic bacteria from freshwater. *Zeitschrift für Naturforschung*, 42, 300–306.

Padmavathy, N. and Vijayaraghavan, R. 2008. Enhanced bioactivity of ZnO nanoparticles—An antimicrobial study. *Science and Technology of Advanced Materials*, 9(3), 035004.

Pal, S., Tak, Y.K., and Song, J.M. 2007. Does the antibacterial activity of silver nanoparticles depend on the shape of the nanoparticle? A study of the gram-negative bacterium *Escherichia coli*. *Applied and Environmental Microbiology*, 73(6), 1712–1720.

Pan, C., Fang, Y., Wu, H., Ahmad, M., Luo, Z., Li, Q., and Zhu, J. 2010. Generating electricity from biofluid with a nanowire-based biofuel cell for self-powered nanodevices. *Advanced Materials*, 22(47), 5388–5392.

Patil, H.B., Borse, S.V., Patil, D.R., Patil, U.K., and Patil, H.M. 2011. Synthesis of silver nanoparticles by microbial method and their characterization. *Archives of Physical Medicine and Rehabilitation*, 2, 153–158.

Perozziello, G., Candeloro, P., Gentile, F., Nicastri, A., Perri, A., Coluccio, M.L., Adamo, A. et al. 2014. Microfluidics and nanotechnology: Towards fully integrated analytical devices for the detection of cancer biomarkers. *RSC Advances*, 4(98), 55590–55598.

Qi, L., Xu, Z., Jiang, X., Hu, C., and Zou, X. 2004. Preparation and antibacterial activity of chitosan nanoparticles. *Carbohydrate Research*, 339(16), 2693–2700.

Ramasami, A.K., Naika, H.R., Nagabhushana, H., Ramakrishnappa, T., Balakrishna, G.R., and Nagaraju, G. 2015. Tapioca starch: An efficient fuel in gel-combustion synthesis of photocatalytically and anti-microbially active ZnO nanoparticles. *Materials Characterization*, 99, 266–276.

Reddy, K.M., Feris, K., Bell, J., Wingett, D.G., Hanley, C., and Punnoose, A. 2007. Selective toxicity of zinc oxide nanoparticles to prokaryotic and eukaryotic systems. *Applied Physics Letters*, 90(21), 213902.

Renn, O. and Roco, M.C. 2006. Nanotechnology and the need for risk governance. *Journal of Nanoparticle Research*, 8(2), 153–191.

Ruparelia, J.P., Chatterjee, A.K., Duttagupta, S.P., and Mukherji, S. 2008. Strain specificity in antimicrobial activity of silver and copper nanoparticles. *Acta Biomaterialia*, 4(3), 707–716.

Sagadevan, S., Savitha, S., and Preethi, R. 2014. Beneficial applications of nanoparticles in medical field—A review. *International Journal of PharmTech Research*, 6(5):1712–1717.

Simon-Deckers, A., Loo, S., Mayne-L'hermite, M., Herlin-Boime, N., Menguy, N., Reynaud, C., and Carrière, M. 2009. Size-, composition- and shape-dependent toxicological impact of metal oxide nanoparticles and carbon nanotubes toward bacteria. *Environmental Science and Technology*, 43(21), 8423–8429.

Sinha, R., Karan, R., Sinha, A., and Khare, S.K. 2011. Interaction and nanotoxic effect of ZnO and Ag nanoparticles on mesophilic and halophilic bacterial cells. *Bioresource Technology*, 102(2), 1516–1520.

Sondi, I. and Salopek-Sondi, B. 2004. Silver nanoparticles as antimicrobial agent: A case study on *E. coli* as a model for Gram-negative bacteria. *Journal of Colloid and Interface Science*, 275(1), 177–182.

Song, F., Li, X., Wang, Q., Liao, L., and Zhang, C. 2015. Nanocomposite hydrogels and their applications in drug delivery and tissue engineering. *Journal of Biomedical Nanotechnology*, 11(1), 40–52.

Stepanov, A.L., Xiao, X., Ren, F., Kavetskyy, T. and Osin, Y.N. 2013. Catalytic and Biological sensitivity of TiO2 and SiO2 matrices with silver nanoparticles created by ion implantation: A review. *Reviews on Advanced Materials Science*, 34, 107–122.

Subbiahdoss, G., Sharifi, S., Grijpma, D.W., Laurent, S., van der Mei, H.C., Mahmoudi, M., and Busscher, H.J. 2012. Magnetic targeting of surface-modified superparamagnetic iron oxide nanoparticles yields antibacterial efficacy against biofilms of gentamicin-resistant staphylococci. *Acta Biomaterialia*, 8(6), 2047–2055.

Taylor, E.N. and Webster, T.J. 2009. The use of superparamagnetic nanoparticles for prosthetic biofilm prevention. *International Journal of Nanomedicine*, 4, 145.

Thirumurugan, T. and Kaur, K. 2013. Biological synthesis and characterization of gold nanoparticles from pomegranate. *International Journal of Future Biotechnology*, 2(2), 1–11.

Thukkaram, M., Sitaram, S., and Subbiahdoss, G. 2014. Antibacterial efficacy of iron-oxide nanoparticles against biofilms on different biomaterial surfaces. *International Journal of Biomaterials*. http://dx.doi.org/10.1155/2014/716080.

Toumey, C. 2010. 35 atoms that changed the nanoworld. *Nature Nanotechnology*, 5, 239–241.

Tran, N., Mir, A., Mallik, D., Sinha, A., Nayar, S., and Webster, T.J. 2010. Bactericidal effect of iron oxide nanoparticles on *Staphylococcus aureus*. *International Journal of Nanomedicine*, 5, 277.

Tsuang, Y.H., Sun, J.S., Huang, Y.C., Lu, C.H., Chang, W.H.S., and Wang, C.C. 2008. Studies of photokilling of bacteria using titanium dioxide nanoparticles. *Artificial Organs*, 32(2), 167–174.

Vasantharaja, D., Ramalingam, V., and Reddy, G.A. 2015. Oral toxic exposure of titanium dioxide nanoparticles on serum biochemical changes in adult male Wistar rats. *Biomedical Research*, 3, 4.

White, B.R., Stackhouse, B.T., and Holcombe, J.A. 2009. Magnetic γ-Fe$_2$O$_3$ nanoparticles coated with poly-L-cysteine for chelation of As (III), Cu (II), Cd (II), Ni (II), Pb (II) and Zn (II). *Journal of Hazardous Materials*, 161(2), 848–853.

Wiley, B., Sun, Y., Mayers, B., and Xia, Y. 2005. Shape-controlled synthesis of metal nanostructures: The case of silver. *Chemistry: A European Journal*, 11, 454–463.

Wu, B., Huang, R., Sahu, M., Feng, X., Biswas, P., and Tang, Y.J. 2010. Bacterial responses to Cu-doped TiO$_2$ nanoparticles. *Science of the Total Environment*, 408(7), 1755–1758.

Xiao, S., Castro, R., Rodrigues, J., Shi, X., and Tomás, H. 2015. PAMAM dendrimer/pDNA functionalized-magnetic iron oxide nanoparticles for gene delivery. *Journal of Biomedical Nanotechnology*, 11(8), 1370–1384.

Xu, K., Huang, J., Ye, Z., Ying, Y., and Li, Y. 2009.Recent development of nano-materials used in DNA biosensors. *Sensors*, 9, 5534–5557.

Xu, S., Shen, Y., Ding, Y., and Wang, Z.L. 2010. Growth and transfer of monolithic horizontal ZnO nanowire superstructures onto flexible substrates. *Advanced Functional Materials*, 20(9), 1493–1497.

Yadav, A., Prasad, V., Kathe, A.A., Raj, S., Yadav, D., Sundaramoorthy, C., and Vigneshwaran, N. 2006. Functional finishing in cotton fabrics using zinc oxide nanoparticles. *Bulletin of Materials Science*, 29(6), 641–645.

Yamanaka, M., Hara, K., and Kudo, J. 2005. Bactericidal actions of a silver ion solution on *Escherichia coli*, studied by energy-filtering transmission electron microscopy and proteomic analysis. *Applied and Environmental Microbiology* 71, 7589–7593.

Yang, Q., Guo, X., Wang, W., Zhang, Y., Xu, S., Lien, D.H., and Wang, Z.L. 2010. Enhancing sensitivity of a single ZnO micro-/nanowire photodetector by piezo-phototronic effect. *ACS Nano*, 4(10), 6285–6291.

Yin, H.T., Zhang, D.G., Wu, X.L., Huang, X.E., and Chen, G. 2013. *In vivo* evaluation of curcumin-loaded nanoparticles in a A549 xenograft mice model. *Asian Pacific Journal of Cancer Prevention*, 14(1), 409–412.

Yoon, K.Y., Byeon, J.H., Park, J.H., and Hwang, J. 2007. Susceptibility constants of *Escherichia coli* and *Bacillus subtilis* to silver and copper nanoparticles. *Science of the Total Environment*, 373(2), 572–575.

chapter twenty-one

Platinum-based anticancer therapeutics and their mechanistic aspects
An overview

Bhaskar Biswas

Contents

21.1 Introduction

A significantly mounting interest in the design of synthetic metal compounds as drugs and diagnostic agents is currently experimental in the area of scientific investigation, appropriately termed medicinal inorganic chemistry. Empirical evidence for the effectiveness of metal-based therapeutics has existed for centuries, and the use of metals and metal complexes in medicine dates back millennia. Investigations in this area focus mostly on the speciation of metals in biological media based on possible interactions of these metal ions with diverse microbials and biomolecules, in an effort to contribute to future development of new antimicrobials, therapeutics, or diagnostic agents. Metallopharmaceuticals used as anticancer agents, metal-mediated antibiotics, antibacterials, antifungals, antivirals, antiparasitics, antiarthritics, antidiabetics, and radio-sensitizing agents appear in therapeutic medicinal inorganic chemistry (Thomson and Orvig, 2003; Reedijk, 2009; Griffith et al., 2012).

History shows that metal-based drugs and remedies have been known and used since very ancient times. For example, silver was employed in the treatment of wounds and ulcers according to the Greek physician Hippocrates, but its antimicrobial properties had probably been recognized long before because it was used to make vessels for storing liquids in pure form. The ancient Egyptians also knew how to sterilize water with copper. The medical use of gold can be dated back to 2500 BC in China. However, the new era of metal-based medicine started almost five decades ago when cisplatin was shown to inhibit cellular division in *Escherichia coli*, thereby leading to the first studies of its antitumor activity in rats and its assessment as one of the most powerful drugs for use against different types of cancer, although many other novel metal-based drugs are promising and they are attracting growing attention in modern clinical medicine (Figure 21.1). Gold salts and arsenic compounds have been in use for decades in the treatment of rheumatoid arthritis and syphilis, respectively, but studies of cisplatin have definitely shifted the attention of researchers to the pool of transition "heavy" metals as potential therapeutic agents (Rosenberg and Vancamp, 1969; Lippard, 1982; Quiroga et al., 2012).

This review will focus on the platinum-based therapeutics with their potential bioactivity and explore the evidences of metal complexes–protein binding relevant to the drug/diagnostic agent's mechanisms of action.

21.2 Role of metal ions in biological system

Bioinorganic chemistry is an interdisciplinary subject at the interface of inorganic chemistry and biology/biochemistry. Medicinal inorganic chemistry includes the study of both nonessential and essential elements with applications to diagnosis and therapies. Bioinorganic chemistry is an extremely dynamic field, and the functional roles of selected biological inorganic elements include charge balance and electrolytic conductivity (Na, K, Cl); structure

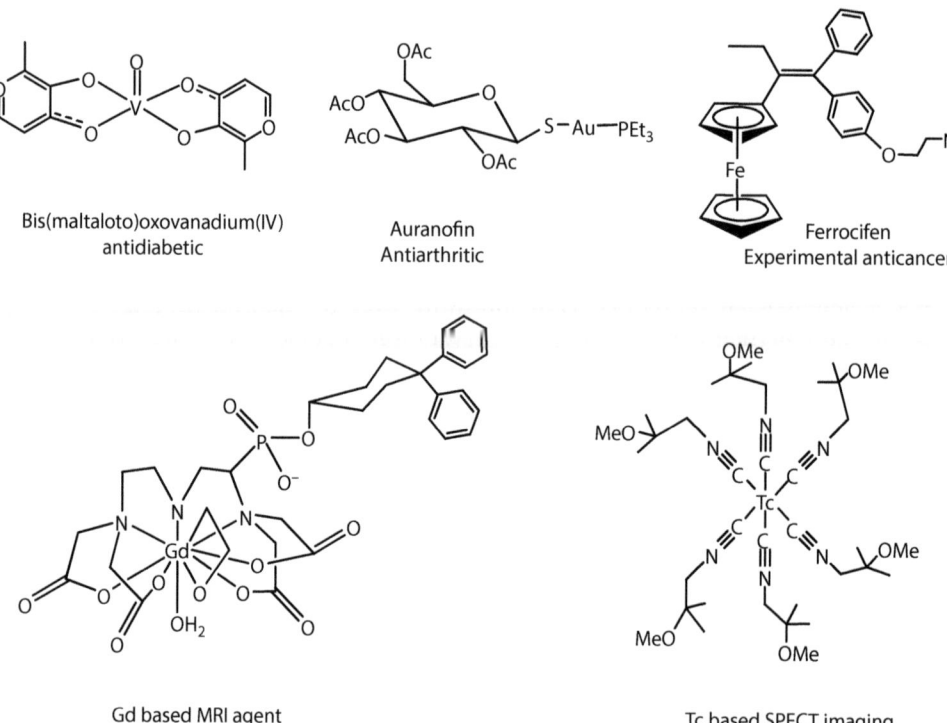

Bis(maltaloto)oxovanadium(IV)
antidiabetic

Auranofin
Antiarthritic

Ferrocifen
Experimental anticancer

Gd based MRI agent

Tc based SPECT imaging

Figure 21.1 Some metal-based potential therapeutics in clinical medicine.

and templating (Ca, Zn, Si, S); signaling (Ca, B, NO); Bronsted acid–base buffering (P, Si, CO); Lewis acid–base catalysis (Zn, Fe, Ni, Mn); electron transfer (Fe, Cu); group transfer such as CH_3, O, and S (V, Fe, Co, Ni, Cu, Mo, W); redox catalysis (V, Mn, Fe, Co, Ni, Cu, W, S, Se); energy storage (H, P, S, Na, K, Fe); and biomineralization (Ca, Mg, Fe, Si, Sr, Cu, P) (Gielen et al., 2005; Norman and Hambley, 2011; Klein and Hambley, 2009).

Lithium carbonate is a major drug for the treatment of manic episodes and for maintenance therapy in bipolar patients. In multicellular organisms, sodium and calcium are extracellular, while potassium and magnesium are largely intracellular (Thompson, 2011). Calcium and magnesium are often metal activators in proteins to which they bind with relatively low affinity. Under appropriate circumstances, these metal ions induce conformational changes in the protein upon binding, and in doing so, they may transmit a signal, for example, the firing of neurons by rapid influx of sodium ions across a cell membrane or the regulation of intracellular functions by calcium-binding proteins such as calmodulin. Bone and teeth are made from calcium phosphate in the form of a mineral hydroxyapatite. Calcium carbonate biominerals such as calcite and aragonite are used for structural support (Alessio, 2011). Nerve cells depend on potassium. Without sodium, our cells can not get the nutrients they need to survive. Sodium also allows our bodies to balance the appropriate amount of water in our blood. Potassium chloride is used to prevent or to treat low blood levels of potassium (hypokalemia). Potassium gluconate is needed for normal functioning of cells, nerve conduction, muscle contraction, kidney function, and acid–base balance. Losartan (1,2-*n*-butyl-4-chloro-1-[*p*-(*o*-1*H*-tetrazol-5-ylphenyl)benzyl]-imidazole-5-methanol monopotassium salt) is a highly selective, orally active, nonpeptide angiotensin II receptor antagonist indicated for the treatment of hypertension (Ochiai, 2008). Ca maintains the cell shape and integrity of membranes, exo- and endocytisis, mytosis, and muscle contraction, causing changes in some proteins/enzymes to modify their function. Calcium is needed by the body for healthy bones, muscles, nervous system, and heart. Calcium carbonate also is used as an antacid to relieve heartburn, acid indigestion, and upset stomach. Calcium citrate, calcium chloride, and calcium gluconate are used to prevent and to treat calcium deficiencies. Mg is needed for the proper growth, formation, and function of our bones and muscles. Mg also controls insulin levels in blood and even helps to prevent depression (Noh et al., 2006).

21.2.1 *Transition metals in biology*

Scientists have shown significant progress in utilization of transition metal complexes as drugs to treat several human diseases like carcinomas, lymphomas, infection control, anti-inflammatory, diabetes, and neurological disorders. Transition metals exhibit different oxidation states and can interact with a number of negatively charged molecules. This activity of transition metals has started the development of metal-based drugs with promising pharmacological application and may offer unique therapeutic opportunities (Gielen and Tiekink, 2005). Vanadium compounds at pharmacological doses display relevant biological actions such as insulin and growth factor mimetic or enhancing effects, as well as osteogenic and cardioprotective activity. On the other hand, depending on the nature of compounds and their concentrations, toxicological actions and adverse side effects may also be shown. The current knowledge and new advances on *in vitro* and *in vivo* effects of inorganic and organically chelated vanadium compounds have been reported (Srivastava and Mehdi, 2005). Vanadium is involved in helping the body convert some foods into energy. It has also been suggested that diabetics may benefit from vanadium when trying to stabilize blood sugar levels. Chromium is a mineral that is essential to humans. Chromium helps regulate sugar levels by interacting with insulin. Chromium picolinate has been used in alternative medicine as an aid to lowering cholesterol or improving the body's use of glucose (sugar). However, chromium supplements are not approved by the FDA for these uses. Manganese is an essential nutrient involved in many chemical processes in the body, including processing of cholesterol, carbohydrates, and protein. It might also be involved in bone formation. A number of iron compounds have been found medically useful. For example, ferrous gluconate, $Fe(C_6H_{11}O_7)_2 \cdot 2H_2O$, and ferric pyrophosphate, $Fe_4(P_2O_7)_xH_2O$, are among the compounds frequently used to treat anemia. Various ferric salts, which act as coagulants, are applied. Iron compounds include hemoglobin, which keeps our blood red. Other essential iron compounds include myoglobin. Iron atoms help join organic molecules (derived from quinoxaline), forming compounds that act as bactericides (killing bacteria) or bacteriostatic agents (preventing bacteria from reproducing) (Tarallo et al., 2010). Ferrous sulfate provides the iron needed by the body to produce red blood cells. Ferrous sulfate comes as regular, coated, and extended-release (long-acting) tablets; regular and extended-release capsules; and oral liquid (syrup, drops, and elixir) to take by mouth. Cobalt is essential in small amount for proper nutrition. Cobalt contained in vitamin B_{12} is important in protein formation and DNA regulation. Cobalt-60, a radioactive isotope, is used as commercial source of high-energy radiation in medicine to destroy cancerous tissue. Cobalt-57 is also used in medicine. It can be used to work out how much vitamin B_{12} is being taken in by the human body. Cobalt-containing drugs are used as cyanide antidote. Cyanide ions will bind to cobalt, which can be supplied in the form of either hydroxocobalamin or dicobalt edetate. Cyanide will also bind to methemoglobin formed after administration of sodium nitrite. Historic uses of copper compounds in medicine are available. Thyroid and immune system health are crucially dependent on copper, and copper deficiency is the most important factor in the development of hyperthyroidism. Women need more copper than men because copper is required for the production of the enzymes that convert progesterone into estrogens. However, more zinc is required by men because zinc helps in the formation of enzyme that converts progesterone into testosterone (Sorenson et al., 1985). Many people wear copper bands to help them with inflammatory disease, such as arthritis. Much of our "dietary" copper actually comes from copper pipes, utensils, and cookware. Copper is important as an electron donor in various biological reactions. Potential benefits of copper gluconate include helping reduce high cholesterol levels in humans, osteoporosis, wound healing, cardiac arrhythmia, hypoglycemia, peripheral vascular disease, osteoarthritis, and rheumatoid arthritis. Evolution has kept stores of copper and iron in excess during the reproductive years because they are so vital to life. But the oxidant damage from these excess stores of metals builds up as we age, and natural selection ceases to act after about age 50 since diseases after that do not contribute to reproductive fitness. Diseases of aging such as Alzheimer's disease, other neurodegenerative diseases, arteriosclerosis, diabetes mellitus, and others may all be contributed to by excess copper and iron. Excess copper along with high-fat diet leads to cognition loss at over three times the normal rate. Inorganic copper in drinking water and in supplements is handled differently than food copper and is therefore more toxic (Dabrowiak, 2010). Zinc is considered to be one of the most important elements of a healthy immune system and is also needed for the growth and repair of tissues throughout our bodies. Zinc oxide topically applied to the skin is used to treat diaper rash, minor burns, severely chapped skin, or other minor skin irritations. Zinc–amino acid chelate provides better absorption (60%–70%), better bioavailability, and better tolerance than zinc salts (Navarro et al., 2010). Gallium mimics ferric iron. Gallium, in the form of intravenously administered gallium nitrate (Ganite®), was approved by the U.S. FDA in 1991 for the treatment of cancer-related hypercalcemia, a disorder that occurs in about 10%–20% of cancer patients. Molybdenum is found in all tissues of the human body but tends to be most concentrated in the liver, kidneys, skin, and bones. Molybdenum is important for transforming sulfur into a usable form in humans. It is required for the proper function of several chemicals in the human body. Metal ions and protein interaction is important regarding binding, stability, and folding. Fe, Cu, and Zn are strongly associated with proteins and form the so-called metalloproteins. For example, ferritin stores iron in the body. Metal cofactors can help catalyze unique chemical reactions and perform specific physiological functions. Fe^{3+}/Fe^{2+} and Cu^{2+}/Cu^+ redox couples play critical roles as cofactors for electron transfer reactions in the catalysis of redox reactions. Some metal ions found deeply buried within proteins interact with the protein and help insure the optimal protein structure and contribute to the stability and appropriate acid–base behavior necessary for the physiological function. For example, the Zn^{2+} ions in Zn fingers are necessary for the adoption of the proper shape of the protein, which allows it to interact with DNA (Tamasi, 2010).

21.3   Platinum drugs in cancer therapy

As cancer remains a major killer in the developed world, a broad spectrum of novel and exciting approaches are being developed and tested. The importance of metal compounds in medicine is undisputed, which can be judged by their use in the treatment of various diseases. In terms of antitumor activity, a wide range of compounds of both transition metal and main group elements have been investigated for efficacy. The existence of a relationship between cancer and metals is widely acknowledged by researchers. Synthetic metal complexes appear to provide a rich platform for the design of novel anticancer drugs.

The documented history of noble metal-based drugs started with the chance discovery of cisplatin (*cis*-diamminedichloroplatinum(II)) (Rosenberg et al., 1965) as one of the most powerful chemotherapeutic agents against ovarian and testicular cancers, although gold salts had been tested against tuberculosis and were employed as antirheumatics in 1929 (Chen, 2006). Since then, a vast library of metal complexes have been synthesized and applied in the pharmacological field, mostly as anticancer agents, but also as anti-inflammatory, antibacterial, antirheumatic, and antimalarial drugs. In particular, they have many applications against tumors, which are based mainly on the strong interactions between metals and DNA. Cisplatin is routinely used for the treatment of testicular and ovarian cancers and is increasingly used against other tumors, such as cervical, bladder, and head/neck tumors (Kelland, 2007).

21.3.1   Mononuclear platinum drugs

Cisplatin contains a square-planar platinum(II) center coordinated to two ammonia ligands and two chloride ligands with a *cis*-ligand conformation. Its activity was discovered by chance in an experiment looking at the effect of electric fields on the growth of bacteria and using platinum electrodes. It is a testament to the tenacity of Rosenberg and his coworkers that the active compound was identified and the chance observation led to such a powerful drug (Rosenberg, 1999). Other few related mononuclear platinum complexes such as carboplatin, nedaplatin, lobaplatin, and oxaliplatin are among the most widely used cancer therapeutic agents (Figure 21.2). Among the platinum drugs, cisplatin was considered as first-generation anticancer drug, while changing of chlorine atoms by ligand produced second-generation anticancer drug, and when ammonium ligands are replaced by different ligands, third-generation drugs are produced.

While cisplatin is among the most effective anticancer agents in the armory of drugs available to cancer clinicians, this broad-spectrum cytotoxic is not without its drawbacks

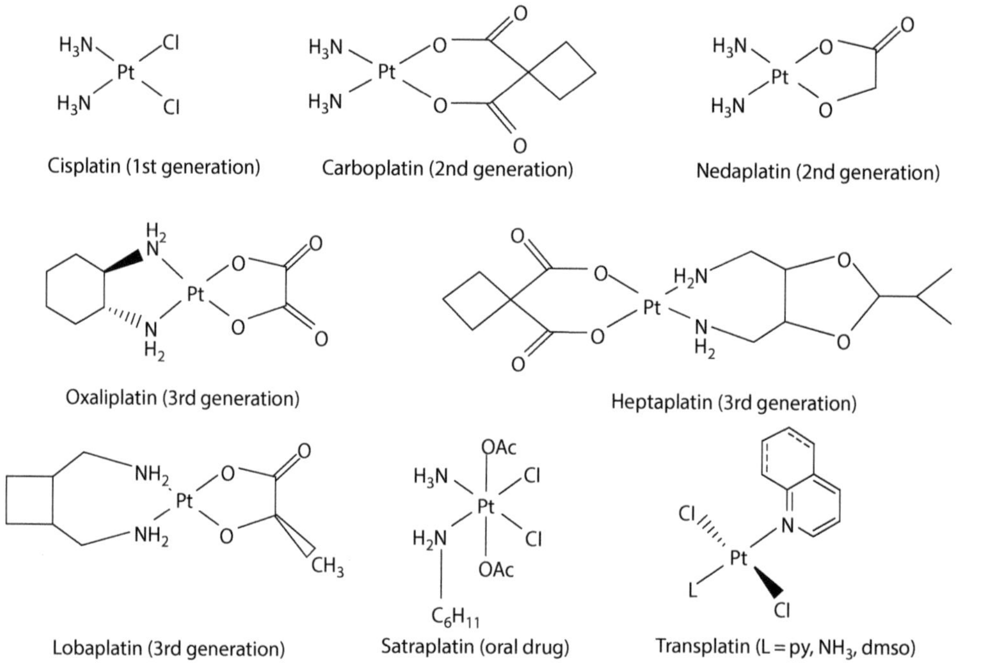

Antimicrobials: Synthetic and natural compounds

Figure 21.2 Structures of the mononuclear platinum drugs.

(Lippert, 1999). Side effects include nephrotoxicity, nausea, vomiting, and loss of sensation in the extremities (though not the hair loss associated with other chemotherapy agents). These are thought to arise through a combination of the nonspecificity of the drug and resulting damage in tissues other than the tumor and platination of the sulfur residues on proteins by the soft platinum(II) center. The change of leaving group does reduce the activity of the agent somewhat, however. While it is as effective in ovarian cancer, it is less potent against testicular, head, and neck cancers. Correspondingly, the side effects are less severe. Consequently, cisplatin has tended to remain the agent of choice, with carboplatin used when there is a clinical need to minimize the platinum drug side effects because of other medical conditions. Alongside cisplatin and carboplatin, three other very similar drugs (Figure 21.2) have appeared, which have been approved for use in specific countries: nedaplatin (Japan; Shionogi and Co. Ltd.), heptaplatin (South Korea; SK Pharma), and lobaplatin (China). Of these, nedaplatin combines the {cis-Pt(NH₃)₂} active fragment with a different bidentate leaving group (and thus is a direct analog of cisplatin and carboplatin), while heptaplatin and lobaplatin link the amines into a bidentate ligand structure and use a dicarboxylate leaving group. Although, no dramatic clinical benefits have been described for these drugs over cisplatin.

Early on, in the studies of cisplatin, it was recognized that the *trans*-isomer (*transplatin*) was inactive, and thereafter, the need for a *cis*-geometry at platinum rapidly became a dogma. However, this dogma (as the other design rules) has more recently been shown to be invalid, and three distinct classes of *trans*-compounds have shown to possess anticancer activity (Natile and Coluccia, 2001): *trans*-compounds containing pyridine ligands (developed by Farrell et al., 1989), *trans*-compounds containing an alkylamine and an isopropylamine (Montero et al., 1999), and *trans*-compounds containing iminoether ligands (developed by Coluccia et al., 1993) (Figure 21.2). These three classes of agents show potencies similar to that of cisplatin and, perhaps more importantly, are active against cis-platin-resistant cell lines. The DNA lesions formed by a *trans*-platinum agent will be inherently different from those formed by a *cis*-agent. Indeed, these *trans*-agents preferentially form monofunctional adducts with the DNA or interstrand cross-links, rather than the 1,2-intrastrand cross-link preferred by cisplatin. Thus, their molecular-level interactions with DNA are different, and hence their activity profiles differ. Despite their promising activity, representatives of these *trans*-platinum(II) classes have not been evaluated in the clinic. Although these platinum(II) cytotoxics are fairly nonspecific in their action and have been in the clinic for three decades, their continuing importance (and the size of the market) is illustrated by the fact that at least seven further platinum drugs are currently in commercial development: satraplatin (GPC Biotech and Pharmion), miriplatin (Dainippon Sumitomo Pharma and Bristol-Myers K.K.), prolindac (Access Pharmaceuticals), BP-C1 (Meabco), cisplatin lipid complex (Transave), aroplatin (Antigenics), and picoplatin (Poinard) (Kelland, 2007).

21.3.2   Di-, tri-, and tetranuclear platinum drugs

The most dramatic example of platinum(II) agents that break the traditional design rules are Farrell's di-, tri-, and tetranuclear platinum compounds (Figure 21.3) in which the metal centers are linked by flexible diamine chains (Farrell, 2004). The first ideas to add a second

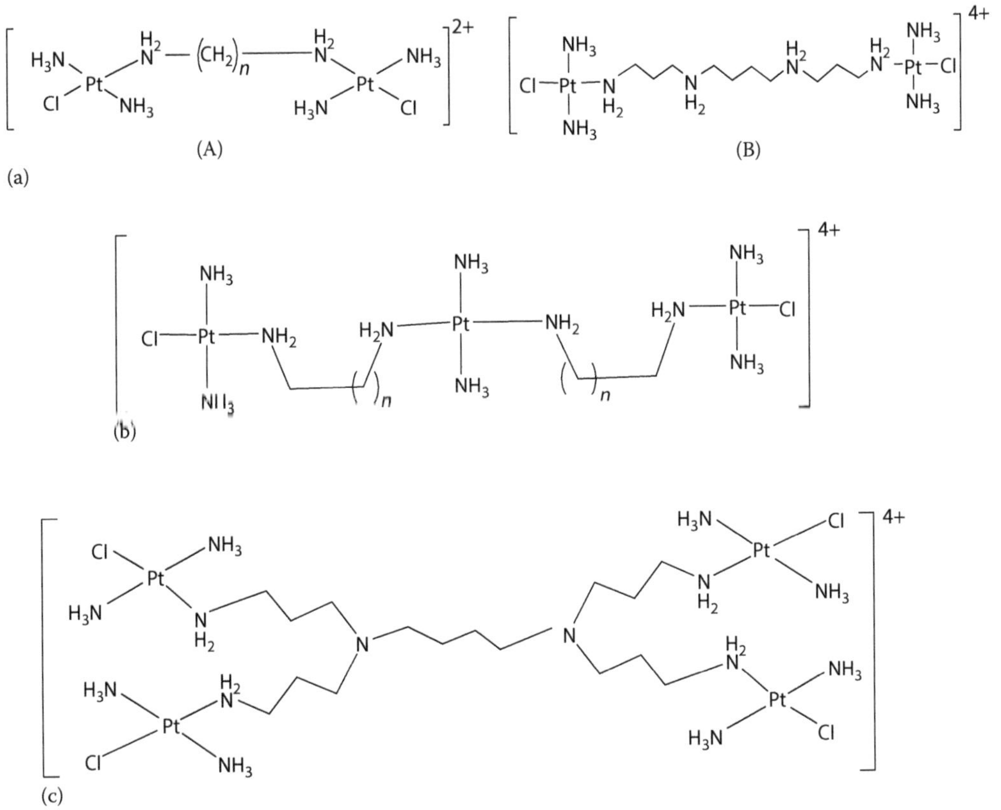

Figure 21.3 (a) Two dinuclear Pt cations (A and B), $n = 6$; (b) trinuclear Pt cation, $n = 5$; and (c) tetranuclear, dendritic-type platinum cation.

metal in an anticancer drug were developed primarily by Farrell et al. (2011). In fact, the observation that nucleobases and nucleotides have more than one binding site already led to the early suggestion of taking an excess Pt in such binding reactions. The work of Lippert's group has subsequently shown that such multiple binding may occur on a relatively large scale (Lippert, 2000), although it seems unlikely that such binding would occur under physiological conditions. Therefore, a search to chemically linked metal ions, by bridging ligands with kinetically stable M–L bonds, is required to study such compounds.

The first dinuclear Pt compounds with flexible linkers from the Farrell group showed a promising activity (Farrell et al., 1995), and quite interesting binding with DNA has been observed, including hairpin folding after binding (Mellish et al., 1997). The success of the dinuclear compounds was soon followed by the trinuclear species, with bifunctional DNA binding (Kabolizadeh et al., 2007). The DNA binding of such compounds is quite interesting and can span long distances, as shown by advanced NMR studies (Zhang et al., 2008).

21.3.3   Hetero-dinuclear platinum drugs

Recently, it is seen that different metal ions with flexible linkers and also a group of dinuclear Pt(II) compounds with rigid linkers have been employed as clinical antitumor agent, and structures are presented in Figure 21.4. Realizing that Cu-phenanthroline compounds can cut DNA in an oxidative way has combined Cu-phenanthroline with several Pt-amines using a variety of spacers, primarily aiming for bifunctionality (de Hoog et al., 2007, 2008; Ozalp-Yaman et al., 2008). In several cases, a high activity was found, and many of the newly prepared compounds were shown to be effective DNA cutting agents. Replacing the Cu part by a fluorescent group in such compounds with relatively flexible spacers had provided strong evidence for the pathway in the cells of Pt compounds (Molenaar et al., 2000). However, changing the Cu–Ru, as studied by Schilden et al. (2004) (Figure 21.4), did not result in any significant anticancer activity, despite DNA binding taking place.

21.3.4   Polymeric platinum drugs

The approach of binding the Pt species for a short time to a polymer (chemically) has been known for some years, and some applications are close to clinical use. As tumors may be hyperpermeable toward macromolecules as a result of compromised vasculature, this "enhanced permeability and retention effect" (the so-called EPR effect) may result in an increased drug concentration within the tumor tissue (Taube, 1952). This is possible when the therapeutic agent is coupled to a macromolecular carrier or may be packed inside a nanosized particle, which after endocytosis will result in drug uptake into the tumor cells (Figure 21.5). The well-known acrylamide-based drug AP5280 with a peptide spacer is given for illustration (Bouma et al., 2002; van Zutphen and Reedijk et al., 2005).

21.3.5   Pt(IV) compounds as anticancer agents

Pt(IV) compounds activated by visible light Pt(IV) complexes are normally less cytotoxic than their divalent counterparts, so they are considered to be prodrugs and they are assumed to provide Pt(II)-active species via a mechanism known as "activation by reduction." New and interesting Pt(IV) prodrugs include those activated by light, which generally have a *trans*-configuration, and they bear two amines (also mixed) (Bednarski et al., 2013; Cubo et al., 2010), two imines (Cubo et al., 2010), two azido ligands (Ruhayel et al., 2011), or a mixed diazido-amine system (Zhao et al., 2013). They are normally inactive or slightly active in the

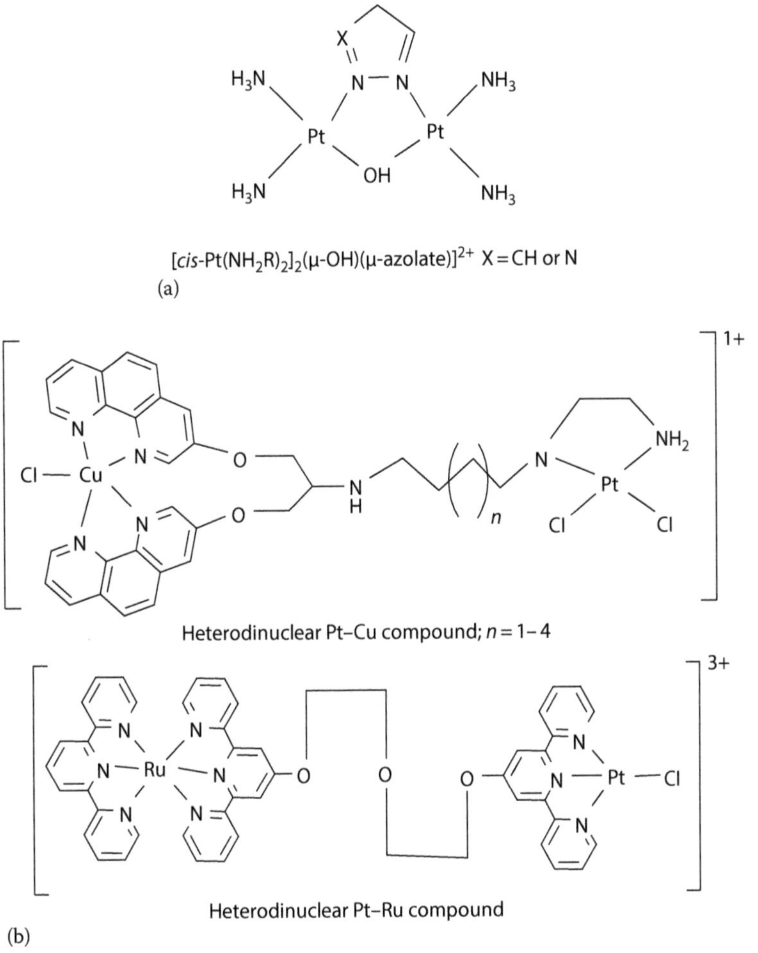

[*cis*-Pt(NH$_2$R)$_2$]$_2$(μ-OH)(μ-azolate)]$^{2+}$ X = CH or N
(a)

Heterodinuclear Pt–Cu compound; *n* = 1–4

Heterodinuclear Pt–Ru compound
(b)

Figure 21.4 (a) Homo- and (b) hetero-dinuclear cationic Pt drugs with flexible and rigid linkers.

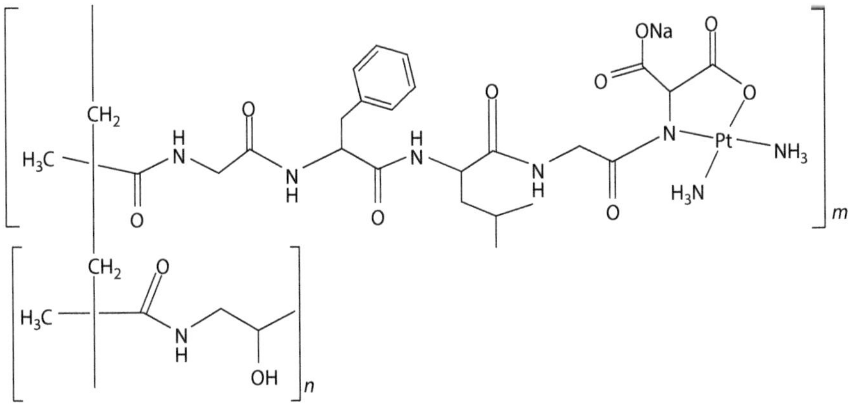

Figure 21.5 Polymeric Pt drug containing amide spacer [AP-5280; *n* > *m*].

dark, but they are selectively activated under UV and/or visible light. Their efficiency is increased by replacing one or two NH_3 ligands with pyridine, which can reach up to 50–65 times that of cisplatin when measured in the same conditions (Zhao et al., 2013).

21.4 Mechanistic aspects for the platinum compounds as anticancer drugs

Differences in coordination geometry, binding preferences according to the hard and soft acids and bases principle, important redox activity, kinetics of ligand exchange reactions, or even the simple capacity of replacement of essential metals form the chemical basis for a diversity of pharmacologically relevant interactions with biomolecules (Kostova, 2010). The metal, its oxidation state, the number and types of coordinated ligands, and the coordination geometry of the complexes can provide a variety of properties. On the other side, the ligands not only control the reactivity of the metal but also play critical roles in determining the nature of interactions involved in the recognition of biological target sites such as DNA, enzymes, and protein receptors. These variables provide enormous potential diversity for the design of metallodrugs (Chaires and Waring, 2001; Lippard, 1982).

The mode of action is now widely accepted to be through the drug's interaction with DNA (Barton, 1994). The compound is administered by injection into the bloodstream and is believed to remain in its neutral state until after it crosses the cell membrane where one or both chlorides are displaced by aqua ligands (the chloride concentration being lower inside than outside the cell) affording cationic compounds. These cationic aqua derivatives react with the bases on DNA, most commonly with the N7 of purine bases (with guanine favored over adenine), which displace the aqua/chlorido ligands. A bifunctional adduct is formed between the {cis-Pt(NH$_3$)$_2$} unit and two adjacent bases on the same strand (the 1,2-intrastrand GG adduct accounts for >70% of all adducts formed; 1,2-intrastrand AG adducts are the next most common at around 20% of all lesions; and 1,3-adducts and monoadducts are much less common). The platinum center is located in the DNA major groove, and the effect of the platinum coordination to two adjacent bases is to bend (kink) the DNA by around 45°, toward the site of platination (Figure 21.6) (Reedijk, 2011). This bent DNA structure is then recognized by nuclear high-mobility group (HMG) proteins that bind and are believed to protect the lesion from DNA repair.

Oxaliplatin and some other second- and third-generation platinum drugs are active against some cisplatin-resistant cancers. The predominant DNA adducts formed by oxaliplatin are 1,2 intrastrand GG adducts analogous to those formed by cisplatin. The DNA lesions formed by a *trans*-platinum agent will be inherently different from those formed by a *cis*-agent. Indeed, these *trans*-agents preferentially form monofunctional adducts with

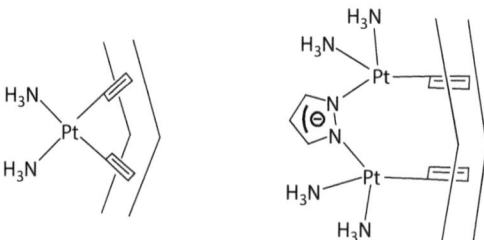

Figure 21.6 Distortion of DNA after binding with cisplatin and dinuclear Pt compounds.

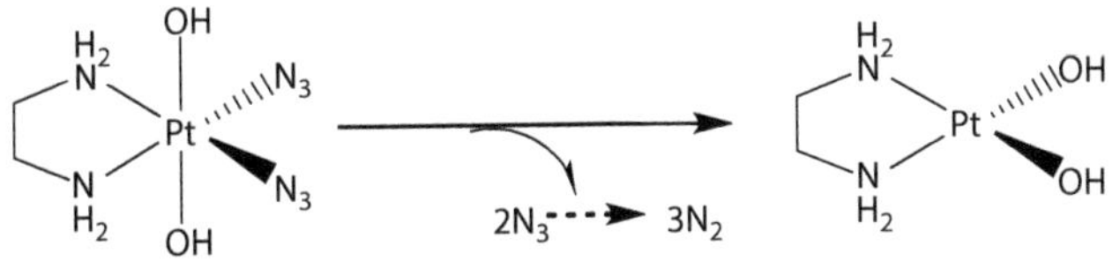

Figure 21.7 Photochemical activation on Pt drug.

the DNA or interstrand cross-links, rather than the 1,2-intrastrand cross-link preferred by cisplatin (Farrer et al., 2009). Thus, their molecular-level interactions with DNA are different, and hence their activity profiles differ. Despite their promising activity, representatives of these *trans*-platinum(II) classes have not been evaluated in the clinic.

The most dramatic example of platinum(II) agents that break the traditional design rules are di-, tri-, and tetranuclear platinum compounds. These platinum complexes mainly form bifunctional long-range DNA adducts (both inter- and intrastrand cross-links). In addition, a DNA conformational change is induced from a right-handed (B) to a left-handed (Z) DNA double helix. This induction of Z-DNA is irreversible and is associated with the cross-linking (Farrell et al., 1995).

With close inspection on homo- and hetero-dinuclear platinum drugs, it is seen that upon loss of the leaving OH group, each Pt would be able to bind at a guanine-N7, which would result in a very small kink of double-stranded DNA when the two G bases would be adjacent. The DNA binding of these dinuclear rigid compounds was as expected, resulting in a very small distortion of the helix only (Zhang et al., 2008), and the activity of the compounds in several cancer cell lines was found to be an order of magnitude better than that of cisplatin (de Hoog et al., 2007). Realizing that Cu-phenanthroline compounds can cut DNA in an oxidative way, de Hoog has combined Cu-phenanthroline with several Pt-amines using a variety of spacers, primarily aiming for bifunctionality. In several cases, a high activity was found, and many of the newly prepared compounds were shown to be effective DNA cutting agents. Changing the Cu–Ru, as studied by van der Schilden et al. (2004), did not result in any significant anticancer activity, despite DNA binding taking place.

It has been demonstrated that light at the appropriate wavelength will cause the dissociation of one or more ligands, thereby giving different active Pt(II) photoproducts, but rarely Pt(IV) species (Figure 21.7). The formation of a cytotoxic azydil radical has also been suggested for complexes containing the azide ion. Their mechanisms of action are still under study, but DNA appears to be the target involved most often, where its damage cannot be recognized by HMGB1 protein, in contrast to cisplatin-type lesions (Zhao et al., 2013). Cell death via nonapoptotic pathways has also been recognized (Hambley et al., 2009).

21.5 Conclusion

The action of metal complexes in living organisms are expected to differ, in general, from the action of nonmetal containing agents and may offer unique research, diagnostic, or therapeutic opportunities. The key to the continuing rapid evolution of the field now lies in its integration with the emerging postgenomic knowledge and technologies from fields, including systems biology, genomics, proteomics, and structural biology. Also underdeveloped, but less directly related to DNA binding, is the control of the toxic side effects; development of the coordination chemistry of Pt compounds with rescue and protective agents (usually S-donor ligands) and especially the reactions of these compounds with other cellular components and their cell wall transport needs serious attention. More directly related to DNA binding of metal compounds is the migration of Pt units along the DNA chain. Finally, it should be mentioned again that some of the new compounds possessing chemical and biological properties related to those of cisplatin might be very active but show weak binding or no binding at all to DNA. Many other metal compounds may be shown to be active in cancer treatment and may primarily interact with other biological targets.

In the long term, "individualized" medical treatments with optimized drugs and clinical regimes tailored to the individual can be envisaged. Harnessing this knowledge and responding to and taking advantage of this new environment require teams committed to integrating chemistry and biology research and knowledge. These advances will revolutionize the field of metallodrugs, taking it far beyond its origins in simple platinum compounds toward sophisticated and modular molecular designs with different and predicted modes of action.

Acknowledgments

B. Biswas gratefully acknowledges the financial support by the Department of Science and Technology (DST), New Delhi, India, under FAST TRACK SCHEME for YOUNG SCIENTIST (No. SB/FT/CS-088/2013 dtd. 21/05/2014). B. Biswas sincerely thanks University Grant Commission, New Delhi, India (F. No. PSW-84/12-13(ERO) dated 05/02/2013) for financial support.

References

Alessio, E. (Ed.). 2011. *Bioinorganic Medicinal Chemistry*. Wiley-VCH, Weinheim, Germany.

Barton, J.K. 1994. Metal/nucleic acid interactions. In *Bioinorganic Chemistry* (Bertini, I., Gray, H.B., Lippard, S.J., and Valentine, J. S., eds.). University Science Books, Mill Valley, CA.

Bednarski, P.J., Korpis, K., Westendorf, A.F., Perfahl, S., and Grünert, R. 2013. Effects of light-activated diazido-PtIV complexes on cancer cells in vitro. *Dalton Trans.* 40:5342–5351.

Bouma, M., Nuijen, B., Stewart, D.R., Rice, J.R., Jansen, B.A.J., Reedijk, J., and Beijnen, J.H. 2002. Stability and compatibility of the investigational polymer-conjugated platinum anticancer agent AP 5280 in infusion systems and its hemolytic potential. *Anti-Cancer Drugs* 13(9):915–924.

Chaires, J.B. and Waring, M.J. 2001. *Drug–Nucleic Acid Interactions*, vol. 340. Academic Press, San Diego, CA.

Chen, H. 2006. *Atlas of Genetic Diagnosis and Counselling*. Humana Press, Totowa, NJ.

Coluccia, M., Nassi, A., Loseto, F., Boccarelli, A., Mariggio, M.A., Giordano, D., and Natile, G. 1993. A trans-platinum complex showing higher antitumor activity than the cis congeners. *J. Med. Chem.* 36:510.

Cubo, L., Pizarro, A.M., Quiroga, A.G., Salassa, L., Navarro-Ranninger, C., and Sadler, P.J. 2010. Photoactivation of *trans* platinum diamine complexes in aqueous solution and effect on reactivity towards nucleotides. *J. Inorg. Biochem.* 104:909–918.

Dabrowiak, J.C. 2010. *Metals in Medicine*. John Wiley & Sons, Hoboken, NJ.

de Hoog, P., Boldron, C., Gamez, P., Sliedregt-Bol, K., Roland, I., Pitié, M., and Reedijk, J. 2007. New approach for the preparation of efficient DNA cleaving agents: Ditopic copper-platinum complexes based on 3-Clip-Phen and cisplatin. *J. Med. Chem.* 50:3148–3155.

de Hoog, P., Pitié, M., Amadei, G., Gamez, P., Meunier, B., Kiss, R., and Reedijk, J. 2008. DNA cleavage and binding selectivity of a heterodinuclear Pt–Cu(3-Clip-Phen) complex. *J. Biol. Inorg. Chem.* 13:575–586.

Farrell, N. et al., 1989. Cytostatic trans-platinum (II) complexes. *J. Med. Chem.* 32:2240.

Farrell, N. 1995. DNA binding and chemistry of dinuclear platinum complexes. *Comments Inorg. Chem.* 16:373–389.

Farrell, N. 2004. Polynuclear platinum drugs. *Metal Ions Biol. Syst.* 41:251–296.

Farrell, N. 2011. Solution studies of dinuclear polyamine-linked platinum-based antitumour complexes. *Dalton Trans.* 40:4147–4154.

Farrer, N.J., Woods, J.A., Munk, V.P., Mackay, F.S., and Sadler, P.J. 2009. Photocytotoxic trans-diam(m)ine platinum(IV) diazido complexes more potent than their cis isomers. *Chem. Res. Toxicol.* 23:413–421.

Gielen, M. and Tiekink, E.R.T. (Eds.). 2005. Metallotherapeutic drugs and metal-based diagnostic agents. *The Use of Metals in Medicine* (Graves, M.G., ed.). Wiley, Chichester, U.K.

Griffith, D.M. et al. 2012. The prevalence of metal-based drugs as therapeutic or diagnostic agents: Beyond platinum. *Dalton Trans.* 41:13239–13257.

Kabolizadeh, P., Ryan, J., and Farrell, N. 2007. Differences in the cellular response and signaling pathways of cisplatin and BBR3464 ([{*trans*-PtCl (NH$_3$)$_2$} 2μ-(*trans*-Pt(NH$_3$)$_2$(H$_2$N(CH$_2$)6-NH$_2$)]$^{4+}$) influenced by copper homeostasis. *Biochem. Pharmacol.* 73:1270–1279.

Kelland, L.R. 2007. The resurgence of platinum-based cancer chemotherapy. *Nat. Rev. Cancer* 7:573–584.

Klein, A. and Hambley, W. 2009. Platinum drug distribution in cancer cells and tumors. *Chem. Rev.* 109:4911–4920.

Kostova, I. 2010. Metal-containing drugs and novel coordination complexes in therapeutic anticancer applications—Part II. *Anti-Cancer Agents Med. Chem.* 10:352–353.

Lippard, S.J. 1982. New chemistry of an old molecule: *cis*-[Pt(NH$_3$)2Cl$_2$). *Science* 218:1075–1082.

Lippert, B. (Ed.). 1999. *Cisplatin, Chemistry and Biochemistry of a Leading Anticancer Drug.* Wiley-VCH, Weinheim, Germany.

Lippert, B. 2000. Multiplicity of metal ion binding patterns to nucleobases. *Coord. Chem. Rev.* 200:487–516.

Mellish, K.J., Qu, Y., Scarsdale, N., and Farrell, N. 1997. Effect of geometric isomerism in dinuclear platinum antitumour complexes on the rate of formation. *Nucleic Acids Res.* 25:1265–1271.

Molenaar, C., Teuben, J.M., Heetebrij, R.J., Tanke, H.J., and Reedijk, J. 2000. New insights in the cellular processing of platinum antitumor compounds, using fluorophore-labeled platinum complexes and digital fluorescence microscopy. *J. Biol. Inorg. Chem.* 5:655–665.

Montero, E.I., Díaz, S., González-Vadillo, A.M., Pérez, J.M., Alonso, C., and Navarro-Ranninger, C. 1999. Preparation and characterization of novel *trans*-[PtCl$_2$(amine)(isopropylamine)] compounds: Cytotoxic activity and apoptosis induction in ras-transformed cells. *J. Med. Chem.* 42:4264.

Natile, G. and Coluccia, M. 2001. Current status of *trans*-platinum compounds in cancer therapy. *Coord. Chem. Rev.* 216–217:383.

Navarro, M., Gabbiani, C., Messori, L., and Gambino, D. 2010. Metal-based drugs for malaria, trypanosomiasis and leishmaniasis: Recent achievements and perspectives. *Drug Discov Today* 15:1070–1078.

Noh, J.Y., Chino, T, and Ito, K. 2006. Treatment with inorganic iodine for Graves' hyperthyroidism. *Nippon Rinsh* 64:2269–2273.

Norman, J.F. and Hambley, T.W. 2011. Targeting strategies for metal-based therapeutics. In Alessio, E. (Ed.). *Bioinorganic Medicinal Chemistry.* Wiley-VCH Verlag GmbH and Co. KGaA, Weinheim, Germany.

Ochiai, E. 2008. *Bioinorganic Chemistry: A Survey, Medical Applications of Inorganic Compounds,* John Wiley & Sons Ltd., England, U.K.

Ozalp-Yaman, Ş., de Hoog, P., Amadei, G., Pitié, M., Gamez, P., Dewelle, J., and Reedijk, J. 2008. Platinated copper (3-clip-phen) complexes as effective DNA-cleaving and cytotoxic agents. *Eur. J. Inorg. Chem.* 14:612–619.

Quiroga, A.G. 2012. Understanding trans platinum complexes as potential antitumor drugs beyond targeting DNA. *J. Inorg. Biochem.* 114:106–112.

Reedijk, J. 2009. Platinum anticancer coordination compounds: Study of DNA binding inspires new drug design. *Eur. J. Inorg. Chem.* 2009:1303–1312.

Reedijk, J. 2011. Increased understanding of platinum anticancer chemistry. *Pure Appl. Chem.* 83:1709–1719.

Rosenberg, B. 1999. In Lippert, B. (Ed.). *Cisplatin, Chemistry and Biochemistry of a Leading Anticancer Drug. Bioinorganic Chemistry: A Survey*, Wiley-VCH, Weinheim, Germany.

Rosenberg, B. and Van Camp, L. 1969. Platinum compounds: A new class of potent antitumour agents. *Nature* 222:385–386.

Rosenberg, B., Van Camp, L., and Krigas, T. 1965. Inhibition of cell division in *Escherichia coli* by electrolysis products from a platinum electrode. *Nature* 205:698–699.

Ruhayel, R.A., Zgani, I., Berners-Price, S.J., and Farrell, N.P. 2011. Solution studies of dinuclear polyamine-linked platinum-based antitumour complexes. *Dalton Trans.* 40:5342–5351.

Schilden, K., Garcia, F., Kooijman, H., Spek, A.L., Haasnoot, J.G., and Reedijk, J. 2004. A highly flexible dinuclear ruthenium (ii)-platinum (ii) complex: Crystal structure and binding to 9-ethylguanine. *Angew. Chem. Int. Ed.* 43:5668–5670.

Sorenson, J.R.J. 1985. Historic uses of copper compounds in medicine. *Trace Elem. Med.*, 2:80–87.

Srivastava, A.K. and Mehdi, M.Z. 2005. Insulino-mimetic and antidiabetic effects of vanadium compounds. *Diabet. Med.* 22:2–13.

Tamasi, G. 2010. Metal-oxicam coordination compounds: Structure, biological activity and strategies for administration. *Open Crystallogr. J.* 3:41–53.

Tarallo, M.B., Urquiola, C., Monge, A., Costa, B.P., Ribeiro, R.R., Costa-Filho, A.J., and Gambino, D. 2010. Design of novel iron compounds as potential therapeutic agents against tuberculosis. *J. Inorg. Biochem.* 104:1164–1170.

Taube, H. 1952. Rates and mechanisms of substitution in inorganic complexes in solution. *Chem. Rev.* 50:69–78.

Thompson, K.H. 2011. Medicinal inorganic chemistry: Metallotherapeutics for chronic diseases. Scott, R.A. (Ed.), *Encyclopedia of Inorganic Chemistry*, John Wiley & Sons, Ltd, Athens, GA.

Thompson, K.H. and Orvig, C. 2003. Boon and bane of metal ions in medicine. *Science* 300:936–939.

Van der Schilden, K., Garcìa, F., Kooijman, H., Spek, A.L., Haasnoot, J.G., and Reedijk, J. 2004. A highly flexible dinuclear ruthenium(II)–platinum(II) complex: Crystal structure and binding to 9-ethylguanine. *Angew. Chem. Int. Ed.* 43:5668–5670.

van Rijt, S.H. and Sadler, P.J. 2009. Current applications and future potential for bioinorganic chemistry in the development of anticancer drugs. *Drug Discov. Today* 14:1089–1097.

van Zutphen, S. and Reedijk, J. 2005. Targeting platinum anti-tumour drugs: Overview of strategies employed to reduce systemic toxicity. *Coord. Chem. Rev.* 249:2845–2853.

Wang, X. 2010. Fresh platinum complexes with promising antitumor activity. *Anti-Cancer Agents Med. Chem.* 10:396–411.

Zhang, J., Thomas, D.S., Berners-Price, S.J., and Farrell, N. 2008. Effects of geometric isomerism and anions on the kinetics and mechanism of the stepwise formation of long-range DNA interstrand cross-links by dinuclear platinum antitumor complexes. *Chem. Eur. J.* 14(21):6391–6405.

Zhao, Y., Woods, J.A., Farrer, N.J., Robinson, K.S., Pracharova, J., Kasparkova, J., and Sadler, P.J. 2013. Diazido mixed-amine platinum(IV) anticancer complexes activatable by visible-light form novel DNA adducts. *Chem. A Eur. J.* 19(29):9578–9591.

section five

Narrow-spectrum antimicrobials

chapter twenty-two

Marine actinobacteria as potential drug storehouses
A future perspective on antituberculosis compounds

N. Tamilselvan, Ernest David, Dharumadurai Dhanasekaran, and Kumar Saurav

Contents

22.1 Introduction

For billions of years, certain bacteria and fungi have produced chemical substances to protect them from attack from other microorganisms. Those used in clinical medicine today are referred to as "antibiotics" or "antimicrobial agents." As a survival mechanism, other microbes have developed mechanisms for resisting the toxic effect of antimicrobials. "Antimicrobial resistance" is thus an ancient phenomenon encoded on resistance genes passed down through microbial lineages. Susceptible strains can become resistant either through mutations in existing genes or by acquiring a resistance gene from another organism that is already resistant. This is the first step in the emergence of "new resistance." Although for most organisms the sudden appearance of new resistance is rare, this is not the case for all pathogens. For example, in patients with tuberculosis (TB) or HIV infection, new mutations in susceptible strains can occur within a patient, especially when therapy is suboptimal. The emergence of resistant strains during therapy greatly increases the risk of a poor clinical outcome, including death. Consequently, effective treatment is absolutely critical to avoid the development of resistance during treatment. Inadequate infection control in hospitals and the lack of effective public health measures to control infections in the community are factors that are contributing to this problem, particularly in developing country settings where the burden of infection is high and the choices for therapy are limited.

The discovery of new drugs for systemic infections is a major challenge in infectious disease research. Despite some major advancement toward drug discovery, the continuing increase in the incidence of infections together with the gradual rise in resistance highlights the need to find novel compounds with divergent mechanisms of action. Hence, considerable research is being directed to isolate new compounds with defined mechanisms of action that can serve as templates for further medicinal chemistry modifications.

22.1.1   Natural products

Natural products are chemical compounds derived from living organisms, for example, plants, animals, and microorganisms. They can be defined as chemical compounds isolated or derived from primary or rather secondary metabolism of organisms concerned naturally. Nature acts as a prominent reservoir for new and novel therapeutics. By employing sophisticated techniques under various screening programs, the rate of discovery of natural compounds exceeds one million so far, out of which 22,500 biologically active compounds that have been extracted are from microbes: 45% are produced by actinobacteria, 38% by fungi, and 17% by unicellular bacteria (Demain and Sanchez, 2009). Members of the class Actinobacteria especially *Streptomyces* spp. have long been recognized as prolific sources of useful bioactive metabolites, providing more than 85% of the naturally occurring antibiotics discovered to date and continuing as a rich source of new bioactive metabolites (Berdy, 2005). Unfortunately, the emergence of drug-resistant pathogens and the increase in diseases affecting the immune system have greatly intensified the need to investigate new bioactive metabolites for potential pharmaceutical and industrial applications. Over the past 75 years, natural product derived compounds have led to the discovery of many drugs to treat numerous human diseases (Grabley, 1999).

22.1.2   Marine microorganisms as a source of natural drugs

The oceans cover more than 70% of the Earth's surface, and little is known about the microbial diversity of marine sediments, which is an inexhaustible resource that has not been fully exploited. Marine extremophiles serves as valuable natural resource for novel products such as antibiotics, antitumor agents, and other therapeutic substances (Amador et al., 2003). Natural products isolated from microorganisms have been the source of most of the antibiotics developed to date for various diseases. Some of them are erythromycin, vancomycin, and streptomycin, which are used for treating bacterial infections (Kelecom, 2002). Microbial natural products are very important not only for their potent therapeutic properties but also for the fact that they possess pharmacokinetic properties required for the clinical development. Marine microbes are particularly attractive because (1) they have not been extensively exploited as their terrestrial counterparts and (2) they have high potency required for bioactive compounds to be effective in the marine environment, due to the diluting effect of seawater. Most of the marine compounds that have been successfully screened and structurally elucidated so far originated from microorganisms, especially from bacteria. It has been shown that marine bacteria produce bioactive substances that are different from known compounds from terrestrial bacteria (Fenical, 1993). It is estimated that less than 1% of potentially useful chemicals from marine environment have been screened so far, with microbial products representing approximately 1% of the total number. Exploration of microbial secondary metabolites has led to the discovery of hundreds of biologically active compounds that possess antibiotic, antitumor, and other pharmacological activities that are currently being used for the treatment of various diseases in humans, animals, and plants. Natural products with antibiotic activity that come from bacteria or fungi are almost always products of secondary metabolic pathways. The focus on the physiology and the potential of bioactive substances of noncultivable marine microorganisms are of current problem and pose a great challenge to researchers to cultivate and isolate novel secondary metabolites for therapeutic applications.

22.2 Marine actinomycetes: a new source for bioactive metabolites

Actinobacteria or actinomycetes are a group of gram-positive bacteria with high G + C ratio. Actinobacteria are widely distributed in terrestrial and aquatic ecosystems, especially in soil, where they play a crucial role in the recycling of refractory biomaterials by decomposing complex mixtures of polymers in dead plant, animal, and fungal materials. They are important in soil biodegradation and humus formation by recycling the nutrients associated with recalcitrant polymers such as keratin, lignocelluloses, and chitin. They also produce several volatile substances like geosmin responsible of the characteristic "wet earth odor." They also exhibit diverse physiological and metabolic properties, such as the production of extracellular enzymes (Olano et al., 2009). In 1940, Waksman discovered actinomycin from soil bacteria. He was awarded with a Nobel Prize for his discovery in 1952. Since then, hundreds of naturally occurring antibiotics have been discovered in actinomycetes, especially from the genus *Streptomyces* (Stackebrandt et al., 1997). Marine actinomycetes exhibit very different 16S rDNA sequences compared to their terrestrial counterparts. As a result, marine actinomycetes produce many novel metabolites that may be biologically active and a potential source of new anti-infective drugs (Lam, 2006). The order Actinomycetales, commonly called as actinomycetes, represent one of the most studied and exploited classes of bacteria for their ability to make a wide range of biologically active metabolites (Ikeda et al., 2003). The actinobacteria play an important role among the marine bacterial communities, because of its diversity and ability to produce novel chemical compounds of high commercial value (Amador et al., 2003, Hopwood, 2007)

Though more than 50% of the microbial antibiotics discovered so far originate from actinomycetes, only *Streptomyces* and *Micromonospora* account for most of these compounds (Berdy, 2005). It was long been known that actinomycetes can be recovered from the sea, and more recently, it was isolated from the deepest known ocean trenches (Pathom-Aree et al., 2006). It was reported that common soil-inhabiting actinomycetes can be readily recovered from marine samples (especially samples collected close to land), but the existence of distinct marine populations was clear in 1984 with the taxonomic description of the first marine actinomycete, *Rhodococcus marinonascens* (Helmke and Weyland, 1984). More recently, at least three marine genera (Maldonado et al., 2005) have been described, including the genus *Salinispora*, which requires seawater and sodium for growth. Sampling for marine actinomycetes began in the late 1960s, but in 2005, the first seawater-obligate marine actinomycete (*Salinispora* spp.) was described (Maldonado et al., 2005). The cultivation of *Salinispora* spp. was originally reported in 1991 (Jensen et al., 1991), but at that time, their taxonomic significance was unaware. Approximately 10 years later, after the application of DNA sequence-based methods to infer evolutionary (phylogenetic) relationships, it became clear that these bacteria are distinct from their terrestrial counterparts (Mincer et al., 2002). Inspired by this finding and by the unusual metabolites observed during initial chemical studies of *Salinispora* spp., studies were broadened to other marine-derived actinomycetes, and several strains were cultured within at least six actinomycete families, including many of them seem to represent new taxa. These efforts have corroborated the existence of marine actinomycetes, and subsequent chemical studies have shown that these strains are also an excellent source of unprecedented secondary metabolites.

Marine actinomycetes have been a rich source of unique and biologically active metabolites. Although heavily studied over the past three decades, actinomycetes continue to prove themselves as reliable sources of novel bioactive compounds. Among the well-characterized pharmaceutically relevant microorganisms, actinomycetes remain a major source of novel, therapeutically relevant natural products. They are the providers of novel structures with unique pharmacological activities that continue to be discovered and reported from all types of natural product sources. The majority of these compounds demonstrate one or more bioactivities that could provide value to human medicine. Analytical methods help us in rapid elucidation of structures and make natural products valuable components of modern drug discovery.

Occurrence of distinct rare genera in the marine ecosystem has been reported with the taxonomic description of the first marine actinomycetes *Rhodococcus marinonascens* (Helmke and Weyland, 1984). Actinomycetes have also been isolated from free swimming as well as sessile marine vertebrates and invertebrates (Ward and Bora, 2006). Unusual actinomycetes belonging to Micrococceae, Dermatophilaceae, and Gordoniaceae have been isolated from sponges (Lam, 2006). Tetrodotoxin-producing actinomycete has been isolated from puffer fish ovaries (Wu et al., 2005). The organism was found to be most closely related to *Nocardiopsis dassonvillei*. Researchers are finding new genera from marine environments on a regular basis and discovering new metabolite producers never reported earlier. Actinomycetes genera identified by cultural and molecular techniques from different marine ecological niches include *Actinomadura, Actinosynnema, Amycolatopsis, Arthrobacter, Blastococcus, Brachybacterium, Corynebacterium, Dietzia, Frankia, Frigoribacterium, Geodermatophilus, Gordonia, Kitasatospora, Micromonospora, Micrococcus, Microbacterium, Mycobacterium, Nocardioides, Nocardiopsis, Nonomuraea, Pseudonocardia, Rhodococcus, Saccharopolyspora, Salinispora, Serinicoccus, Solwaraspora, Streptomyces, Streptosporangium, Tsukamurella, Turicella, Verrucosispora,* and *Williamsia* (Ward and Bora, 2006). In spite of improvements being made in the cultural methods for the isolation of rare marine actinomycetes, many of these organisms still remain uncultivable and have to be identified by using molecular techniques. Metagenomic methods are useful for characterizing microbes that cannot be cultivated and can also be used to isolate their genes (Tringe et al., 2005).

22.1.2 Antibacterial

The α-pyrones are six-membered cyclic unsaturated esters that share chemical and physical properties reminiscent to alkene and aromatic compound. These compounds occur abundantly in naturally occurring molecules and are responsible for vast range of biological activities including antibacterial, antifungal, neurotoxic, and phytotoxic properties (Dickinson, 1993; McGlacken and Fairlamb, 2005) and plant growth-regulating (Kobayashi et al., 1994; Tsuchiya et al., 1997) antitumor (Suzuki et al., 1997; Kondoh et al., 1999), and HIV protease-inhibiting activities (Thaisrivongs et al., 1996; Turner et al., 1998; Chen et al., 2007).

Several α-pyrones, for example, fusapyrones (Altomare et al., 2000), gibepyrones A–F (Barrero et al., 1993), herbarins A–B (Jadulco et al., 2002), styrylpyrones (Rossi et al., 1997; Kamperdick et al., 2002), nocapyrones E–J (Fu et al., 2011a), actinopyrones A–C (Yano et al., 1986), pironetins (Kobayashi et al., 1994), and germicidins A–D (Petersen et al., 1993), have been previously reported with biological activities. For example, nocapyrones E–G (**1a–1c**) and germicidins A–D (**2–5**) were isolated from the actinomycetes *Streptomyces viridochromogenes* NRRL B-1551 and *Nocardiopsis dassonvillei*, respectively, which possess moderate activity against *Bacillus subtilis*.

Pseudonocardians A–C (Li et al., 2011) (**6–8**), three new diazaanthraquinone derivatives, were isolated from the strain SCSIO 01299, a marine actinomycete member of the genus *Pseudonocardia*, collected from deep-sea sediment of the South China Sea, and poses antibacterial activities on *Staphylococcus aureus* ATCC 29213, *Enterococcus faecalis* ATCC 29212, and *Bacillus thuringensis* SCSIO BT01, with minimum inhibitory concentration (MIC) values of 1–4 μg mL^{-1}.

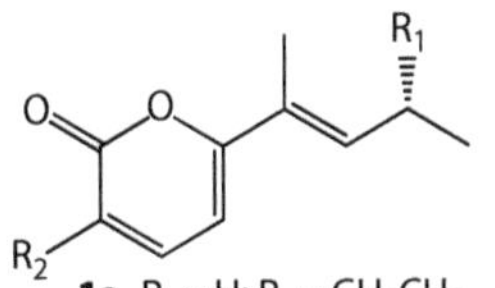

1a. $R_1=H$; $R_2=CH_2CH_3$
1b. $R_1=H$; $R_2=CH_3$
1c. $R_1=OH$; $R_2=CH_2CH_3$

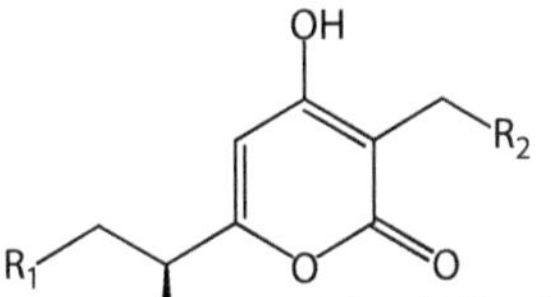

2. $R_1=H$, $R_2=H$
3. $R_1=H$, $R_2=CH_3$
4. $R_1=CH_3$, $R_2=H$
5. $R_1=CH_3$, $R_2=CH_3$

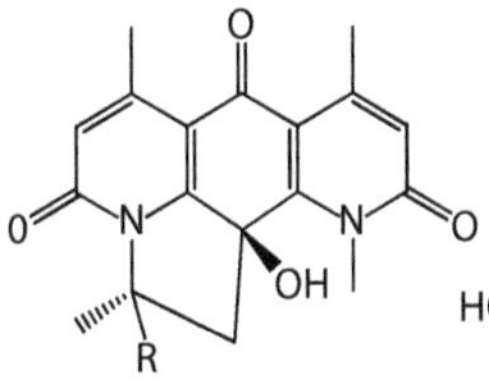

6. $R=CH_3CH_2$
7. $R=CH_3$

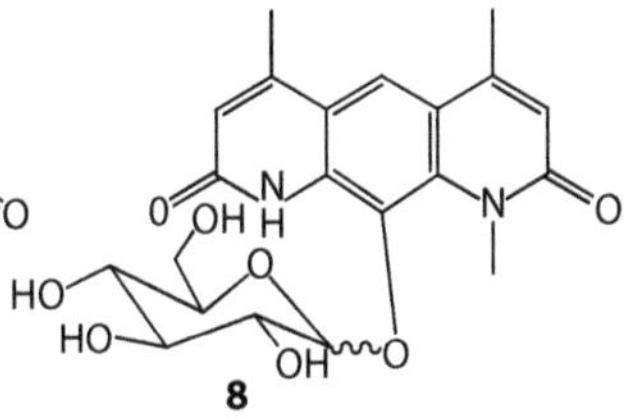

8

Recent advancement led to the discovery of several novel secondary metabolites to circumvent the problem of resistant pathogens, which are no longer susceptible to the currently used drugs. An endophytic *Streptomyces* sp. provided the indolosesquiterpenes, xiamycin B (**9**), indosespene (**10**), and sespenine (**11**) and the known xiamycin A, which had moderate to strong activity against several bacteria, including MRSA and vancomycin-resistant *E. faecalis* (Ding et al., 2011).

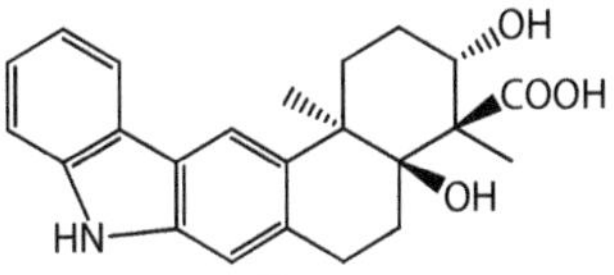

9

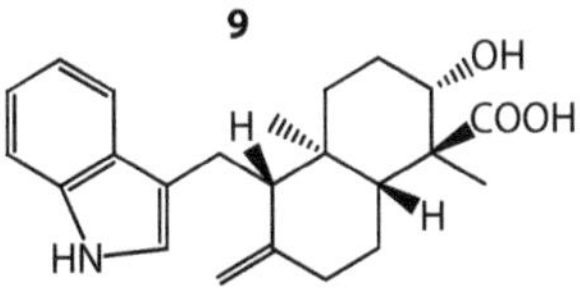

10

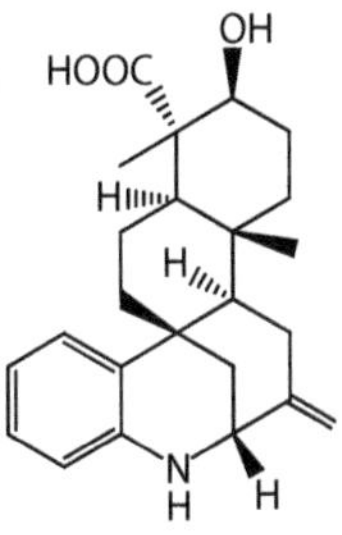

11

A series of chlorinated bisindole pyrroles, lynamicins A–E, was discovered from marine actinomycete, NPS12745, which was isolated from marine sediment collected off the coast of San Diego, California, in which it was shown that these substances possess broad-spectrum activity against both gram-positive and gram-negative organisms. Significantly, compounds **12–16** were active against drug-resistant pathogens such as methicillin-resistant *Staphylococcus aureus* and vancomycin-resistant *Enterococcus faecium* (McArthur et al., 2008).

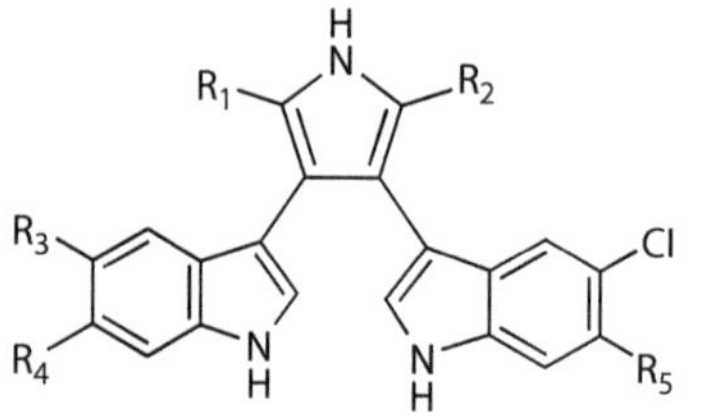

12. $R_1 = R_4 = R_5 = H$, $R_2 = COOMe$, $R_3 = Cl$
13. $R_1 = R_5 = H$, $R_2 = COOMe$, $R_3 = R_4 = Cl$
14. $R_1 = R_2 = H$, $R_3 = R_4 = R_5 = Cl$
15. $R_1 = R_2 = COOMe$, $R_3 = Cl$, $R_4 = R_5 = H$
16. $R_1 = R_2 = COOMe$, $R_3 = R_4 = R_5 = H$

22.2.2 Antifungal

Daryamides (**17**) are cytotoxic and antifungal polyketides isolated from culture broth of a *Streptomyces* strain, CNQ-085. These bioactive compounds have been shown to exhibit moderate cytotoxicity against the human colon carcinoma cell line HCT-116 and moderate antifungal activities against *Candida albicans* (Asolkar et al., 2006). Similarly, chandrananimycins (**18–19**), isolated from marine *Actinomadura* sp. MO48, have been shown to exhibit antibacterial, anticancer, and antifungal activities (Maskey et al., 2003).

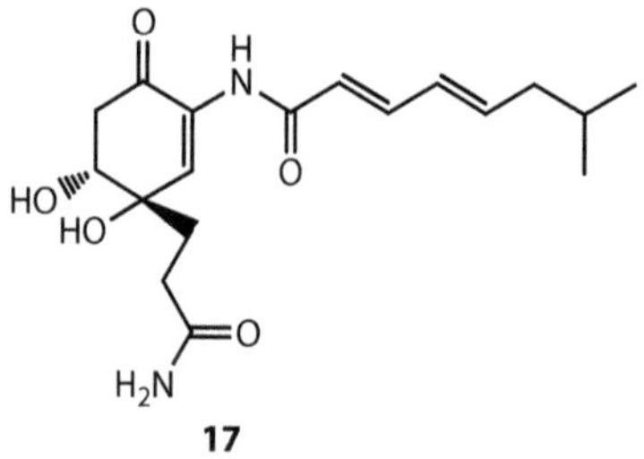

17

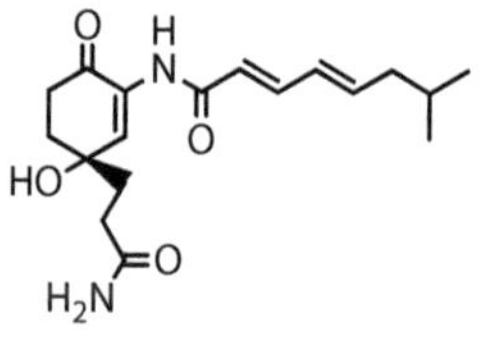

18

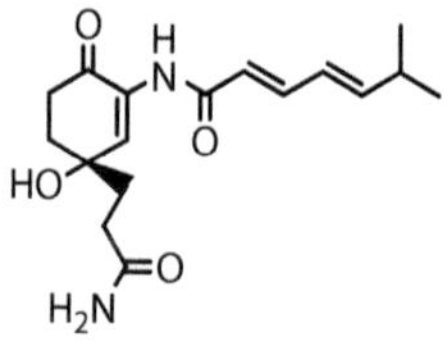

19

The antibiotics antimycin A19 (**20**) and A20 (**21**) were isolated from *Streptomyces antibioticus* with potent activity against *C. albicans* (Xu et al., 2011).

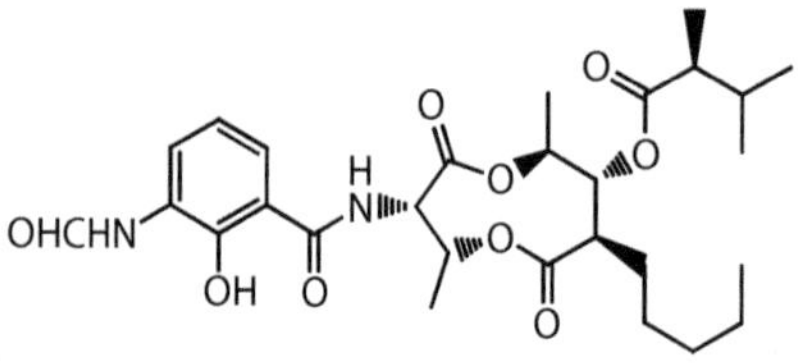

20

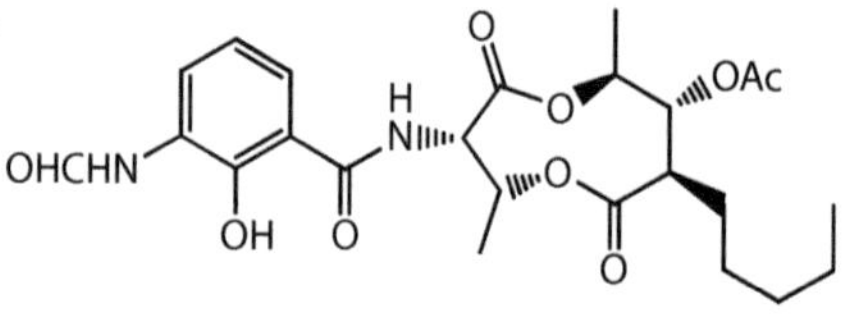

21

22.2.3 Anticancer

Five bipyridine alkaloids named caerulomycins F–J (**22–26**), along with a phenylpyridine alkaloid caerulomycins K (**27**), were discovered from *Actinoalloteichus cyanogriseus* WH1-2216-6 with cytotoxic effect on cancerous cell lines HL-60, K562, KB, and A549.

Caerulomycin analogues also exhibited potential antimicrobial activity at MIC ranging from 9.7 to 38.6 μM (Fu et al., 2011b).

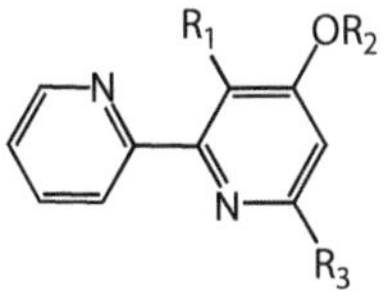

22. R$_1$=H, R$_2$=Me, R$_3$=CH$_2$OH
23. R$_1$=OMe, R$_2$=Me, R$_3$=CH$_2$OH
24. R$_1$=R$_2$=H, R$_3$=CHNOH
25. R$_1$=H, R$_2$=Me, R$_3$=CONHOMe
26. R$_1$=R$_2$=H, R$_3$=CH$_2$NHCOMe

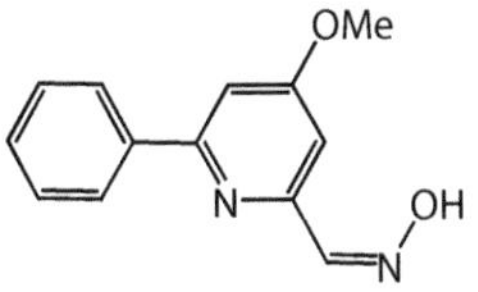

27

A PCR-based screening approach led to the identification of a deep-sea-derived *Streptomyces* sp. SCSIO 03032 capable of producing new bisindole alkaloids spiroindimicins A–D (**28–31**). Spiroindimicins B–D (**29–31**) with a [5,5] spiroring exhibited moderate cytotoxicities against several cancer cell lines (Zhang et al., 2012).

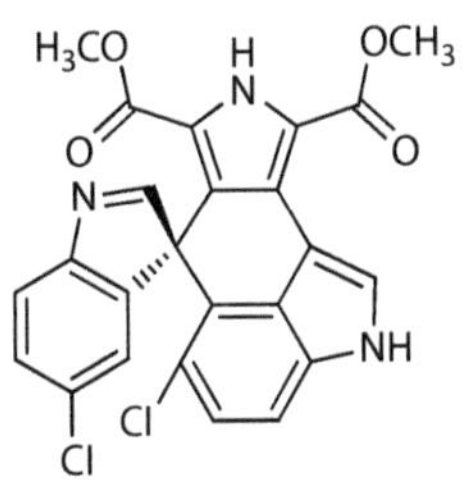

28

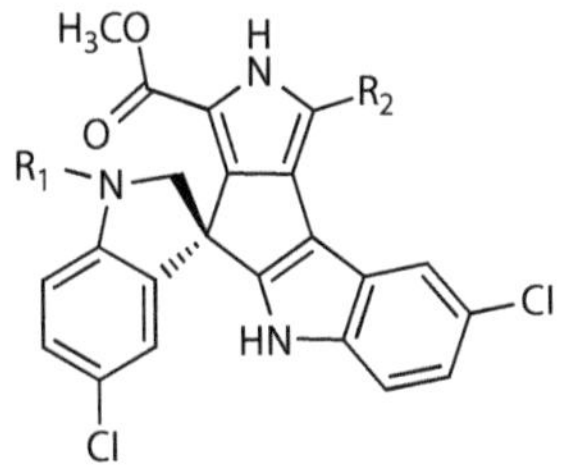

29. R$_1$=CH$_3$, R$_2$=H
30. R$_1$=H, R$_2$=H
31. R$_1$=CH$_3$, R$_2$=COOCH$_3$

Levantilides A (**32**) and B (**33**) are isolated from deep marine sediment derived strain *Micromonospora* sp. with moderate antitumor activity against several cell lines (Gartner et al., 2011).

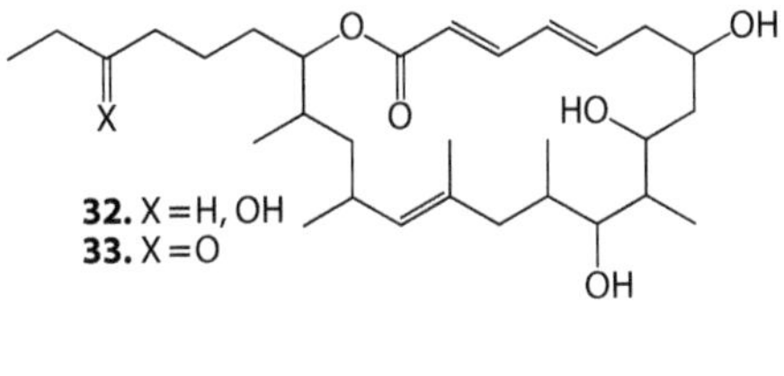

32. X=H, OH
33. X=O

Saliniquinones A–F (Murphy et al., 2010) (**34–39**), anthraquinone-γ-pyrones, and arenjimycin (Asolkar et al., 2010) (**40**) isolated from *Salinospora arenicola* exhibit potent inhibitory activity against HCT-116 cell line.

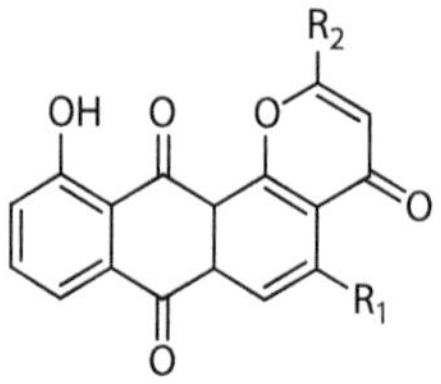

34. R₁ =CH₂OH, R₂=
35. R₁ =CH₂OH, R₂=
36. R₁ =CH₂OH, R₂=
37. R₁ =CH₂OH, R₂=
38. R₁ =Me, R₂=
39. R₁ =CH₂OH, R₂=

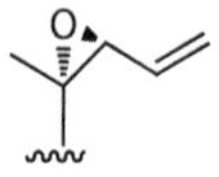

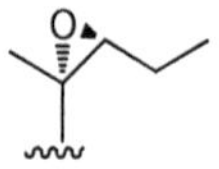

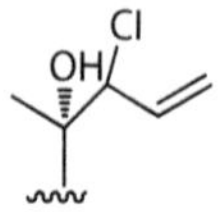

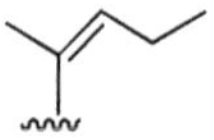

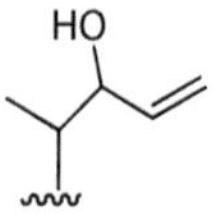

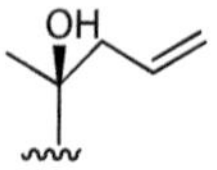

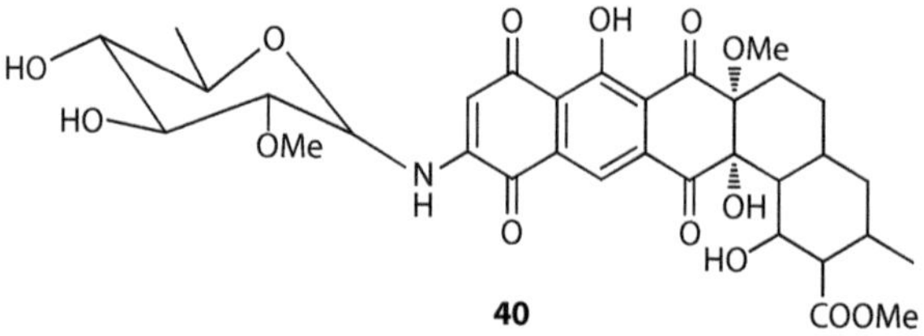

40

Marinomycins A–D (**41–44**), which are unusual macrodiolides composed of dimeric 2-hydroxy-6-alkenyl-benzoic acid lactones with conjugated tetraene-pentahydroxy polyketide chains, produced by *Marinispora* sp. CNQ-140, are isolated from a sediment sample collected at a depth of 56 m offshore of La Jolla, California. These compounds inhibit cancer cell proliferation with an average LC_{50} of 0.2–2.7 μM against the NCI's 60 cancer cell line panel. Marinomycin A showed significant tissue-type selectivity being more active against human melanoma cell lines LOX IMVI, M14, SK-MEL-2, SK-MEL-5, UACC-257, and UACC-62 (Kwon et al., 2006).

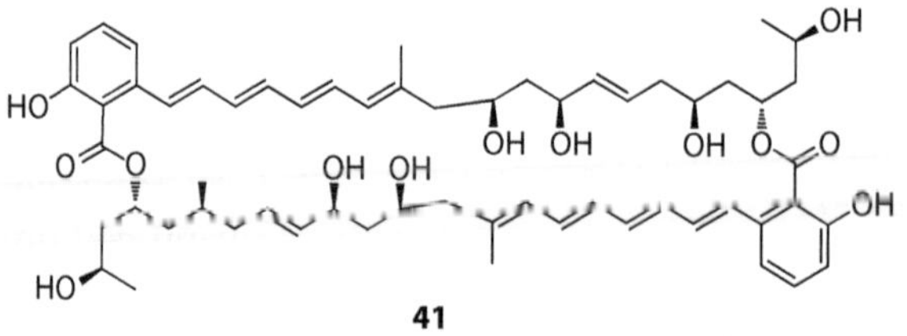

41

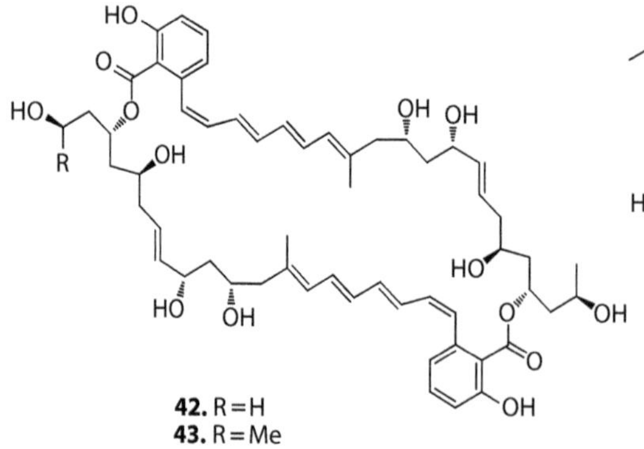

42. R=H
43. R=Me

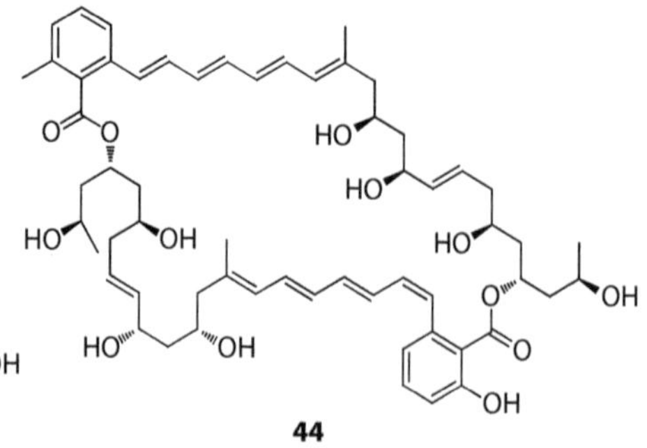

44

The manumycins constitute a class of compounds with antibiotic, cytotoxic, and other biological activities. It has been reported that manumycin A (**45**) and its analogues inhibit Ras farnesyltransferase and the growth of Ki-ras-activated murine fibrosarcoma in mice (Kouchi et al., 1999).

Piericidins C7 (**46**) and C8 (**47**), produced by *Streptomyces* sp. YM14-060, were isolated from an unidentified ascidian collected at Iwayama Bay, Palau Island. The biological activity of piericidins was examined using rat glial cells transformed with the adenovirus E1A gene (RG-E1A-7), Neuro-2a mouse neuroblastoma cells, C6 rat glioma cells, and 3Y1 rat normal fibroblast. Piericidins C7 and C8 showed selective cytotoxicity against RG-E1A-7 cells (IC_{50} of 1.5 and 0.45 nM, respectively) and inhibited the growth of Neuro-2a cells (IC_{50} of 0.83 and 0.21 nM, respectively) without cytotoxic cell death. On the other hand, C6 rat glioma cells and 3Y1 rat normal fibroblast were not affected by piericidins (Hayakawa et al., 2007). Various other antitumors from marine actinomycetes are tabulated in Table 22.1.

Table 22.1 Marine actinomycete compounds as antitumors

Source	Compound	Activity
Streptomyces strain CNQ-085	Daryamides	Antitumor
Streptomyces sp. FMA	Streptocarbazole A	Antitumor
Streptomyces chibaensis AUBN1/7T	Resistoflavin	Antitumor
Streptomyces sp. 04DH10	Streptochlorine	Antitumor
Streptomyces sp. BOSC-022A	Barmumycin	Antitumor
Streptomyces sp. B6007	Caprolactone	Antitumor
Streptomyces aureoverticillatus NPS001583	Aureoverticillactam	Antitumor
Streptomyces strain CNH990	Marmycin A	Antitumor
Streptomyces strain BL-49-58-005	Indoles	Antitumor
Streptomyces strain Mei37	Mansouramycin	Antitumor
Marinospora sp. CNQ-140	Marinomycins	Antitumor
Actinoalloteichus cyanogriseus WH1-2216-6	Caerulomycins	Antitumor
Streptomyces sp. MDG-04-17-069	Tartrolon	Antitumor

22.2.4 Antimalarial

Malaria is an arthropod-borne disease prevalent in many developing countries. Antimalarial drugs are mainly chemically synthesized and extracted from plant source, but recently, the development of drug-resistant strain possess a great challenge. The source of novel compounds from marine sources yielded potent antimalarial compounds from actinomycetes. B-carboline alkaloids, marinacarbolines A–D (48–51), showed potent activity against *Plasmodium falciparum* at varying inhibitory concentrations from 1.9 to 36.03 μM (Huang et al., 2011).

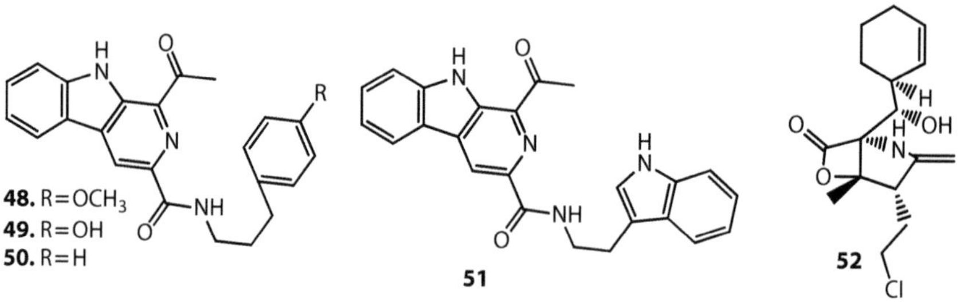

48. R=OCH₃
49. R=OH
50. R=H
51
52

Secondary metabolite salinosporamide A (52) isolated from *Salinispora tropica* showed protective action against the malaria parasite in the erythrocytic stage by inhibiting the 20S proteasome.

22.2.5 Enzyme inhibitory

Benzoxacystol (53–54), a 1,4-benzoxazine-type metabolite obtained from *Streptomyces griseus* (deep-sea sediment, Canary Basin), was an inhibitor of glycogen synthase kinase 3b, in addition to displaying weak antiproliferative activity against mouse fibroblast cells (Nachtigall et al., 2011).

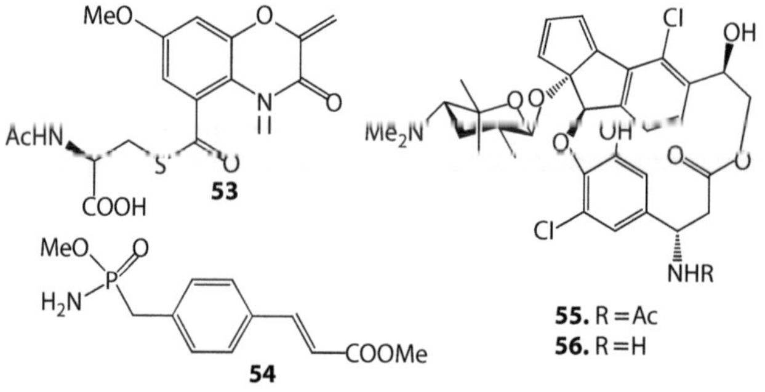

53
54
55. R =Ac
56. R =H

Fijiolides A (55) and B (56) were isolated from a *Nocardiopsis* sp. (sediment, Beqa Lagoon, Fiji). Fijiolide A 17 was a potent inhibitor of the TNF-α-induced transcription factor NFkB and induced quinone reductase activity, while fijiolide B 18 inhibited NFkB to a lesser extent and had no effect on quinone reductase activity.

Screening of marine actinobacterial strains for the presence of hydroxy-3-methylglutaryl-CoA reductase, a key enzyme in the mevalonate (MVA) pathway, identified a *Streptomyces* sp., the culture of which yielded three new phenazine-derived isoprenoids, JBIR-46–48 (**57–59**) (Izumikawa et al., 2010).

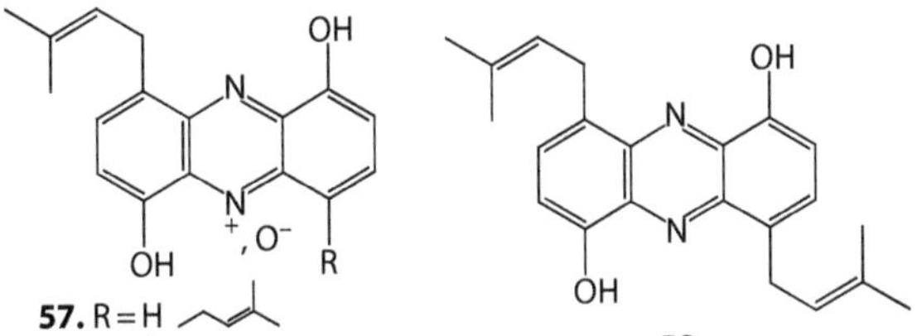

57. R=H
58. R=§

59

22.2.6 *Miscellaneous*

Tumescenamides A (**60**) and B (**61**) are cyclic peptides from *Streptomyces tumescens* (sediment, Big Drop-off, Republic of Palau). Tumescenamide A induced reporter gene expression under the control of the insulin-degrading enzyme promoter, suggesting promise as a potential treatment for Alzheimer's disease (Motohashi et al., 2010a).

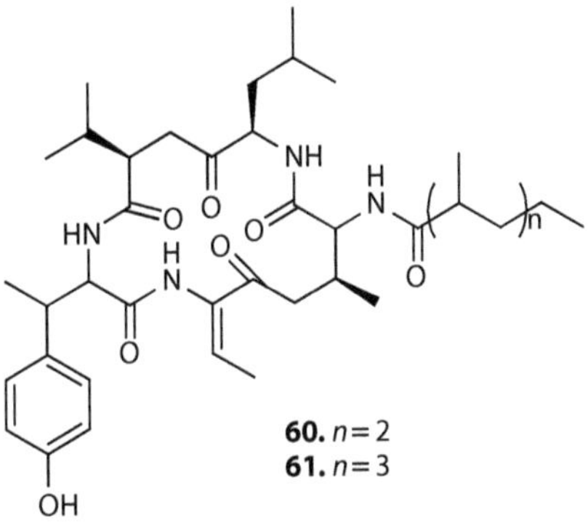

60. $n = 2$
61. $n = 3$

Three new species of *Streptomyces* sp. (sponge *Haliclona* sp., Tateyama City, Chiba, Japan) have been studied. The first species yielded two chlorinated indole-containing tetrapeptides, JBIR-34 (**62**) and JBIR-35 (**63**), both with weak DPPH activity (Motohashi et al., 2010b).

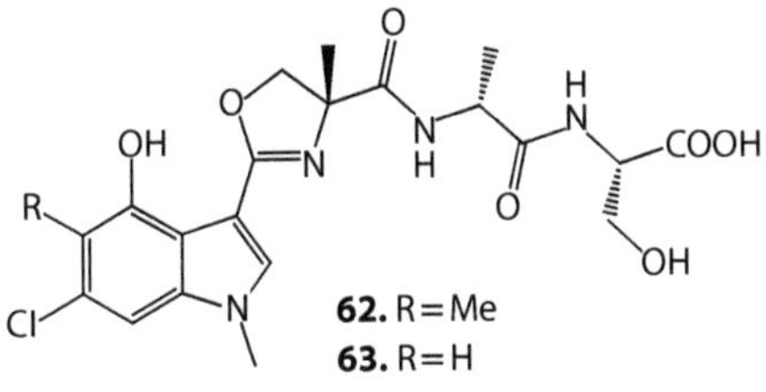

62. R=Me
63. R=H

22.3 *Antituberculosis compound from marine organisms*

TB has been known to humans since ancient days. TB occurred as an endemic among animals long before, which later affected humans. The presence of spinal TB was observed in the fossil bones dated around 8000 BC (Ayyazian, 1993; Basel, 1998; Gregory, 2015).

The confirmed presence of TB in humans was noted in the deformities of the skeletal and muscular parts of the Egyptian mummies of around 2400 BC (Andreas et al., 1997). Nevertheless, it could not be determined whether the disease was due to *Mycobacterium bovis* or *Mycobacterium tuberculosis*. Scientific investigation for the evolutionary origins of the *M. tuberculosis* complex has concluded that the most recent common ancestor was a human-specific pathogen, which underwent population congestion. Analysis of myco-bacterial interspersed repetitive units has allowed dating of the bottleneck to approximately 40,000 years ago, which corresponds to the period subsequent to the expansion of *Homo sapiens* out of Africa. This analysis of mycobacterial interspersed repetitive units also dated the *Mycobacterium bovis* lineage as dispersing approximately 6000 years ago, which may be linked to animal domestication and early farming (Wirth et al., 2008). Prior to this disease, it has been called by numerous names including consumption (because of the severe weight loss and the way the infection appeared to *consume* the patient), phthisis pulmonalis, scrofula, Pott's disease, and the white plague (because of the extreme pallor seen among those infected). In the 1700s and early 1800s, TB prevalence peaked in Western Europe and the United States and was undoubtedly the largest cause of death. For the past 100–200 years, it had spread toward Eastern Europe, Asia, Africa, and South America (Bloom and Murray, 1992). For the past 15,000–20,000 years, *M. tuberculosis* existed. It has been found in relics from ancient Egypt, India, and China. The effectiveness of oral thera-pies was developed in the 1950s that paved a way of optimism, leading to think that TB would soon be a thing of the past. Even today, the majority of the world's population has been exposed and infected with TB, with over 90% in the developing countries, less com-pared to developed countries. More than eight million new cases are recorded each year worldwide, and more than two million people are dying from it. Most of the death cases are recorded from countries that cannot offer modern drug therapy or even simple mod-ern conveniences. Frightening also is the specter of antibiotic-resistant TB, borne of inad-equate treatment and rendering cure difficult and prohibitively expensive. The control of TB remains a global health issue. In the nineteenth century, TB was known as "the captain of all men of death"—and so it remains (Table 22.2).

Marine natural products (MNP) are the best source of chemical diversity structures with promising biological activities. The chemical novelties of those compounds possess novel mechanisms of action. Parental compounds from marine sources are metabolized into fewer compound through metabolism. The use of these resultant compounds as drugs may reduce unwanted effects to host functions. The chemical and structural character-ization of marine natural products is not completed until the drug development process begins. At that time only the therapeutic values are evaluated. However, full assessment of pharmacological significance is not studied until after drugs reach the market. After this major technological process, it's easy to find out the active lead compound present in the parental compound that leads to designing the drug. In earlier days, there was no perfect structure for the active metabolites used for the treatment of infectious dis-eases; meanwhile, these active metabolites were available in the market. This major tech-nological advance allows for drug repurposing or repositioning of the existing drugs in the market to improve the efficacy of available drugs to treat drug resistant pathogens. A biological transformation method is used to design the active metabolites as drug with suitable pharmacological process such as physiochemical, pharmacodynamics and pharmacokinetic properties. Apart from many anti-TB compounds isolated from marine

Table 22.2 Anti-TB compound isolated from marine organisms

S. No.	Isolated compound	Organism	References
1.	Pseudopteroxazole	*Pseudopterogorgia elisabethae*	Rodriguez (1999)
2.	Massetolide A and viscosin	*Pseudomonas* sp.	Said et al. (2002)
3.	Caprazamycin B	*Streptomyces* sp MK730–62F2	Igarashi et al. (2003)
4.	Hexapeptides brunsvicamides	*Tychonema* sp.	Muller et al. (2006)
5.	Sansanmycin	*Streptomyces* sp.	Xie et al. (2007, 2008)
6.	Halicyclamine A	*Haliclona* sp.	Arai et al. (2008)
7.	Trichoderins	*Trichoderma* sp.	Pruksakom et al. (2010)
8.	4-Deoxybostrycin	*Nigrospora* sp.	Wang et al. (2013)
9.	Caulerpin	*Caulerpa* sp.	Chay et al. (2014)
10.	Kahalalide	*Elysia rufescens*	
11.	Lanosterol, nephalsterol	*Nepthea* sp.	Iguchi et al. (1989)
12.	Cyanthiwigins	*Myrmekiodermastyx*	El Sayed et al. (2000)
13.	Manadomanzamines	*Acanthostrongylophora* sp.	Peng et al. (2003)
14.	Saringosterol	*Lessonia nigrescens*	Wachter et al. (2001)
15.	Ingenamine	*Pachychalina* sp.	De Oliveira et al. (2004)
16.	Araguspongine C	*Xestospongia exigua*	Orabi et al. (2002)
17.	Ecteinascidins	*Ecteinascidia thurstoni*	Suwanborirux et al. (2002)
18.	12-Deacetoxyscalarin 19-acetate, manolide 25-acetate	*Brachiaster* sp.	Wonganuchitm et al. (2004)

organisms (Table 22.2), eight marine drugs have been approved by the FDA or EMEA, namely cephalosporin C, cytarabine (Ara-C), vidrabine (Ara-A), zincontide (Prialt), omega-3-acid ethyl esters (Lovaza), ET-743 (Yondelis), E7389 (Halaven), and brentuximabvendotin (SGN-35) (Mayer et al., 2011; Gerwick and Moore, 2012). Some of the MNP were in clinical trials such as soblidotin (TZT 1027), tasidotin, synthadotin (ILX-651), bryostatin 1, hemiasterilin (E7974) and pseudopterosin have been completed, a compound (plitidepsin) in phase III trials, six compounds DMXBAG (GTS-21), plinabulin (NPI 2358), PM00104, elisidepsin, PM01183, CDX-11 are in phase II trials, and four compounds such as marizomib, PM060184, SGN-75 and ASG-5ME are in phase I trials and have been processed with multitudinous marine natural products being investigated preclinical as clinical candidates (Liu, 2012). Unfortunately there were no active drugs from natural products available for Mycobacterium tuberculosis; this makes the new opportunity for medicinal chemists to find the drugs which could overcome drug resistance.

22.3.1   Antitubercular compounds produced by actinomycetes

Streptomycin was the first drug introduced in the market in 1944 for the treatment of TB, but almost immediately after the introduction of this drug, many patients were showing resistance to this antibiotic (Youmans et al., 1946; Pyle et al., 1947; Medical Research Council, 1948). Later, para-aminosalicylate (PAS) was introduced into the market, which overcame the emergence of drug-resistant TB strains (Medical Research Council, 1950). Isoniazide, also known as isonicotinylhydrazine or INH was introduced, and the treatment with isoniazid and streptomycin was even more effective than PAS. To date, many drugs are available, which are classified into two categories: first-line therapy and second-line therapy. First-line therapy includes five medications: isoniazid, pyrazinamide (analog of nicotinamide), ethambutol [(S,S)-2,2(ethylenediamine)di-1-butanol], rifampicin (lipophilic ansamycine), and streptomycin (aminocyclitol glycoside) (Goldberger, 1988). Second-line therapy includes cycloserine, capreomycin, fluoroquinolone, ethionamide, PAS, thioacet-azone, rifabutin, clofazimine, and macrolide, which are prescribed for drug-resistant cases (Iseman, 1993). Multidrug-resistant TB (MDR-TB), defined as resistant to at least isonia-zid and rifampin, was documented in nearly every country from 1994 to 2000 by WHO International Union against Tuberculosis and Surveillance Project (Gupta and Espinal, 2003). The ability of Mtb to remain dormant or persistent within host cells for many years with the potential to be activated allows the bacterium to escape the activated immune system of the host (Meena and Rajini, 2010). The latent or persistent TB infection is a major obstacle in the cure and prevention of TB. Reactivation of latent TB is a high-risk factor for disease development particularly in immunocompromised individuals such as those coinfected with HIV (Koul et al., 2011). However, there is an urgent to discover novel anti-biotics against drug resistance in *M. tuberculosis*. The two natural antibiotics, lasalocid and monensin, are isolated mainly from fungal mycelium of *Streptomyces cinnamonensis* (acti-nomycetes). These antibiotics belong to the family of the polycyclic carboxylic polyethers. Monensin and lasalocid and their metal complexes with Tl(I), La(III), and Gd(III) are active against *M. tuberculosis*. These ionophores could be considered as potential antitubercular agents for the future discovery of new drug design.

22.3.2   Mechanism of action of available drugs against TB bacteria

There are many compounds with antimicrobial activity as antimicrobial agents. The antimicrobial agents presented here will be grouped into three major classes, based on the target microbes: (1) antibacterials, (2) antimycobacterials, and (3) antifungals. Here, we are summarizing the current information about working models for the mechanism of action of antimycobacterial agents. The mechanism of action is categorized into the following types: inhibition of cell wall synthesis, interference with membrane integrity, inhibition of nucleic acid synthesis, protein synthesis, inhibition of synthesis of small molecules, and some other unknown effects. The existing drug therapy may avail this kind of mechanism; then, only the drug must be effective to kill the pathogenic *M. tuberculosis*. There are two types of drug therapy currently available for treating TB, that is, front-line therapy and second-line therapy. Front-line therapy treats the TB-affected person with the combination of different drugs such as rifampicin, isoniazid, ethambutol, and pyrazinamide for 2 months period, followed by rifampicin and isoniazid for an additional of 4 months. Rifampicin inhibits RNA polymerase, and the combination of isoniazid and ethambutol collapses the cell wall biosynthesis, but if these drugs are used singly, there is not much effect. Pyrazinamide specifically stops the replication of TB bacteria (Bai et al., 2011). Second-line therapies specifically for MDR-TB strains include fluoroquinolones (nalidixic acid, levofloxacin, ciprofloxacin, ofloxacin, sparfloxacin, moxifloxacin, and gatifloxacin), bedaquilines (TMC207), and aminoglycosides (kanamycin, streptomycin, spectinomycin, gentamycin, and hygromycin B). Isoniazid (INH) enters the mycobacterial cell by passive diffusion: movement of biochemicals and other atomic or molecular substances across the cell membrane without need of energy. INH is not toxic to the bacterial cell, but it acts as a prodrug and is activated by the enzyme Kat G, a multifunctional catalase–peroxidase. These compounds are potent inhibitors of critical enzymes in the biosynthesis of cell wall lipids and nucleic acids. Some INH-derived reactive species, such as nitric oxide, have a direct role in inhibiting mycobacterial metabolic enzymes (Timmins and Deretic, 2006). The mechanism of action of ethambutol is not completely understood. It inhibits arabinosyl transferase enzymes that are involved in the biosynthesis of arabinoglycan and lipoarabinomannan, which are essential elements within the mycobacterial cell wall (Belanger et al., 1996). Pyrazinamide's exact mechanism is not well understood, and it acts as a prodrug that is converted into the active form, pyrazionic acid, by the bacterial nicotinamidase/pyrazinamidase. However, it appears in the disruption of membrane energetic and inhibited cytoplasmic membrane function in *M. tuberculosis* (Zhang et al., 2003). Rifampicin involves in the inhibition of DNA-dependent RNA synthesis, caused by strong binding to the β subunit of the DNA-dependent RNA polymerase of prokaryotes, with a binding constant in the range of 10⁻⁹ M (Floss and Yu, 2005). Fluoroquinolones have been used as bactericidal by inhibiting its DNA gyrase at concentrations within achievable serum levels (Leysen et al., 1989; De Souza et al., 2006; Kubendiran et al., 2006; Sriram et al., 2006; Shandil et al., 2007). Bedaquiline (TMC207) inhibits the mycobacterial ATP synthase; this drug was recently approved as new drug for MDR-TB. Aminoglycosides disrupt protein synthesis by targeting the bacterial 30S ribosomal subunit. Due to the drug-resistance problem caused by the bacteria, it is necessary to develop new drugs that should inhibit the specific targets; this might be different from those currently used drugs. New drugs should inhibit the target present in bacteria, not in the human host.

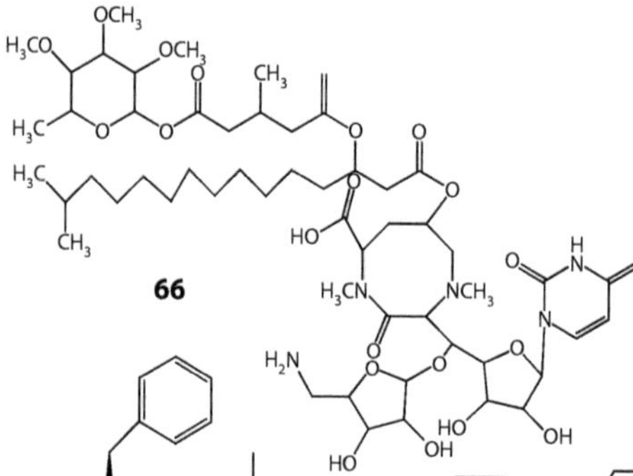

66

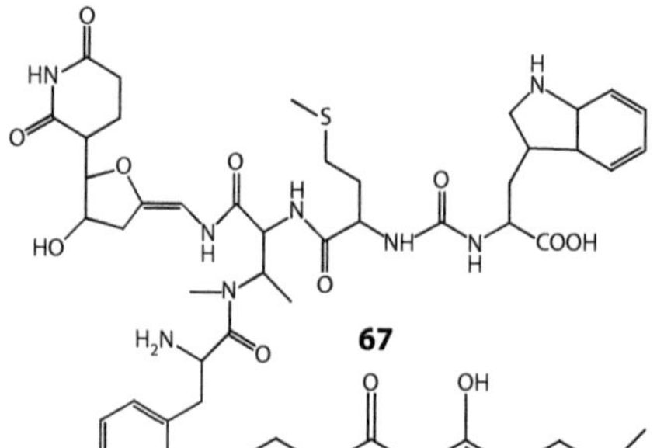

67

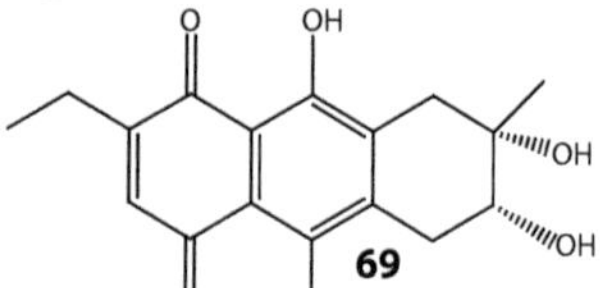

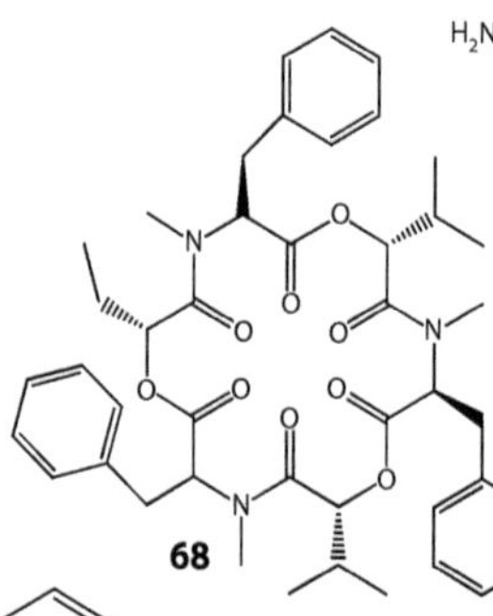

68

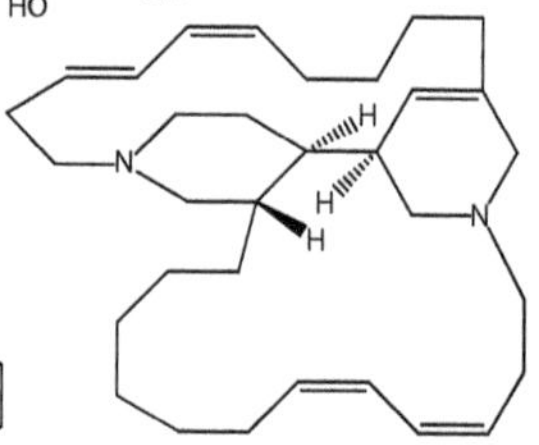

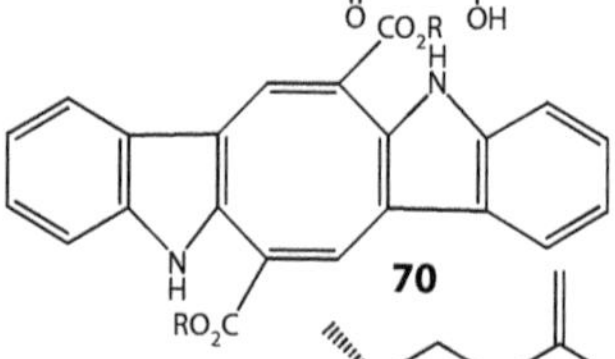

69

70

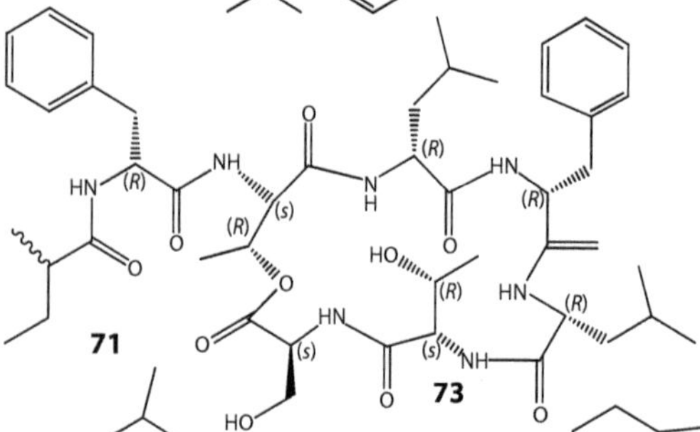

71

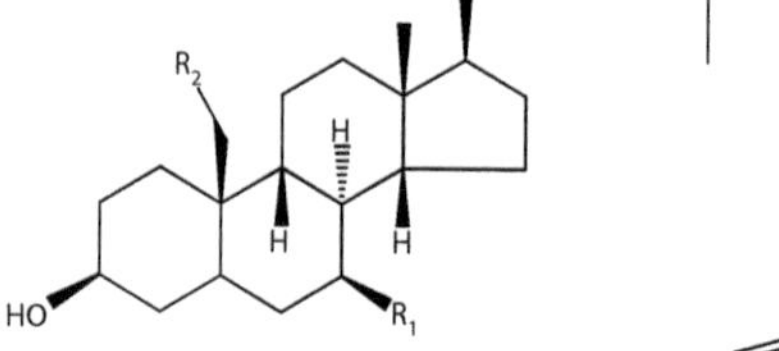

71. R1 = H, R2 = OH
72. R1 = OH, R2 = OH

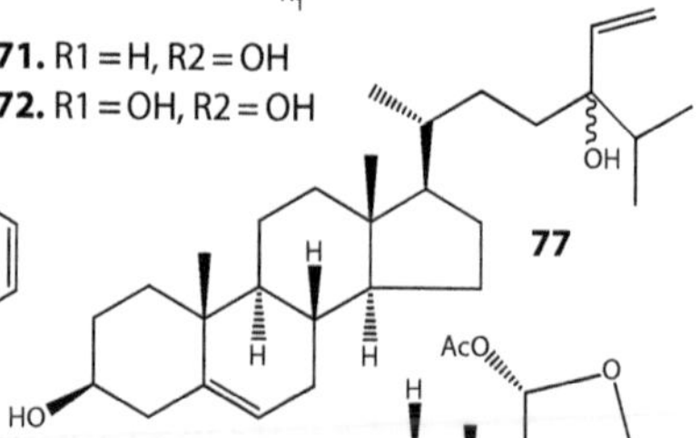

73

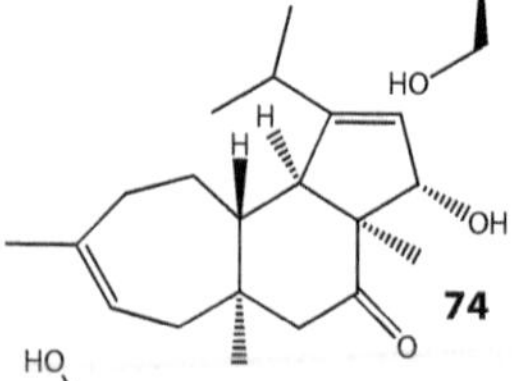

74

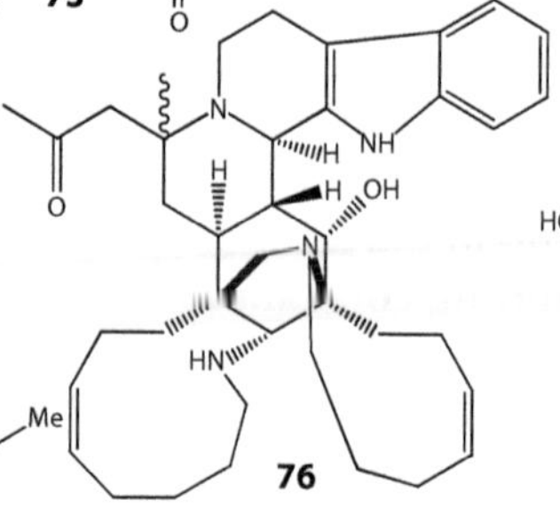

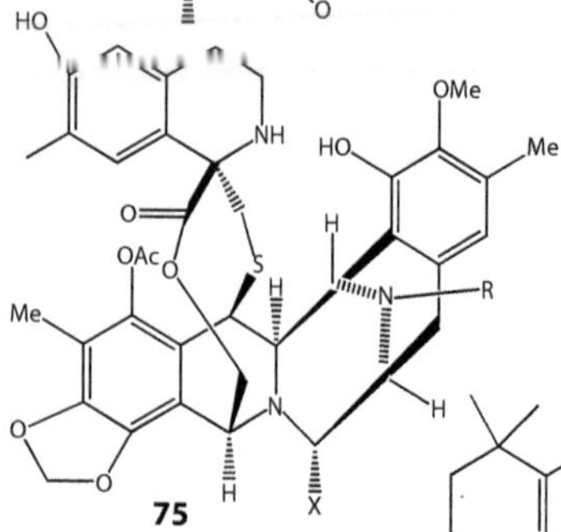

75

76

77

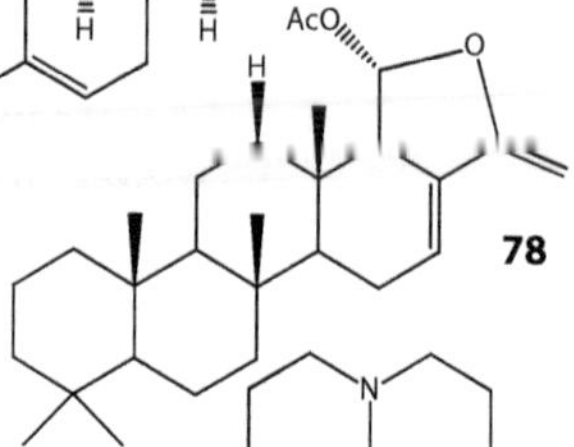

78

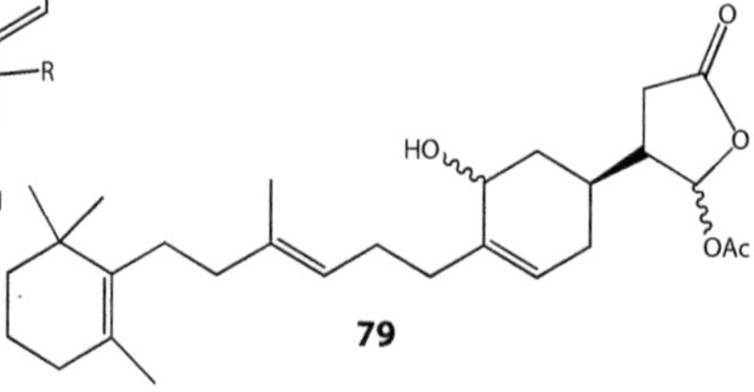

79

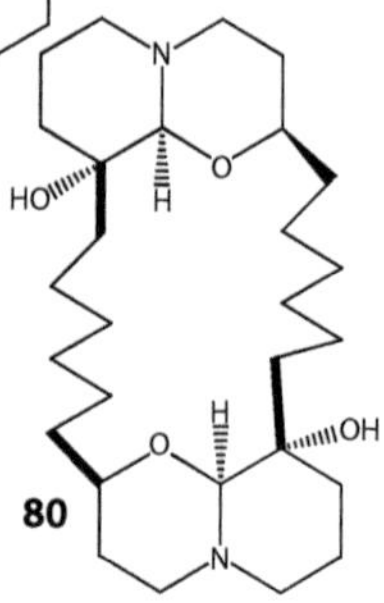

80

22.3.3 Marine-derived active lead compound as TB drug discovery

Pseudopteroxazole is a marine-derived compound that was isolated from the Indian gorgonian coral, *Pseudopterogorgia elisabethae*. It has been used to test against *M. tuberculosis in vitro*; pseudopteroxazole (**63**) was found to possess higher (97%) antimycobacterial activity (Rodriguez, 1999). However, attempts at *in vivo* efficacy studies have not yet been studied. If this compound is found to be efficient *in vivo*, then there is a possibility of medicinal chemistry to design a novel drug for the treatment of TB. Massetolide A (**64**) and viscosin (**65**) compound was isolated from *Pseudomonas* sp., which was symbiotically associated with marine alga, used as a lead compound to test against *M. tuberculosis*, and it was found to be effective in minimum level of dosage (Said et al., 2002). Caprazamycin B (**66**) compounds were isolated from *Streptomyces* sp., which were tested *in vitro* against MDR-TB. Majority of these compounds were isolated from marine species listed in Table 22.1, and their structures were also listed (**66–80**). Though most of the works were done through *in vitro* studies, there is no evidence for these isolated compounds to be tested *in vivo*. Sometimes, *in vitro* testing results will not reappear during the experiment conducted *in vivo*. This is due to metabolism: if compounds with potent *in vitro* activity are rapidly and extensively metabolized into inactive products, this may lead the isolated compound to a failed activity *in vivo*. Studies need to be conducted to determine whether the natural compounds exert a bactericidal and bacteriostatic effect against *M. tuberculosis*. Whole-cell assays and target-based assays are normally used in high-throughput screening (HTS). Target-based assay is more directional, but some natural products with good inhibition of pure enzymes may also show a limited inhibitory effect on Mtb cells due to their limited membrane permeability (Dhiman et al., 2005; Henriksson et al., 2007). In such cases, proper modification of the natural compound will be proposed to overcome the problem. In order to combine the advantages of the traditional whole-cell assays and the target-inhibiting assays, target-based whole-cell assays were developed in recent years to make HTS process more facile and efficient.

References

Altomare, C., Perrone, G., Zonno, M.C., Evidente, A., Pengue, R., Fanti, F., and Polonelli, L. 2000. Biological characterization of fusapyrone and deoxyfusapyrone, two bioactive secondary metabolites of *Fusarium semitectum*. *J. Nat. Prod.* 63(8): 1131–1135.

Amundson, M.L., Jimeno, J., Paz-Ares, L., Cortes-Funes, H., and Hidalgo, M. 2003. Progress in the development and acquisition of anticancer agents from marine sources. *Ann. Oncol.* 14(11): 1607–1615.

Andreas, G.N, Christian, J.H., Albert, Z., Ulrike, S., and Hjalmar, G.H. 1997. Molecular evidence for tuberculosis in an ancient Egyptian mummy. *The Lancet* 350(9088): 1404.

Arai, M., Sobou, M., Vicheze, C., Baughn, A., Hashizume, H., Pruksakorn, P., Ishida, S., Matsumoto, M., and Jacobs, W.R. Jr, Kobayashi M. 2008. Halicyclamine A, a marine spongean alkaloid as a lead for anti-tuberculosis agent. *Bioorgan. Med. Chem.* 16: 6732–6736.

Asolkar, R.N., Jensen, P.R., Kauffman, C.A., and Fenical, W. 2006. Daryamides A-C, weakly cytotoxic polyketides from a marine-derived actinomycete of the genus *Streptomyces* strain CNQ-085. *J. Nat. Prod.* 69(12): 1756–1759.

Asolkar, R.N., Kirkland, T.N., Jensen, P.R., and Fenical, W. 2010. Arenimycin, an antibiotic effective against rifampin- and methicillin-resistant *Staphylococcus aureus* from the marine actinomycete *Salinispora arenicola*. *J. Antibiot.* 63(1): 37–39.

Ayyazian, L.F. 1993. History of tuberculosis. In: Reichman, L.B. and Hershfield, E.S. (Eds.), *Tuberculosis*. Dekker, New York.

Bai, X., Chen, Y., Chen, W., Lei, H., and Shi, G. 2011. Volatile constituents, inorganic elements and primary screening of bioactivity of black coral cigarette holders. *Marine Drugs* 9: 863–878.

Barrero, A.F., Oltra, J.E., Herrador, M.M., Cabrera, E., Sanchez, J.F., Quílez, J.F., and Reyes, J.F. 1993. Gibepyrones: α-pyrones from Gibberella fujikuroi. *Tetrahedron* 49(1): 141–150.

Basel, H.H. 1998. History of tuberculosis. *Respiration* 65: 5–15.

Belanger, A.E., Besra, G.S., Ford, M.E., Mikusova, K., Belisle, J.T., Brennan, P.J., and Inamine, J.M. 1996.The embAB genes of *Mycobacterium avium* encode an arabinosyl transferase involved in cell wall arabinan biosynthesis that is the target for the antimycobacterial drug ethambutol. *Proc. Natl. Acad. Sci.* 93: 11919–11924.

Berdy, J. 2005. Bioactive microbial metabolites. *J. Antibiot.* 58(1): 1–26.

Bloom, B.R. and Murray, C.J. 1992. Tuberculosis: Commentary on re-emergent killer. *Science* 257: 1055–1064.

Chay, C.I.C., Cansino, R.G., Pinzón, C.I.E., Torres-Ochoa, R.O., and Martínez, R. 2014. Synthesis and Anti-tuberculosis activity of the marine natural product caulerpin and its analogues. *Mar. Drugs* 12(4): 1757–1772.

Chen, L., Sabo, J.P., Philip, E., Mao, Y., Norris, S.H., MacGregor, T.R., and Valdez, H. 2007. Steady-state disposition of the nonpeptidic protease inhibitor tipranavir when coadministered with ritonavir. *Antimicrob. Agents. Chemother.* 51(7): 2436–2444.

De Oliveira, J.H.H.L., Grube, A., Kock, M., Berlinck, R.G., Macedo, M.L., Ferreira, A.G., and Hajdu, E. 2004. Ingenamine G and cyclostellettamines G–I, K, and L from the new brazilian species of marine sponge *Pachychalina* sp. *J. Nat. Prod.* 67: 1685–1689.

De Souza, M.V., Vasconcelos, T.R., de Almeida, M.V., and Cardoso, S.H. 2006. Fluoroquinolones: An important class of antibiotics against tuberculosis. *Curr. Med. Chem.* 13: 455–463.

Demain, A.L. and Sanchez, S. 2009. Microbial drug discovery: 80 years of progress. *J. Antibiot.* 62(1): 5–16.

Dhiman, R.K., Schaeffer, M.L., Bailey, A.M., Testa, C.A., Scherman, H., and Crick, D.C. 2005. 1-Deoxy-D-xylulose 5-phosphate reductoisomerase (Ispc) from *Mycobacterium tuberculosis*: towards understanding mycobacterial resistance to fosmidomycin. *J. Bacteriol.* 187: 8395–8402.

Dickinson, J.M. 1993. Microbial pyran-2-ones and dihydropyran-2-ones. *Nat. Prod. Rep.* 10(1): 71–98.

Ding, L., Maier, A., Fiebig, H.H., Lin, W.H., and Hertweck, C. 2011. A family of multicyclic indolosesquiterpenes from a bacterial endophyte. *Organ. Biomol. Chem.* 9(11): 4029–4031.

El Sayed, K.A., Bartyzel, P., Shen, X., Perry, T.L., Zjawiony, J.K., and Hamann, M.T. 2000. Marine natural products as antituberculosis agents. *Tetrahedron* 56: 949–953.

Fenical, W. 1993. Chemical studies of marine bacteria: Developing a new resource. *Chem. Rev.* 93(5): 1673–1683.

Floss, H.G. and Yu, T.W. 2005. Rifamycin-mode of action, resistance, and biosynthesis. *Chem. Rev.* 105: 621–632.

Fu, P., Liu, P., Qu, H., Wang, Y., Chen, D., Wang, H., and Zhu, W. 2011a. Alpha-pyrones and diketopiperazine derivatives from the marine-derived actinomycete *Nocardiopsis dassonvillei* HR10-5. *J. Nat. Prod.* 74(10): 2219–2223.

Fu, P., Wang, S., Hong, K., Li, X., Liu, P., Wang, Y, and Zhu, W. 2011b. Cytotoxic bipyridines from the marine-derived actinomycete *Actinoalloteichus cyanogriseus* WH1-2216-6. *J. Nat. Prod.* 74(8): 1751–1756.

Gärtner, A., Ohlendorf, B., Schulz, D., Zinecker, H., Wiese, J., and Imhoff, J.F. 2011. Levantilides A and B, 20-membered macrolides from a *Micromonospora* strain isolated from the Mediterranean deep sea sediment. *Mar. Drugs* 9(1): 98–108.

Gerwick, W.H., Moore, B.S. 2012. Lessons from the past and charting the future of marine natural products drug discovery and chemical biology. *Chemistry & Biology* 19: 85–98.

Goldberger, M.J. 1988. Antituberculous agents. *Med. Clin. North Am.* 72: 661–668.

Grabley, S.T.R. (ed.). 1999. The impact of natural products on drug discovery. In: *Drug Discovery from Nature*. Berlin Heidelberg: Springer-Verlag, pp. 1–37.

Gregory, M.R. 2015. Tuberculosis retrenched at Saranac Lake: A herald for contemporary hospitals. *News-Medical.net—An AZoNetwork Site*.

Gupta, R. and Espinal, M. 2003. Stop TB Working Group on DOTSPlus for MDR-TB. A prioritized research agenda for DOTS-Plus for multidrug-resistant tuberculosis (MDRTB). *Int. J. Tuberc. Lung Dis.* 7: 410–414.

Hayakawa, Y., Shirasaki, S., Shiba, S., Kawasaki, T., Matsuo, Y., Adachi, K., and Shizuri, Y. 2007. Piericidins C7 and C8, new cytotoxic antibiotics produced by a marine *Streptomyces* sp. *J. Antibiot. (Tokyo)* 60(3): 196–200.

Helmke, E. and Weyland, H. 1984. *Rhodococcus marinonascens* sp. nov., an actinomycete from the sea. *Int. J. Syst. Bacteriol.* 34(2): 127–138.

Henriksson, L.M., Unge, T., Carlsson, J., Aqvist, J., Mowbray, S.L., and Jones, T.A. 2007. Structures of *Mycobacterium tuberculosis* 1-deoxy-D-xylulose-5-phosphate reductoisomerase provide new insights into catalysis. *J. Biol. Chem.* 282: 19905–19916.

Hopewood, D.A. 2007. Therapeutic treasures from the deep. *Nat. Chem. Biol.* 3(8): 457–458.

Huang, H., Yao, Y., He, Z., Yang, T., Ma, J., Tian, X., and Ju, J. 2011. Antimalarial carboline and indol-actam alkaloids from *Marinactinospora thermotolerans*, a deep sea isolate. *J. Nat. Prod.* 74(10): 2122–2127.

Igarashi, M., Nakagawa, N., Doi, N., Hattori, S., Naganawa H., and Hamada, M. 2003. Caprazamycin B, a novel anti-tuberculosis antibiotic, from *Streptomyces* sp. *J. Antibiot.* 56: 580–583.

Iguchi, K., Saitoh, S., and Yamada, Y. 1989. Novel 19-oxygenated sterols from the Okinawan soft coral *Litophyton viridis. Chem. Pharm. Bull.* 37: 2553–2554.

Ikeda, H., Ishikawa, J., Hanamoto, A., Shinose, M., Kikuchi, H., Shiba, T., and Omura, S. 2003. Complete genome sequence and comparative analysis of the industrial microorganism *Streptomyces avermitilis. Nat. Biotechnol.* 21(5): 526–531.

Iseman, M.D. 1993. Treatment of multidrug-resistant tuberculosis. *N. Engl. J. Med.* 329: 784–791.

Izumikawa, M., Khan, S.T., Takagi, M., and Shin-ya, K. 2010. Sponge-derived *Streptomyces* producing isoprenoids via the mevalonate pathway. *J. Nat. Prod.* 73(2): 208–212.

Jadulco, R., Brauers, G., Edrada, R.A., Ebel, R., Wray, V., Sudarsono, S., and Proksch, P. 2002. New metabolites from sponge-derived fungi *Curvularia lunata* and *Cladosporium herbarum. J. Nat. Prod.* 65(5): 730–733.

Jensen, P.R., Dwight, R.Y.A.N., and Fenical, W. 1991. Distribution of actinomycetes in near-shore tropical marine sediments. *Appl. Environ. Microbiol.* 57(4): 1102–1108.

Kamperdick, C., Van, N.H., and Van Sung, T. 2002. Constituents from *Miliusa balansae* (Annonaceae). *Phytochemistry* 61(8): 991–994.

Kelecom, A. 2002. Secondary metabolites from marine microorganisms. *An. Acad. Bras. Cienc.* 74(1): 151–170.

Kobayashi, S., Tsuchiya, K., Kurokawa, T., Nakagawa, T., Shimada, N., and Iitaka, Y. 1994. Pironetin, a novel plant growth regulator produced by *Streptomyces* sp. NK10958. II. Structural elucidation. *J. Antibiot.* 47(6): 703–707.

Kondoh, M., Usui, T., Nishikiori, T., Mayumi, T., and Osada, H. 1999. Apoptosis induction via microtubule disassembly by an antitumour compound, pironetin. *Biochem. J.* 340(Part 2): 411–416.

Konishi, H., Nakamura, K., Fushimi, K., Sakaguchi, M., Miyazaki, M., Ohe, T., and Namba, M. 1999. Manumycin A, inhibitor of ras farnesyltransferase, inhibits proliferation and migration of rat vascular smooth muscle cells. *Biochem. Biophys. Res. Commun.* 264(3): 915–920.

Koul, A., Arnoult, E., Lounis, N., Guillemont, J., and Andries, K. 2011. The challenge of new drug discovery for tuberculosis. *Nature* 469: 483–490.

Kubendiran, G., Paramasivan, C.N., Sulochana, S., and Mitchison, D.A. 2006. Moxifloxacin and gati-floxacin in an acid model of persistent *Mycobacterium tuberculosis. J. Chemother.* 18: 617–623.

Kwon, H.C., Kauffman, C.A., Jensen, P.R., and Fenical, W. 2006. Marinomycins A–D, antitumor-antibiotics of a new structure class from a marine actinomycete of the recently discovered genus "*Marinispora*". *J. Am. Chem. Soc.* 128(5): 1622–1632.

Lam, K.S. 2006. Discovery of novel metabolites from marine actinomycetes. *Curr. Opin. Microbiol.* 9(3): 245–251.

Leysen, D.C., Haemers, A., and Pattyn, S.R. 1989. Mycobacteria and the new quinolones. *Antimicrob. Agents Chemother.* 33: 1–5.

Li, S., Tian, X., Niu, S., Zhang, W., Chen, Y., Zhang, H., and Zhang, C. 2011. Pseudonocardians A-C, new diazaanthraquinone derivatives from a deep-sea actinomycete *Pseudonocardia* sp. SCSIO 01299. *Mar. Drugs* 9(8): 1428–1439.

Liu, Y. 2012. Renaissance of marine natural product drug discovery and development. *J. Marine. Sci. Res. Development* 2: 2.

Maldonado, L.A., Fenical, W., Jensen, P.R., Kauffman, C.A., Mincer, T.J., Ward, A.C., and Goodfellow, M. 2005. *Salinispora arenicola* gen. nov., sp. nov. and *Salinispora tropica* sp. nov., obligate marine actinomycetes belonging to the family Micromonosporaceae. *Int. J. Syst. Evol. Microbiol.* 55(Part 5): 1759–1766.

Maskey, R.P., Li, F.C., Qin, S., Fiebig, H.H., and Laatsch, H. 2003. Chandrananimycins A approximately C: Production of novel anticancer antibiotics from a marine *Actinomadura* sp. isolate M048 by variation of medium composition and growth conditions. *J. Antibiot.* 56(7): 622–629.

Mayer, A.M.S., Rodriguez, A.D., Berlinck, R.G.S., and Fusetani, N. 2011 Marine pharmacology in 2007–8: Marine compounds with antibacterial, anticoagulant, antifungal, anti-inflammatory, antimalarial, antiprotozoal, antituberculosis, and antiviral activities; affecting the immune and nervous system, and other miscellaneous mechanisms of action. *Comp. Biochem. Physiol. C-Toxicol. Pharmacol.* 153: 191–222.

McArthur, K.A., Mitchell, S.S., Tsueng, G., Rheingold, A., White, D.J., Grodberg, J., and Potts, B.C. 2008. Lynamicins a–e, chlorinated bisindole pyrrole antibiotics from a novel marine *Actinomycete*. *J. Nat. Prod.* 71(10): 1732–1737.

McGlacken, G.P. and Fairlamb, I.J.S. 2005. 2-Pyrone natural products and mimetics: Isolation, characterisation and biological activity. *Nat. Prod. Rep.* 22(3): 369–385.

Medical Research Council. 1948. Streptomycin treatment of pulmonary tuberculosis. *British Medical Journal* 2: 769 –782.

Meena, L.S., Rajni. 2010. Survival mechanisms of pathogenic *Mycobacterium tuberculosis* H37Rv. *FEBS* 277: 2416– 2427.

Mincer, T.J., Jensen, P.R., Kauffman, C.A., and Fenical, W. 2002. Widespread and persistent populations of a major new marine actinomycete taxon in ocean sediments. *Appl. Environ. Microbiol.* 68(10): 5005–5011.

Motohashi, K., Takagi, M., and Shin-ya, K. 2010b. Tetrapeptides possessing a unique skeleton, JBIR-34 and JBIR-35, isolated from a sponge-derived actinomycete, *Streptomyces* sp. Sp080513GE-23. *J. Nat. Prod.* 73(2): 226–228.

Motohashi, K., Toda, T., Sue, M., Furihata, K., Shizuri, Y., Matsuo, Y., and Seto, H. 2010a. Isolation and structure elucidation of tumescenamides A and B, two peptides produced by *Streptomyces tumescens* YM23-260. *J. Antibiot.* 63(9): 549–552.

Muller, D., Krick, A., Keluraus, S. 2006. Brunsvicamides A–C: sponge related cyanobacterial peptides with *Mycobacterium tuberculosis* protein tyrosine phosphatase inhibitory activity. *J. Med. Chem.* 49: 4871–4878.

Murphy, B.T., Narender, T., Kauffman, C.A., Woolery, M., Jensen, P.R., and Fenical, W. 2010. Saliniquinones A–F, new members of the highly cytotoxic anthraquinone-γ-pyrones from the marine actinomycete *Salinispora arenicola*. *Aust. J. Chem.* 63(6): 929–934.

Nachtigall, J., Schneider, K., Bruntner, C., Bull, A.T., Goodfellow, M., Zinecker, H., and Fiedler, H.P. 2011. Benzoxacystol, a benzoxazine-type enzyme inhibitor from the deep-sea strain *Streptomyces* sp. NTK 935. *J. Antibiot. (Tokyo)* 64(6): 453–457.

Olano, C., Méndez, C., and Salas, J.A. 2009. Antitumor compounds from actinomycetes: From gene clusters to new derivatives by combinatorial biosynthesis. *Nat. Prod. Rep.* 26(5): 628–660.

Orabi, K.Y., El Sayed, K.A., Hamann, M.T., Dunbar, D.C., Al Said, M.S., Higa, T., and Kelly, M. 2002. Araguspongines K and L, new bioactive bis-1-oxaquinolizidine N-oxide alkaloids from Red Sea specimens of *Xestospongia exigua*. *J. Nat. Prod.* 65(12): 1782–1785.

Pathom-Aree, W., Stach, J.E., Ward, A.C., Horikoshi, K., Bull, A.T., and Goodfellow, M. 2006. Diversity of actinomycetes isolated from Challenger Deep sediment (10,898 m) from the Mariana Trench. *Extremophiles* 10(3): 181–189.

Peng, J., Hu, J.F., Kazi, A.B., Li, Z., Avery, M., Peraud, O., Hill, R.T. et al. 2003. Manadomanzamines A and B: A novel alkaloid ring system with potent activity against mycobacteria and HIV-1. *J. Am. Chem. Soc.* 125(44): 13382–13386.

Petersen, F., Zähner, H., Metzger, J.W., Freund, S., and Hummel, R.P. 1993. Germicidin, an autoregulative germination inhibitor of *Streptomyces viridochromogenes* NRRL B-1551. *J. Antibiot.* 46(7): 1126–1138.

Pruksakorn, P., Arai, M., Kotoku, N., Vilcheze, C., Baughn, A.D., Moodley, P., Jacobs, W.R. Jr, and Kobayashi, M. 2010. Trichoderins, novel aminolipopeptides from a marine sponge derived *Trichoderma* sp., are active against dormant mycobacteria. *Bioorg. Med. Chem. Lett.* 20: 3658–3663.

Pyle, M.M. 1947. Relative numbers of resistant tubercle bacilli in sputa of patients before and during treatment with streptomycin. *Proc. Staff Meet. Mayo Clin.* 22: 465–472.

Rodriguez, A.D., Ramirez, C. Rodriguez II, Gonzalez, E. 1999. Novel antimycobacterialbenzoxazole alkaloids, from the west Indian sea whip *Pseudopterogorgiaelisabethae. Org. Lett.* 1: 527–530.

Rossi, M.H., Yoshida, M., and Maia, J.G.S. 1997. Neolignans, styrylpyrones and flavonoids from an *Aniba* species. *Phytochemistry* 45(6): 1263–1269.

Said, O., Khalil, K., Fulder, S., Azaizeh, M. 2002. Ethnopharmacological survey of medicinal herbs in Israel, the Golan Heights and the West Bank region. *J. Ethnopharmacol.* 83: 251–265.

Shandil, R.K., Jayaram, R., Kaur, P., Gaonkar, S., Suresh, B.L., Mahesh, B.N., Jayashree, R., Nandi, V., Bharath, S., and Balasubramanian, V. 2007. Moxifloxacin, ofloxacin, sparfloxacin, and ciprofloxacin against *Mycobacterium tuberculosis*: Evaluation of in vitro and pharmacodynamic indices that best predict in vivo efficacy. *Antimicrob. Agents Chemother.* 51: 576–582.

Sriram, D., Bal, T.R., Yogeeswari, P., Radha, D.R., and Nagaraja, V. 2006. Evaluation of antimycobacterial and DNA gyrase inhibition of fluoroquinolone derivatives. *J. Gen. Appl. Microbiol.* 52: 195–200.

Stackebrandt, E., Rainey, F.A., and Ward-Rainey, N.L. 1997. Proposal for a new hierarchic classification system, *Actinobacteria classis* nov. *Int. J. Syst. Bacteriol.* 47(2): 479–491.

Suwanborirux, K., Charupant, K., Amnuoypol, S., Pummangura, S., Kubo, A., and Saito, N. 2002. Ecteinascidins 770 and 786 from the Thai tunicate *Ecteinascidia thurstoni. J. Nat. Prod.* 65(6): 935–937.

Suzuki, K., Kuwahara, A., Yoshida, H., Fujita, S., Nishikiori, T., and Nakagawa, T. 1997. NF00659A1, A2, A3, B1 and B2, novel antitumor antibiotics produced by *Aspergillus* sp. NF 00659. I. Taxonomy, fermentation, isolation and biological activities. *J. Antibiot.* 50(4): 314–317.

Thaisrivongs, S., Romero, D.L., Tommasi, R.A., Janakiraman, M.N., Strohbach, J.W., Turner, S.R., and Watenpaugh, K.D. 1996. Structure-based design of HIV protease inhibitors: 5,6-dihydro-4-hydroxy-2-pyrones as effective, nonpeptidic inhibitors. *J. Med. Chem.* 39(23): 4630–4642.

Timmins, G.S. and Deretic, V. 2006. Mechanisms of action of isoniazid. *Mol. Microbiol.* 62: 1220–1227.

Tringe, S.G., Von Mering, C., Kobayashi, A., Salamov, A.A., Chen, K., Chang, H.W., and Rubin, E.M. 2005. Comparative metagenomics of microbial communities. *Science* 308(5721): 554–557.

Tsuchiya, K., Kobayashi, S., Nishikiori, T., Nakagawa, T., and Tatsuta, K. 1997. NK10958P, a novel plant growth regulator produced by *Streptomyces* sp. *J. Antibiot.* 50(3): 259–260.

Turner, S.R., Strohbach, J.W., Tommasi, R.A., Aristoff, P.A., Johnson, P.D., Skulnick, H.I., and Thaisrivongs, S. 1998. Tipranavir (PNU-140690): A potent, orally bioavailable nonpeptidic HIV protease inhibitor of the 5,6-dihydro-4-hydroxy-2-pyrone sulfonamide class â. *J. Med. Chem.* 41(18): 3467–3476.

Wachter, G.A., Franzblau, S.G., Montenegro, G., Hoffmann, J.J., Maiese, W.M., and Timmermann, B.N. 2001. Inhibition of *Mycobacterium tuberculosis* growth by saringosterol from *Lessonia nigrescens. J. Nat. Prod.* 64:1463–1464.

Wang, C., Wang, J., Huang, Y., Chen, H., Li, Y., Zhong, L., Chen, Y. et al. 2013. Anti-mycobacterial activity of marine fungus-derived 4-deoxybostrycin and nigrosporin. *Molecules* 18(2): 1728–1740.

Ward, A.C. and N. Bora 2006. Diversity and biogeography of marine actinobacteria. *Curr. Opin. Microbiol.* 9(3): 279–286.

Wirth, T., Hildebrand, F., Allix-Béguec, C., Wolbeling, F., Kubica, T., Kremer, K., and Niemann, S. 2008. Origin, spread and demography of the *Mycobacterium tuberculosis* complex. *PLoS Pathog.* 4(9): e1000160.

Wonganuchitmeta, S., Yuenyongsawad, S., Keawpradub, N., and Plubrukarn, A. 2004. Antitubercular sesterterpenes from the Thai sponge *Brachiaster* sp. *J. Nat. Prod.* 67(10): 1767–1770.

Wu, Z., Xie, L., Xia, G., Zhang, J., Nie, Y., Hu, J., and Zhang, R. 2005. A new tetrodotoxin-producing actinomycete, *Nocardiopsis* dassonvillei, isolated from the ovaries of puffer fish *Fugu rubripes. Toxicon* 45(7): 851–859.

Xie, Y.Y., Chen, R.X., Si, S.Y., Sun, C.H., Xu, H.Z. 2007. A new nucleosidyl-peptide antibiotic, sansan-mycin. *J. Antibiot.* 60: 158–161.

Xie, Y.Y., Xu, H.Z., Si, S.Y., Sun, C.H., and Chen, R.X. 2008. Sansanmycins B and C, new components of sansanmycins. *J. Antibiot.* 61: 237–240.

Xu, L.-Y., Quan, X.-S., Wang, C., Sheng, H.-F., Zhou, G.-X., Lin, B.-R., Jiang, R.-W., and Yao, X.-S. 2011. Antimycins A19 and A20, two new antimycins produced by marine actinomycete *Streptomyces antibioticus* H74-18. *J. Antibiot. (Tokyo)* 64(10): 661–665.

Yano, K., Yokoi, K., Sato, J., Oono, J., Kouda, T., Ogawa, Y., and Nakashima, T. 1986. Actinopyrones A, B and C, new physiologically active substances. II. Physico-chemical properties and chemical structures. *J. Antibiot.* 39(1): 38–43.

Youmans, G.P., Williston, E.H., Feldman, W.H., and Hinshaw, H.C. (1946) Increase in resistance of tubercle bacilli to streptomycin: a preliminary report. *Proc. Staff Meet. Mayo Clin.* 21: 126–127.

Zhang, W., Liu, Z., Li, S., Yang, T., Zhang, Q., Ma, L., and Zhang, C. 2012. Spiroindimicins A-D: New bisindole alkaloids from a deep-sea-derived actinomycete. *Org. Lett.* 14(13): 3364–3367.

Zhang, Y., Wade, M.M., Scorpio, A., Zhang, H., and Sun, Z. 2003. Mode of action of pyrazinamide: Disruption of *Mycobacterium tuberculosis* membrane transport and energetics by pyrazinoic acid. *J. Antimicrob. Chemother.* 52: 790–795.

chapter twenty-three

Antiprotozoal agents derived from natural soil and aquatic actinobacteria
Fighting one microbe with another

Diana R. Cundell and Michael P. Piechoski

Contents

23.1 Introduction

Human protozoan infections are common in developing nations, where economic resources for their treatment are Spartan at best, but are an issue around the world (Andrews et al., 2014). A list of the most common protozoa, their transmission methods, and their mortality is shown in Table 23.1. Protozoa are transmitted via two main methods: some participate in the life cycle of insects and are acquired through blood meals (insect vector–borne diseases), and others are acquired through ingestion of contaminated food or water (Fletcher et al., 2012; Andrews et al., 2014).

Insect vector–borne diseases such as malaria, sleeping sickness, and Chagas' disease are typically associated with tropical and subtropical climates due to the limitations in the temperature survival range of the vector as well as the protozoan itself (Beck-Johnson et al., 2013). Typical in this respect is the parasite–vector relationship between *Plasmodium*, the agent of malaria, and *Anopheles*, the mosquito, which transmits it (Beck-Johnson et al., 2013). Studies by Beck-Johnson et al. (2013) suggest that the baseline for persistence of the *Anopheles* mosquito is 20°C–26°C, that is, the minimum temperature at which juvenile insects can grow into adults to generate a significant population. The authors' research also predicted that in order for mosquitoes to be long lived, temperatures needed to be in the 20°C–30°C range (Beck-Johnson et al., 2013). These observations correspond with decreased longevity of *Anopheles* at temperatures above 32°C and in the East Africa highlands where temperatures are cool. Similar temperature range data (20°C–26°C) have been seen for the vector of *Trypanosoma brucei*, the tsetse fly (Moore et al., 2011). In contrast, the

Table 23.1 Global impact of common human protozoal parasites

Protozoa	Annual rates	Spread by	Annual deaths
Cryptosporidium parvum[a–c]	20% of diarrhea in developing countries[b]	Contaminated food and water	30%–50% of deaths in children under 5 years[c]
Entamoeba histolytica	500 million[b]	Contaminated food and water	100,000[b]
Giardia lamblia (giardiasis)	20%–30% prevalence in developing countries	Contaminated food and water	Unknown
Leishmania species[a]	12 million	Phlebotamine fly	52,000 (2010)
Plasmodium[a] species (malaria)	207 million[d]	*Anopheles* mosquito	627,000 (2012)
Toxoplasma gondii[a] (toxoplasmosis)	25%–30% of the world's population are infected	Zoonosis or mother to child	May contribute to epilepsy[e] or schizophrenia[f]
Trypanasoma brucei[a] (HAT)	7000–8000 reported[g]	Tsetse fly	9,000 (2010)
T. cruzi[a] (Chagas')	8–10 million[a]	Reduviid bug	10,000–20,000 (2010)[a,h]

Abbreviations: HAT, Human African trypanosomiasis (sleeping sickness).
[a] Andrews et al. (2014).
[b] Fletcher et al. (2012).
[c] Snelling et al. (2007).
[d] World Health Organization (2013).
[e] Flegr et al., (2002).
[f] Torrey et al. (2007).
[g] Mumba et al. (2011).
[h] Kirchhoff (2011).

reduviid (triatomine) bug, the agent of Chagas' disease, requires slightly higher temperatures but similarly possesses a very narrow optimal range of 26°C–29°C (Lazzari, 1991).

In contrast, although many of the insect vector–associated conditions are primarily seen in developing nations, there is strong evidence that gastrointestinal protozoa are carried asymptomatically across the globe but to a lesser degree in developed versus developing nations (Fletcher et al., 2012). A prime example of differential carriage is seen in *Cryptosporidium parvum*, which is the most common causes of gastrointestinal illness in the world (Fletcher et al., 2012). While *C. parvum* has a asymptomatic carriage rate of 0.1% among people in developed countries (Fletcher et al., 2012), typically 10%–30% of those in developing nations carry the microbe (Current and Garcia, 1991). The number of individuals with antibodies against *C. parvum*, that is, demonstrating previous exposure (seroprevalence rate), ranges from 25% to 35% in industrialized countries, up to 68%–88% in Russia (Egorov et al., 2004), and 95% in South America (Casemore et al., 1997). Seroprevalence rates increase with age (Egorov et al., 2004) and are higher in those working with animals, for example, dairy farmers (Lengerich et al., 1993), and in day-care centers (Kuhls et al., 1994).

As with other infectious diseases, the approach to reducing the incidence of protozoan infection has been via a classical two-pronged attack, namely prevention of infection and effective treatment. However, unlike other infectious diseases, protozoa have not been as simple to combat by traditional methods and to understand why requires a fuller understanding of the microbes themselves. In terms of prevention, the first issue is that several of the successful pathogenic protozoa have exploited transmission by insect vectors and require sophisticated eradication techniques in order to reduce infection rates

(Alphey et al., 2010). Furthermore, prevention by vaccination has been stymied since protozoa, unlike bacteria and the majority of viruses, have the capacity to rapidly alter their surface antigens and thus exist in many forms (Gurarie and McKenzie, 2006; Baral, 2010). In terms of effective treatment of protozoa, the main issue is that since they are precursors and very similar to animal cells, the therapeutics that kill them or prevent their reproduction typically also damage human tissues (Andrews et al., 2014).

Given the issues described earlier, it is apparent that much of the focus remains on the development of novel antiprotozoal therapies, which display lower toxicity and greater specificity against the parasites they are designed to eliminate. Such compounds would be expected to be the products of moist soil and water bacteria, which share their ecosystems and wish to avoid predation (Fenical and Jensen, 2006). Indeed, important in this respect are likely to be *Actinobacteria*, known antimicrobial-producing organisms that are dominant in most soil and aquatic ecosystems (Hogan, 2010). *Actinobacteria* have been shown to be less likely to be eaten by protozoa that are bacterial predators (Fenical and Jensen, 2006), and it is of interest that the parent compound of the antiprotozoal drug metronidazole was originally isolated from the actinomycete *Streptomyces eurocidus* (Osato et al., 1955). To fully understand the difficulties in controlling protozoal infections, the next two sections deal with the current status of eradication and treatment strategies. The final sections then explore the currently available literature on natural antiprotozoal agents from soil and aquatic *Actinobacteria* as well as suggesting further directions for drug discovery for this urgently needed field of infectious disease.

23.2   Hurdles in protozoan eradication: vector removal and vaccine development strategies

For those protozoa that are acquired through contaminated food or water, there are a number of practices that can help in limiting disease acquisition, including safe cooking practices, hand washing, and boiling or chlorination of water (Fletcher et al., 2012). In 2012, only 67% of the world's population had access to potable water and that statistic is not expected to improve by 2015 according to the World Health Organization (2012). In addition, two of the three major protozoal causes of gastrointestinal infection, namely, *C. parvum* and *G. lamblia*, produce small cysts resistant to the effects of chlorination and are easily able to enter municipal water supplies (Carmena et al., 2007). Indeed, in a study by Carmena and colleagues (2007), *Cryptosporidium* cysts were found in 15.4%–63.5% of various raw surface water samples, 30.8% of treated water from small treatment facilities, and 26.8% of chlorinated tap water. Hardy *Giardia* cysts were found in 26.9%–92.3% of various raw surface water samples, 19.2% of treated water from small treatment facilities, and 26.8% of chlorinated tap water (Carmena et al., 2007). Thus, contact of food with contaminated water used for irrigation or food preparation or contact with sewage can then spread protozoa further into the human food chain (Fletcher et al., 2012).

To improve potable water, bios and filtration has been used as a low-cost method to purify water in a number of developing countries, most notably Kenya (Tiwari et al., 2009). Hazard Analysis Critical Control Point (HACCP) principles, which identify and address areas where water contamination can occur, have been applied in some countries as a means of providing potable water free of microbiological health hazards (Gunnarsdottir and Gissurarson, 2008). Slow sand filtration, solar technology, and membrane technology are effective methods for reducing pollution from rainwater to be used as a safe drinking water supply (Helmreich and Horn, 2009).

For insect eradication, insecticides have had some success in controlling these agents, especially triatomine bugs that spread Chagas' disease (Schofield and Dias, 1999). The Southern Cone Initiative, mounted in six South American countries (Argentina, Bolivia, Brazil, Chile, Paraguay, and Uruguay) in 1991, sought to keep triatomine bugs out of the dwellings of residents by using housing with less crevasses to eliminate hiding spaces and spraying insecticide (Schofield and Dias, 1999). These measures, together with screening all blood donors for infection, reduced Chagas' disease rates by over 70% in these countries (Schofield and Dias, 1999).

An additional approach to insect vector reduction has been the sterile insect technique (SIT) in which irradiated, sterile male insects are bred and released (Alphey et al., 2010). Since the sterile males are unable to have offspring with wild females, they reduce the insect population over a series of seasons (Alphey et al., 2010). This strategy has recently proven very successful in reducing the tsetse fly population in Senegal by 99% (Dicko et al., 2014). Mosquito control has also been assisted by using the SIT program, and on the surface, this seems a more effective method than spraying of insecticides from crop planes (Alphey et al., 2010).

There are, however, two downsides to the SIT program. The first is that it is expensive to run and therefore not cost-effective for most developing nations (Ansari et al., 1977; Dame et al., 1981). Production costs for sterile males of the *Anopheles albimanus* MACHO mosquito strain in El Salvador in 1979 were estimated at US$156 per million (Dame et al., 1981) and for *Aedes aegypti* and *Culex pipiens fatigans* mosquito pupae at US$58 and US$50, respectively, per million in India in 1975 (Ansari et al., 1977). These numbers do not sound a great deal but when translated into the millions of insects needed make the options very costly. Secondly, for insects where juveniles are also damaging, such as the reduviid (triatomine) bugs that spread Chagas' disease, this technique has been less effective (Alphey et al., 2010). Thirdly, simply removing the insect vector may not be enough as some experts believe that if this occurred, others would merely take their place as carriers of disease (Fang, 2010). Finally, cross-breeding between insect species might also increase the area in which they can survive (Lounibos et al., 2002). Indeed, a study by Lounibos et al. (2002) demonstrated that *Aedes aegypti* mosquitoes could be fertilized by another species, *A. albopictus* (Asian tiger), which can survive much cooler temperatures. If we add in cross-breeding to the equation, these results suggest that vector control is only part of the solution to protozoal disease.

A second strike that has spelt doom for many infections has been the use of vaccines, which have delivered numerous bacteria and viruses into the ash can of history. Despite years of trying, however, vaccine development against protozoa still remains in its infancy, with only two vaccines showing at least partial efficacy (Dumonteil et al., 2012; Walsh, 2013). The first is a vaccine against childhood malaria that has shown a 25% efficacy against infection in early clinical trials (Walsh, 2013). The second is a sophisticated, recombinant vaccine that appears to slow the progression of early Chagas' disease (caused by *Trypanosoma cruzi*) in patients who have either indeterminate (those who have symptoms but have not produced antibodies) or determinate (those with both symptoms and antibodies) disease (Dumonteil et al., 2012). A *G. lamblia* vaccine has been available for dogs since 2000, but no effective human vaccine has yet followed (Olsen et al., 2000). Finally, it is possible that a live attenuated vaccine may be made for *L. donovani* using gene knockout–created versions of the pathogen (Dey et al., 2013). This vaccine, which involves mutation of the p27 gene in the protozoan, has recently been shown to be effective in murine models (Dey et al., 2013), but its efficacy in humans remains to be established. No other vaccines for protozoa exist or are under development.

In the absence of successful preventative methods, research into treatment methods has increased (Andrews et al., 2014). Currently used antiprotozoal pharmaceuticals include those originally developed for the treatment of protozoa and those developed for other conditions, including cancer treatment, have been successfully redirected for protozoal therapy. The focus of the next section is to delve into these strategies further with a view to underlying weaknesses that might be rectified by future drug development.

23.2.1   Hurdles in protozoan treatment: current strategies and drug repurposing

Current treatment methods and their limitations for the treatment of the eight main global protozoal diseases are shown in Table 23.2. Issues with drug treatment of protozoa can be grouped into several criteria, namely resistant strain emergence, toxicity of the drugs, and contraindicated patients. Resistant strains are a particular problem for the treatment of malaria (Andrews et al., 2014) as well as trypanosomal diseases (Kirchhoff, 2011; Odero, 2013) and are thought to result from the significant antigenic variation demonstrated by these species (Gurarie and McKenzie, 2006; Chitanga et al., 2011). Indeed, in the case of malaria, studies by Gurarie and McKenzie (2006) have suggested that in a typical infection, the protozoa present will vary in growth profile, antigenic presentation, and consequently drug sensitivity/resistance. Selective pressure provided by the drug being administered will then sway the distribution in favor of the typically slower growing drug-resistant strains and eliminate the faster growing drug-sensitive ones (Gurarie and McKenzie, 2006).

Table 23.2 Treatment strategies for major protozoal diseases

Protozoa	Treatments	Issues with treatment
C. parvum	Nitazoxanide (3–14 days)[a]	AIDS patients need a higher dosage[a]
E. histolytica	Metronidazole (mild) paromomycin or iodoquinol (invasive) (7 days)[a]	Issues with differential penetration of drugs to reach tissue ameba[b]
G. lamblia	Metronidazole (3 days), tinidazole (single dose), or albendazole (5 days)[a]	Treatment failures and relapses occur probably indicative of resistant strains[c]
Leishmania	Pentamidine/Amp B (IV; 28 days)[d]	Variable efficacy and toxicity[e]
Plasmodium	Gold standard are ACTs (3 days)[f]	Resistant strains emerging[d,g]
Toxoplasma gondii	Pyrimethamine, combined with various drugs (4–6 weeks)[h]	Adverse toxic effects and concerns about efficacy[h]
Trypanosoma brucei (HAT)	Early = suramin (IV; 21 days) or pentamidine (IV; 14–21 days). Late = melarsoprol (10 days) or eflornithine (4 weeks)[i]	Pentamidine is very toxic, melarsoprol-resistant strains now very common (~30% of species)[i]
T. cruzi	Nifurtimox, benznidazole (years)[j]	Toxicity, resistance[j]

Abbreviations: ACTs, Artemisinin combination therapies; Amp B, amphoteracin B; HAT, Human African trypanosomiasis (sleeping sickness); IV, Intravenous.

[a] Fletcher et al. (2012).
[b] Espinosa et al. (2012).
[c] Busatti et al. (2009).
[d] Andrews et al. (2014).
[e] Croft and Coombs (2003).
[f] Perez-Jorge (2014).
[g] Ariey et al. (2014).
[h] Hökelek (2014).
[i] Odero (2013).
[j] Kirchhoff (2011).

Trypanosome resistance is an even more complex phenomenon that seems tied to their immunological diversity (Baral, 2010). These protozoa possess a unique capacity to alter their variant specific glycoproteins (VSGs) into procyclic acidic repetitive proteins (PARPs, or procyclins) and then later back to VSG (Baral, 2010). Apart from hampering vaccine development, this also means that trypanosomes develop drug resistance that is not driven by selective pressure but instead by favorable, random mutations (Chitanga et al., 2011).

Studies by Chitanga et al. (2011) examined the incidence of diminazene aceturate (DA) resistance in a trypanosome strain related to human African trypanosomiasis (HAT) namely *T. congolese*. DA is typically given to livestock to treat trypanosome infections, and although natural populations of *T. congolese* also infect wildlife, these would not be expected to demonstrate resistance to it due to lack of exposure (Chitanga et al., 2011). Chitanga et al. (2011) were surprised to learn that 33 of the 34 strains (97.1%) they isolated from wildlife demonstrated either full resistance (12/34; 35.2%) or partial resistance (21/34; 61.8%) to DA, suggesting that protozoan resistance was a random mutational event, that is, not driven by selective pressure.

Apart from resistance, the second major issue with treating protozoal infections has been the toxicity of the drugs being employed (Kirchhoff, 2011; Croft and Coombs, 2003; Odero, 2013; Andrews et al., 2014). The treatment of late-stage infections with *T. brucei* is a case in point, with melarsoprol, a derivative of arsenic, demonstrating severe side effects including heart failure and death (Brun et al., 2011). Even in the "safer" forms of antiprotozoal medications currently on the market, there are significant contraindications for many of them (Table 23.2). Treatment failures whether due to protozoan resistance (Andrews et al., 2014) or adverse effects are also common in nifurtimox (Coura et al., 1997; Jackson et al., 2010) and benznidazole treatments (Coura et al., 1997; Pinazo et al., 2010). Indeed, in a recent study by Pinazo et al. (2010), performed in an area with good clinical care (Spain), adverse events from benznidazole treatment of Chagas' disease in adults were of the order of 9% of patients studied. The most common issues were severe skin rash (urticaria) and gastrointestinal distress (Pinazo et al., 2010). Nifurtimox produced even higher numbers of adverse events, according to a study by Jackson et al. (2010). These authors observed a 97.5% rate of adverse events, which were often gastrointestinal distress (35.1%) or neurological (27.5%) (Jackson et al., 2010). Sudden critical reactions were also seen in 6.4% of the patients, necessitating immediate withdrawal from the study (Jackson et al., 2010). Jackson et al. (2010) also made a very worrying observation, namely that their *"results call into question the safety of large-scale use of nifurtimox among adult patients in settings with difficult access to medical care, such as poor rural areas of South America"* and that *"Easily accessible medical care and close clinical follow-up appear to be prerequisite to nifurtimox treatment."*

In treating protozoan infections, it has also become apparent that combination therapies seem to be optimal to avoid the development of resistance in the organism (Burrows et al., 2013; Andrews et al., 2014). This has been particularly important in the treatment of diseases like malaria, where the original quinine and then chloroquine–primaquine combinations resulted in rapid development of resistance (Burrows et al., 2013). Nifurtimox–eflornithine combination therapy (NCET) has also recently become a treatment option for HAT, but sadly this is only effective against the gambiense form of the disease and not the rapidly fatal rhodiense form (Priotto et al., 2007, 2009).

Given that new combinations and therapies needed to be developed, pharmaceutical companies have had to make some choices. Although new drug discovery is appealing, cost (around US$800 million) and time (one to two decades) of drug development have been major obstacles and have limited development of significant numbers of novel antimicrobials from nature. In contrast, the use of high-throughput targeted searches of existing drug mechanisms has led to their being put to different uses (Andrews et al., 2014), termed repurposing. Repurposing an existing product has many benefits for both the developer and the consumer other than time to market, namely extant safety data and information about side effects (Andrews et al., 2014). Since many protozoan infections are not associated with significant revenue for companies investing in them, they have therefore become the preferred route to identify new antiprotozoal strategies.

At least nine protozoan drugs exist that were originally developed for use in other areas, most notably chemotherapy and infectious disease treatment (Andrews et al., 2014). Such repurposing can occur because the drug is found to have multiple efficacies, as with doxycycline (Tan et al., 2011) and amphotericin B (Ostrovsky-Ziechner et al., 2003), or because the drug was not effective as a chemotherapeutic agent, as with miltefosine (Dorlo et al., 2012). Like other extant antimicrobial drugs, some repurposed drugs are of natural origin including paromomycin (Wiwanitkit, 2012), semisynthetic like doxycycline (Pickens and Tang, 2010; Tan et al., 2011), or synthetic like miltefosine (Croft and Engel, 2006) and sulfadiazine (Leport et al., 1988). A list of some commonly used protozoal treatments and their origins is shown in Table 23.3.

Drug repurposing has been facilitated by an understanding of protozoan biology as well as a definition of the molecular targets for the therapeutics themselves (Nwaka and Hudson, 2006). Studies have shown that *T. brucei* and *T. cruzi* together with malaria and leishmania express adenylate cyclase, phosphodiesterase (PDE), and kinase enzymes,

Table 23.3 Origin and usage of drugs repurposed for protozoal treatment regimens

Protozoan	Therapeutic	Origin	Original uses
Leishmaniasis	Miltefosine[a]	Synthetic[a]	Breast cancer[a]
Plasmodium	Clindamycin[b]	Natural; *Streptomyces lincolnensis*[b]	Antibiotic[b]
	Doxycycline[c,d]	Semi-synthetic; oxytetracycline parent molecule from *S. rimosus*[c]	Antibiotic[c,d]
	Sulfamethoxazole[f]	Synthetic; derived from dye P[e]	Antibiotic[e]
	Trimethoprim[f]	Synthetic[g]	Antibiotic[e]
T. gondii	Spiramycin[h]	Natural; *S. ambofaciens*[i]	Antibiotic[i]
T. brucei	Amphotericin B[j]	Natural; *S. nodosus*[k]	Antifungal[j]
	Eflornithine[l]	Synthetic[m]	Antitumor[n] and hirsutism[o]
	Paromomycin[p]	Natural; *S. rimosus*[p]	Antibiotic[p]

Source: Modified from Andrews, K.T. et al., *Int. J. Parasitol. Drugs Drug Resist.*, 4, 95, 2014.

[a] Croft and Engel (2006).
[b] Spízek and Rezanka (2004).
[c] Pickens and Tang (2010).
[d] Tan et al. (2011).
[e] Lesch (2007).
[f] Manyando et al. (2013).
[g] Eliopoulos and Huovinen (2001).
[h] Couvreur et al. (1988).
[i] Karray et al. (2007).
[j] Ostrovsky-Zeichner et al. (2003).
[k] Caffrey et al. (2001).
[l] Kennedy (2013).
[m] Metcalf et al. (1978).
[n] Gerner and Mayskens (2004).
[o] Wolf et al. (2007).
[p] Wiwanitkit (2012).

which have slightly differing structures to those of humans (Parsons et al., 2005). Drugs antagonistic to these three target enzymes have been used with much success in tumor therapy (Andrews et al., 2014), where they are effective against rapidly multiplying cells, a situation also seen in malaria, trypanosome, and leishmania infections (Andrews et al., 2014). To date, four classes of chemotherapeutic drugs have been explored for their ability to suppress protozoa, namely histone deacetylase (HDAC) enzyme inhibitors (Kelly et al., 2012; Subathdrage et al., 2012), PDE inhibitors (De Koning et al., 2012), carbonic anhydrase inhibitors (Krungrkai et al., 2008), and tyrosine kinase inhibitors (Patel et al., 2013).

HDAC are a group of therapeutics that have received general acceptance for the treatment of T cell lymphoma and have an effect on the proteins that regulate cell proliferation and differentiation in mammalian cells (Dokmanovic et al., 2007). These agents are currently being tested as to their potential efficacy in the treatment of malaria (Subathdrage et al., 2012) and HAT (Kelly et al., 2012) with promising *in vitro* results already being reported at predicted human serum physiological levels (micromolar range). Recent studies by De Koning et al. (2012) demonstrated that the use of the PDE inhibitor tetrahydrophthalazinone, which they termed compound A (Cpd A), was able to significantly affect the *in vitro* growth of *T. brucei*. Their results showed that *T. brucei* proliferation was inhibited immediately and that protozoa died within 3 days (De Koning et al., 2012). De Koning et al. (2012) also observed that Cpd A prevented cell division of the trypanosome, resulting in multinucleated, multiflagellated cells that eventually lysed (De Koning et al., 2012).

Krungrai et al. (2008) tested a library of carbonic anhydrase inhibitors for efficacy in inhibiting the growth of *P. falciparum* in cell cultures. They observed one compound to be optimal in this respect, the sulfonamide derivative 4-(3,4-dichlorophenylureido) thioureido-benzenesulfonamide, which was inhibitory when used at concentrations as low as 0.18 µM (Krungrai et al., 2008). Finally, the anilinoquinazoline drugs lapatinib and canertinib, which are potent synthetic tyrosine kinase inhibitors, have also been shown to be effective against *T. cruzi* at micromolar levels (estimated dose 50 µM) through blocking the parasite's uptake of transferrin and effectively starving it (Patel et al., 2013). Quinazoline-based drugs have at their heart a dicyclic benzene plus pyrimidine ring and have had numerous uses, most notably antiviral and antibacterial (Wang and Gao, 2013). This latter class of drugs is of interest since they are also structurally similar to the powerful fluoroquinolone antibiotics, which include ciprofloxacin (Cipro) and levafloxacin (Levaquin) (Wang and Gao, 2013).

Despite the advantages of repurposing extant drugs, this strategy is not, however, without risks, and these mainly surround the usage of synthetic agents. The first is that the rapidity with which protozoa develop resistance toward artificial compounds cannot be predicted. The second concerns the reason for repurposing, that is, whether the drugs have significant side effects and therefore require patient monitoring. This second issue may make usage nonviable in developing countries, which is unfortunately where most protozoan infections occur. Third, natural antimicrobials and their derivatives typically possess better bioavailability and capacity to bind to cellular targets than their synthetic counterparts, again lending support for new therapeutics to come from natural sources (Wright, 2010). Thus, despite the considerable time and costs, it still behooves pharmaceutical companies to invest and investigate new sources for antiprotozoal drugs.

As with antibiotics, it appears that many of these will come from the common terrestrial and aquatic *Actinobacteria* species, with the dominant genus of interest being *Streptomyces* (Raja and Prabakarana, 2011). Actively feeding and reproducing protozoa have a need for water, although they can exist in cyst form in dry, desert soils (Darby et al., 2006). Since they are common to all types of water environments and moist soils, it is to be expected that *Actinobacteria* isolated from these locations will secrete antiprotozoals to avoid being grazed on (Fenical and Jensen, 2006). The next section deals with novel antiprotozoal drugs that have been identified from naturally occurring microbes in these environments.

23.3   New sources of antiprotozoal drugs

Bioprospecting for antibiotic agents began with the first discoveries of the soil-dwelling *Streptomyces* species by Selman Waksman in 1952 (Waksman and Lechevalier, 1962) and have continued apace ever since (Raja and Prabakarana, 2011). Actinobacteria are a substantial and diverse group of microbes able to retain dominance in extreme habitats (Kieft, 1991; Kurapova et al., 2012) as well as those with rich microenvironments (Piechoski et al., 2012). They are the source of 80% of the world's currently available antimicrobial drugs (Raja and Prabakarana, 2011), and this usefulness has not even been fully explored yet. Given that they are also the source of the majority of repurposed antiprotozoal drugs (Table 23.3), many scientists are now investigating the capacity of terrestrial and aquatic *Actinobacteria* to generate novel therapeutic agents. The next two sections explore the state-of-the-art knowledge on novel antiprotozoal agents gleaned from these two types of environments.

23.3.1   Terrestrial and endophytic microbes

Several currently available antiprotozoal drugs have been derived from at least four *Streptomyces* species, with *S. lincolensis* (Ryu et al., 2005) and *S. rimosus* (Duggar, 1948) being isolated from cultivated, temperate soils in Lincoln, Nebraska and Columbia, Missouri, respectively. The remaining two have come from diverse tropical soils including China (*S. ambofaciens*; Pinnert-Sindico, 1954) and the Orinoco Basin of Venezuela (*S. nodosus*; Donovick et al., 1954). Although these sound very disparate places, it is important to note that each would be expected to boast a rich microbial ecosystem in which competition for organic nutrients would be intense (Hawksworth and Colwell, 1992). Surviving such locations would need a "competitive edge," such as that provided by antimicrobials, and dominant microbes would therefore be expected to possess this capability. It is not unsurprising that *Streptomyces* species constitute 40% of all soil bacteria (Boone et al., 2001), and under dry and alkaline conditions, these dominate the terrestrial microbial population (Vetsigian et al., 2011).

Although soil and water are primary *Streptomyces* habitats (Ward and Bora 2006; Mokrane et al., 2013), they can also be found in hay (Horn et al., 2012) such as endophytes associating with plants (Ezra et al., 2004; Castillo et al., 2006) and in the gastrointestinal tracts of insects such as termites (Maheswarappa et al., 2013) and wasps (Oh et al., 2011). Each of these locations has its own series of *Actinobacteria* and very different antimicrobial abilities, and several have potential antiprotozoal activities as shown in Table 23.4.

At least two of the newly discovered antiprotozoal agents from terrestrial *Streptomyces*, pyrocoll and prodigiosin, contain pyrolle rings (Dieter et al., 2003; Maheswarappa et al., 2013). Pyrocoll's pyrolle ring is attached to a diketopiperazine core (Dieter et al., 2003), whereas prodigiosin possesses a tripyrrole structure (Maheswarappa et al., 2013), but for both, the pyrolle ring structure is at the heart of their activities (Genes et al., 2011). The importance of the pyrolle ring as an antimicrobial entity has been known for over a decade (Bijev et al., 2000, 2002) and underlined by the more recent studies of Varshney et al. (2014). Early studies by Bijev et al. (2000, 2002) synthesized a total of 12 novel pyrolle ring-containing cephalosporin derivatives with antibacterial activity comparable to cefalexin

Table 23.4 Novel antiprotozoal agents from terrestrial *Actinobacteria*

Location/Soil Type	Species	Agent	Protozoal target
Consett, County Durham United Kingdom (alkaline, black soil)[a]	*Streptomyces* AK409	Pyrocoll (pyrrole-2-carboxylic acid)	*Plasmodium* and *Trypanosoma* species
Yogyakarta, Indonesia (monsoon climate, volcanic-mix soil)[b]	*Streptomyces* a-WM-JG-16.2	Jogyamycin (analog of pactamycin)[c]	*Plasmodium* species
Manu region, Peru on *Monstera* plant species[d]	Endophytic *Streptomyces* (MSU-2110)	Coronomycin (functionalized peptide)	*Plasmodium* species
Northern Territory, Australia on *Kennedia* (snakevine) plants[e]	Endophytic *Streptomyces* NRRL 3052	Munumbicins	*Plasmodium* species
Soil termite gut[f]	Unknown *Streptomyces*	Prodigiosin (tripyrrole)	*T. cruzi* (Chagas' disease)[g]
Tamil Nadu, India (saltpan)[h]	*Streptomyces* VITVSK1	Quinone derivative	*Plasmodium* species

[a] Dieter et al. (2003).
[b] Iwatsuki et al. (2012).
[c] Hanessian et al. (2013).
[d] Ezra et al. (2004).
[e] Castillo et al. (2006).
[f] Maheswarappa et al. (2013).
[g] Genes et al. (2011).
[h] Gopal et al. (2013).

(Bijev et al., 2000, 2002). This group also observed that those cephalosphorins in which the beta-free position and ester groups remained accessible, that is, uncoupled, were typically the most antimicrobial (Bijev et al., 2000, 2002). Varshney et al. (2014) confirmed these observations and additionally observed increased antimicrobial potency with substituted 2,5-dimethyl pyrrole derivatives and substituted 1,3-benzoxazine-4-one derivatives. Indeed, a patent for the use of diaryl piperidyl pyrrole derivatives as antiprotozoal agents was issued in 2006 (Biftu et al., 2006).

Interestingly, prodigiosin possesses all these qualities since it is a 4-methoxy-5-[(Z)-(5-methyl-4-pentyl-2H-pyrrol-2-ylidene) methyl]-1H,1′H-2,2′-bipyrrole (Khanafari et al., 2006). Furthermore, prodigiosin has been shown to be effective in killing not only bacteria but also the protozoan *T. cruzi* (Genes et al., 2011). Genes et al. (2011) determined that the death of *T. cruzi* occurred through the induction of apoptotic death of their mitochondria, in a similar manner to that of cancer cells.

Although prodigiosin and the pyrolle-ringed derivatives that could be made from it have potential, the question at this point may be why are some *Actinobacteria* making these compounds? In the case of prodigiosin, this synthesis may be a successful evolutionary strategy related to continued survival of the triatomine-associated *Streptomyces* during *T. cruzi* infection (Genes et al., 2011). Early studies by Azumbuja et al. (2004) demonstrated that ingestion of a blood meal containing *T. cruzi* caused an increase in triatomine midgut bacteria populations 10,000 fold within a few hours. The authors showed that this increase was specific to *Serratia* and *Actinobacteria* species and was associated with a decline in viability in *T. cruzi* parasites, presumably through the antiprotozoal actions of prodigiosin (Azambuja et al., 2004). Such *Actinobacteria*–insect mutualism has been previously associated with the manufacture of dentigerumycin, a selective antifungal in fungus-farming ants (Oh et al., 2011), and sceliphrolactam, from *Streptomyces* living in wasp midguts (Oh et al., 2011).

In addition to mutualism with insects, some *Streptomyces* have established endophytic relationships with plants, which may or may not be beneficial (Ezra et al., 2004; Castillo et al., 2006). Since the plant–bacterial interaction is a specific one, this is clearly another potential source of unique antimicrobials. At least two useful antimalarial agents have been isolated to date, the munumbicins in 2002 (Castillo et al., 2006) and the coronamycins in 2004 (Ezra et al., 2004). Both appear to be peptide antibiotics with strong efficacy against the malaria parasite but have not been further developed since their discoveries a decade ago.

Bioprospecting yields new compounds but may frequently also uncover modifications of known antimicrobials. This has certainly been the case for jogyamycin, which appears to be a structural analog of pactamycin (Iwatsuki et al., 2012). Pactamycin was first isolated from *S. pactum* in 1961 and appeared to have great potential as an antimicrobial agent until its toxicity emerged (Hanessian et al., 2013). Structure–activity analysis of pactamycin has recently led to the generation of successful synthetic and semisynthetic antiprotozoal agents, which have revolved around modifications of the 6-methylsalicylic acid (6-MSA) ester moiety (Hanessian et al., 2013). Interestingly, the natural product jogyamycin, which is a de-6-methylsalicylyl-7-deoxypactamycin congener of pactamycin, is more potent than the parent molecule against trypanosomes and possesses less toxicity (Iwatsuki et al., 2012; Hanessian et al., 2013). Finally, the discovery of the quinone derivative being secreted by saltpan *Streptomyces* (Gopal et al., 2013) may open up possibilities for this class of agents to be used in malarial treatment. Although structurally separate from quinine and its derivatives, which have been traditionally used for malaria, quinones and their substitutions have been long known to possess broad-spectrum antimicrobial activity (Valderrama et al., 1999). They have also proven particularly useful in the treatment of cancer; for example, the semisynthetic drug adriamycin (doxorubicin) is an anthracycline that contains a quinone ring at the heart of its activities (Weiss, 1992). Given that many of the repurposed drugs being used for antiprotozoal therapy seem to have originated as chemotherapeutics (Table 23.2), quinones and their derivatives may also have a role to play in the treatment of both types of conditions.

23.3.2   Aquatic microbes

The marine environment is the largest ecosystem on the planet and probably possesses the greatest biodiversity in the world (Felczykowska et al., 2012). Actinobacteria species abound in marine environments, comprising about 10% of the species, are harvested from marine aggregates and sediments (Ward and Bora, 2006). Some of these species may be unique to ocean environments (Jensen et al., 2005). For example, *Salinispora* (Maldonado et al., 2005) and *Marinispora* (Jensen et al., 2005), which require seawater for growth, have distinct marine chemotype signatures.

Although able to exist independently of vertebrate and invertebrate life, *Actinobacteria* are often found in association with ocean corals (Lam, 2006). Coral-dwelling species include members of the *Streptomyces*, *Gordoniacae*, and *Actinokineospora* genera and are often unique to each type of cnidarian (Lam, 2006). As with terrestrial locations, those *Actinobacteria* able to live in extreme environments, including those of the Arctic and Antarctic oceans, in the deep oceans, or in seas with high levels of pollutants, are poorly understood since their culture is problematic (Barone et al., 2014). Indeed, it has been estimated that less than 0.1% of all microbes in the oceans today have been discovered so far (Simon and Daniel, 2009). Given the issues with isolation and culture of these important microbes, many investigators have turned to the use of metagenomics (Stach et al., 2003; Mincer et al., 2005). Metagenomic methods, through which genes are identified via high-throughput DNA sequencing technology and cloned into culturable vectors, may well be one major method to allow the natural products they code for to be identified and harvested (Tringe et al., 2005). In recent years, the discovery of several antiprotozoal agents using both classical isolation and metagenomic techniques has been seen, and the characteristics of those of note are shown in Table 23.5.

Some of the agents originating from marine *Actinobacteria* have already been discovered and utilized. A prime example of this is echinomycin, which is also made by the terrestrial species *S. lansiliensis* (Steinerová et al., 1987) and *S. echinatus* (Onnis et al., 2009). Echinomycin was studied as a potential chemotherapeutic agent, when it was found to inhibit hypoxia-inducible factors (HIFs), but clinical trials were halted in the 1980s due to lack of efficacy (Onnis et al., 2009). The observations of Espinosa et al. (2012) that marine *Streptomyces*-isolated echinomycin inhibited the multiplication of *Entamoeba histolytica* suggest that we may again see a chemotherapeutic being repurposed as an antiprotozoal. Similarly, tirandamycin, also originally identified from two terrestrial *Streptomyces* species (Rahman et al., 2010), has now been shown to possess both antiamoebic (Espinosa et al., 2012) and antinematodal activities (Yu et al., 2011), suggesting that this drug may also be repurposed.

Others are novel antimalarial agents, including specific protease inhibitors (Karthik et al., 2014), the polyketide mollemycin (Raju et al., 2014), and the trioxacarcin gutingimycin (Manivasagan et al., 2014). Marine *Actinobacteria* have long been known to produce a

Table 23.5 Antiprotozoal agents from marine *Actinobacteria*

Location	Species	Agent(s)	Target	Protozoa affected
Fishers Island Sound, United States[a]	Novel *Streptomyces* UR-F11 and URI-F39	Echinomycin, tirandamycin	DNA replication RNA polymerase	*E. histolytica*
Nicobar, Indian Ocean[b]	Unknown *Actinobacteria*	Unknown	Protease inhibition	*Plasmodium* species
South Molle Island, Queensland[c]	Novel *Streptomyces* sp. (CMB-M0244)	Mollemycin (polyketide)	Unknown	*Plasmodium* species
Götingen, Sweden[d,e]	*Streptomyces* species B8652	Gutingimycin (trioxacarcin)	Unknown	*Plasmodium* species[e]
Sponge-Associated Actinobacteria				
Mediterranean[f]	Unknown	Valinomycin staurosporine	Ion transport	*Leishmania, T. brucei*
	Streptomyces	Protein kinase C		
		Butenolide	Proapoptotic[g]	
Red sea[h]	*Actinokineospora*	Actinosporins	Unknown	*T. brucei*[h]

[a] Espinosa et al. (2012).
[b] Karthik et al. (2014).
[c] Raju et al. (2014).
[d] Maskey et al. (2004).
[e] Manivasagan et al. (2014).
[f] Pimento-Elardo et al. (2010).
[g] Zhang et al. (2011).
[h] Abdelmohsen et al. (2014).

variety of important enzymes, including protease inhibitors (Karthik et al., 2014). Although originally used in laboratory studies of enzyme kinetics, these are now being explored for their potential as pharmacological agents (Drag and Salvesen, 2010). Antimalarial agents functioning as highly specific protease inhibitors have also been recently recovered from marine *Actinobacteria* (Karthik et al., 2014). These potential therapeutics appear to produce no liver or spleen toxicity in a murine model (Swiss albino mice), again reinforcing their potential as future antimalarial agents (Karthik et al., 2014).

Further, these as yet unknown *Streptomyces* species have the advantage of also being able to synthesize gold nanoparticles, which exhibit significant inhibitory effects on *Plasmodium* multiplication (Karthik et al., 2013). In a murine model, animals infected with a malarial model strain (*P. berghei*) demonstrated delayed onset of parasitemia together with increased survival 8 days postinfection (85% in treated compared with 50% in untreated mice) (Karthik et al., 2013). Angiotensin-converting enzyme (ACE) and HIV-protease-inhibitor drugs have already achieved widespread usage, and it would seem, therefore, that marine *Actinobacteria* have still much untapped potential in this area.

The antimalarial compounds mollemycin and gutingimycin were only discovered and/or ascribed this property recently and currently have no established mode of action against the protozoan (Manivasagan et al., 2014; Raju et al., 2014). Mollemycin has been classified as a polyketide, but since this is a large group of *Streptomyces*-derived agents with diverse effects and targets (Gomes et al., 2013), the delineation does not narrow down its mode of action. One might speculate that, given its effects, mollemycin is related to type II polyketide chemotherapeutic drugs, which include doxorubicin (Gomes et al., 2013). If this is the case, mollemycin exerts its effects by interfering with the enzyme topoisomerase II, thereby impairing DNA uncoiling and ultimately cell replication (Pommier et al., 2010). Thus, mollemycin has the potential to become a chemopreventative or antimalarial drug or both.

Gutingimycin is a trioxacarcin, which is a class of compounds now being carefully scrutinized for chemotherapeutic usage in various types of tumors (Magauer et al., 2013). Given that trioxacarcins are also able to prevent cell proliferation, it is not therefore unexpected that gutingimycin can also prevent protozoan multiplication. With an invention patent just launched for trioxocarcin A and its derivatives (Gruen-Wollny et al., 2014), it remains to be seen whether the trioxacarcins and their future derivatives will in turn be developed or repurposed for use as antimalarial drugs.

Several antiprotozoal compounds have been identified from Mediterranean coral-associated *Actinobacteria* including valinomycin, staurosporine, and butenolide (Pimento-Elardo et al., 2010). None of these agents are novel with staurosporine, for example, being discovered in 1977 (Omura et al., 1977) and currently in use for the treatment of neuroblastoma (Mukthavaram et al., 2013). This discovery that several species of *Actinobacteria* synthesizing the same antimicrobial is not unexpected; valinomycin, for example, is produced by at least 11 separate species of terrestrial and now aquatic *Streptomyces* species (Matter et al., 2009). Instead, because Pimento-Elardo et al. (2010) screened for aquatic microbes producing antitrypanosomal and leishmanial compounds, they may become repurposed drugs.

In contrast, the actinosporins from Red Sea coral *Actinobacteria* possess an activity that is novel, highly effective, and specific against *T. brucei* (Abdelmohsen et al., 2014). Finally, studies by Sosovele et al. (2012) have also identified as yet unknown *Actinobacteria* species from Dar Es Salaam, Tanzanian mangrove swamps capable of inhibiting the multiplication of *Plasmodium falciparum in vitro*. Whether these prove to be novel compounds or new properties ascribed to extant therapeutics still remains to be clarified.

23.4   Conclusions and further directions

Protozoan infections are a significant cause of morbidity and mortality, particularly in developing nations (Table 23.1). Unfortunately, at least three of these types of infections, namely HAT, Chagas' disease, and leishmaniasis, fall under the parameters of neglected tropical diseases (NTDs), which have been largely ignored in terms of novel drug development and discovery (Fevre et al., 2008). The reason for this is one of simple economics; people in poorer countries cannot afford to pay for the drugs, and it has, therefore, not been in the interest of pharmaceutical companies to develop them (Dimitri, 2012).

New initiatives in the area of protozoal drug development are primarily partnerships that combine industry, research, and government agencies, which combine their abilities to create and market cheap, effective, and available drugs (Bompart et al., 2011). This began with the Sanofi-Aventis and Drugs for Neglected Diseases partnership in 2004 (Bompart et al., 2011). Sanofi-Aventis was able to formulate a nonpatented, fixed-dose combination of the antimalarial drugs artesunate and amodiaquine and made them relatively inexpensive, less than US$1 for an adult and 50 cents for a child's full treatment (Bompart et al., 2011). In addition, Sanofi-Aventis also partnered with local experts in the countries where the antimalarial drugs were being administered in order to fully educate patients and healthcare professionals about all forms of community malarial prevention (Bompart et al., 2011). Since then, the combined efforts of pharmaceutical companies and the public sector have resulted in several new treatments being developed for NTDs, including an oral pill for HAT currently in late-stage clinical trials (Tran, 2014). Only in the year 2013, pharmaceutical companies donated 1.4 billion NTD treatments to countries that desperately need them, and each dollar invested is working to make US$10 of pharmaceuticals (Tran, 2014).

These liaisons address the question of financial backing for antiprotozoal drug development and how they will be appropriately administered, but still does not examine where new, nonrepurposed drugs are going to come from. As can be seen from this chapter, there are many viable candidates and, clearly, many more out there still waiting to be identified. Metabolomics has been touted as an optimal screening method for antimicrobials since it is highly cost-effective and provides rapid answers as to the potential utility of novel agents (Creek and Barrett, 2014). Two types of studies are typically employed, targeted and untargeted (Creek and Barrett, 2014). In targeted studies, the investigators are looking for novel molecules that affect specific pathways or sites in the microbe (Creek and Barrett, 2014). For example, in malaria, there are a series of highly conserved and unique serine proteases not found in other parasites or in humans and antagonists of these enzymes have great potential as selective, specific drugs (Alam, 2014).

In reality, most studies are untargeted in nature, being primarily concerned with determining that a potential agent can kill/prevent the multiplication of the protozoan target (Creek and Barrett, 2014). Untargeted metabolomics may also uncover the true mode of action of a drug as was seen in studies of the effective antiprotozoal drug eflornithine (Vincent et al., 2012). Eflornithine, being a repurposed drug, has long been known to affect human tumor cells by inhibiting ornithine decarboxylase (ODC) activity and subsequent polyamine accumulation (Grishin et al., 1999). Metabolomic studies by Vincent et al. (2012) confirmed its efficacy against ODC when they observed an increase in ornithine and decrease in putrescine levels in *T. brucei* protozoa. Interestingly, they also observed that eflornithine downregulated spermidine levels, and results indicate a targeted activity on the polyamine pathway of the parasite. The polyamine pathway in trypanosomes is both essential and unique, combining polyamines and glutathione to generate trypanothione, which protects the protozoan against oxidative stress (Vincent et al., 2012). Thus, in addition to identifying the mode of action for eflornithine, these studies have also uncovered a new target for future drug development, namely those targeting the trypanothione pathway. Metabolomics will be assisted by understanding of the biochemical pathways within protozoa, and this is expected to further unearth specific proteins as potential drug targets (Creek and Barrett, 2014).

Once discovered, there is a second major issue to any drug development from natural products, which is the ability to generate sufficient quantities. One strategy has been to use metagenomics, through which natural organisms producing an antimicrobial compound can be identified and matched with suitable candidate hosts to manufacture these agents (Gomes et al., 2013). Unfortunately, *Streptomyces* species have been shown to be unlike any other bacteria in that they possess a set of preferred codons that runs throughout their genome (Baltz, 2006; Peirú et al., 2008). This has meant that for other *Streptomyces* antimicrobial products, novel strategies have to be employed for gene insertion and cloning; the use of mutant species of *Escherichia coli* will allow for these genes to be inserted (Baltz, 2006; Peirú et al., 2008). Two types of mutation have been successful in this regard (Baltz, 2006; Peirú et al., 2008). The first is to modify the codons present in the *Streptomyces* gene sequence to match those found most abundantly in *E. coli*, thus allowing the host machinery to remain fully functional (Baltz, 2006). The second consists of changes throughout the *E. coli* genome that will increase the occurrence of the rare codons seen in *Streptomyces* (Peirú et al., 2008). To date, both have proven successful in generating genetically modified host bacteria able to generate more *Streptomyces* antimicrobials than the parent organism (Baltz, 2006; Peirú et al., 2008).

Given that the financial, production, and dissemination aspects of antiprotozoal discovery have successful strategies in place, it may now be anticipated that new and repurposed drugs should be appearing over the next decade. Actinobacteria have long been the primary sources of these agents, responsible for at least 10,000 bioactive compounds currently in use (Raja and Prabakarana, 2011). Their diversity and untapped potential, especially in unique and marine environments, should continue to provide novel and effective antimicrobial agents to replace those of the past. The literature is full of partially or fully characterized antiprotozoal agents being secreted from these ubiquitous soil and water microbes, with some being described only recently (Tables 23.4 and 23.5). Whether in their native form or as synthetic or semisynthetic versions, it is likely that these drugs represent the future of antiprotozoal drugs able to combat the diseases that currently afflict millions on our planet. The need for effective antiprotozoals is certainly there and is likely to be met by agents made by Actinobacteria species, the successful mainstay sources of our antimicrobials for the past 62 years.

References

Abdelmohsen, U.R., Cheng, C., Viegelmann, C., Zhang, T., Grkovic, T., Ahmed, S., and Edrada-Ebel, R. 2014. Dereplication strategies for targeted isolation of new antitrypanosomal actinosporins A and B from a marine sponge associated-*Actinokineospora* sp. EG49. *Mar. Drugs*, 12, 1220–1244.

Alam, A. 2014. Serine proteases of malaria parasite *Plasmodium falciparum*: Potential as antimalarial drug targets. *Interdiscip. Perspect. Infect. Dis.*, 2014, 453186.

Alphey, L., Benedict, M., Bellini, R., Clark, G.G., Dame, D.A., Service, M.W., and Dobson, S.L. 2010. Sterile-insect methods for control of mosquito-borne diseases. *Vector Borne Zoonotic Dis.*, 10(3), 295–311.

Andrews, K.T., Fisher, G., and Skinner-Adams, T.S. 2014. Drug repurposing and human parasitic protozoan diseases. *Int. J. Parasitol. Drugs Drug Resist.*, 4, 95–111.

Ansari, M.A., Singh, K.R., Brooks, G.D., Malhotra, P.R., and Vaidyanathan, V. 1977. The development of procedures and techniques for mass rearing of *Aedes aegypti*. *Ind. J. Med. Res.*, 65(Suppl.), 91–99.

Ariey, F., Witkowski, B., Amaratunga, C., Beghain, J., Langlois, A.C., Khim, N., and Ménard, D. 2014. A molecular marker of artemisinin resistant *Plasmodium falciparum* malaria. *Nature* 505(7481), 50–55.

Azambuja, P., Feder, D., and Garcia, E.S. 2004. Isolation of *Serratia marcescens* in the midgut of *Rhodnius prolixus*: Impact on the establishment of the parasite *Trypanosoma cruzi* in the vector. *Exp. Parasitol.*, 107(1–2), 89–96.

Baltz, R.H. 2006. Molecular engineering approaches to peptide, polyketide and other antibiotics. *Nat. Biotechnol.*, 24, 1533–1540.

Baral, T.N. 2010. Immunobiology of African trypanosomes; need for alternative interventions. *J. Biomed. Biotechnol.*, 2010, 389153.

Barone, R., De Santi, C., Palma Esposito, F., Tedesco, P., Galati, F., Visone, M., and De Pascale, D. 2014. Marine metagenomics, a valuable tool for enzymes and bioactive compounds discovery. *Front. Mar. Sci.*, 1, 38.

Beck-Johnson, L.M., Nelson, W.A., Paaijmans, K.P., Read, A.F., Thomas, M.B., and Bjørnstad, O.N. 2013. The effect of temperature on anopheles mosquito population dynamics and the potential for malaria transmission. *PLoS One*, 8(11), e79276.

Biftu, T., Feng, D.D., Liang, G.B., Ponpipom, M.M., Qian, X., Fisher, M.H., and Wyvratt, M.J. 2006. Diaryl piperidyl pyrrole derivatives as anti-protozoal agents, EP 1278520 B1.

Bijev, A., Radev, I., and Borisova, Y. 2000. Synthesis and antibacterial activity of new cephalosporines containing a pyrrole ring in the N-acyl chain. *Pharmazie*, 55(8), 568–571.

Bijev, A.T., Prodanova, P.P., and Nankov, A.N. 2002. Synthesis of new 1*H*-1-pyrrolylcarboxamides by comparative N-aceylation. *Comptes Rendus de l'Academie Bulgare des Sciences*, 55, 9–49.

Bompart, F., Kiechel, J.R., Sebbag, R., and Peccoul, B. 2011. Innovative public-private partnerships to maximize the delivery of anti-malarial medicines: Lessons learned from the ASAQ Winthrop experience. *Malar. J.*, 10, 143.

Boone, R., Castenholtz, R., and Garrity, G. 2001. *Bergey's Manual of Systematic Bacteriology*. Springer-Verlag, New York, vol. 1, pp. 163–164.

Brun, R., Don, R., Jacobs, R.T., Wang, M.Z., and Barrett, M.P. 2011. Development of novel drugs for human African trypanosomiasis. *Future Microbiol.*, 6(6), 677–691.

Burrows, J.N., van Huijsduijnen, R.H., Möhrle, J.J., Oeuvray, C., and Wells, T.N. 2013. Designing the next generation of medicines for malaria control and eradication. *Malar. J.*, 12, 187.

Busatti, H.G., Santos, J.F., and Gomes, M.A. 2009. The old and new therapeutic approaches to the treatment of giardiasis: Where are we? *Biologics*, 3, 273–287.

Caffrey, P., Lynch, S., Flood, E., Finnan, S., and Oliynyk, M. 2001. Amphotericin biosynthesis in *Streptomyces nodosus*: Deductions from analysis of polyketide synthase and late genes. *Chem. Biol.*, 8 (7), 713–723.

Carmena, D., Aguinagalde, X., Zigorraga, C., Fernández-Crespo, J.C., and Ocio, J.A. Presence of Giardia cysts and *Cryptosporidium oocysts* in drinking water supplies in northern Spain. *J. Appl. Microbiol.*, 102, 619–629.

Casemore, D.P., Wright, S.E., and Coop, R.L. 1997. Cryptosporidiosis—Human and animalepidemiology. In Fayer, R. ed., *Cryptosporidium and Cryptosporidiosis*. CRC Press, Boca Raton, FL, pp. 65–92.

Castillo, U.F., Strobel, G.A., Mullenberg, K., Condron, M.M., Teplow, D.B., Folgiano, V., and Jensen, J. 2006. Munumbicins E-4 and E-5: Novel broad-spectrum antibiotics from *Streptomyces* NRRL 3052. *FEMS Microbiol. Lett.*, 255(2), 296–300.

Chitanga, S., Marcotty, T., Namangala, B., Van den Bossche, P., Van Den Abbeele, J., and Delespaux, V. 2011. High prevalence of drug resistance in animal trypanosomes without a history of drug exposure. *PLoS Negl. Trop. Dis.*, 5(12), e1454.

Choi, D.B. et al. 2005. Recovery and purification of lincomycin from the culture broth of *Streptomyces lincolnensis*. *J. Ind. Eng. Chem.* 11(6), 932–937.

Couvreur, J., Desmonts, G., and Thulliez, P. 1988. Prophylaxis of congenital toxoplasmosis. Effects of spiramycin on placental infection. *J. Antimicrob. Chemother.*, 22(Suppl. B), 193–200.

Croft, S.L. and Coombs, G.H. 2003. Leishmaniasis—Current chemotherapy and recent advances in the search for novel drugs. *Trends Parasitol.*, 19(11), 502–508.

Croft, S.L. and Engel, J. 2006. Miltefosine—Discovery of the antileishmanial activity of phospholipid derivatives. *Trans. R Soc. Trop. Med. Hyg.*, 100(Suppl. 1), S4–S8.

Coura, J.R., De Abreu, L.L., Willcox, H.P., and Petana, W. 1997. Comparative controlled study on the use of benznidazole, nifurtimox and placebo, in the chronic form of Chagas' disease, in a field area with interrupted transmission. I. Preliminary evaluation. *Rev. Soc. Bras. Med. Trop.*, 30(2), 139–144.

Creek, D.J. and Barrett, M.P. 2014. Determination of anti-protozoal drug mechanisms by metabolomics approaches. *Parasitology*, 141(1), 83–92.

Current, W.L. and Garcia, L.S. 1991. Cryptosporidiosis. *Clin. Microbiol. Rev.*, 4(3), 225–258.

Dame, D., Lowe, R., and Williamson D. 1981. Assessment of released sterile *Anopheles albimanus* and *Glossina morsitans morsitans*. In: Kyoto, R.P., Kitzmiller, J., Kanda, T., (eds.). *Cytogenetics and Genetics of Vectors. Proc XVI Internat. Cong. Entomol.* New York: Elsevier Science, pp. 231–248.

Darby, B.J., Housman, D.C., Zaki, A.M., Shamout, Y., Adl, S.M., Belnap, J., and Neher, D.A. 2006. Effects of altered temperature and precipitation on desert protozoa associated with biological soil crusts. *J. Eukaryot. Microbiol.*, 53(6), 507–514.

De Koning, H.P., Gould, M.K., Sterk, G.J., Tenor, H., Kunz, S., Luginbuehl, E., and Seebeck, T. 2012. Pharmacological validation of *Trypanosoma brucei* phosphodiesterases as novel drug targets. *J. Infect. Dis.*, 206(2), 229–237.

Dey, R., Dagur, P.K., Selvapandiyan, A., McCoy, J.P., Salotra, P., Duncan, R., and Nakhasi, H.L. 2013. Live attenuated *Leishmania donovani* p27 gene knockout parasites are nonpathogenic and elicit long term effective immunity in BALB/c mice. *J. Immunol.*, 190(5), 2138–2149.

Dicko, A.H., Lancelot, R., Seck, M.T., Guerrini, L., Sall, B., Lo, M., and Bouyer, J. 2014. Using species distribution models to optimize vector control in the framework of the tsetse eradication campaign in Senegal. *Proc. Natl. Acad. Sci.*, 111(28), 10149.

Dieter, A., Hamm, A., Fiedler, H. P., Goodfellow, M., Muller, W. E., Brun, R., and Bringmann, G. 2003. Pyrocoll, an antibiotic, antiparasitic and antitumor compound produced by a novel alkaliphilic *Streptomyces* strain. *J. Antibiot.*, 56(7), 639–646.

Dimitri, N. 2012. R and D investments for neglected diseases can be sensitive to the economic goal of pharmaceutical companies. *Drug Discov. Today*, 17, 818–823, 2012.

Dokmanovic, M., Clarke, C., and Marks, P.A. 2007. Histone deacetylase inhibitors: Overview and perspectives. *Mol. Cancer Res.*, 5, 981–989.

Donovick, R., Gold, W., Pagano, J.F., and Stout, H.A. 1954. Amphotericins A and B, antifungal antibiotics produced by a streptomycete. I. In vitro studies. *Antibiot. Ann.*, 3, 579–586.

Dorlo, T.P., Balasegaram, M., Beijnen, J.H., and de Vries, P.J. 2012. Miltefosine: A review of its pharmacology and therapeutic efficacy in the treatment of leishmaniasis. *J. Antimicrob. Chemother.*, 67(11), 2576–2597.

Drag, M. and Salvesen, G.S. 2010. Emerging principles in protease-based drug discovery. *Nat. Rev. Drug Discov.*, 9, 690–701.

Duggar, B.M. 1948. Aureomycin: A product of the continuing search for new antibiotics. *Ann. N. Y. Acad. Sci.*, 51, 177–181.

Dumonteil, E., Bottazzi, M.E., Zhan, B., Heffernan, M.J., Jones, K., Valenzuela, J.G., and Hotez, P.J. 2012. Accelerating the development of a therapeutic vaccine for human Chagas disease: Rationale and prospects. *Expert Rev. Vaccines*, 11(3), 1044–1055.

Egorov, A., Frost, F., Muller, T., Naumova, E., Tereschenko, A., and Ford, T. 2004. Serological evidence of *Cryptosporidium* infections in a Russian city and evaluation of risk factors for infections. *Ann. Epidemiol.*, 14, 129–136.

Eliopoulos, G.M. and Huovinen, P. 2001. Resistance to trimethoprim-ulfamethoxazole. *Clin. Infect. Dis.*, 32(11), 1608–1614.

Espinosa, A., Socha, A.M., Ryke, E., and Rowley, D.C. 2012. Antiamoebic properties of the actinomycete metabolites echinomycin A and tirandamycin A. *Parasitol. Res.*, 111(6), 2473–2477.

Ezra, D., Castillo, U.F., Strobel, G.A., Hess, W.M., Porter, H., Jensen, J.B., and Yaver, D. 2004. Coronamycins, peptide antibiotics produced by a verticillate *Streptomyces* sp. (MSU-2110) endophytic on *Monstera* sp. *Micobiology*, 150, 785–793.

Fang, J. 2010. A world without mosquitoes. *Nature*, 466, 432–434.

Felczykowska, A., Bloch, S.K., Nejman-Falenczyk, B., and Baranska, S. 2012. Metagenomic approach in the investigation of new bioactive compounds in the marine environment. *Acta Biochim. Pol.*, 59, 501–505, 2012.

Fenical, W. and Jensen, P. R. 2006. Developing a new resource for drug discovery: Marine actinomycete bacteria. *Nat. Chem. Biol.*, 2,666–673.

Fevre, E.M., Wissmann, B.V., Welburn, S.C., and Lutumba, P. 2008. The burden of human African trypanosomiasis. *PLoS Neglect. Trop. Dis.*, 2(12), e333.

Flegr, J., Havlícek, J., Kodym, P., Malý, M., and Smahel, Z. 2002. Increased incidence of traffic accidents in subjects with latent toxoplasmosis: A retrospective case-control study. *BMC Infect. Dis.*, 2, 11.

Fletcher, S.M., Stark, D., Harkness, J., and Ellis J. 2012. Enteric protozoa in the developed world: A public health perspective. *Clin. Microbiol. Rev.*, 25(3), 420–449.

Genes, C., Baquero, E., Echeverri, F., Maya, J.D., and Triana, O. 2011. Mitochondrial dysfunction in *Trypanosoma cruzi*: The role of *Serratia marcescens* prodigiosin in the alternative treatment of Chagas disease. *Parasit. Vectors*, 4(1), 66.

Gerner, E.W. and Meyskens F.L. 2004. Polyamines and cancer: Old molecules, new understanding. *Nat. Rev. Cancer*, 4(10), 781–792.

Gomes, E.S., Schuch, V., and Lemos, E.G.D.M. 2013. Biotechnology of polyketides: New breath of life for the novel antibiotic genetic pathways discovery through metagenomics. *Braz. J. Microbiol.*, 44(4), 1007–1034.

Gopal, J.V., Subashini, E., and Kannabiran, K. 2013. Extraction of quinone derivative from *Streptomyces* sp. VITVSK1 isolated from Cheyyur saltpan, Tamilnadu, India. *J. Kor. Soc. Appl. Biol. Chem.*, 56(4), 361–367.

Grishin, N.V., Osterman, A.L., Brooks, H.B., Phillips, M.A., and Goldsmith, E.J. 1999. X-ray structure of ornithine decarboxylase from *Trypanosoma brucei*: The native structure and the structure in complex with α-difluoromethylornithine. *Biochemistry*, 38, 15174–15184.

Gruen-Wollny, I., Hansske, F., Helmke, E., Kayser, O., Laatsch, H., and Maskey, R.P. 2014. Trioxacarcins and their use against infections. Patent WO 2005080549 A2.

Gunnarsdottir, M.J. and Gissurarson L.R. 2008. HACCP and water safety plans in Icelandic water supply: Preliminary evaluation of experience. *J. Water Health*, 6, 377–382.

Gurarie D. and McKenzie, F.E. 2006. Dynamics of immune response and drug resistance in malaria infection. *Malar. J.*, 5, 86.

Hanessian, S., Vakiti, R.R., Chattopadhyay, A.K., Dorich, S., and Lavallée, C. 2013. Probing functional diversity in pactamycin toward antibiotic, antitumor, and anti-protozoal activity. *Bioorgan. Med. Chem.*, 21(7), 1775–1786.

Hawksworth, D.L. and Colwell, R.R. 1992. Biodiversity amongst microorganisms and its relevance. *Biodivers. Conserv.*, 1, 221–345.

Helmreich, B. and Horn, H. 2009. Opportunities in rainwater harvesting. *Desalination*, 248, 118–124.

Hogan, C M. 2010. Bacteria. In Draggan, S. and Cleveland, C.J., eds., *Encyclopedia of Earth*. National Council for Science and the Environment, Washington, DC.

Hökelek, M. 2014. Toxoplasmosis medication. Updated September 8.

Horn, S., Vaaje-Kolstad, G., Westereng, B., and Eijsink, V. 2012. Novel enzymes for the degradation of cellulose. *Biotechnol. Biofuels*, 5(45), 1–12.

Iwatsuki, M., Nishihara-Tsukashima, A., Ishiyama, A., Namatame, M., Watanabe, Y., Handasah, S., and Ōmura, S. 2012. Jogyamycin, a new anti-protozoal aminocyclopentitol antibiotic, produced by *Streptomyces* sp. a-WM-JG-16.2. *J. Antibiot.*, 65, 169–171.

Jackson, Y., Alirol, E., Getaz, L., Wolff, H., Combescure, C., and Chappuis, F. 2010. Tolerance and safety of nifurtimox in patients with chronic Chagas disease. *Clin. Infect. Dis.*, 51(10), e69–e75.

Jensen, P.R., Mincer, T.J., Williams, P.G., and Fenical, W. 2005. Marine actinomycete diversity and natural product discovery. *Antonie Van Leeuwenhoek*, 87, 43–48.

Karray, F., Darbon, E., Oestreicher, N., Dominguez, H., Tuphile, K., Gagnat, J., and Pernodet, J.L. 2007. Organization of the biosynthetic gene cluster for the macrolide antibiotic spiramycin in *Streptomyces ambofaciens*. *Microbiology*, 153(12), 4111–4122.

Karthik, L., Kumar, G., Keswani, T., Bhattacharyya, A., Chandar, S.S., and Rao, K.B. 2014. Protease inhibitors from marine actinobacteria as a potential source for antimalarial compound. *PLoS One*, 9(3), e90972.

Karthik, L., Kumar, G., Keswani, T., Bhattacharyya, A., Reddy, B.P., and Rao, K.B. 2013. Marine actinobacterial mediated gold nanoparticles synthesis and their antimalarial activity. *Nanomed. Nanotechnol., Biol. Med.*, 9(7), 951–960.

Kelly, J.M., Taylor, M.C., Horn, D., Loza, E., Kalvinsh, I., and Björkling, F. 2012. Inhibitors of human histone deacetylase with potent activity against the African trypanosome *Trypanosoma brucei*. *Bioorg. Med. Chem. Lett.*, 22(5), 1886–1890.

Kennedy, P.G. 2013. Clinical features, diagnosis, and treatment of human African trypanosomiasis (sleeping sickness). *Lancet Neurol.*, 12(2), 186–194.

Khanafari, A., Assadi, M.M., and Fakar, F.A. 2006. Review of prodigiosin, pigmentation is *Serratia marssescens*. *J. Biol. Sci.*, 1, 1–13.

Kieft, T.L. 1991. Soil microbiology in reclamation of arid and semi-arid lands. In Skujins, J., ed., *Semiarid Lands and Deserts: Soil Resource and Reclamation*. Marcel Dekker, New York, pp. 209–256.

Kirchhoff, L.V. 2011. Chagas disease (American trypanosomiasis) diagnosis and treatment. Updated June 3, 2011. Accessed September 22, 2014, http://emedicine.medscape.com/article/214581-overview#a0101.

Krungkrai, J., Krungkrai, S.R., and Supuran, C.T. 2008. Carbonic anhydrase inhibitors: Inhibition of *Plasmodium falciparum* carbonic anhydrase with aromatic/heterocyclic sulfonamides-in vitro and in vivo studies. *Bioorg. Med. Chem. Lett.*, 18(20), 5466–5471.

Kuhls, T.L., Mosier, D.A., Crawford, D.L., and Griffis, J. 1994. Seroprevalence of cryptosporidial antibodies during infancy, childhood, and adolescence. *Clin. Infect. Dis.*, 18(5), 731–735.

Kurapova, A.I., Zenova, G.M., Sudnitsyn, I.I., Kizilova, A.K., Manucharova, N.A., Norovsuren, Z., and Zvyagintsev, D.G. 2012. Thermotolerant and thermophilic actinomycetes from soils of Mongolia desert steppe zone. *Microbiology*, 81, 105–116.

Lam, K.S. 2006. Discovery of novel metabolites from marine actinomycetes. *Curr. Opin. Microbiol.*, 2, 9(3), 245–251.

Lazzari, C.R. 1991. Temperature preference in *Triatoma infestans* (Hemiptera: Reduviidae). *Bull. Entomol. Res.*, 81, 273–276.

Lengerich, E.J., Addiss, D.G., Marx, J.J., Ungar, B.L., and Juranek, D.D. 1993. Increased exposure to cryptosporidia among dairy farmers in Wisconsin. *J. Infect. Dis.*, 167(5), 1252–1255.

Leport, C., Raffi, F., Matheron, S., Katlama, C., Regnier, B., Saimot, A.G., and Vilde, J.L. 1988. Treatment of central nervous system toxoplasmosis with pyrimethamine/sulfadiazine combination in 35 patients with the acquired immunodeficiency syndrome. Efficacy of long-term continuous therapy. *Am. J. Med.*, 84(1), 94–100.

Lesch, J.E. 2007. The first miracle drugs: how the sulfa drugs transformed medicine. Chapter 3: Prontosil. p. 51. Oxford University Press.

Lounibos, L.P., Suárez, S., Menéndez, T., Nishimura, N., Escher, R.L., O'Connell, S.M., and Rey, J.R. 2002. Does temperature affect the outcome of larval competition between *Aedes aegypti* and *Aedes albopictus*? *J. Vector Ecol.*, 27(1), 86–95.

Magauer, T., Smaltz, D.J., and Myers, A.G. 2013. Component-based syntheses of trioxacarcin A, DC-45-A1 and structural analogues. *Nat. Chem.*, 5(10), 886–893.

Maheswarappa, G., Kavitha, D., Vijayarani, K., and Kumanan, K. 2013. Prodigiosin as anticancer drug produced from bacteria of termite gut. *Ind. J. Basic Appl. Med. Res.*, 3(1), 257–266.

Maldonado, L.A., Fenical, W., Jensen, P.R., Kauffman, C.A., Mincer, T.J., Ward, A.C., and Goodfellow, M. 2005. *Salinispora arenicola* gen. nov., sp. nov. and *Salinispora tropica* sp. nov., obligate marine actinomycetes belonging to the family Micromonosporaceae. *Int. J. Syst. Evol.Microbiol.*, 55, 1759–1766, 2005.

Manivasagan, P., Venkatesan, J., Sivakumar, K., and Kim, S.K. 2014. Pharmaceutically active secondary metabolites of marine actinobacteria. *Microbiol. Res.*, 169(4), 262–278.

Manyando, C., Njunju, E.M., D'Alessandro, U., and Van Geertruyden, J.P. 2003. Safety and efficacy of co-trimoxazole for treatment and prevention of *Plasmodium falciparum* malaria: A systematic review. *PLoS One*, 8(2), e56916.

Maskey, R.P., Sevvana, M., Usón, I., Helmke, E., and Laatsch, H. 2004. Gutingimycin: A highly complex metabolite from a marine streptomycete. *Angew. Chem. Int. Ed.*, 43(10), 1281–1283.

Matter, A.M., Hoot, S.B., Anderson, P.D., Neves, S.S., and Cheng, Y.Q. 2009. Valinomycin biosynthetic gene cluster in *Streptomyces*: Conservation, ecology and evolution. *PLoS One*, 4(9), e7194.

Metcalf, B.W., Bey, P., Danzin, C., Jung, M. J., Casara, P., and Vevert, J.P. 1978. Catalytic irreversible inhibition of mammalian ornithine decarboxylase (EC.4.1.1.17) by substrate and product analogs. *J. Am. Chem. Soc.*, 100, 2551–2553.

Mincer, T.J., Fenical, W., and Jensen P.R. 2005. Culture-dependent and culture-independent diversity within the obligate marine actinomycete genus *Salinispora*. *Appl. Environ. Microbiol.*, 71(11), 7019–7028.

Mokrane, S., Bouras, N., Sabaou, N., and Mathieu, F. 2013. Actinomycetes from saline and non-saline soils of Saharan palm groves: Taxonomy, ecology and antagonistic properties. *Afr. J. Microbiol. Res.*, 7(20), 2167–2178.

Moore, S., Shrestha, S., Tomlinson, K.W., and Vuong, H. 2011. Predicting the effect of climate change on African trypanosomiasis: Integrating epidemiology with parasite and vector biology. *J. R. Soc. Interf.*, 9, 817–830.

Mukthavaram, R., Jiang, P., Saklecha, R., Simberg, D., Bharati, I.S., Nomura, N., and Kesari, S.2013. High-efficiency liposomal encapsulation of a tyrosine kinase inhibitor leads to improved in vivo toxicity and tumor response profile. *Int. J. Nanomed. (Dovepress)*, 8(1), 3991–4006.

Mumba, D., Bohorquez, E., Messina, J., Kande, V., Taylor, S.M., Tshefu, A.K., and Meshnick, S.R. 2011. Prevalence of human African trypanosomiasis in the Democratic Republic of the Congo. *PLoS Negl. Trop. Dis.*, 5(8), e1246.

Nwaka, S. and Hudson, A. 2006. Innovative lead discovery strategies for tropical diseases. *Nat. Rev. Drug Discov.*, 5(11), 941–955.

Odero, R.O. 2013. African trypanasomiasis medication. Updated December 13, 2013. Accessed September 23, 2014, http://emedicine.medscape.com/article/228613-medication.

Oh, D.C., Poulsen, M., Currie, C.R., and Clardy, J. 2011. Sceliphrolactam, a polyene macrocyclic lactam from a wasp-associated *Streptomyces* sp. *Org. Lett.*, 13(4), 752–755.

Olsen, M.E., Ceri, H., and Morck, D.W. 2000. Giardia vaccination. *Parasitol. Today*, 16(5), 213–217.

Omura, S., Iwai, Y., Hirano, A., Nakagawa, A., Awaya, J., Tsuchiya, H., and Asuma, R. 1977. A new alkaloid AM-2282 of *Streptomyces* origin taxonomy, fermentation, isolation and preliminary characterization. *J. Antibiot.*, 30(4), 275–282.

Onnis, B., Rapisarda, A., and Melillo, G. 2009. Development of HIF-inhibitors for cancer therapy. *J. Cell. Mol. Med.*, 13, 2780–2786.

Osato, T., Ueda, M., Fukuyama, S., Yagishita, K., Okami, Y., and Umezawa, H. 1955. Production of tertiomycin (a new antibiotic substance), azomycin and eurocidin by *S. eurocidus*. *J. Antibiot.*, 8, 105–109.

Ostrosky-Zeichner, L., Marr, K.A., Rex, J.H., Cohen, S.H. 2003. Amphotericin B: time for a new "gold standard". *Clin. Infect. Dis.* 37(3), 415–425.

Parsons, M., Worthey, E.A., Ward, P.M., and Mottram, J.C. 2005. Comparative analysis of the kinomes of three pathogenic trypanosomatids: *Leishmania major*, *Trypanosoma brucei* and *Trypanosoma cruzi*. *BMC Genomics*, 6, 127.

Patel, G., Karver, C.E., Behera, R., Guyett, P.J., Sullenberger, C., Edwards, P., and Pollastri, M.P. 2013. Kinase scaffold repurposing for neglected disease drug discovery: Discovery of an efficacious, lapatinib-derived lead compound for trypanosomiasis. *J. Med. Chem.*, 56(10), 3820–3832.

Peirú, S., Rodríguez, E., Menzella, H.G., Carney, J.R., and Gramajo, H. 2008. Metabolically engineered *Escherichia coli* for efficient production of glycosylated natural products. *Microb. Biotechnol.*, 1, 476–486, 2008.

Perez-Jorge, E.V. 2014. Malaria medication. Updated March 14, 2014. Accessed September 23, 2014, http://emedicine.medscape.com/article/221134-medication.

Pickens, L.B. and Tang, Y. 2010. Oxytetracycline biosynthesis. *J. Biol. Chem.*, 285, 27509–27515.

Piechoski, M.P., Cundell, D.R., Bower, A.H., and Porter, J.R. 2012. Bioassay and antibiotic activity of jamaican actinomyces isolates. *J. Young Investigators*. October 2012. Accessed January 13, 2014. http://www.jyi.org/issue/bioassay-and-antibiotic-activity-of-jamaican-actinomycetes-isolates/.

Pimentel-Elardo, S.M., Kozytska, S., Bugni, T.S., Ireland, C.M., Moll, H., and Hentschel, U. 2010. Anti-parasitic compounds from *Streptomyces* sp. strains isolated from Mediterranean sponges. *Mar. Drugs*, 8(2), 373–380.

Pinazo, M.J., Muñoz, J., Posada, E., López-Chejade, P., Gállego, M., Ayala, E., and Gascon, J. 2010. Tolerance of benznidazole in treatment of Chagas' disease in adults. *Antimicrob. Agents Chemother.*, 54(11), 4896–4899.

Pinnert-Sindico, S. 1954. Une nouvelle espe' ce de *Streptomyces* productrice d'antibiotiques: *Streptomyces ambofaciens* n. sp. caracte' res culturaux. *Ann. Inst. Pasteur (Paris)*, 87, 702–707.

Pommier, Y., Leo, E., Zhang, H., and Marchand, C. 2010. DNA topoisomerases and their poisoning by anticancer and antibacterial drugs. *Chem. Biol.*, 17(5), 421–433.

Priotto, G., Kasparian, S., Mutombo, W., Ngouama, D., Ghorashian, S., Arnold, U., and Kande, V. 2009. Nifurtimox-eflornithine combination therapy for second-stage African *Trypanosoma brucei* gambiense trypanosomiasis: A multicentre, randomised, phase III, non-inferiority trial. *Lancet*, 374(9683), 56–64.

Priotto, G., Kasparian, S., Ngouama, D., Ghorashian, S., Arnold, U., Ghabri, S., and Karunakara, U. 2007. Nifurtimox-eflornithine combination therapy for second-stage *Trypanosoma brucei* gambiense sleeping sickness: A randomized clinical trial in Congo. *Clin. Infect. Dis.*, 45(11), 1435–1442.

Rahman, H., Austin, B., Mitchell, W.J., Morris, P.C., Jamieson, D.J., Adams, D.R., and Schweizer, M. 2010. Novel anti-infective compounds from marine bacteria. *Marine Drugs*, 8(3), 498–518.

Raja, A. and Prabakarana, P. 2011. Actinomycetes and drug-an overview. *Am. J. Drug Discov. Dev.*, 1, 75–84.

Raju, R., Khalil, Z.G., Piggott, A.M., Blumenthal, A., Gardiner, D.L., Skinner-Adams, T.S., and Capon, R.J. 2014. Mollemycin A: An antimalarial and antibacterial glyco-hexadepsipeptide-polyketide from an Australian marine-derived *Streptomyces* sp. (CMB-M0244). *Org. Lett.*, 16(6), 1716–1719.

Ryu, S.R., Choi, O.Y., Yin, P., and Kwun, K. H. 2005. Recovery and purification of lincomycin from the culture broth of *Streptomyces lincolnensis*. *J. Ind. Eng. Chem.*, 11(6), 932–937.

Schofield, C.J. and Dias, J.C.P. 1999. The southern cone initiative against Chagas disease. *Adv. Parasitol.*, 42, 1–27.

Simon, C. and Daniel, R. 2009. Achievements and new knowledge unraveled by metagenomic approaches. *Appl. Microbiol. Biotechnol.*, 85(2), 265–276.

Snelling, W.J., Xiao, L., Ortega-Pierres, G., Lowery, C.J., Moore, J.E., Rao, J.R., and Dooley, J.S. 2007. Cryptosporidiosis in developing countries. *The Journal of Infection in Developing Countries*, 1(03), 242–256.

Sosovele, M.E., Bergmann, B., Lyimo, T.J., Hosea, K.M., and Mueller, B.I. 2012. Antimalarial activity of marine Actinomycetes isolated from Dar es Salaam mangrove sediments. *Int. Res. Biol. Sci.*, 2(4), 177–181.

Spížek, J. and Řezanka, T. 2004. Lincomycin, cultivation of producing strains and biosynthesis. *Appl. Microbiol. Biotechnol.*, 63(5), 510–519.

Stach, J.E., Maldonado, L.A., Ward, A.C., Goodfellow, M., and Bull, A.T. 2003. New primers for the class Actinobacteria: Application to marine and terrestrial environments. *Environ. Microbiol.*, 5(10), 828–841.

Steinerova, N., Lipavská, H., Stajner, K., Čáslavská, J., Blumauerová, M., Cudlín, J., and Vank, Z. 1987. Production of quinomycin A in *Streptomyces lasaliensis*. *Folia Microbiol.*, 32(1), 1–5.

Sumanadasa, S.D., Goodman, C.D., Lucke, A.J., Skinner-Adams, T., Sahama, I., Haque, A., and Andrews, K.T. 2012. Antimalarial activity of the anticancer histone deacetylase inhibitor SB939. *Antimicrob. Agents Chemother.*, 56(7), 3849–3856.

Tan, K.R., Magill, A.J., Parise, M.E., and Arguin, P.M. 2011. Doxycycline for malaria chemoprophylaxis and treatment: Report from the CDC expert meeting on malaria chemoprophylaxis. *Am. J. Trop. Med. Hyg.*, 84(4), 517–531.

Tiwari, S.S.K., Schmidt, W.P., Darby, J., Kariuki, Z.G., and Jenkins, M.W. 2009. Intermittent slow sand filtration for preventing diarrhoea among children in Kenyan households using unimproved water sources: Randomized controlled trial. *Trop. Med. Int. Health*, 14(11), 1374–1382.

Torrey, E.F., Bartko, J.J., Lun, Z.R., and Yolken, R.H. 2007. Antibodies to *Toxoplasma gondii* in patients with schizophrenia: A meta-analysis. *Schizophr. Bull.*, 33(3), 729–736.

Tran, M.T. 2014. Global partners are taking the "Neglect" out of "Neglected Tropical Diseases". Uniting to Combat Neglected Tropical Diseases. April 1, 2014, Press Release. Accessed October 8, 2014, http://unitingtocombatntds.org/news/global-partners-are-taking-neglect-out-neglected-tropical-diseases.

Tringe, S.G., Von Mering, C., Kobayashi, A., Salamov, A.A., Chen, K., Chang, H.W., and Rubin, E.M. 2005. Comparative metagenomics of microbial communities. *Science*, 308(5721), 554–557.

Valderrama, J., Fournet, A., Valderrama, C., Bastias, S., Astudillo, C., Rojas de Arias, A., and Yaluff, G. 1999. Synthesis and in vitro antiprotozoal activity of thiophene ring-containing quinones. *Chemical and Pharmaceutical Bulletin*, 47(9), 1221–1226.

Varshney, H., Ahmad, A., Rauf, A., Husain, F.M., and Ahmad, I. 2014. Synthesis and antimicrobial evaluation of fatty chain substituted 2,5-dimethyl pyrrole and 1,3-benzoxazin-4-one derivatives. *J. Saudi Chem. Soc.* May 9, 2014. Accessed October 6, 2014, http://www.sciencedirect.com/science/article/pii/S1319610314000647.

Vetsigian, K., Jajoo, R., and Kishony, R. 2011. Structure and evolution of *Streptomyces* interaction networks in soil and in silico. *PLoS Biol.*, 9(10), e1001184.

Vincent, I.M., Creek, D.J., Burgess, K., Woods, D.J., Burchmore, R.J., and Barrett, M.P. 2012. Untargeted metabolomics reveals a lack of synergy between nifurtimox and eflornithine against *Trypanosoma brucei*. *PLoS Negl. Trop. Dis.*, 6(5), e1618.

Waksman, S.A. and Lechevalier, H.A. 1962. *The Actinomycetes, Volume III: Antibiotics of Actinomycetes*. Baltimore, The Williams and Wilkins Company.

Walsh, F. Malaria vaccine trial shows promise. BBC News Health, October 10, 2013.

Wang, D. and Gao, F. 2013. Quinazoline derivatives: Synthesis and bioactivities. *Chem. Cent. J.*, 7, 95.

Ward, A.C. and Bora, N. 2006. Diversity and biogeography of marine actinobacteria. *Curr. Opin. Microbiol.*, 9(3), 279–286.

Weiss, R.B. 1992. The anthracyclines: Will we ever find a better doxorubicin? *Semin. Oncol.*, 19(6), 670–686.

Wiwanitkit, V. 2012. Interest in paromomycin for the treatment of visceral leishmaniasis (kala-azar). *Therap. Clin. Risk Manage.*, 8, 323.

Wolf, J.E., Shander, D., Huber, F., Jackson, J., Lin, C.S., Mathes, B.M., and Schrode, K. 2007. Randomized, double-blind clinical evaluation of the efficacy and safety of topical eflornithine HCl 13.9% cream in the treatment of women with facial hair. *Int. J. Dermatol.*, 46(1), 94–98.

World Health Organization Report. 2012. Global costs and benefits of drinking-water supply and sanitation interventions to reach the MDG target and universal coverage. Accessed September 22, 2014.

World Health Organization. 2013. Malaria report. Accessed September 22, 2014. http://www.who.int/malaria/publications/world_malaria_report_2013/wmr2013_no_profiles.pdf.

Wright, G.D. 2010. Q and A: Antibiotic resistance: Where does it come from and what can we do about it? *BMC Biol.*, 8(1), 123.

Yu, Z., Vodanovic-Jankovic, S., Ledeboer, N., Huang, S.X., Rajski, S.R., Kron, M., and Shen, B. 2011. Tirandamycins from *Streptomyces* sp. 17944 inhibiting the parasite *Brugia malayi* asparagine tRNA synthetase. *Org. Lett.*, 13(8), 2034–2037.

Zhang, Y.F., Xiao, K., Chandramouli, K.H., Xu, Y., Pan, K., Wang, W.X., and Qian, P.Y. 2011. Acute toxicity of the antifouling compound butenolide in non-target organisms. *PLoS One*, 6(8), e23803.

chapter twenty-four

Bioactive compounds from actinomycetes and their antiviral properties
Present trends and future prospectives

Avilala Janardhan, Arthala Praveen Kumar, and Golla Narasimha

Contents

24.1 Introduction

The search for bioactive compounds in nature is a multistep procedure that begins with the selection of suitable sources and then the biological, chemical, or physical interactions of metabolites with test systems that are then qualitatively or quantitatively evaluated (Omura, 1992). This is called screening. Natural products are the organic molecules derived from primary or secondary metabolism of living organisms such as microorganisms. Among them, 50%–60% are produced by plants (alkaloids, flavonoids, terpenoids, steroids, and carbohydrates), and 5% have a microbial origin (Berdy, 2005). The natural products from plants (Han et al., 2007, Huang et al., 2008; Huo et al., 2008), fungi (Krohn et al., 2001; Lin et al., 2002; Wu et al., 2004; Gao et al., 2007), bacteria (Lin et al., 2008), and actinomycetes (Xie et al., 2006; Tang et al., 2007) are the most anti-infectious, anticancer, antibacterial, antiviral, anti-inflammatory, antimalarial, and antidiabetic drugs on the market today. The increasing role in the production of natural products such as antibiotics and other drugs for treatment of serious diseases has been dramatic, but the development of resistance in pathogens and tumor cells has become a major problem and requires much research effort to screen it. The basic premises of a screening program are as follows: (1) drugs operate in a dose–response manner and produce toxicity in higher doses; (2) each class of drug has a characteristic dose–response profile; (3) for the majority of drugs, route of administration produces only a quantitative change in action; (4) absolute potency is not of major importance in therapeutics; and (5) it is possible to predict usefulness and toxicity of a new compound by utilizing a dose–response spectra library of various prototype drugs. The criteria of a good screening program are that it should be simple, economical, reliable, able to pick up new unexpected or unique activity, unbiased, and comprehensive (Irwin, 1962; Lucas and Lewis, 1944; Taylor et al., 1952; Laurence and Bacharach, 1964; Turner, 1965; Mantegazza and Piccinni, 1966; Turner and Hoborn, 1971; Dhawan and Srimal, 1984, 1992; Kamboj and Dhawan, 1989).

Viruses cause many important diseases in humans, with viral-induced emerging and reemerging infectious diseases representing a major health threat to the public. In addition, viruses can also infect livestock and marine species, causing huge losses of many vertebrate food species. Effective control of viral infection and disease has remained an unachieved goal, due to the virus' intracellular replicative nature and readily mutating genome, as well as the limited availability of antiviral drugs and measures. In relation to infectious diseases, the exploration of the marine environment represents a promising strategy in the search for active compounds, whereas there is a need for new medicines, due to the appearance of resistance to available treatments in many microorganisms, specifically concerning antiviral activities. Among the microorganisms, the actinomycetes are the gram-positive bacteria belonging to the order Actinomycetales, which play a significant role in the production of new metabolites (Goodfellow et al., 1988; Demain, 1995). Especially, the *Streptomyces* and *Micromonospora* strains have proven to produce novel antibiotics (Omura et al., 2001; Watve et al., 2001; Bentlley et al., 2002). The screening of microbial natural products leads to the discovery of novel chemicals for the development of new therapeutic agents (Bull et al., 2000). So, it is necessary to continue the screening for new metabolites and evaluate the potential of less-known and new bacterial strains so that the new and improved compounds for future use against drug-resistant bacteria or for chemical modification purposes may be developed (Kurtboke, 2005).

24.2   Antiviral activity

Some compounds have been used for testing antiviral activity in our laboratory. Marine antiviral agents (MAVAs) (Fujioka and Loh, 1996) can be used for the biological control of human enteropathogenic virus contamination and disease transmission in sewage-polluted waters, as chemotherapy for viral diseases of humans and lower animals, as well as the biological control of viral diseases of marine animals. The seeding of MAVAs under natural conditions, or when marine mammals are kept in captivity for various uses, could control viral disease transmission within these select populations. It is clear that the marine environment will play a vital role in the future development and trials of anti infective drugs. The purpose of this study was to establish an *in vitro* model to screen marine extracts for antiviral activity and to evaluate some marine extracts for their antiviral potential, with a long-term goal of discovering new marine compounds to be used as potential antiviral drug candidates.

Viruses cause many diseases in animals; effective control of viral diseases and infections has remained an unachieved goal due to virus intracellular replicative nature and readily mutating genome. Due to the limited availability of antiviral drugs, the use of natural products as drugs was well established. Different studies were conducted to determine the effectiveness of the natural products. Ager (1984) had done experiments on 25 isolates of actinomycetes, which were given as feed for cultured shrimp and tested for their ability to reduce the white spot syndrome virus (WSSV) infection in shrimp. Among these 25 isolates, 6 isolates have shown to be most potential. The pentalactones are extracted from the fermentation broth of *Streptomyces* sp. M-2719 has been reported to be active against several DNA viruses (Kumar et al., 2006). Researchers have reported that guanine-7-N-oxide produced by *Streptococcus* sp. was found to inhibit *in vitro* replication of fish herpes virus, rhabdovirus, and infectious pancreatic necrovirus (Nakagawa et al., 1985).

The antibiotic SF 2487 from *Actinomadura* sp. was found to exhibit antiviral activity against influenza virus *in vitro* (Hasobe et al., 1985). A *Streptomyces* sp. isolated from Brazilian tropical forest soil possessed antiviral activity against herpes simplex virus type 1 (HSV-1) on HEP-2 cells (Hatsu et al., 1990). An antibiotic enriching from *Streptomyces lavendulae* showed inhibition of influenza A and influenza B virus *in vitro* (Sacrament et al., 2004). Current antiviral drugs comprise of over 40 compounds that have been officially approved for clinical use. Among these drugs, half of them were used to treat HIV infections (Bhakuni et al., 1990; Schaeffer and Krylov, 2000; Tziveleka et al., 2003; Mayer and Hamann, 2005). MAVAs were used for biological control of human enteropathogenic virus contamination and disease transmission in sewage-polluted water (Fujioka and Loh, 1996).

MAVAs represent a significantly unique natural marine resource whose multipotential uses include the following applications: (1) One is the biological control of human enteropathogenic virus contamination and disease transmission in sewage-polluted waters. This application would be particularly important to communities that utilize the coastal waters for recreational activities and for food industries (e.g., fish, shellfish), as well as to those regions of the country, such as Hawaii, where the loss of these marine resources would have a devastating effect on the lifestyle and economy of the people. (2) The other one is the chemotherapy of viral diseases of humans and lower animals. To be of practical use, it is imperative that MAVAs are isolated from pure cultures, identified, and characterized. Their spectrum and mechanism of antiviral activity should also be clearly established. Their active principle and moieties should be identified and chemically characterized in order to facilitate application of biotechnological methods for increased yields and cost-effective production.

Currently, it appears that there have been only a few compounds derived from marine actinobacteria with antiviral activity. Benzastatin C (**56**), a 3-chloro-tetrahydroquinolone alkaloid obtained from *Streptomyces nitrosporeus*, showed antiviral activity in a dose-dependent manner with EC50 values of 1.92, 0.53, and 1.99 g/mL against HSV-1, HSV-2, and vesicular stomatitis virus, respectively (Lee et al., 2007). Kumar et al. (2006) reported the antiviral property of a marine *Streptomyces* against WSSV in penaeid shrimp. WSSV infection can cause cumulative mortality up to 100% within 3–10 days, thereby causing considerable economic loss to the shrimp farmers.

24.3 Screening methods

It is suggested that the potential antiviral agents must be screened in a living cell or animal host. Testing for antiviral activity is usually performed in cell culture or embryonated chicken eggs and animal models. *In vitro* antiviral testing using cell cultures involves the virus of interest and a primary or permanent cell line that can support its multiplication. The cells are infected with the virus, or already viral-infected cell lines are exposed to the extracted compound. If the compound has antiviral activity, the multiplication of the virus will be inhibited, which will be evident from the morphology of the cell monolayer. It is important to assess the toxic effect of the test substance on cells at each dilution. This can be done by examining the uninfected cell monolayers exposed to the extracted compound only. From the observed ED50 and LD50 of the compound, its therapeutic index is calculated. Several viral targets are studied to estimate the antiviral effect of compound in a cell culture system. Some of these are viral DNA polymerase activity, ribonucleotide diphosphate reductase, mRNA polyadenylation and RNA-dependent RNA polymerase, terminal deoxynucleotidyl transferase, thymidine kinase, uracil-DNA glycolase, d-UTPase, and reverse transcriptase. Testing for antiviral activity in chicken eggs is very simple.

Here, prophylactic and therapeutic assays may be carried with different test substances since a wide choice of routes and timing of application of both virus and antiviral agents is possible. There are three main routes by which the bioactive compound could be administered into embryonated eggs: allantoic cavity inoculation, amniotic cavity inoculation, and chorioallantoic membrane inoculation. The virus and the compound may be given through different routes; it depends on the type of virus and the compound. The test substance can be given before, along with, or after virus infection. Testing in animal models has relatively the maximum predictive value among the various methods employed for detecting antiviral activity. Testing in these model systems can identify both antiviral activity and antiviral agents. The ideal animal model should have three features: (1) use of a human virus with minimal alteration by adaptation; (2) use of the natural route of infection and size of inoculum as in humans; and (3) similarity of infection, pathogenesis, host response, drug metabolism, and drug toxicity. Animal models exist for both local and systemic virus infections. Antiviral activity of a test substance can also be assessed by titrating the virus in blood and other target organs. The details of these models are described (Bhakuni et al., 1990).

24.4   Viral entry inhibition assay

Cells at exponential growth phase were harvested and seeded into multiwell plates at densities that would allow the formation of an approximately 90% cell monolayer overnight. Marine extracts were diluted with serum-free medium to twice the effective safe concentrations, as determined by the cytotoxicity tests. A 250 μL solution of each extract at twice the maximum nontoxic concentration (e.g., 200 μg/mL for those found to be nontoxic at 100 μg/mL) was mixed with an equal volume of the virus dilution. Positive controls were made by mixing 250 μL of virus dilution with 250 μL of serum-free medium with 0.2% DMSO, in order to yield a final DMSO concentration of 0.1%. The 500 μL virus/extract mixtures were preincubated for 1 h, along with controls, and then assayed for viral infectivity using the optimized plaque assay protocols. Antiviral effect of each extract was categorized as having no meaningful inhibition (<20%), slight inhibition (≥20%), moderate inhibition (≥50%), or high inhibition (≥80%).

24.5   Viral replication inhibition assay

Test cells were seeded into TC 12.5 cm² flasks at a density that would allow the formation of an approximately 90% monolayer the next day. Marine extracts were diluted with a medium containing 5% serum to their safe and effective concentrations. The medium was completely aspirated from the flasks, and then the cell monolayer was briefly washed with Dulbecco's Phosphate Buffered Saline (DPBS) buffer, before infection with test virus at a multiplicity of infection of 0.1. Following a 1 h viral adsorption, all medium in the flask was removed and the flasks were washed twice with DPBS; infected cultures were incubated with 2.5 mL/flask of diluted extract. Two flasks were tested per extract, and these cultures were allowed to incubate for 3 days. Pictures were taken every 12 h using an inverted microscope equipped with a camera, starting at time zero, in order to track the progression of viral-induced CPE. To track viral progression, 200 μL samples of medium were taken from each flask, every 12 h, and stored at −20°C until the end of the experiment. The viral titers of these samples were later determined by standard plaque assay, as previously described. Test extracts shown to produce a visually noticeable reduction in CPE, as well as a reduction in viral titer, were considered for further characterization.

24.6 Anticancer activity

Cancer still remains one of the most serious human health problems, and breast cancer is the second most universal cause of cancer death in women (Ravikumar et al., 2010). Therapeutic methods for cancer treatment are surgery, radiotherapy, immunotherapy, and chemotherapy (Gillet et al., 2007), and these techniques are individually useful in particular situations, and when combined, they offer a more efficient treatment for tumor. Many of the antitumor compounds from marine drugs are derived from marine actinobacteria, and these metabolites play an important role in identification of pharmaceutical compounds (Ravikumar et al., 2012). Currently, it appears that there have been only a few studies focusing on finding bioactive compounds derived from marine actinobacteria to be used as anticancer agents as well as agents against infectious organisms. Pure active compounds extracted from the marine actinobacterium *Salinispora tropica* have shown inhibitory effects in many malignant cell types (Prudhomme et al., 2008). In particular, salinosporamide A is a novel rare bicyclic beta-lactone gamma-lactam isolated from an obligate marine actinobacterium, *S. tropica* (Feling et al., 2003; Jensen et al., 2007). Salinosporamide A is an orally active proteasome inhibitor that induces apoptosis in multiple myeloma cells with mechanisms distinct from the commercial proteasome inhibitor anticancer drug bortezomib (Chauhan et al., 2005). The first anticancer chemical compound for cancer treatment is produced from obligate marine actinnobacterium (Fenical et al., 2009). Prudhomme et al. (2008) tested salinosporamide A for its utility as an anticancer and antimalarial drug. It was shown to have inhibitory activity against parasite development *in vitro* (*Plasmodium falciparum*) and *in vivo* (*Plasmodium yoelii*). The exact mode by which salinosporamide A inhibits *Plasmodium erythrocytic* development is unknown; however, it is likely due to the inhibition of the proteasome complex. It is interesting to note that chloroquine-resistant strains are still sensitive to salinosporamide A. Targeting the proteasome system has a huge therapeutic implication as it can restrain growth and survival of most cell types (Prudhomme et al., 2008). These attributes, taken with the fact that it is already in phase I clinical trials as an antitumor agent, make it an excellent candidate for alternative therapies, such as antibacterial, antiparasitic, antifungal, or antiviral treatments. Caprolactones are new antibiotics isolated from *Streptomyces* sp. showing moderate phytotoxicity and promising activity against cancer cells with concomitant low general cytotoxicity (Stritzke et al., 2004).

References

Ager, Jr., A.L. 1984. Rodent malaria models. In: Peters, W., Ed., *Antimalarial Drugs*, Springer, Berlin, Germany, pp. 225–264.

Bentley, S.D., Chater, K.F., Cerdeno-Tarragfa, A.M., Challis, G.L., Thompson, N.R., James, K.D., Harris, D.E., Quail, M.A., Kieser, H., and Harper, D. 2002. Complete genome sequence of the model actinomycete *Streptomyces coelicolor* A3(2). *Nature*, 417, 141–147.

Berdy, J. 2005. Bioactive microbial metabolites, a personal view. *J. Antibiot.*, 58, 1–26.

Bhakuni, D.S., Goel, A.K., Jain, S., Mehrotra, B.N., and Srimal, R.C. 1990. Screening of Indian plants for biological activity: Part XIV. *Ind. J. Exp. Biol.*, 28, 619.

Bull, A.T., Ward, A.C., and Goodfellow, M. 2000. Search and discovery strategies for biotechnology: The paradigm shift. *Microbiol. Mol. Biol. Rev.*, 64, 573–606.

Chauhan, D., Catley, L., Li, G., Podar, K., Hideshima, T., Velankar, M., and Anderson, K.C. 2005. A novel orally active proteasome inhibitor induces apoptosis in multiple myeloma cells with mechanisms distinct from Bortezomib. *Cancer Cell*, 8(5), 407–419.

Demain, A.L. 1995. Why do microorganisms produce antimicrobials? In: Hunter, P.A., Darby, G.K. and Russell, N.J., Eds., *Fifty Years of Antimicrobials: Prospective and Future Trends—Symposium 53*, Society of General Microbiology, Cambridge University Press, Cambridge, pp. 205–228.

Antimicrobials: Synthetic and natural compounds

Dhawan, B.N. and Srimal, R.C. 1984. In the use of pharmacological techniques or the evaluation of natural products, *UNESCO-CDRI, Worshop on the Pharmacological Techniques Used for Evaluation of Natural Products*, Lucknow, India, 159pp

Dhawan, B.N. and Srimal, R.C. 1992. In *The Use of Pharmacological Techniques for the Study of Natural Products*, UNESCO-CDRI, Lucknow, India.

Feling, R.H., Buchanan, G.O., Mincer, T.J., Kauffman, C.A., Jensen, P.R., and Fenical, W. 2003. Salinosporamide A: A highly cytotoxic proteasome inhibitor from a novel microbial source, a marine bacterium of the new genus *Salinospora*. *Angew. Chem. Int. Ed.*, 42(3), 355–357.

Fenical, W., Jensen, P.R., Palladino, M.A., Lam, K.S., Lloyd, G.K., and Potts, B.C. 2009. Discovery and development of the anticancer agent salinosporamide A (NPI-0052). *Bioorg. Med. Chem.*, 17(6), 2175–2180.

Fujioka, R.S., Loh, P.C. Characterization of the microbiological quality of water in Mamala Bay. Project MB-7 In Mamala Bay, Final Report. 1996;1.

Gao, H., Hong, K., Zhang, X., Liu, H.W., Wang, N.L., Zhang, L., and Yao, X.S. 2007. Polyhydroxylated sterols and new sterol fatty esters from the mangrove fungus *Aspergillus awamori* exhibiting potent cytotoxic activity. *Helv. Chim. Acta*, 90, 1165–1178.

Gillet, J.-P., Efferth, T., and Remacle, J. 2007. Chemotherapy-induced resistance by ATP binding cassette transporter genes. *Biochim. Biophys. Acta Rev. Cancer*, 1775(2), 237–262.

Goodfellow, M., Williams, S.T., and Mordarski, M. 1988. *Actinomycetes in Biotechnology*, vol. 12, pp. 73–74. Academic Press, London, U.K.

Han, L., Huang, X.S., Sattler, I., Fu, H.Z., Grabley, S., and Lin, W.H. 2007. Two new constituents from mangrove *Bruguiera gymnorrhiza*. *J. Asian Nat. Prod. Res.*, 9, 327–331.

Hasobe, M., Saneyoshi, M., and Isono, K. 1985. Antiviral activity and its mechanism of guanine 7-N oxide on DNA and RNA viruses derived from Salmonid. *J. Antibiot.*, 38, 1581–1587.

Hatsu, M., Sasaki, T., Miyadoh, S., Watabe, H., Takeuchi, Y., Kodama, Y., Orikasa, Y. et al. 1990. SF2487, a new polyether antibiotic produced by Actinomadura. *J. Antibiot.*, 43(3), 259–266.

Huang, H., Lv, J., Hu, Y., Fang, Z., Zhang, K., and Bao, S. 2008. *Micromonospora rifamycinica* sp. nov., a novel actinomycete from mangrove sediment. *Int. J. Syst. Evol. Microbiol.*, 58, 17–20.

Huo, C., Liang, H., Tu, G., Zhao, Y., and Lin, W. 2008. A new 5, 11-epoxymegastigmane glucoside from *Acanthus ilicifolius*. *Nat. Prod. Res.*, 22, 896–900.

Irwin, S. 1962. Drug screening and evaluative procedures current approaches do not provide the information needed for properly predicting drug effects in man. *Science*, 136(3511), 123–128.

Jensen, P.R., Williams, P.G., Oh, D.C., Zeigler, L., and Fenical, W. 2007. Species-specific secondary metabolite production in marine actinomycetes of the genus *Salinispora*. *Appl. Environ. Microbiol.*, 73(4), 1146–1152.

Kamboj, V.P. and Dhawan, B.N. 1989. *Contraceptive Research Today and Tomorrow*. Indian Council of Medical Research, New Delhi, India, pp. 115–125.

Krohn, K., Steingröver, K., and Zsila, F. 2001. Five unique compounds: Xyloketals from the mangrove fungus *Xylaria* sp. from the South China Sea coast. *J. Org. Chem.*, 66, 6252–6256.

Kumar, S.S., Philip, R., and Achuthankutty, C. 2006. Antiviral property of marine actinomycetes against white spot syndrome virus in penaeid shrimps. *Curr. Sci.*, 91(6), 807–811.

Kurtboke, D.I. 2005. Actinophages as indicators of actinomycete taxa in marine environments. *Antonie Van Leeuwenhoek*, 1, 19–28.

Laurence, D.R. and Bacharach, A.L. 1964. *Evaluation of Drug Activities: Pharmacometrics*. Academic Press, London, U.K., vol. 1, pp. 135–166.

Lee, J.-G., Yoo, I.-D., and Kim, W.-G. 2007. Differential antiviral activity of benzastatin C and its dechlorinated derivative from *Streptomyces nitrosporeus*. *Biol. Pharm. Bull.*, 30(4), 795–797.

Lin, Y.C., Wu, X.Y., Deng, Z.J., Wang, J. Zhou, S.N., Vrijmoed, L.L.P., and Jones, E.B.G. 2002. The metabolites of the mangrove fungus *Verruculina enalia* No. 2606 from a salt lake in the Bahamas. *Phytochemistry*, 59, 469–471.

Lin, Z., Zhu, T., Fang, Y., and Gu, Q. 2008. 1H and 13C NMR assignments of two new indolic enamide diastereomers from a mangrove endophytic fungus *Aspergillus* sp. *Magn. Reson. Chem.*, 46, 1212–1216.

Lucas, E.H. and Lewis, R.W. 1944. Antibacterial substances in organs of higher plants. *Science*, 100, 597–599.

Mantegazza, P. and Piccini, F. (eds.) 1966. Methods in drug evaluation. *Proceedings of the International Symposium, Milan 1965*. North-Holland, Amsterdam, the Netherlands, 1966, pp. 548–573.

Mayer, A.M. and Hamann, M.T. 2005. Marine pharmacology in 2001–2002: Marine compounds with anthelmintic, antibacterial, anticoagulant, antidiabetic, antifungal, anti-inflammatory, antimalarial, antiplatelet, antiprotozoal, antituberculosis, and antiviral activities; affecting the cardiovascular, immune and nervous systems and other miscellaneous mechanisms of action. *Comp. Biochem. Physiol. Part C*, 140 (3–4), 265–286.

Nakagawa, A., Tomoda, H., Hao, V. M., Iwai, Y., and Omura, S. 1985. Antiviral activities of pentalenolactones. *J. Antibiot.*, 8, 1114–1115.

Omura, S. 1992. *The Search for Bioactive Compounds from Microorganisms*, 1st edn. Springer, Berlin, Germany.

Omura, S., Ikeda, H., Ishikawa, J., Hanamoto, A., Takahashi, C., Shinose, M., Takahashi, Y. et al. 2001. Genome sequence of an industrial microorganism *Streptomyces avermitilis*: Deducing the ability of producing secondary metabolites. *Proc. Natl. Acad. Sci.*, 98, 12215–12220.

Prudhomme, J., McDaniel, E., Ponts, N., Bertani, S., Fenical, W., and Jensen, P. 2008. Marine actinomycetes: A new source of compounds against the human malaria parasite. *PLoS ONE*, 3(6), e2335.

Ravikumar, S., Gnanadesigan, M., Saravanan, A., Monisha, N., Brindha, V., and Muthumari, S. 2012. Antagonistic properties of seagrass associated *Streptomyces* sp., RAUACT-1: A . *Asian Pac. J. Trop. Med.*, 5(11), 887–890.

Ravikumar, S., Gnanadesigan, M., Thajuddin, N., Chakkaravarthi, V., and Banerjee, B. 2010. Anticancer property of sponge associated actinomycetes along Palk Strait. *J. Pharm. Res.*, 3(10), 2415–2417.

Sacrament, O.D.R., Coelho, R.R.R., Wigg, M.D., Linhares, L.F.T.D., Santos, M.G.M.D., Semedo, D.S.L., and DaSilva, A.J.R. 2004. Antimicrobial and antiviral activities of an actinomycete (*Streptomyces* sp.) isolated from a Brazilian tropical forest soil. *World J. Microbiol. Biotechnol.*, 20(3), 225–229.

Schaeffer, D.J. and Krylov, V.S. 2000. Anti-HIV activity of extracts and compounds from algae and cyanobacteria. *Ecotoxicol. Environ. Saf.*, 45, 208–227.

Stritzke, K., Schulz, S., Laatsch, H., Helmke, E., and Beil, W. 2004. Novel caprolactones from a marine *Streptomycete*. *J. Nat. Prod.*, 67(3), 395–401.

Tang, J.S., Gao, H., Hong, K., Yu, Y., Jiang, M.M., Lin, H.P., Ye, W.C., and Yao, X.S. 2007. Complete assignments of 1H and 13C NMR spectral data of nine surfactin isomers. *Magn. Reson. Chem.*, 45, 792–796.

Taylor, A., McKenna, G.F., and Burlage, H.M. 1952. Cancerchemotherapy experiments with plant extracts. *Tox. Rep. Biol. Med.*, 10, 1062.

Turner, R.A. 1965. Analgesics. In *Screening Methods in Pharmacology*. Academic Press, New York, pp. 113–116.

Turner, R.A. and Hoborn, P. 1971. In *Screening Methods in Pharmacology*, Academic Press, New York, 2nd edition.

Tziveleka, L.A., Vagias, C., and Roussis, V. 2003. Natural products with anti-HIV activity from marine organisms. *Curr. Top. Med. Chem.*, 3, 1512–1535.

Watve, M.G., Tickoo, R., Jog, M.M., and Bhole, B.D. 2001. How many antibiotics are produced by the genus *Streptomyces*. *Arch. Microbiol.*, 176, 386–390.

Wu, J., Xiao, Q., Huang, J., Xiao, Z., Qi, S., Li, Q., Zhang, S., and Xyloccensin, O. 2004. Unique 8,9,30-phragmalin ortho esters from *Xylocarpus granatum*. *Org. Lett.*, 6, 1841–1844.

Xie, X.C., Mei, W.L., Zhao, Y.X., Hong, K., and Dai, H.F. 2006. A new degraded sesquiterpene from marine actinomycete *Streptomyces* sp. 0616208. *Chinese Chem. Lett.*, 17, 1463–1465.

chapter twenty-five

Novel antidermatophytic drug candidates from nature

Didem Deliorman Orhan and Nilüfer Orhan

Contents

25.1 Introduction

Historically, natural products have been used since ancient times traditionally for treatment of many diseases. Chemistry and bioactivity studies ongoing for ages enabled scientists to discover thousands of active molecules from a diverse range of chemical structures from plants, fungi, lichens, and microorganisms. Additionally, a huge number of natural compounds have become models for semisynthetic or synthetic compounds and are used by the modern pharmaceutical industry. Hence, nature is a rich source for new drug candidates to cure various diseases.

Over 300 million people suffer from many fungal diseases, including superficial, systemic, cutaneous, subcutaneous, and systemic mycoses, every year (Hean et al., 2011; Global Action Fund for Fungal Infections, 2014). Among all, superficial mycoses are the most frequent fungal diseases throughout the world (Avelar Pires et al., 2014).

Currently, several synthetic antifungal drugs used in the treatment of superficial mycoses exhibit toxic side effects, including headaches, skin hypersensitivity, hepatic toxicity, and gastrointestinal disturbances (Bang et al., 2000). Therefore, many researchers have been focusing on the discovery of safe and effective antidermatophytic drugs from natural resources. For this purpose, the screening of the plants used in traditional medicine is an approach to drug discovery.

25.2 Superficial fungal infections

Superficial fungal pathogens usually infect the outer layer of keratinized tissues such as skin, hair, and nails and cause dermatophytosis, candidiasis, and *Malassezia* infections.

25.2.1 Candidiasis

Candidiasis is a fungal infection caused by the genus *Candida*, which could be found in vagina, mouth, oropharynx, and gastrointestinal tract in many people. *Candida albicans* is the first most common cause of 70%–80% of *Candida* infections. There are many types of these infections such as mucosal candidiasis, candidemia, systemic candidiasis, vulvovaginal candidiasis, invasive candidiasis, cutaneous candidiasis, and oropharyngeal candidiasis. Today, amphotericin B, triazoles, echinocandins, and flucytosine are the treatment options used for *Candida* infections (Ho and Cheng, 2010).

25.2.2 Malassezia *infections*

Malassezia-like lipophilic yeasts (*M. furfur*, *M. sympodialis*, and *M. pachydermatis*), which are frequently found on human skin as commensals, cause *Malassezia* infections. *M. furfur* is especially responsible for several skin diseases, including seborrheic dermatitis, folliculitis, confluent and reticulated papillomatosis, and pityriasis versicolor. First-line treatment of such infections is commonly with topical agents (imidazole antifungals). Oral antifungal agents such as ketoconazole, itraconazole, and griseofulvin are preferred in the treatment of systemic and serious *Malassezia* infections. Prior to use these oral antifungal drugs, liver function and blood tests of patients should be performed (Ho and Cheng, 2010).

25.2.3 Dermatophyte infections

Fungi known as dermatophytes are from *Microsporum*, *Trichophyton*, and *Epidermophyton* genera. These organisms are capable of causing fungal infections in keratinized tissues (skin, hair, and nails) of both human and animals (Kishore et al., 1996). Systemic or topical antifungal drugs could be preferred in the treatment of dermatophyte infections. First-line therapy of dermatophyte infections is topical agents as in the treatment of *Malassezia* infections. Dermatophytes are colonized in stratum corneum layer of the epidermis and cause inflammatory changes in this layer. Topical antifungal drugs diffuse into skin and inhibit the development and growth of fungus. Therefore, these drugs should be capable of binding to stratum corneum cells. The vast majority of patients with dermatophyte infections do not respond to topical drug therapy. In those cases, systemic antifungal drugs or combination therapy should be chosen.

According to clinical appearances, dermatophytosis is categorized into the following nine major groups:

1. *Tinea barbae*: Tinea barbae is one of the most common dermatophyte infections especially among males. It is a rare infection involving the bearded areas of the face and neck that is closely similar to tinea capitis. The transmission of the disease occurs via direct contact with an infected animal. Typical clinical appearances of this infection are severe pustular eruption, deep inflammatory plaques, or noninflammatory superficial patches with invasion of the hair shaft (Sabota et al., 1996; Baran et al., 2004).

2. *Tinea capitis*: Tinea capitis is a dermatophyte infection of eyebrows, eyelashes, scalp hair follicles, and the surrounding skin that is most common in children. Typical symptoms are pustules, large inflammatory swellings (kerion), patchy and scaly alopecia, broken-off hairs, and swollen lymph nodes. This infection is transmitted from personal belongings (combs, bedding, towel, etc.) of children with tinea capitis to other healthy children. Also, house pets, such as kittens and puppies, can spread tinea capitis (Higgins et al., 2000; Rebollo et al., 2008; Moriarty et al., 2012).

3. *Tinea corporis*: Tinea corporis infection occurs in all body parts except for scalp, beard area, feet, groin, and palms. Common clinical appearances are one or more round or oval erythematous scaly skin (ringworm-like lesions). People may be contaminated with dermatophytes causing tinea corporis from clothing, combs, pool surfaces, shower floors, and walls. Tinea corporis has been implicated in several other skin disorders such as nummular eczema, psoriasis, annular erythemas, and pityriasis rosea. It mainly affects children but can occur in people of all ages (Jacyk, 2004; Karakoca et al., 2010).

4. *Tinea cruris* (groin): It is an acute to chronic infection of the groin area, genitals, pubic area, perineum, and perianal skin. Clinical features are a pruritic erythematous rash with an active scaly palpable edge within pustules or vesicles. This infection usually occurs predominantly in adult men. In diabetic patients, obese individuals, and excessive sweating people, the risk of developing tinea cruris is high (Ho and Cheng, 2010).

5. *Tinea favosa* (favus): It is a chronic dermatophyte infection of the scalp and glabrous skin, generally caused by *Trichophyton schoenleinii*. This infection passes from human to human and is most common in Africa and Eurasia. Tinea favosa, also known as favus, is characterized by the presence of scutula (yellowish, cup-shaped crusts) and severe alopecia (favosa) (Khaled et al., 2007; Anane and Chtourou, 2013).

6. *Tinea imbricata*: It is a kind of tinea corporis that is caused by *Trichophyton concentricum*. Contamination is usually by direct contact between family members sharing household items. This infection is characterized by various scaly papulosquamous plaques arranged in concentric rings and appears on skin and scalp. Men and women are affected equally by tinea imbricata infection (Satter, 2009).

7. *Tinea manuum*: Tinea manuum, fungal infection of the hand, is commonly caused by *Trichophyton rubrum* but also associated with *Trichophyton tonsurans* and also known as "the two-feet-one-hand syndrome." This infection often begins to develop on the hands after the onset of tinea pedis. Scaly lesions on palmar

surface, crops of tiny blisters especially on the sides of the fingers and palm, itching, burning, and ringlike appearance of the infection on the skin are typically observed (Ho and Cheng, 2010).

8. *Tinea pedis*: Tinea pedis is also known as athlete's foot and one of the most common types of dermatophyte infections. The strains (*Epidermophyton floccosum, Microsporum gypseum, Trichophyton mentagrophytes, T. rubrum, T. tonsurans*) of dermatophyte causing the infection thrive in moist environments. Therefore, athlete's foot is generally picked up from swimming pools, showers, or locker rooms. Commonly affected areas include the feet, especially the soles and toe webs. Symptoms and signs of this infection are itchiness, redness, fine silvery white flakes on slightly erythematous skin, and sometimes blistering or cracking of skin (Weitzman and Summerbell, 1995; Ayatollahi Mousavi et al., 2009; American Orthopaedic Foot & Ankle Society, 2013).

9. *Tinea unguium*: Tinea unguium caused by *E. floccosum, M. gypseum, T. mentagrophytes, T. rubrum,* and *T. tonsurans* is also known as onychomycosis (Ayatollahi Mousavi et al., 2009). Infection can occur in nail tissues of hands or feet. It is usually picked up in damp areas such as public gyms, showers, or swimming pools and can be spread from human to human. It is characterized by thickening of the nail, discoloration, and destructive and flaky lesions in nails. Symptoms are similar to many conditions such as those of psoriasis, onychogryphosis, and lichen planus (Weitzman and Summerbell, 1995). Thus, diagnosis of the disease should be confirmed by laboratory examinations before the treatment.

25.3   Treatment of dermatophyte infections

Drugs used in the treatment of dermatophyte infections can be classified into six categories (Del Palacio et al., 2000; Gupta and Cooper 2008; Dias et al., 2013). Clinical diagnosis and treatment options of dermatophyte infections and dermatophyte species also are given in Table 25.1.

1. *Azoles*
   a. *Imidazoles*: clotrimazole (cream, spray, lotion and solutions, vaginal tablet), eberconazole (cream), econazole (water miscible cream and lotion), ketoconazole (cream, shampoo, tablet), luliconazole (cream), miconazole (oral gel, powder, ointment, cream, solution, spray, lotion), oxiconazole (cream and lotion), sertaconazole (cream), sulconazole (cream and solution), tioconazole (cream, nail solution)
   b. *Triazoles*: itraconazole (capsule, liquid preparations), fluconazole (capsule, tablet, powder and intravenous infusion)
2. *Allylamines*: terbinafine (tablet, solution, and cream), naftifine (cream)
3. *Benzylamines*: butenafine (cream)
4. *Hydroxypyridone*: ciclopirox (cream, gel, lotion, solution, and nail lacquer)
5. *Morpholine derivatives*: amorolfine (cream and nail lacquer)
6. *Other*: spiro-benzo[b]furan derivative griseofulvin (tablet, spray, and cream)

Clinical diagnosis and treatment options of dermatophyte infections and dermatophyte species (Odom, 1993; Del Palacio et al., 2000; Asticcioli et al., 2008; Gupta and Cooper, 2008, Ayatollahi Mousavi et al., 2009; Moodahadu-Bangera et al., 2012; Dicle and Özkesici, 2013).

Table 25.1 Treatment options for dermatophytes

Clinical diagnosis	Species of dermatophytes	Treatment options
Tinea barbae	Mg, Tm, Tve	Itraconazole, terbinafine
Tinea capitis	Ma, Mc, Md, Mf, Mfu, Mg, Mn, Tm, Tmi, Tsc, Tso, Tt, Tv, Tve, Ty	Fluconazole, griseofulvin, itraconazole, terbinafine
Tinea corporis	Ef, Mc, Mg, Tm, Tr, Tt, Tv, Tve	Amorolfine, butenafine, clotrimazole, eberconazole, econazole, fluconazole, griseofulvin, itraconazole, ketoconazole, miconazole, naftifine, oxiconazole, sertaconazole, sulconazole, terbinafine
Tinea cruris	Ef, Mg, Tr	Butenafine, clotrimazole, eberconazole, econazole, fluconazole, griseofulvin, itraconazole, ketoconazole, miconazole, naftifine, oxiconazole, sertaconazole, sulconazole, terbinafine
Tinea favosa (favus)	Tsc	Griseofulvin
Tinea imbricata	Tc	Terbinafine
Tinea manuum	Tr, Tt	Itraconazole, terbinafine
Tinea pedis	Ef, Mg, Tm, Tr, Tt	Amorolfine, clotrimazole, econazole, fluconazole, griseofulvin, itraconazole, ketaconazole, miconazole, naftifine, oxiconazole, sertaconazole, sulconazole terbinafine
Tinea unguium	Ef, Mg, Tm, Tr, Tt	Amorolfine, fluconazole, griseofulvin itraconazole, ketaconazole, terbinafine, tioconazole

Ef, *Epidermophyton floccosum;* Ma, *Microsporum audouinii;* Mc, *Microsporum canis;* Md, *Microsporum distortum;* Mf, *Microsporum ferrugineum;* Mfu, *Microsporum fulvum;* Mg, *Microsporum gypseum;* Mn, *Microsporum nanum;* Tc, *Trichophyton concentricum;* Tm, *Trichophyton mentagrophytes;* Tmi, *Trichophyton megninii;* Tr, *Trichophyton rubrum;* Tsc, *Trichophyton schoenleinii;* Tso, *Trichophyton soudanense;* Tt, *Trichophyton tonsurans;* Tv, *Trichophyton violaceum;* Tve, *Trichophyton verrucosum;* Ty, *Trichophyton yaoudei.*

25.4   Studies on antidermatophytic activity of natural products and plant extracts

In today's world, the incidence of antifungal infections in humans is increasing day by day. Although there are a number of antifungal preparations, their effectiveness is limited because of many reasons such as side effects, toxicity, arise of resistant strains, and lack of oral and parenteral preparations due to solubility problems. Thus, there is a great desire for the discovery of new antifungal and antidermatophytic drug candidates that are soluble, broad spectrum, and having new mechanism of actions (Vila et al., 2013). Natural products are the most important choices in antidermatophytic activity researches. Many different techniques are used for the determination of antifungal activity of natural products, and these methods are described in detail in numerous reviews (Rios et al., 1998, Jacob and Walker, 2005; Cos et al., 2006; Engelmeier and Hadacek, 2006; Das et al., 2010). Mainly extracts, essential oils, and different secondary metabolite groups of medicinal plants and also isolated compounds were studied for their antifungal activity.

25.4.1 Plant extracts

According to an extensive literature survey, it is known that there is an extremely high number of studies on antidermatophytic activities of medicinal plants. Thus, the summary of these studies is organized in Table 25.2 to be clear and easily understood. Only plants having antidermatophytic activity are mentioned in the table; therefore, inactive plants are not included. Latin names of 142 different plants, their families, parts of the plants that have antidermatophytic activity, type of the extract, affected dermatophyte type, and data of the references are given.

25.4.2 Secondary metabolites

Many secondary plant metabolites having different chemical structures are reported to have antidermatophytic activities. However, most of the antidermatophytic plants, secondary metabolites, or pure compounds have only been tested in *in vitro* experimental models. Thus, their effects, adverse effects, and toxic effects in humans still remain unknown. Despite these facts, natural compounds might be the new precursors to design highly effective drug candidates. A series of molecules from several chemical groups such as terpenes, flavonoids, fatty acids, polyphenols, and alkaloids have been described to have antidermatophytic activities (Dorman and Deans, 2000; Abad et al., 2007; Arif et al., 2009; Lang and Buchbauer, 2012). In this part, examples of active compounds belonging to different secondary metabolite groups are summarized.

25.4.2.1 Essential oils and terpenes

Essential oils have shown several biological activities such as antimicrobial, antiparasitical, insecticidal, anti-inflammatory, wound healing, and antioxidant. Many studies have reported the antifungal activity of essential oils against dermatophytes. In this connection, the antifungal essential oils from the Lamiaceae and Asteraceae families are very popular.

The essential oil obtained by water distillation from the aerial parts of *Thymus pulegioides* (Lamiaceae) growing from Portugal was investigated against *Microsporum canis* FF1, *M. gypseum* FF3, *T. rubrum* FF5, *T. mentagrophytes* FF7, and *E. floccosum* FF9. For *T. rubrum*, essential oil at subinhibitory concentration (0.08 μg/mL) decreased ergosterol content by around 70%. Ergosterol plays a fundamental role in maintaining the integrity and function of the yeast cell membrane. In this report, antifungal activity of *T. pulegioides* essential oil has been ascribed to the presence of high contents of carvacrol and thymol (Pinto et al., 2006).

Effect of an essential oil mixture consisting of *Thymus serpyllum*, *Origanum vulgare*, and *Rosmarinus officinalis* (Lamiaceae) in *Prunus dulcis* oil on dermatophyte infection in sheep caused by *T. mentagrophytes* was evaluated. For this purpose, the mixture and its main components (thymol, carvacrol, 1,8-cineole, α-pinene, *p*-cymene, and γ-terpinene) were tested against fungal clinical isolate. As a result, it was thought that the essential oil mixture would be effective for limiting fungal growth (Mugnaini et al., 2013).

Antifungal activity of the essential oil obtained by water distillation from flowers of *Matricaria recutita* (Asteraceae) was tested on dermatophytes such as *M. canis* PFCC 50691, *M. gypseum* PFCC 50701, *T. rubrum* PFCC 51431, *T. tonsurans* PFCC88-1352, and *T. mentagrophytes* PFCC 50541 using microbioassay technique. The growth of all tested dermatophytes was inhibited to varying degrees by the essential oil. Gas chromatography–mass spectroscopy (GC-MS) analysis exhibited that the main

Table 25.2 Active plant extracts on dermatophytes

Plant name	Family	Plant part	Active extract	Fungi	Reference
Acacia erioloba Edgew.	Fabaceae	Leaf	Dichloromethane/methanol	Tm	Mabona et al. (2013)
Acalypha indica L.	Euphorbiaceae	Leaf	Hexane, ethanol, methanol	Ef, Tm, Tr, Tt	Ponnusamy et al. (2010)
Acalypha manniana Müll. Arg.	Euphorbiaceae	Leaf	Ethyl acetate, methanol	Mg, Te, Tm	Noumedem et al. (2013)
Acanthospermum australe Kuntze	Asteraceae	Aerial parts	Dichloromethane, methanol, water	Mg, Tm	Portillo et al. (2001)
Acanthospermum hispidum DC.	Asteraceae	Aerial parts	Dichloromethane	Mg, Tm	Portillo et al. (2001)
Acokanthera schimperi (A. DC.) Schweinf.	Apocynaceae	Leaf	Methanol	Tm	Tadeg et al. (2005)
Aegle marmelos (L.) Correa	Rutaceae	Leaf	Ethanol, methanol, water	Ef, Mc,Tm	Balakumar et al. (2011)
Allium ascalonicum	Liliaceae	Bulb	Water	Mg, Tm, Tr	Amin and Kapadnis (2005)
Allium cepa L.	Liliaceae	Bulb	Water	Ef, Mc, Mg, Tm, Tr, Ts	Amin and Kapadnis (2005), Shams-Ghahfarokhi et al. (2004, 2006)
Allium sativum L.	Liliaceae	Bulb	Water	Ef, Mc, Mg, Tm, Tr	Amin and Kapadnis (2005), Shams-Ghahfarokhi et al. (2006)
Aloe arborescens Mill.	Xanthorrhoeaceae	Leaf	Dichloromethane/methanol	Tm	Mabona et al. (2013)
Alpinia galanga (L.) Willd.	Zingiberaceae	Rhizome	Chloroform, ethanol	Mg, Tl	Khattak et al. (2005), Phongpaichit et al. (2005)
Anagallis arvensis L.	Primulaceae	Aerial parts	Water	Mc, Tm, Tv	Ali-Shtayeh and Abu Ghdeib (1999)
Anchusa strigosa (Soland.)	Boraginaceae	Aerial parts	Water	Tv	Ali-Shtayeh and Abu Ghdeib (1999)
Andira inermis H. B. & K.	Fabaceae	Bark	Dichloromethane, methanol	Mg, Tm	Freixa et al. (1998)
Andira surinamensis (Boudt.) Splitz	Fabaceae	Bark	Dichloromethane, methanol	Mg, Tm	Freixa et al. (1998)
Annona cherimola Mill.	Annonaceae	Seed	Methanol	Tm, Tr,	García et al. (2003)

(Continued)

Table 25.2 (Continued) Active plant extracts on dermatophytes

Plant name	Family	Plant part	Active extract	Fungi	Reference
Anogeissus leiocarpus (DC.) Guill. & Perr. (L.)	Combretaceae	Bark, leaf, root	Chloroform, ethanol, ethyl acetate, methanol, water	Ma, Mg, Mn, Tm, Tr	Batawila et al. (2005), Mann et al. (2008)
Aristea ecklonii Baker.	Iridaceae	Leaf, root	Dichloromethane/methanol	Tm	Mabona et al. (2013)
Asclepia curassavica L.	Asclepiadaceae	Leaf	Hexane, methanol	Tr	García et al. (2003)
Asphodelus microcarpus Salzm. et Viv.	Liliaceae	Aerial parts	Water	Mc, Tv	Ali-Shtayeh and Abu Ghdeib (1999)
Asphodelus luteus L.	Liliaceae	Aerial parts	Water	Tm, Tv	Ali-Shtayeh and Abu Ghdeib (1999)
Azadirachta indica (Neem)	Meliaceae	Seed	5% DMSO	Mn, Tm, Tr	Natarajan et al. (2003)
Baccharis articulata Pers.	Asteraceae	Aerial parts	Dichloromethane	Mg, Tm	Portillo et al. (2001)
Bixa orellana L.	Bixaceae	Leaf	Dichloromethane, methanol	Mg, Tm, Tr	Freixa et al. (1998), García et al. (2003)
Blepharocalyx tweediei (Hook. et Arn.) Berg.	Myrtaceae	Leaf, seed	Dichloromethane, methanol	Mg, Tm	Freixa et al. (1998)
Boesenbergia pandurata (Robx.) Schltr.	Zingiberaceae	Rhizome	Chloroform	Mg	Phongpaichit et al. (2005)
Calycophyllum multiflorum Griseb.	Rubiaceae	Bark	Dichloromethane, water	Mg, Tm	Portillo et al. (2001)
Capparis spinosa L.	Capparidaceae	Aerial parts	Water	Mc, Tm, Tv	Ali-Shtayeh and Abu Ghdeib (1999)
Cassia alata L.	Fabaceae	Leaf	Hexane, ethanol, methanol	Ef, Tm, Tr, Ts, Tt	Ponnusamy et al. (2010)
Cassia fistula L.	Fabaceae	Flower	Chloroform, hexane, ethyl acetate, methanol, water	Ef, Tm, Tr, Ts, Ss	Duraipandiyan and Ignacimuthu et al. (2007)
Cassia tora L.	Fabaceae	Leaf	Chloroform, ethanol	Ef, Tm	Rath and Mohanty (2013)
Chenopodium ambrosioides Bert. ex Steud.	Chenopodiaceae	Leaf	Dichloromethane/methanol	Tm	Mabona et al. (2013)
Cicca acida Merr.	Euphorbiaceae	Root	Methanol	Mg, Tm	Chauhan et al. (2012)
Cichorium intybus L.	Asteraceae	Root	Acetone	Tt	Mares et al. (2005)
Clematis cirrhosa L.	Ranunculaceae	Aerial parts	Water	Tv	Ali-Shtayeh and Abu Ghdeib (1999)

(Continued)

Table 25.2 (Continued) Active plant extracts on dermatophytes

Plant name	Family	Plant part	Active extract	Fungi	Reference
Clerodendrum inerme (L.) Gaertn.	Verbenaceae	Leaf, stem	Ethyl acetate, hexane	Ef, Tm, Tr, Tt	Anitha and Kannan (2006)
Clerodendrum phlomidis L.f.	Verbenaceae	Leaf, stem	Ethyl acetate, hexane	Ef, Tm, Tr, Tt	Anitha and Kannan (2006)
Combretum fragrans F. Hoffm.	Combretaceae	Leaf	Ethanol	Mn, Tm, Tr	Batawila et al. (2005)
Croton gacilipes Baill.	Euphorbiaceae	Leaf	Dichloromethane	Mg, Tm	Portillo et al. (2001)
Croton urucurana Baill.	Euphorbiaceae	Leaf	Dichloromethane, water	Mg, Tm	Portillo et al. (2001)
Croton zehntneri Pax & K. Hoffm.	Euphorbiaceae	Leaf	Dichloromethane	Mg, Tm	Freixa et al. (1998)
Curcuma longa L.	Zingiberaceae	Rhizome	Ethanol	Ef, Tl, Tm	Khattak et al. (2005), Rath and Mohanty (2013)
Dicoma anomala Sond.	Asteraceae	Tuber	Dichloromethane/ methanol	Tm	Mabona et al. (2013)
Diospyros mespiliformis Hochst. ex A.D.C.	Ebenaceae	Leaf	Dichloromethane/ methanol	Mc, Tm	Mabona et al. (2013)
Dodonaea angustifolia L.f.	Sapindaceae	Leaf	Dichloromethane/ methanol	Tm	Mabona et al. (2013)
Elephantorrhiza elephantina (Burch.) Skeels	Fabaceae	Rhizome, root	Dichloromethane/ methanol	Mc	Mabona et al. (2013)
Erythrina christi-galli L.	Fabaceae	Bark	Dichloromethane	Mg, Tm	Portillo et al. (2001)
Eucalyptus camaldulensis Dehnh.	Myrtaceae	Leaf	Methanol	Ef, Tm, Tsc	Falahati et al. (2005)
Eupatorium aschenborniarum Schauer	Asteraceae	Leaf	Hexane, methanol	Tm,Tr	García et al. (2003)
Eupatorium buniifolium Hook & Arn.	Asteraceae	Aerial parts	Methanol	Mg, Tr, Tm	Muschietti et al. (2005)
Ficus natalensis Hochst.	Moraceae	Leaf	Dichloromethane/ methanol	Tm	Mabona et al. (2013)

(Continued)

Table 25.2 (Continued) Active plant extracts on dermatophytes

Plant name	Family	Plant part	Active extract	Fungi	Reference
Ficus sur Forssk.	Moraceae	Leaf	Dichloromethane/ methanol	Tm	Mabona et al. (2013)
Gallesia integrifolia (Spreng.) Harms	Phytolaccaceae	Bark	Dichloromethane, methanol	Mg, Tm	Freixa et al. (1998)
Galphimia glauca Cav.	Malpighiaceae	Aerial parts	Hexane, methanol	Tm,Tr	García et al. (2003)
Geophila repens (L.) I.M. Johnston	Rubiaceae	Whole parts	Dichloromethane, methanol, water	Mg, Tm	Portillo et al. (2001)
Gentianella nitida Griseb.	Gentianaceae	Whole parts	Methanol	Mg, Tm	Rojas et al. (2004)
Gunnera perpensa L.	Gunneraceae	Leaf	Dichloromethane/ methanol	Tm	Mabona et al. (2013)
Harpephyllum caffrum Bernh. ex Krauss	Anacardiaceae	Bark	Dichloromethane/ methanol	Tm	Mabona et al. (2013)
Hedyosmum anisodorum C. A. Todzia	Chloranthaceae	Leaf	Dichloromethane	Mg, Tm	Freixa et al. (1998)
Heterotheca inuloides Cass.	Asteraceae	Flower	Dichloromethane, methanol	Mg, Tm	Freixa et al. (1998)
Hura crepitans L.	Euphorbiaceae	Bark	Dichloromethane, methanol	Tm	Freixa et al. (1998)
Hymenaea martiana Hayne	Fabaceae	Bark, Trunk	Butanol, ethanol, methanol	Mc, Tm, Tr	Machado de Souza et al. (2009)
Hypericum ternum A. Sit. Hil.	Hypericaceae	Aerial parts	Chloroform, petroleum ether	Ef, Mc, Mg, Tm, Tr	Fenner et al. (2005)
Inula viscosa (L.) Aiton	Asteraceae	Aerial parts	Water	Mc, Mg, Tv	Ali-Shtayeh and Abu Ghdeib (1999)
Ixora coccinia L.	Rubiaceae	Aerial parts	Methanol, water	Mg, Tm	Chauhan et al. (2012)
Jacaranda mimosifolia D. Don	Bignoniaceae	Aerial Parts	Methanol	Mc	Muschietti et al. (2005)
Juglans regia L.	Juglandaceae	Aerial parts	Water	Mc, Mg, Tv	Ali-Shtayeh and Abu Ghdeib (1999)
Lannea discolor Engl.	Anacardiaceae	Leaf	Dichloromethane/ methanol	Tm	Mabona et al. (2013)

(Continued)

Table 5.2 (Continued) Active plant extracts on dermatophytes

Plant name	Family	Plant part	Active extract	Fungi	Reference
Lantana rugosa Thunb.	Verbenaceae	Leaf	Dichloromethane/methanol	Tm	Mabona et al. (2013)
Lawsonia inermis L.	Lythraceae	Leaf	Hexane, ethanol, methanol	Ef, Tm, Tr, Ts, Tt	Ponnusamy et al. (2010)
Lippia adoensis Hochst. ex Walp.	Verbenaceae	Leaf	Chloroform, methanol, petroleum ether	Tm	Tadeg et al. (2005)
Lippia integrifolia (Griseb.) Hieron.	Verbenaceae	Aerial Parts	Methanol	Mc, Mg, Ef, Tr, Tm	Muschietti et al. (2005)
Lithrea molleoides (Vell.) Engl.	Anacardiaceae	Aerial Parts	Methanol	Mc, Mg, Ef, Tr, Tm	Muschietti et al. (2005)
Lysiloma acapulcensis (Kunth) Benth.	Fabaceae	Bark	Methanol	Tm,Tr	García et al. (2003)
Malva parviflora L.	Malvaceae	Leaf, root	Dichloromethane/methanol, methanol	Tm	Mabona et al. (2013), Tadeg et al. (2005)
Mansoa alliaceae (Lam.) A. G.	Bignoniaceae	Leaf	Dichloromethane, methanol	Mg, Tm	Freixa et al. (1998)
Maytenus ilicifolia Mart. ex Reiss	Celastraceae	Stem	Dichloromethane, methanol	Mg, Tm	Portillo et al. (2001)
Melianthus comosus Vahl.	Melianthaceae	Leaf	Dichloromethane/methanol, methanol	Mc, Tm	Mabona et al. (2013)
Melianthus major L.	Melianthaceae	Leaf	Dichloromethane/methanol	Mc, Tm	Mabona et al. (2013)
Mentha longifolia Huds,	Lamiaceae	Leaf	Dichloromethane/methanol	Tm	Mabona et al. (2013)
Micromeria nervosa (Desf.) Benth	Lamiaceae	Aerial parts	Water	Tm, Tv	Ali-Shtayeh and Abu Ghdeib (1999)
Ocimum gratissimum L.	Lamiaceae	Leaves	Hexane	Mc, Mg, Tm, Tr	Silva et al. (2005)
Ocimum micranthum Willd.	Lamiaceae	Aerial parts	Dichloromethane	Mg, Tm	Freixa et al. (1998)

(Continued)

Table 25.2 (Continued) Active plant extracts on dermatophytes

Plant name	Family	Plant part	Active extract	Fungi	Reference
Olinia rochetiana A. Juss.	Oliniaceae	Leaf	Acetone, chloroform, methanol, petroleum ether	Tm	Tadeg et al. (2005)
Parietaria diffusa Mert. & W.D.J. Koch	Urticaceae	Aerial parts	Water	Mc, Tv	Ali-Shtayeh and Abu Ghdeib (1999)
Paronychia argentea Lam.	Caryophyllaceae	Aerial parts	Water	Mc, Tm, Tv	Ali-Shtayeh and Abu Ghdeib (1999)
Peltophorum pterocarpum (DC.) K. Heyne	Fabaceae		Methanol	Mg, Tm	Chauhan et al. (2012)
Persea laevigata H. B. Et K.	Lauraceae	Bark	Dichloromethane, methanol	Mg, Tm	Freixa et al. (1998)
Persicaria glabra Mill.	Polygonaceae		Methanol	Mg, Tm	Chauhan et al. (2012)
Phagnalon rupestre (L.) DC.	Asteraceae	Aerial parts	Water	Mc, Tv	Ali-Shtayeh and Abu Ghdeib (1999)
Phyllanthus amarus Schum. & Thonn.	Euphorbiaceae	Aerial parts	Chloroform	Mg	Agrawal et al. (2004)
Piper betle L.	Piperaceae	Leaf	Ethanol	Mc, Mg, Tm	Trakranrungsie et al. (2008)
Piper elongatum C. DC.	Piperaceae	Leaf	Dichloromethane, methanol	Mg, Tm	Freixa et al. (1998)
Piper fulvescens C. DC.	Piperaceae	Leaf	Dichloromethane	Mg, Tm	Freixa et al. (1998)
Piper solmsianum C. DC. var. *solmsianum*	Piperaceae	Leaf	Dichloromethane, hexane, methanol	Ef, Mc, Mg, Tm, Tr	De Campos et al. (2005)
Pistacia lentiscus L.	Anacardiaceae	Aerial parts	Water	Mc, Tm, Tv	Ali-Shtayeh and Abu Ghdeib (1999)
Pistia stratiotes L.	Araceae	Leaf	Methanol	Ef, Mn, Mg, Tm, Tr	Premkumar and Shyamsundar (2005)
Pittosporum viridiflorum Sims.	Pittosporaceae	Leaf	Dichloromethane/methanol	Tm	Mabona et al. (2013)
Plumbago europaea L.	Plumbaginaceae	Aerial parts	Water	Mc, Tm, Tv	Ali-Shtayeh and Abu Ghdeib (1999)
Polygonum hydropiperoides Michx.	Polygonaceae	Leaf	Dichloromethane	Mg, Tm	Freixa et al. (1998)
Potalia amara Aubl.	Loganiaceae	Leaf, stem	Dichloromethane, methanol	Mg, Tm	Freixa et al. (1998)

(Continued)

Table 25.2 (Continued) Active plant extracts on dermatophytes

Plant name	Family	Plant part	Active extract	Fungi	Reference
Pterocaulon alopecuroides (Lam.) D.C.	Asteraceae	Aerial parts	Dichloromethane, hexane, methanol	Mg, Tr, Tm	Stein et al. (2005)
Pterocaulon balansae Chodat.	Asteraceae	Aerial parts	Dichloromethane, hexane	Mg, Tr, Tm	Stein et al. (2005)
Pterocaulon polystachyum D.C.	Asteraceae	Aerial parts	Hexane, methanol	Mg, Tr, Tm	Stein et al. (2005)
Pterospermum suberifolium (L.) Lam.	Sterculariaceae		Methanol	Mg, Tm	Chauhan et al. (2012)
Punica granatum L.	Lythraceae	Fruit rind	Methanol	Ef, Tm, Tr, Ts, Tt	Ponnusamy et al. (2010)
Retama raetam (Forssk.) Webb	Papilionaceae	Aerial parts	Water	Mc, Tv	Ali-Shtayeh and Abu Ghdeib (1999)
Ruscus aquleatus L.	Liliaceae	Aerial parts	Water	Mc, Tm, Tv	Ali-Shtayeh and Abu Ghdeib (1999)
Ruta chalepensis L.	Rutaceae	Aerial parts	Water	Mc, Tv	Ali-Shtayeh and Abu Ghdeib (1999)
Salvia fruticosa Mill.	Lamiaceae	Aerial parts	Water	Mc, Tv	Ali-Shtayeh and Abu Ghdeib (1999)
Sebastiania brasiliensis Spreng.	Euphorbiaceae	Aerial parts	Methanol	Mc, Mg, Ef, Tr, Tm	Muschietti et al. (2005)
Sebastiania commersoniana (Baill.) L. B. Sm. & B.J. Downs	Euphorbiaceae	Aerial parts	Methanol	Mc, Mg, Ef, Tr, Tm	Muschietti et al. (2005)
Sedum oxypetalum HBK.	Crassulaceae	Leaf	Hexane, methanol	Tm,Tr	García et al. (2003)
Senecio angulifolius DC.	Asteraceae	Leaf	Methanol	Tm	García et al. (2003)
Senecio grisebachii Baker	Asteraceae	Flower	Dichloromethane, methanol, water	Mg, Tm	Portillo et al. (2001)
Senecio inaequidens DC.	Asteraceae	Aerial parts	Chloroform, hexane, methanol	Mg, Tt	Loizzo et al. (2004)
Senecio vulgaris L.	Asteraceae	Aerial parts	Chloroform, hexane, methanol	Mg, Tt	Loizzo et al. (2004)
Solanum nigrum L.	Solanaceae	Aerial parts	Water	Tv	Ali-Shtayeh and Abu Ghdeib (1999)
Spilanthes cavla DC.	Asteraceae	Root	Petroleum ether, water	Tm	Rai et al. (2004)

(Continued)

Table 25.2 (Continued) Active plant extracts on dermatophytes

Plant name	Family	Plant part	Active extract	Fungi	Reference
Tabebuia avellanedae Griseb.	Bignoniaceae	Bark	Dichloromethane, methanol, water	Mg, Tm	Portillo et al. (2001)
Terminalia avicennioides Guill. & Perr.	Combretaceae	Root	Ethanol, ethyl acetate, Methanol, water	Ma, Tr	Mann et al. (2008)
Terminalia brachystemma Welw. ex Hiern	Combretaceae	Leaf	Acetone, dichloromethane, hexane, methanol	Mc	Masoko et al. (2005)
Terminalia elliptica Willd.	Combretaceae		Methanol, water	Mg, Tm	Chauhan et al. (2012)
Terminalia gazensis Bak.f.	Combretaceae	Leaf	Acetone, dichloromethane, hexane, methanol	Mc	Masoko et al. (2005)
Terminalia glaucescens Planh. ex Benth.	Combretaceae	Leaf, root	Ethanol	Mg, Mn, Tm, Tr	Batawila et al. (2005)
Terminalia laxiflora Engl. et Diels	Combretaceae	Leaf, root	Ethanol	Mn, Tm, Tr	Batawila et al. (2005)
Terminalia macroptera Guill. et Perr.	Combretaceae	Bark, leaf, root	Ethanol	Mn, Tm, Tr	Batawila et al. (2005)
Terminalia mollis Laws.	Combretaceae	Leaf	Acetone, dichloromethane, hexane, methanol	Mc	Masoko et al. (2005)
Terminalia prunioides M.A. Lawson	Combretaceae	Leaf	Acetone, dichloromethane, hexane, methanol	Mc	Masoko et al. (2005)
Terminalia sabesica Engl. & Diels	Combretaceae	Leaf	Acetone, dichloromethane, hexane, methanol	Mc	Masoko et al. (2005)
Terminalia sericea Burch ex DC.	Combretaceae	Leaf, root	Acetone, dichloromethane, dichloromethane/methanol, hexane, methanol	Mc, Tm	Masoko et al. (2005), Mabona et al. (2013)

(Continued)

Table 25.2 (Continued) Active plant extracts on dermatophytes

Plant name	Family	Plant part	Active extract	Fungi	Reference
Terminalia trifolia (Griseb.) Lillo	Combretaceae	Aerial parts	Methanol	Mg, Tr, Tm	Muschietti et al. (2005)
Thespesia populnea (L.) Sol.	Malvaceae	Leaf	Chloroform, hexane, ethanol, methanol	Ef, Tm, Tr, Ts, Tt	Ponnusamy et al. (2010)
Thevetia nerifolia Juss.	Apocynaceae	Leaf	Methanol	Mg, Tm, Tr, Tt	Singh and Vidyasagar (2014)
Trachyspermum ammi Sprague	Apiaceae		Methanol	Mg, Tm	Chauhan et al. (2012)
Vernonia tweedieana Baker	Asteraceae	Root	Dichloromethane	Mg, Tm	Portillo et al. (2001)
Warburgia salutaris (G.Bertol.) Chiov.	Canellaceae	Bark, leaf	Dichloromethane/methanol	Tm	Mabona et al. (2013)
Wrightia tinctoria R. Br.	Apocynaceae	Leaf, seed	Chloroform, hexane, ethanol, methanol	Ef, Tm, Tr, Ts, Tt	Ponnusamy et al. (2010)
Xanthosoma sagittifolium L. Scott.	Araceae	Leaf, stalk, root	Water	Tr	Schmourlo et al. (2005)
Xylosma longifolium Clos.	Flacourtiaceae	Bark, leaf	Chloroform, methanol, petroleum ether	Mc, Mg, Tr	Devi et al. (2013)
Zizyphus spina-christi (L.) Desf.	Rhamnaceae	Aerial parts	Water	Tm, Tv	Ali-Shtayeh and Abu Ghdeib (1999)

Ef, *Epidermophyton floccosum*; Ma, *Microsporum audouinii*; Mc, *Microsporum canis*; Mg, *Microsporum gypseum*; Mn, *Microsporum nanum*; Te, *Trichophyton equinum*; Tl, *Trichophyton longifusus*; Tm, *Trichophyton mentagrophytes*; Tr, *Trichophyton rubrum*; Ts, *Trichophyton simii*; Tsc, *Trichophyton schoenleinii*; Tt, *Trichophyton tonsurans*; Tv, *Trichophyton violaceum*; Ss, *Scopulariopsis* sp.

compounds of the essential oil were chamazulene (61.3%), isopropyl hexadecanoate (12.7%), *trans-trans*-farnesol (6.9%), and *E*-β-farnesol (5.2%) (Jamalian et al., 2012).

Varying polarity extracts of the roots of *Vernonanthura tweedieana* (Asteraceae) used for the treatment of skin diseases in Paraguay were screened in order to evaluate their antifungal activity against *M. gypseum* and *T. mentagrophytes*. One active sesquiterpene, identified as 6-cinnamoyloxy-1-hydroxyeudesm-4-en-3-one, against *T. mentagrophytes* was isolated from the antifungal dichloromethane extract through bioassay-guided fractionation procedures (Portillo et al., 2005).

Berry essential oil of *Juniperus communis* sp. *alpina* and leaf and berry essential oils of *Juniperus turbinata* and *Juniperus oxycedrus* sp. *oxycedrus* (Cupressaceae) were tested against *T. rubrum* FF5, *T. mentagrophytes* FF7, *M. canis* FF1, and *M. gypseum* FF3 using a macrodilution method. In this study, *J. oxycedrus* sp. *oxycedrus* leaf oil was found to be effective against all dermatophyte strains tested. One of the main components of the essential oil is δ-3-carene, which has potent antidermatophyte activity (Cavaleiro et al., 2006).

Melaleuca alternifolia (Thymelaeaceae) essential oil, also known as tea tree oil or melaleuca oil, has been used externally as a traditional medicine for various conditions including acne, athlete's foot, nail fungus, wounds, oral candidiasis, cold sores, and skin lesions. Antifungal activity of purchased tea tree oil samples in Australia was investigated against seven dermatophyte strains. Tea tree oil was found to have both its inhibitory and fungicidal effects using *in vitro* susceptibility and time-kill assays (Hammer et al., 2002).

Antidermatophyte activity of *Lonicera japonica* (Caprifoliaceae) leaf essential oil was studied against three strains of *M. canis*, three strains of *T. rubrum*, and two strains of *T. mentagrophytes*. The essential oil (1000 ppm) exhibited moderate antifungal activity (55.1%–70.3%) against all dermatophyte strains tested except for *T. mentagrophytes* KCTC 6085 and also displayed inhibitory effect on spore germination of all tested microorganisms. The findings showed that *L. japonica* leaf essential oil could be a therapeutic alternative for dermatophyte infections (Rahman et al., 2014).

Fruit essential oil of *Pimpinella anisum* (Apiaceae) is utilized as expectorant, carminative, flavoring, and spice. Effect of the essential oil was evaluated *in vitro* on four clinical isolates of dermatophytes (*T. rubrum*, *T. mentagrophytes*, *M. canis*, and *M. gypseum*). Anise oil displayed potent inhibitory activity against tested fungi with minimum inhibitory concentration (MIC) lower than 0.78% (v/v) (Kosalec et al., 2005).

Natural coniferous resin (Norway spruce resin) obtained from *Picea abies* (Pinaceae) has been known to have strong antifungal activity. Sipponen et al. (2012) examined the clinical efficacy of Norway spruce resin for topical treatment of onychomycosis. Thirty-seven patients with onychomycosis caused by *T. rubrum* and *T. mentagrophytes* used topical resin lacquer therapy for 9 months. The 14 patients who completed treatment considered the topical resin treatment to be effective.

The activity of natural essence of bergamot (*Citrus bergamia*, Rutaceae) was tested on 92 clinical isolates of dermatophytes (*Trichophyton*, *Epidermophyton*, and *Microsporum* species) using *in vitro* susceptibility assays. Bergamot oil exhibited activity against all tested fungi with MICs ranging from 0.156% to 2.5%. Therefore, this natural essence could be recommended as potential source for the topical treatment of dermatophyte infections (Sanguinetti et al., 2007).

The leaf essential oils of *Ageratum houstonianum*, *Chenopodium ambrosioides*, *Citrus medica*, *Corymbia citriodora*, *Hyptis suaveolens*, *Melaleuca leucadendron*, *Murraya koenigii*, *Ocimum sanctum*, *Solidago canadensis*, and *Tagetes erecta* were investigated for their fungitoxicity against *Microsporum audouinii* and *T. mentagrophytes*. Only *C. ambrosioides* oil displayed antimycotic activity against all dermatophyte strains tested. *Chenopodium* ointment prepared with petroleum jelly was applied to guinea pigs with ringworm infection for 15 days. The infection was completely treated at the end of the treatment. Chenopodium oil contains mainly ascaridole, and this compound was thought to be responsible for the antimycotic activity of the oil (Kishore et al., 1996).

Beikert et al. (2012) examined the efficacy of 6% coriander oil (Apiaceae) in unguentum leniens in the treatment of tinea pedis patients. This cream exhibited a considerable improvement in the clinical signs of tinea pedis. Also, this medication was well tolerated by patients.

Antidermatophyte activity of several Iranian medicinal plant essential oils (*Artemisia sieberi*, *Cuminum cyminum*, *Foeniculum vulgare*, *Heracleum persicum*, *Mentha spicata*, *Nigella sativa*, *Rosmarinus officinalis*, *Zataria multiflora*, and *Ziziphora clinopodioides*) was studied against *T. mentagrophytes*, *T. rubrum*, *E. floccosum*, *M. gypseum*, and *M. canis* using the broth microdilution technique. The highest activity was exerted by *A. sieberi*, having a lower MIC against dermatophytes than other plant essential oils (Khosravi et al., 2013).

The essential oil obtained from rhizomes of *Homalomena aromatica* (Araliaceae), which is used in the treatment of skin infections and joint pains in India, was assayed for antifungal activity against *T. rubrum*, *T. mentagrophytes*, *Microsporum fulvum*, and *M. gypseum*. The rhizome oil significantly inhibited the growth of all tested strains of dermatophytes. The composition of active essential oil was analyzed by GC–MS, and linalool, terpene-4-ol, δ-cadinene, and T-muurolol were determined as main components (Policegoudra et al., 2012).

The essential oils obtained by hydrodistillation from the rhizomes of nine Zingiberaceae species (*Zingiber officinale*, *Zingiber zerumbet*, *Curcuma aeruginosa*, *Curcuma mangga*, *Curcuma xanthorrhiza*, *Kaempferia galanga*, *Alpinia galanga*, and *Boesenbergia pandurata*) were evaluated for their antifungal activities against five dermatophytes (*T. mentagrophytes*, *T. rubrum*, *M. canis*, *Microsporum nanum*, and *E. floccosum*). Among the samples tested, only *B. pandurata* essential oil exhibited the inhibitory activity on the growth of all fungi. Camphor, geraniol, 1,8-cineole, methyl cinnamate, and camphene were identified as main components in the essential oil. Previous studies reported that the essential oils containing camphor have antifungal activity against some dermatophyte strains such as *T. mentagrophytes* (Jantan et al., 2003).

Five volatile column fractions of *Cupressus lusitanica* (Cupressaceae) leaf hexane extract were screened for their antidermatophytic activity using the agar dilution method against five reference dermatophyte strains (*M. audouinii*, *Microsporum langeronii*, *M. canis*, *T. rubrum*, and *T. tonsurans*). The chemical composition of the column fractions with the highest activity was analyzed by GC–MS, and the main components were identified as α-pinene, *epi*-bicyclosesquiphellandrene, pimaric acid, kaurenoic acid, and 8-β-hydroxysandracopimarane (Kuiate et al., 2006).

Triterpenoids (friedelin, β-amyrin acetate, betulinic acid, and lupeol) isolated from the EtOAc extract of the stem bark of *Syzygium jambos* (Myrtaceae) used in traditional medicine for the treatment of toothache, mouth sores, cough, and as a wound dressing in Cameroon were investigated for their antifungal activity against *M. audouinii*, *T. mentagrophytes*, and *Trichophyton soudanense*. Betulinic acid and friedelolactone have been found to exhibit strong antidermatophytic activity against *T. soudanense* and *T. mentagrophytes* (Kuiate et al., 2007).

Maytenin and pristimerin isolated from the bark of the roots of *Maytenus ilicifolia* (Celastraceae) are quinonemethide triterpenoid compounds, and their antifungal activity was tested against *T. rubrum* and *T. mentagrophytes*. Maytenin showed more strong antidermatophytic activity than pristimerin on *T. rubrum* clinical isolate (Gullo et al., 2012).

25.4.2.2 Phenolic compounds

In recent years, many studies were conducted on the antidermatophytic activity of phenolic compounds isolated from various plant species. These phenolic compounds are mainly flavonoids, coumarins, anthraquinones, tannins, phenolic acids, and their derivatives.

Flavonoids constitute one of the largest groups of phenolics that are widely distributed in the plant kingdom. Antifungal compounds from the bark of the Peruvian plant *Swartzia polyphylla* DC have been isolated as biochanin A and dihydrobiochanin A. These two isoflavones have been found to be active against *T. mentagrophytes* and *M. gypseum* (Rojas et al., 2006). Eight flavonoids have been isolated from the stem bark of *Erythrina burtii*. The flavanones sigmoidin B 4'-methylether, the pterocarpans calopocarpin and neorautenol, and the isoflavanone bidwillon A have been identified as the active components against *T. mentagrophytes* and *M. gypseum* (Yenesew et al., 2005). Additionally, biochanin A has been isolated from the bark of *Andira surinamensis* (Boudt.) Splitz and has evidenced antifungal activity especially against yeasts and dermatophytes (Lock de Ugaz et al., 1991).

Antidermatophytic activity of different extracts of *Psoralea corylifolia* seeds has been examined on *T. rubrum*, *T. mentagrophytes*, *E. floccosum*, and *M. gypseum* by the disk diffusion method. A 4'-metoxyflavone was isolated as the active compound of the active methanol extract. The activity of compound has been tested by tube dilution method showing MICs of 62.5 µg/mL for *T.* and *T. rubrum* and 125 µg/mL for other dermatophytes (Rajendra Prasad et al., 2004).

Apigenin has been isolated as one of the antifungal principles of *Terminalia chebula* stem extracts by Singh et al. (2014). After that, *in vivo* antifungal effect of apigenin oilments (2.5 and 5 mg/g) was tested on mice, which were experimentally induced with *T. mentagrophytes*. High doses of apigenin oilment led to complete recovery from the infection on the 12th day of the experiment. In this study, apigenin gave encouraging results in the topical treatment of dermatophytosis in mice.

Antimicrobial and antifungal activities of 26 *Eucalyptus* species have been investigated, and *E. globulus*, *E. maculata*, and *E. viminalis* were found to have strong antifungal activity against *T. mentagrophytes*. After isolation and structure elucidation processes, three flavonoid compounds were identified as 2',6'-dihydroxy-3'-methyl-4'-methoxy-dihydrochalcone, eucalyptin, and 8-desmethyl eucalyptin. All these flavonoids commonly exhibited significant inhibitory activity against *T. mentagrophytes* with MIC ranging from 1 to 31 mg/L (Takahashi et al., 2004).

According to ethnopharmacological use of *Zuccagnia punctata*, two chalcones (2',4'-dihydroxy-3'-methoxychalcone and 2',4'-dihydroxychalcone) were isolated as the active compounds after bioactivity-guided fractionation. Both chalcones have shown very strong activities against several clinical strains of the dermatophytes *T. mentagrophytes* and *T. rubrum* with MIC values in the range of 1.9–15.6 µg/mL and minimum fungucidal concentration (MFC) values between 1.9 and 7.8 µg/mL (Svetaz et al., 2007).

Asafoetida is a resinous mixture with a smell similar to garlic, which is obtained by drying the exudates of different *Ferula* species. Asafoetida collected from various *Ferula* species were tested for their antifungal activity, and *F. foetida* was found to be the most active one. Nine prenylated coumarins were isolated from this sample and were tested against *M. gypseum* and *Trichophyton interdigitale*. Four of the compounds exhibited strong antifungal activity against the dermatophytes with

5,8-dihydroxyumbelliprenin being most active with an MIC of 10 mM, the positive control miconazole having an MIC of 0.5 mM (Houghton et al., 2006a).

Antifungal activity of dichloromethane extract from leaves of *Piper fulvescens* was examined by using an agar overlay bioautographic method. Activity-guided fractionation of the extract has led to the isolation of three antifungal neolignans named as conocarpan, eupomatenoid 5, and eupomatenoid 6. Conocarpan showed the widest activity, whereas eupomatenoid 6 was the most active against dermatophytes *M. gypseum* and *T. mentagrophytes* (Freixa et al., 2001). In another study, *in vitro* antidermatophytic activity of extracts from leaves of another *Piper* species (*P. regnellii*) was investigated by microdilution methods. The hydroalcoholic extract of leaves presented a strong activity against *T. mentagrophytes*, *T. rubrum*, *M. canis*, and *M. gypseum* with MICs of 15.62, 15.62, 15.62, and 62.5 µg/mL, respectively. After bioactivity-guided fractionation studies, two neolignans (eupomatenoid 3 and eupomatenoid 5) were isolated from the column fractions of chloroform subextract. The pure compounds showed strong activity on *T. rubrum* with MICs of 50 and 6.2 µg/mL, respectively (Koroishi et al., 2008).

Methanolic extract of *Magnolia obovata* stem bark has been found to have antifungal activity against *T. mentagrophytes*, and further studies have revealed that two neolignan compounds (magnolol, honokiol) were responsible for the antidermatophytic activity of the stem bark extract. These two neolignans have shown significant inhibitory activities against *M. gypseum*, *T. mentagrophytes*, and *E. floccosum* with MICs between 25 and 100 µg/mL (Bang et al., 2000).

Lopes et al. (2012) have studied antifungal activities of purified phlorotannin extracts from three brown seaweeds (*Cystoseira nodicaulis*, *Cystoseira usneoides*, and *Fucus spiralis*). The purified phlorotannin extracts possessed both fungistatic and fungicidal activities against yeast and dermatophytes. *E. floccosum* and *T. rubrum* were the most susceptible species among dermatophytes.

Antimicrobial effect of phytoalexin resveratrol on dermatophytes and bacterial pathogens of the skin was investigated by Chan (2002). Antifungal activity was tested on *T. mentagrophytes*, *T. tonsurans*, *T. rubrum*, *E. floccosum*, and *M. gypseum*. The growth of dermatophytes was inhibited at 25–50 µg/mL of resveratrol.

The methanol extract and subextracts of the aerial parts of *Geophila repens* were tested against many dermatophytes. The butanol subextract was found to be the most active one against *Cryptococcus neoformans*, *M. gypseum*, and *T. mentagrophytes*. After fractionation by successive column chromatography, an ester of caffeic acid and maleic anhydride identified as maleic anhydride caffeate was determined as the active principle (Vila et al., 2013).

Zacchino et al. (1999) investigated *in vitro* antidermatophytic effect of 34 arylpropanoids and related compounds isolated from different plants against *E. floccosum*, *M. canis*, *M. gypseum*, *T. mentagrophytes*, and *T. rubrum* by agar dilution method. Alpha-halopropiophenones displayed a broad spectrum of activities against dermatophytes with MICs between 0.5 and >50 µg/mL. Additionally, keto, alcohol, and alpha-haloketo propyl derivatives of naphthalene and phenanthrene possessed high antidermatophytic activity. Also, phenanthryl derivatives were found to be more active (MICs: 3–20 µg/mL) than naphthyl derivatives (MICs 3–50 µg/mL).

25.4.2.3 Saponins

Triterpene and steroidal saponins have also been isolated as antidermatophytic constituents of some medicinal plants. These studies concern mainly species of the Solanaceae family. A spirostanol saponin and three saponins have been isolated from the leaves of *Solanum hispidum*. All the isolated compounds have revealed antimycotic activity. The most active compound was a spirostanol derivative, with IC50 values of 25 µg/mL against both *T. mentagrophytes* and *T. rubrum* (Gonzales et al., 2004). Alvarez et al. (2001) have isolated a steroidal saponin named SC-1 from the leaves of *Solanum chrysotrichum*. Its structure has been characterized as 3-*O*-{β-quinovopyranosyl(1 → 6)-β-glucopyranosyl(1 → 6)-β-glucopyranosyl}chlorogenin, and it has possessed straight fungitoxic activity against the dermatophyte *T. mentagrophytes* (MIC = 40 µg/mL; nystatin MIC = 10 µg/mL). Additionally, five spirostan saponins and two sterol glycosides have been isolated from the leaves of the same plant having anti-dermatophytic activity. The most active compound is one of the spirostan derivatives and had low MIC values (12.5 µg/mL) against *T. mentagrophytes* and *T. rubrum* (Zamilpa et al., 2002).

Saponins isolated from roots and aerial parts of *Medicago sativa*, *Medicago murex*, *Medicago arabica*, and *Medicago hybrida* were reported to be active against three dermato-phytic fungi *M. gypseum*, *T. interdigitale*, and *T. tonsurans* (MIC < 0.09 mM). *T. tonsurans* was the most sensitive dermatophyte. Activities of glycosides were higher than the activities of aglycones and monodesmosidic glycosides of medicagenic acid were the most active compounds (Houghton et al., 2006b).

25.4.2.4   Fatty acids

According to a recent study, fatty acids have antifungal activity mainly due to their capac-ity to disrupt cell membranes. They are able to interfere with cell membrane structure, displacing phospholipids and increasing its permeability. Undecylenic acid is a good example of a semisynthetic antifungal compound mainly used in the treatment of super-ficial mycoses. It is prepared from ricinoleic acid obtained from the seed oil of *Ricinus communis* (Vila et al., 2013).

A number of fatty acids were isolated from the bark of *Calycophyllum spruceanum* var. *multiflorum* by bioguided fractionation, and their structure has been elucidated as 6-hexadecinoic acid, 6-heptadecinoic acid, 6-octadecinoic acid, 6-nonadecinoic acid and 6-eicosinoic acid, palmitic acid, heptadecanoic acid, and stearic acid. The mixture of these fatty acids was found to be active against the dermatophytes *M. gypseum* and *T. mentagro-phytes* with MIC and MFC values of 0.25 µg/mL, lower than those of the reference drugs nystatin and amphotericin B (Vila et al., 2013).

25.4.2.5   Other compounds

Two piperidine alkaloids and named as haloxylines A and B have been isolated from *Haloxylon salicornicum*. Both alkaloids showed moderate to potent antifungal activities against *Trichophyton longifusus* and *M. canis* (Ferheen et al., 2005).

In vitro antifungal activity of hexane, methanol, ethanol, and chloroform extracts of some Indian plants has been studied by Ponnusamy et al. (2010). Active chloroform extract of *Wrightia tinctoria* leaves has been fractionated, and the indole compound indiru-bin was isolated as the active principle. It exhibited activity against dermatophytes such as *E. floccosum* (MIC = 6.25 µg/mL), *T. rubrum* and *T. tonsurans* (MIC = 25 µg/mL), and *Trichophyton simii* (MIC = 50 µg/mL).

After antifungal assay–directed fractionation of methanolic extract of *Eleutherine americana* bulbs, a naphthoquinone derivative was isolated from the *n*-hexane-soluble fraction. The compound was determined as eleutherin, and it was found to have antider-matophytic activity against *T. mentagrophytes* in agar diffusion assay (Kusuma et al., 2010).

A study of the antimicrobial compounds from *Moneses uniflora* resulted in the isolation of naphthoquinone derivatives named 8-chlorochimaphilin, together with chimaphilin and 3-hydroxychimaphilin having antifungal activity against *M. gypseum* with MIC values 12.5 and 25 μg/mL (Saxena et al., 1996).

The leaves of *Allamanda cathartica* have been used to isolate an iridoid called plumieride. It has been tested against dermatophytes *E. floccosum* and *M. gypseum* by a modified paper disk technique. Plumieride possessed significant fungitoxicity by inhibiting the growth of both test dermatophytes completely (Tiwari et al., 2002).

Antidermatophytic activity of ether extract of *Nigella sativa* and its active principle thymoquinone have been tested against different dermatophyte species (*E. floccosum, M. canis, T. rubrum, T. mentagrophytes,* and *T. interdigitale*) by agar diffusion method. The MICs of ether extract of the plant and thymoquinone have been between 10 and 40 and 0.125 and 0.25 mg/mL, respectively. The MICs of thymoquinone have been less for *E. floccosum* and *M. canis* than for the *Trichophyton* species (Aljabre et al., 2005).

Antidermatophytic activities of crude steroidal glycoside extract and two spirostanol glucosides (yuccaloeside B and C) isolated from *Yucca gloriosa* were tested by using agar dilution method on *T. rubrum, T. mentagrophytes, T. soudanense, M. canis, M. gypseum,* and *E. floccosum.* The MICs of yuccaloeside B and C were found to be between 0.78 and 12.5 μg/mL (Favel et al., 2005).

25.5   Conclusion

Although the incidence of dermatophytic infections in humans is increasing day by day, there are only a few therapeutic options for treatment because of side effects and toxicity of medicines, drug resistance, etc. Beside synthetic drugs, traditional medicines and natural products are widely used for fungal infections all over the world. Today, nearly 1000 plants have been reported for their antifungal activities on different strains; however, only 100–200 of them have possessed antidermatophytic activity. Antidermatophytic activity screening studies on plant extracts are almost common, but even so, only a few scientists have resumed their studies to isolate and determine the active principles.

Additionally, the *in vivo* antidermatophytic activity studies of the *in vitro* active compounds have to be tested on animal and humans to find their effectiveness, toxicity, and dose–response relationship.

In this chapter, the studies on essential oils, terpenoids, phenolic compounds, saponins, other natural compounds, and plant extracts with antidermatophytic activity were reviewed. Further studies are needed to enlighten structure–activity relationship of these secondary metabolites and also to determine their active and safe doses in humans

References

Abad, M.J., Ansuategui, M., and Bermejo, P. 2007. Active antifungal substances from natural sources. *Arkivoc* 7:116–145.

Agrawal, A., Srivastava, S., Srivastava, J.N., and Srivasava, M.M. 2004. Evaluation of inhibitory effect of the plant *Phyllanthus amarus* against dermatophytic fungi *Microsporum gypseum. Biomed. Environ. Sci.* 17:359–365.

Ali-Shtayeh, M.S. and Abu Ghdeib, S.I. 1999. Antifungal activity of plant extracts against dermatophytes. *Mycoses* 42:665–672.

Aljabre, S.H.M., Randhawa, M.A., Akhtar, N., Alakloby, O.M., Alqurashi, A.M., and Aldossary, A. 2005. Antidermatophytic activity of ether extract of *Nigella sativa* and its active principle, thymoquinone. *J. Ethnopharmacol.* 101:116–119.

Alvarez, L., Perez, M.C., Villarreal, M.L., and Navarro, V. 2001. SC-1, An antimycotic spirostan saponin from *Solanum chrysotrichum. Planta Med.* 67:372–374.

American Orthopaedic Foot & Ankle Society. 2015. Athlete's foot.

Amin, M. and Kapadnis, B.P. 2005. Heat stable antimicrobial activity of *Allium ascalonicum* against bacteria and fungi. *Indian J. Exp. Biol.* 43:751–754.

Anane, S. and Chtourou, O. 2013. Tinea capitis favosa misdiagnosed as tinea amiantacea. *Med. Mycol. Case Rep.* 2:29–31.

Anitha, R. and Kannan, P. 2006. Antifungal activity of *Clerodendrum inerme* (L.) and *Clerodendrum phlomidis* (L.). *Turk. J. Biol.* 30:139–142.

Arif, T., Bhosale, J.D., Kumar, N., Mandal, T.K., Bendre, R.S., Lavekar, G.S., and Dabur, R. 2009. Natural products-antifungal agents derived from plants. *J. Asian Nat. Prod. Res.* 11:621–637.

Asticcioli, S., Di Silverio, A., Sacco, L., Fusi, I., Vincenti, L., and Romero, E. 2008. Dermatophyte infections in patients attending a tertiary care hospital in northern Italy. *New Microbiol.* 31:543–548.

Avelar Pires, C.A., Monteiro Lobato, A., Oliveira Carneiro, F.R., Satos da Cruz, N.F., Oliveira de Sousa, P., and Darwich Mendes, A.M. 2014. Clinical, epidemiological, and therapeutic profile of dermatophytosis. *Ann. Bras. Dermatol.* 89:259–274.

Ayatollahi Mousavi, S.A., Salari Sardoui, S., and Shamsadini, S. 2009. A first case of tinea imbricate from Iran. *Jundishapur J. Microbiol.* 2:71–74.

Balakumar, S., Rajan, S., Thirunalasundari, T., and Jeeva, S. 2011. Antifungal activity of *Aegle marmelos* (L.) Correa (Rutaceae) leaf extract on dermatophytes. *Asian Pac. J. Trop. Biomed.* 1(4):309–312.

Bang, K.H., Kim, Y.K., Min, B.S., Na, M.K., Rhee, Y.H., Lee, J.P., and Bae, K.H. 2000. Antifungal activity of magnolol and hinokiol. *Arch. Pharm. Res.* 23:46–49.

Baran, W., Szepietowski, J.C., and Schwartz, R.A. 2004. Tinea barbae. *Acta Dermatoven APA* 13:91–94.

Batawila, K., Kokou, K., Koumaglo, K., Gbeassor, M., De Foucault, B., Bouchet, P., and Akpagana, K. 2005. Antifungal activities of five Combretaceae used in Togolese traditional medicine. *Fitoterapia* 76:264–268.

Beikert, F.C., Anastasiadou, Z., Fritzen, B., Frank, U., and Augustin, M. 2013. Topical treatment of tinea pedis using 6% coriander oil in unguentum leniens: A randomized, controlled, comparative pilot study. *Dermatology* 226(1):47–51.

Cavaleiro, C., Pinto, E., Gonçalves, M.J., and Salgueiro, L. 2006. Antifungal activity of *Juniperus* essential oils against dermatophyte, *Aspergillus* and *Candida* strains. *J. Appl. Microbiol.* 100:1333–1338.

Chan, M.M.-Y. 2002. Antimicrobial effects of resveratrol on dermatophytes and bacterial pathogens of the skin. *Biochem. Pharmacol.* 63:99–104.

Chauhan, V.S., Suthar, A., Naik, V., and Salkar, K. 2012. Screening of plants for antidermatophyte activity. *Int. J. Pharm. Sci. Res.* 3:1502–1506.

Cos, P., Vlietinck, A., Vanden Berghe, D., and Maes, L. 2006. Anti-infective potential of natural products: How to develop a stronger in vitro "proof-of-concept". *J. Ethnopharmacol.* 106:290–302.

Das, K., Tiwari, R.K.S., and Shrivastava, D.K. 2010. Techniques for evaluation of medicinal plant products as antimicrobial agent: Current methods and future trends. *J. Med. Plants Res.* 4:104 111.

De Campos, M.P., Cechinel Filho, V., Da Silva, R.Z., Yunes, R A., Zacchino, S., Juarez, S., and Bella Cruz, A. 2005. Evaluation of antifungal activity of *Piper solmsianum* C. DC. var. *solmsianum* (Piperaceae). *Biol. Pharm. Bull.* 28:1527–1530.

Del Palacio, A., Garau, M., Gonzalez-Escalada, A., and Calvo, M.T. 2000. Trends in the treatment of dermatophytosis. *Rev. Iberoam. Micol.* 16:148–158.

Devi, W.R., Singh, S.D.B., and Singh, C.B. 2013. Antioxidant and anti-dermatophytic properties leaf and stem bark of *Xylosma longifolium* Clos. *BMC Complement. Altern. Med.* 13:155.

Dias, M.F.R.G., Quaresma-Santos, M.V.P., Schectman, R.C., Bernardes-Filho, F., Da Fonseca Amorim, A.G., and Azulay, D.R. 2013. Treatment of superficial mycoses: Review-part II. *Ann. Bras. Dermatol.* 88:937–944.

Dicle, Ö. and Özkesici, B. 2013. Tinea kapitis. *Turk. J. Dermatol.* 7:1–8.

Dorman, H.J.D. and Deans, S.G. 2000. Antimicrobial agents from plants: Antibacterial activity of plant volatile oils. *J. Appl. Microbiol.* 88:308–316.

Duraipandiyan, V. and Ignacimuthu, S. 2007. Antibacterial and antifungal activity of *Cassia fistula* L.: An ethnomedicinal plant. *J. Ethnopharmacol.* 112(3):590–594.

Engelmeier, D. and Hadacek, F. 2006. Antifungal natural products: Assays and applications. In: *Advances in Phytomedicine Series.* Vol. III: *Naturally Occurring Bioactive Compounds*, eds. M. Rai and M.C. Carpinella, pp. 23–43. Amsterdam, the Netherlands: Elsevier.

Falahati, M., Tabrizib, N.O., and Jahaniani, F. 2005. Antidermatophyte activities of *Eucalyptus camaldulensis* in comparison with griseofulvin. *Iran. J. Pharmacol. Ther.* 4:80–83.

Favel, A., Kemertelidze, E., Benidze, M., Fallague, K., and Regli, P. 2005. Antifungal activity of steroidal glycosides from *Yucca gloriosa* L. *Phytother. Res.* 19:158–161.

Fenner, R., Sortino, M., Rates, S.K., Dall'Agnol, R., Ferraz, A., Bernardi, A.P., and Zacchino, S. 2005. Antifungal activity of some Brazilian *Hypericum* species. *Phytomedicine* 12:236–240.

Ferheen, S., Ahmed, E., Afza, N., Malik, A., Shah, M.R., Nawaz, S.A., and Choudhary, M.I. 2005. Haloxylines a and b, antifungal and cholinesterase inhibiting piperidine alkaloids from *Haloxylon salicornicum*. *Chem. Pharm. Bull.* 53:570–572.

Freixa, B., Vila, R., Ferro, E.A., Adzet, T., and Cañigueral, S. 2001. Antifungal principles from *Piper fulvescens*. *Planta Med.* 67:873–875.

Freixa, B., Vila, R., Vargaz, L., Lozano, N., Adzet, T., and Cañigueral, S. 1998. Screening for antifungal activity of nineteen Latin American plants. *Phytother. Res.* 12:427–430.

Garcıa, V.N., Gonzalez, A., Fuentes, M., Aviles, M., Rios, M.Y., Zepeda, G., and Rojas, M.G. 2003. Antifungal activities of nine traditional Mexican medicinal plants. *J. Ethnopharmacol.* 87(1):85–88.

Global Action Fund for Fungal Infections. 2015. http://www.gaffi.org/why/fungal-disease-frequency.

Gonzales, M., Zamilpa, A., Marquina, S., Navarro, V., and Alvarez, L. 2004. Antimycotic spirostanol saponins from *Solanum hispidum* leaves and their structure-activity relationships. *J. Nat. Prod.* 67:938.

Gullo, F.P., Sardi, J.C., Santos, V.A., Sangalli-Leite, F., Pitangui, N.S., Rossi, S.A., de Paula, E.S.A.C., Soares, L.A., Silva, J.F., Oliveira, H.C., Furlan, M., Silva, D.H., Bolzani, V.S., Mendes-Giannini, M.J., and Fusco-Almeida, A.M. 2012. Antifungal activity of maytenin and pristimerin. *e-CAM* 1–6.

Gupta, A.K. and Cooper, E.A. 2008. Update in antifungal therapy of dermatophytosis. *Mycopathologia* 166:353–367.

Hammer, K.A., Carson, C.F., and Riley, T.V. 2002. *In vitro* activity of *Melaleuca alternifolia* (tea tree) oil against dermatophytes and other filamentous fungi. *J. Antimicrob. Chemother.* 50:195–199.

Hean, C.C., Rahim, I.N.A.B.A., Harn, G.L., Halimi, N.A.B.A., Hamzah, M.H.B., Aisyah, T.A.F., and Said, N.F.B. 2011. Fungal disease and therapy. *Webmed Central Pharm. Sci.* 2(12):WMC002693.

Higgins, E.M., Fuller, L.C., and Smith, C.H. 2000. Guidelines for the management of tinea capitis. *Br. J. Dermatol.* 143:53–58.

Ho, K. and Cheng, T. 2010. Common superficial fungal infections—A short review. *The Hong Kong Med. Diary* 15:23–27.

Houghton, P., Patel, N., Jurzysta, M., Biely, Z., and Cheung, C. 2006b. Antidermatophyte activity of *Medicago* extracts and contained saponins and their structure-activity relationships. *Phytother. Res.* 20:1061–1066.

Houghton, P.J., Ismail, K.M., Maxia, L., and Appendino, G. 2006a. Antidermatophytic prenylated coumarins from *asafetida*. *Planta Med.* 72:S-008.

Jacob, M.R. and Walker, L.A. 2005. Natural products and antifungal drug discovery. In: *Methods in Molecular Medicine*, Vol. 118: *Antifungal Agents: Methods and Protocols*, eds. E.J. Ernst, and P.D. Rogers. Totowa, NJ: Humana Press Inc.

Jacyk, W.K. 2004. Four common infectious skin condition: Tinea corporis, pityriasis versicolor, scabies, larva migrans. *S. Afr. Fam. Pract.* 46:13–16.

Jamalian, A., Shams-Ghahfarokhi, M., Jaimand, K., Pashootan, N., Amani, A., and Razzaghi-Abyaneh, M. 2012. Chemical composition and antifungal activity of *Matricaria recutita* flower essential oil against medically important dermatophytes and soil-borne pathogens. *J. Mycol. Med.* 22:308–315.

Jantan, I., Yassin, M.S.M., Chin, C.B., Chen, L.L., and Sim, N.L. 2003. Antifungal activity of the essential oils of nine Zingiberaceae species. *Pharm. Biol.* 41:392–397.

Karakoca, Y., Endoğru, E., Turgut Erdemir, A., Kiremitçi, Ü., Gürel, M.S., and Güçin, Z. 2010. Generalized inflammatory tinea corporis. *J. Turk. Acad. Dermatol.* 4:1–3.

Khaled, A., Mbarek, L.B., and Kharfi, M. 2007. Tinea capitis favosa due to *Trichophyton schoenleinii*. *Acta Dermatoven. APA* 16:34–36.

Khattak, S., Saeed, P., Ullah, H., Ahmad, W., and Ahmad, M. 2005. Biological effects of indigenous medicinal plants *Curcuma longa* and *Alpinia galanga*. *Fitoterapia* 76:254–257.

Khosravi, R.A., Shokri, H., Farahnejat, Z., Chalangari, R., and Katalin, M. 2013. Antimycotic efficacy of Iranian medicinal plants towards dermatophytes obtained from patients with dermatophytosis. *Chin. J. Nat. Med.* 11:43–48.

Kishore, N., Chansouria, J.P.N., and Dubey, N.K. 1996. Antidermatophytic action of the essential oil of *Chenopodium ambrosioides* and an ointment prepared from it. *Phytother. Res.* 10:453–455.

Koroishi, A.M., Foss S.R., Cortez, D.A.G., Ueda-Nakamura, T., Nakamurad, C.V., and Filho, B.P.D. 2008. *In vitro* antifungal activity of extracts and neolignans from *Piper regnellii* against dermatophytes. *J. Ethnopharmacol.* 117:270–277.

Kosalec, I., Pepeljnjak, S., and Kustrak, D. 2005. Antifungal activity of fluid extract and essential oil from anise fruits (*Pimpinella anisum* L., Apiaceae). *Acta Pharm.* 55:377–385.

Kuiate, J.R., Bessière, J.M., Vilarem, G., and Amvam Zollo, P.H. 2006. Chemical composition and antidermatophytic properties of the essential oils from leaves, flowers and fruits of *Cupressus lusitanica* mill from Cameroon. *Flavour Frag. J.* 21:693–697.

Kuiate, J.R., Mouokeu, S., Wabo, H.K., and Tane, P. 2007. Antidermatophytic triterpenoids from *Syzygium jambos* (L.) Alston (Myrtaceae). *Phytother. Res.* 21:149–152.

Kusuma, I.W., Arung, E.T., Rosamah, E., Purwatiningsih, S., Kuspradini, H., Astuti, J., and Shimizu, K. 2010. Antidermatophyte and antimelanogenesis compound from *Eleutherine americana* grown in Indonesia. *J. Nat. Med.* 64:223–226.

Lang, G. and Buchbauer, G. 2012. A review on recent research results (2008–2010) on essential oils as antimicrobials and antifungals. A review. *Flavour Frag. J.* 27:13–39.

Lock de Ugaz, O., Costa, J., Sanchez, L., Ubillas Sanchez, R.P., and Tempesta, M.S. 1991. Flavonoids from *Andira inermis*. *Fitoterapia* 62:89–90.

Loizzo, M. R., Statti, G. A., Tundis, R., Conforti, F., Bonesi, M., Autelitano, G., and Menichini, F. 2004. Antibacterial and antifungal activity of *Senecio inaequidens* DC and *Senecio vulgaris* L. *Phytother. Res.* 18:777–779.

Lopes, G., Sousa, C., and Silva, L.R. 2012. Can phlorotannins purified extracts constitute a novel pharmacological alternative for microbial infections with associated inflammatory conditions? *PLoS One* 7(2):e31145.

Mabona, U., Viljoen, A., Shikanga, E., Marston, A., and Vuuren, S.V. 2013. Antimicrobial activity of southern African medicinal plants with dermatological relevance: From an ethnopharmacological screening approach, to combination studies and the isolation of a bioactive compound. *J. Ethnopharmacol.* 148:45–55.

Machado de Souza, A.C., Kato, L., Conceição da Silva, C., Cidade, A.F., Alves de Oliveira, C.M., and Silva, M.R.R. 2009. Antimicrobial activity of *Hymenaea martiana* towards dermatophytes and *Cryptococcus neoformans*. *Mycoses* 53:500–503.

Mann, A., Banso, A., and Clifford, L.C. 2008. An antifungal property of crude plant extracts from *Anogeissus leiocarpus* and *Terminalia avicennioides*. *Tanzan. J. Health Res.* 10:34–38.

Mares, D., Romagnoli, C., Tosi, B., Andreotti, E., Chillemi, G., and Poli, F. 2005. Chicory extracts from *Cichorium intybus* L. as potential antifungals. *Mycopathologia* 160:85–92.

Masoko, P., Picard, J., and Eloff, J.N. 2005. Antifungal activities of six South African *Terminalia* species (Combretaceae). *J. Ethnopharmacol.* 99:301–308.

Moodahadu-Bangera, L.S., Martis, J., Mittal, R., Krishnankutty, B., Kumar, N., Bellary, S., and Rao, P.K. 2012. Eberconazole-pharmacological and clinical review. *Indian J. Dermatol.* 78:217–222.

Moriarty, B., Hay, R., and Morris-Jones, R. 2012. The diagnosis and management of tinea. *Br. Med. J.* 345:1–10.

Mugnaini, L., Nardoni, S., Pistelli, L., Leonardi, M., Giuliotti, L., Benvenuti, M.N., and Mancianti, F. 2013. A herbal antifungal formulation of *Thymus serpillum*, *Origanum vulgare* and *Rosmarinus officinalis* for treating ovine dermatopytosis due to *Tricophyton mentagrophytes*. *Mycoses* 56:333–337.

Muschietti, L., Derita, M., Sülsen, V., de Dios Muñoz, J., Ferraro, G., Zacchino, S., and Martino, V. 2005. *In vitro* antifungal assay of traditional Argentine medicinal plants. *J. Ethnopharmacol.* 102:233–238.

Natarajan, V., Venugopal, P.V., and Menon, T. 2003. Effect of *Azadirachta indica* (Neem) on the growth pattern of dermatophytes. *Indian J. Med. Microbiol.* 21:98–101.

Noumedem, J.A.K., Tamoko, J.D., Teke G.N., Momo, R.C.D., Kuete, V., and Kuiate J.R. 2013. Phytochemical analysis, antimicrobial and radical-scavenging properties of *Acalypha manniana* leaves. *Springer Plus* 2:503.

Odom, R. 1993. Pathophysiology of dermatophyte infections. *J. Am. Acad. Dermatol.* 28:2–7.

Phongpaichit, S., Subhadhirasakul, S., and Wattanapirosakul, C. 2005. Antifungal activities of extracts from Thai medicinal plants against opportunistic fungal pathogens associated with AIDS patients. *Mycoses* 48:333–338.

Pinto, E., Pina-Vaz, C., Salgueiro, L., Gonçalves, M.J., Costa-de-Oliveira, S., Cavaleiro, C., and Martinez-de-Oliveira, J. 2006. Antifungal activity of the essential oil of *Thymus pulegioides* on *Candida*, *Aspergillus* and dermatophyte species. *J. Med. Microbiol.* 55:1367–1373.

Policegoudra, R.S., Goswami, S., Aradhya, S.M., Chatterjee, S., Datta, S., Sivaswamy, R., and Singh, L. 2012. Bioactive constituents of *Homalomena aromatica* essential oil and its antifungal activity against dermatophytes and yeasts. *J. Mycol. Med.* 22:83–87.

Ponnusamy, K., Petchiammal, C., Mohankumar, R., and Hopper, W. 2010. *In vitro* antifungal activity of indirubin isolated from a South Indian ethnomedicinal plant *Wrightia tinctoria* R. Br. *J. Ethnopharmacol.* 132:349–354.

Portillo, A., Vila, R., Freixa, B., Adzet, T., and Cañigueral, S. 2001. Antifungal activity of Paraguayan plants used in traditional medicine. *J. Ethnopharmacol.* 6:93–98.

Portillo, A., Vila, R., Freixa, B., Ferro, E., Parella, T., Casanova, J., and Cañigueral, S. 2005. Antifungal sesquiterpene from the root of *Vernonanthura tweedieana*. *J. Ethnopharmacol.* 97:49–52.

Premkumar, V.G. and Shyamsundar, D. 2005. Antidermatophytic activity of *Pistia stratiotes*. *Ind. J. Pharmacol.* 37:126–128.

Rahman, A., Al-Reza, S.M., Siddiqui, S.A., Chang, T., and Kang, S.C. 2014. Antifungal potential of essential oil and ethanol extracts of *Lonicera japonica* Thunb against dermatophytes. *EXCLI J.* 13:427–436.

Rai, M.K., Varma, A., and Pandev, A.K. 2004. Antifungal potential of *Spilanthes calva* after inoculation of *Piriformospora indica*. *Mycoses* 47:479–481.

Rajendra Prasad, N., Anandi, C., Balasubramanian, S., and Pugalendi, K.V. 2004. Antidermatophytic activity of extracts from *Psoralea corylifolia* (Fabaceae) correlated with the presence of a flavonoid compound. *J. Ethnopharmacol.* 91:21–24.

Rath, S. and Mohanty, R.C. 2013. Antifungal screening of *Curcuma longa* and *Cassia tora* on dermatophytes. *Int. J. Life Sci. Pharma Res.* 2(4):88–94.

Rebollo, N., López-Barcenas, A.P., and Arenas, R. 2008. Tinea capitis. *Actas Dermosifiliogr.* 99:91–100.

Ríos, J.L., Recio, M.C., and Villar, A. 1998. Screening methods for natural products with antimicrobial activity: A review of the literature. *J. Ethnopharmacol.* 23:127–149.

Rojas, R., Bustamante, B., Ventosilla, P., Fernádez, I., Caviedes, L., Gilman, R.H., and Hammond, G.B. 2006. Larvicidal, antimycobacterial and antifungal compounds from the bark of the peruvian plant *Swartzia polyphylla* DC. *Chem. Pharm. Bull.* 54:278–279.

Rojas, R., Doroteo, V., Bustamante, B., Davel, J, and Lock, O. 2004. Antimicrobial and free radical scavenging activity of *Gentianella nitida*. *Fitoterapia* 75:754–757.

Sabota, J., Brodell, R., Rutecki, G.W., and Hoppes, W.L. 1996. Severe tinea barbae due to *Trichophyton verrucosum* infection in dairy farmers. *Clin. Infect. Dis.* 23:1308–1310.

Sanguinetti, M., Posteraro, B., Romano, L., Battaglia, F., Lopizzo, T., De Carolis, E., and Fadda, G. 2007. *In vitro* activity of *Citrus bergamia* (bergamot) oil against clinical isolates of dermatophytes. *J. Antimicrob. Chemother.* 59:305–308.

Satter, E.K. 2009. Tinea imbricata. *Cutis* 83:188–191.

Saxena, G. Farmer, S.W., Hancock, R.E.W., and Towers, G.H.N. 1996. Chlorochimaphilin: A new antibiotic from *Moneses uniflora*. *J. Nat. Prod.* 59:62–65.

Schmourlo, G., Mendonça-Filho, R.R., Alviano, C.S., and Costa, S.C. 2005. Screening of antifungal agents using ethanol precipitation and bioautography of medicinal and food plants. *J. Ethnopharmacol.* 96:563–568.

Shams-Ghahfarakhi, M., Goodarzi, M., Abvaneh, M.R., Al Tiraihi, T., and Sevedipovr, G. 2004. Morphological evidences for onion-induced growth inhibition of *Trichophyton rubrum* and *Trichophyton mentagrophytes*. *Fitoterapia* 75:645–655.

Shams-Ghahfarokhi, M., Shokoohamiri, M.R., Amirrajab, N., Moghadasi, B., Ghajari, A., Zeini, F., and Razzaghi-Abyaneh, M. 2006. *In vitro* antifungal activities of *Allium cepa, Allium sativum* and ketoconazole against some pathogenic yeasts and dermatophytes. *Fitoterapia* 77:321–332.

Silva, M.R.R., Oliveira, J.G., Fernandes, O.F.L., Passos, X.S., Costa, C.R., Souza, L.K. H., and Paula, J.R. 2005. Antifungal activity of *Ocimum gratissimum* towards dermatophytes. *Mycoses* 48:172–175.

Singh, G., Kumar, P., and Joshi, C. 2014. Treatment of dermatophytosis by a new antifungal agent 'apigenin'. *Mycoses* 57:497–506.

Singh, S. and Vidyasagar, G.M. 2014. *In vitro* antidermatophytic and preliminary phytochemical studies of petroleum ether and inter-polar methanolic leaf extracts of *Thevetia nerrifolia*. *J. Appl. Pharma. Sci.* 4(3):81–85.

Sipponen, P., Sipponen, A., Lohi, J., Soini, M., Tapanainen, R., and Jokinen, J.J. 2012. Natural coniferous resin lacquer in treatment of toenail onychomycosis: An observational study. *Mycoses* 56:289–296.

Stein, A.C., Sortino, M., Avancini, C., Zacchino, S., and Von Poser, G. 2005. Ethnoveterinary medicine in the search for antimicrobial agents: Antifungal activity of some species of *Pterocaulon* (Asteraceae). *J. Ethnopharmacol.* 99:211–214.

Svetaz, L., Agüero, M.B., Alvarez, S., Luna, L., Feresin, G., Derita, M., and Zacchino, S. 2007. Antifungal activity of *Zuccagnia punctata* Cav: Evidence for the mechanism of action. *Planta Med.* 73:1074–1080.

Tadeg, H., Mohammed, E., Asres, K., and Gebre-Mariam T. 2005. Antimicrobial activities of some selected traditional Ethiopian medicinal plants used in the treatment of skin disorders. *J. Ethnopharmacol.* 100:168–175.

Takahashi, T., Kokubo, R., and Sakaino, M. 2004. Antimicrobial activities of eucalyptus leaf extracts and flavonoids from *Eucalyptus maculata*. *Lett. Appl. Microbiol.* 39:60–64.

Tiwari, T.N., Pandey, V.B., and Dubey, N.K. 2002. Plumieride from *Allamanda cathartica* as an antidermatophytic agent. *Phytother. Res.* 16:393–394.

Trakranrungsie, N., Chatchawanchonteera, A., and Khunkitti, W. 2008. Ethnoveterinary study for antidermatophytic activity of *Piper betle, Alpinia galanga* and *Allium ascalonicum* extracts *in vitro*. *Res. Vet. Sci.* 84:80–84.

Vila, R., Freixa, B., and Cañigueral, S. 2013. Antifungal compounds from plants. In: *Recent Advances in Pharmaceutical Sciences III*, eds. D. Muñoz-Torrero, A. Cortés, and E.L. Mariño, pp. 23–43. Kerala, India: Transworld Research Network.

Weitzman, I. and Summerbell, R.C. 1995. The dermatophytes. *Clin. Microbiol. Rev.* 8:240–259.

Yenesew, A., Derese, S., Midiwo, J.O., Bii, C.C., Heydenreich, M., and Peter, M.G. 2005. Antimicrobial flavonoids from the stem bark of *Erythrina burttii*. *Fitoterapia* 76:469–472.

Zacchino, S.A., López, S.N., Pezzenati, G.D., Furlán, R.L., Santecchia, C.B., Muñoz, L., and Enriz, R.D. 1999. *In vitro* evaluation of antifungal properties of phenylpropanoids and related compounds acting against dermatophytes. *J. Nat. Prod.* 62:1353–1357.

Zamilpa, A., Tortoriello, J., Navarro, V., Delgado, G., and Alvarez, L. 2002. Five new steroidal saponins from *Solanum chrysotrichum* leaves and their antimycotic activity. *J. Nat. Prod.* 65:1815–1819.

Index

Milton Keynes UK
Ingram Content Group UK Ltd.
UKHW052026071024
449327UK00027B/2438